012122372

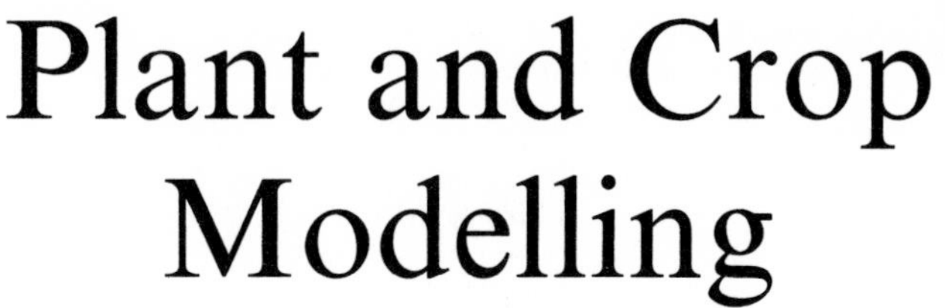

Plant and Crop Modelling

A Mathematical Approach to Plant and Crop Physiology

John H. M. Thornley

Agricultural and Food Research Council
Institute for Grassland and Animal Production
Hurley, Maidenhead, Berkshire

and

Ian R. Johnson

Department of Agronomy and Soil Science
University of New England
Armidale, New South Wales

CLARENDON PRESS · OXFORD
1990

Oxford University Press, Walton Street, Oxford OX2 6DP
Oxford New York Toronto
Delhi Bombay Calcutta Madras Karachi
Petaling Jaya Singapore Hong Kong Tokyo
Nairobi Dar es Salaam Cape Town
Melbourne Auckland
and associated companies in
Berlin Ibadan

Oxford is a trade mark of Oxford University Press

Published in the United States
by Oxford University Press, New York

British Library Cataloguing in Publication Data
Thornley, John H. M.
Plant and crop modelling.
1. Plants. Physiology, Mathematical models
I. Title II. Johnson, Ian
581.1'0724
ISBN 0 19 854160 0

Library of Congress Cataloging in Publication Data
Thornley, J. H. M.
Plant and crop modelling: a mathematical approach to plant and crop physiology / John H. M. Thornley and Ian R. Johnson.
Includes bibliographical references.
1. Plant physiology—Mathematical models. 2. Crops—Physiology—Mathematical models. I. Johnson, Ian R. II. Title.
QK711.2.T483 1990 581.1'01'5118—dc20 89-16249
CIP

ISBN 0 19 854160 0

Set by Asco Trade Typesetting Ltd., Hong Kong
Printed in Great Britain by
Bookcraft Ltd., Midsomer Norton, Avon

'Truth comes out of error more easily than out of confusion'
Francis Bacon (1561–1626)

'There is nothing so practical as a good theory'
Emanuel Kant (1724–1804)

'Things should be made as simple as possible, but no simpler'
Albert Einstein (1879–1955)

Preface

This book sets out to be a textbook: for instance, there are exercises, and there are outline solutions. It is a textbook of physiology, although not of the traditional type. We attempt to show plant scientists how to express their physiological ideas mathematically, and how to deduce the appropriate quantitative conclusions, which can then be compared with experiment. This is an area where, arguably, a higher level of expertise would enable progress to be more rapid.

The universe, including our particular corner of it, functions in much the same way today as it did a thousand or a million years ago. Presumably the mechanisms, whether physical, chemical, or biochemical, have hardly changed. What has very definitely changed is our understanding of how it all works. Science is nothing but conjecture, and the viability of these conjectures is continually assessed and reassessed against the physical world by means of experimentation. For this process to be effective, it is essential that current experimental technology is matched by the precision of our concepts and theories. These concepts and theories as to how and why everything happens span a continuum from the qualitative to the highly quantitative. The matching of experiment with theory may mean using mathematics for formulating and expressing our ideas. One of the aims of this book is to provide plant scientists with the techniques and methodology for doing this.

It is sometimes said that contemporary science is characterized by the increasing importance of convergent patterns of research. That is, the merging of different disciplines, and the integration of knowledge and approaches, are becoming more and more relevant. There are now many examples of interdisciplinary studies: mathematicians and physicists work together in the search for a unified field theory to span space and time from the subatomic to the cosmological scales; in molecular biology, techniques such as recombinant DNA are being harnessed, combined with chemical engineering, to yield products on an industrial scale; in non-linear systems dynamics, topics such as dissipative systems, limit cycles, biological clocks, strange attractors, chaos, and turbulence are moving close together; in the nervous system, studies of the brain make use of cybernetics and biochemistry; and new subjects are being formed all the time. Computers are a common element in most of these developments: they enable us, by means of models, to examine the consequences of complexity. The increasing importance of modelling in plant science research can be regarded as a local manifestation of more global movements in research methodology.

Biology in general, and the plant sciences in particular, can be distinguished from the physical sciences by the complexity of their hierarchical structure. For instance, the plant scientist may be concerned with responses and mechanisms at the crop, plant, organ, tissue, or cell levels of structure. While research into (say) photosynthesis at the cell level may provide knowledge that is of value in its own right, that value may be greatly enhanced if the cell-level understanding can be related to plant and crop responses; for this purpose, a model is required. Mathematics provides a powerful means of integrating several items of knowledge at one level to describe responses at a higher organizational level. Such integration may be valuable for the purposes of research, leading to understanding, or for the purposes of crop management when the models have a practical predictive value. As these methods become standard equipment in the tool-kit of the plant scientist, they constitute a springboard, both enabling and constraining the next advances to be made. We are convinced that the integration of molecular genetics, biochemistry, physiology and whole-organism behaviour will be a major and productive area of science for a considerable period of time. The benefits of this enterprise are likely to be as inconceivable to us now, as today's state of the art was beyond the imagination of a visionary of 50 years ago.

This book is designed for possible self-study, since there is little teaching available in this area; it is suitable for research and advisory workers, graduates, and students in the plant sciences who are interested in these developments. While the main thrust of the book is concerned with the area of plant and crop physiology, the material covered in Part I is of general relevance to biological modelling, and similarly Chapters 12, 18, 19 may be more widely applicable. The level of mathematics required is mostly elementary, comprising calculus, differential equations, and some algebra; we have attempted not to take too much for granted, and to give step-by-step accounts of most topics (details may be covered in an exercise, and the exercises with the worked solutions are sometimes an important part of the text). Fortunately, the computer revolution has removed much of the need for the heroic mathematical and numerical analyses that were common last century and in the first half of this century. The emphasis is now firmly on the biological ideas, their mathematical expression, and the interpretation of the solutions in terms of mechanism; the numerical techniques of problem solution are often relatively straightforward.

The scope of the book is best revealed by the list of contents. We suggest that Part I (General topics) is required reading, as it is concerned with subjects of wide relevance to modelling; in Part II (Plant and crop physiology), the reader may wish to select topics of particular interest; Part III (Plant morphology) stands somewhat alone, but is included because we feel that it is only a matter of time before morphological features are included more explicitly in plant and crop models, as well as the problems themselves possessing considerable intrinsic interest. The book should not be regarded as a set of recipes which can simply be assembled when modelling a particular plant or crop, but rather as suggesting possible conceptual approaches which will almost certainly need modification and extension when applied to a particular problem. In constructing a model of a specific crop, it is essential that the submodels used are thoroughly understood and are appropriate. While we have tried to cover the important topics, inevitably our own experiences have affected the subjects chosen and the depth of treatment. It has not proved practical to define a single set of symbols for the whole book; however, this should not give rise to confusion. We have deliberately avoided long or exhaustive lists or references, but have attempted to include a few particularly helpful or in some way typical citations for each chapter.

We wish to thank John Prescott, former Director of the Institute for Grassland and Animal Production, for encouraging us to begin this project. It is not possible to list all our many friends and colleagues from whose knowledge and ideas we have benefited over the years; however, we are very conscious of the debt we owe and we express our gratitude for their contributions. The support of David Sweeney and his team in computing matters has been invaluable. The patient explanations by Tony Parsons of many physiological topics have been of great assistance, and we have also been much helped by discussions with Merv Ludlow. Some of the material we describe has been developed, in joint work or otherwise, by Roger Brugge and Jim France; we are indebted to them for allowing us to make use of their efforts, and also for many fruitful past collaborations. Graham Jones and Jeff Moorby have kindly read and commented on Chapter 12 and Chapter 19 respectively; their contribution has been much appreciated. We also thank Pam Berridge, who has typed and edited much of the manuscript. It would have been difficult to finish this book had one of us (JHMT) not been able to spend a year in the Department of Agronomy and Soil Science at the University of New England. This was made possible by the hospitality of John Lovett and his colleagues in Armidale, and generous support from the University of New England and the Stapledon Memorial Trust.

Ruskin said that anyone who expects perfection from a work of art knows nothing about works of art. Much the same can be said about works of science, which share more with works of art than is sometimes admitted. We ask readers to inform us of notable errors and omissions, and indeed to send us any other comments.

Hurley and Armidale J.H.M.T.
January 1989 I.R.J.

Contents

Part I General topics

Part I
General topics

1
Dynamic modelling

1.1 Introduction

Science is concerned with prediction. We can predict only by virtue of having models or conceptual schemes of the world. The models we use today are those conjectures that have best survived the unremitting criticism and scepticism that are an integral part of the scientific process. There are (it is assumed) no new processes under the sun, but our models of these processes have allowed mankind to transform its corner of the universe. Science is commonsense; it is also an unpredictable, fascinating, and thoroughly human activity. These are simple truths, self-evident to the practising scientist, but they need frequent repetition, especially at a time when the scientific community is increasingly assailed by politicians or other careerists who, while enjoying the fruits of science, would deny its methods (Thornley and Doyle 1984).

Plant and crop modelling has, broadly speaking, two aims. The first, and here we nail our flag clearly to the mast, is to increase knowledge in this area of science. Historically, increasing knowledge has led to unpredicted (in detail) and enormous benefits to the human race, and there is no reason why this should not continue. The second aim might be called 'applied' and 'strategic' in today's terminology, and is directed at the solution of currently perceived problems in the short to medium term. Given present agricultural and horticultural practice, one can envisage many uses for models of plant and crop growth, which could increase efficiency, improve the environment, and generally contribute positively to life. There is then, no difficulty in defending the practice of the techniques, ideas, and approaches which we are about to expound.

Scientific knowledge is not only about observational data, but also about having a theory (or hypothesis, or conceptual scheme, or model) that corresponds to the data. It is the continual interaction between hypothesis (how we think things work) and observational data (how they actually do work) that leads to progress. With the passage of time, measurements become more accurate and more extensive; similarly we are continually widening the scope of our theories and demanding more accurate predictions from them. When comparing theory with experiment, we attempt to connect the theory to nature at as many points as possible and as precisely as possible. As a branch of science progresses from the qualitative to the quantitative, one day it may be expected to reach the point where the connections between theory and experiment are most efficiently made using the language of mathematics. It is to be emphasized that the ideas and hypotheses of the theory are not contributed by mathematics. Mathematics is used as a tool or language, enabling biological scientists to express their ideas so that quantitative prediction is possible, and these predictions are then compared with observational data.

Agricultural and horticultural practice is based partly on tradition, partly on scientific knowledge, and partly on conjecture or guesswork. By tradition we mean an inherited folk-wisdom or set of customs where things are done because it is known that they work to a certain degree, but it is not understood why, or whether better results might be obtained by doing things a little differently. The formal knowledge of science can give a rationale for decision-taking when problems fall within the scope of current knowledge. Conjecture is also needed because, from time to time, a novel situation arises, there is no guidance from current knowledge on what to do, and yet a decision has to be taken. One purpose of agricultural research is to increase the knowledge-based component of agricultural decision-taking at the expense of the other two components. Increased knowledge does not necessarily lead to higher efficiency, but it may uncover more efficient options. With present agricultural practices, and other things remaining equal, the current efficiency of production provides a baseline from which it is only possible to move forwards. Equally important is the fact that increases in scientific knowledge allow a more rational response when other things do not remain equal, when the environment in which the farmer operates, natural or man-made (if this dubious distinction is permitted), changes.

Mathematical models can contribute to both of the aims discussed in the second paragraph of this section; that is, enlarging knowledge and helping with practical applications. Not only can models encapsulate knowledge, but, suitably programmed for the increasingly ubiquitous computer, they can also make this knowledge accessible to and usable by the non-expert. While the research worker delves ever more deeply into the minutiae and mechanisms of phenomena, it seems that technical developments will continue to make this detailed knowledge ever more easily available to the non-specialist farmer and farm adviser.

As with many things in this good life, there are models and models. Mostly, this book is concerned with dynamic deterministic models: dynamic models predict how a system unfolds with the passage of time—the time course of events; deterministic models make definite predictions (e.g. on 1 July the dry matter per unit area of the wheat crop will be 1 kg m^{-2}) without any associated probability distribution. Even dynamic deterministic models come in three types, demonstrating yet again the richness of science and the diversity of approaches possible. We call these types teleonomic, empirical, and mechanistic, although some would choose a different terminology. In terms of the organizational hierarchy of levels to be discussed later, teleonomic models look (mostly) upwards to higher levels, empirical models examine a single level, and mechanistic models look downwards, considering a level in relation to lower levels. Teleonomic models are sometimes called teleological or goal seeking. Empirical models belong to the category associated with curve fitting, regression, and applying mathematical formulae directly to observational data, usually without being constrained by scientific principles or any knowledge of mechanism. Mechanistic models are reductionist, concerned with mechanism, and integrative; they contribute understanding, and are sometimes called explanatory. In any given investigation, the objectives of the enterprise should determine what modelling approach, if any, can be used. It is therefore important to understand how these different types of model relate to each other and to the structure of the problem, and this is the main concern of the next section.

De Wit (1970) gives an excellent early account of concepts in the crop modelling area, which are also discussed by Thornley (1976, 1980).

1.2 Hierarchical systems

Biology, including plant biology, is notable for its many organizational levels. Whereas in physics and chemistry one travels more or less directly from atomic and molecular behaviour to that of liquids and solids, in biology there are several intervening organizational entities. It is the existence of the different levels of organization that gives rise to the great diversity of the biological world. For the plant sciences, a typical scheme for the hierarchy of organizational levels is as follows.

Level	Description of level	
...	...	
$i + 1$	crop	
i	plant	
$i - 1$	organs	
...	tissues	(1.1a)
...	cells	
...	organelles	
...	macromolecules	
...	molecules and atoms	

The levels that are of principal interest to this book are labelled $i + 1$, i, and $i - 1$. Using this diagram, we shall pinpoint the differences in viewpoint associated with the empirical, mechanistic, and teleonomic approaches to modelling, but first we discuss the principal properties of a hierarchical system.

Hierarchical systems have several important properties.

1. Each level has its own language, which is unique to that level. For example, the terms crop yield, leaf area, or whole-plant dry mass have little meaning at the cell or organelle levels.
2. Each level is an integration of items from lower levels. The response of the system at level i can be related to the responses at lower levels. This is scientific reductionism, and leads to mechanistic models.
3. Successful operation of a given level requires lower levels to function properly, but not vice versa. For example, if a cup is smashed to small pieces, it will no longer function as a cup, although the molecular interactions are hardly altered.

4. The higher levels provide the constraints, boundary values, and driving functions, including any inputs and outputs, to the lower levels.
5. On descending to a lower level, generally both the spatial and temporal scales become smaller; this corresponds to smaller physical size and to faster processes at the lower levels.

1.2.1 *Empirical models*

Empirical models are essentially direct descriptions of observational data, but they can, none the less, be exceedingly useful. The well-known saying 'red sky at night, shepherd's delight' has helped in the planning of many harvesting activities and family picnics. The tables that describe the tides round our coastlines are constructed by totally empirical methods. In an empirical model, any mathematical relationships that are written down are usually unconstrained by physical laws such as that of energy conservation or the laws of thermodynamics, by biological information, or by any knowledge of the structure of the system. The empirical modeller attempts to describe level i behaviour (observational data) in terms of level i attributes alone, without regard to any biological theory. The approach is primarily one of examining the data, deciding on an equation or set of equations, and fitting these to the data. Essentially, an empirical model re-represents the data, perhaps more conveniently, and no new information is acquired.

In Fig. 1.1 a simple example of an empirical mathematical model is given. The observational data shown give the response of crop yield Y to the level of fertilizer N applied. Often such data can be fitted by a three-parameter rectangular hyperbola

$$Y = Y_{\max} \frac{(N_s + N)}{K + (N_s + N)} \tag{1.1b}$$

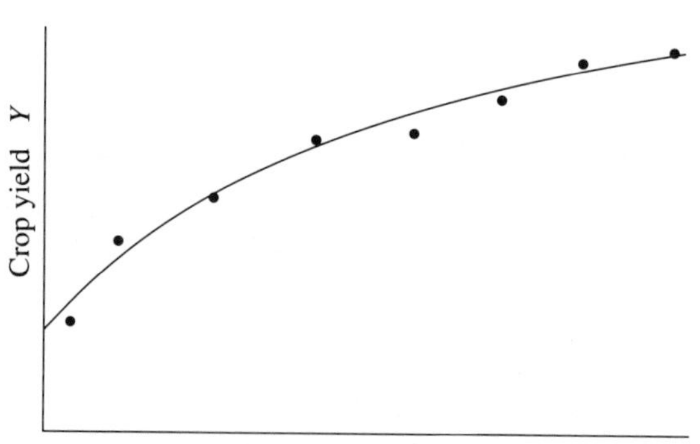

Fig. 1.1. A simple empirical model of the response of crop yield Y to nitrogen fertilizer level N. ●, observational data; —, eqn (1.1b).

where Y_{max} is the maximum value of yield Y at high N values, N_s is the effective residual soil nitrogen which gives a yield at $N = 0$, and K is a parameter determining the initial slope of the $Y : N$ response at $N_s + N = 0$. Also shown in Fig. 1.1 is the curve obtained by fitting eqn (1.1b) to the observational data and adjusting the parameters.

The important point to recognize is that the fitted curve (or empirical model) contains no information beyond the original data; it says nothing about the mechanisms that give rise to the response (for which a mechanistic model of the system would be needed), or why the response is as it is in terms of possible goal-oriented behaviour of the plant (we use 'why?' in the 'for what purpose?' sense); for the latter a teleonomic model would be appropriate. Empirical models can provide a powerful means of summarizing data and interpolation, and may provide a practical tool for the farmer or farm adviser. Traditionally much agricultural research has been of a descriptive nature, and a great deal of essential groundwork has been done using empirical models. However, our thesis is that the needs of the subject are changing, and it is now timely to seek mechanistic explanations and an understanding of these responses by means of models that integrate the underlying mechanisms.

1.2.2 *Mechanistic models*

The main concern of this book is with dynamic deterministic models that are concerned with mechanism and can lead to an understanding of the ith level (1.1a) that is based on component processes at the $(i - 1)$th level, and possibly at lower levels. The mechanistic modeller attempts to construct a description of level i behaviour which has some extra content of mechanism, understanding or explanation at lower levels. Mechanistic modelling is 'hard' science, and it follows the traditional reductionist method that has been so very successful in the physical sciences, molecular biology, and biochemistry. As shown in Fig. 1.2, in contrast with the empirical modeller who proceeds directly to the whole-plant variables that are of interest and may connect these in whatever way seems best to fit the data, the mechanistic modeller goes round a relatively circuitous route: under analysis and reduction he breaks the system down into components and assigns processes and properties to these components; this introduces extra variables at the $(i - 1)$th level, and additional observational data are generally also available at the $(i - 1)$th level; finally, it is by the integration of the set of equations that define the system that the responses at the whole-plant level are synthesized.

A mechanistic model of responses at a certain hierarchical level is always far more complex than an empirical model; it will generally fit the data at the ith level less well because it has many constraints built into its structure by means of the assumptions of the model. However, its content is far richer in that it applies to a greater range of phenomena, relating them to each other. Because

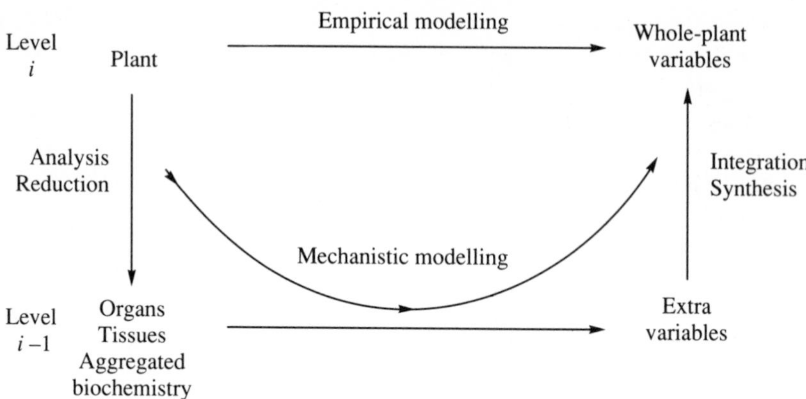

Fig. 1.2. Mechanistic and empirical modelling.

of this, a mechanistic model always offers more possibilities for manipulating and improving the system.

Some have questioned whether integration of knowledge at the $(i - 1)$th level can indeed give rise to new knowledge at the ith level. Is the whole more than the sum of its parts? Are there 'emergent' properties? There is, in our view, no doubt about the answer to this question. Shakespeare was more creative than the proverbial monkey at a typewriter because of the way in which he assembled familiar words and phrases into new patterns. The same is true in music. In the sciences, the juxtaposition of well-known components to give rise to new results (or 'emergent' properties) has occurred again and again. The mathematician Turing showed that diffusion combined with chemical reaction could generate patterns—a result that has been most valuable in studies of morphogenesis (Chapter 19). The kinetic theory of gases with all its ramifications is a consequence of a few very simple assumptions at the molecular level. The theories of the atmosphere are becoming steadily more successful and yet, at base level, nothing has changed. The special theory of relativity is a consequence of attributing a constant velocity of light to all vacuum reference frames—this seems innocuous enough, and yet the results of the ensuing analysis are stupendous. The whole is more than the sum of the parts, and yet it is explainable in terms of the parts and how they interact. The post-war progress in molecular biology and biochemistry indicates that a reductionist programme is applicable to all areas of biology, even including value systems and religion (Monod 1972). An explanation of the responses, behaviour, and mechanisms of an organism does not mean that the particular combination of mechanisms that gives rise to the observed responses and behaviour could have been predicted. Even with deterministic equations, the prediction of a detailed time course may be totally unreliable, owing for instance to sensitivity to initial conditions, as with chaotic systems. The range of future possibilities may be so great that the particular time

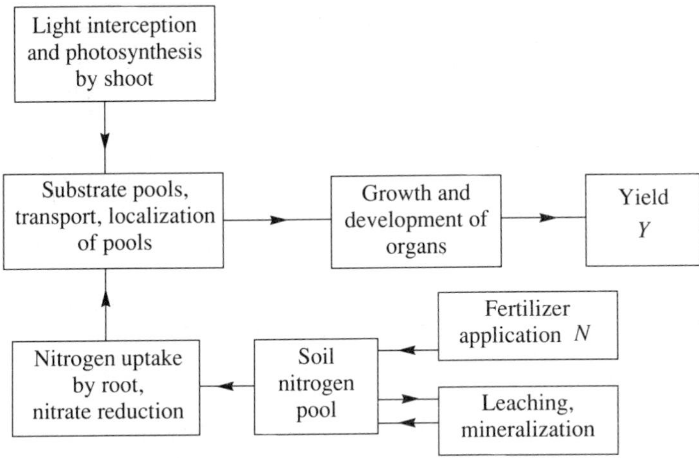

Fig. 1.3. Elements of a possible mechanistic model of response of crop yield Y to fertilizer level N.

course followed is unpredictable, and yet it is explainable in terms of established mechanisms.

In Fig. 1.1 and eqn (1.1b), an elementary empirical model of crop yield response to fertilizer is given. In Fig. 1.3, a possible mechanistic approach to this problem is shown. There are usually several ways in which a system might be analysed into components. There are no hard and fast rules. One attempts to incorporate the 'important' components in the model, to make 'reasonable' assumptions, and to strike the right balance between complexity, tractability, and realism commensurate with the overall objectives. If possible, elegance is also part of the structure. Modelling is partly an art form, although hopefully the end result is of scientific value. Metaphysics has been the precursor to much good science. Conjecture (or guesswork) is needed because there are many areas of ignorance. Individual scientists will always assign different weights to the various ingredients in the mix; hence the variety of crop models that exists. In biology, most models are simplifications and are therefore wrong at some level of detail (this is less true of models in physics and chemistry). Therefore critics can always legitimize their reasons for rejecting a model. A model should be criticized mostly in relation to the stated objectives. The strengths of the model, which should enable the modelling objectives to be met, ought to be emphasized more than the weaknesses, which should not detract appreciably from the model's performance in relation to the stated modelling objectives. However, a different investigator with different objectives might find those weaknesses unacceptable.

Frequently, people will talk about the 'complexity' of a model, and yet there is no consensus about what is meant by this term. Possible measures of complexity might be the number of state variables (p. 15), the number of parameters, the topology of the system diagram (as in Fig. 1.3) with perhaps the number of closed loops or cycles, the level of mathematics required (partial differential equations, perturbation theory or topology may be regarded as 'complex' or 'difficult'), or even the computing power needed to generate solutions. Generally, in a 'good' model, only significant parameters are retained—those that have an appreciable effect on the solutions or the scope of the model. Thus we choose the number of parameters as the best measure of model complexity.

Sometimes, it is said that a situation is 'too complex' to model, or that not enough is known about the system to build a model. Yet often, we model because it is the only way of grappling with complexity, or we model in order to define what we do not know about the problem. Complex situations with uncertain outcomes, which are common in agriculture and biology, are ripe for modelling.

Models can also be used to calculate physiologically meaningful parameters which are not accessible to direct observation. For example, plant and root respiration parameters can be obtained by analysis (Chapter 11, Sections 11.3 and 11.5). In physics and chemistry, there is a well-established tradition of finding quantities such as the mass of the electron, or the activation energies of reactions, by indirect methods using a model.

Let us assume we have a model (empirical or mechanistic) with n parameters. The model makes predictions for the m values of some observable quantity Y, Y_i, $i = 1, 2, \ldots, m$. These predictions can be compared with the corresponding observational data y_i, $i = 1, 2, \ldots, m$. Regarding the n parameters as all adjustable and as having been adjusted to give the best fit to the observational data, the estimated mean square prediction error (MSPE) can be written as

$$\text{MSPE} = \sum_{i=1}^{m} \frac{(y_i - Y_i)^2}{m - n} \tag{1.2a}$$

Figure 1.4 shows schematically how we might expect the MSPE for empirical and mechanistic models to vary with the complexity parameter n. The empirical model (EM) is applied to a given set of observational data at the organizational level i (1.1a); as the complexity n of the model (as measured by the number of parameters) is increased, the observational data base m remains the same. Eventually the data are being over-fitted, and the MSPE, after initially decreasing, increases again. The situation with the mechanistic model (MM) is rather different. To begin with, for low numbers of parameters and with roughly the same data base, MM will always give a worse fit because it is constrained by the structure and assumptions of the model, whereas EM is 'free'. As model complexity increases (with increasing n), the observational data base increases; there are data at the level of the assumptions of the model, and the model can now be tested at the level of the assumptions ($i - 1$) as well as at the level of its predictions (i). Thus the MSPE for MM can decrease with increasing complexity, but may now conceivably approach an asymptote. It may be argued that in comparing empirical with mechanistic models in this way, like is not being compared with like, and this is correct since the scope of the mechanistic model is allowed to

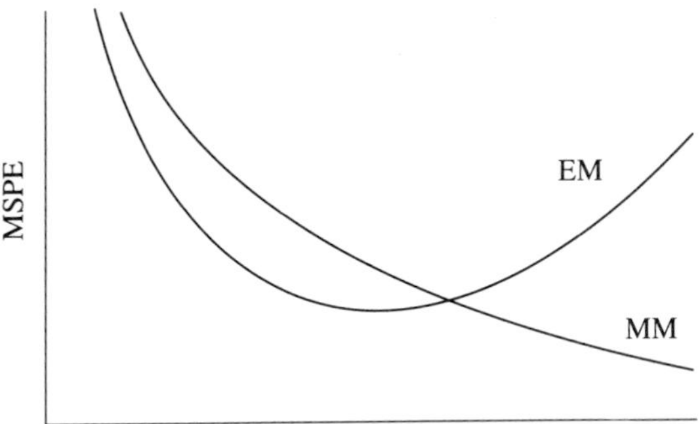

Fig. 1.4. Relative predictive ability of empirical (EM) and mechanistic (MM) models. The hypothetical dependence of the mean square prediction error (MSPE) on model complexity, represented by the number of parameters n, is shown.

increase with increasing complexity, whereas this is not the case for the empirical model. However, this is generally the manner in which empirical and mechanistic models are developed and applied.

1.2.3 *Teleonomic models*

From time to time in this book, a third class of model is used or referred to—teleonomic models. Such models may be relatively unfamiliar to the plant scientist, and although they only play a minor role currently in plant and crop modelling, this could expand, and it is important that the teleonomic viewpoint be appreciated and understood (Monod 1972) including the relationship between teleonomic models, empirical models, and mechanistic models.

Teleonomic models are applicable to goal-directed behaviour, and are formulated explicitly in terms of goals. These models may refer to a single level, say the ith level; they may provide a teleonomic interpretation of an otherwise empirical model; they may also refer responses at the ith level to the constraints provided by the $(i + 1)$th level (p. 5). The goals of level i can be viewed as the requirements imposed by level $i + 1$. It is the higher-level constraints which, via evolutionary pressures, can select out combinations of the lower-level mechanisms of biochemistry, which may lead to apparently goal-directed behaviour at level i. We give two examples of possible teleonomy in plant growth and development (see also Chapter 7, Section 7.6, p. 193).

Dry-matter partitioning is an aspect of most plant and crop growth models, and it can be approached by the modeller in various ways. For root : shoot partitioning all three kinds of model exist: empirical, teleonomic and mechanistic. Chapter 13 considers the topic in detail, and here we give a brief qualitative discussion to illustrate general modelling issues. The empirical approach is to measure (say) root : shoot partitioning for a variety of situations of both environment and plant. These observational data provide the basis for an empirical model of root : shoot partitioning. However, there have been various attempts to interpret these measured responses in terms of goal-directed behaviour of the plant. An early suggestion was that the root : shoot ratio adjusts so as to maintain a constant carbon : nitrogen ratio in the plant. Another suggestion is that the root : shoot ratio and the carbon : nitrogen ratio both adjust so that they are proportional to each other. The observational data are not of sufficient quality to discriminate unambiguously between these alternative teleonomic interpretations, but nevertheless such goal-directed models have been incorporated, with a degree of success, into some crop-growth models. However, we can go beyond a direct teleonomic interpretation of observational data, and attempt to construct a teleonomic model of root : shoot partitioning by hypothesis. For example, it can be assumed that plants have evolved controls so that, when the plant is faced with a given environment, new material is allocated to the shoot and to the root in such a way that the plant attains an optimal specific growth rate in that given environment. This then leads to a teleonomic partitioning model. A goal-oriented model at the $(i - 1)$th level (the organ level) has been constructed by assuming that, at the ith level (the plant level), the plant has certain 'requirements' such as optimizing the plant specific growth rate; it is of course assumed that the diversity of possible mechanisms at the lower levels is able to satisfy these requirements. The teleonomic model can be written directly in terms of the lower-level mechanisms, giving a mechanistic model which is more complicated than the teleonomic representation but is 'objective' in its operation (as are the laws of physics and chemistry).

Phyllotaxis is the second example we consider (Chapter 19). It is especially interesting because it appears to be a case where mechanistic models of the problem and teleonomic models deduced from the 'requirements' of the plant both lead to Fibonacci arrangements for leaves around a stem or for seeds in a flower-head. It is an observational fact that leaves in plants and seeds in flower-heads are often positioned so that they are separated from adjacent members by the Fibonacci angle. The teleonomic view of this is as follows. It is assumed that the plant 'requires' optimum interception of light from the azimuth by its leaves; this assumption leads to the Fibonacci angle. Also, if it is assumed that the plant 'requires' optimum packing of seeds in flower-heads (thus maximizing the number of seeds), an analysis again leads to the Fibonacci angle. Thus, in this case, the Fibonacci model is derivable from an assumption about the goals of the plant. There are two mechanistic approaches to modelling phyllotaxis—giving rise to models that in both cases generate successive primordia (leaves or seeds). One is a 'contact' theory, and works with the 'next available space'; the second is a diffusion–reaction model involving a postulated morphogen. Both are able to generate Fibonacci patterns of primordia. Thus, in this case, mechanistic considerations lead to patterns that also satisfy teleonomic requirements.

1.3 Research models and applications models

The question that is addressed in this section is this: are there important differences between models constructed for an applications purpose and models used in research? The term 'application' means here application in agricultural production. Models of plant and crop growth are, potentially at least, capable of being used by farmers, farm advisers, horticulturalists, and even enthusiastic amateurs. A good crop model that encompasses response to environment and also controllable inputs such as fertilizer levels and irrigation could provide a valuable tool for crop management.

The answer to the question posed in the first sentence above is an emphatic 'yes'. Although in time research models may evolve into applications models, when they achieve the latter state, they may cease to be of value in research. With a research model, aimed at improving understanding, much progress can be made with a model that fails to describe correctly what is happening. Part of the value (and the enjoyment) of modelling is that it is possible to speculate inexpensively and try out various ideas. Sometimes, after a perhaps long and arduous progression round and round the cycle of conjecture, prediction, and observation, one arrives at a set of hypotheses that are compatible with current observational data.

However, if an applications model is to be successful (which means being used), it must give predictions and guidance that are demonstrably better, in some way, than existing practice. All models, whether oriented towards research or applications, are based on a mix of observational data, currently accepted knowledge, and conjecture. While it may be desirable for a research model to have a high proportion of conjecture, an applications model must clearly be as firmly founded as possible on data and knowledge that are relatively secure. If an applications model were to fail, this could put into jeopardy the livelihood of many people. The failure of a research model has no such consequences.

Table 1.1 summarizes the principal differences that usually exist between research models and applications models. Since empirical models are related directly to observational data, and are often mathematically and computationally simple, applications models for farm management purposes frequently have a high content of empiricism. As increasing user sophistication demands a wider range of applicability and scope of these models, it may well be that more mechanistic models prove to be the best way forward, although such models must always be carefully evaluated under the conditions where they might be applied. Sheer empiricism can require almost infinite experimental resources in complex input-output situations. Also, empiricism on the ith level alone may be very inefficient in ignoring the ever-widening base of physiological and biochemical knowledge at the lower levels. The possible benefits of application to the problems of agriculture and the environment provide a good reason for continuing with basic and strategic research. While crop management may continue to use refined empirical models for some years, we believe that con-

Table 1.1. Research models and applications models: a comparison of the principal differences

	Research	Applications
Hypotheses	Speculative	Well-accepted
Connections to observational data	Tenuous (often)	Good
Accuracy of predictions	Variable	Good
Scope	Wide	Restricted
Complexity	Complex	Simple
Type	Mechanistic	Empirical

tinued progress in the area depends on pursuing a long-term strategy of developing more mechanistic models, although such models may often be simplified for practical application. A mechanistic model must always uncover more options than an empirical model. Until such models have been successfully constructed and evaluated, there will remain areas of ignorance that neither science nor the practitioners of agriculture can afford to leave undisturbed.

1.4 Mathematical models: objectives and contributions

There are many roles for modelling work in biology and in the plant and crop sciences. When undertaking a modelling project, the most important need is for a clearly defined and realistic set of objectives. It would be wasteful to construct a complex mechanistic model where an empirical approach would better meet the requirements. Since objectives are linked to potential contributions, we list some of the possible objectives and potential benefits associated with mathematical modelling. The abbreviations EM for empirical model and MM for mechanistic model are used.

1. Hypotheses expressed in mathematical terms can provide a quantitative description and mechanistic understanding of a biological system (MM).
2. A model requires a completely defined conceptual framework, and this may pinpoint areas where knowledge is lacking, and perhaps stimulate new ideas and experimental approaches (MM).
3. A mathematical model, especially if implemented in an easy-to-use computer program, may provide an excellent recipe by which recent research knowledge is made available to the farm manager or adviser (EM mostly, some MM).
4. Agro-economic models may highlight the benefits of new crop management techniques suggested by recent research, thereby stimulating the adoption of more efficient production methods (MM, EM).
5. Modelling may lead to less *ad hoc* experimentation, as models may make it possible to design experiments to answer particular questions, or to discriminate between alternative mechanisms (MM, EM).

6. In a system with several components, a model provides a means of bringing together knowledge about the parts, giving an integrated view of whole-system behaviour (MM).
7. Modelling can provide strategic and tactical support to a research programme, motivating scientists and encouraging collaboration (MM).
8. A model may provide a powerful means of summarizing data, and also a method for interpolation and cautious extrapolation (EM mostly, MM).
9. Observational data are becoming more precise, but also more expensive to obtain; a mathematical model may be able to make more complete use of such data (EM, MM).
10. The predictive power of a successful model can be used in many ways. For instance a model can be used to answer 'what if...?' questions. What are the consequences on crop production of halving the maintenance requirements of plant tissue? What are the effects on crop yields of changing within-plant transport resistances? However, it should be remembered that the answers given by a model are, in a sense, built into it by hypothesis. Thus a model can be used to stimulate thought, but it may be dangerous to use a model to manage a research and development programme (MM).

Any given model is only likely to contribute under two or three of these ten points. However, this list indicates the many possible reasons for undertaking modelling work.

1.5 Deterministic dynamic differential equation models

It is assumed that the state of the system under investigation at time t is defined by q variables $X_1, X_2, \ldots, X_q$; these q variables are called state variables. The q state variables are independent; that is, it is not possible to derive one of the state variables, X_1 say, from a knowledge of the values of the other state variables. The state variables represent properties or attributes of the system being considered (such as dry matter, number of cells, leaf area, starch content, etc.). The choice of state variables is the first and most important assumption that the modeller makes. The scope of the model is defined by its state variables.

The next step is to construct the q first-order ordinary differential equations that describe how the q state variables change with time t. These can be written formally as

$$\frac{\mathrm{d}X_1}{\mathrm{d}t} = f_1(X_1, X_2, \ldots, X_q; P; E)$$

$$\frac{\mathrm{d}X_2}{\mathrm{d}t} = f_2(X_1, X_2, \ldots, X_q; P; E) \qquad (1.3a)$$

$$\frac{\mathrm{d}X_q}{\mathrm{d}t} = f_q(X_1, X_2, \ldots, X_q; P; E).$$

The $f_1, f_2, \ldots, f_q$ denote functions of the state variables, of a number of parameters which are indicated by P, and of environmental quantities denoted by E. Equations (1.3a) are called 'rate-state' equations, and they show how the rates of change of the system state variables depend explicitly on the current values of the state variables. In plant and crop models it is helpful to indicate explicitly the presence of parameters and environmental quantities in eqns (1.3a) by P and E.

Writing $f_1(X_1, X_2, \ldots, X_q; P; E)$ does not mean that the function f_1 must contain all the state variables, parameters, and environmental quantities; the rates of change of most state variables will not depend directly upon the environment, and it would be usual for the rate of change of a given state variable to depend only upon two or three other state variables. Denoting a single state variable by X, we can write its rate-state equation as

$$\frac{\mathrm{d}X}{\mathrm{d}t} = \text{inputs} - \text{outputs}; \tag{1.3b}$$

the inputs are the terms that contribute positively to the rate of change of X, and the outputs contribute negatively. Each term on the right-hand side of eqn (1.3b) gives the rate of a process. In the plant sciences, all processes are either transport or chemical conversion. For example, eqn (1.3b) could take the form

$$\frac{\mathrm{d}X}{\mathrm{d}t} = c - \frac{kX}{K + X} + g(a - X); \tag{1.3c}$$

c stands for a constant production of X, the second term on the right for a Michaelis–Menten-like utilization of X (Chapter 2, p. 51) where k and K are parameters, and the last term stands for a diffusive-like transport of X (Chapter 4, p. 95) where g and a are parameters.

1.5.1 *Explicit time dependence*

Suppose that one of the rate-state equations (1.3a) takes the form

$$\frac{\mathrm{d}X}{\mathrm{d}t} = aX(b - t) \tag{1.4a}$$

where a and b are constants. Here, the time variable t appears explicitly on the right-hand side of the equation. In eqns (1.3a) as written the time variable t does not appear explicitly on the right-hand side of the equation (excluding its possible appearance in the environmental specifications in E). Generally speaking, it is not sound scientific methodology to include t directly in the rate-state equations. The system is completely specified by the set of state variables and does not 'know' what the time is, although one or more state variables may effectively be keeping track of time. However, the elimination of one or more state variables may lead to a reduced set of rate-state equations with explicit time dependence; thus explicit time dependence can be viewed as representing hidden state variables. Consequently, although the use of explicit time dependence in the rate-state equations can be viewed as a device that is sometimes convenient and can usefully simplify the model, it may impose an undesirable external constraint on the dynamics of the system.

From eqn (1.4a), we could expand the set of state variables by defining a second state variable Y by

$$Y = b - t, \tag{1.4b}$$

from which t is easily eliminated by a single differentiation, giving

$$\frac{dY}{dt} = -1 \qquad \text{with } Y = b \text{ at } t = 0. \tag{1.4c}$$

The time-independent rate-state equations are now

$$\frac{dX}{dt} = aXY \qquad \text{and} \qquad \frac{dY}{dt} = -1. \tag{1.4d}$$

Equation (1.4a) can be integrated quite easily, since the variables are separable, to give

$$X = X_0 \exp\left[\frac{at(2b - t)}{2}\right] \tag{1.4e}$$

where $X = X_0$ at $t = 0$. This function is known as the exponential quadratic (Chapter 2, eqn (2.15d), Fig. 2.12a, pp. 69, 70; Chapter 3, eqn (3.7d), p. 87), and is sometimes employed in plant-growth analysis. As used in plant-growth analysis, eqns (1.4a) and (1.4e) are rightly regarded as 'empirical' or 'curve-fitting' approaches. However, eqns (1.4d) may permit a more mechanistic biological interpretation (namely, that there is some component of the growth machinery that is steadily being depleted (see Chapter 2, Fig. 2.10, p. 66)).

The Gompertz growth equation in differential form is

$$\frac{dW}{dt} = \mu_0 We^{-Dt} \qquad \text{with } W = W_0 \text{ at } t = 0, \tag{1.4f}$$

where W is the state variable (denoting dry matter) with initial value W_0, and μ_0 and D are constants (Chapter 3, eqn (3.5d), p. 80, and Fig. 3.4). The problem of putting eqn (1.4f) into a two-state-variable formulation without explicit time dependence is posed in Exercise 1.1.

1.5.2 *Memory functions and delays*

Consider a two-state-variable problem with state variables X and Y, as in eqns (1.4d). The differential equation for X can be written

$$\frac{dX}{dt} = f_1(X, Y), \tag{1.5a}$$

where f_1 denotes some function. Sometimes we wish to make the current value of $\mathrm{d}X/\mathrm{d}t$ dependent upon the value of a variable at some time in the past, say a time period τ ago. This remembered variable could itself be X affecting $\mathrm{d}X/\mathrm{d}t$, so that eqn (1.5a) becomes (we use f_i to denote a function)

$$\frac{\mathrm{d}X}{\mathrm{d}t}(t) = f_2[X(t), Y(t), X(t-\tau)]. \tag{1.5b}$$

Alternatively, the remembered variable affecting $\mathrm{d}X/\mathrm{d}t$ might be Y, giving

$$\frac{\mathrm{d}X}{\mathrm{d}t}(t) = f_3[X(t), Y(t), Y(t-\tau)]. \tag{1.5c}$$

For either of the above equations, we can denote the remembered variable as Z, which is computed from past values of one or more of the state variables of the system, and eqns (1.5b) and (1.5c) can be written simply as (omitting Y)

$$\frac{\mathrm{d}X}{\mathrm{d}t} = f(X, Z). \tag{1.5d}$$

The remembered variable Z is in general obtained by an integral computed over the entire past history of the system up to the present:

$$Z = \int_0^t g(X)\,\mathrm{d}t', \tag{1.5e}$$

where g denotes the memory function and t' is a dummy time variable.

The Dirac delta function $\delta(x)$, where x is a variable, can be thought of as a very sharp spike of unit area at the origin $x = 0$, which is zero everywhere else. It is defined by the equation (x' is a dummy variable)

$$f(x) = \int f(x')\delta(x - x')\,\mathrm{d}x'. \tag{1.5f}$$

The delta function $\delta(x)$ just selects out the value of f where the spike occurs, i.e. at $x' = x$. When the Dirac delta function is used, the discrete memory function which remembers the value of state variable X at time period τ ago is given by

$$Z = X(t - \tau) = \int_0^t \delta(t - \tau - t')X(t')\,\mathrm{d}t'. \tag{1.5g}$$

Instead of (and perhaps more realistic than) remembering a value of X at some instant in the past, one might remember some weighted average of past values. Equation (1.5g) is replaced by

$$Z(t) = \int_0^t w(t - t')X(t')\,\mathrm{d}t', \tag{1.5h}$$

where w is a weighting function. (See Exercise 1.2.)

In the last section, in the discussion of explicit time dependence, it was said that the system cannot 'know' what the time is, and a representation with more state variables can always be found which does not have an explicit time dependence. It is similar here for 'remembered' variables. The only way in which a system remembers a variable is through other variables, and so, again, a representation can always be constructed which does not have remembered variables but which may be considerably more complicated. For example, eqn (1.5g) can be represented by a large number of intermediate variables I_i, as in the scheme

$$X \rightarrow I_1 \rightarrow I_2 \rightarrow \cdots \rightarrow Z. \tag{1.5i}$$

This compartmental scheme, which can be regarded as a delay line or a pipe with plug flow, can be equivalent to a discrete delay (see Chapter 7, Section 7.3.4, p. 177, and eqn (7.7d) *et seq.*).

1.6 Numerical integration

For all but the very simplest models, eqns (1.3a) can only be integrated numerically. This requires specification of the initial values X_i at time $t = 0$, the parameters P and the environment E; the results of the numerical integration are the values of the state variables X_i at any subsequent time t. Computing technology has transformed numerical methods and now provides the facility to solve complex problems. There are many excellent textbooks on the subject, and here we introduce a few of the simpler methods and point out some of the pitfalls.

We begin by simplifying eqns (1.3a) and considering the single equation

$$\frac{\mathrm{d}x}{\mathrm{d}t} = f(x, t) \tag{1.6a}$$

with a single state variable x; f denotes a function of x and time t, and the parameters P and environment E are not shown explicitly. The process of numerical integration can be shown as

$$x(t) \rightarrow x(t + \Delta t). \tag{1.6b}$$

Given the value of x at time t, and given eqn (1.6a) so that the rate of change x can be calculated, how do we calculate the value of x at time $t + \Delta t$, where Δt denotes an increment in the time variable? A prescription of the type of (1.6b) can be applied iteratively, starting from time $t = 0$ and proceeding to any time t.

1.6.1 *Euler's method*

This is the simplest method of all. It is called a first-order method because it takes account of terms of order Δt, and the error is of order $(\Delta t)^2$. The rate of change of the state variable x at time t is by definition given by

$$\frac{\mathrm{d}x}{\mathrm{d}t} = \lim_{\Delta t \to 0} \left[\frac{x(t + \Delta t) - x(t)}{\Delta t} \right]. \tag{1.7a}$$

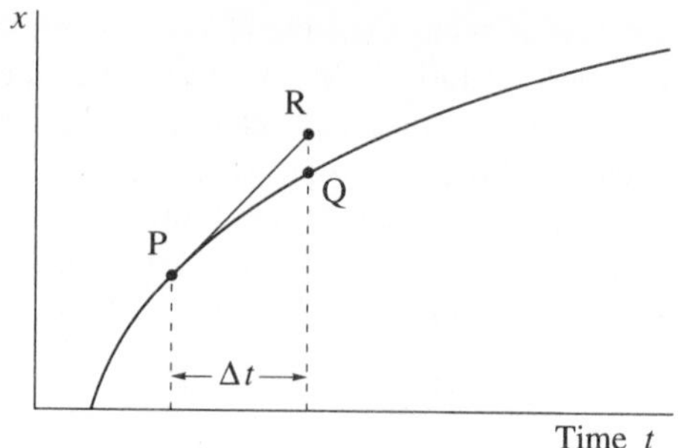

Fig. 1.5. Numerical integration by Euler's method. The curve shows the true solution to eqn (1.6a) passing through P and Q. PR is the tangent to the true solution at point P. R represents the solution predicted by Euler's method for time $t + \Delta t$, which is in error by RQ.

Substituting eqn (1.6a) into (1.7a) gives

$$f[x(t), t] = \lim_{\Delta t \to 0} \left[\frac{x(t + \Delta t) - x(t)}{\Delta t} \right]. \tag{1.7b}$$

For small values of Δt, it is approximately true that

$$f[x(t), t] = \frac{x(t + \Delta t) - x(t)}{\Delta t}. \tag{1.7c}$$

Rearranging this equation gives Euler's formula

$$x(t + \Delta t) = x(t) + \Delta t\, f[x(t), t]. \tag{1.7d}$$

An illustration of the operation of Euler's method over a single interval Δt is given in Fig. 1.5. Given the value of x at time $t = 0$, this equation can be applied iteratively to give values for x at times $t = \Delta t, 2\Delta t, 3\Delta t, \ldots$. (See Exercise 1.3.)

A first-order method such as Euler's method gives a second-order error of order $(\Delta t)^2$. This can be demonstrated by considering the Taylor series expansion of $x(t + \Delta t)$. This is

$$x(t + \Delta t) = x(t) + \Delta t \frac{\mathrm{d}x}{\mathrm{d}t} + \frac{1}{2!}(\Delta t)^2 \frac{\mathrm{d}^2 x}{\mathrm{d}t^2} + \frac{1}{3!}(\Delta t)^3 \frac{\mathrm{d}^3 x}{\mathrm{d}t^3} + \cdots. \tag{1.7e}$$

On comparing eqn (1.7e) with eqn (1.7d) and noting that $f[x(t), t] = \mathrm{d}x/\mathrm{d}t$, the error ε in Euler's formula (eqn (1.7d)) is given by

$$\varepsilon = \frac{1}{2}(\Delta t)^2 \frac{\mathrm{d}^2 x}{\mathrm{d}t^2} + \text{higher-order terms}. \tag{1.7f}$$

This error is known as the truncation error. For a straight line, the second and higher derivatives are zero and Euler's formula gives an exact result. Halving the integration interval Δt reduces the error in eqn (1.7f) by a factor of 4. However, we now need to take twice as many steps to cover the same period of time, so that the resultant error per unit of time is reduced by 2.

Despite the availability of more sophisticated methods, Euler's method is of great practical value. Because of its simplicity, it is easily checked out in complete detail, and one can feel sure of just what is going on in a calculation.

Oscillations as an artefact of integration Using Euler's method, we next show how instability and oscillations can arise. Assume that eqn (1.6a) takes the form

$$\frac{dx}{dt} = -kx \qquad \text{with } x = 1 \text{ at } t = 0. \tag{1.7g}$$

k is a constant. Integration gives the exact analytical solution

$$x = e^{-kt}. \tag{1.7h}$$

Next we generate an approximate numerical solution to eqn (1.7g) using Euler's formula of eqn (1.7d) with $k = \frac{1}{2}$ and $\Delta t = 1$, to give

t	0	1	2	3	4	5	....
x	1	1/2	1/4	1/8	1/16	1/32	....

(1.7i)

However, now assume that $k = 3$. Applying eqn (1.7d) gives

t	0	1	2	3	4	5	....
x	1	−2	4	−8	16	−32	....

(1.7j)

For eqn (1.7g) it can be shown that

$k\,\Delta t < 1$ gives asymptotic stability

$1 < k\,\Delta t < 2$ gives oscillations of decreasing amplitude (1.7k)

$2 < k\,\Delta t$ gives oscillations of increasing amplitude.

With smaller values of Δt, the stability of the numerical solution is increased; the truncation error of eqn (1.7f) is also reduced. However, computing time increases because more iterations are required and there is an increased possibility of rounding errors which result from the fact that the computer can only store values to a certain number of decimal digits. The Δt used is a compromise between these conflicting criteria (Section 1.6.6, p. 30). Usually, to reduce the truncation error, it is more efficient to use a higher-order method than to reduce Δt. However, some problems are not easily solved with higher-order methods, as these may require the calculation of the derivatives within the time interval Δt.

1.6.2 *Trapezoidal method*

Euler's formula of eqn (1.7d) can be regarded as the first two terms of the Taylor series of eqn (1.7e). Occasionally, it is possible to calculate higher derivatives analytically and use these directly in the Taylor series to obtain a better approximation. Thus, given

$$\frac{dx}{dx} = f(x, t), \tag{1.8a}$$

by analytic differentiation the second derivative is

$$\frac{d^2x}{dt^2} = g(x, t), \tag{1.8b}$$

and then the first three terms of the Taylor series of eqn (1.7e) are calculable, giving

$$x(t + \Delta t) = x(t) + \Delta t\, f(x, t) + \tfrac{1}{2}(\Delta t)^2 g(x, t). \tag{1.8c}$$

This second-order equation has a third-order error which is the fourth term on the right-hand side of eqn (1.7e).

More usually, it is not practical to calculate analytically using algebra the higher derivatives of the Taylor series expansion of eqn (1.7e). Essentially, the higher-order numerical integration methods operate by evaluating the first derivatives using eqn (1.8a) at different points within the integration interval, and the higher derivatives are obtained from the way in which the first derivatives change. We consider here the simplest of the second-order methods, which is known as the trapezoidal method.

Consider the second derivatives as defined by differentiating eqn (1.7a) to give (with $\dot{x} \equiv dx/dt$)

$$\frac{d^2x}{dt^2} = \lim_{\Delta t \to 0} \left[\frac{\dot{x}(t + \Delta t) - \dot{x}(t)}{\Delta t}\right]. \tag{1.8d}$$

For small Δt, therefore,

$$\frac{d^2x}{dt^2} = \frac{\dot{x}(t + \Delta t) - \dot{x}(t)}{\Delta t}. \tag{1.8e}$$

As $f(x, t) = dx/dt$, eqn (1.8e) can be written

$$\frac{d^2x}{dt^2} = \frac{f[x(t + \Delta t), t + \Delta t] - f(x, t)}{\Delta t}. \tag{1.8f}$$

Euler's method of eqn (1.7d) provides a first-order estimate x_1 of $x(t + \Delta t)$, giving

$$x_1 = x(t + \Delta t) = x(t) + \Delta t\, f(x, t). \tag{1.8g}$$

Next, substitute x_1 for $x(t + \Delta t)$ in eqn (1.8f) to give

$$\frac{d^2x}{dt^2} = \frac{f(x_1, t + \Delta t) - f(x, t)}{\Delta t}. \tag{1.8h}$$

Now take the first three terms of the Taylor series of eqn (1.7e) and use eqns

(1.8a) and (1.8h) to substitute for the first and second derivatives, giving

$$x(t + \Delta t) = x(t) + \Delta t\, f(x, t) + \frac{1}{2}(\Delta t)^2 \left[\frac{f(x_1, t + \Delta t) - f(x, t)}{\Delta t}\right], \qquad (1.8i)$$

where the higher-order terms are omitted. On simplifying this equation, the trapezoidal method is obtained:

$$x(t + \Delta t) = x(t) + \tfrac{1}{2}\Delta t[f(x_1, t + \Delta t) + f(x, t)]. \qquad (1.8j)$$

This method is a simple example of a 'predictor–corrector' method: Euler's method is used to predict $x_1 = x(t + \Delta t)$ (eqn (1.8g)), and then eqn (1.8j) is used to improve or correct this first estimate of $x(t + \Delta t)$. The trapezoidal method is second order, so that the error ε is third order. It can be shown that

$$\varepsilon = -\frac{1}{12}(\Delta t)^3 \frac{\mathrm{d}^3 x}{\mathrm{d}t^3} + \text{higher-order terms.} \qquad (1.8k)$$

This truncation error is smaller than that given by Euler's method (eqn (1.7f)). (See Exercise 1.4.)

1.6.3 *Higher-order methods*

Several higher-order methods are available, such as Simpson's rule, Runge–Kutta methods, and Milne's method. These involve calculating the first derivative within or across the interval. Just as in the trapezoidal method, where in eqn (1.8d) two values of the first derivative allow the second derivative to be estimated, three values of the first derivative allow the third derivative to be estimated, and four values of the first derivative allow the fourth derivative to be estimated. With the latter, the error (see eqn (1.7e)) is then fifth order.

The fourth-order Runge–Kutta method is outlined as an example of a higher-order method. Given the function

$$\frac{\mathrm{d}x}{\mathrm{d}t} = f(x, t)$$

and given x at time t, at time $t + \Delta t$ this algorithm takes the form

$$x + \Delta x = x + \tfrac{1}{6}(\Delta x_1 + 2\Delta x_2 + 2\Delta x_3 + \Delta x_4), \qquad (1.8l)$$

where

$$\begin{aligned}
\Delta x_1 &= \Delta t\, f(x, t)\\
\Delta x_2 &= \Delta t\, f(x + \tfrac{1}{2}\Delta x_1, t + \tfrac{1}{2}\Delta t)\\
\Delta x_3 &= \Delta t\, f(x + \tfrac{1}{2}\Delta x_2, t + \tfrac{1}{2}\Delta t)\\
\Delta x_4 &= \Delta t\, f(x + \Delta x_3, t + \Delta t).
\end{aligned}$$

The truncation error ε is of order $(\Delta t)^5$.

1.6.4 *Variable-step methods*

Variable-step methods are popular because they enable a numerical integration to be performed efficiently in terms of the length of the time step Δt and computer time. The idea is that the time step is maximized subject to the truncation error being less than a certain value. Calculation of the error requires an extra function evaluation. The technique is outlined with respect to Euler's formula.

Euler's method gives (eqn (1.7d))

$$\Delta x = \Delta t\, f(x, t). \tag{1.8m}$$

Therefore, from eqn (1.7f) for the truncation error ε and eqn (1.8f) for the second derivative,

$$\varepsilon = \tfrac{1}{2}(\Delta t)[f(x + \Delta x, t + \Delta t) - f(x, t)]. \tag{1.8n}$$

The relative error is thus given by

$$\frac{\varepsilon}{\Delta x} = \frac{1}{2}\left[\frac{f(x + \Delta x, t + \Delta t) - f(x, t)}{f(x, t)}\right]. \tag{1.8o}$$

The integration time step Δt is then increased until the relative error is just less than a chosen value, which may depend on the word length of the computer being used.

1.6.5 *Stiff equations*

The constant k in eqn (1.7g) is often described as a rate constant. The higher the value of k, the more rapid is the rate of change of x. k has dimensions of $(\text{time})^{-1}$; thus $1/k$ has dimensions of time, and is sometimes referred to as the relaxation time of the equation. In eqn (1.7g), x falls to $1/e$ of its original value in a time interval of $1/k$.

Stiff equations occur when the rate constants in a system of differential equations differ markedly. The equations then have solutions whose relaxation times are very different. In many areas of biology, including the plant sciences, there may be fast biochemical processes which take milliseconds or less for their completion, at the macromolecular and cellular levels processes may take seconds or even minutes, and at the organ level (for instance, root : shoot ratios) a response may take several days. An integration interval of Δt that suits the slow processes in a model may cause unstable oscillations of increasing amplitude with the fast processes, as in (1.7j). A short time interval that gives stable integration with the fast processes may be very expensive to run over the total time period of interest. It may also be difficult to restrict the range of rate constants in a model without distorting or discarding important components of physiology or biochemistry.

Consider the process (or chemical reaction)

$$z \xrightarrow{k} \cdots, \tag{1.9a}$$

where z is a state variable and k is a rate constant. This leads to $dz/dt = -kz$, which, with $z = z_0$ at time $t = 0$, gives

$$z = z_0 e^{-kt}. \tag{1.9b}$$

Now replace (1.9a) by a two-state variable scheme

$$y \underset{k_2}{\overset{k_1}{\rightleftharpoons}} x \overset{k_3}{\rightarrow} \cdots, \tag{1.9c}$$

with state variables x and y, and rate constants k_1, k_2, and k_3. The differential equations of this system are

$$\begin{aligned} \frac{dx}{dt} &= -(k_2 + k_3)x + k_1 y \\ \frac{dy}{dt} &= -k_1 y + k_2 x. \end{aligned} \tag{1.9d}$$

Consider the case where $k_1 = k_2 = 10$, so that x and y are in comparatively rapid equilibrium with each other, and $k_3 = 1$, so that x is converted relatively slowly into other products. Some solutions, starting from several different initial values, are shown in Fig. 1.6. The solution of interest is usually the slowly developing solution, as determined by k_3 in Fig. 1.6, and in this solution x and y are in quasi-equilibrium with each other. The short spurs in Fig. 1.6 illustrate the rapid process that occurs when the system point (x, y) is started at values of x and y which are not in quasi-equilibrium: there is rapid movement due to k_1 and k_2 to re-establish the quasi-equilibrium.

As an aside, it is interesting to consider the consequences of reversing the direction of time in eqns (1.9d) and Fig. 1.6; that is, replace t by $-t$ in eqns (1.9d),

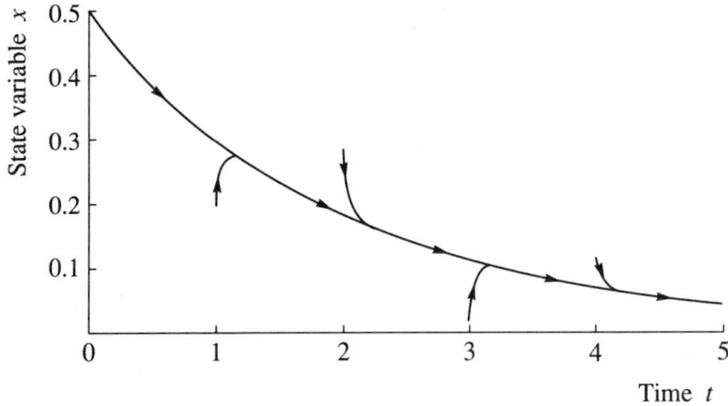

Fig. 1.6. Stiff equations. Results were obtained by integration of eqns (1.9d) with $k_1 = k_2 = 10$, $k_3 = 1$, and $\Delta t = 0.0005$, and using the fourth-order Runge–Kutta method. The 'slow' solution is calculated from initial values of $x = y = 0.5$; the 'fast' solutions shown by the short spurs were obtained by displacing the system point from the slow solution but maintaining $x + y$ at the same value.

so that the system point moves from right to left in Fig. 1.6. It will be clear that the path traced out by the system point is so sensitive to the initial values (now at $t = 5$ in Fig. 1.6) that effectively the system is indeterminate. This is reminiscent of the behaviour of the so-called chaotic systems (Cvitanovic 1984). As has been pointed out by Popper (1982) and others, even a deterministic system of equations can fail to have a determined outcome; thus nineteenth-century scientific attitudes rooted in classical (pre-quantum-mechanical) physics were in error in assuming that the universe is determined.

There are several techniques which can be applied to stiff equation problems in order to obtain solutions without using the brute-force method of a small integration interval and a lot of computing. In a given situation, one of these may well be applicable.

Combining rapidly exchanging pools Rapidly exchanging pools can sometimes be brought together. In eqns (1.9d), define z by

$$z = x + y \tag{1.9e}$$

and add the two eqns (1.9d) together to give

$$\frac{dz}{dt} = -k_3 x. \tag{1.9f}$$

If we assume that $k_1, k_2 \gg k_3$, then, since x and y are now in quasi-equilibrium under k_1 and k_2, from either of eqns (1.9d)

$$k_1 y = k_2 x. \tag{1.9g}$$

Therefore, eliminating y between eqns (1.9e) and (1.9g),

$$x = \frac{k_1}{k_1 + k_2} z. \tag{1.9h}$$

Substituting for x with eqn (1.9h) in eqn (1.9f)

$$\frac{dz}{dt} = -kz \qquad \text{where} \qquad k = \left(\frac{k_1}{k_1 + k_2}\right) k_3. \tag{1.9i}$$

This is equivalent to (1.9a), and a simplification of the problem has thus been achieved.

Reducing the fast rate constants The particular values ascribed to the fast rate constants, which cause difficulties in the integration, may have little effect on the solutions. In Fig. 1.6 the slow solutions are much the same for k_1 and k_2 values of 10, 100, or 1000. Of course varying the fast rate constants has a great effect on the rapid transients in the solutions. Some of the fluxes in the system are reduced in magnitude by reducing the values of rate constants. This may be important to the investigator. An alternative method that leaves the fluxes unaltered is described below.

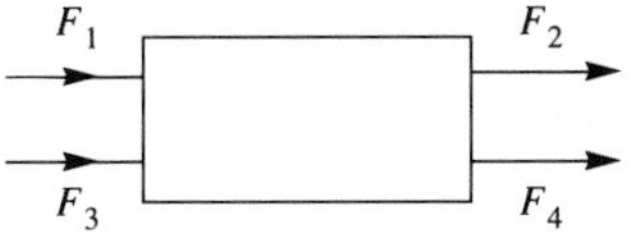

Fig. 1.7. A single compartment with state variable Q of a model is shown; the F_i are fluxes into and out of the compartment.

Increasing pool sizes Consider the 'biochemical' system shown in Fig. 1.7. Q is a state variable denoting a quantity of substance; F_1 and F_3 are fluxes of substance into the pool, and F_2 and F_4 are fluxes out of the pool. The differential equation for Q is

$$\frac{\mathrm{d}Q}{\mathrm{d}t} = F_1 + F_3 - F_2 - F_4. \tag{1.9j}$$

Assume that substance Q is distributed over a volume V, so that the concentration C of Q is given by

$$C = Q/V. \tag{1.9k}$$

For some applications, the steady state solutions of eqn (1.9j) may be of interest, i.e. when $\mathrm{d}Q/\mathrm{d}t = 0$ and $F_1 + F_3 = F_2 + F_4$. The fluxes in general depend on the concentration C, so that the quantity Q and therefore C finds a value where the fluxes are in equilibrium. Sometimes, quite small excursions from the steady state give rise to large fluxes (in relation to the pool size which is the steady state value of Q), so that $\mathrm{d}Q/\mathrm{d}t$ from eqn (1.9j) is large and causes the integration to break down. This type of behaviour usually occurs with small highly labile pools, such as adenosine triphosphate (ATP) or $NADPH_2$. A way round this problem is to inflate the pool size by a multiplier m. Assume that the volume associated with substance Q is

$$V' = mV \tag{1.9l}$$

and, most important, that the initial value ascribed to the pool at time $t = 0$ is changed from $Q(t = 0)$ to

$$Q'(t = 0) = mQ(t = 0). \tag{1.9m}$$

Note that the concentration given by eqn (1.9k) is unaltered when calculated using $C = Q'/V'$. It is assumed that the reactive volumes are unchanged—these are the volumes within which the reactions take place that give rise to the fluxes F_1 etc.—so that the fluxes are unaltered, since these depend on the concentrations and the reactive volumes. Thus $\mathrm{d}Q/\mathrm{d}t$ calculated from eqn (1.9j) remains the same, but now this rate of change is applied to updating Q' which is m times larger. Integration problems as in (1.7j) are less likely to arise when a rate of change of 10 is applied to a pool of 30 than when it is applied to a pool of 3, as for $m = 10$ with $Q(t = 0) = 3$.

The rate at which the steady state is approached is reduced by factor m, and so the transient behaviour is changed, but in the steady state the correct solutions are still obtained.

Elimination of labile pools In eqns (1.9e) and (1.9i) above we described how rapidly exchanging pools could be combined to give a simpler and more stable representation of the system. Another technique which could be applied to small labile pools is to assume they are vanishingly small and eliminate them from the system. Thus, referring to Fig. 1.7 and eqn (1.9j), it is assumed that

$$F_1 + F_3 = F_2 + F_4. \tag{1.9n}$$

The residual problem is that some of the fluxes may depend on the concentration C. How is this to be dealt with as Q and C are now undetermined? There are two cases to be considered.

1. All fluxes in the system are assumed independent of C: one of the fluxes in eqn (1.9n) (F_2 say) can adjust freely so that eqn (1.9n) is satisfied, and the other fluxes in the equation are determined elsewhere. This same type of balance can be obtained if it is assumed that two or more of the fluxes in eqn (1.9n) can adjust freely but in a fixed ratio: for example, $F_2 + F_4 = (1 + \lambda)F_2$, λ is assigned a value, and F_2 adjusts freely.
2. Some priority scheme is assumed. For example, it is assumed that F_4 is zero and F_2 adjusts freely up to a certain limit, after which F_2 remains constant and F_4 takes up the slack. In plant models it may (for instance) be assumed that first the apex, then the young growing leaves, and lastly the roots take up the consumption of carbohydrate, and this assumption may enable the carbohydrate pool to be eliminated from a plant model.

Special algorithms Recently a number of algorithms have been developed for the stiff differential equation problem. Essentially, they enable step lengths to be used that are considerably greater than the shortest relaxation times in the system. However, there is a computational price to pay in that a state transition matrix must be constructed and inverted. The most important point to be realized is that these algorithms are of no benefit if there are high-frequency components in the driving functions. Thus, if we drive a plant or crop model with rapidly changing light flux densities which follow the fluctuating light levels through the day, stiff algorithms do not enable us to take daily time steps and hence save computing time. It is then necessary to use the brute-force method of a sufficiently small time step and adequate computing power.

1.6.6 *Choice of integration method*

This is a matter of obtaining sufficient accuracy without using excessive computer time. Five fixed-step methods are compared in Table 1.2. The equation integrated was

Table 1.2. Comparison of integration errors (truncation + rounding) and execution times (relative units) for five different fixed-step methods for eqn (1.9o)

	Integration method									
	Euler		Adams		Trapezoidal		Simpson's		Runge–Kutta	
Δt	Time	Error (%)	Time	Error (%)	Time	Error (%)	Time	Error (%)	Time	Error (%)
0.1	1	−99	1	−32	1	−14	2	−11	2	−0.03
0.01	3	−39	3	−1	4	−0.3	5	−0.3	7	−0.2
0.001	17	−6	19	−2	29	−2	43	−2	58	−2
0.0001	160	−15	180	−15	290	−15	410	−15	570	−15

Δt is the integration interval; a word length of 32 bits was employed. The error was computed using eqn (1.9p).

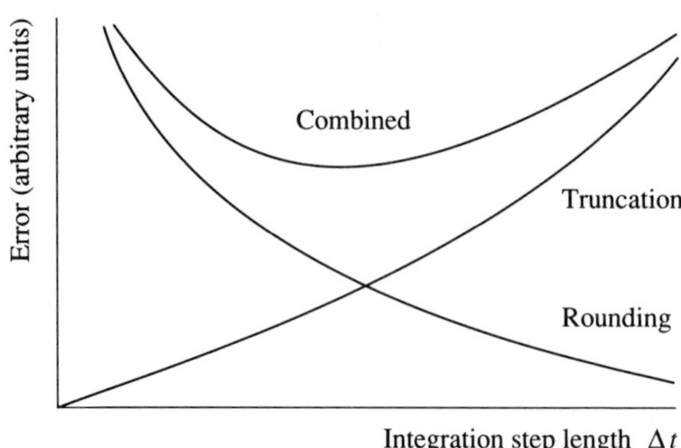

Fig. 1.8. Effect of the integration step length Δt on truncation errors and rounding errors in numerical integration by computer.

$$\frac{\mathrm{d}x}{\mathrm{d}t} = x \qquad \text{with } x = 1 \text{ at } t = 0. \tag{1.9o}$$

At time $t = 100$, the numerical result $x(t = 100$, numerical) is compared with the analytical result $x(t = 100$, analytical) = exp(100), and a percentage error is calculated as follows:

$$\text{error (\%)} = \frac{100[x(t = 100, \text{numerical}) - x(t = 100, \text{analytical})]}{x(t = 100, \text{analytical})} \tag{1.9p}$$

It can be seen from Table 1.2 that (for this particular problem and for the computer used) the best error : execution time ratio is given by the high-order Runge–Kutta method with the largest time step of $\Delta t = 0.1$. Rounding errors very quickly become important as Δt is decreased. The compromise between truncation errors and rounding errors is shown schematically in Fig. 1.8.

While the fourth-order Runge–Kutta method is well suited to many problems, it may sometimes give bogus results and there are some problems where it is inapplicable; these arise in crop modelling in particular. Euler's method has an advantage of being so transparent that it is easily checked out in detail. The higher-order methods can, each in its own way, exhibit unstable behaviour and generate spurious results. We advocate that the results of a numerical integration should be viewed sceptically, at least until they have been generated using a range of step lengths and if possible two or three different integration methods.

The plant and crop modeller has a special problem which can compel the use of Euler's method in preference to any higher-order methods, which all involve evaluating the first derivatives more than once per time step as in eqns (1.8j) and (1.8l). This arises because the equations of the model (eqns (1.3a)) are often driven by environmental variables denoted by E. These may be quantities such as temperature, radiation, or rainfall, which are not available as continuous variables but only as averages over some time interval. For instance, the radiation receipt may be 10 MJ m^{-2} on day 3 of a simulation and 5 MJ m^{-2} on day 4; similarly the mean temperatures on successive days may be 22 °C and 17 °C. These discontinuities cause discontinuities in the first derivatives in eqns (1.3a) on moving from one time interval to the next. It may not be obvious how the first derivatives should be calculated within or at the two ends of a time interval. The method that many plant and crop modellers use is to measure all the environmental variables with respect to a common time interval (or refer them to a common time interval), and then to use this time interval for numerical integration using Euler's method. Many modelling languages (ACSL, CSMP; see Section 1.7.3, p. 34) allow a choice of several integration methods.

1.7 Evaluation of models

The term evaluation is used here in the wider scientific sense—as part of the general methodology of science—and, potentially at least, evaluation may be concerned with any aspect of a model. It must be stressed that the evaluation of a model is not a wholly objective process. It may be connected with properties of the model such as its utility, simplicity, elegance, economy, plausibility, goodness of fit, and appropriateness to objectives. Researchers will always attach different weights to these items, and so it is unsurprising that models in the same problem area are differently ranked by different people. Some modellers speak of model verification and validation, although this suggests a lack of understanding of the 'working hypothesis' status of a scientific theory. Popper's (1958) thesis, now widely accepted, is that theories can only be falsified, and so the term 'validation' must be presumed to mean a failed attempt at falsification. Most crop and plant models have a short life, and they are soon discarded in favour of other, usually more general, models. It is even mistaken to talk about 'validation' for an applied model with a clear practical objective, such as the efficient application of nitrogen fertilizer to grassland. All that needs to be demonstrated is that the proposed model produces better results than the current model (current practice) in a defined set of circumstances; this is sometimes referred to as the 'champion–challenger' approach. From the farmer's viewpoint, it may not matter too much that the model may be based on (what some would regard as) mistaken assumptions.

Perhaps it hardly needs to be said, but any model should be thoroughly tested for methodological correctness: the mathematical equations must correctly represent the stated biological assumptions; both assumptions and equations must be self-consistent, with the equations being dimensionally homogeneous; any algebra or analysis must be free from error; any computer code must be correct. The modelling literature is not free from such failings. Indeed, in large and complex models, it can be so difficult not to make errors that it is better to assume that errors will be made and to adopt a self-checking method of working that traps the errors as or shortly after they are made; it may be almost impossible to find them later. In contrast with model evaluation, model testing is an objective process which has the result 'true' or 'false'.

Both evaluation and testing should be applied continuously throughout a modelling project right from the beginning. Final evaluation of a model depends upon being sure of the model's methodological correctness.

Most modelling projects can be considered in five parts.

1. The objectives of the research (p. 14): it is common for objectives to change somewhat as the project proceeds, and this is legitimate because science is intrinsically unpredictable.
2. Definition of the structure of the model: this includes the type of description (levels of aggregation), often a diagram such as Fig. 1.3, and the biological assumptions.
3. Representing 2 mathematically, and performing any subsequent analysis.
4. Solving the mathematical problem resulting from 3, usually by means of a computer program.
5. Examining and interpreting the model predictions, especially in relation to 1; attempting to simulate experimental data if suitable data are available.

Evaluation and testing are carried out at each step. The steps do overlap and interact, and it is usual to move backwards and forwards around these five items.

1.7.1 *Structure of the model*

A mathematical model represents a set of biological assumptions which are always a simplification of reality. Inevitably there are colleagues and critics who are unwilling to accept the level of simplification chosen: the model is either 'too complex' or is 'over-simplified'. Although mathematics is the servant of the sciences (for current purposes), there is little double that the assumptions of model builders are often constrained by their ability to express those assumptions mathematically. A good appreciation of the biological state of the art is essential. Much of the skill and art of modelling then lies in deciding which details can be ignored, what approximations are reasonable and appropriate, and in striking that fruitful compromise which leads to progress. Procrustean assumptions may well be needed, and colleagues who see their research area brusquely set aside may be much offended. There are no objective rules for going about this process. Conjecture is used, but it should be informed conjecture; however, it is inevitably based on the experience, skills, and judgement of the modeller.

Sometimes one sees modellers attempting biological modelling, but keeping the biology of the problem at arm's length. It rarely works: the collaborating biologist may fail to grasp what is being attempted in the mathematical representation, and the modeller may be unable to assess and rank the biological possibilities. The result is likely to be work of little value, and frustration for the concerned parties.

1.7.2 *Mathematical equations*

The accurate translation of biological ideas into mathematics requires mathematical fluency, a wide familiarity with mathematical possibilities, and a sound understanding of the biological ideas being translated. Clearly this requires either highly numerate biologists or close collaboration between biologists and people with the requisite mathematical skills. Some simple guidelines, which reduce the possibility of error and help in detecting errors, can be followed to accomplish this.

The first step is to define the symbols. It is worth giving this careful thought, since equations are much easier to read, understand, and check if similar symbols are used for similar quantities, and if the similar symbols have the same units. For example, rate constants with dimensions time^{-1} may be denoted by k_1, k_2, ...,. In a plant model the components of dry mass may be shown by W_l, W_s, and W_r for the leaf, stem, and root. Where there is a consensus in the literature about the use of certain symbols for certain quantities, the traditional symbols should be used unless there are good reasons for doing otherwise. The use of computer language notation such as FORTRAN in mathematical analysis (or in scientific papers) is, in our view, mistaken. It is less efficient, less readable, and less easily checked than the more conventional mathematical notation that has evolved over many centuries using the Latin and Greek alphabets, with upper or lower case, and subscripts and superscripts as needed. Computer notation, while still quite primitive, is rapidly evolving. Work presented in such notation may quickly become inaccessible. Indeed, many journals do not allow computer notation to be used within the body of a scientific communication.

The second step is to check dimensions. In an equation, each term must have the same dimensions as all the others. For this purpose, a symbol table should be constructed which has a verbal definition of each symbol used and also its dimensions. Sometimes it is also helpful to work out the dimensions of groups of symbols that often occur together. A single system of units (preferably SI) should be used throughout the model, even when these are not the customary units. This avoids troublesome conversion factors for quantities like grams to kilograms, or cubic metres to litres, which can very easily give rise to errors.

A third step is to check for mathematical consistency and completeness. There must be enough equations to define the problem, but the problem must not be over-defined. For instance, for a dynamic model with three state variables, three difference equations or differential equations are required. For a simple static problem with five variables, five equations are needed.

A fourth useful check is for biological consistency and completeness at the whole-system level. In many plant models (e.g. Chapter 16, Section 16.5, p. 464), carbon (C) and nitrogen (N) accounting can be carried out. We can write

$$\frac{\mathrm{d}}{\mathrm{d}t}(\text{total C or N in system}) = \text{system inputs} - \text{system outputs}. \quad (1.10a)$$

Internal transfers T_{ij}, say from pool i to pool j, occur twice in the mathematical equations of the model: positively in the differential equation for pool j, and negatively in that for pool i. Summing the equations should give cancellation of all internal transfers. For instance, however complicated a plant model is, some of its equations should sum to

$$\text{gross photosynthetic rate} = \text{growth rate} + \text{respiration rate} + \text{senescence rate}, \qquad (1.10b)$$

where these quantities are expressed in the same units.

1.7.3 *Solving the equations of the model*

This process is usually carried out by computer. If possible, one should choose to use a modern portable language that lends itself to well-structured self-documenting programs. Variable names should be chosen with care, and should correspond to the mathematical/biological variables. The general principle to be followed is that mistakes will be made, so that programs are written so that the mistakes are easily located and corrected.

There are now some good non-procedural modelling languages available, such as Continuous Systems Modelling Program (CSMP) (Speckhart and Green 1976) and Advanced Computer Simulation Language (ACSL) (Mitchell and Gauthier Associates 1987); ACSL has been used to solve most of the dynamic models described in this book. In a non-procedural program, the program statements can be written in any order; during the compilation process, the statements are put in an executable order. To program in a non-procedural language is a liberating experience that must be experienced to be appreciated. Using such languages is easy and quick (compared with using FORTRAN), and the program can be structured according to the biology of the problem, which increases program readability enormously. While these languages are not suitable for all problems, they are ideal for dynamic deterministic models of the type of eqns (1.3a).

Where possible, self-consistency checks of the type in eqns (1.10a) and (1.10b) should be written into the program; these may pin-point programming errors or mathematical errors in model formulation. In the early runs of a program, it is often worthwhile to print out every left-hand-side quantity in the program, and sometimes errors can be located by performing a detailed check on a hand calculator working direct from the mathematical equations (not from the programmed version of these equations).

Checks should also be applied against the possibility of integration errors due to an inappropriate integration method or to too large an integration interval. The results of running the program should be reasonably stable against variation in integration method and interval. Some machines have rather a short word-length, and rounding errors can be soon encountered if very short integration intervals are used (p. 30).

1.7.4 *Comparison of model with experiment: model fitting*

After the model has been carefully tested, and is free from mathematical, computational, and numerical errors, its predictions then truly reflect the assumptions on which it is based. It is essential to evaluate a model first by examining its qualitative behaviour. If this is satisfactory, then one can proceed to a direct comparison of the model's predictions with observational data, if suitable data are available, which is often not the case. When comparing a model's predictions with observational data, a measure of goodness of fit is required, and frequently some of the parameters of the model can be estimated by optimizing the goodness of fit. This last process is called fitting, or sometimes tuning or calibrating, a model.

Fitting a model generally means adjusting some of the parameter values (shown by P in eqns (1.3a)) and perhaps some of the initial values also ($X_i(t = 0)$, $i = 1, 2, \ldots, q$) so that the predictions of the model more closely resemble the observational data; this adjustment process does not alter the structure or basic equations of the model. To some modellers this approach is unacceptable—they feel that parameters should be available from independent investigations, possibly at the level of the assumptions. This view may neglect the practical objectives and empirical content which are associated with many models. Fitting can be an important part of model evaluation, and the simple procedure outlined below has been found to be of practical value.

Consider the case where a single attribute y of the system (dry mass say) is measured at m time points (t is the time variable), to give a set of m number pairs:

$$(y_1, t_1), (y_2, t_2), \ldots, (y_m, t_m). \tag{1.11a}$$

We further assume, for simplicity, that each y_i is a mean over any replicates that are taken at the ith time point.

The state variables of the model are X_k, $k = 1, 2, \ldots, q$ (see eqns (1.3a)), and running the model on the computer predicts a value for X_k at any time point t_i. We write this as

$$X_k(t_i; P; E) \tag{1.11b}$$

to emphasize the fact that the predicted values depend on the values assigned to the parameters P. In (1.11b) the parameters P include the initial values $X_k(t = 0)$, and from now on when we refer to parameters, we include initial values in with the parameters. One of the state variables of (1.11b) may correspond directly to the experimentally measured quantity y, or it may be necessary to derive an auxiliary variable that does correspond to y. (State variables may be root, stem and leaf dry masses; from these the total dry matter is calculated (this is an auxiliary variable), and it is the predicted value of total dry matter that can be compared with experiment.) Thus the predicted variable that corresponds to observation y_i is denoted by Y_i, and the model gives a set of predicted values, one at each of the m time points t_i:

$$(Y_1, t_1), (Y_2, t_2), \ldots, (Y_m, t_m). \tag{1.11c}$$

The sets (1.11a) and (1.11c) can now be compared; a perfect fit would give $y_i = Y_i$ at every time point. Since the model is deterministic, the Y_i are obtained without any probability distribution. The environment in which the observational data were taken is the same as the environment E used within the model; if two environmental treatments were used, then there would be two sets of observational data and two sets of predicted data.

Calculation of a residual A residual r_i can be calculated according to

$$r_i = y_i - Y_i \qquad \text{or} \qquad r_i = \ln\left(\frac{y_i}{Y_i}\right), \tag{1.11d}$$

or by using some other measure. In the plant sciences the second of eqns (11.1d) is usually appropriate, since this weights data points with the same proportional error equally. Thus a 10 per cent error in a predicted dry mass of 0.001 kg is of the same importance as a 10 per cent error in a predicted dry mass of 1 kg. Equations (1.11d) can be summed to give a residual sum of squares

$$R = \sum_{i=1}^{m} g_i r_i^2, \tag{1.11e}$$

using a weighting factor g_i if required. The residual sum of squares R is a measure of the lack of fit of the model and it depends on the parameter values, so that we can write

$$R \equiv R(P). \tag{1.11f}$$

Consider the simplified case where the model has just a single parameter P. Best fit is obtained by adjusting P so that R is a minimum, giving

$$\frac{dR}{dP} = 0 \qquad \text{and} \qquad \frac{d^2R}{dP^2} > 0. \tag{1.11g}$$

Examples of a sensitive parameter (curve A) and an insensitive parameter (curve B) are illustrated in Fig. 1.9. In curve A the value of d^2R/dP^2 is large, and this defines the value of P for best fit more narrowly. It is desirable that the residual R should be reasonably sensitive to all the parameters of the model, and if there are some parameters to which R is completely insensitive, this can indicate areas in which the model might be simplified. It must be remembered that such a result depends totally on the nature of the observational data against which the model is measured.

If truncation and rounding errors (which are always present but are usually insignificant) are ignored, the predicted values $Y_i = 1, 2, \ldots, m$ are computed without error; however, the experimental data y_i, $i = 1, 2, \ldots, m$ are subject to error, and this puts a lower limit on the value of R that can be achieved by parameter adjustment. The residual sum of squares R can be divided into two

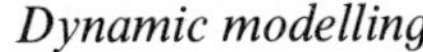

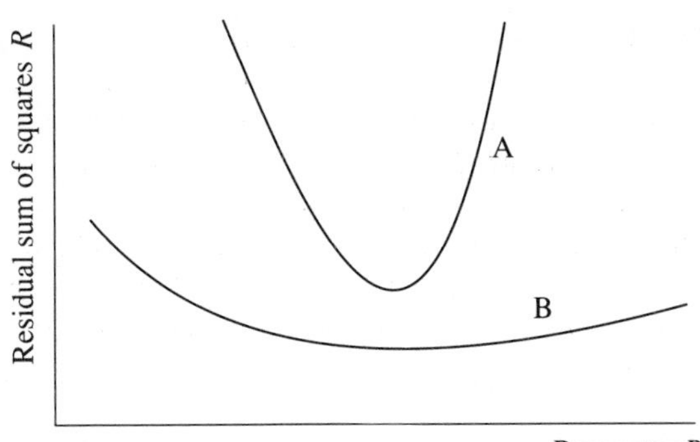

Fig. 1.9. Sensitive and insensitive model parameters: curve A, the residual sum of squares R varies rapidly as the parameter P changes, and the value of P is therefore critical to the fit obtained with the model; curve B, the value of P has little effect on R, so that fitting the model is not able to define P closely and the value assigned to P is not critical to the model.

components:

$$R = R_l + R_e \tag{1.11h}$$

where R_l is the part of the residual due to the lack of fit of the model and R_e is due to error in the experimental data. R_e has an expected value of

$$R_e = (m - n)\sigma^2, \tag{1.11i}$$

where m is the number of data points, n is the number of parameters in P which are adjusted, and σ^2 is the error variance. The parameters P only affect R_l, but if R_e is very large (adding a large constant term to the curves in Fig. 1.9) R may be rendered much less sensitive to the P values. If the experiment resulting in the data $y_1, y_2, \ldots, y_m$ can be replicated, then the error term R_e, may be known. Otherwise, an upper limit an R_e and also on σ^2 is given by the minimum value of R obtained by adjusting the parameters P.

Suppose that instead of the single set of data as in (1.11a), we have two sets, perhaps plant dry mass and leaf area, giving

$$W_i, i = 1, \ldots, m_W,$$

and (1.11j)

$$A_i, i = 1, \ldots, m_A.$$

Using eqns (1.11d) and (1.11e), we can calculate two residual sums of squares, that with respect to dry mass R_W and that with respect to leaf area R_A. It is necessary to combine these in order to carry out parameter adjustment; this can be achieved using

$$R = \frac{R_W}{\sigma_W^2} + \frac{R_A}{\sigma_A^2} \tag{1.11k}$$

where the σ_W^2 and the σ_A^2 are the respective error variances. The parameters can now be adjusted with respect to the combined residual R. If the error variances are not known, then clearly eqn (1.11k) cannot be constructed, although for fitting it is still necessary to combine the residuals in some way. If the second of eqns (1.11d) is used to calculate the residuals r_i, then R_W and R_A are independent of the dimensions of W or A and of any scaling factors. If it is further assumed that dry mass W and area A have the same coefficient of variation, which is often a reasonable assumption, then the combined residual can be written

$$R = R_W + R_A, \tag{1.11l}$$

and the parameters can be adjusted so that R in eqn (1.11l) is minimized.

The goodness of fit can be estimated by comparing the residual sum of squares due to lack of fit of the model (R_l in eqn (1.11h)) with the error residual sum of squares R_e using the F test (see Exercise 1.5). An estimate of the error term is only available if a replicated experiment has been performed; if this is not the case, only a qualitative and subjective assessment of the fit of the model to the data may be possible.

Confidence intervals for fitted parameters The dependence of the residual sum of squares R on the n parameters P_j, $j = 1, 2, \ldots, n$ can be expressed by

$$R = R(P_1, P_2, \ldots, P_n). \tag{1.11m}$$

Computer methods are generally used to search for the minimum of R with respect to the P_j using an optimization procedure. The function R, whose minimum one is trying to find, is known as the objective function. The gradient vector is the set of first partial derivatives of R with respect to the parameters P_j, $i = 1, 2, \ldots$ n, namely $\partial R/\partial P_1, \partial R/\partial P_2, \ldots, \partial R/\partial P_n$. At a minimum,

$$\frac{\partial R}{\partial P_1} = \frac{\partial R}{\partial P_2} = \cdots = \frac{\partial R}{\partial P_n} = 0. \tag{1.11n}$$

The matrix of second partial derivatives of the objective function is called the Hessian matrix, and its elements take the form

$$H_{jj'} = \frac{\partial^2 R}{\partial P_j \partial P_{j'}}. \tag{1.11o}$$

At the minimum all the eigenvalues of the Hessian matrix must be positive. Essentially this means that the second derivatives with respect to the parameters are positive, and R increases in whatever direction one moves from the point defined by $\partial R/\partial P_j = 0$, $j = 1, 2, \ldots, n$. The matrix in (1.11o) is obtained by numerical differentiation, for which computer procedures are available.

Let the matrix G with elements $G_{jj'}$ be the inverse of the Hessian matrix H with elements $H_{jj'}$. This can be represented symbolically by

$$G = H^{-1}, \tag{1.11p}$$

although it is usual to use a computer procedure to obtain the inverse matrix.

The number of degrees of freedom is denoted by ν. If there are m data points and n adjustable parameters, ν is given by

$$\nu = m - n, \tag{1.11q}$$

although if there are two data sets as in (1.11j), ν is given by

$$\nu = m_W + m_A - n.$$

An unbiased estimate of the variance V of parameter P_j is

$$V(P_j) = \frac{R}{\nu} G_{jj}, \tag{1.11r}$$

and an unbiased estimate of the covariance C of parameters P_j and $P_{j'}$ is

$$C(P_j, P_{j'}) = \frac{R}{\nu} G_{jj'}. \tag{1.11s}$$

If P_j^* is the true value of the parameter P_j, then the $100(1 - \beta)$ per cent confidence interval of P_j^* is

$$P_j \pm [V(P_j)]^{1/2} t_{\beta,\nu} \tag{1.11t}$$

where $t_{\beta,\nu}$ is the 100β percentage point of the t distribution with ν degrees of freedom (France and Thornley 1984, pp. 282–283). (See Exercise 1.5.)

1.7.5 *Sensitivity analysis*

Sensitivity with respect to observational data Consider a model with a single adjustable parameter P which has been fitted to a data set by minimizing a residual sum of squares R with ν degrees of freedom (eqns (1.11d), (1.11e), and (1.11q)). The variance $V(P)$ of P is given by (eqns (1.11o), (1.11p), and (1.11r))

$$V(P) = \frac{R}{\nu} \frac{1}{\partial^2 R/\partial P^2}. \tag{1.12a}$$

To compare the effects of different parameters on model performance, a dimensionless quantity is required that is independent of the absolute value of a parameter; the variance $V(P)$ of P has the same units as P^2 and is therefore not acceptable. The coefficient of variation $\mathrm{CV}(P)$ of V is such a quantity, and is defined by

$$\mathrm{CV}(P) = \frac{[V(P)]^{1/2}}{P}. \tag{1.12b}$$

In Fig. 1.9, curve A has a high value of curvature $\partial^2 R/\partial P^2$, giving reduced values of $V(P)$ and $\mathrm{CV}(P)$; a low value of $\mathrm{CV}(P)$ indicates that the model is sensitive to the parameter P. Similarly, curve B in Fig. 1.9, which denotes an insensitive parameter, has a low value of curvature and a higher value of $\mathrm{CV}(P)$.

Now consider a model with several parameters P_j, all of which affect the performance of the model. Analogous to eqn (1.12b), the coefficient of variation of P_j can be calculated by

$$\mathrm{CV}(P_j) = \frac{[V(P_j)]^{1/2}}{P_j}, \tag{1.12c}$$

where the variance $V(P_j)$ of P_j is obtained from eqn (1.11r). The coefficients of variation can be used to rank the parameters—a low value of $\mathrm{CV}(P_j)$ denotes that parameter P_j has a considerable effect on the fit of the model to that particular data set, and vice versa. It must be emphasized that the results of a sensitivity analysis depend entirely on the data set being used, and different data sets can give very different results. There is, therefore, nothing 'objective' or absolute about the method, which may say more about the data set than about the model. Nevertheless, the technique can be useful, especially when preparing a model for applications purposes which may require defining the precision required of the parameters of the model and specifying its range of validity (by this we mean a certain specified level of predictive accuracy). Experience with plant and crop models suggests that values of $\mathrm{CV}(P_j)$ in the range 0.05–0.2 are reasonable.

Ranking the parameters by means of a sensitivity analysis may indicate ways in which a model might be simplified. A high value of $\mathrm{CV}(P_j)$ means that the parameters P_j has little effect on the predictions of the model. It may be possible to remove such a parameter from the model. However, there may be good reasons, biological or physiological, for retaining the parameter, even though it has little effect on the model's predictive performance.

Sensitivity with respect to model predictions So far we have considered sensitivity analysis in terms of the effects of a parameter P_j on the goodness of fit as measured by a residual. It may be more pertinent to evaluate the sensitivity of some quantity Q to the parameter P_j; we denote this by $S(Q, P_j)$. For example, we may be concerned with the yield of a crop at maturity or at a particular time during the growing season. Suppose that a model has been evaluated and is judged satisfactory. Some of the parameters of the model will be physiological/genetic and some will be environmental. Some of the environmental parameters, such as the timing and quantity of fertilizer applications or irrigation, may be within the control of management. Objectives for programmes of plant breeding, genetic manipulation, or management priorities can be formulated more effectively if it is possible to rank the parameters P_j according to their effects on (say) yield Y. A dimensionless measure of the sensitivity of yield Y to a parameter P_j is

$$S(Y, P_j) = \frac{\partial Y}{\partial P_j}\frac{P_j}{Y} \approx \frac{\delta Y}{Y}\frac{P_j}{\delta P_j}. \tag{1.12d}$$

δP_j denotes a small finite change in the parameter P_j, and δY is the change that this produces in Y. For computing the sensitivity $S(Y, P_j)$ of Y and P_j, a 5 per cent parameter increment is usually sufficient, so that $\delta P_j/P_j = 0.05$. If a 5 per cent change in P_j gives a 5 per cent change in Y, then $S(Y, P_j) = 1$. Parameters with $S(Y, P_j) > 1$ have larger effects on yield, and vice versa.

An example of the use of a model to rank parameters is given by Thornley, Hurd, and Pooley (1981, Table 1); they rank the parameters of a leaf-growth model in terms of their marginal contribution to the carbon budget of the plant.

Finally, it should be noted that the sensitivity of a model, as defined by eqn (1.12d), is very similar to the economist's concepts of demand elasticities and cost elasticities. If the price of some good is p and x is the number of goods sold per unit time, then a demand elasticity E can be defined by

$$E = \frac{\delta x/x}{\delta p/p} = \frac{\delta(\ln x)}{\delta(\ln p)}. \tag{1.12e}$$

The prefix δ denotes a small increment in the variable. Comparison of eqns (1.12d) and (1.12e) shows that they are identical in structure.

1.8 The presentation of models

The continued health and progress of science depends upon the existence of open channels of communication. Publication is the means by which scientists put their efforts before the scientific community, which is then free to ignore, refute, modify, or applaud the contribution. However, even the brief history of science reveals many examples of attempts by the 'scientific establishment' to suppress innovative contributions; examples of such censorship include theories of continental drift and the origin of the solar system. Fortunately the pluralistic and fragmented nature of science has thwarted such approaches, although the inefficiencies caused by 'establishment' attitudes should not be underestimated (Lock 1986).

The modeller of plant and crop growth, and indeed of wider agricultural problems, will doubtless have run up against some of the problems that arise in attempting to publish modelling papers. Our reason for writing this section is to help modellers fight the battles with which they may be faced. Just as the weak administrator likes to construct and rely upon an extensive set of rules, rather than have some guiding principles and then treat each problem on its merits, so some journals like to decide a quite detailed 'policy', which can then be used as a blunt instrument to decide what is and what is not acceptable to the journal. Of course, it will always be difficult for any journal to strike the right balance between material that is not worth publishing and encouraging the growing points of the subject, which will often seem heretical or of little value to many and will inevitably be difficult to assess (Lock 1986).

Some biological journals will not accept papers that are 'purely theoretical' and some will not accept papers that are 'purely observational', even though these papers may be highly relevant to the field of scholarship of the journal. However, to exclude purely theoretical papers (papers without direct reference to observational data) is really as unwise as to do the converse, i.e. to exclude purely observational papers which have no interpretative content, mathematical or otherwise. Not only does this indicate a misunderstanding of the very nature of science and its components of observation, speculation (or hypothesis), and deduction (perhaps by means of a mathematical model), but in other areas of science one can readily see that all these different types of science, including the extremes, have made profound contributions. Examples of purely theoretical work, sometimes untested and untestable for many years, include Einstein's theory of general relativity, Maxwell's electromagnetic theory, and the prediction of the neutrino; all these were enormously stimulating to the subject. Examples of pure observation are early work in atomic and molecular spectroscopy, superconductivity, and antibiotic activity. It would have been a great loss if these contributions had not won publication, and one could speculate endlessly on what might be missing from current science as a result of editorial policy. Of course many papers lie at neither extreme, and they contain a mix of observational data and interpretation, sometimes by means of a model.

The referees and editors of some biological journals sometimes demand of modelling papers the satisfaction of criteria that are seldom applied to other papers, and they may also require the meeting of objectives of their own choosing (such as an extensive 'validation'; see p. 31). It should be remembered that science progresses mostly by quite modest steps, and this is true of theoretical work as well as experiment. Small but useful contributions are often noticed far away, where further work may be stimulated. Just as experiments may lead to further experiments, so may models lead to other models, experiments may stimulate models, and models may give rise to directed experimentation. Further, one may construct different models of the same system for different purposes, and confronting a model with experimental data may or may not be needed to meet the modelling objectives. Although the desirability of comparing models with experimental data is self-evident, anyone who has attempted to do this will be aware of (a) the difficulties of finding suitable data, (b) the operational problems of comparing complex mechanistic models covering two or more hierarchical levels with data, and (c) the inconclusiveness of such procedures even when carried out. Indeed, in some areas, such as modelling the geochemical carbon cycle or the 'nuclear winter' scenarios, comparison of theory with experiment may be impossible, yet there is little doubt about the value of these models. It seems to be of more scientific value to examine qualitatively the trends and patterns predicted by the model; the topology of the responses is of greater significance than precise numerical values, at least in the current state of the art. Perhaps the most important thing is that the objectives stated by the author(s) should be given close scrutiny, and one should assess to what extent these objectives are legitimate and are subsequently met.

1.8.1 *Requirements for publication*

It seems that editorial policy and authors should be concerned with five principal items:

(i) clarity;
(ii) economy;
(iii) methodological correctness;
(iv) not a trivial repetition of already published work;
(v) accuracy.

Referees and editors who pass judgements that go beyond these five points are assuming an authority that belongs properly to the scientific community as a whole. Work that satisfies the above criteria should be published so that evaluation by the scientific community can proceed unhindered.

Clarity is essential if the work is to be understood, and if it cannot be understood, at least by a few who can communicate it further, it will not have any value. Readability and clarity can usually be assessed by workers in the same general area, and a specialist in the particular topic is not necessarily required; indeed, specialists may be so concerned with what is being said, that they find it difficult to evaluate how it is being said.

Economy is needed simply because journal space is expensive. Clarity and economy may sometimes be in conflict, but more often they go together.

Methodological correctness is usually easy to assess. In many areas of both mathematics and experimental technique there is a wide acceptance of a basic methodology, and in these areas the work should be free from error. For the mathematical modeller, this might include algebra, calculus, and numerical analysis; for an experimentalist this could include the measurement of temperature and dry mass, and many chemical techniques.

The fourth item—that the work should not be a trivial repetition of work that has already been published—is self-explanatory. All work stands on what has gone before, and to give continuity, comprehensibility, and context, there will always be some repetition. However, there must be some non-trivial aspect of the newly reported work which is different from previous work. This may be interpreted as a requirement for 'originality', although what is original may be a subject for much debate; for instance, a synthesis of existing concepts may be unoriginal at one level, but may lead to novel insights. Modelling is often about the integration of ideas—the whole is more than the sum of the parts, although it is explainable in terms of those parts and how they fit together (Section 1.2.2, p. 8).

With regard to the fifth and last item, accuracy, authors should realize that if they allow minor errors of referencing, style, tabulation, equation layout, and typing to appear too frequently in the final manuscript, this must cast substantial doubts on the thoroughness and correctness of the work, and greatly reduces its overall credibility.

There is no wholely objective way of scoring a piece of work with respect to these attributes or of combining the attributes into a single score. One manuscript may be rather obscure but highly original, whereas another might present little new at the ideas level but may be a lucid and accessible exposition of a difficult topic; the originality of the latter lies in its lucidity. Many different types of contribution play a valuable role in furthering science. The scientific community has its own way of sorting out the wheat from the chaff. It is possible that a wider adoption of 'open refereeing' would be beneficial. This would encourage referees to be more objective in their criticisms. It would not be possible to express opinions or prejudices under the shelter of anonymity; in science (and indeed elsewhere) there should be no place for secret judgements. For most of us, it is a continual struggle to regard our views and attitudes as 'working hypotheses', but open refereeing could help greatly in this respect.

Exercises

1.1. The Gompertz growth equation in differential form is

$$\frac{dW}{dt} = \mu_0 W e^{-Dt} \qquad \text{with } W = W_0 \text{ at time } t = 0.$$

W is a state variable denoting dry matter with initial value W_0; μ_0 and D are constants. Write this single differential equation with explicit time dependence as two differential equations for two state variables. Suggest a biological interpretation.

1.2. The compartmental scheme $X \to Z \to \cdots$ where the rate constants into and out of compartment Z are both k has the differential equation $dZ/dt = k(X - Z)$. Show that the integral equation

$$Z(t) = k \int_{-\infty}^{t} X(t') \exp[-k(t - t')] \, dt'$$

is equivalent to the differential equation above. This is equivalent to a remembered variable as in eqn (1.5h), with the weighting function equal to $k \exp[-k(t - t')]$.

1.3. Use Euler's method of eqn (1.7d) to integrate numerically the differential equation $dx/dt = -x$ with $x = 1$ at time $t = 0$. Use a time interval $\Delta t = 0.1$, and perform the calculation over three time steps up to $t = 0.3$. Also check your results against the analytical solution.

1.4. Use the second-order trapezoidal method of eqn (1.8j) to rework the integration in Exercise 1.3. Note the greatly increased accuracy.

1.5. A model has been fitted to 84 data points by adjusting four parameters to minimize the log residual sum of squares R (eqns (1.11d) and (1.11e)), obtaining $R(\text{minimum}) = 0.8$. Calculate the mean residual sum of squares and estimate the average relative error (or lack of fit) between prediction and observation. Assume that the error residual (total) is found to be 0.4 with 50 degrees of freedom. Is the model giving an acceptable fit to the data at the 10 per cent probability level?

1.6. Suppose that the fractional carbon content f_C of plant tissue is defined by the equation $W_C = f_C W$, where W_C (kilograms of carbon) and W (kilograms of total dry matter) denote the masses of carbon and total plant dry matter respectively. Derive the units of f_C. Can these units be simplified? What are the units of leaf area index (LAI) and can these be simplified?

2
Some subjects of general importance

2.1 Introduction

In this chapter we consider some topics which are of general importance in the plant sciences, and are useful for much of the subsequent material of the book. We begin by considering the basic system of units and conversion factors. This is an area where confusion can often arise, but can easily be avoided. We then consider some of the principles of enzyme kinetics, since it is possible to derive from enzyme-kinetic considerations several functions which are of considerable utility to the modeller, as they are mathematically well behaved and have biologically interpretable parameters. An added advantage of understanding the basic concepts of enzyme kinetics is that it gives some insight into the underlying biochemical processes involved in plant and crop modelling. The final topic covered is cell division and organ growth, which is relevant to the difficult area of differentiation and development, and again provides useful background knowledge for modelling plant and crop processes.

2.2 Units and conversion factors

Confusion that arises from the choice of units can cause unnecessary problems in plant and crop modelling, and in science in general. For example, single-leaf photosynthesis may have units mg CO_2 (m^2 leaf)$^{-1}$ s^{-1}, and crop yield may have units kg (dry matter) ha^{-1}. The mixture milligrams and kilograms for mass, and square metres and hectares for area within the model invites error and confusion. It is important to adhere to a consistent set of units throughout any model. Any conversions to what may seem more appropriate units should be made at the end of, or preferably outside, the model. By doing so, the model is independent of conversion factors, and the modeller is in no doubt as to the units of quantities within the model. Another important advantage of this approach is that it facilitates the essential check for dimensional consistency of equations which should always be done.

In this book, we use the now almost universally accepted International System of Units (SI) (Royal Society 1975), with one exception (discussed below) in regard to the mole as the amount of substance. The basic units for mass, length, and time, are kilogram (kg), metre (m), and second (s). Derived units for quantities such as energy (joule (J)), pressure (pascal (Pa)), and force (newton (N)) are all defined in terms of these base units. For dynamic models of crop growth over the growing cycle of the crop the day may be the natural description of time and in such cases will be used, where

$$1 \text{ day} = 86\,400 \text{ s.} \tag{2.1a}$$

2.2.1 *Relative molecular mass and kilogram mole*

We shall first define these units as they will be used in this book, and then relate them to the SI definitions. In so doing we shall highlight the weaknesses of that system.

A kilogram mole (kg mol) is defined as that amount of a substance which contains N_A units, or entities of the substance, where N_A is Avogadro's number. Choose a reference substance where the mass of one entity is m_r kg. By definition

$$N_A m_r = r \text{ kg (kg mol)}^{-1}, \tag{2.2a}$$

where r is the mass of 1 kg mol of the reference substance. In (2.2a), either N_A or r can be assigned an arbitrary numerical value. We select a reference substance with known m_r, define r, and then derive the corresponding N_A. Following convention, the reference substance is taken to be carbon 12, and r is defined as 12 kg (kg mol)$^{-1}$, so that

$$N_A m_{^{12}\text{C}} = 12 \text{ kg (kg mol)}^{-1}, \tag{2.2b}$$

and this defines N_A as

$$N_A = 6.022 \times 10^{26} \text{ (kg mol)}^{-1}. \tag{2.2c}$$

It follows from these equations that the kilogram mole is defined as the amount of substance of a system which contains as many elementary entities as there are in 12 kg of carbon 12; that is, the mass of 1 kg mol of carbon 12 is 12 kg. Now consider a substance with n entities, each of mass m kg. The total mass M of the substance is

$$M = nm \text{ kg.} \tag{2.2d}$$

The quantity Q of the substance, measured in kilogram moles, is

$$Q = \frac{n}{N_A} \text{ kg mol.} \tag{2.2e}$$

We can now define the *molar mass* or *relative molecular mass* μ as

$$\mu = \frac{M}{Q} \text{ kg (kg mol)}^{-1}, \tag{2.2f}$$

which, using eqns (2.2d) and (2.2e), is equivalent to

$$\mu = N_A m \text{ kg (kg mol)}^{-1}. \tag{2.2g}$$

Eliminating N_A, using eqn (2.2a), we can write this as

$$\mu = \frac{m}{m_r/r} \text{ kg (kg mol)}^{-1}. \tag{2.2h}$$

Now, if carbon 12 is used as the reference, then this states that the relative molecular mass is the ratio of the average mass per molecule of the natural isotopic composition of the elements to 1/12 of the mass of an atom of carbon 12. It is important to note here the factor 1/12 actually has units kg mol kg^{-1} (eqns (2.2a) and (2.2b)) and the relative molecular mass has dimensions kg (kg mol)$^{-1}$. The relative molecular mass of any substance can now be calculated. For example, $\mu_{CO_2} = 44.0098$, which means that a kilogram mole of CO_2 is 44.0098 kg of CO_2 (the value is not exactly 44 since there are other isotopes apart from carbon 12 and oxygen 16).

In these definitions, carbon 12 has been used as the reference. However, the choice of carbon 12 is arbitrary, and oxygen 16, hydrogen 1, or any other elemental isotope could be used. This follows from Avogadro's law that a unit volume at standard temperature and pressure contains the same number of molecules, regardless of what those molecules are. For example, if oxygen 16 were used then, since

$$\frac{m_{^{16}O}}{m_{^{12}C}} = \frac{16}{12}, \tag{2.2i}$$

eqn (2.2b) would become

$$N_A m_{^{16}O} = 16 \text{ kg (kg mol)}^{-1}, \tag{2.2j}$$

and the values for the relative molecular mass for any substances would be unchanged.

It is not always appreciated that relative molecular mass is not dimensionless, but has dimensions kg (kg mol)$^{-1}$, although this is apparent from the above equations. The impression scientists generally have is that a mole is the molecular weight in grams; converted to the present units, a kilogram mole is the relative molecular mass in kilograms. This statement is a special case of eqn (2.2f) which can be rewritten

$$M = \mu Q, \tag{2.2k}$$

so that if $Q = 1$ kg mol, then M is *numerically* equal to μ.

Now consider the SI definition of the mole and the reasons why we feel it to be inadequate. The SI definition of the mole is the amount of substance of a system which contains as many elementary entities as there are atoms in 0.012 kg of carbon 12 (Royal Society 1975, p. 22). The SI definition of relative molecular mass is the ratio of the average mass per atom (molecule) of the natural isotopic composition of an element (the elements) to 1/12 of the mass of an atom of the nuclide ^{12}C (Royal Society 1975, p. 15). In this definition, it follows that the factor 1/12 has units mol g^{-1} which is equivalent to kg mol kg^{-1}. There is an obvious inconsistency here, in that the gram appears, and this is due to the SI definition of the mole. It follows that the relative molecular mass of carbon 12 is 0.012 kg (mol)$^{-1}$, which is clearly unsatisfactory. Another example is, say, oxygen, which according to this definition has a relative molecular mass of 0.032 kg (mol)$^{-1}$ (or, more accurately, 0.0319988 kg (mol)$^{-1}$), although most scientists would give the value as 32. If the kilogram mole is used, as defined above, then the units are entirely consistent; for example, the relative molecular mass of oxygen is 31.9988 kg (kg mol)$^{-1}$. A further point to note is that, according to our set of definitions, Avogadro's number N_A is a factor of 10^3 greater than as defined by the SI system (Royal Society 1975, p. 44).

We use the term relative molecular mass, which is that recommended by the Royal Society (1975) and replaces the traditional term *molecular weight*. Relative molecular mass is preferred as it gives quite an accurate indication of the definition of the term. The term molecular weight is inappropriate and should be discarded; a weight is a force and has dimensions of newtons (N). *Molar mass* would be acceptable, although it is better to use only one term.

2.2.2 *Concentration*

The concentration C of a substance, is given in units of kilogram moles of a substance per cubic metre (kg mol m^{-3}). This is 10^3 times the SI definition. Note that the unit of 1 kg mol m^{-3} is exactly equivalent to the older unit of 1 gram molecule per litre, which can be convenient.

From the gas laws, the concentration of any gas is

$$C = \frac{P}{RT} \tag{2.3a}$$

where P is the pressure (Pa), T is the temperature (K), and R is the gas constant (8314 J K^{-1}(kg mol)$^{-1}$). Thus at normal temperature (273.15 K) and pressure (101 325.0 Pa, equivalent to 0.76 mHg), the concentration of any gas is

$$C = 0.044\,618 \text{ kg mol m}^{-3}. \tag{2.3b}$$

At arbitrary temperature and pressure, therefore, the concentration is

$$C = \frac{273.15}{T}\frac{P}{101\,325.0}0.044\,618 \text{ kg mol m}^{-3}. \tag{2.3c}$$

We also use the term concentration in another sense in crop models to define the substrate status of the plant or crop. This is discussed below.

2.2.3 *Density*

The density ρ of a substance is the number of kilograms of the substance per cubic metre (kg m^{-3}). This is exactly the same as the SI unit. The quantities density, relative molecular mass, and concentration are related by the equation

$$\text{density} = \text{relative molecular mass} \times \text{concentration}. \tag{2.3d}$$

2.2.4 *Carbon dioxide concentrations and densities*

The concentration and density of CO_2 are obtained directly from eqns (2.3c) and (2.3d). Plant scientists generally refer to CO_2 concentrations in units of parts per million (ppm). Parts per million, defined as volume parts per million, is related to concentration and density, as defined above, by

$$C = \frac{\text{ppm}}{10^6} \frac{273.15}{T} \frac{P}{101\,325.0} 0.044\,618 \text{ kg mol CO}_2 \text{ m}^{-3} \tag{2.4a}$$

and

$$\rho = \frac{\text{ppm}}{10^6} \frac{273.15}{T} \frac{P}{101\,325.0} 1.963\,6 \text{ kg CO}_2 \text{ m}^{-3}, \tag{2.4b}$$

where the relative molecular mass μ_{CO_2} of CO_2 is 44.009 8 kg (kg mol)$^{-1}$.

The term parts per million has two deficiencies. First, it is not clear whether the definition is kilograms per million kilograms or molecules per million molecules, although traditionally it is taken to be the latter (as in the above equations) and this problem can, in part, be overcome by using the term volume parts per million (vpm). The second, and more serious, problem is that photosynthesis depends upon the absolute number of CO_2 molecules per unit volume, and not just the proportion of CO_2 molecules in the air. In any model, therefore, it is more appropriate to use the definition of concentration above. There are times when parts per million may be useful as it is quite easy to visualize—for example, when talking of the general increase in atmospheric CO_2 levels during this century from around 300 to 340 ppm—but in mathematical models it should be avoided.

2.2.5 *Plant composition*

The simplest means of defining plant composition is to consider plant dry matter. Although this may be quite sufficient for many purposes, it soon becomes necessary when modelling plant processes to incorporate the metabolic function of the various components within the plant. For many purposes it is helpful to separate the plant into substrate and structure. Structure comprises the cell wall material—mainly cellulose and hemicellulose—and the 'machinery' of the plant—protein. The remainder of the plant is taken to be substrate and this includes labile compounds such as glucose and amino acids. Clearly this is a simplification. However, it is the logical step from considering plant dry matter alone and, as is apparent from many of the models considered in this book, does permit considerable progress in plant and crop modelling.

The structural dry matter will generally be denoted by W_G and substrate by W_S, although other subscripts will be introduced to define, for example, shoot and root dry matter, or carbon and nitrogen substrate. A useful technique for defining the plant status with respect to a particular substrate is to look at the substrate concentration, which differs from the definition of concentration given above, and is given by

$$W_S/W_G. \tag{2.5a}$$

The dimensions of concentration in this context are kg substrate (kg structure)$^{-1}$.

2.2.6 *Water potential*

Water potential an important variable when considering the water status of a plant or crop. Any physical system will attempt to minimize its potential energy, and this is the basis for the understanding of many physical problems. For example, if a chain is suspended from two points (not necessarily at the same height), then the profile of the chain can be derived by calculating the shape required to minimize the potential energy (the curve is the well-known catenary given by the hyperbolic cosine function). In an equivalent way, in any system which involves water, the water will flow so as to minimize its potential energy. Later in the book (Chapter 14 and 15) we look at transpiration by a crop canopy and crop water use, and we need to use the concept of the water potential, which is the energy of the water in the crop, soil, or air.

An adequate definition of water potential for most plant and crop studies is the amount of energy required to transport a unit mass of pure water from a reference state to its position in the system. Water potential is therefore measured relative to some reference height, which is usually taken to be ground level. Water potential can be considered, in simple terms, as the energy per unit mass of water, and has units of joules per kilogram ($J\ kg^{-1}$).

Other units often used are the pascal (Pa) (or more commonly the kPa) and the bar, which are pressure units. The pascal is the SI unit of pressure, and has dimensions of force per unit area ($N\ m^{-2}$) which is equivalent to $J\ m^{-3}$. Consequently, if water potential is defined in pascals, this is energy per unit volume rather than per unit mass. Since a given mass of water can occupy a different volume, depending on the temperature and pressure, it is more appropriate to use $J\ kg^{-1}$. The density of water depends on temperature and takes its maximum value of 1000 $kg\ m^{-3}$ at 4 °C. Using this value, $J\ kg^{-1}$ and the pascal are related by

$$1\ J\ kg^{-1} \equiv 10^3\ Pa = 1\ kPa. \tag{2.6a}$$

The bar is not an SI unit, but is a unit of pressure in the cgs (centimetre-gram-second) system which preceded the SI system. The bar is defined as

$$1\ bar = 10^5\ Pa, \tag{2.6b}$$

so that

$$1\ J\ kg^{-1} \equiv 0.01\ bar. \tag{2.6c}$$

There are perhaps three main reasons why the bar is still retained in some quarters, despite its not being an SI unit. The first is that 1 bar is approximately equal to standard atmospheric pressure, which is 101 325 Pa = 1.013 25 bar. The second is that plant water potentials generally lie in the range 0 to −20 bar, which is an easy range to work with. The final reason is that, in practice, water potential is often measured by measuring a pressure exerted by the water in the system. In all branches of science, quantities are measured both directly and indirectly, and there is little justification for ascribing units on this basis.

2.3 Useful responses derived from enzyme kinetics

We now consider some of the basic ideas of enzyme kinetics both as an introduction to the subject and in order to derive several useful equations for the plant and crop modeller. The equations that are derived can be used both as direct models for the reaction types considered and also in a more qualitative sense to represent aggregated processes. For example, the biochemistry of protein synthesis from sugars and amino acids is complex but, for some modelling purposes, it may be appropriate to represent this by using a bi-substrate Michaelis–Menten equation. The reaction schemes and equations considered below cover all the equations of this type that are used in this book. For a more complete discussion of the topic, including a historical perspective of its development, the interested reader should consult Dixon and Webb (1979).

2.3.1 *Michaelis–Menten equation (rectangular hyperbola)*

The most widely applied model of enzyme kinetics is the Michaelis–Menten equation. The reaction scheme is

$$\mathrm{E} + \mathrm{S} \underset{k_{-1}}{\overset{k_{+1}}{\rightleftharpoons}} \mathrm{ES} \overset{k_{+2}}{\rightarrow} \mathrm{E} + \mathrm{P}, \qquad (2.7a)$$

where E, S, and P indicate the enzyme, substrate, and products of the reaction respectively; k_{+1}, k_{-1}, and k_{+2} are the rate constants for the reactions, where the plus and minus signs refer to forward and reverse reactions. ES denotes the enzyme-substrate complex. In the steady state the concentration of ES is constant, so that the rate of production of ES must equal the rate of degradation of ES:

$$k_{+1}[\mathrm{E}][\mathrm{S}] = (k_{-1} + k_{+2})[\mathrm{ES}] \qquad (2.7b)$$

where the square brackets denote concentrations. If E_0 is the total concentration of the enzyme present, which does not vary with time, then

$$E_0 = [\mathrm{E}] + [\mathrm{ES}] \qquad (2.7c)$$

(that is, the total number of kilogram moles of enzyme is constant). Combining eqns (2.7b) and (2.7c) and rearranging leads to

$$[\mathrm{ES}] = \frac{k_{+1}[\mathrm{S}]E_0}{k_{+1}[\mathrm{S}] + k_{-1} + k_{+2}}. \qquad (2.7d)$$

This gives the concentration of substrate molecules which are combined with (adsorbed onto the surface of) the enzyme molecules in terms of the substrate concentration [S] and the total enzyme concentration E_0. A similar equation regarding the adsorption of gas molecules onto solid surfaces, which is known as Langmuir's isotherm, was obtained many years prior to the work of Michaelis and Menten.

If the speed of the steady-state reaction is v, then

$$v = k_{+2}[\mathrm{ES}] = \frac{k_{+1}k_{+2}[\mathrm{S}]E_0}{k_{+1}[\mathrm{S}] + k_{-1} + k_{+2}}, \tag{2.7e}$$

which can be written

$$v = \frac{v_{\mathrm{m}}[\mathrm{S}]}{K + [\mathrm{S}]} \tag{2.7f}$$

where

$$v_{\mathrm{m}} = k_{+2}E_0 \qquad \text{and} \qquad K = \frac{k_{-1} + k_{+2}}{k_{+1}}. \tag{2.7g}$$

v_{m} is the maximum speed of the reaction which occurs when all the active sites on the enzyme molecules are occupied by substrate molecules. K is known as the Michaelis–Menten constant, and is the value of the substrate concentration for half-maximal speed $v = \frac{1}{2}v_{\mathrm{m}}$. Equation (2.7f) is illustrated in Fig. 2.1. If $k_{+2} \ll k_{-1}$, then $K \approx k_{-1}/k_{+1}$ and the first stage of the reaction is virtually in equilibrium. In this case eqn (2.7d) becomes

$$\frac{[\mathrm{ES}]}{E_0} = \frac{[\mathrm{S}]}{[\mathrm{S}] + K}, \tag{2.7h}$$

and K can be regarded as a binding constant (or dissociation constant) in that it indicates the proportion of enzyme E that is bound to the substrate S; thus a low value of K means that the enzyme has a high affinity for the substrate, and the reaction saturates at low concentrations of substrate.

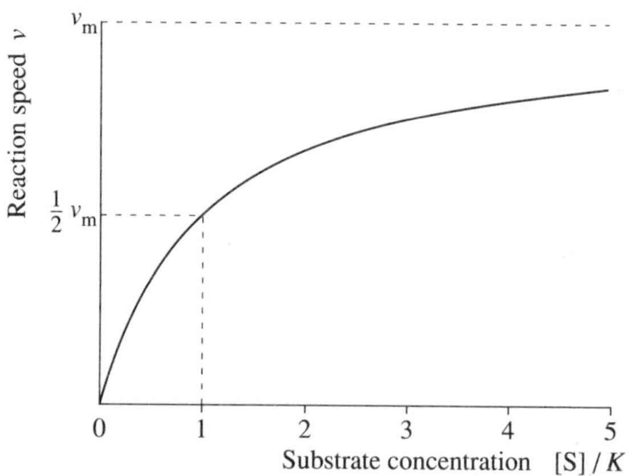

Fig. 2.1. The Michaelis–Menten equation (2.7f) for the reaction speed v as a function of substrate concentration [S]. There is an asymptote at $v = v_{\mathrm{m}}$, and $v = \frac{1}{2}v_{\mathrm{m}}$ when $[\mathrm{S}] = K$.

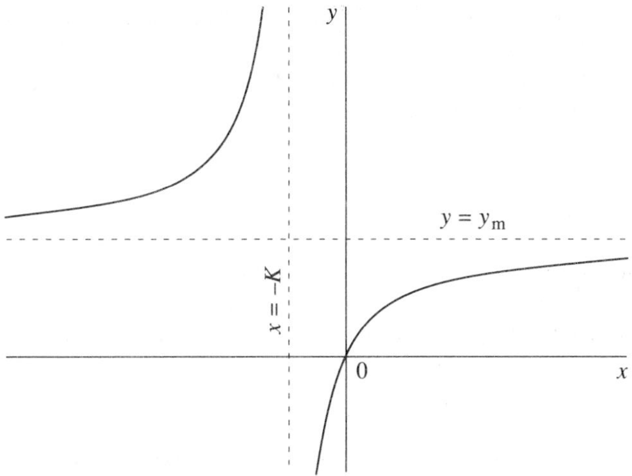

Fig. 2.2. The rectangular hyperbola (eqn (2.7i)). The asymptotes $y = y_m$ and $x = -K$ are shown.

Equation (2.7f) is known as a rectangular hyperbola. The name is apparent from the illustration in Fig. 2.2 where the equation

$$y = \frac{y_m x}{K + x} \tag{2.7i}$$

is presented. There are two branches to the curve, both approaching asymptotes defined by the lines

$$y = y_m \quad \text{and} \quad x = -K. \tag{2.7j}$$

These two asymptotes are at right angles to each other, and this is the basis for the name rectangular hyperbola. Equation (2.7f) is a particular case of the curve for the range $x > 0$.

An alternative formulation for the rectangular hyperbola is

$$y = \frac{\alpha x y_m}{\alpha x + y_m}, \tag{2.7k}$$

where it is readily shown that the constant α is the initial slope of the curve:

$$\frac{dy}{dx}(x = 0) = \alpha. \tag{2.7l}$$

The two forms presented for the rectangular hyperbola (eqns (2.7f) and (2.7k)) are mathematically equivalent in that K and α are related by

$$K = y_m/\alpha. \tag{2.7m}$$

The choice of which form to use will depend on the nature of the problem. For example, if it is convenient to prescribe the substrate level at which the reaction speed is half its maximum, then eqn (2.7f) is appropriate, whereas if it is more appropriate to incorporate the reaction rate at low substrate levels, then eqn (2.7k) might be preferred. Both forms are used throughout this book.

2.3.2 *Threshold response equations*

The rectangular hyperbola does not have a point of inflexion. However, many biological processes show this type of behaviour. An equation of this form can be derived by modifying the Michaelis–Menten reaction scheme (eqn (2.7a)) to

$$E + 2S \underset{k_{-1}}{\overset{k_{+1}}{\rightleftharpoons}} ES_2 \overset{k_{+2}}{\rightarrow} E + P. \tag{2.8a}$$

In this case molecules of substrate S can combine with the enzyme E at two sites. Using the approach of the previous section, it can be shown that the speed of this reaction is given by (Exercise 2.1)

$$v = \frac{v_m[S]^2}{K^2 + [S]^2} \tag{2.8b}$$

where v_m and K are given by eqns (2.7g). This equation is analogous to eqn (2.7f) and is illustrated in Fig. 2.3: the initial slope is zero, the asymptote is $v = v_m$, $v = \frac{1}{2}v_m$ when $S = K$, and there is a point of inflexion at $S/K = 1/\sqrt{3}$ (Exercise 2.1).

Equation (2.8a) can be generalized to

$$E + nS \underset{k_{-1}}{\overset{k_{+1}}{\rightleftharpoons}} ES_n \overset{k_{+2}}{\rightarrow} E + P \tag{2.8c}$$

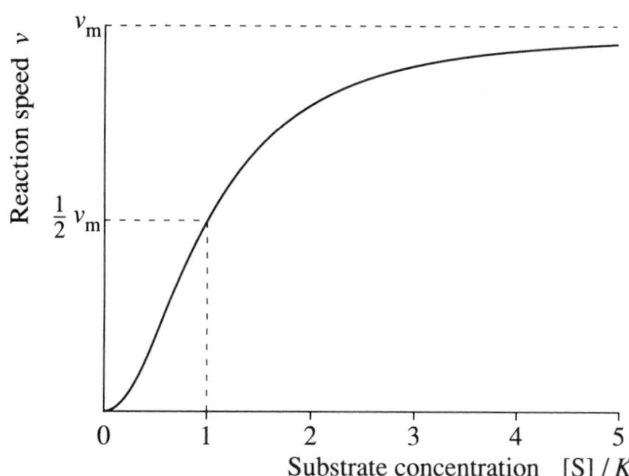

Fig. 2.3. The sigmoidal response curve (eqn (2.8b)) for the reaction speed v as a function of substrate concentration [S]. As for the Michaelis–Menten equation, there is an asymptote at $v = v_m$, and $v = \frac{1}{2}v_m$ when [S] $= K$.

where n is a positive integer. According to this scheme, there are n active sites per enzyme molecule for the substrate. Proceeding as above generates the family of curves given by (Exercise 2.2)

$$v = \frac{v_m[S]^n}{K^n + [S]^n}. \tag{2.8d}$$

Again, v_m and K are given by eqns (2.7g) and $v = \frac{1}{2}v_m$ when $S = K$. The equation for v is illustrated in Fig. 2.4 for several values of n: when $n = 1$, the curve is the Michaelis–Menten rectangular hyperbola (eqn (2.7f)) as illustrated in Fig. 2.1 and there is no point of inflexion; for $n = 2$, the curve is the same as that illustrated in Fig. 2.3 and there is weak sigmoid behaviour; as n increases the sigmoid behaviour becomes more and more pronounced, until in the limit $n \to \infty$ a step function is obtained. For all values of n ($n \geqslant 1$) the asymptote is v_m. For $n \geqslant 2$ the initial slope is zero, and there is a point of inflexion at (Exercise 2.2)

$$\frac{[S]}{K} = \left(\frac{n-1}{n+1}\right)^{1/n}. \tag{2.8e}$$

Although eqn (2.8d) has been derived from the reaction of eqn (2.8c), for some applications it can be regarded as being a useful empirical equation and there is no reason why n need be integral.

Equation (2.8d) and Fig. 2.4 show a type of 'switch-on' behaviour where the sharpness of the switching characteristic depends on the value of n. A similar relation can be constructed to represent 'switch-off' behaviour, and has the form

$$v = \frac{v_m K^n}{K^n + [S]^n}. \tag{2.8f}$$

This equation is illustrated in Fig. 2.5. v_m now occurs at $[S] = 0$, and v

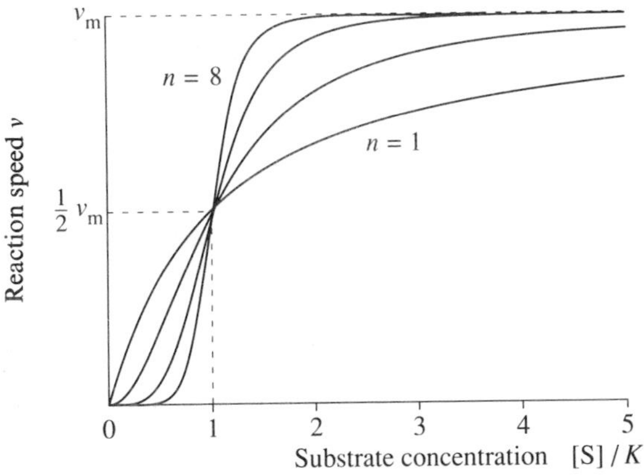

Fig. 2.4. Sigmoidal 'switch-on' response (eqn (2.8d)): $n = 1$ and $n = 8$ are indicated, and the intermediate curves are for $n = 2$ and $n = 4$. All curves have an asymptote at $v = v_m$, and $v = \frac{1}{2}v_m$ when $[S] = K$.

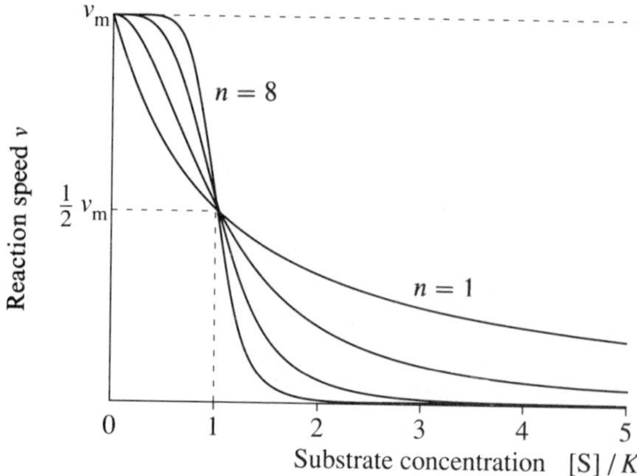

Fig. 2.5. Sigmoidal 'switch-off' response (eqn (2.8f)): $n = 1$ and $n = 8$ are indicated, and the intermediate curves are for $n = 2$ and $n = 4$. For all curves $v = v_m$ when $[S] = 0$, and $v = \frac{1}{2}v_m$ when $[S] = K$.

approaches zero as [S] increases for all values of n ($n \geqslant 1$). The point of inflexion is still given by eqn (2.8e) (Exercise 2.2).

2.3.3 *Non-rectangular hyperbola*

The non-rectangular hyperbola is a useful generalization of the rectangular hyperbola as it is a more versatile curve. One way of deriving it is to assume that there is an enzyme–substrate reaction which can be described by a rectangular hyperbola, but that now the substrate has to diffuse across some boundary from an external region to the site of the reaction. In this case, the aim is to derive an expression for the speed of the reaction in terms of the external substrate concentration. As an example, this type of scheme might be regarded as representing a reaction that takes place within the root system of a plant where the substrate has to diffuse across the root cell membranes. The speed of the reaction is

$$v = \frac{v_m[S_i]}{K + [S_i]}, \tag{2.9a}$$

where the subscript i indicates the internal substrate. It is assumed that the substrate diffuses across some boundary to the site of the reaction, and that the system is in the steady state. This means that $[S_i]$ is constant so that the rate of utilization of substrate at the site of the reaction must equal the rate of diffusion across the boundary, and hence

$$v = \frac{[S] - [S_i]}{r}, \tag{2.9b}$$

where [S] is the external substrate concentration and r is a resistance. Using eqns

(2.9a) and (2.9b) to eliminate $[S_i]$ lead to the quadratic equation

$$v^2 - v\left(\frac{K + [S]}{r} + v_m\right) + \frac{v_m[S]}{r} = 0, \tag{2.9c}$$

which can be factorized to give

$$\left(v - \frac{[S] + K}{r}\right)(v - v_m) = \frac{Kv_m}{r}. \tag{2.9d}$$

This defines a pair of curves with asymptotes

$$v = \frac{[S] + K}{r} \quad \text{and} \quad v = v_m. \tag{2.9e}$$

[S] is only physiologically defined for $[S] \geqslant 0$ and v in the range $0 \leqslant v \leqslant v_m$. However, it is instructive to look at the solutions to eqn (2.9d) over the whole range of values of [S] and v, and so the curves given by eqn (2.9d) are presented in Fig. 2.6 where the full curve represents the physiologically realistic part of the solution while the broken curves denote the unrealistic solutions.

The non-rectangular hyperbola can be written in other ways. The formulation given by eqns (2.9c) and (2.9d) is useful because it has been derived from a simple model, although it may not always be the most convenient. An alternative is to write (replace [S] by x, v by y, v_m by y_m, $1/r$ by α/θ, and K by $y_m(1 - \theta)/\alpha$)

$$\theta y^2 - (\alpha x + y_m)y + \alpha x y_m = 0. \tag{2.9f}$$

For $\theta = 0$ this reduces to a simple rectangular hyperbola given by

$$y = \frac{\alpha x y_m}{\alpha x + y_m}, \tag{2.9g}$$

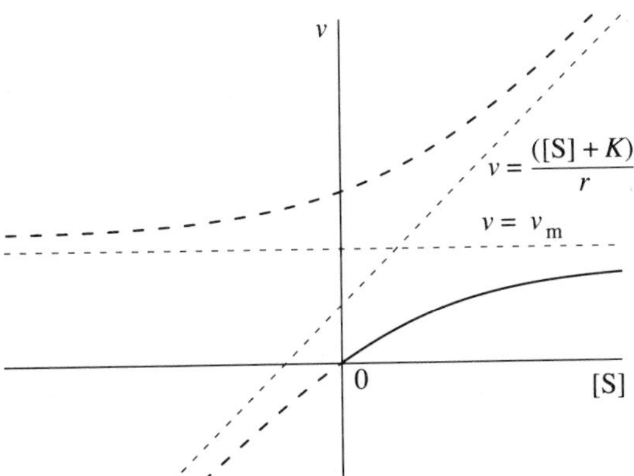

Fig. 2.6. The non-rectangular hyperbola given by the solutions to the quadratic equation (2.9c). The asymptotes $v = v_m$ and $v = ([S] + K)/r$ are indicated. The broken curves indicate the physiologically unrealistic solutions.

and for $\theta = 1$ it factorizes to

$$(y - \alpha x)(y - y_m) = 0 \tag{2.9h}$$

so that y is given by the two straight lines

$$y = \begin{cases} \alpha x & x \leqslant y_m/\alpha \\ y_m & x > y_m/\alpha. \end{cases} \tag{2.9i}$$

For θ lying in the range $0 < \theta < 1$, the physiologically realistic solution for y is the lower root of eqn (2.9f) which is

$$y = \frac{1}{2\theta}\{\alpha x + y_m - [(\alpha x + y_m)^2 - 4\theta\alpha x y_m]^{1/2}\}. \tag{2.9j}$$

It can be shown that in the limit $\theta \to 0$ eqn (2.9j) reduces to eqn (2.9g) and for $\theta = 1$ it becomes eqn (2.9i) (Exercise 2.3). For all values of θ in the physiologically sensible range $0 \leqslant \theta \leqslant 1$, the initial slope of the curve is

$$\frac{dy}{dx}(x = 0) = \alpha \tag{2.9k}$$

and the asymptote is (Exercise 2.3)

$$y(x \to \infty) = y_m. \tag{2.9l}$$

The non-rectangular hyperbola for several values of θ in the range $0 \leqslant \theta \leqslant 1$ is illustrated in Fig. 2.7.

The extra parameter θ in the non-rectangular hyperbola gives more control over the response than is the case for the rectangular hyperbola. In general, when looking at responses which increase without a point of inflexion to an asymptote, there are three basic features of the curve. The first is the initial slope, the second

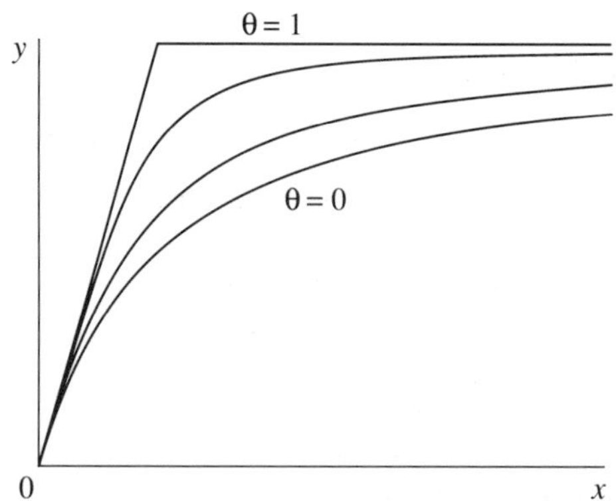

Fig. 2.7. The non-rectangular hyperbola (eqn (2.9j)): $\theta = 0$ and $\theta = 1$ are indicated and the intermediate curves are for $\theta = 0.5$ and $\theta = 0.9$. When $\theta = 0$ eqn (2.9j) reduces to the rectangular hyperbola (eqn (2.9g)) and for $\theta = 1$ it becomes two straight lines (eqn (2.9i)).

is the asymptote, and the third is the sharpness of the response i.e. how rapidly it approaches the asymptote. With the two-parameter rectangular hyperbola there is only control over two of these features. With the non-rectangular hyperbola, however, the extra parameter gives control over all three aspects of the response, which is often useful.

2.3.4 *Bi-substrate Michaelis–Menten equation*

Consider now a reaction depending on two substrates. We make the simplifying assumption that all reactions are in equilibrium, and that the order of the reactions is random. The system can be represented by the reaction scheme

$$\mathrm{E} + \mathrm{A} \rightleftharpoons \mathrm{EA} \tag{2.10a}$$

$$\mathrm{E} + \mathrm{B} \rightleftharpoons \mathrm{EB} \tag{2.10b}$$

$$\mathrm{EA} + \mathrm{B} \rightleftharpoons \mathrm{EAB} \tag{2.10c}$$

$$\mathrm{EB} + \mathrm{A} \rightleftharpoons \mathrm{EAB} \tag{2.10d}$$

where A and B are the substrates. The equilibrium equations corresponding to these reactions are

$$[\mathrm{E}][\mathrm{A}] = K_{\mathrm{A}}[\mathrm{EA}] \tag{2.10e}$$

$$[\mathrm{E}][\mathrm{B}] = K_{\mathrm{B}}[\mathrm{EB}] \tag{2.10f}$$

$$[\mathrm{EA}][\mathrm{B}] = K'_{\mathrm{B}}[\mathrm{EAB}] \tag{2.10g}$$

$$[\mathrm{EB}][\mathrm{A}] = K'_{\mathrm{A}}[\mathrm{EAB}]. \tag{2.10h}$$

K_{A}, K_{B}, K'_{B}, and K'_{A} are the binding constants for the equilibrium reactions. It is assumed that the product of the scheme is formed by the reaction

$$\mathrm{EAB} \xrightarrow{k} \mathrm{E} + \mathrm{P}, \tag{2.10i}$$

so that the speed of formation of P is

$$v = k[\mathrm{EAB}]. \tag{2.10j}$$

The aim is to derive an expression for v in terms of [A] and [B]. Equations (2.10e)–(2.10h) must therefore be used to eliminate [E], [EA], and [EB] to give [EAB] from which v is given by eqn (2.10j). However, since the total concentration of enzyme is constant

$$E_0 = [\mathrm{E}] + [\mathrm{EA}] + [\mathrm{EB}] + [\mathrm{EAB}], \tag{2.10k}$$

there are five equations for the four unknowns [E], [EA], [EB], and [EAB]. For the equations to be consistent, it can be shown that (Exercise 2.4)

$$K_{\mathrm{A}}K'_{\mathrm{B}} = K_{\mathrm{B}}K'_{\mathrm{A}} = K_{\mathrm{AB}}, \text{ say.} \tag{2.10l}$$

[EAB] can now be derived (Exercise 2.4) and v is given by

$$v = \frac{v_m}{1 + K'_B/[B] + K'_A/[A] + K_{AB}/[A][B]}, \tag{2.10m}$$

where

$$v_m = kE_0. \tag{2.10n}$$

Equation (2.10m) describes the speed of the reaction in terms of the substrate levels. If the concentration of either substrate is kept constant, then the response to variation in the other substrate is a rectangular hyperbola. For example, if v is plotted as a function of [A], the initial slope of the curve is

$$\frac{v_m B}{K'_A[B] + K_{AB}} \tag{2.10o}$$

and the asymptote is

$$\frac{v_m[B]}{[B] + K'_B}. \tag{2.10p}$$

Similar expressions apply when v is expressed in terms of [B].

For the special case where each substrate can only combine with the enzyme at a specific site on the enzyme and this is unaffected by the other substrate,

$$K_A = K'_A \qquad \text{and} \qquad K_B = K'_B, \tag{2.10q}$$

and eqn (2.10m) (using eqn (2.10l)) simplifies to

$$v = \frac{v_m}{(1 + K_A/[A])(1 + K_B/[B])}. \tag{2.10r}$$

This type of equation is very useful for representing processes that depend on the supply of two substrates, and is simpler than eqn (2.10m).

2.3.5 *Inhibitors*

The final equations to be considered are the generalization of the Michaelis–Menten theory, which led to the rectangular hyperbola eqn (2.7f), to include inhibitors. Inhibitors which manifest themselves as increasing the K parameter without altering v_m in the Michaelis–Menten equation are termed competitive; those which do not influence K but result in a decrease in v_m are known as non-competitive. There are several other forms of inhibitors which involve combinations of these effects, and these are discussed by Dixon and Webb (1979). We restrict attention to the two simplest forms known as fully competitive and fully non-competitive inhibitors. The two equations derived below (eqns (2.11h) and (2.12j)) provide useful means of representing reactions where a substrate is required but the reaction is inhibited by other components. As for the bi-substrate Michaelis–Menten equation, it is assumed that all reactions are in equilibrium.

Fully competitive inhibitor The system is represented by the reaction schemes

$$\mathrm{E} + \mathrm{S} \rightleftharpoons \mathrm{ES}, \tag{2.11a}$$

$$\mathrm{E} + \mathrm{I} \rightleftharpoons \mathrm{EI}, \tag{2.11b}$$

where S is the substrate and I the inhibitor. It is apparent from this reaction scheme that if an enzyme molecule combines with a molecule of the inhibitor it is completely unavailable to the substrate. The equilibrium equations for these reactions are

$$[\mathrm{E}][\mathrm{S}] = K_{\mathrm{S}}[\mathrm{ES}] \tag{2.11c}$$

and

$$[\mathrm{E}][\mathrm{I}] = K_{\mathrm{I}}[\mathrm{EI}], \tag{2.11d}$$

where K_S and K_I are the binding constants for the equilibrium reactions. As for the previous schemes, the product P of the reaction is formed by

$$\mathrm{ES} \xrightarrow{k} \mathrm{E} + \mathrm{P}, \tag{2.11e}$$

and hence the speed of formation of P is

$$v = k[\mathrm{ES}]. \tag{2.11f}$$

Again it is assumed that the total concentration of enzyme E_0, is constant, so that

$$E_0 = [\mathrm{E}] + [\mathrm{ES}] + [\mathrm{EI}]. \tag{2.11g}$$

Eliminating [E] and [EI] from eqns (2.11c), (2.11d), and (2.11g) to obtain [ES] and substituting in eqn (2.11f) gives (Exercise 2.5)

$$v = \frac{v_{\mathrm{m}}}{1 + (K_{\mathrm{S}}/[\mathrm{S}])(1 + [\mathrm{I}]/K_{\mathrm{I}})}, \tag{2.11h}$$

where v_m, given by

$$v_{\mathrm{m}} = kE_0, \tag{2.11i}$$

is the maximum speed of the reaction and occurs in the limit

$$\frac{K_{\mathrm{S}}}{[\mathrm{S}]}\left(1 + \frac{[\mathrm{I}]}{K_{\mathrm{S}}}\right) \to 0 \tag{2.11j}$$

which requires

$$\frac{[\mathrm{S}]}{K_{\mathrm{S}}} \gg 1 + \frac{[\mathrm{I}]}{K_{\mathrm{I}}}. \tag{2.11k}$$

This implies that the substrate is non-limiting and is in such supply that it occupies virtually all the sites on the enzyme, thus preventing any effect of the inhibitor. It is clear, therefore, that eqn (2.11h) corresponds to the Michaelis–Menten equation but with

$$K = K_{\mathrm{S}}\left(1 + \frac{[\mathrm{I}]}{K_{\mathrm{I}}}\right). \tag{2.11l}$$

Fully non-competitive inhibitor For the fully competitive inhibitor, we saw that the enzyme could combine with either the substrate or the inhibitor. If it combined with the latter, then it was completely unavailable to the substrate for the formation of products. The type of inhibitor known as fully non-competitive is one where an intermediate complex [ESI] may be formed, and where the affinity of enzyme for either the substrate or the inhibitor is unaffected by the presence of the other. The reaction scheme is now

$$\mathrm{E + S \rightleftharpoons ES} \tag{2.12a}$$

$$\mathrm{E + I \rightleftharpoons EI} \tag{2.12b}$$

$$\mathrm{ES + I \rightleftharpoons ESI} \tag{2.12c}$$

$$\mathrm{EI + S \rightleftharpoons ESI}. \tag{2.12d}$$

The equilibrium equations are

$$[\mathrm{E}][\mathrm{S}] = K_{\mathrm{S}}[\mathrm{ES}] \tag{2.12e}$$

$$[\mathrm{E}][\mathrm{I}] = K_{\mathrm{I}}[\mathrm{EI}] \tag{2.12f}$$

$$[\mathrm{EI}][\mathrm{S}] = K_{\mathrm{S}}[\mathrm{ESI}] \tag{2.12g}$$

$$[\mathrm{ES}][\mathrm{I}] = K_{\mathrm{I}}[\mathrm{ESI}], \tag{2.12h}$$

where there are only two binding constants since, as stated above, the affinity of the enzyme for either S or I is independent of the presence of the other. As for the fully competitive inhibitor in the previous section, the formation of the product P of the reaction is given by eqn (2.11e), so that eqn (2.11f) defines the speed of formation of P. Equations (2.12e)–(2.12h) must now be used to eliminate [EI] and [ESI] in order to obtain [ES] and hence the reaction speed v. Although there are four equations and only three unknowns, the equations are consistent since dividing either eqn (2.12e) by eqn (2.12f) or eqn (2.12g) by eqn (2.12h) gives

$$\frac{[\mathrm{S}]}{[\mathrm{I}]} = \frac{K_{\mathrm{S}}}{K_{\mathrm{I}}}\frac{[\mathrm{ES}]}{[\mathrm{EI}]}. \tag{2.12i}$$

Proceeding as above, it can be shown (Exercise 2.5) that the reaction speed v is given by

$$v = \frac{v_{\mathrm{m}}}{(1 + K_{\mathrm{S}}/[\mathrm{S}])(1 + [\mathrm{I}]/K_{\mathrm{I}})}. \tag{2.12j}$$

The effect of the inhibitor in this case is to reduce the speed of the reaction by a factor of $1 + [\mathrm{I}]/K_{\mathrm{I}}$, which is unaffected by the level of [S]. Also,

$$\text{as } [S] \to \infty,\ v \to v_{\mathrm{m}}^{*} = \frac{v_{\mathrm{m}}}{1 + [\mathrm{I}]/K_{\mathrm{I}}}, \tag{2.12k}$$

where v_{m}^{*} is the maximum speed of the reaction when the inhibitor is present. K_{S}

is now equivalent to K in eqn (2.7f) for the Michaelis–Menten equation in that

$$\text{when } [\mathrm{S}] = K, v = \tfrac{1}{2}v_{\mathrm{m}}^{*}, \tag{2.12l}$$

so that eqn (2.12j) is equivalent to the Michaelis–Menten equation (2.7f) with v_{m} replaced by v_{m}^{*}.

2.4 Cell division and organ growth

We now turn our attention to the growth of an organ in terms of the cell division within that organ. Since this book is primarily concerned with processes at the plant and crop levels in the hierarchical structure discussed in Chapter 1, this topic is not of central importance, but it is useful to have some understanding of this lower level in the hierarchy. We restrict attention to the dynamics of cell division and do not consider cell growth in itself. However, it is important to remember that cell division and growth are different processes, which may occur separately or together. For further discussion of cell growth see Thornley (1981).

2.4.1 *Cell division of a purely meristematic culture*

Consider a culture comprising only meristematic (dividing) cells. It is assumed that the cells divide by binary fission, where τ_{d} is the time interval between divisions, so that the maximum age of any cell is τ_{d}, and this is taken to be constant for all cells. This assumption that τ_{d} is constant for all cells can be relaxed, and the theory has been presented by Powell (1956). Let the equilibrium age distribution of the cells be

$$\phi(\tau), \text{ where } 0 \leqslant \tau \leqslant \tau_{\mathrm{d}} \text{ and } \int_0^{\tau_{\mathrm{d}}} \phi(\tau)\,\mathrm{d}\tau = 1, \tag{2.13a}$$

where ϕ has units of time^{-1}. It is assumed that this is independent of time so that for any time t the proportion of cells of a given age will be constant. If M is the cell number at time t, then the number of cells with ages lying in the interval τ to $\tau + \mathrm{d}\tau$ is

$$M(t)\phi(\tau)\,\mathrm{d}\tau. \tag{2.13b}$$

The number of cells that divide in a time interval $\mathrm{d}t$ is

$$M(t)\phi(\tau_{\mathrm{d}})\,\mathrm{d}t. \tag{2.13c}$$

With binary fission and assuming that all the progeny cells are meristematic, the increment in cell number is

$$\mathrm{d}M = M\phi(\tau_{\mathrm{d}})\,\mathrm{d}t. \tag{2.13d}$$

If we define the cell number growth constant ν (time^{-1}) by

$$\nu = \phi(\tau_{\mathrm{d}}), \tag{2.13e}$$

then

$$\frac{dM}{dt} = vM \tag{2.13f}$$

and

$$M = M_0 e^{vt} \tag{2.13g}$$

where M_0 is the cell number at time $t = 0$. This is the usual equation for exponential growth of cell numbers. Now, since τ_d is the time between cell divisions, exactly all the cells present at time t will have doubled by time $t + \tau_d$, so that

$$M(t + \tau_d) = 2M(t) \tag{2.13h}$$

which, substituted in eqn (2.13g), leads to

$$v = \frac{\ln 2}{\tau_d}. \tag{2.13i}$$

Equations (2.13g) and (2.13i) define the cell population $M(t)$ in terms of the initial cell number M_0 and the time interval τ_d between cell divisions. $M(t)$ is illustrated in Fig. 2.8.

Now consider the age distribution function $\phi(\tau)$ (eqn (2.13a)). The number of cells at time t of age τ per unit age interval is

$$M(t)\phi(\tau). \tag{2.13j}$$

After an incremental increase $d\tau$ in time, the number of cells of age $\tau + d\tau$ per

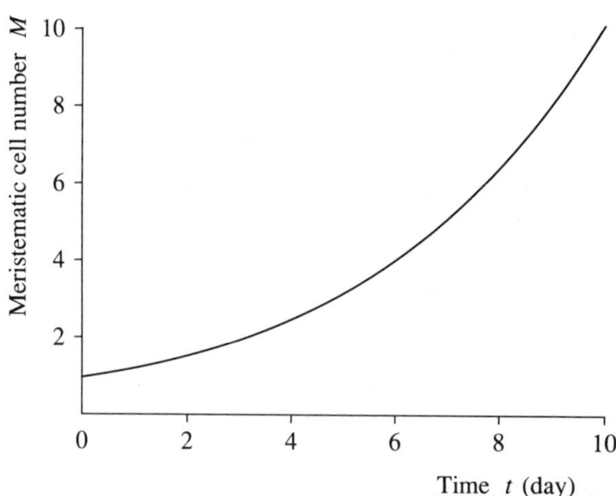

Fig. 2.8. Meristematic cell number $M(t)$ for exponential growth (eqns (2.13g) and (2.13i)). The initial value M_0 is arbitrarily taken to be unity, and $\tau_d = 3$ days (so that $v = 0.23$ day^{-1}).

unit age interval is

$$M(t + \mathrm{d}\tau)\phi(\tau + \mathrm{d}\tau). \tag{2.13k}$$

However, the cells of age τ at time t are exactly the same cells of age $\tau + \mathrm{d}\tau$ at time $t + \mathrm{d}\tau$, so that (2.13j) and (2.13k) are equal, i.e.

$$M(t)\phi(\tau) = M(t + \mathrm{d}\tau)\phi(\tau + \mathrm{d}\tau). \tag{2.13l}$$

To proceed, we use the Taylor series which is defined as follows:

$$f(x + h) = f(x) + hf'(x) + \frac{h^2}{2!}f''(x)\ldots, \tag{2.13m}$$

where the primes denote derivatives with respect to x and $n! = n(n-1)(n-2)\ldots 3.2.1$, which is referred to as n factorial. Expanding the terms on the right-hand side of eqn (2.13l) using the Taylor series gives, neglecting terms of order $\mathrm{d}\tau^2$ and higher,

$$M(t)\phi(\tau) = \left[M(t) + \mathrm{d}\tau\frac{\mathrm{d}M}{\mathrm{d}t}(t)\right]\left[\phi(\tau) + \mathrm{d}\tau\frac{\mathrm{d}\phi}{\mathrm{d}\tau}(\tau)\right], \tag{2.13n}$$

which reduces to (again neglecting the terms in $\mathrm{d}\tau^2$)

$$\frac{\mathrm{d}\phi}{\mathrm{d}\tau} = -\phi\frac{1}{M}\frac{\mathrm{d}M}{\mathrm{d}t}. \tag{2.13o}$$

Substituting from eqn (2.13g) and integrating gives

$$\phi(\tau) = \phi_0 \mathrm{e}^{-\nu\tau}, \tag{2.13p}$$

where ϕ_0 denotes $\phi(\tau = 0)$. From this equation it can be seen that young cells predominate. Indeed, setting $\tau = \tau_\mathrm{d}$ and using eqn (2.13i), the physiologically obvious result

$$\phi(\tau_\mathrm{d}) = \tfrac{1}{2}\phi_0 \tag{2.13q}$$

is obtained; that is, the number of cells about to divide is half the number that have just divided. ϕ_0 is derived by noting that

$$M(t) = \int_0^{\tau_\mathrm{d}} M(t)\phi(\tau)\,\mathrm{d}\tau \tag{2.13r}$$

and hence

$$\int_0^{\tau_\mathrm{d}} \phi(\tau)\,\mathrm{d}\tau = 1, \tag{2.13s}$$

from which it immediately follows that

$$\phi_0 = 2\nu. \tag{2.13t}$$

$\phi(\tau)$ is illustrated in Fig. 2.9.

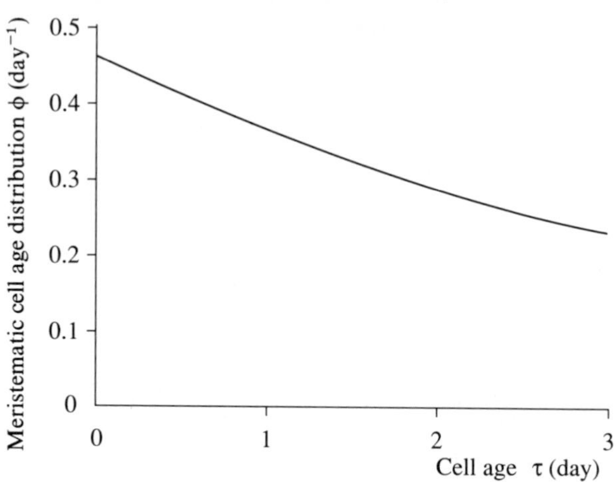

Fig. 2.9. Age distribution $\phi(\tau)$ of meristematic cells, corresponding to Fig. 2.8, and given by eqns (2.13p) and (2.13t). Note that $\phi(\tau_d) = \frac{1}{2}\phi_0$ (recall that $\tau_d = 3$ days).

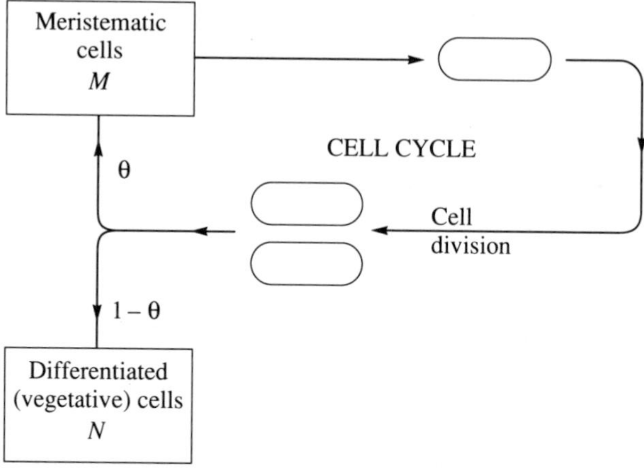

Fig. 2.10. Scheme for meristematic cell division and the production of differentiating cells.

2.4.2 *Cell division and its cessation*

Consider now the case where only a proportion θ, $0 \leqslant \theta \leqslant 1$, of newly divided cells continues to divide and the remainder are removed from the meristematic system. (This is a different θ from that of (2.9f).) This could arise for several reasons. Some of the newly divided cells may simply not be viable. Alternatively, some of these cells may be committed to a vegetative pathway of development which excludes cell division. This is likely to follow a period of purely meristematic growth. For convenience, we shall refer to non-meristematic cells as differentiating cells. This scheme, adapted from Thornley (1981), is illustrated in Fig. 2.10. The differential equation for M is now

$$\frac{dM}{dt} = \lambda \nu M, \tag{2.14a}$$

where the parameter λ accounts for the fact that not all cells continue to divide. λ is constrained by

$$\lambda \leqslant 1 \tag{2.14b}$$

since M cannot exceed the value for a purely meristematic culture. For constant λ, which corresponds to a constant value of θ, eqn (2.14a) integrates to

$$M(t) = M_0 e^{\lambda \nu t}. \tag{2.14c}$$

This equation, which is analogous to eqn (2.13g), involves the new parameter λ. To derive λ it is necessary to consider the age distribution function $\phi(\tau)$. Integrating eqn (2.13o) for $\phi(\tau)$, combined with eqn (2.14c), gives

$$\phi(\tau) = \phi_0 e^{-\lambda \nu \tau} \tag{2.14d}$$

for the age distribution of the cells, which is analogous to eqn (2.13p). Clearly, ϕ_0 will be affected by the proportion of cells that become non-meristematic, and so both λ and ϕ_0 must now be derived.

Applying the constraint (2.13s), ϕ_0 can be calculated as (Exercise 2.6)

$$\phi_0 = \frac{\lambda \nu}{1 - e^{-\lambda \ln 2}}. \tag{2.14e}$$

λ is derived as follows. Meristematic cells of age τ_d are continually dividing and producing cells of age zero. When the cells divide, a fraction θ of these newly formed cells remain meristematic, while the remainder are no longer meristematic. Thus

$$\phi_0 = 2\theta \phi(\tau_d), \tag{2.14f}$$

from which it can be shown that (Exercise 2.6)

$$\lambda = 1 + \frac{\ln \theta}{\ln 2}. \tag{2.14g}$$

The total number M of meristematic cells is therefore (eqn (2.14c))

$$M(t) = M_0 \exp\left[\nu\left(1 + \frac{\ln \theta}{\ln 2}\right)t\right], \tag{2.14h}$$

and the age distribution is

$$\phi(\tau) = \phi_0 \exp\left[-\nu\left(1 + \frac{\ln \theta}{\ln 2}\right)\tau\right]. \tag{2.14i}$$

$M(t)$ and $\phi(\tau)$ are illustrated in Fig. 2.11. For $\theta = 1$ the population is purely meristematic and the solutions for $M(t)$ and $\phi(\tau)$ are identical with those of the previous section. If $\frac{1}{2} < \theta < 1$, M is an increasing function and young cells predominate. When $\theta = \frac{1}{2}$, $\lambda = 0$ so that $M(t) = M_0$ and $\phi(\tau) = \phi_0$, which

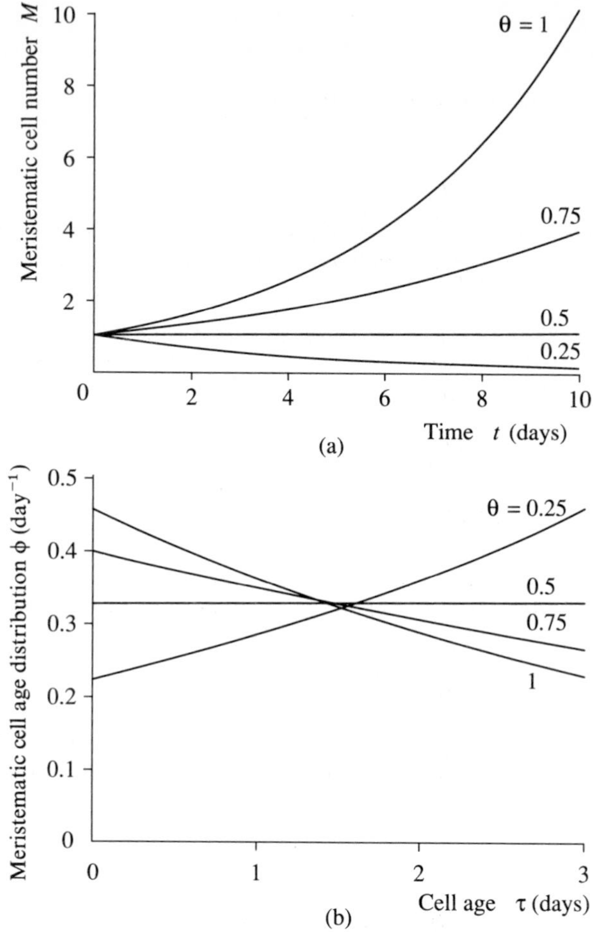

Fig. 2.11. (a) Meristematic cell number $M(t)$ as given by eqn (2.14h) for θ values of 1, 0.75, 0.5, and 0.25 as indicated ($M_0 = 1$ and $\tau_d = 3$ days, so that $\nu = 0.23$ day^{-1}); (b) corresponding age distribution function $\phi(\tau)$ (eqn (2.14i)) (note that when $\theta = 0.5$, $\lambda = 0$, and $\phi = \phi_0 = 1/\tau_d$ is constant).

means that the meristematic cell population and its age distribution are constant. For $0 < \theta < \frac{1}{2}$, $\lambda < 0$ and in this case $M(t)$ is a decreasing function with old cells predominating.

It should be noted that eqn (2.14e) for ϕ_0 is indeterminate when $\lambda = 0$. However, in this case $\phi(\tau)$ can be derived directly from the differential equation (2.13o) to be

$$\phi(\tau) = \phi_0, \tag{2.14j}$$

and from the constraint (2.13s), ϕ_0 is readily shown to be

$$\phi_0 = 1/\tau_d. \tag{2.14k}$$

It can be shown (Exercise 2.6) from eqn (2.14e) that

$$\lim_{\lambda \to 0} \phi_0 = 1/\tau_d, \tag{2.14l}$$

which is consistent with eqn (2.14k).

So far θ has been taken to be constant, although this will not be the case in a determined organ, such as a leaf, where meristematic activity eventually ceases. Following Thornley (1981) it is simply assumed that the time course of θ is given by

$$\frac{d\theta}{dt} = -D\theta, \tag{2.15a}$$

and taking D (which has dimensions of time^{-1}) to be constant, this has solution

$$\theta = e^{-Dt}. \tag{2.15b}$$

Substituting in eqn (2.14g) for λ gives

$$\lambda = 1 - \frac{Dt}{\ln 2}, \tag{2.15c}$$

and hence, integrating eqns (2.14a) and (2.13o), $M(t)$ and $\phi(t, \tau)$ become

$$M(t) = M_0 \exp\left[v\left(1 - \frac{Dt}{2 \ln 2}\right)t\right] \tag{2.15d}$$

and

$$\phi(t, \tau) = \phi_0 \exp\left[-v\left(1 - \frac{Dt}{\ln 2}\right)\tau\right]. \tag{2.15e}$$

Equation (2.15d) for $M(t)$ is generally referred to as an exponential quadratic function. It is symmetric about a maximum value given by

$$M(t_m) = M_0 \exp(\tfrac{1}{2} v t_m) \tag{2.15f}$$

which occurs at time

$$t_m = \frac{\ln 2}{D}. \tag{2.15g}$$

It is convenient to express $M(t)$ and $\phi(\tau)$ in terms of t_m rather than D, in which case eqns (2.15d) and (2.15e) become

$$M(t) = M_0 \exp\left[v\left(1 - \frac{t}{2t_m}\right)t\right] \tag{2.15h}$$

and

$$\phi(t, \tau) = \phi_0 \exp\left[-\nu\left(1 - \frac{t}{t_m}\right)\tau\right]. \tag{2.15i}$$

Note that ϕ_0 is still defined by eqn (2.14e) but is now a function of time t, through the dependence of λ on t in eqn (2.15c), and is given by

$$\phi_0 = \frac{(1 - t/t_m)\nu}{1 - \frac{1}{2}\exp[(t/t_m)\ln 2]}. \tag{2.15j}$$

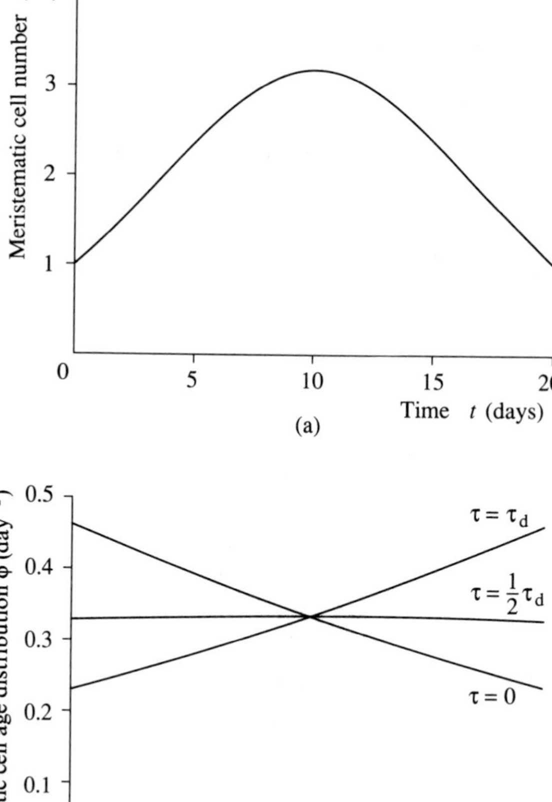

Fig. 2.12. (a) Meristematic cell number $M(t)$ as given by eqn (2.15h) with $M_0 = 1$ and $\tau_d = 3$ days, so that $\nu = 0.23$ day^{-1} and $t_m = 10$ days; (b) corresponding age distribution function $\phi(t, \tau)$ (eqn (2.15i)) of cells of age 0, $\frac{1}{2}\tau_d$, and τ_d in response to time t.

For $t = t_m$, $\lambda = 0$ and ϕ_0 defined by eqn (2.14e) is indeterminate as discussed above. In this case, eqn (2.14k) again applies, i.e. $\phi_0 = 1/\tau_d$. When computing ϕ care should be taken to avoid any problems of this nature. $M(t)$ and the corresponding age distribution $\phi(t, \tau)$, for $\tau = 0, \frac{1}{2}\tau_d, \tau_d$, are illustrated in Figs 2.12. It can be seen from Fig. 2.12b that initially young cells predominate whereas, as would be expected, this is reversed as time progresses.

2.4.3 *Production of differentiating cells*

The analysis of the previous section gives rise to non-meristematic or differentiating cells. If there are no differentiating cells and the population is purely meristematic, then the rate of production of these cells is given by eqn (2.13f). In the situation where differentiating cells are produced, eqn (2.14a) defines the rate of production of meristematic cells. The rate of production of differentiating cells is therefore the difference between eqns (2.13f) and (2.14a) which, denoting these cells by N and using eqn (2.14g) for λ, is given by

$$\frac{\mathrm{d}N}{\mathrm{d}t} = -\frac{\ln\theta}{\ln 2}\nu M. \tag{2.16a}$$

Substituting for θ, M, and D from eqns (2.15b), (2.15h), and (2.15g) gives

$$\frac{\mathrm{d}N}{\mathrm{d}t} = \left(\frac{\nu M_0}{t_m}\right) t \exp\left[\nu\left(1 - \frac{t}{2t_m}\right)t\right]. \tag{2.16b}$$

At time $t = 0$ there are no differentiating cells, so that $N(t = 0) = 0$. It is an interesting and challenging mathematical problem to show that eqn (2.16b) can be integrated to give (Exercise 2.7)

$$\frac{N}{M_0} = 1 - \exp\left[\nu\left(1 - \frac{t}{2t_m}\right)t\right] + \left(\frac{\pi\nu t_m}{2}\right)^{1/2} \exp\left(\frac{1}{2}\nu t_m\right)$$
$$\times \left\{\operatorname{erf}\left(\frac{1}{2}\nu t_m\right)^{1/2} + \operatorname{sgn}(t - t_m)\operatorname{erf}\left[\left(\frac{\nu}{2t_m}\right)^{1/2} |t - t_m|\right]\right\}, \tag{2.16c}$$

where the error function is defined by

$$\operatorname{erf}(z) = \frac{2}{\sqrt{\pi}} \int_0^z \exp(-x^2)\,\mathrm{d}x, \tag{2.16d}$$

the sign function by

$$\operatorname{sgn}(z) = \begin{cases} \text{sign of } z & z \neq 0, \\ 0 & z = 0, \end{cases} \tag{2.16e}$$

and

$$|z| = \text{modulus of } z. \tag{2.16f}$$

In practice, it is perhaps easier to integrate eqn (2.16b) numerically rather than use the analytical solution (2.16c). The ideal language for doing so is ACSL, which

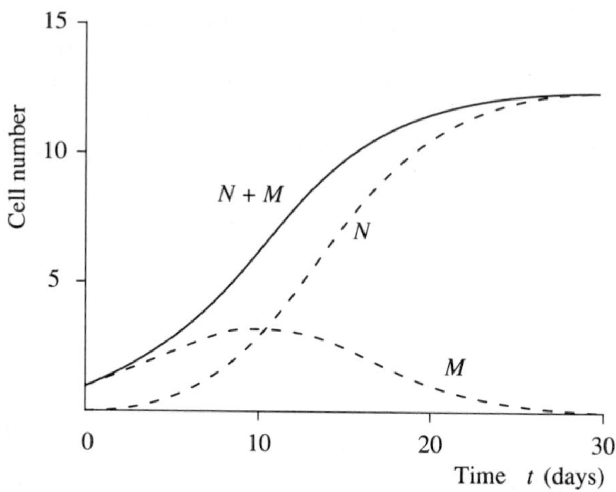

Fig. 2.13. Meristematic cell number $M(t)$, differentiating cell number $N(t)$ (broken curves as indicated), and total cell number $N + M$ (full curve) for the model of Section 2.4.3. $M(t)$ is given by eqn (2.15h) and $N(t)$ by eqn (2.16c). Parameter values are $M_0 = 1$, $\nu = 0.23\ \text{day}^{-1}$, and $t_m = 10$ day.

was discussed in Chapter 1. The solutions for M (identical with that of Fig. 2.12a), N, and the total cell number $M + N$ are illustrated in Fig. 2.13, where it can be seen that the growth of the total cell number follows the expected sigmoidal pattern.

Exercises

2.1. (a) Derive eqn (2.8b) for the reaction scheme of eqn (2.8a).
(b) Show that eqn (2.8b) has a point of inflexion at $S/K = 1/\sqrt{3}$.

2.2. (a) Derive eqn (2.8d) for the reaction scheme of eqn (2.8c).
(b) Show that eqn (2.8d) has a point of inflexion given by eqn (2.8e).
(c) Show that eqn (2.8f) also has a point of inflexion given by eqn (2.8e).

2.3. (a) Show that in the limit $\theta \to 0$ eqn (2.9j) (the non-rectangular hyperbola) reduces to eqn (2.9g).
(b) Show that eqn (2.9j) is equivalent to eqn (2.9i) when $\theta = 1$.
(c) Show that the initial slope of the non-rectangular hyperbola (eqn (2.9j)) is α (eqn (2.9k)).
(d) Show that the asymptote of the non-rectangular hyperbola (eqn (2.9j)) is y_m (eqn (2.9l)).
Hint. For part (a) use the binomial expansion in the form

$$(1 + x)^n = 1 + nx + \frac{n(n-1)}{2!}x + \frac{n(n-1)(n-2)}{3!}x^3 + \cdots \quad |x| < 1 \text{ and } n \neq -1, \quad \text{(E2.3a)}$$

where

$$m! = m(m-1)(m-2)\ldots 3.2.1 \quad \text{(E2.3b)}$$

and is termed m factorial. Note that, although not required here, when $n = -1$ the series expansion is

$$(1+x)^{-1} = 1 - x + x^2 - x^3 + \cdots. \qquad \text{(E2.3c)}$$

2.4. (a) Derive the constraint (2.10l) for the bi-substrate Michaelis–Menten reaction scheme.

(b) Derive eqn (2.10m) for the speed of the bi-substrate Michaelis–Menten reaction scheme.

2.5. (a) Derive eqn (2.11h) for the speed of the Michaelis–Menten reaction with a fully competitive inhibitor.

(b) Derive eqn (2.12j) for the speed of the Michaelis–Menten equation with a fully non-competitive inhibitor.

2.6. (a) Show that ϕ_0 is given by eqn (2.14e) for the scheme illustrated in Fig. 2.10, as discussed in Section 2.4.2.

(b) Derive eqn (2.14f) for λ for this model.

(c) Show that in the limit $\lambda \to 0$, ϕ_0 as defined by eqn (2.14f) reduces to (2.14l). To do so, use the series expansion for e^x:

$$e^x = 1 + x + \frac{x^2}{2!} + \frac{x^3}{3!} + \cdots. \qquad \text{(E2.6a)}$$

2.7. Derive eqn (2.16c) for the time course of the number of differentiating cells for the model presented in Section 2.4.3. This is quite a difficult problem.

3
Plant growth functions

3.1 Introduction

Growth functions are widely applied in many branches of the biological sciences, often in the analysis of the time course of experimental data in plant and animal studies (Hunt 1982, France and Thornley 1984). The term growth function generally denotes an analytical function which can be written as a single equation. Thus a growth function for the time course of dry mass W is

$$W = f(t) \tag{3.1a}$$

where t is time.

Growth functions are empirical models, as described in Chapter 1. Their value is mainly as descriptions and summaries of experimental data, or for use as sub-models in larger mechanistic models. Since they are empirical, the principal test of growth functions is the statistical accuracy of the fit to data. However, it is not a difficult task to construct an equation of the form (3.1) which goes through all the data points, although the contribution of such a model is questionable. Clearly, it is important to put growth functions in their proper perspective.

We propose two basic criteria which a growth function must satisfy if it is to have any significant value in plant and crop modeling. First it should be derived from a differential equation for $\mathrm{d}W/\mathrm{d}t$, and second the parameters in this equation should, in some way, be biologically meaningful—accepting that there is no absolute standard by which this can be defined. The second point will become clear as we proceed. The rationale for working from a differential equation is that cumulative dry mass is, by definition, the integral of growth rate which, in turn, will vary according to substrate supply, environmental conditions, developmental rates and so on.

Thus (3.1a) is derived from a differential equation of the form

$$\frac{\mathrm{d}W}{\mathrm{d}t} = g(W, t) \tag{3.1b}$$

which defines the growth rate in terms of the current dry mass W and time t. This choice of formulation is physiologically reasonable since plants or crops of different dry mass might be expected to have different growth rates, and ontogenetic development will depend (in some way) on time. W is therefore obtained from the equation

$$\int_{W_0}^{W} \mathrm{d}W' = \int_0^t g(W', t')\,\mathrm{d}t', \tag{3.1c}$$

where W_0 is the dry mass at time $t = 0$ and the primes denote the dummy variables of integration.

In the illustrations that follow, plots of W and $\ln W$ will be presented for all the growth curves discussed. Owing to the many different growth equations applied to experimental data in recent years, we select only those which we feel make a useful contribution to the analysis of plant growth data. In the discussion at the end of the chapter, we present our arguments as to why we believe that some commonly applied approaches have limited value.

3.2 Simple exponential growth

The most basic growth function is that which describes simple exponential growth. The assumptions are that the quantity of growth machinery is proportional to dry mass W, that the growth machinery works at a constant rate, and that growth is irreversible. The governing differential equation is

$$\frac{\mathrm{d}W}{\mathrm{d}t} = \mu W \tag{3.2a}$$

where μ is the specific growth rate; μ depends on the proportion of W that constitutes growth machinery, and also on the efficiency or speed with which this machinery can operate. Integrating (3.2a) gives

$$W = W_0 \mathrm{e}^{\mu t}. \tag{3.2b}$$

Equation (3.2b) is illustrated in Fig. 3.1. On the semi-logarithmic plot the growth curve is linear, since from eqn (3.2b)

$$\ln W = \ln W_0 + \mu t. \tag{3.2c}$$

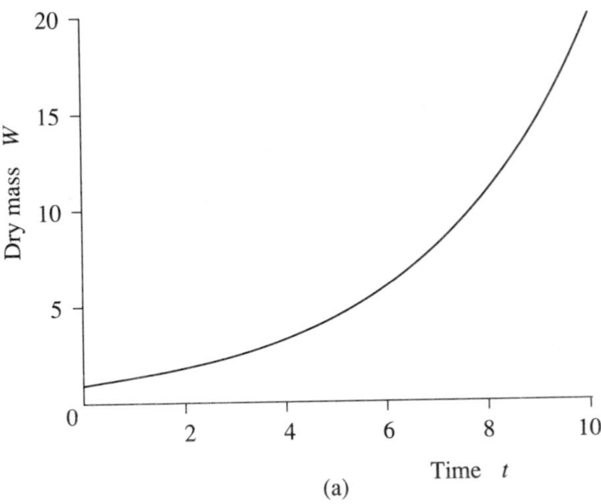

Fig. 3.1. Simple exponential growth as described by eqns (3.2b) and (3.2c): W is the dry mass and t is time, both in arbitrary units; parameter values are $W_0 = 1$ and $\mu = 0.3$.

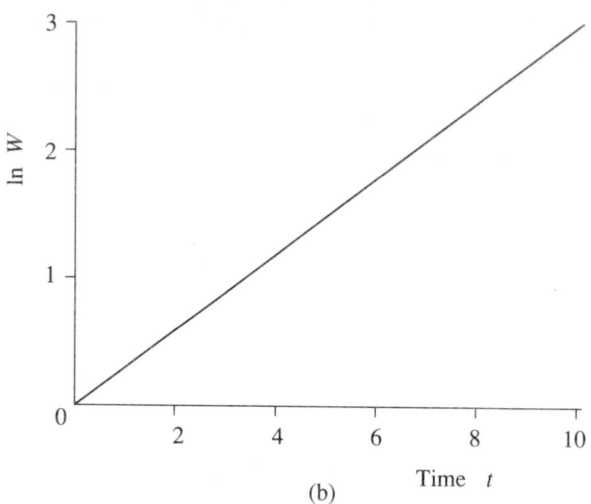

(b)

Fig. 3.1 (*continued*)

This model is defined independently of substrate availability, which implies the assumption of 'non-limiting' substrate, or that substrate supply is proportional to dry mass. Consequently the model can apply when 'sources' are much stronger than 'sinks', or in the early stages of growth when there is negligible shading so that substrate supply is approximately proportional to plant dry mass and the plant is vegetative. The other extreme is to relate growth solely to the available substrate.

3.3 The monomolecular equation

This equation describes the progress of a simple irreversible first-order chemical reaction. The assumptions are that the quantity of growth machinery is constant and independent of dry mass W, that this machinery works at a rate proportional to the substrate level S, and that growth is irreversible. Thus we can write

$$\frac{\mathrm{d}W}{\mathrm{d}t} = kS, \tag{3.3a}$$

where k is a constant and indicates the proportion of S utilized per unit time. It is assumed that there is no net gain or loss from the system, so that

$$W + S = W_0 + S_0 = W_f = \text{constant} \tag{3.3b}$$

where W_0 and S_0 are the initial values of W and S at time $t = 0$, and W_f is the final value of W approached as $t \rightarrow \infty$. (Note that all of the substrate is used up, so that $S \rightarrow 0$ as $t \rightarrow \infty$.) Equation (3.3a) can therefore be written as

$$\frac{\mathrm{d}W}{\mathrm{d}t} = k(W_f - W), \tag{3.3c}$$

which is readily integrated from $t = 0$ to give (Exercise 3.1)

$$W = W_f - (W_f - W_0)e^{-kt}. \tag{3.3d}$$

Equation (3.3d) is the monomolecular equation, which is also known as the Mitscherlich equation. For the simple case where $W_0 = 0$, it reduces to

$$W = W_f(1 - e^{-kt}). \tag{3.3e}$$

Equation (3.3e) is illustrated in Fig. 3.2. The growth rate decreases continually and there is no point of inflexion (Exercise 3.1).

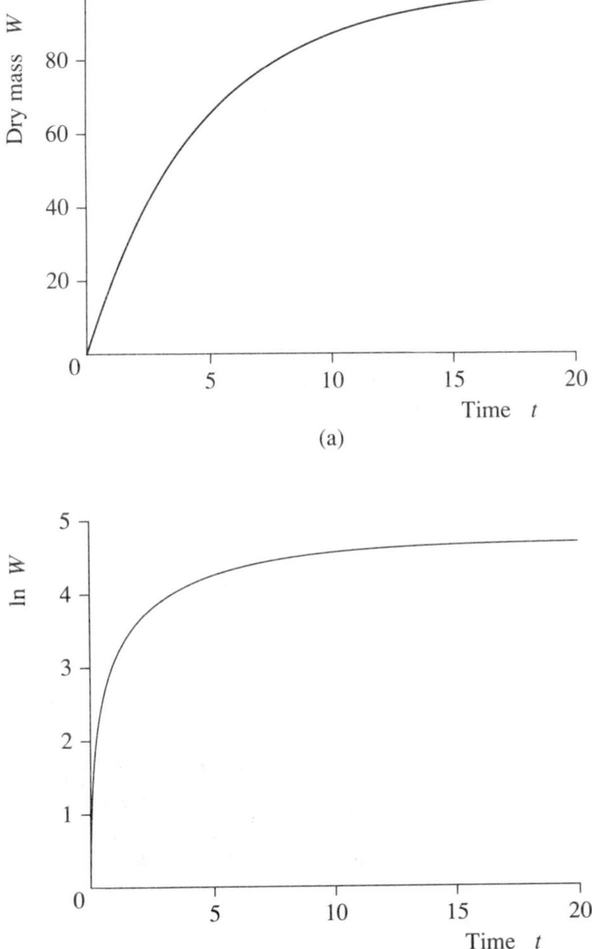

Fig. 3.2. Monomolecular growth equation as described by eqn (3.3d): W is the dry mass and t is time, both in arbitrary units; Parameter values are $W_0 = 1$, $W_f = 100$, and $k = 0.2$.

3.4 The logistic growth equation

The two growth equations described so far considered the extreme situations where growth depends only on the quantity of growth machinery and not on substrate availability—the simple exponential equation—and where the growth rate depends only on the substrate level and not on the plant dry mass—the monomolecular equation. Clearly, neither of these situations is likely to be particularly realistic in practice for extended periods of time. The logistic equation is derived from a composite assumption: the quantity of growth machinery is proportional to dry mass W; this growth machinery works at a rate proportional to the amount of substrate S; growth is irreversible. The governing differential equation is now

$$\frac{dW}{dt} = k'WS, \tag{3.4a}$$

where k' is a constant and, again, S is the substrate level. Substituting from eqn (3.3b) for S, this becomes

$$\frac{dW}{dt} = k'W(W_f - W). \tag{3.4b}$$

It is convenient to work with a constant μ, defined by

$$\mu = k'W_f, \tag{3.4c}$$

so that

$$\frac{dW}{dt} = \mu W\left(1 - \frac{W}{W_f}\right). \tag{3.4d}$$

Equation (3.4d) is readily integrated (Exercise 3.2) to give

$$W = \frac{W_0 W_f}{W_0 + (W_f - W_0)e^{-\mu t}} \tag{3.4e}$$

During the early stages of growth, where t is small and

$$\frac{W}{W_f} \ll 1, \tag{3.4f}$$

eqn (3.4e) approximates to

$$W \approx W_0 e^{\mu t}. \tag{3.4g}$$

This is the same as eqn (3.2b), so that during the early stages of growth the plant is growing exponentially with specific growth rate μ. This also follows directly from eqn (3.4d).

Differentiating eqn (3.4d) we obtain

$$\frac{1}{\mu}\frac{d^2W}{dt^2} = \frac{dW}{dt}\left(1 - \frac{2W}{W_f}\right). \tag{3.4h}$$

so that there is a point of inflexion at

$$W = \tfrac{1}{2}W_f, \tag{3.4i}$$

which, substituting in eqn (3.4e), occurs at $t = t^*$ where

$$t^* = \frac{1}{\mu}\ln\left(\frac{W_f - W_0}{W_0}\right). \tag{3.4j}$$

The growth curve is illustrated in Fig. 3.3, and it can be seen that the response is a smooth sigmoid.

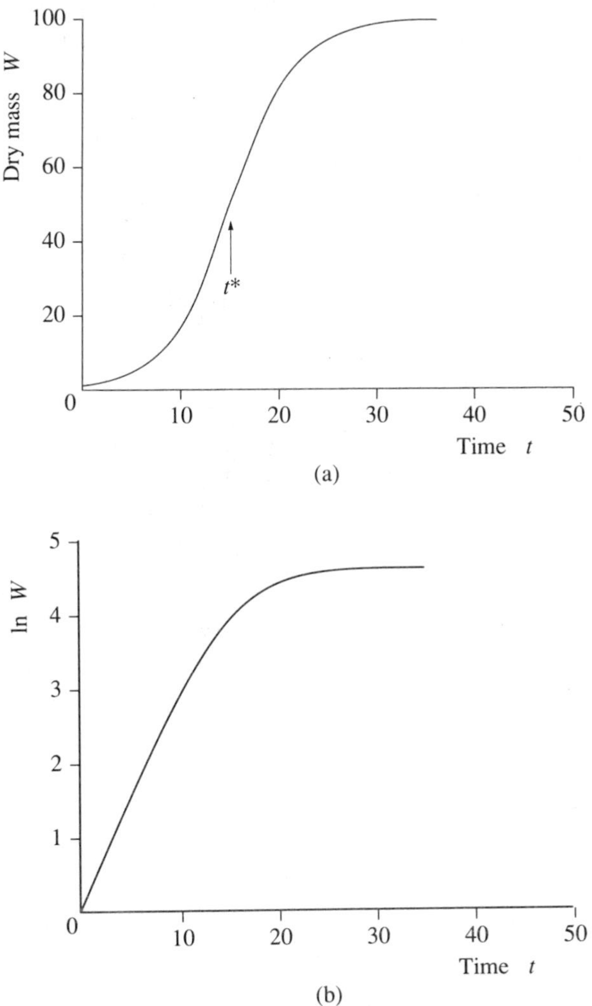

Fig. 3.3. Logistic growth equation as described by eqn (3.4e): W is the dry mass and t is time, both in arbitrary units; parameter values are $W_0 = 1$, $W_f = 100$, and $\mu = 0.3$. The time t^* of the point of inflexion (eqn (3.4j)) is indicated.

A good initial assessment as to the possibility of a set of data being well described by the logistic equation is to look at the semi-logarithmic plot of the data. If there is a relatively long linear phase, as in Fig. 3.3b, then it is likely that the curve will be appropriate.

3.5 The Gompertz growth equation

In deriving the logistic equation it is assumed that the rate of the autocatalytic growth is modified by substrate availability, according to eqn (3.4a). The Gompertz equation can be derived by making the alternative assumptions that the substrate is non-limiting so that the growth machinery is always saturated with substrate, the quantity of growth machinery is proportional to the dry mass W with a constant of proportionality μ, and the effectiveness of the growth machinery decays with time according to first-order kinetics, giving exponential decay (see eqn (1.4f) and Exercise 1.1)). This decay can be viewed as due to degradation (possibly of enzymes), or senescence, or development and differentiation. There are alternative sets of assumptions that give rise to the Gompertz equation. Formalizing the above, therefore, we obtain

$$\frac{\mathrm{d}W}{\mathrm{d}t} = \mu W, \tag{3.5a}$$

as in eqn (3.2a), but now the specific growth rate parameter μ is no longer constant but is governed by

$$\frac{\mathrm{d}\mu}{\mathrm{d}t} = -D\mu, \tag{3.5b}$$

where D is a parameter describing the decay in the specific growth rate. Integrating eqn (3.5b) gives

$$\mu = \mu_0 \mathrm{e}^{-Dt}, \tag{3.5c}$$

where μ_0 is the value of μ at time $t = 0$. Combining eqns (3.5b) and (3.5c), we obtain the differential equation for the Gompertz equation

$$\frac{\mathrm{d}W}{\mathrm{d}t} = \mu_0 W \mathrm{e}^{-Dt}, \tag{3.5d}$$

which integrates to (Exercise 3.2)

$$W = W_0 \exp\left[\frac{\mu_0(1 - \mathrm{e}^{-Dt})}{D}\right]. \tag{3.5e}$$

For small values of t

$$1 - \mathrm{e}^{-Dt} \approx Dt, \tag{3.5f}$$

so that growth is exponential with

$$W \approx W_0 \exp(\mu_0 t). \tag{3.5g}$$

As $t \to \infty$, an asymptote at $W = W_f$ is approached, where

$$W_f = W_0 \exp\left(\frac{\mu_0}{D}\right). \tag{3.5h}$$

Differentiating eqn (3.5d) gives

$$\frac{1}{\mu_0}\frac{d^2W}{dt^2} = \frac{dW}{dt}e^{-Dt} - DWe^{-Dt}. \tag{3.5i}$$

Equating this to zero and substituting for dW/dt from eqn (3.5d), we find a point of inflexion at time t^*, where

$$t^* = \frac{1}{D}\ln\left(\frac{\mu_0}{D}\right) \quad \text{and} \quad W(t = t^*) = \frac{W_f}{e}. \tag{3.5j}$$

The differential equation from which the Gompertz equation is derived here involves both the state variable W and the time variable t. By substituting from eqn (3.5e) for $\exp(-Dt)$, eqn (3.5d) can be written independently of t and so in the form 'rate is a function of state' (eqn (1.3a)), giving

$$\frac{dW}{dt} = \mu_0 W\left[1 - \frac{D}{\mu_0}\ln\left(\frac{W}{W_0}\right)\right]. \tag{3.5k}$$

In turn, either D or μ_0 can be eliminated from this equation by using eqn (3.5h), so that the Gompertz equation can also be written in the forms

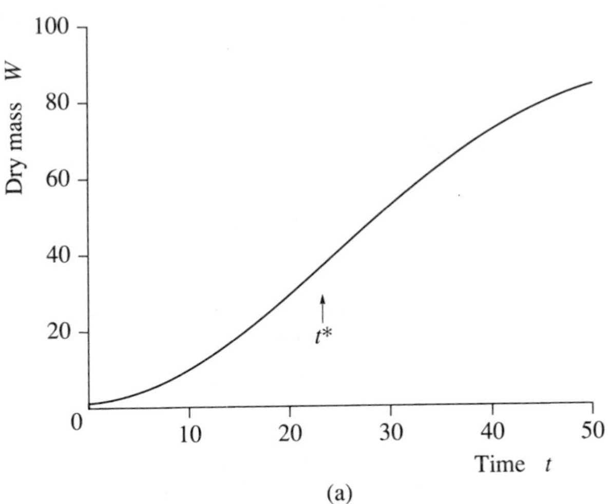

Fig. 3.4. Gompertz growth equation as described by eqn (3.5e): W is the dry mass and t is time, both in arbitrary units; parameter values are $W_0 = 1$, $W_f = 100$, and $\mu_0 = 0.3$. The time t^* of the point of inflexion (eqn (3.5j)) is indicated.

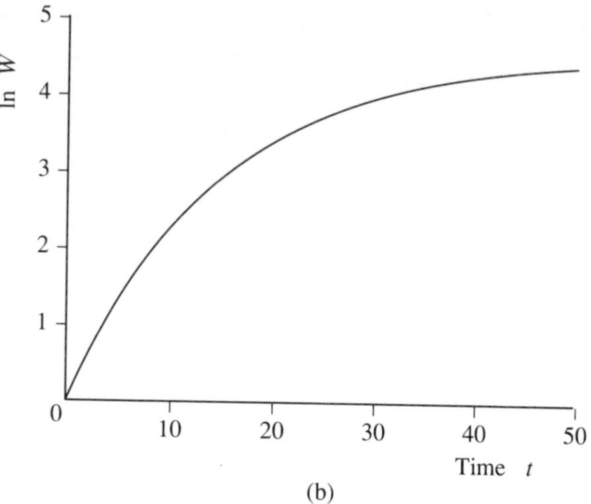

(b)

Fig. 3.4 (*continued*)

$$\frac{dW}{dt} = \mu_0 W\left[\frac{\ln(W_f/W)}{\ln(W_f/W_0)}\right] = DW\ln\left(\frac{W_f}{W}\right). \tag{3.5l}$$

The Gompertz equation is illustrated in Fig. 3.4. Compared with the logistic equation it has the same number of parameters (three), but the inflexion point now occurs at 1/e (0.37) times the final mass rather than half the final mass (eqns (3.4i) and (3.5j)). Comparision of Figs 3.3 and 3.4 (presented in Fig. 3.5 below where the Chanter growth curve, which is a composite of the logistic and Gompertz curves, is discussed) shows that, for the same initial and final masses and initial specific growth rate, the Gompertz equation shows a slower approach to the asymptote, with a longer linear period about the inflexion. The semi-logarithmic plot is less linear for the Gompertz curve than for the logistic curve, and this can be used as an initial indication as to how appropriate the Gompertz curve is likely to be when analyzing data.

3.6 The Chanter growth equation

Chanter (1976, Appendix 2; discussed by France and Thornley 1984) devised a growth function which is a hybrid of the logistic and Gompertz equations, and whose parameters can be similarly interpreted. By analogy with eqns (3.4d) and (3.5d), an equation

$$\frac{dW}{dt} = \mu W\left(1 - \frac{W}{B}\right)e^{-Dt} \tag{3.6a}$$

can be constructed, where μ, B, and D are constants. Thus in this equation the specific growth rate is modified by two factors: the first, $1 - W/B$, depends linearly on substrate level as in the logistic equation (eqn (3.4d)); the second,

$\exp(-Dt)$, depends on the passage of time, and can be interpreted as differentiation, development, or senescence, as in the Gompertz equation (eqn (3.5d)). The logistic and Gompertz equations are contained in eqn (3.6a) as special cases: with $D = 0$, it becomes the logistic equation and $W_f = B$; as $B \to \infty$, it reduces to the Gompertz equation. As for the previous growth curves, μ is the specific growth rate for early growth (Exercise 3.3).

Integrating eqn (3.6a) gives (Exercise 3.3)

$$W = \frac{W_0 B}{W_0 + (\mathrm{B} - W_0)\exp\{-[\mu(1 - \mathrm{e}^{-Dt})/\mathrm{D}]\}}. \tag{3.6b}$$

At time $t = 0$ the initial dry mass is given by $W = W_0$. However, the final dry mass, obtained as $t \to 0$, is given by a more complex expression:

$$W_f = \frac{W_0 B}{W_0 + (B - W_0)\mathrm{e}^{-\mu/D}}. \tag{3.6c}$$

This equation can also be written as

$$B = W_f \frac{(\mathrm{e}^{\mu/D} - 1)}{(\mathrm{e}^{\mu/D} - W_f/W_0)}, \tag{3.6d}$$

which allows us to derive the following constraints on W_f in terms of the fitted parameters B and D. First, since B must be positive but can approach infinity through positive values,

$$W_f \leqslant W_0 \mathrm{e}^{\mu/D}, \tag{3.6e}$$

where, when the equality holds, $B \to \infty$ and the Chanter equation reduces to the Gompertz equation. Second, since $W_f > W_0$, it follows that

$$W_f \leqslant B, \tag{3.6f}$$

and in this case the equality holds when $D \to 0$ and the Chanter equation reduces to the logistic equation.

The Chanter equation is best illustrated by keeping W_0, W_f, and μ constant and allowing B or D to vary over the range allowed by (3.6e) and (3.6f); B and D are not independent, but are related by (3.6d). Note that from the inequality (3.6e) and the fact that D must be positive for physiologically realistic solutions, D lies in the range

$$0 \leqslant D \leqslant \frac{\mu}{\ln(W_f/W_0)}. \tag{3.6g}$$

In Fig. 3.5 the Chanter equation (eqn (3.6b)) is illustrated for fixed values of W_0, W_f, and μ, where D varies over the range of the inequality (3.6g) and B is given by eqn (3.6d). Note, once again, that the lower limit of D corresponds to the logistic equation and the upper limit to the Gompertz equation. It can be seen that there is a family of curves lying between the logistic and Gompertz equations.

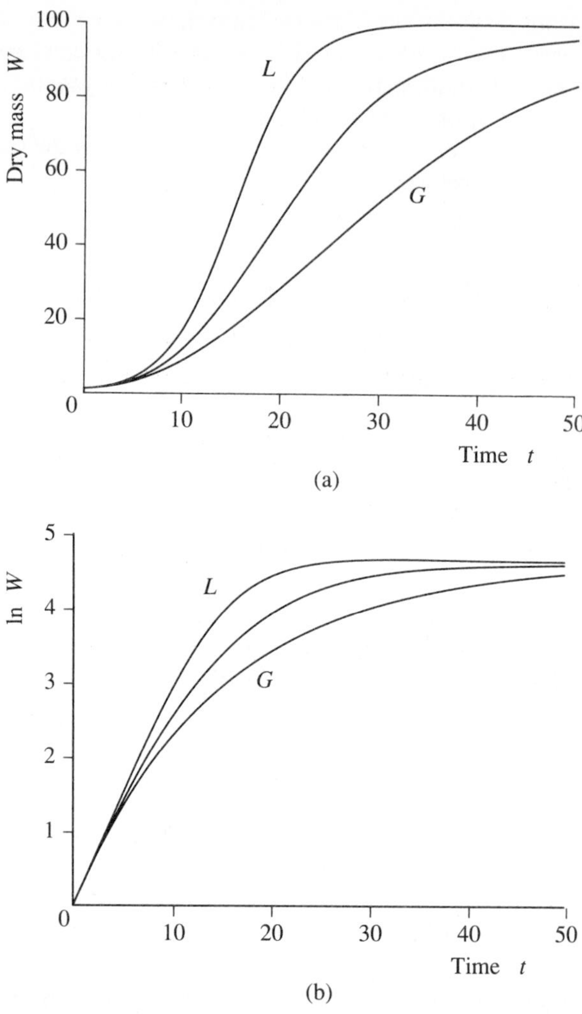

Fig. 3.5. Chanter growth equation as described by eqn (3.6b): W is the dry mass and t is time, both in arbitrary units; parameter values are $W_0 = 1$, $W_f = 100$, and $\mu = 0.3$. The limiting cases of the logistic equation ($D \to 0, B = 100$), denoted by L, and the Gompertz equation ($D = \mu/\ln(W_f/W_0) = 0.065, B \to \infty$), denoted by G, are indicated, and the intermediate curve is for $D = 0.03$. The parameter B is given by eqn (3.6d).

All the curves have a point of inflexion, but this can only be obtained numerically (Exercise 3.3).

The constraint (3.6g) on D is of no real consequence when fitting to data. Only four of the five parameters μ, W_0, W_f, B, and D are independent. Thus, the four fitted parameters may be μ, W_0, B, and D, and W_f is derived from these (eqn (3.6c)). The only constraints that must be specified when fitting the curve are that both B and D are positive.

When considering the Gompertz equation, we showed that the governing differential equation could be expressed independent of the time variable t. This is also possible with the Chanter equation: first eqn (3.6b) can be solved for e^{-Dt} and then combined with eqn (3.6a) to give

$$\frac{dW}{dt} = \mu W\left(\frac{B - W}{B}\right)\left\{1 - \frac{D}{\mu}\ln\left[\left(\frac{W}{B - W}\right)\left(\frac{B - W_0}{W_0}\right)\right]\right\}. \tag{3.6h}$$

3.7 Discussion

We have presented five growth functions in this chapter: simple exponential growth, the monomolecular equation, and the logistic, Gompertz, and Chanter equations, although the logistic and Gompertz equations are contained in the Chanter growth function as special cases. These equations are all relatively simple analytically, have biologically meaningful parameters, and are easy to apply to experimental data. It is unlikely that one of these equations would not provide a useful description for fairly smooth experimental data. If the data are not smooth, we should question the value of using growth functions to analyze such data: it may be more appropriate to study the physiology of the problem in order to understand the behaviour of the data.

The main applications of simple exponential growth are to plants growing in a uniform environment in the early stages of growth, or to plants grown at low temperatures which are substrate saturated. Hunt (1982) has shown that for a limited period of time in early growth plants grow exponentially, and Brouwer (1962) has demonstrated that during this time the shoot dry mass is proportional to the root dry mass. (Although Brouwer did not analyse his data in this way, his data also indicate that the plants were growing exponentially.) Consequently, we can conclude that there is a period of time during which the shoot and root grow exponentially and with the same specific growth rate. We refer to this as 'balanced exponential growth'. It is often of value to examine the behaviour of models of plant and crop processes (e.g. shoot : root partitioning (Chapter 13)) under the assumption of balanced exponential growth. However, care should be taken in the interpretation of the initial dry mass W_0 since during the early stages of growth, or following some environmental perturbation (e.g. a change in nitrogen supply), there may be transients to plant growth prior to balanced exponential growth. In such cases, the initial dry mass W_0 will refer to the value of W once the transients have settled down and, likewise, t will have to be measured from that point. The value in using this assumption is that, for systems of first-order differential equations which have balanced exponential growth solutions, these solutions are independent of the choice of initial conditions so that we can examine the behaviour of the model without any unintended influence from our choice of these initial conditions. Having done so, and being satisfied that the model behaves in the expected manner, we are then in a position to look at more general circumstances. This is discussed further in Chapter 13, where shoot : root partitioning is considered.

The choice between using the monomolecular equation or a sigmoidal equation is obviously governed by the general shape of the data. The model behind the monomolecular equation is very simple in that there is a constant supply of substrate and growth is proportional to the substrate supply and continues until it is exhausted. The sigmoidal responses incorporate the assumptions that growth is autocatalytic and that there is a finite supply of substrate (logistic), or the specific growth rate decays exponentially with time (Gompertz), or both these conditions apply (Chanter). These models do have some mechanistic basis: for example the substrate can be regarded as a finite supply of nitrogen to the soil which is exhausted during the growth of the crop, and temperature could be the main environmental factor influencing the decrease in the specific growth rate. However if, for example, nitrogen is continually applied and the growth data are still sigmoidal, then in this case it may not be clear exactly as to what the finite supply of substrate refers. The individual researcher must decide where the balance between mechanism and empiricism lies.

As we mentioned in Section 3.1, several commonly applied growth functions have been omitted from our discussion, and the reasons for this are now discussed. References are not given to the source of these equations, but the reader may consult France and Thornley (1984, Chapter 5) for further details.

The Richards function is defined by the differential equation

$$\frac{dW}{dt} = \frac{kW(W_f^n - W^n)}{nW_f^n}, \tag{3.7a}$$

where k is a growth constant, W_f is the asymptote of W as $t \to \infty$, and n is an empirical parameter. n is constrained by $n \geqslant -1$, since $n < -1$ is non-physiological and gives infinite growth rates as $W \to 0$. There is no obvious biological interpretation of the parameter n, although with n values of -1, 0, and 1 eqn (3.7a) reduces to the monomolecular, Gompertz, and logistic equations respectively, which makes it an attractive equation to deal with. However, it is our own experience that practical problems are often encountered when fitting to experimental data, in that the equation is very sensitive to variation in the parameter n. Although the Chanter equation only includes the logistic and Gompertz equations as special limiting case, we would recommend its use in preference to the Richards function. If the data are not sigmoidal, then a non-sigmoidal curve such as the monomolecular equation can be used directly. Usually, the more analytically versatile the growth function, the more difficult it may be to apply to experimental data.

Exponential polynomials are equations of the form

$$W = W_0 \exp(a_1 t + a_2 t^2 + a_3 t^3 + \cdots), \tag{3.7b}$$

where $a_1, a_2, \ldots$ are constants. Taking logarithms, eqn (3.7b) becomes

$$\ln W = \ln W_0 + a_1 t + a_2 t^2 + a_3 t^3 + \cdots. \tag{3.7c}$$

The appeal of this equation is clear from eqn (3.7c) in that polynomials are fitted to the logarithm of the dry masses; this may be an attractive statistical procedure, but has little physiological basis. Several authors have argued convincingly that only exponential quadratic functions should be used because of the unrealistic shape of the function with more terms. The exponential quadratic function is

$$W = W_0 \exp(a_1 t - a_2 t^2), \tag{3.7d}$$

which is illustrated in Fig. 3.6. This curve is symmetric about a maximum and

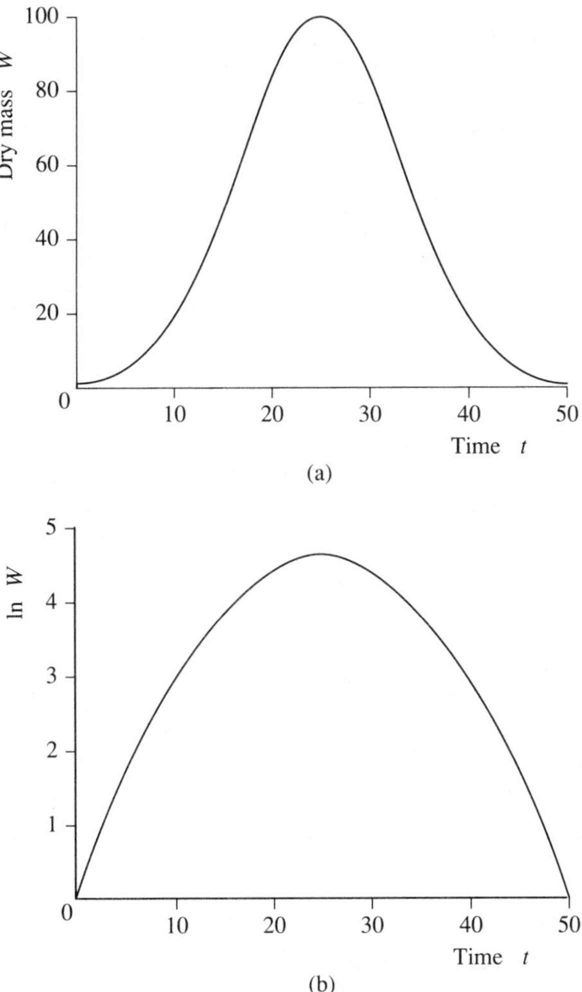

Fig. 3.6. Exponential quadratic equation as described by eqn (3.7d): W is the dry mass and t is time, both in arbitrary units; parameter values are $W_0 = 1$, $a_1 = 0.37$, and $a_2 = 0.0074$. The values of a_1 and a_2 are chosen so that the maximum dry mass is 100 and occurs at $t = 25$.

the usual procedure is to use only that part of the curve up to the maximum. This is clearly unsatisfactory, and again we would recommend the use of the Chanter equation in preference. It should be noted that there is a mechanistic basis to the exponential quadratic equation for meristematic cell growth where the proportion of cells which remain meristematic decreases with time, so that the meristematic cell number (and therefore mass) rises and then falls with time. The remaining cells differentiate, and the whole organ has sigmoidal growth. This model is discussed in Chapter 2 (p. 69).

The final approach is that of allometry, where the growth of two organs or plant parts, say the shoot and root, denoted by W_1 and W_2 are related by

$$W_1 = aW_2^b \tag{3.7e}$$

where a and b are parameters. Taking logarithms and then differentiating eqn (3.7e) leads to the differential equation

$$\frac{1}{W_1}\frac{\mathrm{d}W_1}{\mathrm{d}t} = b\frac{1}{W_2}\frac{\mathrm{d}W_2}{\mathrm{d}t}, \tag{3.7f}$$

suggesting that allometry originates from observations that the specific growth rate of one plant part is proportional to that of another. While this may, in some cases, be a reasonable empirical observation, there is little evidence to suggest any substantial mechanistic basis to the relationship. An important interpretative problem with allometry is that the dimensions of the parameter a depend on the numerical value of the parameter b. Suppose that W is measured in units of kilograms dry mass, then from eqn (3.7e) it follows that a has dimensions (kilograms dry mass)$^{1-b}$. This precludes comparison of data sets with different values of b, and so imposes considerable limitations on the usefulness of the technique.

To conclude, therefore, many approaches have been proposed for the statistical analysis of plant and crop growth data in recent years, and we feel it necessary to select from these those which are likely to be of most value. The few equations we have considered cover a wide range of responses and are likely to be sufficient for many applications. If these equations prove to be inadequate it may perhaps be appropriate to seek a more mechanistic understanding rather than use detailed and complex statistical techniques.

Exercises

3.1. Derive the monomolecular equation (eqn (3.3d)) from the governing differential equation (eqn (3.3c)). Show that this equation does not have a point of inflexion.

3.2. (a) Derive the logistic equation (eqn (3.4e)) from its governing differential equation (eqn (3.4d)).

(b) Derive the Gompertz equation (eqn (3.5e)) from its governing differential equation (eqn (3.5d)).

3.3. (a) Derive the Chanter equation (eqn (3.6b)) from its governing differential equation (eqn (3.6a)), and show that, during the early stages of growth, the specific growth rate is given by μ.

(b) Derive an equation whose solution will give the inflexion point of the Chanter growth equation.

3.4. Two frequently applied approximations for the specific growth rate of an organ, plant, or crop are

$$\frac{1}{W}\frac{\mathrm{d}W}{\mathrm{d}t}(t) \approx \frac{W_2 - W_1}{[(W_1 + W_2)/2](t_2 - t_1)} \tag{E3.4a}$$

and

$$\frac{1}{W}\frac{\mathrm{d}W}{\mathrm{d}t}(t) \approx \frac{\ln W_2 - \ln W_1}{t_2 - t_1} \tag{E3.4b}$$

where W_1 and W_2 are the dry masses at times t_2 and t_2 respectively and

$$t = \frac{t_1 + t_2}{2}. \tag{E3.4c}$$

(a) Show that eqn (E3.4a) is exact when growth is a linear function of time.

(b) Show that eqn (E3.4b) is exact for simple exponential growth (eqn (3.2b)).

(c) If eqn (E3.4a) is applied to the simple exponential growth equation at times $t \pm \delta$, calculate the associated fractional error in the approximation. Using $\mu = 0.1$, calculate the maximum value of δ, in whole days, such that this error is less than 0.1. In doing so, it may be useful to use the hyperbolic tangent tanh(x) with series expansion given by

$$\tanh(x) = \frac{\mathrm{e}^x - \mathrm{e}^{-x}}{\mathrm{e}^x + \mathrm{e}^{-x}} = x - \frac{x^3}{3} + \frac{2x^5}{15} + \cdots, \qquad \text{for } |x| < 1. \tag{E3.4d}$$

4
Transport processes

4.1 Introduction

In this chapter transport processes are considered—the transport of a single chemical species S from one location L_1 to a second location L_2, without any overall chemical or biochemical conversion occurring. Note that some transport mechanisms, such as facilitated diffusion or active transport, may involve a temporary chemical association with a carrier particle.

The growth and development of all forms of plant life, from forest trees to unicellular organisms, depend upon the transport of many substances to, within, and from the organism. While it is still arguable whether within-plant transport costs significantly affect or limit plant and crop growth, there is little doubt that restrictions in the supplies of carbon, nitrogen, water, and other nutrients are important physiologically and agronomically. Our interest is in the mathematical representation of these phenomena.

4.1.1 *Notation and terminology*

The symbol T is used to denote a transport flux, with units such as kilograms of substance S per day or kilogram moles of substance S per second. T_{ij} is the transport flux from pool i into pool j; T_{0i} and T_{i0} are the fluxes from outside the system into pool i and vice versa. The flux T_{ij} can be expected to depend upon some parameters P and also upon S_i and S_j, the concentrations of substance at the two locations i and j. The flux density J with units of, for instance, kilograms of substance per square metre per day, is related to the flux T by the equation

$$T = AJ \tag{4.1a}$$

where A (m^2) is the cross-sectional area of the transporting pathway.

Transport processes are sometimes described using the terms 'diffusion-like', 'facilitated', 'active', and 'convective flow'. It needs emphasizing that all continuing transport processes, of whatever sort, require inputs of energy in one form or another in order to be sustained. Even diffusion-like processes have their costs. While the more detailed examples given later should clarify matters, some questions that can be asked are as follows. Is the substance S in the same chemical/physical form along the pathway as at the two ends, or is there an association, as with oxygen and haemoglobin? Are there inputs of energy along the pathway or only at the two ends? Does the flux T_{ij} vanish if $S_i = S_j$? Is there some associated secondary flux of material, as with a solution or with migrating carriers (birds, insects, or molecules)?

Consider the equation

$$T_{ij} = k(S_i - S_j) \tag{4.1b}$$

for the transport flux of substance S between locations i and j, where S_i and S_j are the concentrations of S at i and j, and k is a parameter. Note that $T_{ij} = 0$ if $S_i = S_j$. Equation (4.1b) may apply, under certain conditions, to several mechanisms: diffusion, facilitated diffusion, and convective flow. The whole-plant modeller may prefer to view eqn (4.1b) as a phenomenological equation, providing a useful parameterization of the process while saying nothing about mechanism. However, consideration of some possible mechanisms leads to other relationships, such as

$$T_{ij} = kS_i(S_i - S_j) \tag{4.2a}$$

$$T_{ij} = k(S_i^2 - S_j^2) = k(S_i + S_j)(S_i - S_j) \tag{4.2b}$$

$$T_{ij} = k_1 S_i - k_2 S_j. \tag{4.2c}$$

It can be seen that in eqn (4.2c), $T_{ij} \neq 0$ when $S_i = S_j$, and this equation may describe transport up a concentration gradient (it is derived in eqn (4.18i)). Any of eqns (4.1b) and (4.2) may be of value to the crop and plant modeller, even when details of the mechanism are unknown or disputed.

4.2 Diffusion

Many phenomena, in biology, chemistry and physics, obey the diffusion equation (Crank 1975) which describes the process of diffusion mathematically. First we consider diffusion in one dimension (Fig. 4.1a). Let J (kg m^{-2} s^{-1}) be the flux

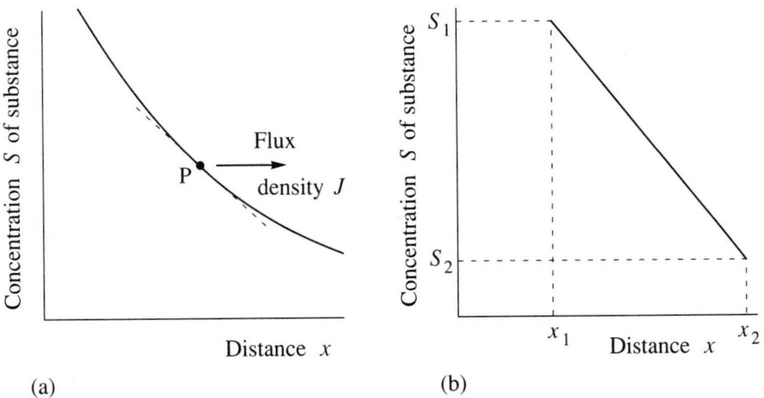

Fig. 4.1. Diffusion: (a) illustration of eqn (4.3) (the broken line at point P is a tangent to the concentration : distance curve and shows the concentration gradient at P; the flux density J is in the direction down the concentration gradient); (b) a constant concentration gradient, as in eqns (4.4).

density of substance S in the positive x direction. If S denotes the concentration of substance S ($\mathrm{kg\,m^{-3}}$), Fick's law of diffusion states that

$$J = -D\frac{\partial S}{\partial x}, \tag{4.3}$$

where D ($\mathrm{m^2\,s^{-1}}$) is the diffusion coefficient and $\partial S/\partial x$ denotes the concentration gradient.

If there are no sources or sinks of substance S along the path, then J is constant and therefore the concentration gradient is constant (Fig. 4.1b). Using an obvious notation, the concentration S can be written

$$S = S_1 + \frac{S_2 - S_1}{x_2 - x_1}(x - x_1), \tag{4.4a}$$

giving

$$\frac{\partial S}{\partial x} = \frac{S_2 - S_1}{x_2 - x_1}. \tag{4.4b}$$

Substituting in eqn (4.3), therefore, we have

$$J = -D\frac{S_2 - S_1}{x_2 - x_1}. \tag{4.4c}$$

This is positive for the values shown in Fig. 4.1b, indicating a flux density in the positive x direction.

4.2.1 *Derivation of the diffusion equation*

Next, diffusion down a uniform cylinder of constant cross-section A ($\mathrm{m^2}$) is considered (Fig. 4.2). At a given point in time, the concentration of the diffusing substance S only varies with the coordinate x along the length of the cylinder. An element of cylinder lying between x and $x + \Delta x$ is shown in Fig. 4.2; this element has the volume

$$\Delta V = A\,\Delta x. \tag{4.5a}$$

The flux into the element is AJ_x and the flux out of it is $AJ_{x+\Delta x}$ ($\mathrm{kg\,s^{-1}}$). Assume that the substance S can be created or destroyed with a source/sink activity of σ ($\mathrm{kg\,m^{-3}\,s^{-1}}$). A positive value of σ indicates that substance is created, and vice versa. The amount of substance in the volume element is $S\,\Delta V$. The conservation equation is

rate of change of quantity of substance = inflow − outflow + source strength (4.5b)

which mathematically becomes

$$\frac{\partial}{\partial t}(S\,\Delta V) = A(J_x - J_{x+\Delta x}) + \sigma\,\Delta V. \tag{4.5c}$$

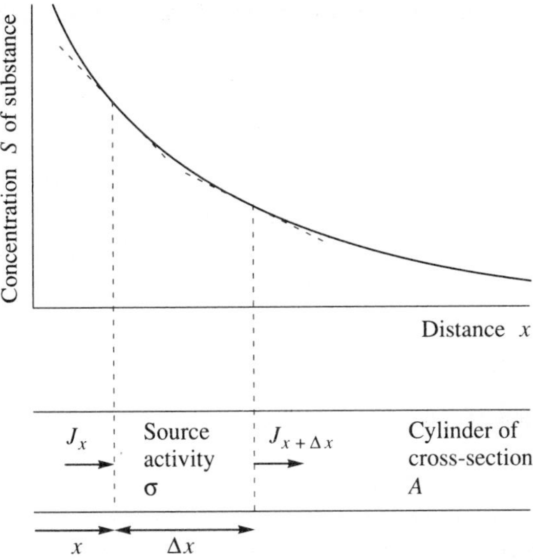

Fig. 4.2. Diffusion along a uniform cylinder of cross-section A. An element of length Δx is considered, with flux densities into and out of the element of J_x and $J_{x+\Delta x}$. See text and eqns (4.5).

Approximately, to order Δx,

$$J_{x+\Delta x} = J_x + \frac{\partial J_x}{\partial x}\Delta x, \tag{4.5d}$$

which, with eqn (4.3), gives

$$J_x - J_{x+\Delta x} = \Delta x \frac{\partial}{\partial x}\left(D\frac{\partial S}{\partial x}\right). \tag{4.5e}$$

Substituting in eqn (4.5c), using eqn (4.5a), and simplifying, therefore, we obtain

$$\frac{\partial S}{\partial t} = \frac{\partial}{\partial x}\left(D\frac{\partial S}{\partial x}\right) + \sigma. \tag{4.5f}$$

If the diffusion coefficient D is assumed to be constant, this equation takes the more familiar form

$$\frac{\partial S}{\partial t} = D\frac{\partial^2 S}{\partial x^2} + \sigma. \tag{4.5g}$$

4.2.2 *Some steady state solutions of the diffusion equation in one dimension*

Constant concentrations at the pathway ends: no sources/sinks This is the simple introductory example discussed in eqns (4.4). Boundary values are $x = x_1, S = S_1$

and $x = x_2$, $S = S_2$. As $\partial S/\partial t = 0$ in the steady state and $\sigma = 0$, from eqn (4.5g)

$$\frac{d^2S}{dx^2} = 0. \tag{4.6a}$$

Thus

$$\frac{dS}{dx} = \text{constant} = \frac{S_2 - S_1}{x_2 - x_1} \tag{4.6b}$$

and

$$S = S_1 + \frac{S_2 - S_1}{x_2 - x_1}(x - x_1). \tag{4.6c}$$

For a cross-section of area A, the transport flux from x_1 to x_2 is

$$T_{12} = AJ = -AD\frac{S_2 - S_1}{x_2 - x_1}. \tag{4.6d}$$

This result is the same as that given earlier in eqns (4.4). See Exercise 4.1.

First-order sink: constant concentration of S at $x = 0$ A first-order sink means that the source activity σ can be written

$$\sigma = -kS \tag{4.7a}$$

where k is a constant (s^{-1}). Boundary values are $x = 0$, $S = S_0$ and $x \to \infty$, $S \to 0$. Equation (4.5g) becomes

$$0 = D\frac{d^2S}{dx^2} - kS. \tag{4.7b}$$

With

$$\alpha = \left(\frac{k}{D}\right)^{1/2}, \tag{4.7c}$$

the general solution to eqn (4.7b) is

$$S = C_1 e^{\alpha x} + C_2 e^{-\alpha x}, \tag{4.7d}$$

where C_1 and C_2 are constants. We now put in the boundary values $C_1 = 0$ (to keep the solution finite) and $C_2 = S_0$. Hence

$$S = S_0 e^{-\alpha x}. \tag{4.7e}$$

The flux that must be provided at the origin ($x = 0$) to maintain the concentration constant at S_0 is

$$-AD\frac{dS}{dx}(x = 0) = ADS_0\alpha = AS_0(kD)^{1/2}. \tag{4.7f}$$

See Exercise 4.2. See also Chapter 7, Section 7.4, p. 182.

4.2.3 *Resistance representation of diffusion*

It may sometimes be useful to think in terms of a diffusion resistance r_{12} (m^{-3} s or m^{-3} day) between pathway ends. Equation (4.6d) can be written

$$T_{12} = \frac{S_1 - S_2}{r_{12}} \tag{4.8a}$$

where

$$r_{12} = \frac{x_2 - x_1}{AD}. \tag{4.8b}$$

Some workers prefer to use a conductance g ($m^3 \, s^{-1}$ or $m^3 \, day^{-1}$) related to resistance r by means of

$$g = 1/r. \tag{4.8c}$$

Note that g can be interpreted as a volume flow rate. Equation (4.8a) then becomes

$$T_{12} = g_{12}(S_1 - S_2). \tag{4.8d}$$

4.2.4 *Diffusion resistance plus a biochemical sink*

Consider the scheme defined in Fig. 4.3. The pathway ABC is assumed to have negligible capacity for the storage of substrate, and therefore the transport flux T between A and B is the same as that between B and C (this would also be true given storage at B and making the steady state assumption). Thus

$$T = \frac{S_1 - S_2}{r_d} = \frac{kS_2}{K + S_2} \tag{4.9a}$$

where r_d (m^{-3} day) is a diffusion resistance; k (kg day^{-1}) and K (kg m^{-3}) are parameters defining the substrate response of the biochemical sink for which the Michaelis–Menten equation (p. 52) has been assumed. In terms of the variable resistance r_c of Fig. 4.3, the last term in eqn (4.9a) can be replaced by

$$\frac{S_2}{r_c} \quad \text{where} \quad r_c = \frac{K + S_2}{k}. \tag{4.9b}$$

The elimination of S_2 between the two equations in (4.9a) leads to the quadratic

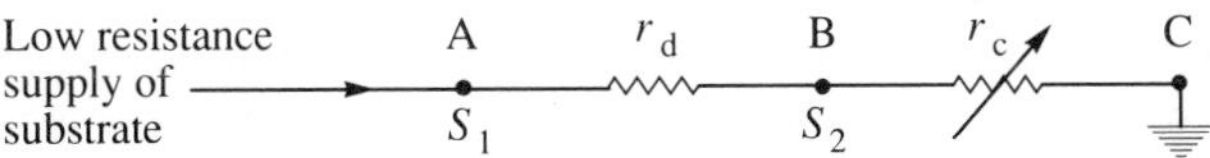

Fig. 4.3. Diffusion resistance plus a biochemical sink. The diffusion process is represented by the resistance r_d (eqn (4.8b)) and the biochemical sink is represented by a variable resistance r_c connected to earth at C. The substrate concentraton at A is S_1 (assumed constant) and at B is S_2.

$$0 = T^2 r_d - T(K + S_1 + kr_d) + kS_1 \tag{4.9c}$$

with the (biologically significant) root

$$T = \frac{(K + S_1 + kr_d) - [(K + S_1 + kr_d)^2 - 4r_d kS_1]^{1/2}}{2r_d}. \tag{4.9d}$$

The other root corresponds to negative values of S_2, which is a consequence of $T > S_1/r_d$ and $S_2 = S_1 - Tr_d$. Equation (4.9c) describes a non-rectangular hyperbola whose properties are discussed in detail on p. 57. Equation (4.9c) can be put in reduced form

$$0 = \left(\frac{T}{k}\right)^2 \frac{kr_d}{K} - \frac{T}{k}\left(1 + \frac{S_1}{K} + \frac{kr_d}{K}\right) + \frac{S_1}{K}, \tag{4.9e}$$

so that plotting T/k against S_1/K results effectively in the shape of the curve's being determined by a single parameter kr_d/K (cf. Fig. 2.7, p. 58). The sharpness of the knee of the curve is easily adjusted: if $r_d = 0$, then the response is simply the Michaelis–Menten response of eqn (2.7f) (p. 52); if $K = 0$, then the response follows the asymptotes ($T = S_1/r_d$ and $T = k$), switching abruptly from one to the other where they cross.

Sometimes the slope of the response is required; this is given by

$$\frac{dT}{dS_1} = \frac{1 - [(K + S_1 + kr_d)^2 - 4r_d kS_1]^{-1/2}[(K + S_1 + kr_d) - 2r_d k]}{2r_d}. \tag{4.9f}$$

It can be shown that the initial response to the substrate is

$$\frac{dT}{dS_1}(S_1 = 0) = \frac{1}{r_d + K/k}. \tag{4.9g}$$

At low values of substrate concentration, the flux is determined by both diffusion resistance and biochemical resistance, as in eqn (4.9g); at high values the flux is limited by the biochemical parameter k. See Exercise 4.3.

4.3 Facilitated diffusion

Diffusion plays a role in many biological phenomena. However, in some cases the rates of transport achieved by pure diffusion are too small and would restrict growth of the organism: the diffusion coefficient may be rather low, coupled sometimes to a very low solubility of the substance in question. Facilitated diffusion is essentially still a diffusion process, but some of the limitations of pure diffusion have been overcome.

The main idea in facilitated diffusion is that carrier molecules, denoted by C, are present in the region PQ (Fig. 4.4) across which transport is to occur. The carrier molecules can combine reversibly with the substrate S, according to

$$\mathrm{S} + \mathrm{C} \underset{k_2}{\overset{k_1}{\rightleftharpoons}} X \tag{4.10a}$$

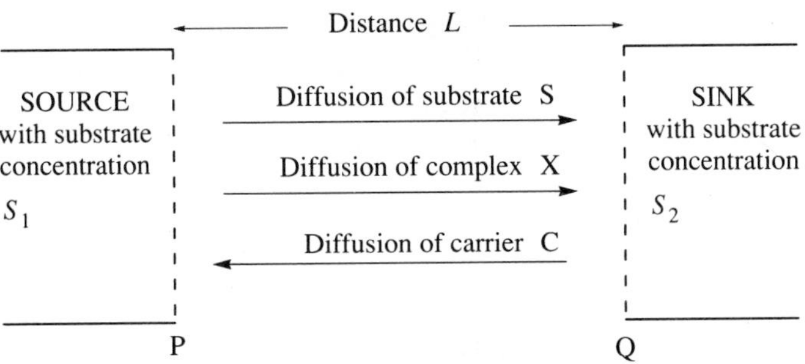

Fig. 4.4. Facilitated diffusion of substrate S from a source, where S is maintained at a concentration S_1, to a sink, where the substrate concentration is S_2. The carrier C can combine reversibly with substrate S to give a carrier–substrate complex X. Neither C nor X can pass through the barriers shown by the broken lines at P and Q; it is assumed that substrate can pass freely through these barriers.

where k_1 and k_2 are rate constants and X denotes the substrate–carrier complex. Effectively, the reaction (4.10a) increases the solubility of the substrate, thereby increasing the concentrations and concentration gradients which drive diffusion. The best-known examples of facilitated diffusion concern oxygen transport in muscle tissue, where the carrier is the protein myoglobin, and in leguminous plants with leghaemoglobin. In these cases both the carrier molecule C and the substrate–carrier molecule X, which are both large molecules compared with oxygen, diffuse more slowly, but the large increase in solubility (over the low solubility of oxygen) enhances the diffusive transport flux considerably. A simplified account of the problem is given here; more extensive treatments can be found in Murray (1977) and Stein (1986, Chapter 4).

The scheme assumed is shown in Fig. 4.4. The substrate concentrations in the source and sink are maintained at S_1 and S_2 (kg mol m^{-3}); the pathway between source and sink is of length L (m) and the cross-sectional area of the pathway is A (m^2). The carrier C and the substrate–carrier complex X are confined to the region PQ by means of semi-permeable barriers, presumably membranes, situated at P and Q. Let D_S, D_C, and D_X (m^2 s^{-1}) be the diffusion coefficients of S, C, and X. If C_t is the total concentration of carrier molecules (both combined and uncombined), then

$$C_t = C + X. \tag{4.10b}$$

With the assumption that

$$D_C = D_X, \tag{4.10c}$$

which is generally reasonable (the carrier molecule C and the substrate–carrier molecule X are of similar size), then in the steady state C_t is constant along the pathway PQ. While this is demonstrated mathematically below, it can be seen intuitively: in Fig. 4.4, the flux of X to the right is equal to the flux of C to the left in the steady state; with eqn (4.10c) the concentration gradients of X and C must be equal and opposite; adding concentration gradients that are equal and opposite, as in eqn (4.10b), gives a constant value of C_t.

From eqn (4.10a), a function σ is defined by

$$\sigma = k_1 SC - k_2 X, \tag{4.10d}$$

which, with eqn (4.10b), gives

$$\sigma = k_1 S(C_t - X) - k_2 X. \tag{4.10e}$$

In the steady state the diffusion eqn (4.5g) becomes, for the three species S, C and X,

$$D_S \frac{d^2 S}{dx^2} = \sigma \tag{4.10f}$$

$$D_X \frac{d^2 X}{dx^2} = -\sigma \tag{4.10g}$$

$$D_C \frac{d^2 C}{dx^2} = \sigma. \tag{4.10h}$$

Adding eqns (4.10g) and (4.10h) with $D_X = D_C$ gives

$$\frac{d^2 X}{dx^2} + \frac{d^2 C}{dx^2} = 0;$$

hence

$$\frac{dX}{dx} + \frac{dC}{dx} = \text{constant.}$$

This constant is zero since it is assumed that no net flux of the carrier species takes place. Thus

$$\frac{dX}{dx} + \frac{dC}{dx} = 0$$

giving

$$C + X = C_t,$$

where C_t is constant throughout the membrane. Adding eqns (4.10f) and (4.10g) gives

$$D_S \frac{d^2 S}{dx^2} + D_X \frac{d^2 X}{dx^2} = 0. \tag{4.10i}$$

Integrating leads to

$$D_S \frac{dS}{dx} + D_C \frac{dX}{dx} = -J \tag{4.10j}$$

where J is a constant of integration and is in fact equal to the total transported flux density of substrate, since substrate is transported by diffusion of S and of the carrier–substrate complex X. Equation (4.10j) can be written

$$D_S \int_{S_1}^{S_2} dS + D_X \int_{X_1}^{X_2} dX = -J \int_0^L dx$$

and hence, on integrating, we obtain

$$J = \frac{1}{L}[D_S(S_1 - S_2) + D_X(X_1 - X_2)]. \tag{4.10k}$$

Multiplying by the cross-sectional area A, the flux T_d of substrate from P to Q due to pure diffusion is, from the first term of eqn (4.10k),

$$T_d = \frac{A}{L} D_S(S_1 - S_2). \tag{4.10l}$$

It is assumed throughout that the semi-permeable barriers at P and Q do not present any resistance to movement of substrate. The flux T_{fd} of substrate from P to Q due to facilitated diffusion is, from the second term of eqn (4.10k),

$$T_{fd} = \frac{A}{L} D_X(X_1 - X_2). \tag{4.10m}$$

At P, (4.10a) becomes

$$S_1 + C_1 \underset{k_2}{\overset{k_1}{\rightleftharpoons}} X_1$$

and, if this is assumed to be in equilibrium, we obtain

$$k_1 S_1 C_1 = k_2 X_1,$$

Defining K (kg mol m^{-3}) by

$$K = k_2/k_1 \tag{4.10n}$$

gives

$$X_1 = C_t \frac{S_1}{K + S_1}. \tag{4.10o}$$

Similarly at Q

$$X_2 = C_t \frac{S_2}{K + S_2}. \tag{4.10p}$$

In terms of the source and sink substrate concentrations, eqn (4.10m) for the facilitated diffusion flux becomes

$$T_{fd} = \frac{A}{L} D_X C_t \left[\frac{K(S_1 - S_2)}{(K + S_1)(K + S_2)} \right]. \tag{4.10q}$$

The quantity in square brackets is dimensionless, and we denote it by y, where

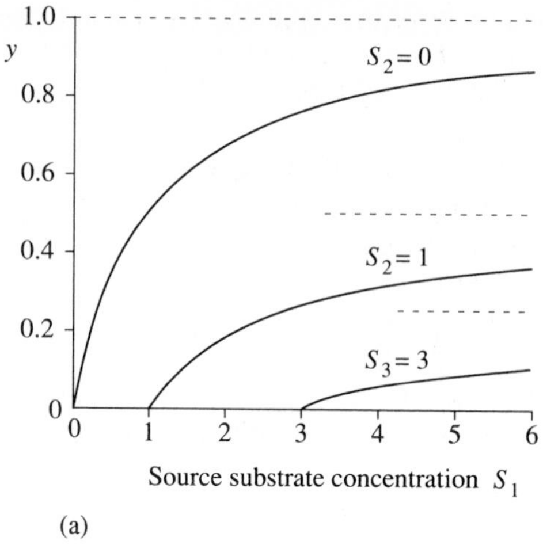

(a)

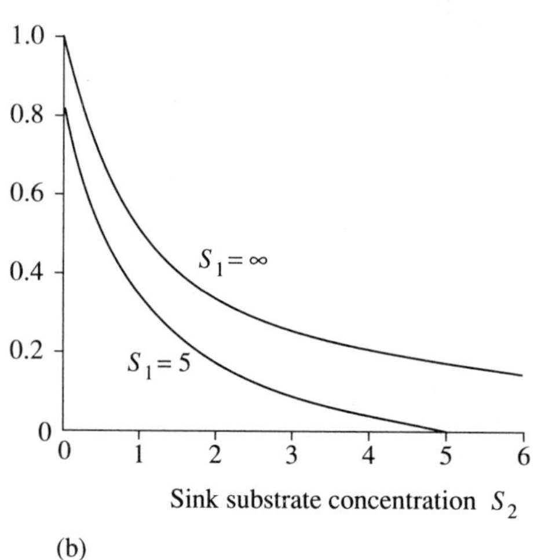

(b)

Fig. 4.5. Response of facilitated diffusion (eqns (4.10q) and (4.10r)) to the source and sink substrate concentrations S_1 and S_2. Equation (4.10r) is drawn for $K = 1$, and various values of S_2 and S_1 as shown. The broken lines indicate asymptotes.

$$y = \frac{K(S_1 - S_2)}{(K + S_1)(K + S_2)}. \tag{4.10r}$$

This expression defines the response of the facilitated diffusion flux T_{fd} to the substrate concentrations of the source and sink, and is illustrated in Fig. 4.5. The ratio of facilitated diffusion (eqn (4.10q)) to ordinary diffusion (eqn (4.10l)) is defined as ϕ, so that

$$\phi = \frac{T_{fd}}{T_d} = \frac{D_X}{D_S} \frac{C_t}{S_1 - S_2} y. \tag{4.10s}$$

Considering the behaviour of eqns (4.10q), (4.10r), and (4.10s), if S_1, $S_2 \ll K$, then

$$y = \frac{S_1 - S_2}{K} \quad \text{and} \quad \phi = \frac{D_X C_t}{D_S K}. \tag{4.10t}$$

In this limit, facilitated diffusion is diffusion like in its response to source and sink substrate concentrations.

A powerful sink will maintain $S_2 \approx 0$, and with $S_2 \ll S_1$ and $S_2 \ll K$,

$$y = \frac{S_1}{K + S_1}. \tag{4.10u}$$

In this case, facilitated diffusion exhibits a Michaelis–Menten response; at high source substrate concentration ($S_1 \to \infty$), $y \to 1$ and

$$T_{fd}(\text{maximum}) = \frac{A}{L} D_X C_t. \tag{4.10v}$$

Differentiation of eqn (4.10r) gives

$$\frac{\partial y}{\partial S_1} = \frac{S_1}{(K + S_1)^2} \quad \text{and} \quad \frac{\partial y}{\partial S_2} = -\frac{S_2}{(K + S_2)^2}. \tag{4.10w}$$

With increasing source concentration S_1, the facilitated flux always increases, but at a decreasing rate; with increasing sink concentration, the facilitated flux always decreases; the response of the facilitated flux to changes in substrate concentration is maximum when S_1 or S_2 is equal to K. This is easily demonstrated by evaluating

$$\frac{\partial^2 y}{\partial S_1^2} = \frac{K - S_1}{(K + S_1)^2} \quad \text{and} \quad \frac{\partial^2 y}{\partial S_2^2} = -\frac{K - S_2}{(K + S_2)^2}, \tag{4.10x}$$

which are zero at $S_1 = K$ and $S_2 = K$ respectively, giving maxima in eqns (4.10w).

Differentiation of eqn (4.10r) with respect to K leads to

$$\frac{dy}{dK} = \frac{(S_1 - S_2)(S_1 S_2 - K^2)}{(K + S_1)^2 (K + S_2)^2}. \tag{4.10y}$$

For given substrate concentrations S_1 and S_2, the flux is a maximum if

$$K^2 = S_1 S_2. \tag{4.10z}$$

Finally, we note that the most obvious consequence of facilitated diffusion is increased transport, as obtained by adding together the two transport fluxes of eqns (4.10l) and (4.10q); the increase is most significant with a high carrier concentration C_t, a reasonable value for the diffusion coefficient D_X of the carrier–substrate complex, a source concentration S_1 much higher than K, so that the carrier at P (Fig. 4.4) is saturated with substrate, and a sink concentration S_2 that is much less than K so that the substrate is easily detached from the carrier complex at Q. A bonus of facilitated transport is the control of the flux that is obtained from the sink end of the system: as soon as the sink concentration $S_2 \approx K$, the flux is rapidly cut back (eqn (4.10q) and Fig. 4.5). See Exercises 4.4 and 4.5.

4.4 Transport across membranes

The transport of ions and of various compounds across membranes, either into or out of plant cells, or into or out of organelles such as chloroplasts and vacuoles, is important to plant growth and development. Some ions of particular concern include nitrate, phosphate, potassium, and sodium; some important substances are glucose and amino acids. While many detailed physical, chemical, and biochemical studies of membranes have been made, both experimentally and from the theoretical standpoint (e.g. see Curran and Schultz 1968, Stein 1986), we describe below two simple models (Thornley, 1976, pp. 61–66) in order to gain some insight into the processes involved and to deduce mathematical expressions that can be used semi-empirically by the plant modeller for processes such as nitrogen uptake by root systems. The schemes discussed have certain features in common with facilitated transport as outlined in the last section and in Fig. 4.4; however, the facilitated transport flux of eqn (4.10q) requires a concentration gradient, and then substrate is transported down the gradient from the high concentration to the low concentration. A characteristic that distinguishes many membrane transport processes is their ability to undertake what is known as 'active' transport—in which substances may move from regions of low concentration to regions of high concentration. This requires an energy input—here it is assumed that adenosine triphosphate (ATP) is coupled into the reaction on the inside of the membrane, so as to make the reaction of substrate release irreversible.

4.4.1 *ATP-driven transport with fast diffusion*

The scheme assumed is shown in Fig. 4.6. Note that the membrane is asymmetrical with respect to the reactions that can take place on its surface. On the outer surface, substrate molecules at concentration S_1 can combine reversibly with carrier C to give a carrier–substrate complex X, according to

$$\mathrm{S_1} + \mathrm{C} \underset{k_2}{\overset{k_1}{\rightleftharpoons}} \mathrm{X}, \tag{4.11a}$$

where k_1 and k_2 are constants. If fast diffusion of carrier C and of carrier–substrate complex X within the membrane is assumed, then the concentrations of C and X are the same on both sides of the membrane. On the inner side of the membrane it is assumed that the release of substrate from the substrate–carrier complex is driven irreversibly by ATP, namely

$$\mathrm{X} + \mathrm{ATP} \overset{k_3}{\rightarrow} \mathrm{S_2} + \mathrm{C} + \mathrm{ADP} + \text{other products} \tag{4.11b}$$

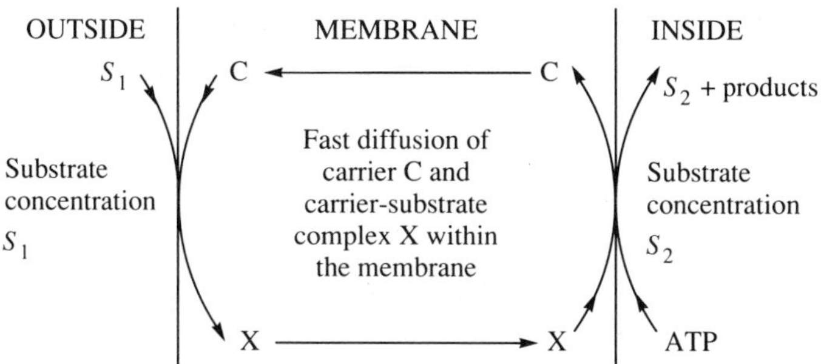

Fig. 4.6. Model of active transport across a membrane. S is the substrate being transported and C is a carrier confined to the membrane that can combine with substrate S on the outer membrane surface to give a carrier–substrate complex X. On the inner membrane surface ATP drives the irreversible release of substrate. Rapid diffusion of C and X within the membrane are assumed. The response deduced for this scheme is given in eqn (4.11f).

where k_3 is a constant. If T (kg mol s^{-1}) is the transport flux across the membrane, which is of area A (m^2), then in the steady state the fluxes across the two boundaries are equal, so that

$$T = A(k_1 C S_1 - k_2 X), \tag{4.11c}$$

and

$$T = A k_3 c_{\text{ATP}} \text{X}, \tag{4.11d}$$

where c_{ATP} is the concentration of ATP. By balancing dimensions in eqns (4.11c) and (4.11d), it can be seen that k_1, k_2, and k_3 are not normal rate constants, but involve surface activities, or volume activities with an active depth; k_1, k_2, and k_3 have units of (kg mol)$^{-1}$ m^4 s^{-1}, m s^{-1}, and (kg mol)$^{-1}$ m^4 s^{-1} respectively.

Let C_t be the total concentration of carrier, where

$$C_t = C + X. \tag{4.11e}$$

This equation is used to substitute for C in eqn (4.11c), and the resulting equation is used to eliminate X from eqn (4.11d), giving

$$T = \frac{A k_1 k_3 C_t S_1 c_{\text{ATP}}}{k_1 S_1 + k_2 + k_3 c_{\text{ATP}}}. \tag{4.11f}$$

This familiar expression describes a transport flux which has a rectangular-hyperbolic dependence on S_1, the substrate concentration outside the membrane, and c_{ATP}, the concentration of ATP inside the membrane. Note that the flux increases without limit as S_1 and c_{ATP} increase; this unrealistic result is a consequence of assuming very rapid diffusion within the membrane—this assumption is modified in the next section. The concentration c_{ATP} of ATP can be simply related to (for instance) the concentration of carbohydrate which can be oxidized, thereby producing ATP.

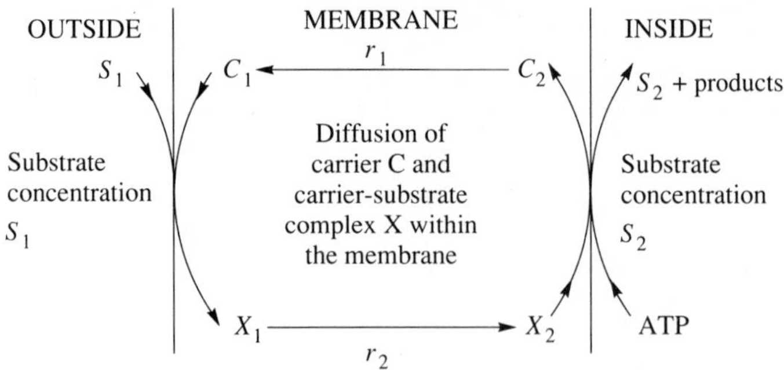

Fig. 4.7. Model of active transport across a membrane. It is assumed that the carrier C and the carrier–substrate complex X diffuse across the membrane with diffusion resistances of r_1 and r_2 as shown. On the inner membrane surface ATP drives the irreversible release of substrate S. The response deduced for this scheme is given in eqn (4.12l).

4.4.2 *ATP-driven transport with a slowly diffusing carrier*

In contrast with the last section, it is now assumed that there are concentration gradients of carrier C and carrier–substrate complex X across the membrane. Using the scheme and definitions shown in Fig. 4.7, we obtain

$$S_1 + C_1 \underset{k_2}{\overset{k_1}{\rightleftharpoons}} X_1 \tag{4.12a}$$

on the outer surface of the membrane and

$$X_2 + ATP \overset{k_3}{\rightarrow} S_2 + C_2 + \text{products} \tag{4.12b}$$

on the inner surface; k_1, k_2, and k_3 are constants. Denoting the transport flux across the membrane by T (kg mol s^{-1}) and the area of the membrane by A (m^2), we obtain in the steady state

$$T = A(k_1 C_1 S_1 - k_2 X_1) \tag{4.12c}$$

$$T = A k_3 c_{ATP} X_2, \tag{4.12d}$$

and diffusion across the membrane gives

$$T = \frac{A(X_1 - X_2)}{r_1} \tag{4.12e}$$

$$T = \frac{A(C_2 - C_1)}{r_2}. \tag{4.12f}$$

The diffusion resistances r_1 and r_2 have units of s m^{-1}, k_1 and k_3 have units of (kg mol)$^{-1}$ m^4 s^{-1}, and k_2 has units of m s^{-1}. The concentrations of carrier C and carrier–substrate complex X are linear across the membrane, and the total concentration C_t of carrier is therefore given by

$$C_t = \tfrac{1}{2}(C_1 + C_2 + X_1 + X_2). \tag{4.12g}$$

X_1, X_2, C_1, and C_2 are eliminated between eqns (4.12c)–(4.12g) to obtain an expression for the transport flux T. From eqns (4.12e) and (4.12f),

$$X_1 = \frac{T}{A} r_1 + X_2 \quad \text{and} \quad C_2 = \frac{T}{A} r_2 + C_1. \tag{4.12h}$$

Substituting these in eqn (4.12g) gives

$$C_t = \frac{1}{2}\left[2C_1 + 2C_2 + \frac{T}{A}(r_1 + r_2)\right]. \tag{4.12i}$$

Eliminating X_1 between eqn (4.12c) and the first of eqns (4.12h) gives

$$k_1 C_1 S_1 = \frac{T}{A}(1 + k_2 r_1) + k_2 X_2. \tag{4.12j}$$

Substituting for C_1 in (4.12i) now gives

$$2C_t k_1 S_1 = 2X_2(k_1 S_1 + k_2) + \frac{T}{A}[k_1 s_1(r_1 + r_2) + 2(1 + k_2 r_1)]. \tag{4.12k}$$

Equation (4.12d) is used to substitute for X_2 in eqn (4.12k), leading to

$$T = \frac{A C_t k_1 k_3 S_1 c_{\mathrm{ATP}}}{k_2 + k_1 S_1 + k_3(1 + k_2 r_1)c_{\mathrm{ATP}} + k_1 k_3 (r_1 + r_2) S_1 c_{\mathrm{ATP}}/2}. \tag{4.12l}$$

This apparently complicated expression is a typical Michaelis–Menten bi-substrate response (p. 60) of the form $xy/(1 + x + y + xy)$; in contrast with eqn (4.11f), the transport flux saturates for high values of S_1 and c_{ATP}. See Exercise 4.6.

4.5 Long-distance transport

Diffusion does not produce fluxes of sufficient magnitude for plant growth (Exercise 4.1) except for very small and simple organisms. Both plants and animals have developed specialized structures and mechanisms for providing transport fluxes of adequate size. If a plant or crop model is going to help us understand the role of transport in determining growth and productivity, it is essential that transport processes are represented explicitly within the model.

Some have argued that within-plant transport occurs so easily that transport resistances do not significantly affect the patterns of crop growth, and therefore transport processes need not be included within a crop growth simulator. This is a viewpoint which we do not share: in most natural systems there is a tendency for the source mechanisms (of substrates), the transport processes, and the sink (or utilization) processes to be in balance, so that each similarly constrains plant growth; further, there is unequivocal evidence that there are gradients of substrate concentrations within plants, and these would not exist for vanishingly small transport resistances.

Some recent physiological accounts of the subject have been given by Peel (1974) and Moorby (1981). Our aim in this section is to analyse some simple model systems in order to extract equations that may be useful in plant and crop modelling. We are more concerned with describing the overall behaviour of the transport processes in terms of those properties of the plant that are of interest than in the fine details of mechanism.

In 1928 Mason and Maskell reported a now classical series of experiments on sugar movement in cotton plants. What they observed in an extensive series of experiments was that sugar movement appeared to obey a diffusion-type equation (eqn (4.4c)), that is, sugar moves down the concentration gradient at a rate proportional to the gradient. Our considerations in this section show that a variety of different mechanisms are able to lead to this pattern of response.

4.5.1 *Convective flow in a simple recycling scheme*

The scheme assumed is shown in Fig. 4.8. The transport flux T_{12} is

$$T_{12} = q(S_1 - S_2). \tag{4.13}$$

This is of the same form as the transport flux due to diffusion in eqn (4.6d). The system is non-polar, without any intrinsic directionality in the system, and the polarity is imposed by the source and the sink. There are two sources of energy required to drive the system: the first maintains the circulation of solution at rate q; the second is the source/sink system which supplies molecules at concentration S_1 and removes them at concentration S_2, so that $S_1 \neq S_2$ and transport of S can occur. In the steady-state, the rate of transport of S is independent of the volume flow rate q, and is set by the rates of loading and unloading, which are equal; S_1 and S_2 adjust themselves so that the rates of loading, transport and unloading are the same.

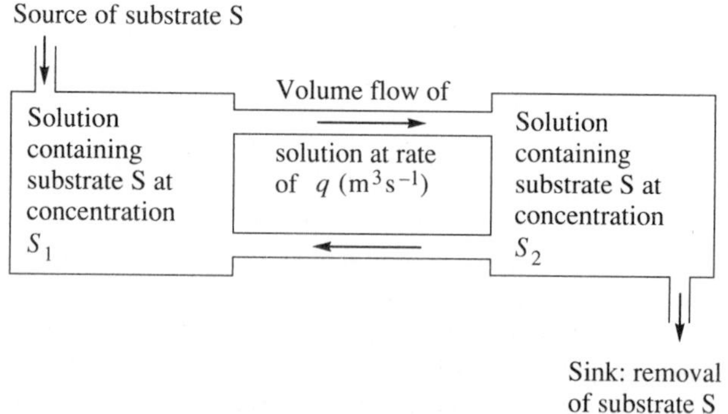

Fig. 4.8. Convective flow with simple recycling of solution.

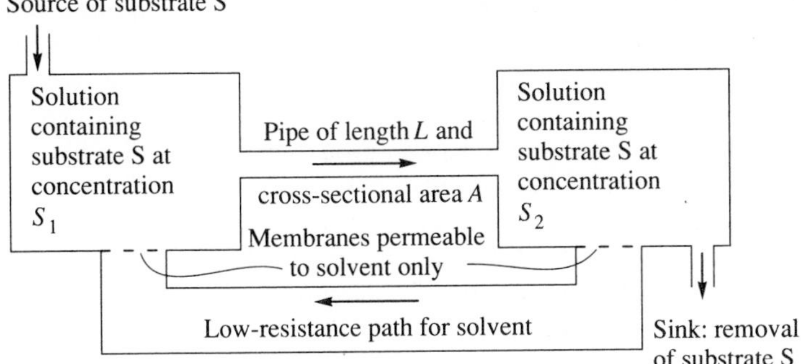

Fig. 4.9. Simple scheme for convective flow driven by osmotic pressure—the Münch hypothesis.

4.5.2 *Convective flow driven by osmotic pressure*

This mechanism was suggested by Münch (1930) to account for the transport of sugars in plants, and it is now widely referred to as the Münch hypothesis. The scheme shown in Fig. 4.9 is assumed. Fig. 4.9 differs from Fig. 4.8 in that there are now two semi-permeable membranes, and there is a low-resistance return path for the solvent.

Taking units of kg mol m^{-3} for the concentrations S_1 and S_2 at the ends of the pathway, we obtain the osmotic pressures p_1 and p_2 (kg m^{-1} s^{-2}) in the two compartments as

$$p_1 = S_1 R\theta \qquad \text{and} \qquad p_2 = S_2 R\theta, \tag{4.14a}$$

where θ is the absolute temperature (K) and $R = 8314.32$ J (kg mol)$^{-1}$ K^{-1} is the gas constant. If non-turbulent flow is assumed, Poiseuille's formula gives the volume flow rate q (m^3 s^{-1}) as

$$q = \frac{(p_1 - p_2)\pi r^4}{8L\eta}, \tag{4.14b}$$

where r (m) is the radius of the pipe (assumed circular) and η is the coefficient of viscosity (kg m^{-1} s^{-1}). Writing $A = \pi r^2$ for the cross-sectional area of the pipe (m^2), substituting for p_1 and p_2 from eqns (4.14a), and multiplying volume flow rate q by source concentration S_1, we obtain the transport flux T (kg mol s^{-1}) as

$$T = \frac{R\theta A^2}{8L\eta\pi} S_1(S_1 - S_2). \tag{4.14c}$$

As for diffusion (eqn (4.6d)), the system is essentially non-directional or non-polar; the polarity is imposed by the source and the sink, which together provide the driving energy for the process, adding molecules to the higher concentration S_1 and removing them from the lower concentration S_2. Comparison of eqn

(4.14c) with eqn (4.6d) shows that the osmotic mechanism is more responsive than a diffusion-like mechanism to the source concentration S_1. In Exercise 4.7, an estimate is made of the transport flux from this mechanism through the wheat peduncle, showing that this process is capable in principle of transporting the quantities of substrates required for plant growth.

A rigorous treatment of this problem is far from trivial and involves many factors that have been ignored here, such as the movement of water out of the sieve tube across to the xylem, the utilization and leakage of sugars from the transport pathway, and the variation in the viscosity of sucrose solution with concentration. The reader is referred to Lang (1978) and Ross and Tyree (1979) for further discussion.

The credibility of the osmotic mechanism as a mechanism for translocation is generally considered to depend on the state of the sieve plate pores, and to what extent these are open. Also, the possibility of non-Newtonian flow, and consequent deviations from Poiseuille's formula, have not been seriously investigated. For example, it is known from studies of the flow of blood (Nubar, 1971) and other solutions that the presence of quite small amounts of protein can give rise to significant deviations from the usual behaviour; such effects give higher flow rates than would be expected using a straightforward calculation, and assume increasing importance the smaller the conducting vessel. In addition, interfacial flow (p. 115) may cause slip at the sieve tube surfaces, further enhancing the transport flux. All these factors tend to improve the acceptability of this hypothesis.

4.5.3 *Convective flow driven by gravity*

This mechanism can produce a small contribution to the movement of sugars within the plant (about 0.1 per cent of that arising from osmotic pressure); it is usually completely ignored in discussions of translocation, but we give a brief account here (from Thornley 1976).

A scheme for considering this is shown in Fig. 4.10. In the pipe P_1P_2 the density of solution of concentration S_1 is ρ_1 (kg m^{-3}), and along $P_3P_4P_5$ it is ρ_2. With the arrangement of source and sink shown $S_1 > S_2$ and $\rho_1 > \rho_2$, and gravity will cause a flow in the direction shown.

For the purpose of making a simple calculation, it is assumed that the density $\rho(S)$ of a solution of concentration S is given by

$$\rho(S) = (1000 + \mu_S S), \tag{4.15a}$$

where μ_S is the relative molecular mass of substrate S, the density of water is 1000 kg m^{-3}, and the concentration S is in units of kg mol m^{-3}. The length of P_1P_2 is L(m) and, if the reservoirs in Fig. 4.10 are assumed to be small, the force per unit area (kg m^{-1} s^{-2}) acting around $P_1P_2P_3P_4P_5$ is

$$Lg(\rho_1 - \rho_2) = Lg\mu_S(S_1 - S_2), \tag{4.15b}$$

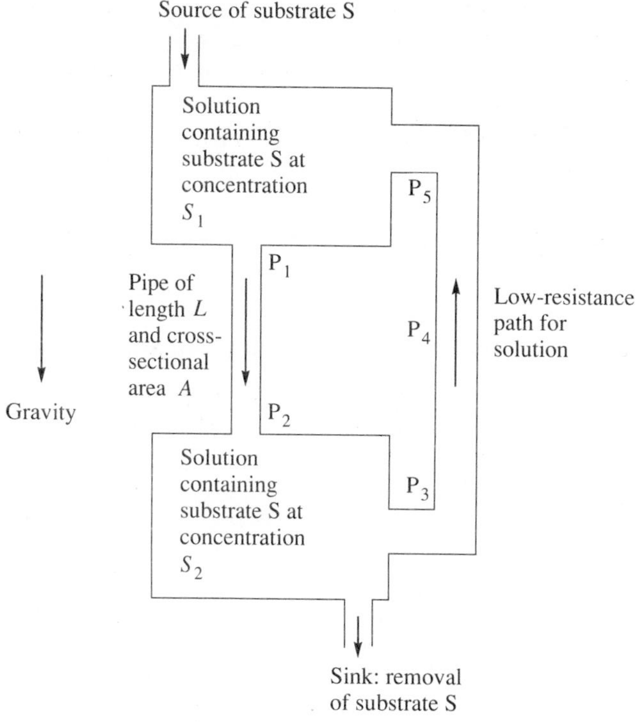

Fig. 4.10. Scheme for convective flow driven by gravity.

where g (m s^{-2}) is the acceleration due to gravity. Assuming non-turbulent flow and applying Poiseuille's formula (cf. eqn (4.14b)), we obtain the volume rate of flow q (m^3 s^{-1}) as

$$q = \frac{(S_1 - S_2)L\mu_S g\pi r^4}{8L\eta}, \tag{4.15c}$$

where it is assumed that the part of the pathway giving resistance to flow is also of length L; as before, r is the radius of the pipe (m) and η is the viscosity (kg m^{-1} s^{-1}). Writing $A = \pi r^2$ for the cross-sectional area of the pipe and multiplying volume flow rate q by concentration S_1, we obtain the transport flux T (kg mol s^{-1}) as

$$T = \frac{Lg\mu_S A^2}{8L\eta\pi} S_1(S_1 - S_2). \tag{4.15d}$$

Comparison of the relative magnitudes of osmotic pressure and gravity can be obtained by comparing eqns (4.15d) and (4.14c), giving the radio

$$\frac{Lg\mu_S}{R\theta}. \tag{4.15e}$$

With $L = 1$ m, $g = 10\,\mathrm{m\,s^{-2}}$, $\mu_S = 342$ (sucrose), $R = 8314\,\mathrm{J\,(kg\,mol)^{-1}\,K^{-1}}$, and $\theta = 293$ K (20 °C), this ratio is 0.0014. The contribution is small, but will be larger for large plants and trees; for a tree with $L = 50$ m, the ratio becomes 0.07 giving a 7% contribution.

Equations (4.15d) and (4.14c) both have the same response to solute concentrations. However, the gravity mechanism is intrinsically directional, in contrast with the osmotic mechanism. Although gravity produces the driving force, the energy is derived from the source and sink, which maintain the concentration and density differences.

4.5.4 *Convective flow driven by temperature gradients*

This is similar to the mechanism discussed in the last section: temperature differences within the plant cause density differences, and gravity acts upon these density differences (Thornley, 1987). Temperature differences may arise from the effects of sunlight and shade, or from differing air and soil temperatures. It is related to the Benard–Rayleigh convective instability (Pippard, 1985, pp. 122–123).

Consider the scheme in Fig. 4.11: the total length of the pathway is $2L$ (m), and the horizontal components are ignored. The area of the conducting pathway is A ($\mathrm{m^2}$) and, if it is assumed that there are n cylindrical tubes of radius r (m),

$$A = n\pi r^2. \tag{4.15f}$$

Given that ΔT (K) is the temperature difference between the two vertical arms of Fig. 4.11, g ($\mathrm{m^2\,s^{-1}}$) is the acceleration due to gravity, and $\mathrm{d}\rho/\mathrm{d}T$ ($\mathrm{kg\,m^{-3}\,K^{-1}}$) is the temperature coefficient of density, the pressure Δp (Pa) exerted round the loop is

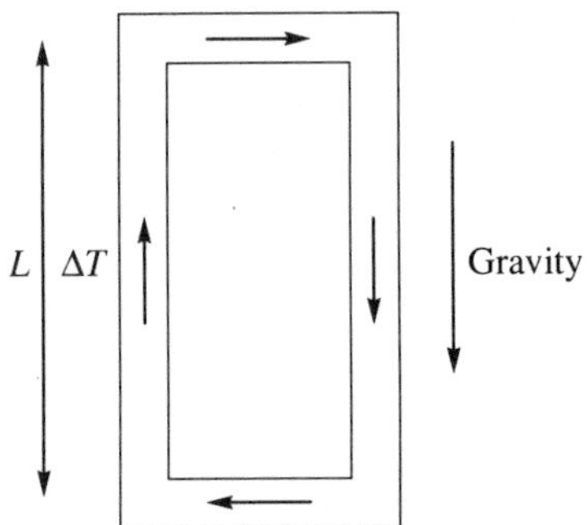

Fig. 4.11. Scheme for temperature-driven transport. The vertical side is of length L, and the horizontal members are assumed to be of negligible length. One of the vertical arms is maintained at a temperature of ΔT K above that of the other vertical arm. Gravity operates as shown, and the transport flux is in the sense indicated.

$$\Delta p = g\left(-\frac{\mathrm{d}\rho}{\mathrm{d}T}\right)L\Delta T. \tag{4.15g}$$

The coefficient $-\mathrm{d}\rho/\mathrm{d}T$ is evaluated at the mean temperature.

If we assume streamline flow (which can be shown to be valid (Thornley, 1987)), Poiseuille's formula gives the volume rate of flow q ($\mathrm{m^3\,s^{-1}}$) as

$$q = \frac{n\Delta p \pi r^4}{8(2L)\eta}, \tag{4.15h}$$

where η (Pa s) is the viscosity. The speed v ($\mathrm{m\,s^{-1}}$) of the convective flow is

$$v = \frac{q}{A} = \frac{\Delta p\, r^2}{(16L\eta)}.$$

which becomes

$$v = \frac{g(-\mathrm{d}\rho/\mathrm{d}T)r^2\Delta T}{16\eta}. \tag{4.15i}$$

We make the following numerical assumptions:

$$g = 10\,\mathrm{m\,s^{-1}} \qquad \Delta T = 1\,\mathrm{K}$$

$$-\frac{\mathrm{d}\rho}{\mathrm{d}T}(\text{water at } 20\,^\circ\mathrm{C}) = 0.21\,\mathrm{kg\,m^{-3}\,K^{-1}}; \tag{4.15j}$$

$$r = 10^{-4}\,\mathrm{m} \qquad \eta\,(\text{water at } 20\,^\circ\mathrm{C}) = 0.001\,\mathrm{Pa\,s}.$$

Any secondary effects of temperature or solutes on viscosity are ignored, the assumption of $\Delta T = 1$ K is conservative, as air–soil temperature differences can be as large as 10 K, and $\mathrm{d}\rho/\mathrm{d}T$ is non-linear, varying from about -0.1 at 10 °C to about $-0.3\,\mathrm{kg\,m^{-3}\,K^{-1}}$ at 30 °C. Then substitution in eqn (4.15i) gives

$$v = 1.25 \times 10^{-6}\,\mathrm{m\,s^{-1}}. \tag{4.15k}$$

This is much smaller than a typical xylem or phloem speed of flow of about $10^{-3}\,\mathrm{ms^{-1}}$, which may vary greatly diurnally. The speed of convective flow per hour is $3600v = 0.5 \times 10^{-2}\,\mathrm{m\,h^{-1}}$, giving a movement of 6 cm over 12 h, or 60 cm with a 10 K temperature difference.

4.5.5 *Convective flow driven by electro-osmosis*

Electro-osmosis belongs to the class of electrokinetic phenomena in which there is a coupling between volume flow, electric potential difference, current flow (of ions), and hydrostatic pressure difference (Katchalsky and Curran, 1965). Spanner (1958, 1970) suggested that electro-osmosis at the sieve plates in the phloem could provide a driving force for the volume movement of solution, which could then circulate as shown in Figs 4.8, 4.9, or 4.10. In simple terms, electro-

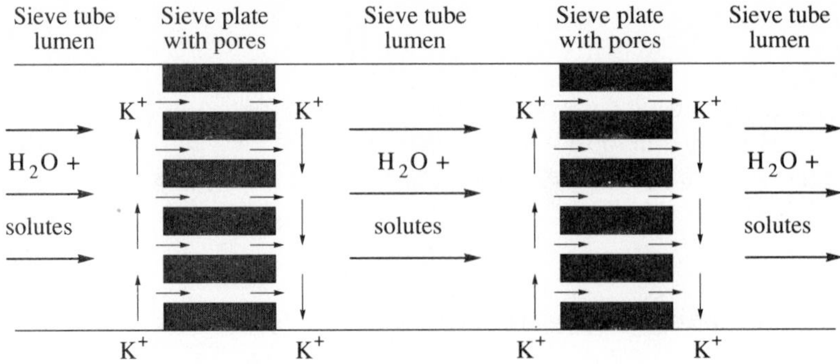

Fig. 4.12. A hypothetical scheme for electro-osmotically driven convective flow in the phloem. A sieve tube crossed by two sieve plates is shown.

osmosis can be envisaged as working as follows (Fig. 4.12). Suppose that an electric field is maintained across a solution which contains, amongst other things, potassium ions K^+: the K^+ ions move down the field gradient towards the negative electrode. There is a 'frictional' force between the K^+ ions and the water molecules, so that the K^+ ions tend to drag water molecules along with them. This produces a volume flow of water, and anything that is dissolved in the water is carried along. Some experimental estimates of electro-osmotic efficiency indicate that 150 molecules of water may be dragged along by each K^+ ion.

Figure 4.12 shows a hypothetical scheme for convective flow driven by electro-osmosis, with the sieve tube crossed by two sieve plates which are traversed by pores. The K^+ ions are shown as cycling round a loop which passes through the sieve plate pores, and here the loop is closed outside the sieve tube. It is assumed that metabolic activity is involved in setting up electric potential differences, and promoting the active movement of the K^+ ions, and it is usually assumed that the charged pores in the sieve plates may play an important role in this.

If j (kg mol K^+ s^{-1}) is the molar flux of potassium through the sieve plate, and j' (kg mol H_2O s^{-1}) is the molar flux of water along the sieve tube, then we can write simply

$$j' = fj \tag{4.16}$$

where f is a coefficient. A rigorous treatment of all the hydraulic and electrical effects, in which account is taken of the length of the sieve plate and also that of the lumen between the sieve plates, would be out of place here, but the interested reader should refer to Spanner (1970). The current consensus is that electro-osmosis is an unlikely candidate for driving sugar transport in plants because of the very high current requirements and other anatomical and theoretical difficulties.

However, it should be noted that in this case the driving force is applied quasi-continuously along the pathway of transport, unlike the mechanisms considered earlier, and the flux could therefore be relatively independent of the pathway length. In order that the system can function continuously, a return pathway for solvent must exist, but this is not shown on Fig. 4.12. If q ($m^3 s^{-1}$) is the volume flow rate of solution and S_1 and S_2 are the sugar concentrations at the source and sink ends of the pathway, with solution of concentration S_2 being returned to the sink, then the transport flux T of sugar will be the same as in eqn (4.13), namely

$$T = q(S_1 - S_2). \tag{4.17a}$$

It is possible that, as the volume flux q depends upon metabolic activity which may depend upon substrate availability along the pathway, we could write

$$q = \frac{c}{2}(S_1 + S_2) \tag{4.17b}$$

where c is a constant, so that the volume flux is just proportional to the average concentration along the pathway. Combining these two equations, we obtain

$$T = \frac{c}{2}(S_1 - S_2)(S_1 + S_2). \tag{4.17c}$$

Thus, quite complicated mechanisms such as electro-osmosis may give rise to relatively simple transport equations such as eqn (4.17c) for possible use in plant and crop models.

4.5.6 *A model combining a diffusion-analogue mechanism with active loading and unloading*

Active transport at the ends of the transport system (loading a unloading) may cause an increase or decrease in the effective substrate concentrations which are driving the long-range transport mechanism. For example, eqns (4.13) and (4.14c) for convective flow are of the type

$$T = \sigma_1(S_1 - S_2) \tag{4.18a}$$

and

$$T = \sigma_2 S_1(S_1 - S_2), \tag{4.18b}$$

where T is the transport flux, σ_1 and σ_2 are constants, and S_1 and S_2 are the substrate concentrations at the source and sink ends of the pathway. Note that a simple increase in the substrate concentrations by a factor of α will change the transport flux by α or α^2. However, eqns (4.18) are still only able to describe transport down a concentration gradient, and there are many plant systems where this is inadequate as uphill movement is occurring. In this section, following Thornley (1977), we combine active transport with an equation of type (4.18a) to obtain an overall phenomenological equation that may have a wider range of application than eqns (4.18).

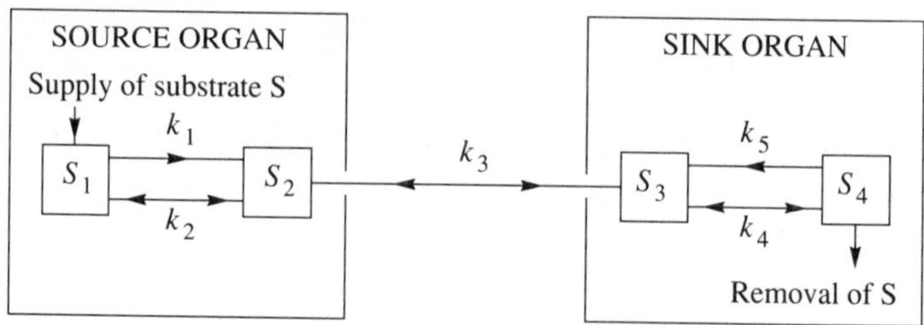

Fig. 4.13. Active and passive processes are combined in a model of substrate transport. S_1, S_2, S_3, and S_4 denote the concentrations of substrate S in the four compartments shown. The first two compartments (S_1, S_2) are in the source organ and the second two (S_3, S_4) are in the sink organ. k_1, k_2, k_3, k_4, and k_5 are rate constants. A single arrow (k_1, k_5) indicates active transport in the direction indicated; double arrows (k_2, k_3, k_4) denote bi-directional diffusion-like transport as in eqn (4.18a).

Using simple models of membrane transport, we showed in eqns (4.12l) and (4.11f) that the substrate dependence of active transport follows a Michaelis–Menten relationship

$$T = \frac{k_1 S}{1 + LS}, \qquad (4.18c)$$

where T is the transport flux, k_1 and L are constants, and S is the substrate concentration outside the membrane. For low levels of substrate, eqn (4.18c) can be approximated by

$$T = k_1 S. \qquad (4.18d)$$

The scheme considered is given in Fig. 4.13, where the notation is defined. Let T_{ij} be the transport flux between any two adjacent compartments i and j; in the steady state the fluxes into and out of a compartment exactly balance. From the first to the second compartment

$$T_{12} = k_1 S_1 + k_2(S_1 - S_2). \qquad (4.18e)$$

Also

$$T_{23} = k_3(S_2 - S_3) \qquad (4.18f)$$

and

$$T_{34} = k_4(S_3 - S_4) - k_5 S_4. \qquad (4.18g)$$

Therefore, putting $T = T_{12} = T_{23} = T_{34}$ and eliminating S_2 and S_3, we obtain

$$T = \frac{(1 + k_1/k_2)S_1 - (1 + k_5/k_4)S_4}{1/k_2 + 1/k_3 + 1/k_4}, \qquad (4.18h)$$

which gives the transport flux T in terms of the substrate concentrations S_1 and S_4 at the ends of the pathway. This expression can be more simply written as

$$T = k_6 S_1 - k_7 S_4, \tag{4.18i}$$

using two derived rate constants k_6 and k_7. k_6 and k_7 are independent of each other, and thus eqn (4.18i) with $k_6 > k_7$ can give uphill transport when $S_1 < S_4$. Equation (4.18i) provides a possible basis for parameterizing a wide range of transport processes.

4.5.7 *Other mechanisms suggested for sugar movement in plants*

A major difficulty in deciding on the mechanism of translocation is the continuing disagreement about the anatomy of the sieve tubes, which is felt by some to cause problems for the pressure flow hypotheses. Thus, in addition to the mechanisms outlined above, a number of other possibilities have been proposed, and some of these are now briefly discussed.

Interfacial movement This quite old proposal has recently been discussed by Sutcliffe and Collins (1975) and Richmond and Wardlaw (1976). The idea is that an interfacial flow of assimilates along surfaces within the sieve tube structure may occur. This requires a compound or compounds able to reduce the surface tension at the interfaces. The result is that there is slip at the sieve tube surfaces which modifies the application of Poiseuille's equation (eqn (4.14b)), enhancing the transport flux.

Peristalsis Aikman and Anderson (1971) examined the possible role of peristalsis in driving solution flow in translocation. In contrast with interfacial movement (which may cause a passive enhancement of transport fluxes), this is an active mechanism with energy being applied along the transport pathway. The energy inputs required appear to be reasonable, although there is no physiological or anatomical evidence in direct support of this hypothesis. If such a mechanism were to apply, simple considerations suggest that equations such as eqn (4.13) or (4.17c) may describe the phenomenon.

Transcellular streaming Over a century ago it was suggested that protoplasmic streaming may be responsible for sieve tube transport. Although high streaming velocities of almost 500 cm h^{-1} have been observed in slime moulds, only much lower velocities of up to 6 cm h^{-1} have been seen in higher plants, and these appear to be insufficient. Also, there is doubt as to whether discrete macroscopic streaming strands really exist in sieve tubes. Canny (1973) and Peel (1974) give a detailed account with references to the original work.

4.6 Phenomenology of transport in plants

A variety of expressions for the transport flux T between two locations where the substrate concentrations are S_1 and S_2 have been derived. In summary six of these can be written as follows:

$$T = \sigma(S_1 - S_2) \tag{4.19a}$$

relates to eqn (4.4c) on diffusion and eqn (4.13) for a recycling scheme (here and in what follows the σs, ks, Ks, k_1, and k_2 are constants). From eqn (4.10q) for facilitated diffusion

$$T = \sigma \frac{S_1 - S_2}{(K + S_1)(K + S_2)}. \tag{4.19b}$$

Active transport gives (eqns (4.11f) and (4.12l))

$$T = \frac{kS_1}{K + S_1}. \tag{4.19c}$$

for a unidirectional flux. In eqn (4.14c) the osmotic pressure flow hypothesis gives rise to

$$T = \sigma S_1(S_1 - S_2). \tag{4.19d}$$

Electro-osmosis and eqn (4.17c) give

$$T = \sigma(S_1 - S_2)(S_1 + S_2). \tag{4.19e}$$

Finally, a combination of active mechanisms with diffusion-analogue transport leads to eqn (4.18i), namely

$$T = k_1 S_1 - k_2 S_2, \tag{4.19f}$$

which permits uphill movement of substrates with appropriate values of k_1 and k_2.

These six equations, and possibly others as well, can be regarded as phenomenological equations, i.e. as equations that can describe the overall response, or the gross phenomenon, although justification at lower levels may be lacking and parameter values may not be derivable from more fundamental considerations. Such equations provide a convenient resting place in the long trek between behaviour at the plant level, and the cellular and molecular details of transport. The specialist in translocation mechanisms will aim to establish these or similar equations and to provide parameter estimates. The crop and whole-plant physiologists can attempt to discern these relationships working at the organ level, although regrettably few investigations of this type have been carried out. For example, Ho and Thornley (1978) examined carbon export from mature tomato leaves in terms of the equation

$$T = \sigma(S_1 - S_2), \tag{4.20a}$$

with units of milligrams of carbon per day for the flux T and grams sucrose per 100 g fresh weight for sucrose concentration, and obtained $\sigma = 403 \pm 65$ mg carbon day^{-1} (g sucrose per 100 g fresh weight)$^{-1}$. In a similar investigation of the import of carbon into tomato fruit (Walker and Thornley, 1977), again using eqn (4.20a), it was found that $\sigma = 307 \pm 59$ mg carbon day^{-1} (g sucrose per 100 g fresh weight)$^{-1}$. There is, we believe, a great need for a similar parameterization of transport flux response equations at the organ level in a wide variety of plant systems.

Exercises

4.1. Calculate the sucrose transport flux (kg day^{-1}) due to diffusion of sucrose in aqueous solution at 25 °C along a 1 m pathway of cross-section 10^{-4} m^2, assuming a zero concentration at the distal end of the pathway and a concentration of 100 kg sucrose m^{-3} (about a 10 per cent solution) at the proximal end. The diffusion constant is $D = 5.2 \times 10^{-10}$ m^2 s^{-1}.

4.2. Figure 4.14 shows a scheme where there is diffusion-like transport with diffusion constant D (m^2 day^{-1}) along a pathway of length L (m) and cross-section A (m^2). Per unit volume of solution, there is a first-order loss of substrate along the pathway at a rate $k_p S$ (kg m^{-3} day^{-1}) where S is the substrate concentration (kg m^{-3}) and k_p is a rate constant (day^{-1}); this could be interpreted as leakage from the pathway, or a substrate requirement for maintenance of the pathway structure, or a simple degradation. At the end of the pathway is a sink, possibly a growing fruit, which utilizes substrate according to first-order kinetics, i.e. at a rate $k_1 S_1$ (kg day^{-1}) where k_1 (m^3 day^{-1}) is a constant.

Find expressions for the fluxes (kg day^{-1}) (a) into the sink, (b) of pathway loss, and (c) provided at $x = 0$ to maintain S_0 constant.

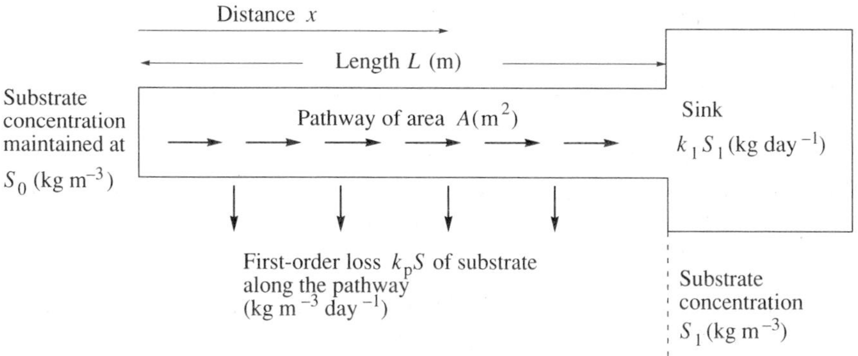

Fig. 4.14. Scheme for Exercise 4.2.

4.3. For a diffusion resistance plus two biochemical sinks in parallel (Fig. 4.15) derive a cubic equation for the transport flux T equivalent to eqn (4.9c). Assume that the substrate responses of the two sinks are $k_1 S/(K_1 + S)$ and $k_2 S/(K_2 + S)$, where S is substrate concentration and k_1, k_2, K_1, K_2 are parameters. Derive expressions for the asymptote and the initial slope of the substrate response.

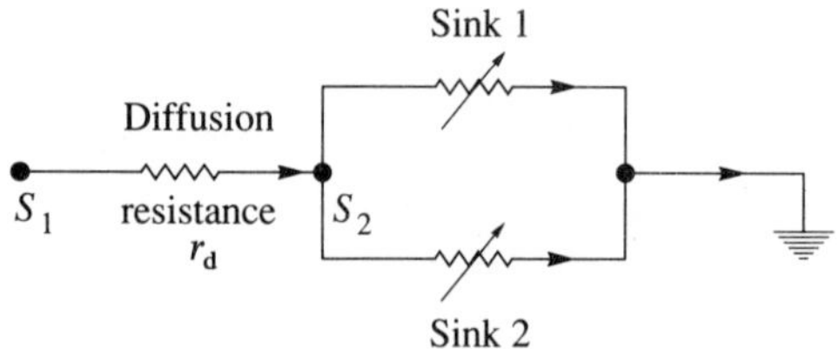

Fig. 4.15. Scheme for Exercise 4.3 showing a diffusion resistance r_d, in series with two biochemical sinks in parallel. S_1 and S_2 indicate substrate concentrations.

4.4. Compare the fluxes arising from ordinary diffusion and facilitated diffusion (eqns (4.10l) and (4.10q)) for oxygen in aqueous solution with myoglobin as a carrier molecule, assuming the following parameters (partly from Murray 1977, p. 55):

A	pathway cross-sectional area	$10^{-4}\,m^2$
C_t	myoglobin concentration	$0.012\,kg\,mol\,m^{-3}$
D_S	diffusion coefficient of oxygen in water	$1.2 \times 10^{-9}\,m^2\,s^{-1}$
D_X	diffusion coefficient of myoglobin	$4.35 \times 10^{-11}\,m^2\,s^{-1}$
K	binding constant	$7.9 \times 10^{-7}\,kg\,mol\,m^{-3}$
k_1	rate constant	$1.4 \times 10^7\,(lg\,mol\,m^{-3})^{-1}\,s^{-1}$
k_2	rate constant	$11\,s^{-1}$
L	path length	$0.01\,m$
S_1	concentration of oxygen at source	$2.6 \times 10^{-4}\,kg\,mol\,m^{-3}$
S_2	concentration of oxygen at sink	$0\,kg\,mol\,m^{-3}$

Verify that, for values of $S_2 \sim K$ (e.g. $S_2 = K, 2K, 4K, 9K$), the facilitated flux is sensitive to S_2 whereas the ordinary diffusive flux is not appreciably affected.

4.5. In facilitated diffusion, assume that (4.10a) for the reversible combination of substrate S with a carrier C is replaced by

$$q\mathrm{S} + \mathrm{C} \underset{k_2}{\overset{k_1}{\rightleftharpoons}} X$$

where q is an integer. Such a scheme may apply for cooperative binding of S to a protein where there are q interacting sites to which S can be attached (p. 54), Derive, following the procedure in the text, an equivalent equation to eqn (4.10q) for the facilitated flux (write $K^q = k_2/k_1$).

If the facilitated process is followed by a biochemical process with rate $kS_2/(K' + S_2)$, with k and K' constants, show graphically that if $K' \lesssim K$, then in the steady state the substrate concentration S_2 at the biochemical sink is largely independent of the rate constant k for the biochemical sink.

4.6. Nitrate uptake by roots is one of the most important processes in crop and plant growth. There is evidence that the rate of this process depends on the nitrogen concentration N_{soil} in the soil solution and the sugar levels in the root system (c_{glu} = concentration of glucose in the plant root) which enable the process to be active, and is inhibited by adequate nitrogen levels N_{plant} within the root.

Derive, based on the models of Sections 4.4.1 and 4.4.2, nitrogen-uptake models where the transport flux depends on the three (state) variables N_{soil}, c_{glu}, and N_{plant}. It is suggested that $S_1 = N_{soil}$, and it is assumed that $c_{ATP} = c_{glu}$ (apart from a constant factor) and that inhibition by N_{plant} is accommodated by taking k_3 of the reactions (4.11b) and (4.12b) equal to $k_{3m}/(1 + N_{plant}/J)$, where J is an inhibition constant (p. 62), and substituting for k_3 in eqns (4.11f) and (4.12l).

4.7. Calculate the sucrose mass flux (kg day^{-1}) through the peduncle into the wheat kernel, assuming that eqn (4.14c) for convective flow driven by osmotic pressure applies. Assume that the source solution is a 10 per cent sucrose solution, the relative molecular mass of sucrose RMM (sucrose) is 342.30 kg (kg mol)$^{-1}$, the sink concentration S_2 is zero, the temperature is 20 °C, the length L of the pathway is 0.1 m, the viscosity η is 1.15×10^{-3} kg s^{-1} m^{-1} (at 20 °C for a 5 per cent sucrose solution), and the pathway consists of 50 phloem bundles with one phloem tube per bundle of cross-sectional area 0.12×10^{-8} m^2.

5
Temperature effects on plant and crop processes

5.1 Introduction

The effects of temperature in biology have long been a source of both confusion and controversy. For example, from the temperature dependence of a complex process it has sometimes been inferred that a single rate-limiting step is operating or, in other cases, that phase transitions are taking place in membrances or other biological complexes. It is not our intention to adjudicate on such matters here, but rather to set out as simply as possible the essentials of our current understanding and to provide the reseacher with the appropriate conceptual and analytical tools so that he can exercise his own judgment. This chapter is based on a review article by Johnson and Thornley (1985).

First let us consider a typical example of the type of response it is wished to describe and understand. To obviate the difficulty of defining a single quantitative measure of plant growth and development, we take our illustration from the microbial world: in a steady state bacterial culture (for instance a chemostat), growth and development occur at the same rate, and this rate is an altogether unambiguous quantity. In Fig. 5.1 the specific growth rate of *Escherichia coli* is shown over a range of temperatures. This has dimensions of time^{-1}, and is analogous to the rate constants of chemical and biological processes. The response is qualitatively typical of the temperature response of many complex processes: an accelerating increase from zero, a linear section, and an optimum, followed by a rapid fall-off.

Of course, not all temperature responses of biological processes are identical and it is necessary to develop a general theory to understand, as well as describe, the various phenomena that are observed. In this chapter we shall examine the effect of temperature on the rate constants of reactions. Since the dimensions of these rate constants depend upon the order of the reaction involved, they will not always be the same. Consequently, in many cases, rate constants will appear with no clearly defined dimensions. However, when applying the techniques to specific problems, these dimensions should be apparent.

5.2 Arrhenius equation and Q_{10}

Consider the chemical reaction

$$\mathrm{S} \xrightarrow{k} \text{products}, \tag{5.1a}$$

where S is some substance and k is a rate constant. Arrhenius developed a simple

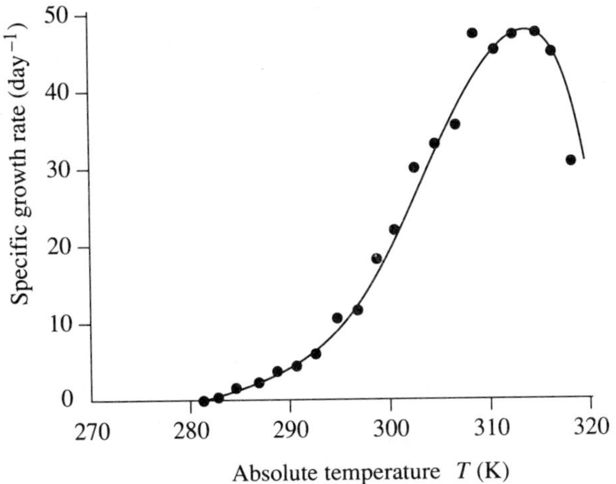

Fig. 5.1. Temperature response of the specific growth rate of *Escherichia coli* (adapted from the data of Ingraham (1958)). The curve has been drawn by eye.

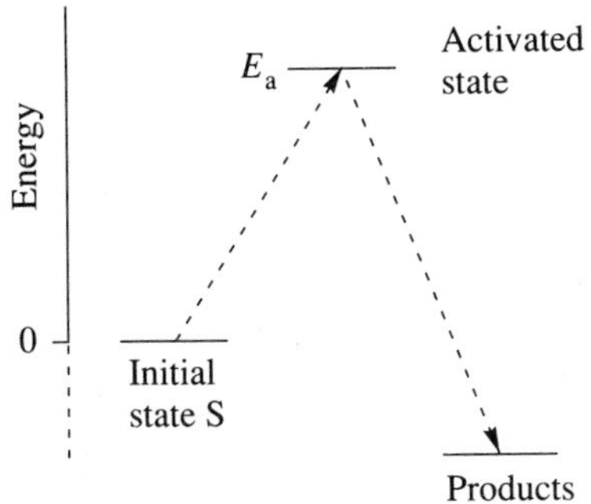

Fig. 5.2. Energy level scheme for the Arrhenius equation (eqn (5.1e)).

and primarily qualitative theory for the rate constant k which has been an important contribution to our efforts to understand the rates of chemical reactions.

The scheme that leads to the reaction (5.1a) is illustrated in Fig. 5.2. This figure is an energy level diagram and the horizontal line marked S denotes the normal state of substance S which is at zero on the energy scale. The idea behind the Arrhenius theory is that there is a higher-energy transient state, called the activated state, marked with energy E_a in the diagram. The reaction S $\rightarrow$ products can only take place from the activated state, and does so irreversibly (note that the reaction scheme illustrated in Fig. 5.2 is exothermic since the energy level of the products is lower than S). Thus the reaction rate depends on the number of molecules of S which are present in the activated state. If the Boltzmann distribution function is assumed, since the reagent molecules are at zero energy level the proportion of the total number of reagent molecules in the activated state is

$$\frac{\exp(-E_a/RT)}{1+\exp(-E_a/RT)}, \tag{5.1b}$$

where T is the absolute temperature, R is the gas constant (for 1 kg mol), and E_a is the activation energy (J (kg mol)$^{-1}$). This can be approximated by (see Exercise 5.1 and Fig. S5.1, p. 593)

$$\exp\left(-\frac{E_a}{RT}\right) \tag{5.1c}$$

since, in practice,

$$\exp\left(-\frac{E_a}{RT}\right) \ll 1. \tag{5.1d}$$

We can therefore write for the rate constant k,

$$k = A\exp\left(-\frac{E_a}{RT}\right) \tag{5.1e}$$

where A is a constant. This equation, attributed to Arrhenius, can be linearized by taking logarithms, giving

$$\ln k = \ln A - \frac{E_a}{RT}. \tag{5.1f}$$

Both forms of the equation are illustrated in Fig. 5.3 for data of Hecht and Conrad (1889) as quoted by Barrow (1961, p. 486). These data relate to the reaction

$$CH_3I + C_2H_5ONa \rightarrow CH_3OC_2H_5 + NaI \tag{5.1g}$$

in ethanol. The parameter values are $E_a/R = 9924$ K and $A = 3.3 \times 10^{11}$ s^{-1} (these values are slightly different from those obtained by Barrow). Typically, for an enzyme–substrate reaction, E_a/R takes values of the order 5000–15 000 K (see, for example, Dixon and Webb 1964, p. 158). Note that, with values of E_a/R in this range and for biologically reasonable values of T, the strong inequality described by (5.1d) is satisfied so that the approximation of eqn (5.1b), given by eqn (5.1c), is valid. (See Exercise 5.1)

A widely applied approach for describing the temperature response of k uses the Q_{10} factor. This empirical approach is based on the observation that a given temperature increment often increases the reaction rate by a constant factor. Formally, this result can be written

$$k = k_r Q_{10}^{[(T-T_r)/10]} \tag{5.1h}$$

where k_r is the rate constant at the reference temperature T_r and Q_{10} is the factor by which the rate constant increases for a temperature increment of 10 K.

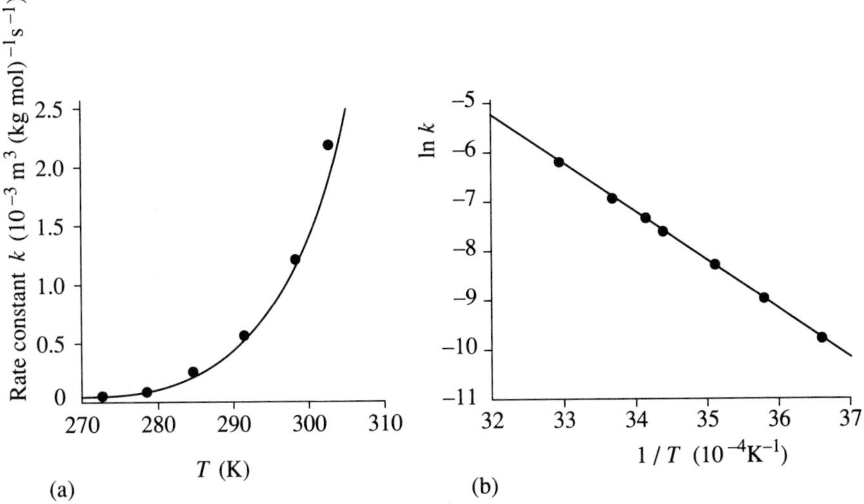

Fig. 5.3. Arrhenius equation for the rate constant k. The data, due to Hecht and Conrad (1889) as quoted by Barrow (1961), are for the reaction in eqn (5.1g). T is the absolute temperature, $A = 3.3 \times 10^{11}\ \mathrm{s}^{-1}$, and $E_a/R = 9924$ K. In (a) the temperature dependence is shown directly, using eqn (5.1e), and in (b) the linearized form given by eqn (5.1f) is shown.

Equation (5.1h) can be linearized by taking logarithms:

$$\ln k = \ln k_r + \left(\frac{T - T_r}{10}\right) \ln Q_{10}. \tag{5.1i}$$

Comparison of eqns (5.1e) and (5.1f) with eqns (5.1h) and (5.1i) shows that there is obviously no direct equivalence between the Arrhenius equation and the Q_{10} equation (although both are three-parameter relationships between k and T). However, by taking a Taylor series expansion of k as given by eqns (5.1e) and (5.1h), it is possible to interpret the parameters of the Q_{10} equation in terms of the Arrhenius equation. The Taylor series expansion of any given rate parameter $k(T)$ about a reference temperature T_r, is given by

$$k(T) = k(T_r) + \Delta T \frac{\mathrm{d}k(T_r)}{\mathrm{d}T} + \frac{(\Delta T)^2}{2} \frac{\mathrm{d}^2 k(T_r)}{\mathrm{d}T^2} + \cdots \tag{5.1j}$$

where

$$\Delta T = T - T_r. \tag{5.1k}$$

The Taylor series expansion of the Arrhenius equation, (5.1e), is

$$k(T) = A \exp\left(-\frac{E_a}{RT_r}\right)\left\{1 + \frac{\Delta T E_a}{RT_r^2} + \frac{(\Delta T)^2}{2}\left[\left(\frac{E_a}{RT_r^2}\right)^2 - \frac{2E_a}{RT_r^3}\right] + \cdots\right\}. \tag{5.1l}$$

The Q_{10} equation can be written as

$$k = k_r \exp\left(\frac{\Delta T \ln Q_{10}}{10}\right) \tag{5.1m}$$

from which the Taylor series up to order $(\Delta T)^2$ is readily obtained:

$$k(T) = k_r\left[1 + \frac{\Delta T \ln Q_{10}}{10} + \frac{(\Delta T)^2}{2}\left(\frac{\ln Q_{10}}{10}\right)^2 + \cdots\right]. \tag{5.1n}$$

Comparing eqns (5.1l) and (5.1n) shows that the Arrhenius equation and the Q_{10} equation can be made to be identical and to have identical slope at the reference temperature T_r by setting

$$k_r = A \exp\left(-\frac{E_a}{RT_r}\right) \tag{5.1o}$$

and

$$Q_{10} = \exp\left(\frac{10E_a}{RT_r^2}\right). \tag{5.1p}$$

An indication as to the accuracy of the two approaches can be obtained by comparing the second derivatives, i.e. the coefficients of $(\Delta T)^2$, at T_r. These are approximatey equal if

$$\frac{2E_a}{RT_r^3} \ll \left(\frac{E_a}{RT_r^2}\right)^2, \tag{5.1q}$$

i.e.

$$T_r \ll \frac{E_a}{2R}. \tag{5.1r}$$

Consequently, if this strong inequality is satisfied the Arrhenius equation and the Q_{10} equation, as well as being equal and having the same slope at T_r, also have similar curvature. It is not unexpected that T_r must be bounded by an upper limit in order for the two approaches to be similar since the Arrhenius equation approaches an asymptote as T becomes large (Fig. S5.1, p. 594), whereas the Q_{10} equation increases unboundedly.

Thus, given the parameters A and E_a of the Arrhenius equation, we choose a reference temperature T_r and then use eqns (5.1o) and (5.1p) to compute k_r and Q_{10}. Alternatively, given T_r, k_r, and Q_{10}, eqns (5.1o) and (5.1p) can be used to give first E_a and then A. Note that Q_{10} and a reference temperature T_r alone are sufficient to define an activation energy E_a. For example, setting $T_r = 290$ K in eqn (5.1p) gives

$$E_a/R = 8410 \ln Q_{10}, \tag{5.1s}$$

in units of kelvins. With $Q_{10} = 1.5$, $E_a/R = 3410$ K, and for $Q_{10} = 2$, $E_a/R = 5829$ K. Note that, in both these examples, eqn (5.1r) is reasonably well satisfied.

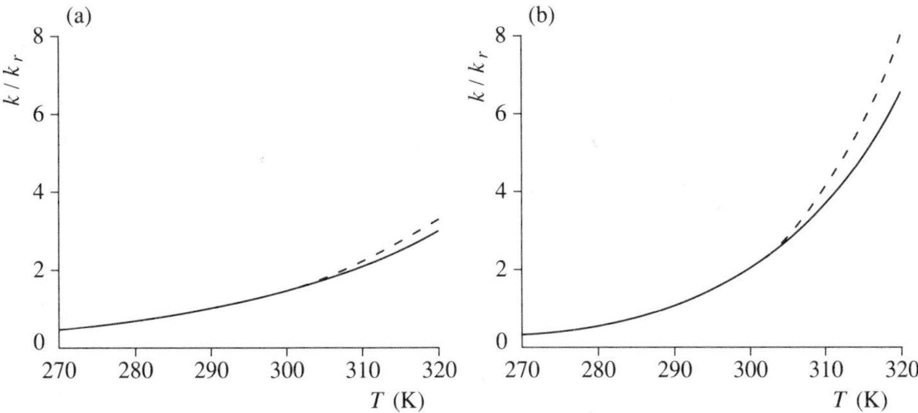

Fig. 5.4. Comparison of the Q_{10} equation (eqn (5.1h)) (– – –) with the Arrhenius equation (eqn (5.1e)) (——) for a reference temperatue $T_r = 290$ K: (a) $Q_{10} = 1.5$; (b) $Q_{10} = 2$.

In Figs. 5.4 these two examples are drawn for T in the range (270, 320) K, and it can be seen that, over the restricted temperature range which is usually of interest to the biologist, the Q_{10} equation is in quite good agreement with the Arrhenius equation. However, it must be emphasized that there is no theoretical justification for using the Q_{10} equation, apart from its convenience, and the approach is purely empirical, whereas the Arrhenius equation does have some theoretical basis.

5.3 Collision theory and transition-state theory of the rate constant

In the preceding discussion of the Arrhenius equation, it was assumed that the parameter A of eqn (5.1e) is a constant and independent of temperature. More detailed consideration of the nature of A hinges upon the details of the reaction mechanism, and different assumptions are possible. The two principal approaches are the collision and transition-state theories, which are discussed in most text-books of physical chemistry (e.g. Barrow, 1961).

The collision theory applies to bimolecular reactions in gases. Considering the reaction

$$S_1 + S_2 \xrightarrow{k} \text{products}, \tag{5.2a}$$

and calculating the number of collisions between molecules of S_1 and S_2 per unit time per unit volume which have the necessary energy to allow the reaction to take place leads to the equation

$$k = BT^{1/2} \exp\left(-\frac{E_a}{RT}\right) \tag{5.2b}$$

for the rate constant k; B is a constant.

The transition-state theory, which is due to Eyring (1935), applies to both gases and solutions. It is assumed that molecules S_1 and S_2 react to establish an

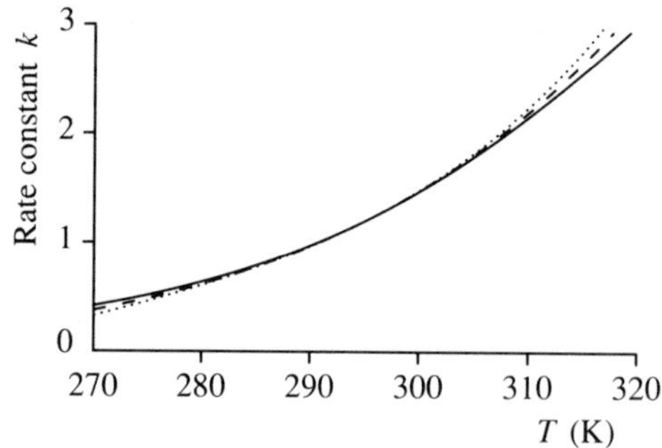

Fig. 5.5. Rate constants according to the Arrhenius equation (——) eqn (5.1e) with $A = 1.28 \times 10^5$), collision theory (---; eqn (5.2b) with $B = 7.51 \times 10^3$), and transition-state theory (..., eqn (5.2d) with $C = 4.41 \times 10^2$). $E_a/R = 3410$ K throughout (corresponding to a Q_{10} value of 1.5). The constants A, B, and C are chosen so that the curves all take a value of unity at $T = 290$ K.

equilibrium with the transition-state (or reacting-state) molecules $(S_1, S_2)^*$, and this transition state reacts further to produce products. The scheme can be described by

$$S_1 + S_2 \rightleftharpoons (S_1, S_2)^* \xrightarrow{k} \text{products}. \tag{5.2c}$$

By calculating the concentration of the transition-state species and the rate at which it breaks up to produce products, we can obtain the expression

$$k = CT \exp\left(-\frac{E_a}{RT}\right) \tag{5.2d}$$

for the rate constant k; C is a constant.

In order to compare the simple Arrhenius equation (5.1e) with eqns (5.2b) and (5.2d) these three equations are drawn in Fig. 5.5 for T in the range (270, 320). The same value of $E_a/R = 3410$ K is used throughout (this gives $Q_{10} = 1.5$ at 290 K with the Arrhenius equation), and the values of A, B, and C are chosen so that $k = 1$ at $T = 290$ K in each case. From Fig. 5.5 it is clear that, over the restricted temperature range of interest to the biologist, the curves are essentially the same. Since at present there are few theoretical grounds for preferring one equation to the other for biological reactions, it seems reasonable to make use of the simpler unmodified Arrhenius equation. It is interesting to note that both eqns (5.2b) and (5.2d) increase without limit as the temperature T increases, so that neither possess an asymptote. Furthermore, while eqn (5.2b) has a point of inflexion, this is not the case for eqn (5.2d). (See Exercise 5.2)

5.4 Models with a temperature optimum

The discussion so far has been based on the temperature response of a single chemical reaction. The plant scientist often needs to describe, either empirically or mechanistically, the type of biological response shown in Fig. 5.1. In this section we consider models in which an enzyme can exist in an equilibrium between two or three different forms or states; these models lead to a temperature response for the rate constant k (Hearon 1952).

In many cases this type of approach still represents a considerable simplification of the actual biological mechanisms involved. However, the general trends may be similar to the simple models. In these situations, the equations developed here can be regarded as useful empirical equations to apply although the appropriate parameter values may not necessarily be biologically plausible. Thus, using these equations, we can examine growth or development or some other rate process of the plant or crop system.

5.4.1 *Equilibrium between two conformational states*

Assume that the growth or development rate arises from the activity of an enzyme that can exist in two forms: an active state and an inactive state. Products are formed from the activated state of the active enzyme, and it is assumed that this reaction has a rate constant k_A which obeys the Arrhenius equation (5.1e). The scheme is illustrated in Fig. 5.6. With increasing temperature, the equilibrium shifts in favour of the inactive state. Let the free energy ΔF of the inactive state relative to the active state be given by

$$\Delta F = \Delta H - T\Delta S, \tag{5.3a}$$

where H is the enthalpy and S is the entropy. According to the Boltzmann distribution function, the fraction f_A of the total number of enzyme molecules in the active state is therefore given by

$$f_A = \frac{1}{1 + \exp(-\Delta F/RT)}. \tag{5.3b}$$

The rate constant k for the production of the products is given by

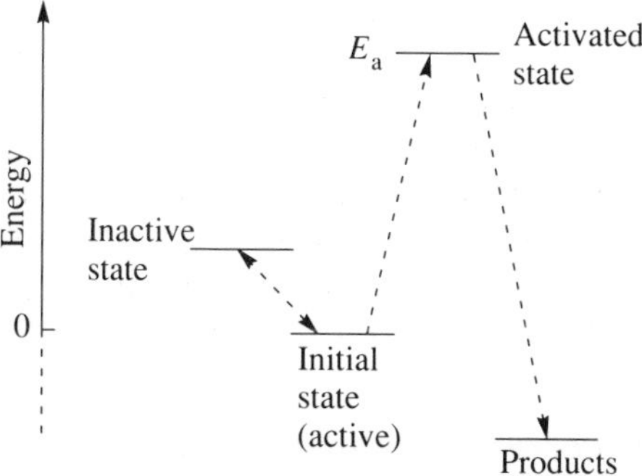

Fig. 5.6. Energy level scheme for the enzyme–substrate reaction where the enzyme can exist in either the active or inactive state.

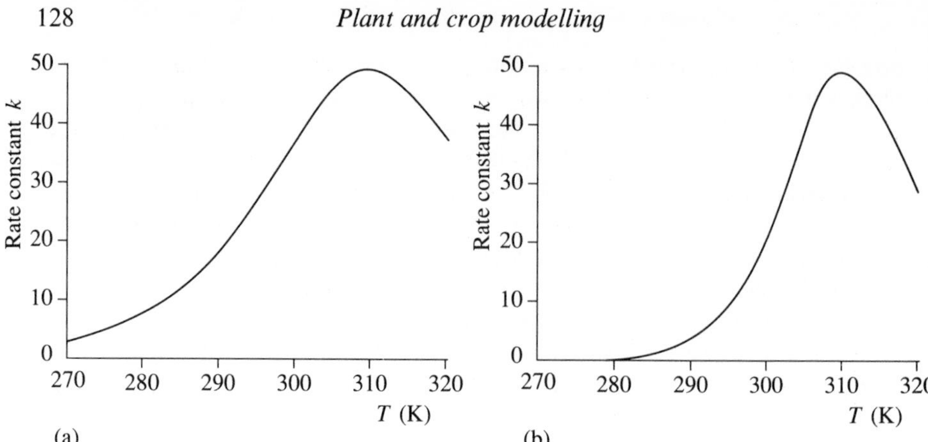

Fig. 5.7. Rate constant k for the case when equilibrium between two species, one active and the other inactive, is assumed (eqn (5.3d)): (a) $A = 3.2 \times 10^{12}$, $E_a/R = 7.5 \times 10^3$ K, $\Delta S/R = 48$, and $\Delta H/R = 15 \times 10^3$ K; (b) $A = 1.3 \times 10^{23}$, $E_a/R = 15 \times 10^3$ K, $\Delta S/R = 81$, and $\Delta H/R = 25 \times 10^3$ K. The parameter values have been chosen so that k is a maximum at $T = 310$ K and takes the value 50 at the maximum.

$$k = f_A k_A \tag{5.3c}$$

which, combining eqns (5.1e), (5.3a), and (5.3b), is

$$k = \frac{A \exp(-E_a/RT)}{1 + \exp(\Delta S/R - \Delta H/RT)}. \tag{5.3d}$$

Note that in practice, when considering enzymes, the free energy ΔF is dominated by the enthalpy ΔH, since the contribution $T\Delta S$ is generally extremely small. Consequently, the free energy ΔF is relatively insensitive to changes in temperature (over the temperature range for which the enzyme exists.). However, the qualitative response of eqn (5.3d) to temperature is comparable with that which is frequently observed in biology, and the scheme from which it is derived can be regarded as a useful conceptual approach. Thus, by selecting parameter values for E_a, ΔH and ΔS (which are perhaps not mechanistically acceptable for very simple reactions), eqn (5.3d) can be used empirically to describe more complex processes. This equation is illustrated in Fig. 5.7 for two sets of parameter values, demonstrating that the curve is versatile and can be applied in a wide variety of circumstances (e.g. to germination, p. 190). The curve shown in Fig. 5.7(b) bears a close resemblance to that shown in Fig. 5.1. It can be seen that the response is sigmoidal up to an optimum, after which it decreases; the parameters have been chosen so that the maximum occurs at $T = 310$ K and $k(310) = 50$. (Exercise 5.3)

5.4.2 *Equilibrium between three conformational states*

The above method can be developed further as follows. Assume now that there are two inactive forms, of which one is preferentially present at low temperatures and the other at high temperatures. Let the free energies of the low- and high-temperature forms relative to the active state be

$$\Delta H_L - T\Delta S_L \qquad \text{and} \qquad \Delta H_H - T\Delta S_H, \tag{5.3e}$$

respectively. The subscripts L and H denote the low- and high-temperature forms. Applying the Boltzmann distribution function again gives the fraction of enzyme molecules in the active state as (analogous to eqn (5.3b))

$$f_A = \frac{1}{1 + \exp(\Delta S_L/R)\exp(-\Delta H_L/RT) + \exp(\Delta S_H/R)\exp(-\Delta H_H/RT)}. \tag{5.3f}$$

When this is combined with the Arrhenius equation (eqn (5.1e)), the rate constant k is given by

$$k = \frac{A\exp(-E_a/RT)}{1 + \exp(\Delta S_L/R)\exp(-\Delta H_L/RT) + \exp(\Delta S_H/R)\exp(-\Delta H_H/RT)}. \tag{5.3g}$$

Equation (5.3g) is similar in structure to eqn (5.3d), but the extra parameters produce a more versatile curve when fitting to data.

A similar expression to eqn (5.3g) is considered by Sharpe and DeMichele (1977, eqn (17)), although they used the Eyring equation (5.2d) rather than the Arrhenius equation which has been used here. Sharpe and DeMichele fit their response equation to the temperature responses for a range of organisms and, although they obtain a good fit to the data, comparison of their Table 2 with Dixon and Webb (1964, pp. 158, 165) indicates that the parameters for best fit, particularly the entropy terms, are not acceptable as thermodynamic parameters. This emphasizes the point that the merits of the approach used in this and the last section seem to be largely empirical, although the analysis provides a tentative basis for adopting these empirical equations.

5.5 Models without a temperature optimum

In this section several simple networks of reactions are considered with a view to attempting to answer the question as to whether overall behaviour can tell us anything useful about underlying responses or mechanisms. The Arrhenius equation is again used throughout.

5.5.1 *Sequential reactions*

Irreversible reaction sequences Consider the irreversible reaction described by

$$\mathrm{S} \xrightarrow{k_1} \mathrm{S}_1 \xrightarrow{k_2} \mathrm{P} \tag{5.4a}$$

where S is the initial substrate, S_1 is an intermediate substrate, P is the reaction product, and k_1 and k_2 are rate constants. In the steady state the rate of increase of P is given by

$$\frac{dP}{dt} = k_1 S = k_2 S_1. \tag{5.4b}$$

Thus, for a given S, the reaction rate depends only on k_1; the magnitude of k_2 and its temperature dependence are irrelevant, as S_1 simply adjusts itself to satisfy eqn (5.4b). If k_1 obeys the Arrhenius equation, then the rate of product formation obeys the same equation with the same activation energy.

This result is easily generalized. For any sequence of irreversible reactions, however long, the rate is determined by the rate of the first step in the sequence. This is because the substrate concentration for any subsequent step is able to rise to a value where that step occurs at the same rate as the first step. Thus the temperature dependence of the overall steady state process is controlled by the temperature dependence of the first reaction.

Reversible reaction sequences A simple extension to eqn (5.4b) which immediately gives different behaviour is, retaining the earlier notation,

$$\mathrm{S} \underset{k_2}{\overset{k_1}{\rightleftharpoons}} \mathrm{S}_1 \overset{k_3}{\rightarrow} \mathrm{P}, \tag{5.4c}$$

where k_1, k_2, and k_3 are rate constants. The difference between this equation and eqn (5.4b) is that the reaction from S to S_1 has been made reversible. In the steady state the rate of change of the intermediate product S_1 is zero, so that

$$Sk_1 = S_1(k_2 + k_3), \tag{5.4d}$$

and the rate of increase in products is

$$\frac{\mathrm{d}P}{\mathrm{d}t} = S_1 k_3 = \frac{Sk_1 k_3}{k_2 + k_3}. \tag{5.4e}$$

After substituting the Arrhenius equation, with an obvious notation, the temperature dependence of the overall process is therefore

$$\frac{\mathrm{d}P}{\mathrm{d}t} = \frac{SA_1A_3 \exp[-(E_1 + E_3)/RT]}{A_2 \exp(-E_2/RT) + A_3 \exp(-E_3/RT)}. \tag{5.4f}$$

An initial impression may be that the Arrhenius plot of this equation is a combination of two straight lines occurring when first one and then the other of the denominator terms dominates. However, this is not the case. The Arrhenius plot of eqn (5.4f) is obtained by taking logarithms to give

$$\ln\left(\frac{\mathrm{d}P}{\mathrm{d}t}\right) = \ln(SA_1A_3) + \ln\left\{\frac{\exp[-(E_1 + E_3)/RT]}{A_2 \exp(-E_2/RT) + A_3 \exp(-E_3/RT)}\right\}. \tag{5.4g}$$

Equation (5.4f) is illustrated in Fig. 5.8(a) and the corresponding Arrhenius plot is shown in Fig. 5.8(b). In these figures $E_2 > E_3$, and the asymptote of the Arrhenius plot can be obtained from eqn (5.4g). This asymptote is also illustrated in Fig. 5.8(b). The responses of the rate constant to temperature are continuous throughout, without any abrupt transitions.

Longer sequences of the type described by eqn (5.4c) are easily written down, e.g.

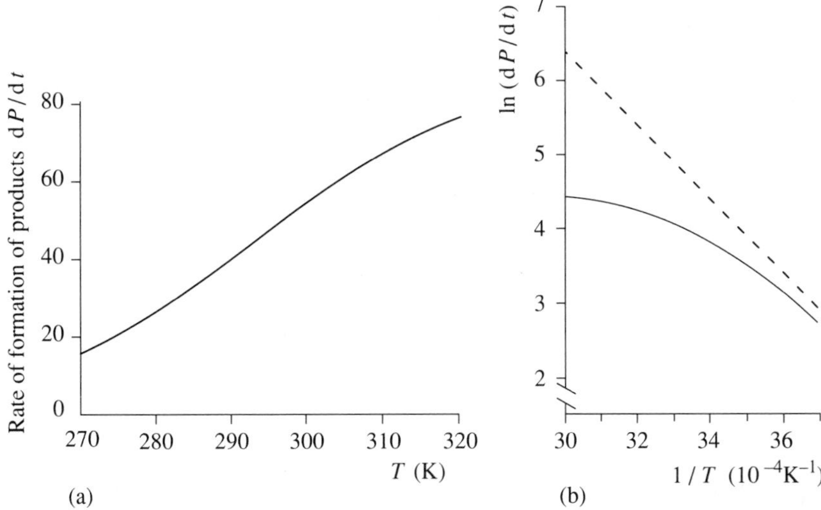

Fig. 5.8. The rate of formation of products (arbitrary units) for the reversible reaction scheme given by eqn (5.4c): (a) direct temperature dependence as given by eqn (5.4f); (b) the Arrhenius plot given by eqn (5.4g). The asymptote is shown by the broken line. The parameter values are $[S] = 0.2$, $A_1 = 10^{10}$, $A_2/A_3 = 2 \times 10^7$, $E_1/R = 5 \times 10^3$ K, and $(E_2 - E_3)/R = 5 \times 10^3$ K.

$$S \underset{k_2}{\overset{k_1}{\rightleftharpoons}} S_1 \underset{k_4}{\overset{k_3}{\rightleftharpoons}} S_2 \xrightarrow{k_5} P. \tag{5.4h}$$

In the steady state the rate of formation of product is given by

$$\frac{dP}{dt} = \frac{Sk_1k_3k_5}{k_2k_4 + k_2k_5 + k_3k_5}. \tag{5.4i}$$

The temperature equation is obtained from eqn (5.4i) by putting

$$k_i = A_i \exp\left(-\frac{E_i}{RT}\right) \tag{5.4j}$$

The behaviour of the Arrhenius plots for the formation of products can be examined by identifying the dominant term in the denominator of eqn (5.4i).

When more terms are included in the reaction scheme, in the manner of eqn (5.4i), there is more control (through the adjustment of parameters) over the response $dP(T)/dt$. For example, in Fig. 5.8(b) the Arrhenius plot approaches the asymptote quite gently, whereas with more terms a sharper bend in the curve may be obtained. Such responses for enzyme activity have been observed (see e.g. Gordon and Flood 1979) and it is likely that in such cases the underlying mechanisms are quite complex.

5.5.2 *Parallel reactions*

An alternative scheme to a sequential reaction is to assume that there are parallel processes, for example

$$\mathrm{S} \overset{k_1}{\underset{k_2}{\rightrightarrows}} \mathrm{P}, \tag{5.4k}$$

giving a rate of formation of product of

$$\frac{\mathrm{d}P}{\mathrm{d}t} = (k_1 + k_2)S = S\left[A_1 \exp\left(-\frac{E_1}{RT}\right) + A_2 \exp\left(-\frac{E_2}{RT}\right)\right]. \tag{5.4l}$$

To investigate the Arrhenius form of eqn (5.4l), assume that

$$E_1 < E_2, \tag{5.4m}$$

so that

$$\ln\left(\frac{\mathrm{d}P}{\mathrm{d}t}\right) = \ln(SA_1) + \ln\left[1 + \frac{A_2}{A_1}\exp\left(-\frac{E_2 - E_1}{RT}\right)\right] - \frac{E_1}{RT}. \tag{5.4n}$$

This is of a similar form to the Arrhenius plots described in the previous section with the difference that, as $1/T$ increases, $\ln(\mathrm{d}[\mathrm{P}]/\mathrm{d}t)$ approaches the asymptote

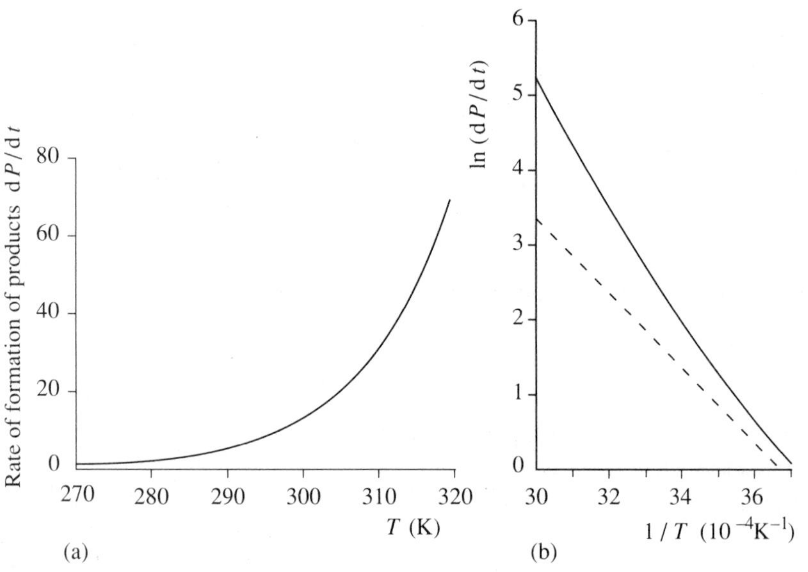

Fig. 5.9. The rate of formation of products (arbitrary units) for the parallel reaction scheme given by eqn (5.4k): (a) direct temperature dependence as given by eqn (5.4l); (b) the Arrhenius plot given by eqn (5.4n). The asymptote is shown by the broken line. The parameter values are $SA_1 = 10^8$, $A_2/A_1 = 2 \times 10^7$, $E_1/R = 5 \times 10^3$ K, and $E_2/R = 10^4$ K.

$$\ln\left(\frac{\mathrm{d}P}{\mathrm{d}t}\right) \approx \ln(SA_1) - \frac{E_1}{RT} \tag{5.4o}$$

from above rather than from below. Equations (5.4l) and (5.4n) are illustrated in Figs 5.9(a) and 5.9(b) respectively, and the asymptote given by eqn (5.4o) is also drawn in Fig. 5.9(b).

5.5.3 *Phase changes*

The term 'chilling injury' refers to tissue damage and plant death caused by temperatures above freezing but below about 15 °C; such injury is of considerable commercial importance. Simple considerations from the Arrhenius equation suggest that, although reduced temperatures would cause plant metabolism to slow down, the plant might be expected to survive but grow at a reduced rate. The survival of a plant is a complex matter, depending upon photosynthesis, nutrition, and water, all of which may be affected by temperature, as well as environmental history, which may lead to cold hardening. There is no single critical temperature below which a plant always perishes and above which survival is assured. Some workers have sought a single basic process which could be responsible, and, indeed, there appears to be a correlation between plant survival temperature and the melting point of its lipids (Lyons and Asmundson 1965). The idea is that there is a critical temperature at which some of the plant membrane lipids undergo an abrupt phase change, and this changes the activation energy of some key reactions (Lyons 1973). This hypothesis is reviewed critically by Wolfe (1978). The relevance of this to the foregoing discussion of reaction kinetics and the Arrhenius equation is that there has been a continuing controversy over the existence or non-existence of discontinuities in the temperature responses of various plant processes, and also as to whether these (if they exist) are compatible with the simple application of the Arrhenius function to different reaction schemes, in series, in parallel, in competition, etc. If such simple application is unable to account for the observational data, then it may be necessary, arguably, to assume that the activation energy is quite suddenly, at or about the critical temperature, switched to a new value. This then leads directly to theoretical temperature responses which exhibit a discontinuity.

In the models described in eqns (5.4a)–(5.4o), the Arrhenius plots are seen to approach a straight line asymptote from either below or above, depending on whether product formation resulted from reactions in series or in parallel. In no case is there any evidence to suggest combinations of linear phases in the Arrhenius plots. Our tentative conclusion, therefore, based on the theoretical evidence described above, is that Arrhenius plots are generally smooth and approach a linear asymptote as $1/T$ increases. However, the nature of the Arrhenius plots illustrated shows that it is possible for experimental data to exhibit, superficially, distinct linear regions. It is often instructive to analyse data on linear plots rather than logarithmic plots which may be misleading, although in some cases where there are large changes in the magnitude of the rate of formation of products, logarithmic plots may be unavoidable.

5.6 Diffusion, viscosity, and translocation

These three phenomena are involved in many aspects of plant growth and development. For example, diffusion in the gas phase is important in CO_2 uptake and water loss by transpiration, diffusion in the liquid phase affects within-cell processes, and liquid phase diffusion and biochemistry combined may be crucial over small distances in morphogenesis (as in the shoot and root apices, and bud formation); viscosity plays a major role in long-distance transport—the translocation of sugars in the phloem, and the movement of water and other nutrients in the xylem. We therefore review the relevant temperature responses.

In the gas phase, for a CO_2–air mixture and for an oxygen–nitrogen mixture the usual binary coefficients of diffusion can be described by (Jost 1960, p. 425)

$$D(CO_2\text{–air}) = c_1 T^{1.97} \tag{5.5a}$$

and

$$D(O_2\text{–}N_2) = c_2 T^{1.79}, \tag{5.5b}$$

where c_1 and c_2 are constants. An increase of 10 K in temperature from 290 to 300 K thus produces increases of 6.9 per cent and 6.3 per cent respectively in the diffusion constants. The temperature coefficients of gaseous diffusion are small compared with those of most chemical or biochemical processes. (Exercise 5.5)

However, diffusion in the liquid phase can have a temperature dependence comparable with that of chemical or biochemical processes. This is sometimes not appreciated. The temperature dependence of diffusion in liquids can be reasonably described by (Jost 1960, p. 469)

$$D = c_3 T^{1/2} \exp\left(-\frac{E}{RT}\right), \tag{5.5c}$$

where c_3 is a constant. This is of the same type as eqn (5.2b) which applies to a chemical or biochemical rate constant; as is shown in Fig. 5.5, this is virtually indistinguishable from the Arrhenius equation (eqn (5.1e)) over a practical temperature range. Consequently, it may be impossible to discriminate between diffusive and metabolic processes by virtue of their temperature response.

Next we consider viscosity, which is one of the most important parameters in translocation. As discussed by Jost (1960, p. 462), diffusion and viscosity are inversely proportional to each other in liquids (although they are directly proportional in gases). Thus, for liquids, it follows from eqn (5.5c) that

$$\eta = \frac{c_4}{T^{1/2}} \exp\left(\frac{E'}{RT}\right), \tag{5.5d}$$

where η is the viscosity, c_4 is a constant, and E' is an activation energy. Comparing this equation with eqn (5.2b), this can vary just as rapidly with temperature as the rates of metabolic processes (Fig. 5.5).

The effect of temperature on translocation can only be discussed with reference to a particular model of the process. If F is the flux of solute (kg mol s^{-1}) between two organs, with solute concentrations S_1 and S_2 (kg mol m^{-3}), then a conductivity σ ($m^3\ s^{-1}$) can be defined by

$$F = (S_1 - S_2)\sigma. \tag{5.5e}$$

As an example we shall consider mass flow cyclosis which is actuated by osmotic pressure. This scheme, usually referred to as the Münch hypothesis, is illustrated in Fig. 4.9 (p. 107). The flux F of solute is given by (cf. eqn (4.14c))

$$F = \frac{RT\pi a^4}{8L\eta} S_1(S_1 - S_2), \tag{5.5f}$$

where a is the radius and L is the length of the conducting tube. For a short path length $L = \Delta x$, eqns (5.5e) and (5.5f) can be combined to give

$$\sigma = \frac{RT\pi a^4}{8\Delta x}\frac{S}{\eta}, \tag{5.5g}$$

where η and S are interpreted as average values of viscosity and solute concentration over the path. Substituting eqn (5.5d) in (5.5g) yields, for the temperature dependence of σ,

$$\sigma = c_5 T^{3/2} \exp\left(-\frac{E'}{RT}\right) \tag{5.5h}$$

where c_5 is independent of temperture. Comparison with eqns (5.1e), (5.2b), and (5.2d), and with Fig. 5.5, shows that this is not very different from the Arrhenius equation, and therefore translocation may not differ greatly from metabolic processes in its temperature dependence.

Many measurements of the temperature response of the translocation flux have been made, often with the hope that the type of response observed would discrimate between the possible mechanisms that might be involved in translocation. Many different treatments are possible, for instance involving various lengths of the conducting pathway, including the source and/or sink in the temperature treatment, or following the time course of events to the new steady state. Also, other environmental factors, including the state of the plant (with respect to development and growth), may vary. All these factors may affect the balance between the source, the sink, and the conducting pathway, and, as discussed in the next section, may be expected to influence the observed response markedly. Unsurprisingly, the observational data do not show a consensus. Helms and Wardlaw (1977) report that in *Nicotiniana* translocation out of the leaf does not change significantly with temperature. Lang (1974), working with a *Nymphoides* petiole, finds Q_{10} factors between unity and infinity, with an average of 1.18. In *Salix*, Watson (1975) finds that cooling 65 cm of stem from 23 to 0 °C reduces the translocation flux by not more than 20 per cent. Geiger and Sovonick (1970) also observed that translocation fluxes are well maintained in cooled systems. The complexity of the responses that can be obtained theoretically from a simple source–transport–sink system is illustrated below, and has been investigated experimentally as well as by means of a model by Thornley, Gifford, and Bremner (1981), who considered sugar movement in a wheat spikelet.

5.7 Temperature response of an integrated source–transport–sink system

Consider the scheme illustrated in Fig. 5.10. This shows a simple hybrid scheme, which will be analysed in order to indicate the complexity of the temperature responses resulting from the interactions between a Michaelis–Menten sink and a transport resistance. S and S_c are the substrate concentrations at the source and sink respectively. The resistance between the source and sink is r, and F is the flux of substrate between the source and sink. The sink resistance is r_c so that, at the sink,

$$F = S_c/r_c \tag{5.6a}$$

and, assuming the sink to be a Michaelis–Menten sink, it follows that

$$r_c = \frac{K + S_c}{k}, \tag{5.6b}$$

where K is a Michaelis–Menten constant and k is a rate constant. Assume also that the transport resistance r behaves according to Ohm's law, so that

$$F = \frac{S - S_c}{r}. \tag{5.6c}$$

The flux F, which may be proportional to the growth of a fruit or other organ, will clearly depend on the substrate concentration S and the three parameters r, k, and K. The problem discussed in this section is to what extent the temperature

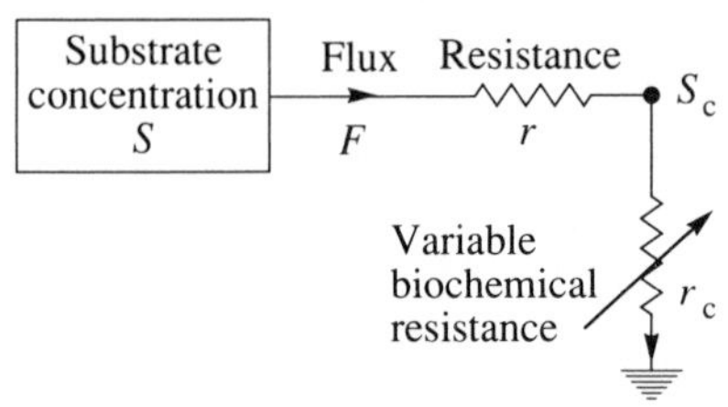

Fig. 5.10. Hybrid scheme shown using the resistance analogy form. S_c is the substrate concentration between resistances r and r_c. See eqns (5.6a)–(5.6c).

response of F will depend on the temperature responses of the underlying parameters, especially as the source concentration S is varied.

S_c and r_c can be eliminated from eqns (5.6a)–(5.6c) to give

$$F^2 r - F(K + S + kr) + kS = 0. \tag{5.6d}$$

The hyperbolic nature of the $F:S$ response is shown by writing this equation in factors

$$(F - k)(Fr - K - S) = kK, \tag{5.6e}$$

which is a non-rectangular hyperbola in F and S. The asymptotes are obtained by equating the two factors on the left-hand side of the equation to zero, i.e.

$$F = k \qquad \text{and} \qquad F = \frac{S + K}{r}. \tag{5.6f}$$

The solution to eqn (5.6d) is given by

$$F = \frac{K + S + kr - [(K + S + kr)^2 - 4rkS]^{1/2}}{2r}, \tag{5.6g}$$

and by specifying the temperature dependence of the parameters

$$r = r(T), \qquad k = k(T), \qquad K = K(T) \tag{5.6h}$$

we can investigate the effect of temperature upon F. Equation (5.6g) is illustrated in Fig. 5.11 for arbitrary units. Also shown are the asymptotes given by eqns (5.6f).

It is instructive to examine the behaviour of $F(S)$ (eqn (5.6g)) in response to various values of the parameters r, k, and K. Consider first the resistance parameter r. From eqn (5.6d) or eqn (5.6g) it can be shown that

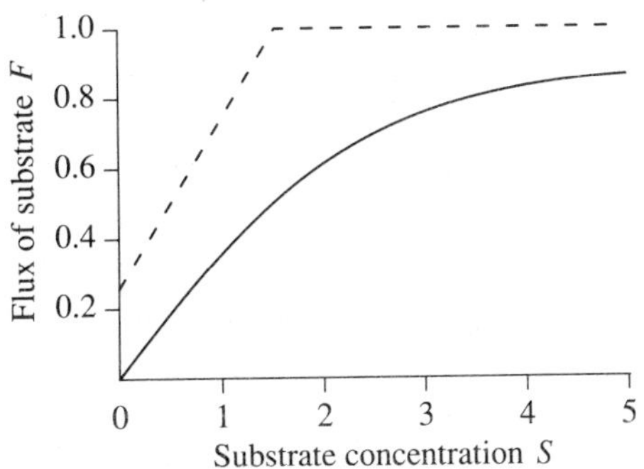

Fig. 5.11. Response of the flux F of substrate between source and sink (arbitrary units) to the substrate concentration S at the source for the simple hybrid scheme illustrated in Fig. 5.10, as given by eqn (5.6g). Broken lines indicate the asymptotes. The parameter values are $k = 1$, $r = 2$, and $K = 0.5$.

$$F \to \frac{kS}{K + S} \quad \text{as} \quad r \to 0 \tag{5.6i}$$

and

$$F \to 0 \quad \text{as} \quad r \to \infty. \tag{5.6j}$$

The first of these limits corresponds to the case when $S = S_c$ and there is no transport resistance between the source and sink, and in the second limit there is infinite resistance and consequently no flux between source and sink.

Now consider the rate parameter k. As above, we can obtain the limiting cases

$$F \to 0 \quad \text{as} \quad k \to 0 \tag{5.6k}$$

and

$$F \to S/r \quad \text{as} \quad k \to \infty. \tag{5.6l}$$

As expected, when the rate constant is zero the corresponding flux is also zero. When k is very large the substrate at the sink is readily utilized, so that $S \gg S_c$ and the flux is limited by the transport resistance between the source and sink.

Proceeding as before, it follows that

$$F \to \begin{cases} S/r, & S \leqslant kr, \\ k, & S > kr, \end{cases} \quad \text{as } K \to 0 \tag{5.6m}$$

and

$$F \to 0 \quad \text{as} \quad K \to \infty. \tag{5.6n}$$

Thus, if K is small, there is a sharp transition between transport-limited behaviour ($F \approx S/r$) and sink-limited behaviour ($F \simeq k$) (cf. the Blackman photosynthetic response of Fig. 9.6 (p. 228) with $\theta = 1$).

Note that the initial slope of the substrate response curve is

$$\frac{\mathrm{d}F}{\mathrm{d}S}(S = 0) = \frac{1}{r + K/k}, \tag{5.6o}$$

so that, for non-zero K, even at the lowest substrate values the response may be a balance between transport and biochemistry—pure transport-limited growth does not occur for non-zero K.

The limiting responses of eqns (5.6i), (5.6l), (5.6m), and (5.6o) all depend differently on the parameters r, k, and K, each of which are functions of temperature. Therefore the temperature response of such systems may be expected to depend on the conditions of the experiment, and it may be quite unrealistic to talk about *the* temperature response of a plant or part of a plant: eqn (5.6l) will show the temperature behaviour of r alone; eqn (5.6m) at high substrate levels will show that of k alone; eqn (5.6i) depends on the combined temperature responses of k and K and is affected by the substrate level S; eqn (5.6o) combines all three parameters with their associated temperature responses.

5.8 Temperature effects at the organ, plant, and crop level

In this section we outline some schemes for analysing and describing the effects of temperature on the growth and development of plants and crops. Rather than concentrate on detailed biochemical processes, our aim is to outline how the basic responses to temperature can be investigated in a manner which is practical at the whole-plant and crop levels.

5.8.1 *Photosynthesis*

Although there has been a substantial research effort on photosynthesis, there is still relatively little quantitative information regarding the temperature response of single leaves to temperature (see also Chapter 9, Section 9.3.3, p. 229). An appropriate curve for describing single-leaf photosynthesis as a function of leaf irradiance I ($W\ m^{-2}$), is the non-rectangular hyperbola (eqn (9.10f), p. 228), and the photosynthetic rate P ($kg\ CO_2\ m^{-2}\ s^{-1}$) is given by the lower root of the equation

$$\theta P^2 - (\alpha I + P_m)P + \alpha I P_m = 0, \qquad (5.7a)$$

i.e.

$$P = \frac{1}{2\theta}\{\alpha I + P_m - [(\alpha I + P_m)^2 - 4\theta\alpha I P_m]^{1/2}\}. \qquad (5.7b)$$

α is the photosynthetic efficiency ($kg\ CO_2\ J^{-1}$), P_m is the rate of photosynthesis at saturating irradiance levels, and θ is a dimensionless parameter with $0 \leqslant \theta \leqslant 1$. For low irradiance levels, eqn (5.7b) is approximated by

$$P \approx \alpha I, \qquad (5.7c)$$

and α therefore represents the initial slope of the response.

Although eqns (5.7a) and (5.7b) have a theoretical basis (pp. 223–8), in practice α, θ, and P_m can be regarded as adjustable parameters when analysing photosynthesis data. Considered as an empirical equation, eqn (5.7b) is appropriate as the three parameters α, θ, and P_m in turn influence the $P(I)$ response at low, medium, and high irradiance levels.

Experimental evidence indicates that θ is largely independent of temperature: for C_4 plants α is independent of temperature, whereas for C_3 plants over the range 14–24 °C, α decreases by about 14 per cent (Ehrlinger and Björkman 1977). However, P_m depends strongly on temperature. Johnson and Thornley (1984) assign a simple linear relationship to P_m:

$$P_m = P_m(T_r)\left(\frac{T - T^*}{T_r - T^*}\right), \qquad (5.7d)$$

where T_r is some reference temperature and T^* is the temperature at which photosynthetic activity ceases. Assuming that θ and α are temperature independent, they derive expressions for instantaneous and daily canopy photosynthesis in fluctuating temperature conditions. Although eqn (5.7d) may well break down at high temperatures, the simple form for a practical temperature range makes it suitable for many purposes.

It is our view that, while there is a great deal of information available concerning many aspects of photosynthesis, an accurate catalogue of responses of the form of eqn (5.7b) with parameter values and their temperature dependences is still lacking.

5.8.2 *Respiration*

Consider the organ, plant, or crop growth scheme illustrated in Fig. 11.2 (p. 270), which was first proposed by Thornley (1977). The model has three state variables: the storage dry mass w_s, the degradable component of the structural mass w_d, and the non-degradable component of the structural mass w_n. The total mass is (see pp. 270–2 for a detailed treatment)

$$w = w_s + w_d + w_n. \tag{5.7e}$$

Fractions of matter in the different components are defined by

$$f_s = w_s/w \quad \text{and} \quad f_d = w_d/w. \tag{5.7f}$$

Respiration is considered in terms of the two components arising from growth and from maintenance. Growth respiration r_g is interpreted as that component of respiration associated with the synthesis of new structures (either degradable or non-degradable), whereas maintenance respiration r_m is that resulting from the resynthesis of degraded structure. Proceeding in this way (cf. eqn (11.5k), p. 271) therefore, we have

$$r_g = \frac{1 - Y_g}{Y_g}(1 - f_s)\frac{1}{w}\frac{dw}{dt} \tag{5.7g}$$

$$r_m = \frac{1 - Y_g}{Y_g} k_d f_d, \tag{5.7h}$$

and the specific respiration rate r is

$$r = r_g + r_m = \frac{1 - Y_g}{Y_g}\left[(1 - f_s)\frac{1}{w}\frac{dw}{dt} + k_d f_d\right]. \tag{5.7i}$$

Thus, by fitting the total specific respiration to the specific growth rate of the organ or plant, we can identify the growth and maintenance components. In deriving these equations it is assumed that the storage fraction f_s does not change with time.

There is, at present, no biochemical evidence to suggest that the growth efficiency Y_g, is temperature dependent. Any small effects observed by experiment are probably due to variation in the term $1 - f_s$ which is generally not considered:

plants grown at high temperatures have lower substrate levels than those grown at lower temperatures. Although the growth respiration is temperature dependent, so is the growth rate, and thus the growth efficiency, which depends on their ratio, may be independent of temperature.

The maintenance respiration, in contrast, is strongly temperature dependent since it is directly related to the enzymatic process of degradation. However, once again, care must be taken to examine the f_d factor for the degradable component in eqns (5.7h) and (5.7i) as this may vary for plants grown in different environments.

5.8.3 *Development and the day-degree hypothesis*

For many years crop scientists have been concerned with the response of developmental rate to temperature. The day-degree hypothesis or temperature summation rule is widely used (see Section 7.2, p. 167, for a wider discussion). The idea is as follows. The mean temperature $\bar{T}_i$ for each day i is measured, and a sum h is formed according to

$$h = \sum_{i=1}^{j} (\bar{T}_i - T_c), \tag{5.7j}$$

which includes only those terms where $\bar{T}_i$ is above some threshold value T_c. When h reaches a particular value, this signifies that a phase in development is complete, and this is generally associated with a biological event which occurs over a short period of time and is readily observed. The day-degree sum h essentially integrates some underlying temperature-sensitive process or processes. For example, for a cereal (Robertson 1968) there are various phases in the development of the plant, and the temperature sum is found to have a certain value for the successful completion of each:

$$\begin{aligned} &h\ (\text{sowing} \rightarrow \text{emergence}) \\ &h\ (\text{emergence} \rightarrow \text{floral initiation}) \\ &h\ (\text{floral initiation} \rightarrow \text{anthesis}) \\ &h\ (\text{anthesis} \rightarrow \text{maturity}). \end{aligned} \tag{5.7k}$$

The temperature threshold T_c may be different for each of these phases.

The approach is based on the notion of a developmental rate k_d, whose response to temperature is approximately linear over a restricted temperature range. Comparison with actual temperature responses as in Fig. 5.12 or Fig. 5.1 suggest that this is not unreasonable, and the method works well in practice. It is implicitly assumed that the organ possesses a developmental clock which is proceeding at the rate k_d. In general, it is to be expected that the development rate k_d may depend on a number of quantities. This can be represented by

$$k_d = f(V, P, E), \tag{5.7l}$$

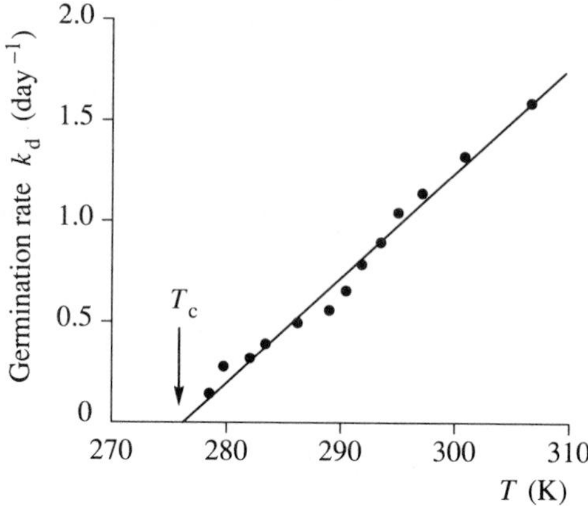

Fig. 5.12. Temperature summation rule: k_d is the developmental rate, T is the temperature, and T_c is a temperature threshold. The line drawn represents a linear approximation to development (cf. eqns (5.7j) and (5.7n)). The data points (●) denote the germination rate of mustard seed (the reciprocal of the time for 50 per cent germination) and are derived from Simon *et al.* (1976).

where f denotes some function of the state variables V, parameters P, and environmental quantities E. The temperature-sum rule works because the most important of the environmental variables is temperature, and the response to temperature is approximately linear. Generalizing eqn (5.7j), we can write

$$h_{ij} = \int_{t_i}^{t_j} k_d \, dt, \tag{5.7m}$$

where t is time and h_{ij} is the 'distance' along the developmental pathway between points i and j. The length of the phase is $t_j - t_i$ and, given t_i (the beginning of the phase), then t_j (the end of the phase) is the time point where the integral gives a particular value for the index h_{ij}. (Exercise 5.6)

More elaborate non-linear forms of eqn (5.7j) of the type

$$h = \sum_{p=1}^{j} (a + b\bar{T}_p + c\bar{T}_p^{\,2} + d\bar{T}_p^{\,3}) \tag{5.7n}$$

have been applied (Robertson 1968) or, generalizing eqn (5.7m) in a similar way,

$$h_{ij} = \int_{t_i}^{t_j} [a + bT(t) + cT(t)^2 + dT(t)^3] \, dt, \tag{5.7o}$$

where a, b, c, and d are constants.

The developmental index approach can be extended in many ways, by adding quadratic and higher terms, cross terms, other environmental variables, and so on. However, the method assumes that development is a scalar (one-dimensional) quantity and can be characterized by a single number. This may not always be a good approximation.

Exercises

5.1. Differentiate the Arrhenius equation for the rate constant k (eqn (5.1e)), i.e.

$$k = A\exp\left(-\frac{E_a}{RT}\right), \tag{E5.1a}$$

twice with respect to temperature T to give dk/dT and d^2k/dT^2. Find the inflexion temperature T_i. Sketch the $k:T$ curve. Apart from a factor, the Arrhenius equation is an approximation of eqn (5.1b); repeat the exercise with eqn (5.1b), and consider the differences between the two graphs you have sketched.

5.2. In Exercise 5.1 it was shown that the Arrhenius equation (E5.1a) possesses a point of inflexion at temperature T_i, and T_i was determined (eqn (S5.1c)). Show that the rate equation obtained from bimolecular collision theory (eqn (5.2b)), i.e.

$$k = BT^{1/2}\exp\left(-\frac{E_a}{RT}\right), \tag{E5.2a}$$

also has an inflexion point, whereas the Eyring equation (eqn (5.2d))

$$k = CT\exp\left(-\frac{E_a}{RT}\right) \tag{E5.2b}$$

does not.

5.3. As shown in Fig. 5.7, eqn (5.3d), which was derived from the two-conformational-state model of Fig. 5.6, has an optimum temperature when the rate constant k is maximum. Rewriting this equation more simply as

$$k = \frac{A\exp(-ax)}{1 + b\exp(-cx)} \tag{E5.3a}$$

where $x = 1/T$ and a, b, and c are constants, find the maximum value of k and the corresponding value of T.

5.4. Combining the reversible reaction scheme of (5.4c) with the parallel reaction scheme of (5.4k) gives a scheme

$$S \underset{k_2}{\overset{k_1}{\rightleftharpoons}} S_1 \underset{k_4}{\overset{k_3}{\rightrightarrows}} P. \tag{E5.4a}$$

For the steady state, find the rate of formation of product P in terms of the rate constants. Assuming tthe Arrhenius eqn (5.1e) for the rate constants, write down an expression for the temperature dependence of the rate of product formation.

5.5. In the gas phase, the diffusion coefficient D obeys an expression of the type (eqns (5.5a) and (5.5b))

$$D = cT^x \tag{E5.5a}$$

where c and x are constants and T is the absolute temperature. By approximating, write this in the form of the Q_{10} equation (eqns (5.1h) and (5.1i) and hence find the Q_{10}s for CO_2 and oxygen from the numerical values in eqns (5.5a) and (5.5b).

5.6. The day-degree approach to development in its simplest form (eqn (5.7j)) assumes that the developmental rate constant k responds linearly to temperature above some base temperature T_0. If an Arrhenius equation (eqn (E5.1a)) is assumed for k, the best straight line approximation to this equation can be written

$$k = c(T - T_0) \tag{E5.6a}$$

where the line is drawn through the inflexion point with the gradient c equal to the maximum slope of the Arrhenius equation (Fig. S5.1, p. 594). Find c and T_0 in terms of the parameters A and E_a of the Arrhenius equation (E5.1a).

6
Biological switches

6.1 Introduction

Several processes in plant and crop physiology exhibit apparent discontinuities at the macroscopic level. These can sometimes be viewed as triggers for a new time course of development, e.g. in seed dormancy and germination, or continued metabolic activity and senescence; in other cases these are seen as bifurcations or branch points, from which alternative pathways of development arise, as in the vegetative or reproductive development of shoot apices (Thornley 1976, pp. 24, 233, 1981, 1987).

It is straightforward to write into a computer program statements such as

$$\text{if } x < K \text{ then do } A, \text{ else do } B, \tag{6.1a}$$

where x is a variable (plant or environment), K is a constant, and A and B represent plant processes. Such statements are generally empirical. Our objective in this chapter is to interpret (6.1a) or similar statements in terms of biochemical or morphological processes. First we need to discuss just what is meant by the terms 'switch, 'trigger', and 'event'.

Consider a system which is described, at time t, by q state variables

$$X_1, X_2, \ldots, X_q. \tag{6.1b}$$

The X_i are scalar quantities, and initially it is assumed that there is no spatial variation attached to the state variables. (This does not exclude the possibility that some of the variables may denote attributes associated with pattern or structure, e.g. internode distance or length-to-breadth ratio.) Using a deterministic approach, we can describe the dynamics of the system by q first-order differential equations

$$\frac{\mathrm{d}X_i}{\mathrm{d}t} = f_i(X_1, \ldots, X_q; E; P) \qquad i = 1, \ldots, q, \tag{6.1c}$$

where the f_i are functions of the state variables X_i, the environment E, and any parameters or constants P.

A switch is sometimes defined as a device that converts a smooth input into a relatively discontinuous output. Whenever the term 'switch' is used, there is an implicit division of the system into two parts: the environment or agent or other subsystem which operates the switch, and the system or subsystem that undergoes the quasi-discontinuous change. It is convenient to consider switches as two types: exogenous switches and endogenous switches. These are illustrated in Fig. 6.1.

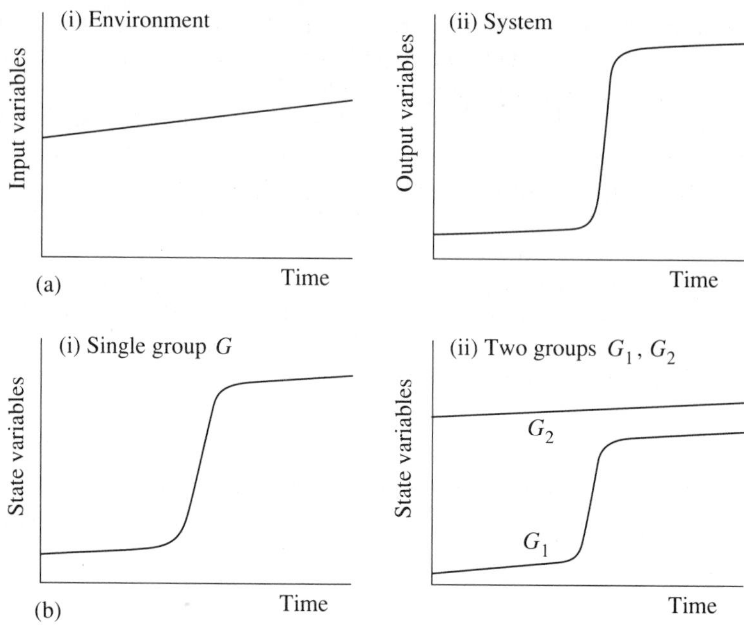

Fig. 6.1. Types of switch: (a) the usual form of switch—an exogenous switch with the environmental input varying slowly and the system output variables showing a sudden change; (b) an endogenous switch in which (i) the whole group G of state variables undergoes a switch response with a sudden change in values or (ii) the state variables can be divided into two groups, G_1 and G_2, of which one is switched (G_1), and the other (G_2) only changes slowly. Given that the two subsystems G_1 and G_2 are coupled, then it may be possible to regard G_2 as providing the smooth switch input (analogous to (a)(i)) and G_1 as the switch output ((a)(ii)).

In an exogenous switch some component of the environment, e.g. E in eqns (6.1c), is the actuator of the switch, and the discontinuity may show itself in some or all of the state variables of the system (Fig. 6.1(a)). Examples of this are germination or non-germination in seeds depending on moisture and temperature, vegetative or reproductive growth depending upon daylength, and leaf senescence and abscission or continued normal leaf function depending on environment.

In an endogenous switch, the switch behaviour may occur in a constant or varying environment regardless of the previse details of the environment; thus the behavioural discontinuity may be part of the intrinsic developmental dynamics of the system (Fig. 6.1(b)). Examples are as follows: some plants switch in an obligate manner from vegetative to reproductive growth, and some plant organs undergo obligate senescence and abscission; primordial initiation can also largely be viewed as endogenously switched. The discontinuity might occur in some or all of the state variables of the system. It may be possible to divide the state variables into two groups, corresponding to a division of the system into two subsystems, one of which (varying slowly) could be viewed as actuating switch behaviour in the other group (Fig. 6.1(b) (ii)).

The role of time-scales is important in considering switches. A so-called switch is viewed as such if it takes place over a small fraction of the time span of interest. For example, if growth and development of the shoot apex over a period of 50 days is being investigated, a change in morphology or the level of some chemical that is substantially complete over 2 or 3 days can reasonably be regarded as a switch, whereas the same changes occurring over 20 days would be regarded differently. Thus switch dynamics begin with all the state variables varying slowly, and then over a short period of time some or all of these variables change rapidly, moving to a new level from which a slow rate of variation is resumed.

6.2 Differential equation representation of a decision process

In this section, the sigmoidal response curves discussed on pp. 54–6 are related to decision statements such as (6.1a), and it is shown how the time course of the process can be represented by ordinary differential equations.

The equation

$$y = y_{\mathrm{m}} \frac{x^n}{x^n + K^n} \tag{6.2a}$$

describes a very useful family of responses; x and y are variables, y_{m} is the asymptotic value of y, and K is a parameter ($y = y_{\mathrm{m}}/2$ at $x = K$). These responses are illustrated in Fig. 6.2(b). If x is regarded as the input variable (changing smoothly) for a decision process, then it is assumed that y is the output variable,

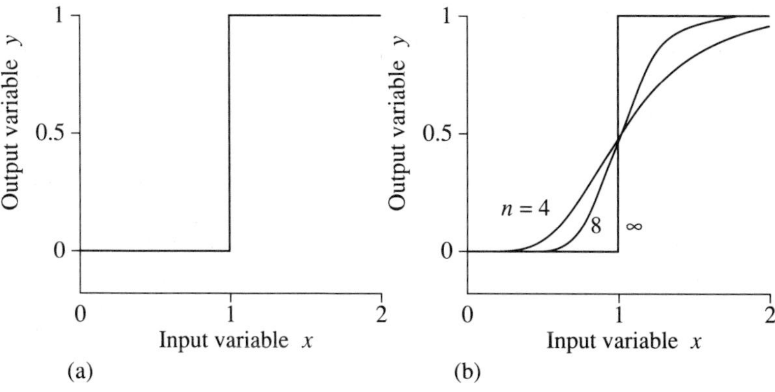

Fig. 6.2. Comparison of (a) eqn (6.1a) with $K = 1$ with (b) an equivalent graded response (eqn (6.2a) with $K = y_{\mathrm{m}} = 1$) for decision processes. $y < 0.5$ corresponds to process A; $y \geqslant 0.5$ corresponds to process B.

$y < y_m/2$ corresponds to process A, $y \geqslant y_m/2$ to process B. This result, which gives a graded equivalence to (6.1a), is shown in Fig. 6.2.

Used in this way, eqn (6.2a) is hardly less empirical than (6.1a), although the biochemical basis for the equation and the fact that it gives a graded response rather than a binary response may make it preferable to the biologist.

The simplest method of putting either of these switching mechanisms on to a time-course basis is to construct a differential equation for the input variable x. For instance, take

$$\frac{dx}{dt} = g \quad \text{and} \quad x = x_0 + gt, \tag{6.2b}$$

where g is a constant (the rate of increase of x) and x_0 is the value of x at $t = 0$. The 'event' can be interpreted as occurring at $x = K$ or $t = (K - x_0)/g$; this is compatible with the temperature-sum (p. 167) or day-degree (p. 141) approaches to development.

We may wish to distinguish between the value of the output decision variable y and the morphological expression of the developmental processes, for which the variable z is used, which may lag behind y. One way of achieving this is to use another differential equation, such as

$$\frac{dz}{dt} = k(y - z), \tag{6.2c}$$

where k is a rate constant. In this formulation z will follow y, at a rate determined by k, giving, for constant y,

$$z = y - (y - z_0)\exp(-kt) \tag{6.2d}$$

where z_0 is the initial value of z. This corresponds to a delay or lag between y and z of $1/k$.

6.3 Single-variable systems with multiple steady states

Consider the single state variable x, whose time dependence is governed by

$$\frac{dx}{dt} = f(x), \tag{6.3a}$$

where f denotes a function of x. In the steady state $dx/dt = 0$, and therefore

$$f(x) = 0. \tag{6.3b}$$

If eqn (6.3a) is to have more than one stable steady state, eqn (6.3b) must possess three roots, i.e. be equivalent to a cubic or higher equation. This can be demonstrated by considering the equation

$$\frac{dx}{dt} = g_m \frac{x}{x + K} - kx, \tag{6.3c}$$

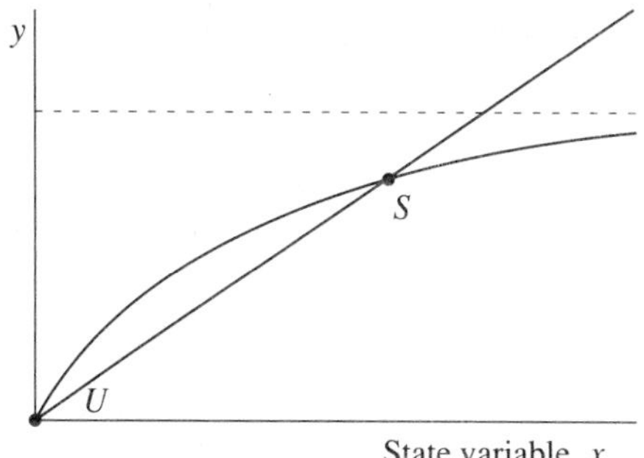

Fig. 6.3. Illustration of the steady state solutions to eqn (6.3c) by the intersection of eqns (6.3e). U is unstable and S is stable. The broken line denotes the asymptote of the rectangular hyperbola at $y = g_m$.

where g_m, K, and k are parameters. Biologically the first positive term can be interpreted as autocatalytic Michaelis–Menten-limited synthesis, and the second term as linear degradation. The equation $dx/dt = 0$ is quadratic in x with the two roots

$$x = 0 \quad \text{and} \quad x = \frac{1}{k}(g_m - kK). \tag{6.3d}$$

It is helpful to consider the intersections between

$$y = g_m \frac{x}{x + K} \quad \text{and} \quad y = kx \tag{6.3e}$$

as shown in Fig. 6.3. The point U is an unstable steady state; this can be seen by considering small values of x, when the synthetic term is greater than the degradative term (for $g_m/K > k$) so that dx/dt is positive and x continues to increase. Similarly S can be seen to be a stable steady state solution. Finally, if $g_m/K < k$, there is just one stable steady state solution at the origin (for $x \geqslant 0$). Do Exercise 6.1.

Now replace eqn (6.3c) by

$$\frac{dx}{dt} = g_m \frac{x^2}{x^2 + K^2} - kx; \tag{6.4a}$$

this equation has the same number of parameters (g_m, K, k), but the Michaelis–Menten term is replaced by a weakly sigmoidal term. The equation $dx/dt = 0$ for the steady state solutions gives rise to the cubic equation

$$g_m x^2 - kx(x^2 + K^2) = 0, \tag{6.4b}$$

which has three roots

$$x = 0, \quad x = \frac{1}{2}\left\{\frac{g_m}{k} \pm \left[\left(\frac{g_m}{k}\right)^2 - 4K^2\right]^{1/2}\right\}. \tag{6.4c}$$

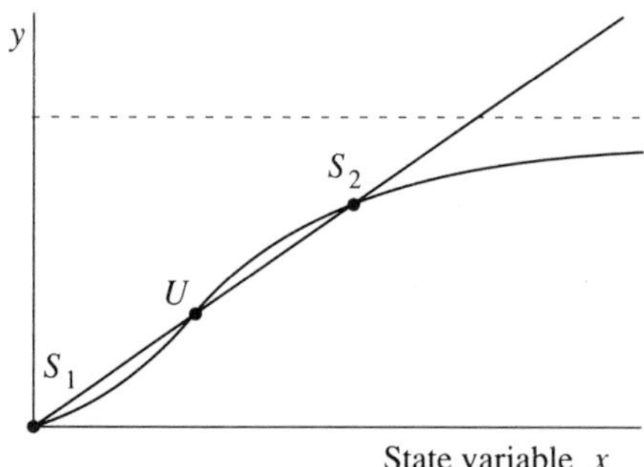

Fig. 6.4. Illustration of the steady state solutions to eqn (6.4a), by the intersection of eqns (6.4d). U is unstable; S_1 and S_2 are stable. The broken line denotes the asymptote of the sigmoidal term in eqn (6.4a).

Again, these are shown graphically in Fig. 6.4 by examining the intersections between

$$y = g_m \frac{x^2}{x^2 + K^2} \quad \text{and} \quad y = kx. \tag{6.4d}$$

It can be seen that there is an unstable steady state U separating two stable steady states at S_1 and S_2.

The introduction of still sharper sigmoidal responses, as in

$$\frac{dx}{dt} = g_m \frac{x^q}{x^q + K^q} - kx, \tag{6.4e}$$

where $q > 2$, does not qualitatively change the features shown in Fig. 6.4, but increases both the stability of the two stable steady states and the instability of the unstable point.

Finally, we consider an equation related to those above, which gives a simple representation of the stability of grazed pastures (this topic has been examined physiologically by Johnson and Parsons (1985)):

$$\frac{dx}{dt} = ax(x_m - x) - g_m \frac{x^q}{x^q + K^q}, \tag{6.5a}$$

where a, x_m, g_m, K, and q are parameters. Suppose that x denotes the standing dry matter in a pasture. The first term on the right-hand side defines a logistic growth curve, and is parabolic in x (Fig. 6.5): at low x only a small part of the available light is intercepted and therefore growth is autocatalytic; at higher x light interception is complete and the processes of maintenance and senescence are an increasing burden until at $x = x_m$ there is no net growth. The second sigmoidal term can be regarded as a typical animal intake function. In Fig. 6.5 the steady states are shown by plotting

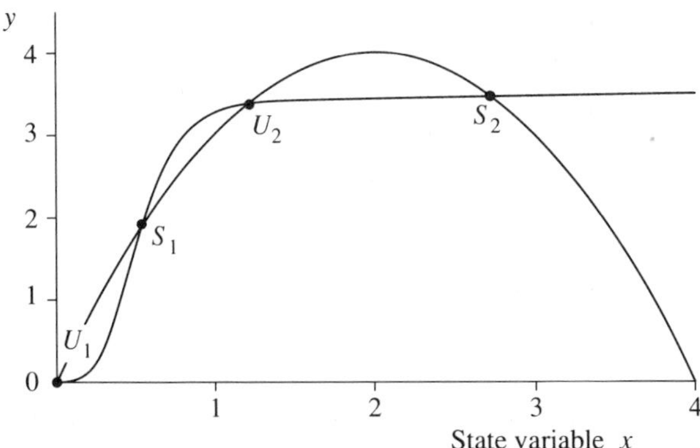

Fig. 6.5. Illustration of the steady state solutions to eqn (6.5a), by the intersection of eqns (6.5b): $a = 1, x_m = 4, g_m = 3.5, K = 0.5$, and $q = 4$. U_1 and U_2 are unstable; S_1 and S_2 are stable.

$$y = ax(x_m - x) \quad \text{and} \quad y = g_m \frac{x^q}{x^q + K^q}. \tag{6.5b}$$

With the parameter values chosen there are four steady states, two of which are stable. Depending upon when stocking commences in the pasture growth cycle, either of these stable steady states, which have different productivities, may be selected. A detailed examination of the grazing problem gives valuable insights into the relations between pasture growth, environment, stocking density, and animal production.

6.4 Two-variable systems with multiple steady states

In this section an account is given of a simple symmetrical two-variable system which can give rise to multiple steady states; the scheme used is biochemical and has been applied by Thornley (1972, 1976) to the problem of vegetative and reproductive growth in plants. The model is represented in Fig. 6.6, and the assumptions that are made and the ensuing mathematical description are now discussed.

6.4.1 *Assumptions*

A pool of substrate with concentration S is available for the synthesis of new plant material—it is assumed that S is maintained at a constant level. As shown in Fig. 6.6, the substrate can be used in two pathways, producing either vegetative plant material by the upper pathway or reproductive material by the lower pathway. While it is helpful in developing the model to assign explicit roles to the compartments as shown in Fig. 6.6, the interpretation or application of the

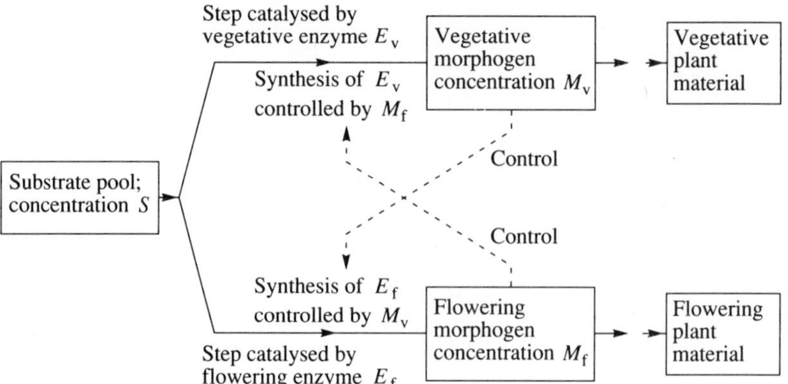

Fig. 6.6. A bi-stable biochemical scheme used for switched pathways of development: —, fluxes of material; - - -, control pathways.

scheme may be quite different. For example, vegetative and reproductive apices are similar biochemically, and the differences are more in the organization of the apex than in its composition.

One of the metabolites associated with the upper vegetative pathway is the vegetative morphogen, which has concentration M_v. It is assumed that the value of M_v is determined by the level of a vegetative enzyme E_v, which catalyses the step preceding M_v according to

$$M_v = a_v E_v, \tag{6.6a}$$

where a_v is a constant, possibly involving the substrate concentration S. Analogously, one of the metabolites of the lower flowering pathway is a flowering morphogen with concentration M_f, where M_f depends on the concentration E_f of a flowering enzyme catalysing the step preceding M_f, with

$$M_f = a_f E_f, \tag{6.6b}$$

where a_f is a constant. Note that in eqns (6.6a) and (6.6b) the morphogen levels respond immediately to the prevailing concentration of the relevant enzyme, implying that the pool sizes of M_v and M_f are small and no time lag is caused in the response due to the pools filling or emptying.

It is assumed that the rate of synthesis of vegetative enzyme E_v is repressed by the flowering morphogen M_f, according to a sigmoidal repression curve of the type shown in Fig. 2.5 (p. 56), so that the gross rate of synthesis of E_v is

$$\frac{k_v}{1 + (M_f/c_f)^n} \tag{6.6c}$$

where k_v and c_f are constants and n is a positive integer constant. Similarly it is supposed that the gross rate of synthesis of the flowering enzyme E_f is repressed by the vegetative morphogen M_v according to

$$\frac{k_f}{1 + (M_v/c_v)^n} \tag{6.6d}$$

where k_f, c_v and n are constants.

It is assumed that the vegetative and flowering enzymes E_v and E_f are both degraded according to first-order kinetics with rate constants k_v' and k_f', so that the rates at which the two enzymes are destroyed are

$$k_v'E_v \quad \text{and} \quad k_f'E_f. \tag{6.6e}$$

Assuming that the volumes associated with the pools are constant, combining eqns (6.6c)–(6.6e) leads directly to the differential equations

$$\begin{aligned}\frac{dE_v}{dt} &= \frac{k_v}{1 + (M_f/c_f)^n} - k_v'E_v \\ \frac{dE_f}{dt} &= \frac{k_f}{1 + (M_v/c_v)^n} - k_f'E_f,\end{aligned} \tag{6.6f}$$

where t is the time variable. Eliminating M_v and M_f from eqns (6.6f) with eqns (6.6a) and (6.6b) gives

$$\begin{aligned}\frac{dE_v}{dt} &= \frac{k_v}{1 + (a_fE_f/c_f)^n} - k_v'E_v \\ \frac{dE_f}{dt} &= \frac{k_f}{1 + (a_vE_v/c_v)^n} - k_f'E_f.\end{aligned} \tag{6.6g}$$

These two equations define the behaviour of the system. However, this is more easily investigated if symmetry is assumed, with

$$k_v = k_f = k, \quad a_v = a_f = a, \quad c_v = c_f = c, \quad k_v' = k_f' = k', \tag{6.6h}$$

and the variables are changed using

$$E_v = \frac{k}{k'}x, \quad E_f = \frac{k}{k'}y, \quad t = \frac{\tau}{k'}, \quad \frac{a}{c} = \frac{k'}{k}\frac{1}{g}, \tag{6.6i}$$

where g is a constant. Equations (6.6g) become

$$\begin{aligned}\frac{dx}{d\tau} &= \frac{1}{1 + (y/g)^n} - x \\ \frac{dy}{d\tau} &= \frac{1}{1 + (x/g)^n} - y.\end{aligned} \tag{6.6j}$$

6.4.2 *Steady state solutions*

In the steady state the time derivatives of eqns (6.6j) are equated to zero, giving two simultaneous equations in x and y:

$$0 = \frac{1}{1 + (y/g)^n} - x$$
$$0 = \frac{1}{1 + (x/g)^n} - y. \tag{6.7a}$$

Equations (6.7a) yield one or three biologically acceptable solutions, depending upon the values of g and n. There are various ways of examining these solutions. y (or x) can be eliminated directly, giving

$$x + g^n(x - 1)\left[1 + \left(\frac{x}{g}\right)^n\right]^n = 0. \tag{6.7b}$$

For $n = 1$, this becomes

$$x^2 + gx - g = 0, \tag{6.7c}$$

giving one positive root at

$$x = \tfrac{1}{2}[(g^2 + 4g)^{1/2} - g]. \tag{6.7d}$$

Thus with $n = 1$ the model cannot describe alternate states or pathways of development. With $n = 2$, eqn (6.7b) gives

$$x^5 - x^4 + 2g^2x^3 - 2g^2x^2 + (g^4 + g^2)x - g^4 = 0. \tag{6.7e}$$

This quintic equation has one or three positive roots for x depending upon the value of g: for $g > 0.5$ there is a single solution; for $g < 0.5$ there are three solutions, of which two are stable and the third is unstable (Exercise 6.2). For higher n the polynomial resulting from eqn (6.7b) becomes quite complicated. Better insights are obtained by using an alternative approach and writing eqns (6.7a) as

$$x = \frac{1}{1 + (y/g)^n}$$
$$y = \frac{1}{1 + (x/g)^n}; \tag{6.7f}$$

these can be plotted and the points of intersection examined. Graphical solutions for various values of the parameters n and g are shown in Fig. 6.7. Some general rules governing the solutions can be discerned from Fig. 6.7 or by analysis.

1. $n = 1$ gives rise to one stable steady state solution.
2. $n > 1$ gives rise to a single solution or to three solutions depending upon the value of g. There is a critical value of g, denoted g_c; for $g < g_c$ there are three solutions; for $g > g_c$ there is one solution; g_c is a function of n (Fig. 6.8).
3. There is always one solution at $x = y$; this is unstable (U) if there are three solutions, and stable (S) if there is only one solution.
4. If $g < 1$, then increasing n moves the solutions towards (1,0), (1,1), and (0,1); if $g > 1$, then increasing n generally moves the solutions towards (1,1), although for some g ($1 < g < 1.12$) increasing n gives more complex results,

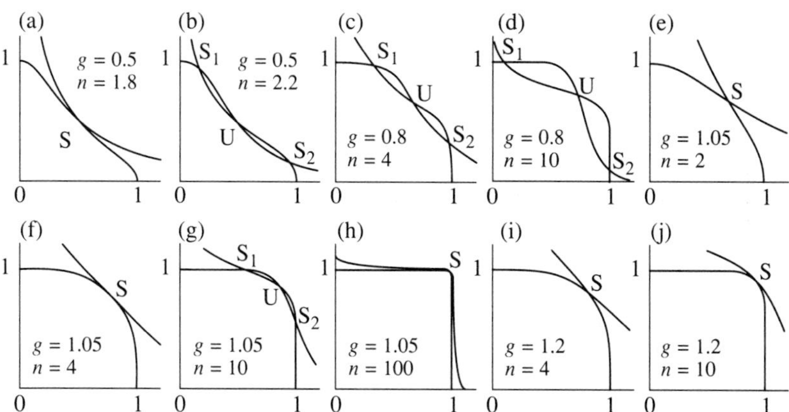

Fig. 6.7. Graphical illustration of the steady state solutions of the symmetrical biochemical switch and eqns (6.7f) with abscissa x and ordinate y. U denotes an unstable steady-state solution, and S, S_1, and S_2 denote stable steady state solutions for various values of g and n.

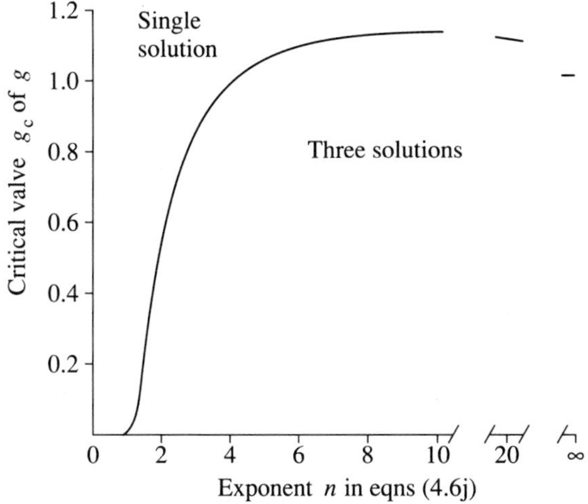

Fig. 6.8. The curve separates the single-solution region from the three-solution region in the $g:n$ parameter space of the symmetrical biochemical switch, with steady state solutions obtained from eqns (6.7f).

passing from one solution to three solutions and back to one solution (Figs 6.7(e)–6.7(h) and 6.8).

5. The $g_c : n$ curve is sketched in Fig. 6.8; this was obtained numerically. For small positive values of $(n - 1)$, $g_c = (n - 1)^2$.

The model of Fig. 6.6 is capable of describing switch behaviour for certain parameter values which give rise to three steady-state solutions U, S_1 and S_2. The symmetrical U solution with $x = y$ is unstable, and a small perturbation

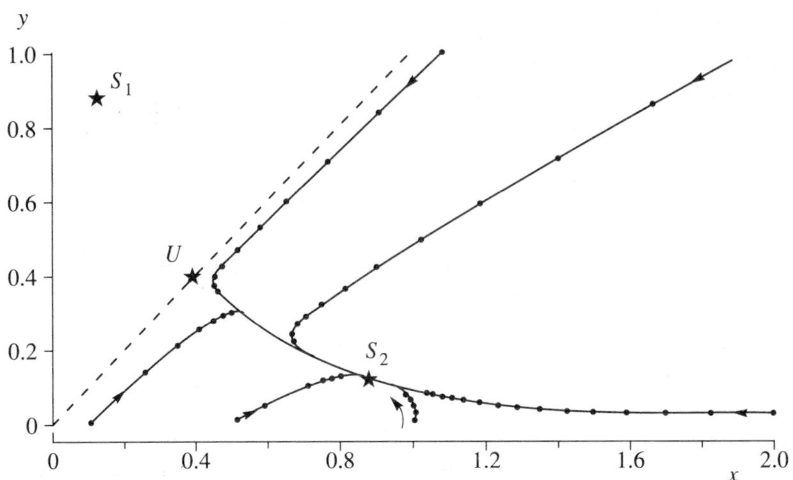

Fig. 6.9. Dynamic solutions of eqns (6.6j) with $n = 2$, $g = 0.333$. S_1 and S_2 are stable (attractor) steady state solutions; U is an unstable steady state solution. The broken line divides the system space into two zones wherein the system point is attracted to S_1 or S_2. Where shown, the dots on the trajectories denote a $\Delta\tau$ of 0.2.

causes the system point to move from U to S_1 or S_2, which may denote alternative pathways of development. It is possible for an external perturbation to flip the system from one stable steady state, S_1 say, to the other, S_2.

6.4.3 *Dynamic behaviour*

This is defined by eqns (6.6j). Clearly the line $x = y$ divides the two zones of development, corresponding to the two stable (attractor) steady state solutions at S_1 and S_2. In Fig. 6.9 some trajectories starting from various initial values of (x, y) are shown. These are all shown finishing at S_2; if some perturbation increments x and/or y so that the line $x = y$ is crossed by the system point, then the new trajectory will finish at S_1. Given the symmetry of eqns (6.6j), Fig. 6.9 can be rotated about the line $x = y$ through 180°, thereby interchanging x and y. It can be seen that there is a tendency for all trajectories to move along a common curve in their final approaches to S_2.

6.4.4 *Application to flowering*

The terms *competence*, *initiation*, and *commitment* are sometimes used in connection with the biological processes of development and differentiation, particularly when a fairly abrupt change in the characteristics of the system are observed.

If a plant is competent to flower, this implies that an appropriate stimulus can set in train the developmental processes leading to flowering, and in due course a flower is formed. If this process is interpreted in terms of Fig. 6.9, the perturbation which arises from the flowering stimulus results in a displacement of the system point in the xy plane. Initiation of flowering then corresponds to moving the system point across the broken line dividing the two zones of development which contain the attractor points S_1 and S_2. It is possible that the system point is so far from the broken line that the largest stimulus that can be applied is not sufficient to initiate flowering; such a plant would not be competent. If a young plant begins life with x, y values that are far from the broken line separating the zones of development, then some time may elapse before the system point comes within striking distance of the zonal dividing line. Alternatively, lack of competence may result from there being only a single allowed growth mode of the system, and the parameters are such that eqns (6.7a) have only one solution. This may relate to the terms 'juvenile phase' and 'ripeness to flower' used by some physiologists. Commitment to flowering implies that the new developmental processes begun after initiation are not reversible. It seems feasible that, shortly after initiation when the system point is still quite close to the zonal dividing line, a suitable stimulus might move the system point back into the vegetative zone and stop the flowering process.

A strict interpretation of the model would suggest that vegetative and reproductive growth are not entirely distinct; for example x is non-zero at point S_1 in Fig. 6.9. This 'escape' synthesis is equivalent to the leakage current of an electrical switch in the off position, or the low level of synthesis of a fully repressed and unneeded enzyme such as β-galactosidase in *Escherichia coli* growing on a glucose medium. The switch action could of course be made sharper and the escape synthesis smaller by taking different parameter values (Fig. 6.7). In some bacterial systems fully repressed and depressed rates of protein synthesis may differ by factors of up to about 50. It is possible that the aggregative properties that give rise to macroscopic structure and morphology are determined more by the balance of components present than by their absolute presence or absence.

Experimental work on flowering indicates that at least two processes may be involved in flower initiation: the first is the production of some stimulating agent by the leaves; the second is the response of the shoot apex to this stimulus. Possibly the ideas of competence and commitment should apply to both of these processes.

There is great variety in nature in the types of flowering response to photoperiod and other stimuli. Some plants will not flower at all unless they are given a suitable stimulus, whereas others will flower eventually in the absence of a stimulus, although a stimulus greatly expedites flowering. It is interesting to consider how the present scheme, with g and n as crucial parameters (Fig. 6.7), could meet these requirements. The parameter g might be regarded as a phenomenological parameter summarizing the effects of many interactions in the real system. g could change with time as the plant develops, and a system with a single allowed state of growth (vegetative) may develop into a system with two alternative growth patterns (vegetative and reproductive). The presence of a changing asymmetry may cause the broken line in Fig. 6.9 to sweep across the system space; the system point will normally move away from an interzonal dividing line, although this line could move more rapidly than the system point can respond, thus causing initiation without any of the usual stimuli.

6.5 Cusp catastrophe model

The contribution of catastrophe theory to biology is disputed (Zahler and Sussmann, 1977), although some workers find that it gives helpful descriptions of some processes (Bellairs, Goodwin, and Mackley, 1977). An account of the simplest catastrophe, the cusp catastrophe, is given here in order to provide an introductory understanding of the topic. Further discussion is given by Wilson and Kirkby (1980, Chapter 10) and Poston and Stewart (1978). The method has been applied more specifically by Thornley and Cockshull (1980) to the problem of the switch from vegetative to reproductive growth in the shoot apex.

A potential function $U_1(x)$ which possesses two minima with respect to the variable x (Exercise 6.3) is

$$U_1(x) = bx^4 - ax^2, \tag{6.8a}$$

where a and b are positive parameters. Adding a term γxy, where y is a second variable and γ is a parameter which may be positive or negative, gives a modified potential function

$$U_2(x) = bx^4 - ax^2 + \gamma yx. \tag{6.8b}$$

$U_2(x)$ is known as the cusp catastrophe function (Poston and Stewart 1978). It can be shown (Exercises 6.4 and 6.5) that there is one minimum if

$$(\gamma y)^2 > \frac{8a^3}{27b}, \tag{6.8c}$$

and there are two minima separated by a maximum if

$$(\gamma y)^2 < \frac{8a^3}{27b}. \tag{6.8d}$$

For biological applications it is convenient to shift the origin of x so that the relevant range of x is at positive x; replacing x in eqn (6.8b) by $x - x_0$, where x_0 is a constant, therefore gives

$$U_3(x) = b(x - x_0)^4 - a(x - x_0)^2 + \gamma y(x - x_0). \tag{6.8e}$$

The potential function U_3 is drawn in Fig. 6.10 for a range of y values. It is assumed that the system point resides at a minimum of this potential function. Thus when $y = -0.08$ the system point is at S; as y increases, there are now two minima, but the system point stays at the second minimum S_2 which is barely different from its original position as S; as y increases further, it is now the turn

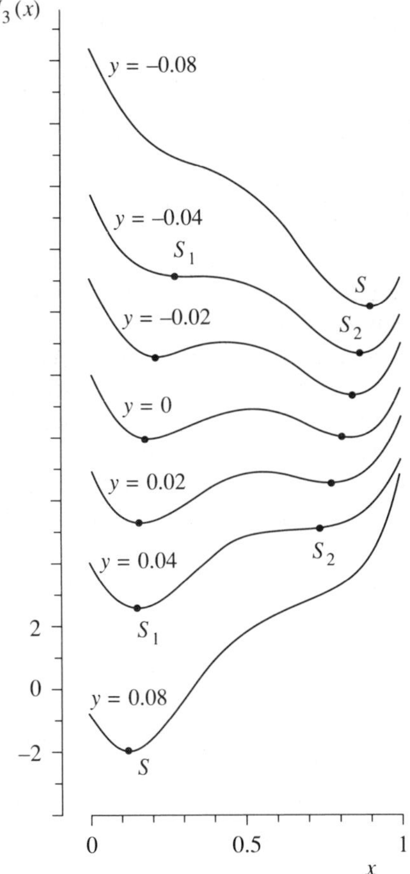

Fig. 6.10. The cusp catastrophe potential function of eqn (6.8e) is drawn for vanens y as shown, with parameter values $a = 0.2$, $b = 1.0$, $\gamma = 1.0$, and $x_0 = 0.5$. The system point (•) is at S for $y = -0.08$, then at S_2, for $y = -0.04, -0.02, \ldots, 0.04$. When y is increased from 0.04 to 0.08, the minimum at S_2 ceases to exist (at $y = 0.0487$ by eqn (6.8c)) and the system point moves catastrophically to S. The ordinates of the curves have been displaced for clarity.

of the second minimum at S_2 to disappear, and the system point moves quickly (catastrophically) across and down to the minimum point at S with $y = 0.08$.

6.5.1 *Application to flowering*

A principal difficulty with catastrophe theory in biology has been in its application to specific problems and in constructing a correspondence between variables in the catastrophe equations and observable quantities. The following account is based on the work of Thornley and Cockshull (1980), and is intended as a tentative indication of a possible approach to these difficult problems.

Equation (6.8e), describing the cusp catastrophe, can be viewed as a phenomenological equation, or equation of state, which relates useful morphological variables to each other. It describes macroscopic behaviour, but ignores mechanism, and may be able to mimic a switch between two modes of apical organization (vegetative and reproductive). This is somewhat analogous to van der Waal's equation for the gas–liquid transition which similarly ignores molecular mechanisms but describes and parameterizes gross behaviour. Anticipating a correspondence between the model described above (Fig. 6.10 and eqn (6.8e)) and the physiology of the shoot apex, we relate increasing y values to increasing apex size and relate x to primordial size. With $y = -0.08$ the point S denotes obligate vegetative growth with a small apex (low y) and large primordia (large x); at intermediate y the two minima S_1 and S_2 correspond to alternative modes of development; finally, with $y = 0.08$ there is obligate reproductive growth with a larger apex (y) and small primordia (x) at the point S.

Let W_p be the dry weight of a primordium just after its initiation and W_a be the dry weight of the shoot apex just before a primordium is initiated. To relate apical development to the foregoing discussion of the cusp catastrophe and eqns (6.8), the variable x is defined by

$$x = W_p/W_a. \tag{6.9a}$$

Clearly x must lie in the range $0 \leqslant x \leqslant 1$. However, it seems biologically reasonable to suppose that a primordium, as an organized entity, possesses a minimum size either absolutely or relative to the size of the apex. It is also assumed that it is not possible for the whole of the apex to be transformed into a primordium. Thus x is constrained by

$$x_{\min} \leqslant x \leqslant x_{\max}, \tag{6.9b}$$

where $x_{\min}$ and $x_{\max}$ limit the range of x. To build this restriction into the potential function $U_3(x)$ of eqn (6.8e), an extra term is added to give

$$U(x) = b(x - x_0)^4 - a(x - x_0)^2 + \gamma y(x - x_0) + \frac{h}{(x_{\max} - x)(x - x_{\min})}, \tag{6.9c}$$

where h is a constant. The last term becomes large when x approaches $x_{\min}$ or $x_{\max}$ so that the minima of $U(x)$ are kept within physiological limits.

The last connection to be made between eqn (6.9c) and the apical system is to relate the variable y to apex size W_a. It is assumed that

$$y = \alpha(W_a - C_a), \tag{6.9d}$$

where C_a is a constant and α is a scale factor. C_a is introduced to shift the origin of y so that positive and negative values of y correspond to positive values of W_a. Hence, by eqn (6.9d), increasing apical size W_a corresponds to increasing values of y, producing the catastrophic behaviour shown in Fig. 6.10. Equations (6.9a) and (6.9d) together relate the two variables x and y of the potential function of eqn (6.9c) to the measurable characteristics W_p and W_a of the apex.

It is assumed that, for a given value of apex size W_a and therefore y, the size of the primordium at initiation (xW_a by eqn (6.9a)) is such that, in $U(x)$ of eqn (6.9c), x is at a minimum point. Thus

$$\frac{\partial U}{\partial x} = 0, \tag{6.9e}$$

and therefore

$$0 = 4b(x - x_0)^3 - 2a(x - x_0) + fy + \frac{h(2x - x_{\min} - x_{\max})}{(x_{\max} - x)^2(x - x_{\min})^2}. \tag{6.9f}$$

This equation can be regarded as defining the relationship between apex size and primordial size at initiation, when initiation occurs. For a given y (apex size), eqn (6.9f) may have one or three roots for x, which defines primordial size. A convenient numerical technique for calculating x at equilibrium, and hence primordial size, is to integrate with respect to time the rate equation

$$\frac{dx}{dt} = \beta\left(-\frac{\partial U}{\partial x}\right), \tag{6.9g}$$

where t is time, and β is a factor which is sufficiently large to ensure that the primordial size is near its equilibrium value, and eqn (6.9e) is closely satisfied. This is equivalent to assuming that the system point is close to one of the minima in Fig. 6.10.

Growth relations in the shoot apex As discussed by Lyndon (1977), there are relationships between certain apical characteristics, such as the intrinsic specific growth rate μ of the apex, the apical and primordial sizes W_a and W_p, the plastochron t_p (the time interval between the times of initiation of successive primordia), and the overall specific growth rate R_w of the apex, which allows for the loss of material from the apex to the new primordia. The intrinsic specific growth rate μ refers to the growth of apical mass in the absence of or between primordial initations (BC in Fig. 6.11); the overall apical specific growth rate R_w refers to a time span covering primordial initiation events such as AC in Fig. 6.11.

Consider the growth curve in Fig. 6.11. By equating overall growth along the broken line AC ($\exp(R_w t_p)$) to the actual growth path AB + BC, we obtain

$$(W_a - W_p)\exp(\mu t_p) = W_a \exp(R_w t_p) \tag{6.9h}$$

Using eqn (6.9a) it follows that

$$R_w = \mu + \frac{1}{t_p}\ln(1 - x). \tag{6.9i}$$

In this equation, the intrinsic specific growth rate μ is regarded as constant (it

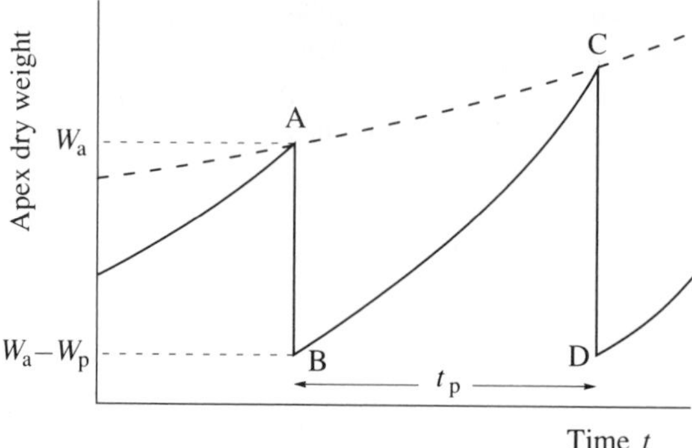

Fig. 6.11. Growth relations in the shoot apex. W_a is the apex dry weight at A just before initiation of a primordium of dry weight W_p. Just after initiation, the apex dry weight falls to $W_a - W_p$ at B. The line BC represents the intrinsic specific growth rate μ of the apex, and the broken line AC represents the overall specific growth rate R_w. The plastochron t_p is the time interval between initiations of successive primordia.

would be possible to make μ depend upon carbohydrate status and/or other substrates), and $x(= W_p/W_a)$ varies as the apex grows and is given by solving eqn (6.9f); the plastochron t_p is still to be discussed.

As W_a measures the size of the apex just before each primordial initiation, it grows according to the growth rate R_w, giving

$$\frac{dW_a}{dt} = R_w W_a. \tag{6.9j}$$

The inhibitor approach to primordial initiation will be followed (Thornley 1975, Charles-Edwards, Cockshull, Horridge, and Thornley 1979). Let I be the concentration of an inhibitor of primordial initiation in the apex, and assume that this is diluted by the intrinsic growth of the apex μ and is also degraded at a rate k_d. When I falls to a level I_c, it is assumed that primordial initiation takes place. There is then instantaneous synthesis of more inhibitor: an amount λW_p is produced where λ is a constant, and this is distributed over the apex and the new primordium. Thus there is an increment in the inhibitor concentration of $\lambda W_p/W_a$ or λx. This will decay again, according to

$$I(t) = (I_c + \lambda x)\exp[-(\mu + k_d)t]. \tag{6.9k}$$

The next primordium arises when $I(t) = I_c$, so that the plastochron t_p is determined by the relation

$$t_p = \frac{1}{\mu + k_d}\ln\left(1 + \frac{\lambda x}{I_c}\right). \tag{6.9l}$$

Summary statement of the model The equations that define the system are gathered together here. The model has two state variables W_p and W_a, although it is convenient to write the model in terms of x and y which are related to W_p and W_a by

$$x = \frac{W_p}{W_a} \quad \text{and} \quad y = \alpha(W_a - C_a). \tag{6.10a}$$

The two auxiliary variables t_p and R_w are obtained from W_p and W_a by means of eqns (6.10a) and

$$\begin{aligned} t_p &= \frac{1}{\mu + k_d} \ln\left(1 + \frac{\lambda x}{I_c}\right) \\ R_w &= \mu + \frac{1}{t_p} \ln(1 - x). \end{aligned} \tag{6.10b}$$

The two rate equations are (eqns (6.9g) and (6.9c))

$$\frac{dx}{dt} = \beta\left[-4b(x - x_0)^3 + 2a(x - x_0) - \gamma y - \frac{h(2x - x_{min} - x_{max})}{(x_{max} - x)^2(x - x_{min})^2}\right] \tag{6.10c}$$

and (eqns (6.9d) and (6.9j))

$$\frac{dy}{dt} = R_w(y + \alpha C_a). \tag{6.10d}$$

Initial values are required for x and y at time $t = 0$, although since β in eqn (6.10c) is large, the initial value of x is not critical; y at $t = 0$ is equivalent to W_a at $t = 0$. Numerical values are required for the parameters α, C_a, μ, k_d, λ, I_c, β, b, a, γ, h, x_0, x_{min}, and x_{man}.

Numerical values and methods The seven parameters b, a, γ, h, x_0, x_{min}, and x_{max} are chosen such that the switch occurs within the allowed range of x values. The only physiological input is that the vegetative primordium can occupy up to 80 per cent of the apex (x_{min}). Thus it is assumed that

$$\begin{aligned} &a = 0.2 \text{ day}^{-1}, \quad b = 1 \text{ day}^{-1}, \quad \gamma = 1 \text{ g}^{-1}\text{day}^{-1}, \quad h = 0.0001 \text{ day}^{-1}, \\ &x_0 = 0.5, \quad x_{min} = 0.05, \quad x_{max} = 0.8. \end{aligned} \tag{6.11a}$$

From Horridge and Cockshull (1979), the volume of the chrysanthemum apex at about the time it becomes reproductive is about 0.01 mm^3 = 10^{-5} cm^3 which, assuming a dry weight:fresh weight ratio of 0.1, is equivalent to 10^{-6} g. From Fig. 6.10 the range of y required to give switching behaviour is known, and C_a, α, and $W_a(t = 0)$ are chosen to cover this range:

$$C_a = 6 \times 10^{-7} \text{ g}, \quad \alpha = 2 \times 10^{-5}, \quad W_a(t = 0) = 3 \times 10^{-7} \text{ g}. \tag{6.11b}$$

The initial value $x(t = 0)$ of x is not important, and it is simply assigned the value

$$x(t = 0) = 0.9x_{max}. \tag{6.11c}$$

The relations between the growth rates μ and R_w and the plastochron t_p in eqns (6.10b) gave some difficulty. Following Lyndon (1977) it was first assumed that the intrinsic specific growth rate μ is 0.8 day^{-1} and that, in the absence of any evidence, $k_d = 0$ day^{-1}. To obtain a plastochron t_p of 3 days then requires (eqn (6.10b))

$$\lambda/I_c = 12.5. \tag{6.11d}$$

However, on running the model, the apex grew far too rapidly, and more realistic physiological behaviour was obtained by taking

$$\mu = 0.6 \text{ day}^{-1}, \quad k_d = 0.2 \text{ day}^{-1}. \tag{6.11e}$$

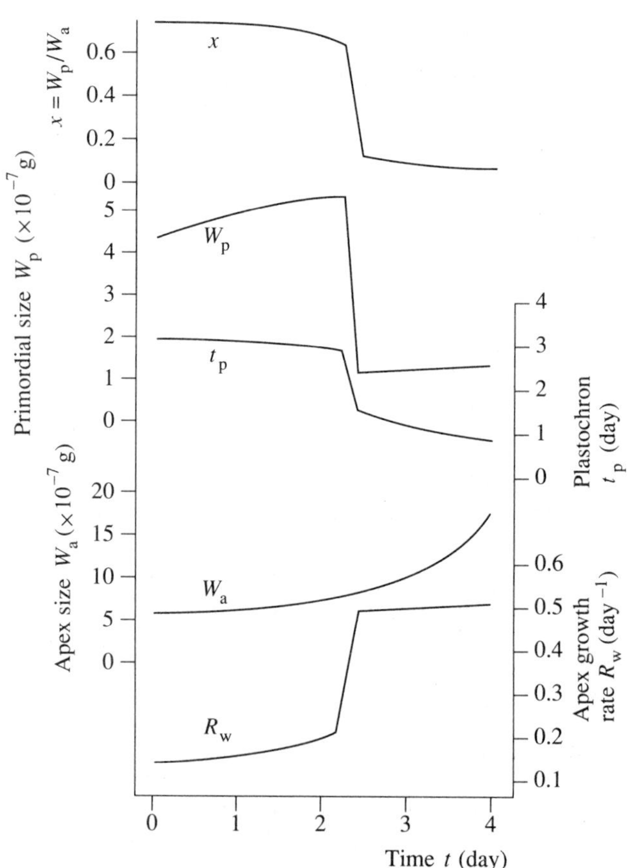

Fig. 6.12. Initiation of reproductive growth in the shoot apex using a cusp catastrope model. W_p is the primordial size at initiation, W_a is the apex size just before primordial initiation, t_p is the time period between successive primordial initiations, and R_w is the overall specific growth rate of the apex (AC in Fig. 6.11).

Trapezoidal (first-order) integration was used with a time interval of Δt. Since the model requires the system point to reside at the bottom of one of the minima, as in Fig. 6.10, β in eqn (6.10c) was made as large as possible and Δt correspondingly small within the two constraints of stability and rounding errors (p. 21). Satisfactory performance was obtained with

$$\beta = 100, \quad \Delta t = 0.001 \text{ day}. \tag{6.11f}$$

Results and discussion The simulated behaviour of the shoot apex is shown in Fig. 6.12. The size of the primordium at initiation and the plastochron both shift suddenly to lower values, and at the same time the apex growth rate increases rapidly. It should be noted that the discontinuity can be made much less abrupt if the parameter β of eqn (6.10c) is made smaller.

Changes in the plastochron at the transition from vegetative to reproductive development are well documented (e.g. Lyndon 1977, Table 1, Horridge and Cockshull 1979), and the experimental findings agree generally with Fig. 6.12.

Miller and Lyndon (1976) find that the growth rate of the apex increases during the first stage of flower formation in *Silene*. The data of Horridge and Cockshull (1979) on chrysanthemum indicate clearly an increase in the growth rate of apical diameter about the time of initiation. There are few measurements of primordial sizes at initiation which can be compared with the present prediction. However, Lyndon (1977, Table 5) gives estimates for different parts of the developing flower (sepals, stamens, petals), and primordial sizes become much smaller as flower development proceeds.

Exercises

6.1. Show, graphically or algebraically, that the equation

$$\frac{dx}{dt} = k_1 \frac{x}{K_1 + x} - k_2 \frac{x}{K_2 + x},$$

where x represents the amount of substance, t is time, and k_1, k_2, K_1, and K_2 are constants, has only one or two steady state solutions. The first term on the right-hand side can be viewed as autocatalytic Michaelis–Menten-limited synthesis, and the second term can be viewed as Michaelis–Menten-limited degradation.

6.2. Prove that eqn (6.7e) with $g = \frac{1}{2}$ has three coincident roots for x at $x = \frac{1}{2}$.

6.3. Show that eqn (6.8a) has two minima and one maximum, and find their coordinates.

6.4. Show, by considering the conditions under which a cubic can have only one real root, that eqns (6.8c) and (6.8d) apply.

6.5. Sketch the functions $U_1(x)$ and $U_2(x)$ of eqns (6.8a) and (6.8b) for $b = 1$, $a = 0$ and 1, $\gamma y = 0$ and 1.

7
Development

7.1 Introduction

It is often useful to consider growth and development as processes that can occur independently, although this is only an approximation. The phasic development or phenology of a crop can be regarded as an ordered sequence of processes, which take place over a period of time, punctuated by more or less discrete events, so that the following table can be drawn up, beginning with sowing:

	Event	*Process*	
1	Sowing	Germination	
2	Emergence	Vegetative growth	
3	Initiation of flowering	Flower development	(7.1a)
4	Anthesis	Fruit growth	
5	Maturity		

Much modelling work on development is based on the idea that it is possible to associate a scalar (single-valued) variable h with each phase, and to use the value of h to denote progression through that phase. The subscript i, with $i = 1, 2, \ldots, 5$, is used in the above example for the discrete events from sowing (1) to maturity (5), as in (7.1a). h_i is the value of the development index during the ith process which begins with the ith event; k_i is a rate constant (day^{-1}) which defines the rate of progress of the ith process. Thus

$$h_i = \int_{t_i}^{t} k_i \, \mathrm{d}t \tag{7.1b}$$

where t is the time variable (day) and t_i is the time of the ith event. h_i increases with time; by definition $h_i(t = t_i) = 0$; also, by assumption, h_i attains the value h_{ij} (a constant) at time $t = t_j$, and it is deemed that the ith phase of development is complete and that event j has occurred.

It is implicitly assumed that the plant, or part of the plant, possesses a developmental clock which proceeds at a rate k_i for the ith phase. (Henceforth we omit the subscript i for simplicity.) Ideally, the internal state variables of the system would be sufficient so that the developmental index h could be calculated directly from them:

$$h = h(\text{internal state variables of system}); \tag{7.1c}$$

however, apart from some tentative efforts in this direction (p. 159), progress in this area has been limited. It has been more usual to assume that the developmental rate constant k, and hence the variable h through eqn (7.1b), depend directly upon the environmental variables, so that

$$h = \int_0^t k(E)\,\mathrm{d}t, \tag{7.1d}$$

where E denotes the set of environmental variables and time $t = 0$ is the beginning of the developmental phase under consideration. Again, h increases monotonically with time, at a rate depending upon the environment E which affects the rate constant k, and when h reaches a particular value, the phase of development has reached completion.

In our discussion of these problems we begin with the simplest models of type (7.1d) which correspond to the temperature-sum concept and generalize these in terms of other environmental variables. This type of model gives predictions of mean development times. For some purposes (germination, maturity, and harvest) the distribution of times is of particular interest, and for this purpose compartment models are often applied. Compartmental approaches are discussed. The effects that diffusion processes may have on the time course of development is analysed for a simple case. A compartmental model of germination is constructed and solved. Finally, a teleonomic model of time of flowering is discussed.

7.2 Empirical threshold models of developmental rate—the temperature-sum rule

Using a chemical analogy, a reaction of the type

$$\text{substrate} \xrightarrow{k} h \tag{7.2a}$$

is zero order if the rate at which h is produced is independent of the substrate level; we write

$$\frac{\mathrm{d}h}{\mathrm{d}t} = k \tag{7.2b}$$

for the rate of increase in the variable h; the value of h indicates the progress that has been through the ith phase of development, which begins with the ith event and terminates with the jth event (7.1a). Therefore, with k constant,

$$h_{ij} = k(t_j - t_i), \tag{7.2c}$$

and this equation determines the time t_j of the jth event. h_{ij} can be set at any level by scaling the rate constant k; sometimes h_{ij} is normalized to unity, but in any case h_{ij} is the threshold or critical value of h that signifies occurrence of the event which terminates this phase of development. The model is completely deterministic, and all the plants in the population complete the process simultaneously, as shown in Fig. 7.1.

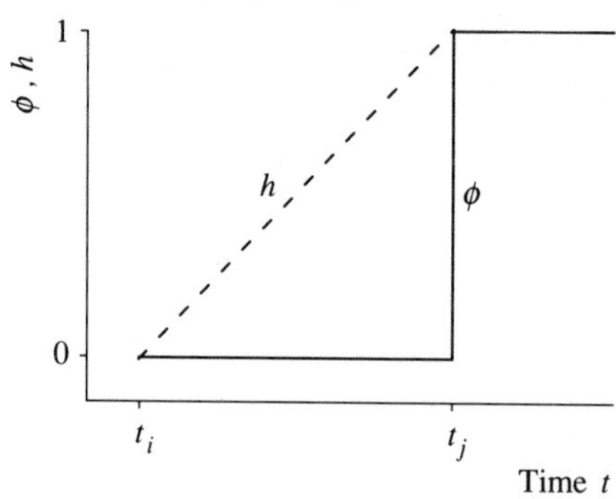

Fig. 7.1. h (– – –) indicates a constant rate of development (k is constant in eqn (7.2b)) terminating when $h = 1$ at time $t = t_j$ with event j. ϕ denotes the proportion of plants that has completed the phase of development at time t.

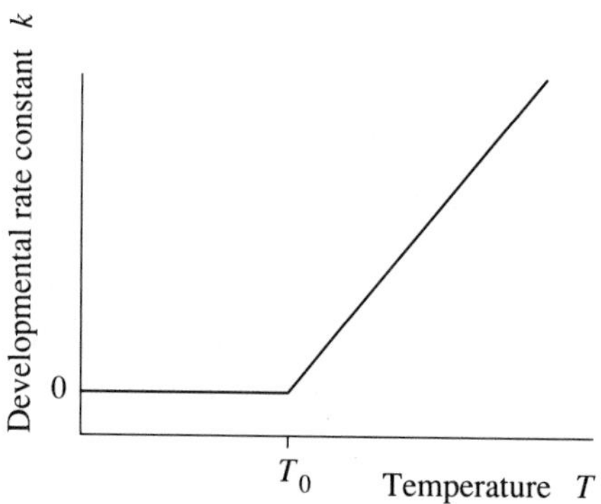

Fig. 7.2. Assumed temperature dependence of the rate constant k which leads to the temperature-sum rule, as in eqn (7.2d). T_0 is the base temperature below which development does not occur.

The temperature-sum method (see also Chapter 5, Section 5.8.3, p. 141) is obtained by assuming that the rate constant k depends on temperature T alone according to

$$k = \begin{cases} k_0(T - T_0) & T \geqslant T_0 \\ 0 & \text{otherwise,} \end{cases} \tag{7.2d}$$

where k_0 is a constant and T_0 is a base temperature below which development does not proceed at all. Equation (7.2d) is illustrated in Fig. 7.2. The integral form of eqn (7.2b) is

$$h = \int k(T)\,\mathrm{d}t, \tag{7.2e}$$

where temperature T can vary with time t. In combining eqns (7.2d) and (7.2e) it is usual to make the approximation that

$$h = \sum_{i=1}^{j} k_0(\bar{T}_i - T_0), \tag{7.2f}$$

where $\bar{T}_i$ is the mean temperature on the ith day and only positive terms ($\bar{T}_i \geqslant T_0$) are included in the sum. To demonstrate how eqn (7.2f) works in practice, consider a period of 25 days (i = 1, 2, ..., 25) with a base temperature of 5 °C. Given the mean daily temperatures and assuming $k_0 = 1$, the following table can be constructed:

i	1	2	3	4	5	6	7	8	9	10	11	12	13	14	15	16	17	18	19	20	21	22	23	24	25
$\bar{T}_i$	2	3	6	7	4	4	3	7	5	6	8	8	7	9	10	7	8	10	12	11	14	13	12	9	9
Δh	0	0	1	2	0	0	0	2	0	1	3	3	2	4	5	2	3	5	7	6	9	8	7	4	4
h	0	0	1	3	3	3	3	5	5	6	9	12	14	18	23	25	28	33	40	46	54	62	69	73	77.

Δh is the increment in h arising from the ith term in (7.2f). For a particular crop and cultivar, it is often observed that approximately a certain value of the sum h, perhaps 70 day °C, corresponds to the termination of a given phase of development. An application of the temperature-sum method may be to consider how well the average temperatures at a given location would suit the developmental aspects of a range of possible crops.

Table 7.1 lists some results for the germination process or the sowing to emergence phase taken from Angus, Cunningham, Moncur and Mackenzie (1981a, p. 373), who found that a linear day-degree system gave a very satisfactory account of their data; these relate to the parameters T_0 and h (day °C) of eqn (7.2f) with $k_0 = 1$. It can be seen that, in general, as the threshold temperature increases, the temperature sum h decreases.

The approach of eqns (7.2d)–(7.2f) can be generalized in a variety of ways (Robertson 1968). Equation (7.2d) can be expanded to give

$$k = \begin{cases} a(T - T_0) + b(T - T_0)^2 + c(T - T_0)^3 & T \geqslant T_0, \\ 0 \quad \text{otherwise} & \end{cases} \tag{7.2g}$$

where a, b, and c are constants. Equation (7.2f) then becomes

$$h = \sum_{i=1}^{j} a(\bar{T}_i - T_0) + b(\bar{T}_i - T_0)^2 + c(\bar{T}_i - T_0)^3, \tag{7.2h}$$

where only positive terms are included in the sum.

Although not usually important in germination and up to emergence, environmental variables other than temperature, such as day length, radiation and vernalization (temperatures below some threshold temperature value), may be important in other phases of development. Following Robertson (1968) and assuming linearity, a more general developmental index can be constructed:

Table 7.1. Temperature-sum method of eqn (7.2f) with $k_0 = 1$ for some of the crops considered by Angus *et al.* (1981a)

Species	T_0 (°C)	h (day °C)
Linseed	1.9 (0.3)	89 (4)
Oats	2.2 (0.3)	91 (3)
Rape	2.6 (0.3)	79 (3)
Wheat	2.6 (0.2)	78 (2)
Barley	2.6 (0.3)	79 (3)
Sunflower	7.9 (0.4)	67 (3)
Maize	9.8 (0.3)	61 (3)
Sorghum	10.6 (0.1)	48 (1)
Mung bean	10.8 (0.2)	50 (2)
Cowpea	11.0 (0.2)	43 (2)
Amaranthus	11.7 (1.0)	32 (8)
Peanut	13.3 (0.2)	76 (6)
Finger millet	13.5 (0.6)	40 (6)

T_0 is the threshold temperature and h is the temperature sum for germination and emergence; standard errors are given in parentheses.

$$h = \sum_{i=1}^{j} [a(\bar{T}_i - T_0) + b(g_i - g_0) + c(J_i - J_0) + d(T_v - \bar{T}_i)] \qquad (7.2i)$$

where a, b, c, and d are constants, $\bar{T}_i$, g_i, and J_i are the mean temperature, day length and radiation receipt on the ith day, T_0, g_0, and J_0 are (lower) threshold values of temperature, day length and radiation, and T_v is an upper (vernalization) temperature threshold ($\bar{T}_i < T_v$ for development to be hastened). Each term only contributes to the sum if it is positive; otherwise it is set to zero. Equation (7.2i) applies to a long-day plant (long days expedite development); for a short-day plant the second term would be written $b(g_0 - g_i)$. Because temperature has the dominant effect on development, it is customary to take $a = 1$ and to scale b, c, and d to give the relative contributions of the other environmental factors to the temperature sum. Soil temperatures and daily maximum or minimum temperatures could also be included in eqn (7.2i) as terms contributing to development.

Equation (7.2i) is not only linear in its four terms, but more importantly it does not contain any interactions between the different environmental factors. Nuttonson (1955) considered what he called the photothermal requirements of wheat in which temperature and radiation effects interact. The idea is that high temperatures are more effective in promoting development if the plant is supplied with high light levels, presumably ensuring plentiful substrate availability. For example, in the tomato plant high-temperature–low-light regimes are positively harmful to aspects of reproductive development.

A photothermal index can be constructed using an equation such as

$$h = \sum_{i=1}^{j} a(\bar{T}_i - T_0) f_J(J_i), \tag{7.2j}$$

where f_J denotes some function of radiation J_i on the ith day. Taking the simplest response form with $f_J = J_i$ gives

$$h = \sum_{i=1}^{j} aJ_i(\bar{T}_i - T_0), \tag{7.2k}$$

and this clearly increases the temperature contributions to the developmental index h with increasing irradiance levels.

In considering the post-emergence development of spring wheat, Angus, Mackenzie, Morton,and Schafer (1981b) examined the effects of both temperature and photoperiod, and concluded that a linear equation, as in eqn (7.2i), was not satisfactory. For emergence to anthesis, where both temperature and photoperiod are important, they proposed using a product formulation

$$h = \sum_{i=1}^{j} f_T(\bar{T}_i) f_g(g_i), \tag{7.2l}$$

where f_T is a function of temperature T and f_g is a function of daylength g, and we have written down the approximation of eqn (7.2e) that uses a time step of 1 day. After considering several possible equations, it was found that a threshold negative exponential function, related to the monomolecular growth equation (p. 76), gave the best results, with

$$f_T(T) = \alpha\{1 - \exp[-\beta(T - T_0)_+]\} \tag{7.2m}$$

$$f_g(g) = \{1 - \exp[-\gamma(g - g_0)_+]\}; \tag{7.2n}$$

α, β, and γ are constants, and the + signs indicate that the quantities in the brackets can only take positive values, otherwise they are set equal to zero, as in eqn (7.2d). Normalizing the sum in eqn (7.2l) to unity for the phase from emergence to anthesis, Angus *et al.* (1981b, p. 275) determined the following parameter values (standard errors in parentheses):

$$\begin{aligned} \alpha &= 0.025\ (0.002)\ \text{day}^{-1} \\ \beta &= 0.15\ (0.03)\ ^\circ\text{C}^{-1} \\ \gamma &= 0.30\ (0.01)\ \text{h}^{-1} \\ \bar{T}_0 &= 3.5\ (0.6)^\circ\text{C} \\ g_0 &= 9.2\ (0.2)\ \text{h}. \end{aligned} \tag{7.2o}$$

The parameters of (7.2o) encompass two developmental phases: vegetative growth and flower development. Angus *et al.* also treated these two phases

Table 7.2. Parameter estimates for the product threshold negative exponential model of eqns (7.2*l*)–(7.2n) for the vegetative and flower development phases in the development of a spring wheat cultivar, taking account of temperature and daylength effects on development, and using daily time steps

	Vegetative growth (emergence to flower initiation)	Flower development (flower initiation to anthesis)
α (day^{-1})	0.076 (0.03)	0.050 (0.014)
β (°C^{-1})	0.23 (0.09)	0.12 (0.07)
γ (h^{-1})	0.09 (0.08)	0.28 (0.09)
$\bar{T}_0$ (°C)	3.3 (0.7)	5.1 (2.1)
g_0 (h)	4.8 (2.7)	9.3 (0.5)

Standard errors in parentheses.
From Angus *et al.* 1981b, p. 278.

separately, i.e. from emergence to flower initiation, and from flower initiation to anthesis. The parameter values for the two-phase process are given in Table 7.2. Despite the evident parameter differences, there was nothing to be gained from using the two-stage process for predicting anthesis.

It is well known that high temperatures can be deleterious to plant growth and inhibit development. A developmental rate constant k that is simply linear in temperature above a base temperature, as in eqn (7.2d) and Fig. 7.2, is not able to describe such harmful effects. Considering the germination of pearl millet over a temperature range 12–47 °C, Garcia-Huidobro, Monteith, and Squire (1982) found it useful to replace eqn (7.2d) by

$$k = \begin{cases} k_0(T - T_0) & T_0 \leqslant T < T_1, \\ k_0(T_2 - T)\dfrac{T_1 - T_0}{T_2 - T_1} & T_1 \leqslant T \leqslant T_2, \\ 0 & \text{otherwise.} \end{cases} \tag{7.2p}$$

k_0 is a constant, often put equal to unity; T_0, T_1, and T_2 are threshold temperatures, which are shown in Fig. 7.3 where eqn (7.2p) is drawn. This response function gives a sharp optimum in k at $T = T_1$ which is not always observed (Labouriau and Osborn, 1984).

7.3 Compartmental models of development

In Chapter 6, biological switches are discussed as a means of representing binary decision processes—a good example of this is the 'choice' between vegetative and reproductive growth in the shoot apex, which may be facultative (optional) in

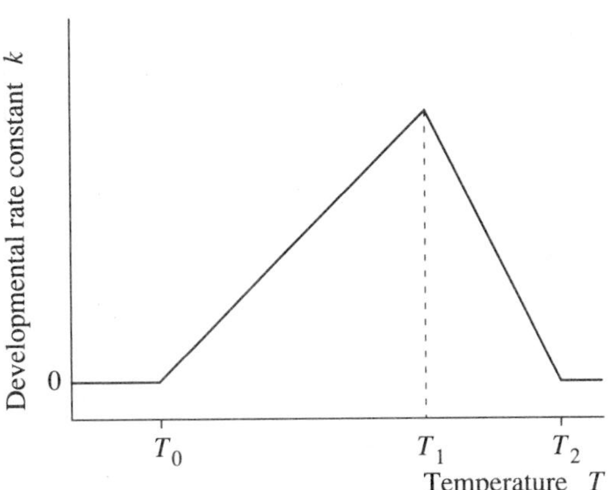

Fig. 7.3. Temperature dependence of developmental rate k showing an optimum at $T = T_1$; T_0 and T_2 are the lower and upper limits of the temperature range for development (see eqn (7.2n)).

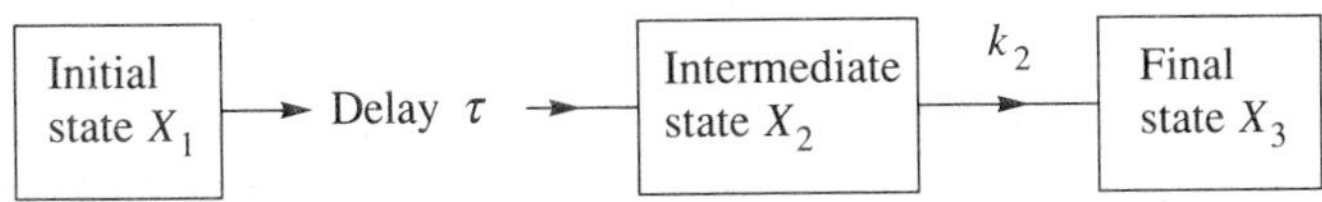

Fig. 7.4. Discrete time delay τ with a first-order process k_2: X_1, X_2, and X_3 represent three possible states of the system.

some plants and obligate in others. In this section we are concerned with the time course of development processes that are obligate, such as germination. Compartmental models (Godfrey, 1983) can be applied to these problems to give insights into the underlying processes and to help interpret the observational data (France, Thornley, Dhanoa, and Siddons 1985).

7.3.1 *Discrete time delay and a first-order process*

The compartmental scheme assumed is shown in Fig. 7.4. After time τ, the contents of the first compartment X_1 are transferred in their entirety to the second compartment X_2, whence a first-order process with rate constant k_2 transfers X_2 into the final state X_3. The equations that describe this are

$$t < \tau, \quad X_1 = X_1(t = 0), \quad X_2 = 0, X_3 = 0 \tag{7.3a}$$

$$t \geqslant \tau, \quad X_1 = 0$$

$$\frac{dX_2}{dt} = -k_2 X_2 \qquad \text{with } X_2(t = \tau) = X_1(0)$$

$$\frac{dX_3}{dt} = k_2 X_2 \qquad \text{with } X_3(t = \tau) = 0. \tag{7.3b}$$

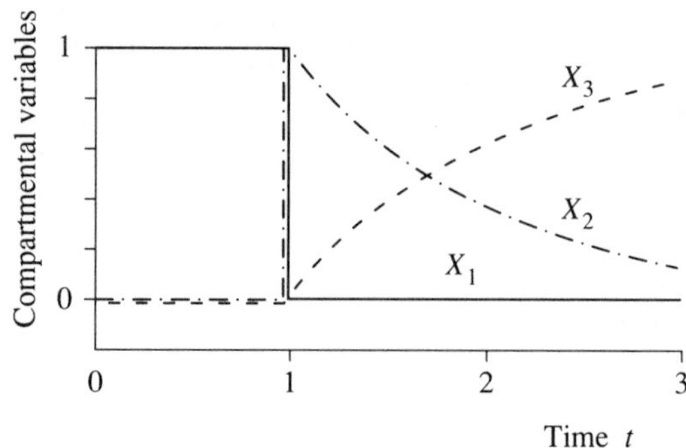

Fig. 7.5. Solutions to the scheme in Fig. 7.4, as given by eqns (7.3b) and (7.3c) with $X_1(0) = 1$, $\tau = 1$, and $k_2 = 1$.

Therefore, solving for $t \geqslant \tau$,

$$\begin{aligned} X_2(t) &= X_1(0)\exp[-k_2(t-\tau)] \\ X_3(t) &= X_1(0)\{1-\exp[-k_2(t-\tau)]\} \end{aligned} \tag{7.3c}$$

The time course of the process is drawn in Fig. 7.5. After time τ has elapsed the dynamics of the process are non-sigmoidal with respect to time. Apart from germination, few developmental time courses have been observed, but for germination, the X_3 time course, interpreted as the cumulative fraction of germinated seeds, closely resembles some data given by Garcia-Huidobro *et al.* (1982, Fig. 3) for pearl millet.

7.3.2 *Discrete time delay and two sequential first-order processes*

Two first-order processes are required in order to obtain a sigmoidal time response, and in Fig. 7.6 these are assumed to occur after a discrete delay τ. The scheme of Fig. 7.6 is described by the following equations:

$$t < \tau, \quad X_1 = X_1(t=0), \quad X_2 = X_3 = X_4 = 0 \tag{7.4a}$$

$$t \geqslant \tau, \quad X_1 = 0$$

$$\begin{aligned} \frac{dX_2}{dt} &= -k_2X_2 \qquad \text{with } X_2(t=\tau) = X_1(0) \\ \frac{dX_3}{dt} &= k_2X_2 - k_3X_3 \qquad \text{with } X_3(t=\tau) = 0. \\ \frac{dX_4}{dt} &= k_3X_3 \qquad \text{with } X_4(t=\tau) = 0. \end{aligned} \tag{7.4b}$$

Therefore, solving for $t \geqslant \tau$ (see Exercise 7.1), with $k_2 \neq k_3$, gives

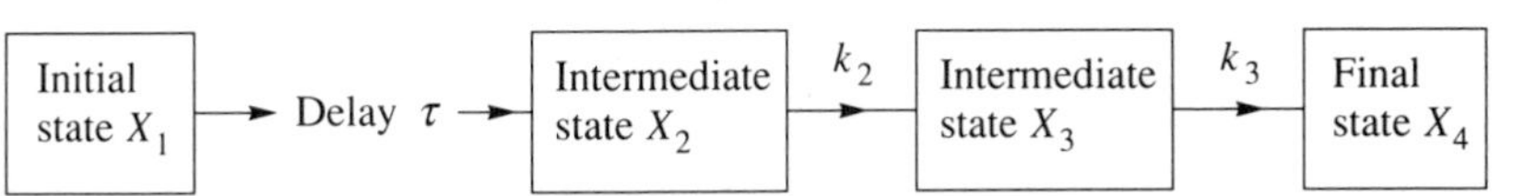

Fig. 7.6. Discrete time delay τ with two first-order processes k_1 and k_2. X_1, X_2, X_3, and X_4 represent four possible states of the system.

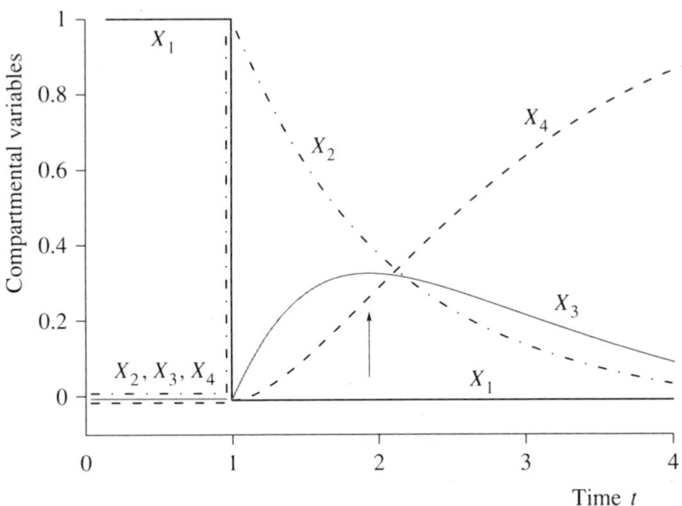

Fig. 7.7. Solutions to the scheme in Fig. 7.6 as given by eqns (7.4b) and (7.4c) with $X_1(0) = 1$, $\tau = 1$, $k_2 = 1$, and $k_3 = 1.2$. The maximum of X_3 and the point of inflexion of X_4 are at $t = 1.91$ by eqn (7.4d) and are indicated by the arrow.

$$\begin{aligned} X_2(t) &= X_1(0)\exp[-k_2(t-\tau)] \\ X_3(t) &= \frac{k_2 X_1(0)}{k_3 - k_2}\{\exp[-k_2(t-\tau)] - \exp[-k_3(t-\tau)]\} \\ X_4(t) &= X_1(0)\left\{1 - \frac{k_3\exp[-k_2(t-\tau)] - k_2\exp[-k_3(t-\tau)]}{k_3 - k_2}\right\}. \end{aligned} \tag{7.4c}$$

The solutions are shown in Fig. 7.7; the final state X_4 is weakly sigmoidal with time, compared with the non-sigmoidal behaviour of the final state X_3 in Fig. 7.5. The time t_i of inflexion of X_4, and the time $t_{\max}$ when X_3 is maximum coincide, and are given by (Exercise 7.2)

$$t_i = t_{\max} = \tau + \frac{\ln(k_3/k_2)}{k_3 - k_2}. \tag{7.4d}$$

The above solutions are valid for $k_2 \neq k_3$; the solutions for $k_2 = k_3$ are considered in Exercise 7.3.

Fig. 7.8. A general linear compartmental model with $n + 1$ compartments $X_1, X_2, \ldots, X_{n+1}$ and n rate constants $k_1, k_2, \ldots, k_n$.

7.3.3 *A general linear compartmental model*

Consider the $(n + 1)$-compartment model shown in Fig. 7.8. The system is described by the equations

$$\frac{dX_1}{dt} = -k_1 X_1, \tag{7.5a}$$

$$\frac{dX_2}{dt} = k_1 X_1 - k_2 X_2, \tag{7.5b}$$

$$\frac{dX_3}{dt} = k_2 X_2 - k_3 X_3, \tag{7.5c}$$

$$\vdots$$

$$\frac{dX_n}{dt} = k_{n-1} X_{n-1} - k_n X_n, \tag{7.5d}$$

$$\frac{dX_{n+1}}{dt} = k_n X_n. \tag{7.5e}$$

At time $t = 0$, $X_1 = 1$ and $X_2 = X_3 = \ldots = X_n = X_{n+1} = 0$. The solutions of eqns (7.5a)–(7.5e) are

$$X_1 = \exp(-k_1 t) \tag{7.5f}$$

$$X_2 = k_1 \left[\frac{\exp(-k_1 t)}{k_2 - k_1} + \frac{\exp(-k_2 t)}{k_1 - k_2} \right] \tag{7.5g}$$

$$X_3 = k_1 k_2 \left[\frac{\exp(-k_1 t)}{(k_2 - k_1)(k_3 - k_1)} + \frac{\exp(-k_2 t)}{(k_1 - k_2)(k_3 - k_2)} + \frac{\exp(-k_3 t)}{(k_1 - k_3)(k_2 - k_3)} \right] \tag{7.5h}$$

$$\vdots$$

$$X_n = (k_1 k_2 k_3 \ldots k_{n-1}) \sum_{i=1}^{n} \frac{\exp(-k_i t)}{\prod_{j=1, j \neq i}^{n} (k_j - k_i)} \tag{7.5i}$$

$$X_{n+1} = 1 - \sum_{i=1}^{n} \left[\left(\prod_{\substack{j=1 \\ j \neq i}}^{n} k_j \right) \exp(-k_i t) \Big/ \prod_{\substack{j=1 \\ j \neq i}}^{n} (k_j - k_i) \right] \tag{7.5j}$$

It is easily verified by direct substitution that eqns (7.5f)–(7.5h) do indeed satisfy eqns (7.5a)–(7.5c); eqn (7.5i) looks more daunting than it really is, and it is suggested that the reader writes down X_4 and X_5 explicitly, in order to grasp the increasing but trivial complexity of the series X_1, X_2, X_3, X_4, ...; the final compartment X_{n+1} is obtained by integration of eqn (7.5e), and, again, expanding (7.5j) for $n = 1, 2, 3$ is a useful exercise (Exercise 7.4). The analytical solutions given above are only valid if the rate constants are distinct, i.e. $k_i \neq k_j$.

7.3.4 *A linear compartmental model leading to the gamma distribution*

Assume that the rate constants in the scheme shown in Fig. 7.8 are all equal to k:

$$k_1 = k_2 = \ldots = k_{n-1} = k_n = k. \tag{7.6a}$$

Instead of eqns (7.5a)–(7.5e), the differential equations become

$$\frac{dX_1}{dt} = -kX_1 \tag{7.6b}$$

$$\frac{dX_2}{dt} = kX_1 - kX_2, \tag{7.6c}$$

$$\frac{dX_3}{dt} = k(X_2 - X_3) \tag{7.6d}$$

$$\vdots$$

$$\frac{dX_n}{dt} = k(X_{n-1} - X_n) \tag{7.6e}$$

$$\frac{dX_{n+1}}{dt} = kX_n. \tag{7.6f}$$

The solution to eqn (7.6b) is (with the initial conditions $t = 0, X_1 = 1, X_2 = X_3 = \cdots = X_n = X_{n+1} = 0$)

$$X_1 = e^{-kt}. \tag{7.6g}$$

Substituting this into (7.6c) leads to

$$\frac{dX_2}{dt} = k(e^{-kt} - X_2),$$

and it is easily verified that a solution to this is

$$X_2 = kte^{-kt}. \tag{7.6h}$$

Therefore, continuing in this manner, we obtain

$$\frac{dX_3}{dt} = k(kte^{-kt} - X_3), \tag{7.6i}$$

and

$$X_3 = \frac{(kt)^2}{2!} e^{-kt}$$

is a solution to eqn (7.6i). In general

$$X_j = \frac{(kt)^{j-1}}{(j-1)!} e^{-kt} \tag{7.6j}$$

satisfies

$$\frac{dX_j}{dt} = k(X_{j-1} - X_j). \tag{7.6k}$$

Finally, from eqn (7.6f), the final compartment is given by

$$X_{n+1} = \int_0^t \frac{k(kt)^{n-1}}{(n-1)!} e^{-kt}\, dt. \tag{7.6l}$$

In terms of the incomplete gamma function ((7.7e) *et seq.*), eqn (7.6l) can be written

$$X_{n+1} = \frac{1}{(n-1)!} \gamma(n, kt). \tag{7.6m}$$

The gamma distribution $f(\tau)$ can be written as

$$f(\tau) = \frac{k(k\tau)^{n-1} e^{-k\tau}}{(n-1)!}, \tag{7.7a}$$

where τ is a dummy variable. Apart from a factor of k, this describes the time course of the jth compartment X_j of eqn (7.6j). It is also identical with the integrand of eqn (7.6l), and has the following properties:

$$\int_0^\infty f(\tau)\, d\tau = 1. \tag{7.7b}$$

The mean $\bar{\tau}$, the mean square $\overline{\tau^2}$, the variance V, and the mode τ_{max}, are given by

$$\bar{\tau} = \frac{n}{k}, \quad \overline{\tau^2} = \frac{n(n+1)}{k^2}$$

$$V = \overline{\tau^2} - \bar{\tau}^2 = \frac{n}{k^2} \quad \tau_{max} = \frac{n-1}{k}. \tag{7.7c}$$

This useful distribution is plotted in Fig. 7.9 (a) for a range of n and k values that maintain a constant value of the mean $\bar{\tau}$. The cumulative gamma distribution $F(\tau)$ is defined by

$$F(\tau) = \int_0^\tau f(\tau)\, d\tau = \int_0^\tau \frac{k(k\tau)^{n-1} e^{-k\tau}}{(n-1)!} d\tau. \tag{7.7d}$$

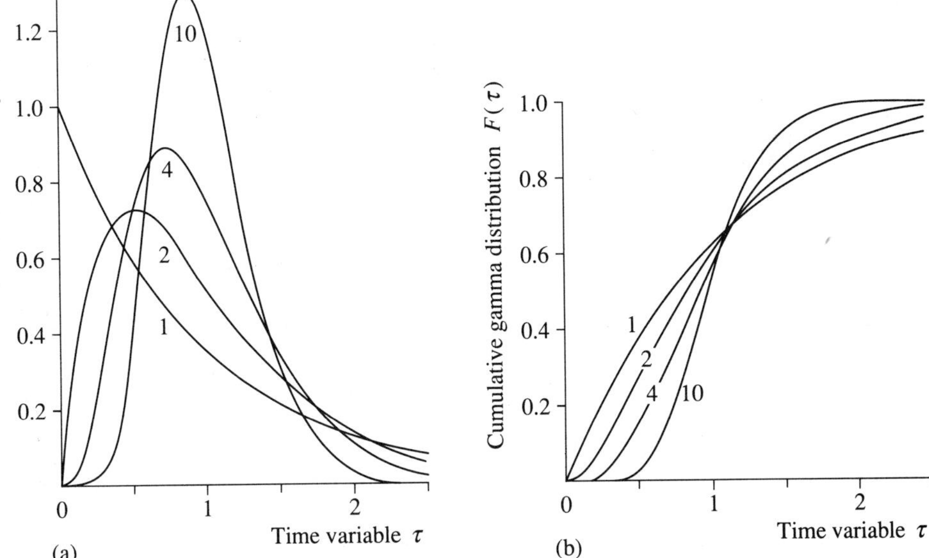

Fig. 7.9. (a) The gamma distribution $f(\tau)$ of eqn (7.7a) and (b) the cumulative gamma distribution $F(\tau)$ of eqn (7.7d). The mean $\bar{\tau}$ of the distribution is unity throughout, so that $n = k$ (eqns (7.7c)). The range of n considered is shown.

As mentioned above, this describes the time course of the terminal compartment X_{n+1}, of eqn (7.6l). $F(\tau)$ is shown in Fig. 7.9 (b). By taking very high values of n and k, a very sharp spike can be generated at $\bar{\tau} = n/k$, and in fact as n and k become infinite this spike approaches a delta function (this is a spike of unit area with infinite height and zero width); in this limit the cumulative gamma distribution $F(\tau)$ becomes a unit step function at $\bar{\tau}$.

In Sections 7.3.1 and 7.3.2 and Figs 7.4 and 7.6 (pp. 173–5), discrete time delays are used. From the foregoing discussion and Figs 7.8 and 7.9, it can be seen that a discrete time delay is equivalent to a sequence of many compartments (large n) with a fast rate constant between adjacent compartments (large k) to give an average time delay of n/k. For some purposes it may be more convenient to use the gamma distribution of Fig. 7.9(a) and eqn (7.7a) to describe a distributed time delay, since this can, in the limit, be used to simulate a simple discrete delay.

Finally, it is worth noting that the cumulative gamma distribution of eqn (7.7d) and the population of the terminal compartment of eqns (7.6l) and (7.6m) are related to the incomplete gamma function $\gamma(n, x)$, which is widely tabulated and for which computer subroutines are available. The incomplete gamma function is defined by

$$\gamma(n, x) = \int_0^x z^{n-1} e^{-z} \, dz. \tag{7.7e}$$

Substituting $k\tau = z$ and $k\mathrm{d}\tau = \mathrm{d}z$ in eqn (7.7d), gives

$$F(\tau) = \frac{\gamma(n, k\tau)}{(n-1)!}. \tag{7.7f}$$

The factorial function $(n - 1)!$ is defined by

$$(n-1)! = \int_0^\infty z^{n-1} \mathrm{e}^{-z} \, \mathrm{d}z. \tag{7.7g}$$

For integer n, this is equal to $(n-1)(n-2) \ldots 3.2.1$. Other useful properties of $\gamma(n, x)$ are

$$\gamma(n, 0) = 0 \tag{7.7h}$$

$$\gamma(n, \infty) = (n-1)! \tag{7.7i}$$

$$\frac{\mathrm{d}\gamma(n, x)}{\mathrm{d}x} = x^{n-1}\mathrm{e}^{-x} \tag{7.7j}$$

$$\gamma(n, x) \approx \frac{x^n}{n} \qquad \text{for } x \ll 1 \tag{7.7k}$$

$$\gamma(n+1, x) = n\gamma(n, x) - x^n \mathrm{e}^{-kt}. \tag{7.7l}$$

7.3.5 *Gamma delay with a first-order process*

In Section 7.3.1 (p. 173) we discuss a discrete delay followed by a first-order process, as shown in Fig. 7.4. In Fig. 7.9 and eqns (7.7a)–(7.7g) it is shown how the gamma distribution, which is itself based on the compartmental scheme of Fig. 7.8 with equal rate constants, can represent a distributed delay or, in the limit of large n and k (the two parameters of the gamma distribution), a discrete delay. Here, we consider the scheme in Fig. 7.10, which represents a gamma delay followed by a first-order process, so that the consequences of the biologically more acceptable distributed delay, and also the consequences of the transition towards a discrete delay, can be examined.

Using eqns (7.6g) and (7.6j), we obtain

$$X_1 = \mathrm{e}^{-kt}$$

$$X_n = \frac{(kt)^{n-1}}{(n-1)!}\mathrm{e}^{-kt}. \tag{7.8a}$$

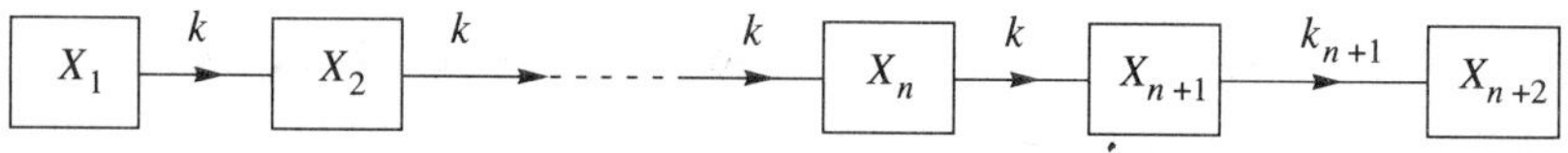

Fig. 7.10. Gamma-distributed delay followed by a first-order process k_{n+1}. The rate constants from $X_1, X_2, \ldots, X_n$ are all equal to k, giving a gamma-distributed arrival in X_{n+1} as in Fig. 7.9(a) and eqn (7.7a). Solutions to this scheme are shown in Fig. 7.11.

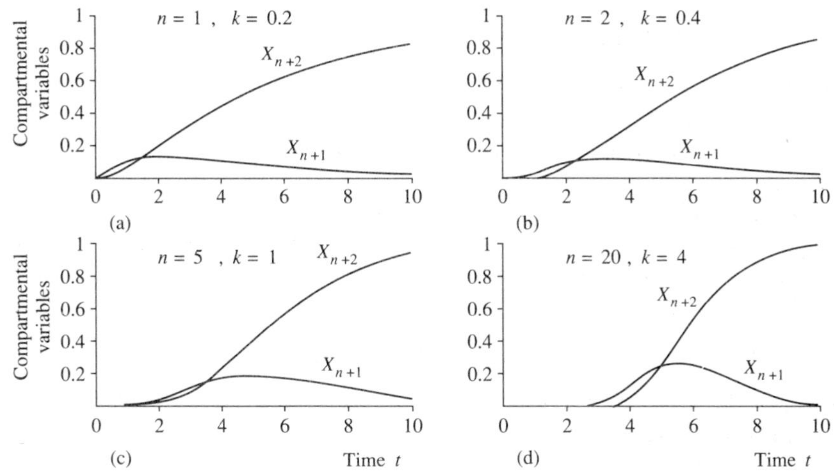

Fig. 7.11. Solutions to the scheme of Fig. 7.10 for $k_{n+1} = 1$ throughout and various parameter values as shown; $n/k = 5$ is constant (eqn (7.7c)). Results were obtained numerically from eqns (7.8b) and (7.8c).

X_{n+1} and X_{n+2} are given by the differential equations

$$\frac{dX_{n+1}}{dt} = kX_n - k_{n+1}X_{n+1} \tag{7.8b}$$

$$\frac{dX_{n+2}}{dt} = k_{n+1}X_{n+1}. \tag{7.8c}$$

Numerical integration of these equations gives the results shown in Fig. 7.11, where a range of n and k values have been examined with n/k maintained constant—this gives the same mean time of arrival in the $(n + 1)$th compartment as n is increased (eqn (7.7c)). In Fig. 7.11(d) with $n = 20$, the response is approaching that produced by a discrete delay of 5 time units, and the rate of arrival in the $(n + 2)$th compartment is dominated by the final rate constant k_{n+1}.

In terms of the incomplete gamma function, it can be shown (Exercise 7.5) that eqns (7.8a)–(7.8c) have solutions

$$X_{n+1} = \exp(-k_{n+1}t)\left(\frac{k}{k - k_{n+1}}\right)^n \frac{1}{(1-n)!}\gamma[n, (k - k_{n+1})t] \tag{7.8d}$$

$$X_{n+2} = \frac{1}{(n-1)!}\left\{\gamma(n, kt) - \exp(-k_{n+1}t)\left(\frac{k}{k - k_{n+1}}\right)^n \gamma[n, (k - k_{n+1})t]\right\}. \tag{7.8e}$$

It can be proved (Exercise 7.6) that, on taking the limit $k_{n+1} \to k$, eqn (7.8d) becomes identical with eqn (7.6j) with $j = n + 1$, and eqn (7.8e) becomes identical with eqn (7.6m) with n replaced by $n + 1$.

7.4 Diffusion and development

The time course of a developmental process can sometimes be limited by diffusion or by a diffusion-like mechanism. For example, the diffusion of water into a seed may be slow enough under certain conditions to determine substantially the time of emergence. A diffusive-like movement of a morphogen, from the shoot apex or from a leaf to the root or some other growing point, may be the determinant of a developmental event. It is therefore worth considering the time course given by diffusion for some simple situations.

7.4.1 *One-dimensional diffusion of a morphogen*

Consider the scheme in Fig. 7.12. The initial distribution of morphogen, of concentration C, is given by

$$\begin{aligned} t \leqslant 0, x \leqslant 0, \quad C &= C_0, \\ x > 0, \quad C &= 0 \end{aligned} \tag{7.9a}$$

where t is time, x is distance, and C_0 is a constant. Equation (7.9a) is depicted by the solid line in Fig. 7.12. The one-dimensional diffusion equation is (p. 93)

$$\frac{\partial C}{\partial t} = D \frac{\partial^2 C}{\partial x^2}, \tag{7.9b}$$

where D is the diffusion constant. With the initial values of eqn (7.9a) and the boundary value of

$$t > 0, \quad x \leqslant 0, \quad C = C_0, \tag{7.9c}$$

the solution for $t > 0$ and $x > 0$ is (Crank 1956, p. 19, eqn (2.25))

$$C = C_0 \left(1 - \frac{2}{\pi^{1/2}} \int_0^{x/2(Dt)^{1/2}} e^{-u^2}\, du \right), \tag{7.9d}$$

where u is a dummy variable. Note that

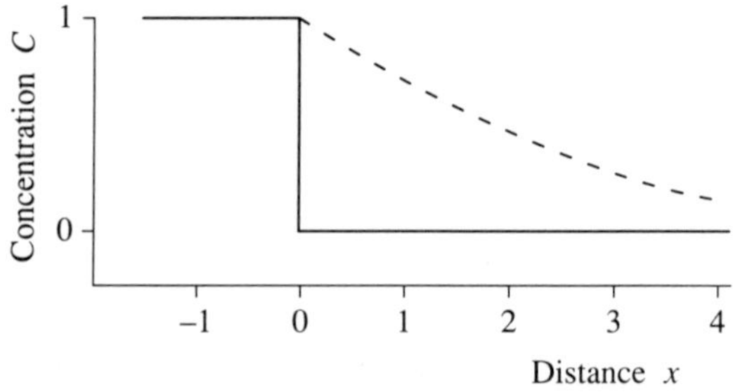

Fig. 7.12. One-dimensional diffusion of a morphogen: —, initial distribution; – – –, distribution at time $t = 0.25$. The results are calculated from eqn (7.9d) with $D = 16$ and $C_0 = 1$.

$$1 = \frac{2}{\pi^{1/2}} \int_0^\infty \exp(-u^2)\, du. \tag{7.9e}$$

It can be shown that eqn (7.9d) does indeed satisfy eqn (7.9b) (Exercise 7.7).

The morphogen distribution resulting from diffusion is also shown in Fig. 7.12; this decreases monotonically with distance, without a point of inflexion.

It is more interesting to consider the time behaviour of the morphogen concentration at a given point in space. If we define

$$y = C/C_0, \tag{7.9f}$$

then, from eqn (7.9d),

$$\frac{\mathrm{d}y}{\mathrm{d}t} = \frac{1}{\pi^{1/2}} \frac{x}{2D^{1/2}} \frac{1}{t^{3/2}} \exp\left(-\frac{x^2}{4Dt}\right). \tag{7.9g}$$

Writing

$$\alpha = \frac{x}{2D^{1/2}}, \tag{7.9h}$$

we can express this equation more succinctly as

$$g(t) = \frac{\mathrm{d}y}{\mathrm{d}t}(t) = \frac{\alpha}{\pi^{1/2}} \frac{1}{t^{3/2}} \exp\left(-\frac{\alpha^2}{t}\right), \tag{7.9i}$$

where the function $g(t)$, the rate of increase of y, is defined. With respect to the time variable t, y always exhibits a point of inflexion. This occurs when $\mathrm{d}g/\mathrm{d}t = 0$, at a time t_i given by

$$t_i = \frac{2\alpha^2}{3}, \tag{7.9j}$$

and the value g_i of g is

$$g_i = \frac{1}{\pi^{1/2}} \frac{3\sqrt{3}}{2\sqrt{2}} \frac{1}{\alpha^2} \mathrm{e}^{-3/2}. \tag{7.9k}$$

We can interpret t_i as the delay or the time required for the morphogen signal to diffuse as far as distance x. A typical time course for the process is shown in Fig. 7.14—this was obtained by integrating eqn (7.9g).

7.4.2 *One-dimensional diffusion of a substance from two sides*

Some of the time required for seed germination may be attributable to the diffusion of water or other substances into the seed. In the last section, diffusion in a single direction, as in Fig. 7.12, was considered. We now discuss the scheme shown in Fig. 7.13, which possibly represents diffusion into both faces of a rather flat-sided seed. The switching character of the time course given by the scheme of Fig. 7.13 is rather more abrupt than that given by Fig. 7.12.

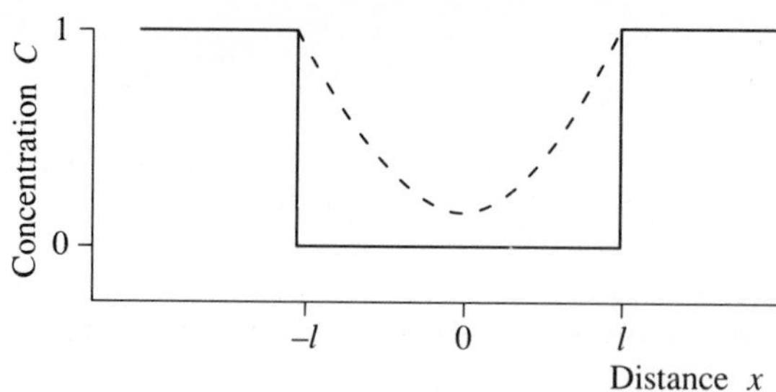

Fig. 7.13. One-dimensional diffusion of a morphogen or a substance from two sides: —, the initial distribution; ---, the distribution at time $t = 0.25$, obtained by solving eqn (7.10b) with $l = 5$, $D = 16$, and $C_0 = 1$.

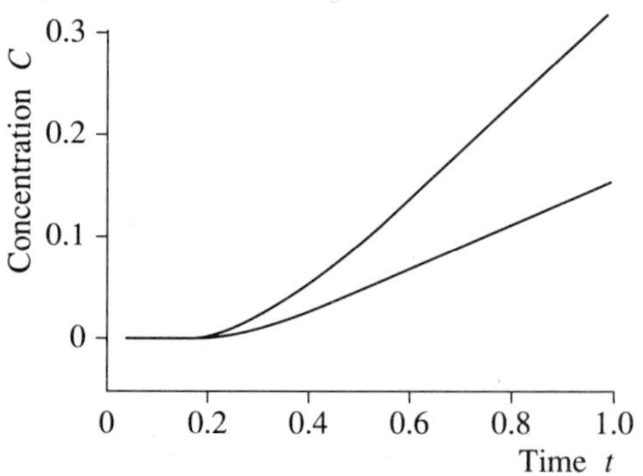

Fig. 7.14. Time course of arrival of a diffusing substance or morphogen: the lower curve represents the one-sided scheme of Fig. 7.12 and eqn (7.9d) with $C_0 = 1, D = 16$, and $x = 8$, giving $\alpha = 1$ by eqn (7.9h), and was calculated by integrating eqn (7.9i); the upper curve represents the two-sided scheme of Fig. 7.13 and eqn (7.10b) with $C_0 = 1$, $D = 16$, $l = 8$, and $x = 0$.

The initial distribution of a substance of concentration C is

$$\begin{aligned} t \leqslant 0, \quad |x| \geqslant l, \quad C &= C_0 \\ |x| < l, \quad C &= 0, \end{aligned} \tag{7.10a}$$

where $2l$ is the thickness of the slab into which diffusion is taking place, C_0 is a constant, and the distance x is measured from the centre of the slab. For $t > 0$ and with $|x| \geqslant l$, $C = C_0$, in the interior of the slab with $|x| < 1$, Crank (1956, p. 22, eqn (2.67)) has shown that C is given by

$$C = C_0 - \frac{4}{\pi} \sum_{n=0}^{\infty} \frac{(-1)^n}{2n+1} \exp\left[\frac{-D(2n+1)^2\pi^2 t}{4l^2}\right] \cos\left[\frac{(2n+1)\pi x}{2l}\right]. \tag{7.10b}$$

A typical time course calculated from eqn (7.10b) is shown in Fig. 7.14, where the results of the scheme in Fig. 7.12 and eqn (7.9d) are also drawn. In both cases, diffusion gives rise to a delay time of about $x^2/4D$ (see eqns (7.9h) and (7.9j)); note that this delay time is proportional to the square of the distance. The temperature response of diffusion may be similar to that of chemical or biochemical processes (p. 134), and diffusion may similarly be a component of schemes leading to the temperature-sum rules discussed earlier in this chapter.

7.5 A compartmental model of seed germination: temperature and time responses

The effects of environment on seed germination are important agriculturally. The topic is relevant to efficient crop establishment, as well as to the stability, longevity, and productivity of swards used in animal production and also the dynamics of weed infestation. The aim of this section is to describe and analyse the simplest compartmental model which seems able to encompass the range of dynamic behaviour and temperature responses that are found in nature (Thornley 1986).

The dynamics of the phenomenon and the time course of germination, and how these respond to temperature, are of particular applied interest. The topic has understandably received a great deal of attention, both experimentally and theoretically. Labouriau (e.g. Labouriau and Osborn 1984) has reported many detailed investigations, as have Hageseth (1978) has reported many detailed investigations, as have Hageseth (1978), Garcia-Huidobro *et al.* (1982), Hsu, Nelson, and Chow (1984), and many others. These authors also report theoretical studies of the problem, as do Goloff and Bazzaz (1975), Hageseth and Joyner (1975), Thornley (1977), and Washitani and Takenaka (1984). However, there is still a need for a simple model which will describe the temperature responses of the principal germination parameters—i.e. germinability and germination rate—and also the time course of germination.

There are no mechanistic mathematical models of seed germination, i.e. models based on a representation of underlying physical, chemical, and physiological processes. Such a model would undoubtely be very complex, and even if enough were known to permit model construction, it would hardly be a practical tool. The highly empirical approaches to the problem (e.g. Hsu *et al.* 1984) do not usually give helpful scientific insights, and it can be difficult to represent environmental effects within the model. Here we have chosen the middle ground of a compartmental representation: such a model is easily constructed, described, and analysed—whether or not the compartments included in the model correspond to physiological entities is very debatable. Nevertheless, the Arrhenius equation approach (p. 120) allows temperature to be incorporated in a straightforward manner: analytical results are possible, facilitating application, and the equations and their solutions are well behaved and bounded.

First the general framework within which germination is considered needs definition. Suppose that S_0 seeds are put out to germinate at time $t = 0$. Let $P(t)$ be the number that have emerged at time t. The germinability G_m is defined as the total fraction of seeds that eventually germinate after waiting a long time, so that

$$G_m = \frac{P(t = \infty)}{S_0}. \tag{7.11a}$$

G_m obviously has a maximum value of unity. The fraction $G(t)$ of the total number of seeds that have germinated at time t is

$$G(t) = \frac{P(t)}{S_0}. \tag{7.11b}$$

It is convenient to define a time t_h as the time required for half the final germination to be achieved (this is preferable to defining t_h in terms of the total number of seeds, since under certain conditions less than half the seeds may eventually germinate). Thus

$$\frac{G(t = t_h)}{G_m} = \frac{1}{2}. \tag{7.11c}$$

The overall rate constant for germination k_{gh} (day^{-1}) is then defined by

$$k_{gh} = 1/t_h. \tag{7.11d}$$

Rather than using the time for half-final germination, some investigators (e.g. Labouriau and Osborn 1984) prefer to calculate the mean germination time $\bar{t}$ of the seeds that germinate, which is defined by

$$\bar{t} = \int_0^\infty t\frac{dP}{dt}dt \bigg/ \int_0^\infty \frac{dP}{dt}dt. \tag{7.11e}$$

Another overall rate constant for germination k_{gt}, can be obtained from the mean germination time $\bar{t}$ as follows:

$$k_{gt} = 1/\bar{t}. \tag{7.11f}$$

In practice, the two germination time (half-final germination and mean germination time) do not usually differ from each other by more than about 20 per cent. As a measure of speed of germination, we prefer the half-final germination time t_h, to the mean germination $\bar{t}$, because it is less affected by a small number of seeds with very long germination times. However, for the compartmental model described below, the mean germination time of eqn (7.11e) can be obtained analytically, whereas the half-final germination time of eqn (7.11c) requires a numerical solution.

The observational phenomena to be modelled are summarized in Fig. 7.15 where the curves are drawn by eye but are related to experimental work by Hageseth (1978), Garcia-Huidobro *et al.* (1982), Hsu *et al.*, and Labouriau and Osborn (1984); many other investigators have reported similar results. A feature to be noted in Fig. 7.15(a) is the changing sigmoidicity in the time-course curve,

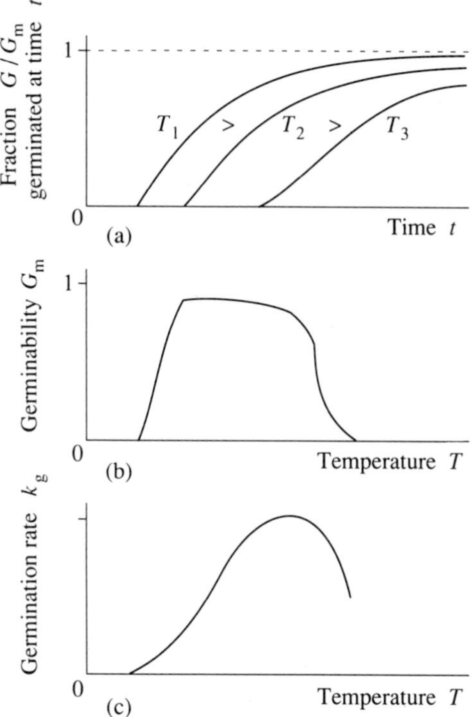

Fig. 7.15. Time and temperature responses occurring in seed germination. In (a), T_1, T_2, and T_3 denote the temperatures at which germination occurred, with $T_1 > T_2 > T_3$.

which increases as the temperature is decreased. The challenge facing the modeller of germination is to construct a scheme that will simulate these time and temperature responses realistically.

7.5.1 *The model*

The four-compartment model applied to this problem is shown in Fig. 7.16. It is assumed that there is a discrete delay τ; this might be interpreted as the time for seed wetting to occur, and could be related to the type of the time delay seen in Fig. 7.14. There then follow two intermediate compartments I_1, and I_2 with first-order rate constants as shown. Finally, the emergence of the young plant is denoted by compartment P. A process k_{1d} giving loss in germinability at low temperatures is shown as branching off the first intermediate state I_1; a process k_{2d} giving high-temperature loss in germinability is shown as branching off the second intermediate state I_2. In any linear compartment model, the order of the intermediate states is immaterial to the overall responses obtained. The discrete delay τ and the four rate constants k_1, k_2, k_{1d}, and k_{2d}, are all assumed to be

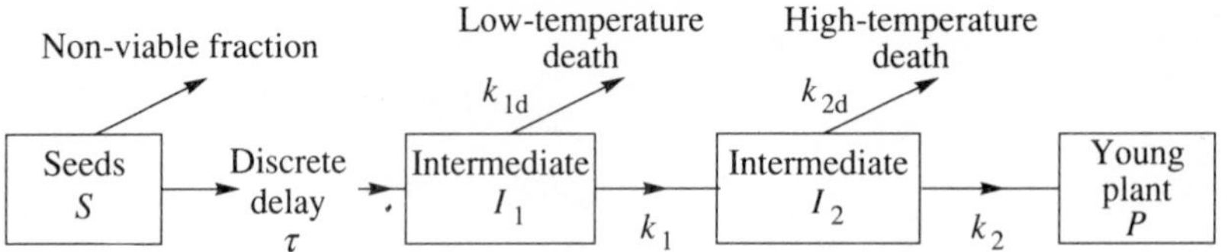

Fig. 7.16. Seed germination model. k_1, k_2, k_{1d}, and k_{2d} are rate constants; k_{1d} is the process that gives low-temperature loss in germinability, and k_{2d} gives high-temperature loss in germinability. A fraction of the seeds is non-viable under all conditions.

temperature dependent. The use of two intermediate compartments I_1 and I_2, makes sigmoidal time responses a possible outcome of the model (see Figs 7.6 and 7.7). A fraction of the seeds may be non-viable under all conditions (Fig. 7.16), even the optimum; this fraction may be negligible, or up to 10 or 20 per cent, depending upon plant type, variety, and the conditions under which the seeds were prepared. It is assumed here that this non-viable fraction is zero, so that the maximum possible germination is 100 per cent. It is easy to relax this assumption.

The initial conditions are

$$t < \tau, \quad S = 1, \quad I_1 = I_2 = P = 0. \tag{7.12a}$$

With the discrete delay assumption, at time $t = \tau$, the contents of compartment S are moved to I_1, so that

$$t = \tau, \quad S = 0, \quad I_1 = 1, \quad I_2 = P = 0. \tag{7.12b}$$

For $t \geqslant \tau$, the differential equations governing the system are

$$\frac{dI_1}{dt} = -k_1 I_1 - k_{1d} I_1, \tag{7.12c}$$

$$\frac{dI_2}{dt} = k_1 I_1 - k_2 I_2 - k_{2d} I_2, \tag{7.12d}$$

$$\frac{dP}{dt} = k_2 I_2. \tag{7.12e}$$

The solution to eqn (7.12c) with (7.12b) is

$$I_1 = \exp[-(k_1 + k_{1d})(t - \tau)]. \tag{7.12f}$$

Substituting this into eqn (7.12d) and integrating gives

$$I_2 = k_1 \frac{\exp[-(k_1 + k_{1d})(t - \tau)] - \exp[-(k_2 + k_{2d})(t - \tau)]}{k_2 + k_{2d} - k_1 - k_{1d}}. \tag{7.12g}$$

Substituting into eqn (7.12e), a further integration leads to

$$P = \frac{k_1}{k_1 + k_{1d}} \frac{k_2}{k_2 + k_{2d}}$$

$$\times \left\{1 - \frac{(k_2 + k_{2d}) \exp[-(k_1 + k_{1d})(t - \tau)] - (k_1 + k_{1d}) \exp[-(k_2 + k_{2d})(t - \tau)]}{k_2 + k_{2d} - k_1 - k_{1d}}\right\}. \tag{7.12h}$$

These last two equations are not valid if

$$k_1 + k_{1d} = k_2 + k_{2d}, \tag{7.12i}$$

in which case eqns (7.12g) and (7.12h) are replaced by (see Exercise 7.3)

$$I_2 = k_1(t - \tau) \exp[-(k_1 + k_{1d})(t - \tau)] \tag{7.12j}$$

$$P = \frac{k_1 k_2}{(k_1 + k_{1d})(k_2 + k_{2d})} \{1 - [1 + (k_1 + k_{1d})(t - \tau)] \exp[-(k_1 + k_{1d})(t - \tau)]\}. \tag{7.12k}$$

Equations (7.12j) and (7.12k) can be compared with eqns (S7.3g) and (S7.3i) (p. 600) obtained from the simpler scheme of Fig. 7.6.

Applying eqn (7.11a) with $S_0 = 1$, gives the germinability G_m as

$$G_m = P(t = \infty). \tag{7.12l}$$

Therefore, with eqn (7.12h) (or eqn (7.12k)),

$$G_m = \frac{k_1}{k_1 + k_{1d}} \frac{k_2}{k_2 + k_{2d}}. \tag{7.12m}$$

This expression is intuitively obvious directly from Fig. 7.16, since any developing seed in the intermediate state I_1 has relative probabilities in the ratio $k_1 : k_{1d}$ of continued development or loss of viability, and similarly for the intermediate state I_2.

The time t_h for half-final germination is defined by

$$P(t = t_h) = \tfrac{1}{2} P(t = \infty), \tag{7.12n}$$

which, with eqn (7.12h), gives the transcendental equation

$$\tfrac{1}{2}(k_2 + k_{2d} - k_1 - k_{1d}) = (k_2 + k_{2d}) \exp\{-(k_1 + k_{1d})(t_h - \tau)\}$$
$$-(k_1 + k_{1d}) \exp\{(k_2 + k_{2d})(t_h - \tau)\}. \tag{7.12o}$$

This equation can be solved numerically to give t_h; the Newton–Raphson method (p. 605) is a convenient technique for finding the solutions. The germination rate k_{gh} is then calculated from eqn (7.11d). (Exercise 7.8)

If eqn (7.12i) holds, then instead of eqn (7.12o), we obtain

$$\tfrac{1}{2} = [1 + (k_1 + k_{1d})(t_h + \tau)] \exp[(k_1 + k_{1d})(t_h - \tau)] \tag{7.12p}$$

from eqns (7.12k) and (7.12n). Thus equation can also be solved numerically for t_h.

The mean germination time $\bar{t}$ can be calculated by applying eqn (7.11e) to eqn (7.12h) to give

$$\bar{t} = \tau + \frac{k_1 + k_{1d} + k_2 + k_{2d}}{(k_1 + k_{1d})(k_2 + k_{2d})}, \tag{7.12q}$$

and this expression is also valid for the case of eqn (7.12i). The germination rate based on the mean germination time k_{gt}, is obtained from eqn (7.11f).

7.5.2 *Temperature effects on the parameters*

Temperature responses are discussed in general terms in Chapter 5. Here we use the simple Arrhenius equation (eqn (5.1e), p. 122) or an equation based on the Arrhenius equation (eqn (5.3d), p. 128) to model the temperature dependence of the five parameters of the model of Fig. 7.16, i.e. k_1, k_{1d}, k_2, k_{2d}, and τ. From the time delay τ, an equivalent rate constant k_0 is defined by

$$k_0 = 1/\tau. \tag{7.13a}$$

The Arrhenius equation is

$$k = A\exp\left(-\frac{B}{T}\right) \tag{7.13b}$$

for the rate constant k in terms of the absolute temperature T and the two further parameters A and B. It is assumed that four of the five rate constants, i.e. k_1, k_{1d}, k_2, k_{2d}, can be written as functions of temperature T directly using eqn (7.13b):

$$k_i = A_i\exp\left(-\frac{B_i}{T}\right), \tag{7.13c}$$

where the subscripts on parameters A and B are $i = 1$, 1d, 2, and 2d.

In order that the germination rate should decline at high temperatures, it is necessary to assume that at least one of the rate constants should rise to a maximum and then decline with further increases in temperature. Based on the expression in eqn (5.3d) derived from the Arrhenius equation and shown in Fig. 5.7 (p. 128), it is assumed that the rate constant k_0 (which determines the discrete delay τ through eqn (7.13a)) is given by

$$k_0 = \frac{A_0\exp(-B_0/T)}{1 + C_0\exp(-D_0/T)}, \tag{7.13d}$$

where A_0, B_0, C_0, and D_0 are the four parameters defining the rate constant k_0.

The advantage of using eqns (7.13c) and (7.13d) for the rate constants of the compartmental model is that these expressions behave in a bounded and biologically reasonable manner; our use of these equations is otherwise totally empirical, and it is not supposed that the parameter values used below truthfully reflect the constructs of the Arrhenius theory such as activation energies, collision frequencies, or transition probabilities.

7.5.3 *Results and discussion*

The 12 parameters of the model are assumed to have the following values:

$$\begin{aligned} &A_0 = 10^{15}, \quad B_0 = 10^4, \quad C_0 = 10^{36}, \quad D_0 = 25\,000 \\ &A_1 = 2.25 \times 10^{61}, \quad B_1 = 40\,000 \\ &A_{1d} = 1330, \quad B_{1d} = 2000 \\ &A_2 = 1330, \quad B_2 = 2000 \\ &A_{2d} = 4.74 \times 10^{56}, \quad B_{2d} = 40\,000. \end{aligned} \tag{7.14a}$$

The values in (7.14a) were estimated in an *ad hoc* manner from a knowledge of the behaviour of eqns (7.13c) and (7.13d) and the range of rate constants required in the compartmental model to give the required germination response.

The resulting rate constants, calculated using eqns (7.13a)–(7.13d), and the principal germination parameters are given for a range of temperatures in Table 7.3. In Fig. 7.17, the responses to temperature of the principal germination parameters, i.e. germinability G_m and germination rate k_{gh}, are shown graphically. The main features are the decline in the germinability at both high and low temperatures below its maximum value close to unity, and the sharp increase in germination rate with temperature to a maximum followed by a decrease at high temperatures. Some of the corresponding time courses are shown in Fig. 7.18,

Table 7.3. Temperature response of the parameters of the model.

T (°C)	τ (days)	k_0 (day^{-1})	k_1 (day^{-1})	k_{1d} (day^{-1})	k_2 (day^{-1})	k_{2d} (day^{-1})	G_m	t_h (days)	k_{gh} (day^{-1})
5	4.1	0.24	0.08	1.00	1.00	2.0×10^{-6}	0.08	5.7	0.17
10	2.2	0.46	1.03	1.14	1.14	2.1×10^{-5}	0.48	3.3	0.48
15	1.2	0.83	12.0	1.29	1.29	2.4×10^{-4}	0.90	1.8	0.55
20	0.7	1.40	128	1.45	1.45	0.0026	0.99	1.2	0.83
25	0.5	1.96	1261	1.62	1.62	0.026	0.98	0.9	1.07
30	0.5	1.87	1.2×10^4	1.81	1.81	0.24	0.89	0.9	1.14
35	0.9	1.18	9.8×10^4	2.02	2.02	2.0	0.50	1.0	0.98
40	1.6	0.61	7.8×10^5	2.24	2.24	15.9	0.12	1.7	0.59

T is the temperature; τ is the discrete delay shown in Fig. 7.16, and is related to the rate constant k_0 by eqn (7.13a), which is calculated using eqn (7.13d); the rate constants k_1, k_{1d}, k_2, and k_{2d} are illustrated in Fig. 7.16, and are calculated using eqn (7.13c). The underlying Arrhenius theory parameters are given in eqns (7.14a). The germination parameters given are the germinability G_m, the half-final germination time t_h and the germination rate k_{gh}, which are evaluated using eqns (7.12m), (7.12o) and (7.12d).

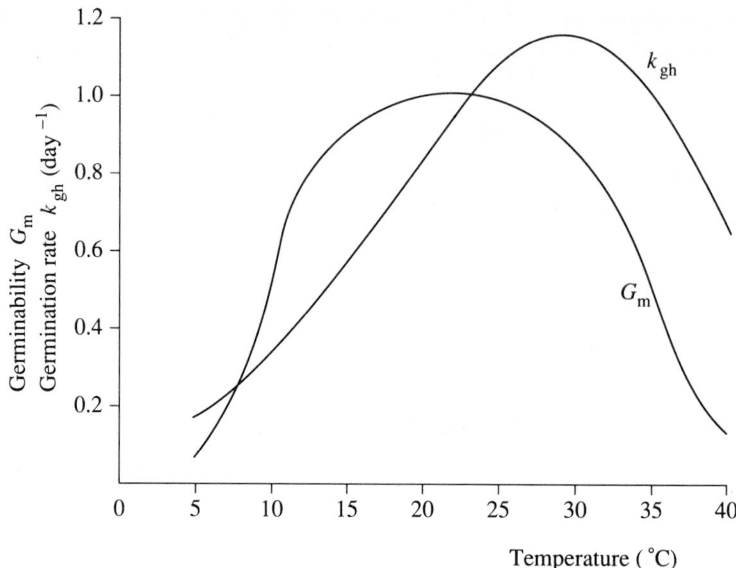

Fig. 7.17. Temperature responses of germinability G_m, and germination rate k_{gh}, for the model of Fig. 7.16 with the parameter values in eqns (7.14a). The germinability was computed using eqns (7.12m) and (7.13c); the germination rate was computed using eqns (7.11d), (7.12o), (7.13a), (7.13c) and (7.13d).

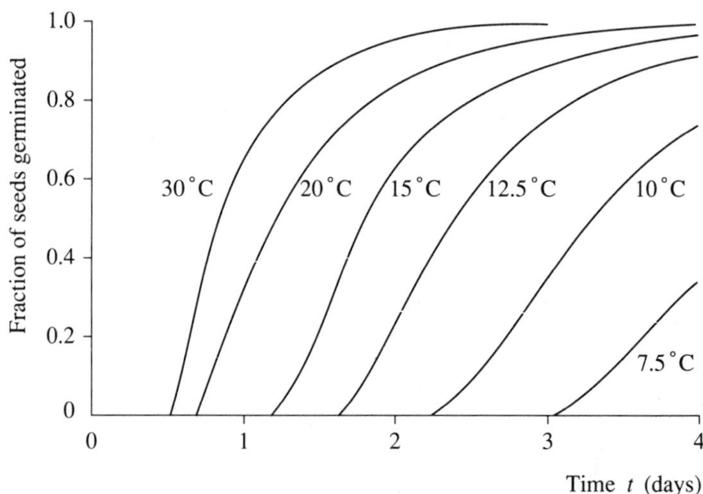

Fig. 7.18. Time course of seed germination for the range of temperatures indicated. The fraction of the germinable seeds is shown (eqns (7.11a) and (7.11b)), so that all curves approach an asymptote of unity regardless of germinability. Parameter values are as in eqns (7.14a). The results were computed using eqns (7.13c), (7.13d), and (7.13a), eqns (7.12h) or (7.12k), and eqns (7.11a) and (7.11b).

where $G(t)/G_m$ is plotted; each curve approaches an asymptote of unity regardless of the germinability. At lower temperatures the curves are somewhat sigmoidal in character; at the higher temperatures, the curves resemble the monomolecular growth function (p. 77) after the discrete time delay.

The model is realistic in its predictions, and although it contains as many as 12 parameters (eqns (7.14a)), it seems unlikely that a phenomenon as varied in its responses as germination can be described by fewer.

7.6 A teleonomic model of time of flowering

In this section we describe a very simple model to illustrate a teleonomic (p. 11) approach to time of flowering.

Consider a plant with mass w_0 at time $t = 0$ growing vegetatively with a constant growth rate of μ. The mass w of the plant at time t is

$$w = w_0 e^{\mu t}. \tag{7.15a}$$

Suppose that flowering occurs at time $t = t_f$, when the growth rate g is given by (differentiating (7.15a) and putting $t = t_f$)

$$g = \mu w_0 \exp(\mu t_f). \tag{7.15b}$$

It is assumed that, after flowering, growth continues at a constant rate (7.15b), and that all growth goes into seed dry matter. Thus, by the end of the growing season at time $t = t_s$, the seed mass w_s is

$$w_s = \mu w_0 (t_s - t_f) \exp(\mu t_f). \tag{7.15c}$$

The later flowering occurs, the larger is t_f and the greater is the vegetative growth rate (7.15b) to support seed growth, but the shorter is the remaining season in which seed growth can occur. Maximizing (7.15c) with respect to t_f gives

$$t_f = t_s - 1/\mu. \tag{7.15d}$$

A large vegetative specific growth rate μ requires a late flowering time, according to this view of the goals of the plant.

Exercises

7.1. For the compartmental scheme in Fig. 7.6, solve the differential eqns (7.4b) to prove eqns (7.4c). Do this for $k_2 \neq k_3$, and to simplify the notation take $\tau = 0$.

7.2. From eqns (7.4c), and taking $\tau = 0$ to simplify the notation, show that X_3 is a maximum at time $t = t_{max}$, and that X_4 has a point of inflexion at time $t = t_i$, where

$$t_i = t_{max} = \frac{\ln(k_3/k_2)}{k_3 - k_2}.$$

7.3. For the compartmental scheme of Fig. 7.6 with $k_2 = k_3 = k$, find the corresponding equation to eqns (7.4b)–(7.4d). *Hint*: there are two ways of attacking this problem; one is to solve eqns (7.4b) with $k_2 = k_3 = k$; the second is to take the existing results in eqns (7.4c) and (7.4d) and to apply the limit $k_2 \to k$ with $k_3 = k$.

7.4. For the general linear compartmental scheme of Fig. 7.8, use eqn (7.5i) to write down X_4 (you may prefer to do this by extending the series X_1, X_2, X_3 in eqns (7.5f)–(7.5h)). Use eqn (7.5e) to integrate eqns (7.5f)–(7.h) to find the solution for the terminal compartment X_2, X_3, X_4 ($n = 1, 2, 3$), and check that these solutions agree with eqn (7.5j) with $n = 1, 2, 3$. Using eqn (7.5j) or otherwise, find X_5 when X_5 is the final compartment.

7.5. Using eqns (7.8a) and (7.8b) derive eqn (7.8d) for X_{n+1}. Substitute this solution into eqn (7.8c) and, using integration by parts, derive eqn (7.8e) for X_{n+2}.

7.6. In eqns (7.8d) and (7.8e) take the limit $k_{n+1} \to k$ and show that these two equations are equivalent to eqns (7.6j) and (7.6m).

7.7. Verify that eqn (7.9d) satisfies the diffusion equation (eqn (7.9b)).

7.8. Apply the Newton–Raphson method to solve eqn (7.12o), giving the time t_{h} for half-final germination and the germination rate k_{gh} (eqns (7.11c) and (7.11d)). Use the parameter values $k_1 = 1$, $k_{1\mathrm{d}} = 0$, $k_2 = 2$, $k_{2\mathrm{d}} = 0$, $\tau = 0$.

Part II
Plant and crop physiology

8
Light relations in canopies

8.1 Introduction

In this chapter we consider the light relations in plant canopies, which include light interception, absorption, and attenuation. This is important for studies of photosynthesis, where the light reactions are driven by solar energy. Solar radiation energy is also a major component of the energy balance of the crop and provides energy for transpiration (see Chapter 14 where we consider the transpiration from a crop canopy).

In crop models, light attenuation in a canopy is generally described by the equation

$$I(l) = I_0 e^{-kl}, \tag{8.1a}$$

where I_0 and $I(l)$ (W (m^2 ground)$^{-1}$) are the irradiances above and within the canopy respectively at cumulative leaf area index l, and k (m^2 ground (m^2 leaf)$^{-1}$) is known as the extinction coefficient. The physical units and definition of irradiance are considered in the next section. However, it should be noted here that in defining these units, we distinguish between leaf area and horizontal area, where the latter is denoted by m^2 ground. This distinction is important since leaf photosynthesis is governed by the irradiance on the leaf surface. The fact that k has dimensions m^2 ground (m^2 leaf)$^{-1}$ is frequently unappreciated: while mathematically this implies that k is dimensionless, physically this is not the case, and we shall see that this is important when calculating the irradiance incident on the leaves within the canopy. Taking logarithms of eqn (8.1a) gives

$$\ln(I/I_0) = -kl, \tag{8.1b}$$

and this form of the equation is often used to fit to experimental data. Equation (8.1a) (or the logarithmic form (8.1b)) has frequently been shown to give a good description of the light attenuation in canopies (e.g. Brown and Blaser 1968, Sheehy and Peacock 1975).

Equation (8.1a) was first applied to plant canopies by Monsi and Saeki (1953), and is sometimes referred to as the Monsi–Saeki equation. It is analogous to the Beer–Lambert law in physics for the light attenuation through a murky medium. As k increases, a greater proportion of light is attenuated for a given amount of leaf area, so that k values for canopies with prostrate leaves are generally observed to be greater than those where the leaves are more erect. For example, typical values for perennial ryegrass and white clover are 0.5 and 0.8 respectively. As we shall see in Section 8.2.1, for canopies with randomly distributed leaves,

k is constrained by

$$0 < k < 1, \tag{8.1c}$$

although for more regular leaf distributions this may not be the case.

The Monsi–Saeki equation forms the basis of our treatment of light interception, attenuation, and absorption in monocultures, mixtures, row crops, and single plants. It is obviously a very simplified approach, and ignores variation due to cloud cover, atmospheric conditions, solar angle, and variation in brightness over the sky, as well as the photosynthetic quality of the solar radiation. However, these details are seldom available for crop models, and there is value in considering the problem at a simplified level which is useful in practice. We also restrict attention to the irradiance incident on the upper surface of the leaves so that, although we account for scattered (or reflected) irradiance, the subsequent component of irradiance incident on the lower surfaces of the leaves is ignored; it is unlikely that this is of great significance.

In presenting the theory, we make a distinction between the irradiance incident on the leaf surfaces, the irradiance intercepted by the leaves within the canopy, the irradiance intercepted by the canopy as a whole, the irradiance absorbed by the leaves, and the irradiance absorbed by the canopy as a whole. It is important to identify these components for the correct interpretation and application of environmental data. In studies of leaf photosynthesis, while the absorbed irradiance is required for a mechanistic understanding of the light reactions of photosynthesis (see Chapter 9), experiments to measure the response of leaf photosynthesis to irradiance often relate to incident irradiance (see Chapter 10, Fig. 10.1, p. 245). When considering leaf or canopy transpiration, it is necessary to define the absorbed irradiance as this provides the energy for water evaporation (see Chapter 14). We also derive expressions for the irradiance which is reflected and transmitted by the canopy.

We present a simple derivation of the Monsi–Saeki equation for a uniform crop canopy, where a uniform canopy is defined as being homogeneous in the horizontal plane. This is then used to consider the light relations in monocultures and mixtures. The problem is obviously more complex for row crops and isolated plants owing to the irregular nature of the leaf distribution. However, we show that the Monsi–Saeki equation can also be applied in these situations. First, however, some important background aspects of light energy are considered.

8.1.1 *Units and terminology of light quantities*

Irradiance In studies of plant and crop physiology, light energy is generally required to be defined as that incident on a unit surface per unit of time. This is referred to as *irradiance*, with units

$$\mathrm{J\ m^{-2}\ s^{-1} \equiv W\ m^{-2}}, \tag{8.2a}$$

where J (joule) is the SI unit of energy and W (watt) is the unit of power. Only the visible component of the sun's energy drives the light reactions of photosynthesis. Under normal conditions the visible component comprises approximately 50 per cent of the solar spectrum, although this may vary between 40 and 60 per cent depending on such factors as the solar elevation and atmospheric composition (e.g. cloud cover) (Ludlow 1983). The visible component drives the light reactions of photosynthesis and is referred to as photosynthetically active radiation (PAR). For transpiration studies, solar radiation provides energy for the evaporation of water, but in this case it is the total spectrum and not just the PAR component. It is therefore important to identify whether the PAR component or total solar radiation is relevant to a particular problem.

Since light comprises photons, which are energy particles, irradiance is often defined in terms of the number of photons incident per unit area per unit time. This number is clearly going to be large, and so the units are mol (photons) m^{-2} s^{-1}, where a mole of photons is Avogadro's number of photons. The attraction of this unit is due to the fact that the light reactions of photosynthesis are related to excitation of electrons by incident photons, as discussed in Chapter 9. However, neither these units nor those of eqn (8.2a) give information about the different photosynthetic activity of the photons, in that a green light photon is less photosynthetically active than a blue or a red. As a consequence, there is no unique conversion factor between W m^{-2} (PAR) and mol (photons) m^{-2} s^{-1} (PAR). Ludlow (1983) quotes 4.6 μmol J^{-1}; i.e. 1 mol (photons) $\equiv 0.22 \times 10^6$ J. Other authors, fully recognizing that there is no exact conversion, have quoted slightly different values (for example, Robson and Sheehy (1981) suggest the factor 1 mol (photons) $\equiv 0.235 \times 10^6$ J). The variation in these conversion factors is unlikely to be of significance in photosynthesis models. We hold no strong preference for either set of units, but choose to work with W m^{-2} because of our reservations about the mole as an SI unit, in that it clearly relates to the gram rather than the kilogram, as discussed in Chapter 2. It would probably introduce some confusion if we used the more appropriate kilogram mole. For further discussion of the units of light energy, see Bell and Rose (1981).

Brightness function The irradiance incident on the horizontal plane depends on the brightness of the sky, which is a combination of the direct solar beam and diffuse light. The brightness function for the sky is defined in terms of spherical polar coordinates (r, θ, ϕ), although it does not involve the r coordinate. Spherical polar coordinates are illustrated in Fig. 8.1. The angle θ defines the inclination to the vertical and ϕ gives the position in the horizontal plane. Note that $0 \leqslant \theta \leqslant \pi/2$ and $0 \leqslant \phi \leqslant 2\pi$. When defining the position of the sun, θ is known as the zenith angle (when $\theta = 0$ the sun is directly overhead) and ϕ is termed the azimuth angle. (The zenith angle is sometimes measured from the horizontal rather than the vertical, although this is not consistent with the mathematical definition of spherical polar coordinates.)

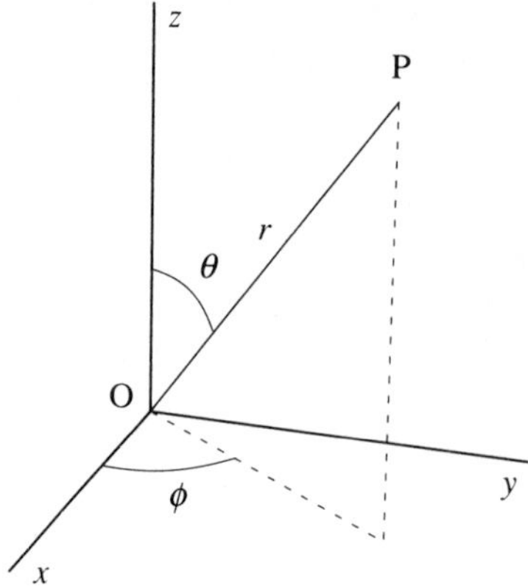

Fig. 8.1. Spherical polar coordinate system (r, θ, ϕ) in relation to the Cartesian system (x, y, z) for a point P relative to origin O. The two systems are related by $x = r \sin\theta \cos\phi$, $y = r \sin\theta \sin\phi$, $z = r\cos\theta$.

The brightness function of the sky is defined as $B(\theta, \phi)$, with units W m^{-2} srad^{-1}. The steradian (srad) is the SI unit for solid angle: a solid angle is the area of the surface of a portion of a sphere divided by the square of the radius of the sphere. Thus the total solid angle of a sphere is 4π srad.

To define the brightness function $B(\theta, \phi)$ and to relate this to the irradiance I_0, requires some knowledge of spherical polar coordinates. We present an outline of the approach here, but the reader unfamiliar with this mathematics should consult a standard text such as Simons (1970). For the spherical polar coordinates (r, θ, ϕ), the unit vector in the r direction, denoted by $\boldsymbol{r}$, can be defined in terms of unit vectors $(\boldsymbol{x}, \boldsymbol{y}, \boldsymbol{z})$ in each of the (x, y, z) directions as

$$\boldsymbol{r} = \sin\theta\cos\phi\boldsymbol{x} + \sin\theta\sin\phi\boldsymbol{y} + \cos\theta\boldsymbol{z}. \tag{8.3a}$$

The projection of the brightness function on any plane whose normal is the unit vector $\boldsymbol{n}$ is

$$B(\theta, \phi)\boldsymbol{r}\cdot\boldsymbol{n}, \tag{8.3b}$$

where $\boldsymbol{r}\cdot\boldsymbol{n}$ is the scalar product of $\boldsymbol{r}$ and $\boldsymbol{n}$. The total irradiance on the plane is

$$I_n = \int B(\theta, \phi)\boldsymbol{r}\cdot\boldsymbol{n}\,\mathrm{d}\Omega, \tag{8.3c}$$

where $\mathrm{d}\Omega$ denotes the element of solid angle and the integration is taken over the total sky. It can be shown that (e.g. Simons 1970)

$$\mathrm{d}\Omega = \sin\theta\,\mathrm{d}\theta\,\mathrm{d}\phi, \tag{8.3d}$$

and so I_n can now be evaluated. In particular, for canopies we are interested in I_0 for a horizontal sensor facing upwards, where $\boldsymbol{n} = (0, 0, 1)$, so that $\boldsymbol{r}\cdot\boldsymbol{n} = \cos\theta$, and hence

$$I_0 = \int_0^{\pi/2} \int_0^{2\pi} B(\theta, \phi) \cos\theta \sin\theta \, d\theta \, d\phi. \tag{8.3e}$$

When considering row crops or individual plants, the horizontal irradiance may also be of importance, and this is evaluated by setting $\boldsymbol{n}$ equal to $(1, 0, 0)$ or $(0, 1, 0)$.

One simple form for the brightness function is to assume that $B(\theta, \phi)$ is independent of ϕ and write

$$B(\theta) = B_0[\mu + (1 - \mu)\cos\theta], \tag{8.4a}$$

where B_0 is the brightness of the sky directly overhead ($\theta = 0$) and the brightness at the horizon is a fraction μ of B_0:

$$B(\theta = \pi/2) = \mu B_0. \tag{8.4b}$$

I_0 can then be derived from eqn (8.3e) as (Exercise 8.1)

$$I_0 = B_0 \pi \frac{\mu + 2}{3}. \tag{8.4c}$$

The two forms most commonly used for $B(\theta)$ are for a uniformly overcast sky, where $\mu = 1$, in which case

$$I_0 = B_0 \pi, \tag{8.4d}$$

and for a standard overcast sky with $\mu = \frac{1}{3}$ so that the sky is three times as bright overhead as on the horizon and

$$I_0 = \tfrac{7}{9} B_0 \pi. \tag{8.4e}$$

In practice the brightness will depend on the solar elevation, even on overcast days. It is relatively straightforward to include the direct solar beam in the brightness function, but with variable cloud cover the function will be complex. Such analysis is beyond the scope of this book.

8.2 Uniform canopies

8.2.1 *Monocultures*

Consider an increment in leaf area index within the canopy of dl m^2 leaf (m^2 ground)$^{-1}$. The leaves will, in general, be inclined to the horizontal; therefore let the projection of the leaves onto the horizontal plane be

$$\zeta \, dl. \tag{8.5a}$$

ζ is related to the leaf inclination to the horizontal ψ by

$$\zeta = \cos\psi. \tag{8.5b}$$

Now the irradiance in the horizontal plane will, in general, not be uniform but will fluctuate. If the leaf increments are randomly distributed in the horizontal plane, then the mean irradiance per unit of projected leaf area will be equal to the mean irradiance in the horizontal plane. In other words, the leaf projections can be regarded as providing a statistically accurate sample of the irradiance in the horizontal plane. Thus the irradiance I_l W (m^2 leaf)$^{-1}$ incident on the leaves within the canopy is

$$I_l = \zeta I. \tag{8.5c}$$

The intercepted irradiance for the increment dl is $I_l\,\mathrm{d}l$. This can be separated into scattered, absorbed, and transmitted components

$$n\mathrm{I}_l\,\mathrm{d}l, \quad (1 - n - m)I_l\,\mathrm{d}l, \quad mI_l\,\mathrm{d}l, \tag{8.5d}$$

where n and m denote the scattered and transmitted fractions respectively. n and m can vary considerably depending on such factors as leaf thickness, water status, and the waxiness of the leaf surface: for PAR they are typically of order 0.1, but for the total spectrum they are generally of order 0.2–0.3. If we assume that the scattered irradiance moves upwards and the transmitted irradiance moves downwards, the irradiance lost from I is the sum of the scattered and absorbed components, so that

$$\mathrm{d}I = -(1 - m)I_l\,\mathrm{d}l. \tag{8.5e}$$

Using eqn (8.5c), this is

$$I = -I(1 - m)\zeta\,\mathrm{d}l. \tag{8.5f}$$

In order to integrate this equation through the depth of canopy, it is necessary to have some information regarding the variation of ζ within the canopy. Here we make the assumption that ζ is constant, although the analysis is possible with ζ depending on l (Exercise 8.2). While this is quite clearly a fairly drastic assumption, the resulting equation has been shown to give a good description of light attenuation in many swards. With ζ constant, therefore, eqn (8.5f) is readily integrated to give

$$I = I_0 \exp[-(1 - m)\zeta l], \tag{8.5g}$$

which is equivalent to eqn (8.1a) with $k = (1 - m)\zeta$ (the reason for these different symbols will become clear).

Now consider non-random leaf distributions in the horizontal plane. Equation (8.5c) will no longer apply, but must be modified to

$$I_l = \xi\zeta I. \tag{8.6a}$$

If the leaves are clumped, so that there is a tendency for leaves to overlap one another, then less light will be intercepted than in the random distribution, so that

$$\xi < 1. \tag{8.6b}$$

Similarly, if the leaves are regularly placed in such a manner as to intercept the brighter portions of the light—as is the case with spiral phyllotaxis patterns which are discussed in Chapter 19—then

$$\xi > 1. \tag{8.6c}$$

Equation (8.5f) now becomes

$$\mathrm{d}I = -(1 - m)\xi\zeta I \,\mathrm{d}l. \tag{8.6d}$$

If we now adopt the assumption than ξ is constant through the depth of the canopy this can be integrated to give

$$I = I_0 \exp[-(1 - m)\xi\zeta l] \tag{8.6e}$$

which is equivalent to eqn (8.1a) with

$$k = (1 - m)\xi\zeta. \tag{8.6f}$$

This provides a rather simplified derivation of eqn (8.1a). Given the number of assumptions made, it is perhaps better to regard eqn (8.1a) as a semi-empirical description of light attenuation by a canopy, where the extinction coefficient k accounts for leaf inclination, leaf transmission, and also whether the leaf distribution is random, regular, or clumped. We can conclude that, for canopies with randomly distributed or clumped leaves, k is constrained by

$$0 < k < 1, \tag{8.6g}$$

but for regularly spaced leaves the lesser constraint

$$k > 0 \tag{8.6h}$$

applies.

The irradiance incident on the leaf surface is, combining eqns (8.6a) and (8.6f),

$$I_1 = \frac{k}{1 - m} I_0 \mathrm{e}^{-kl} = \frac{k}{1 - m} I. \tag{8.6i}$$

The importance of defining the units of k can be seen from this expression: I has units of W (m^2 ground)$^{-1}$ which, multiplied by k with units of m^2 leaf (m^2 ground)$^{-1}$, gives the appropriate units for I_1 as W (m^2 leaf)$^{-1}$.

The irradiance absorbed by the leaves is (analogous to eqn (8.5d))

$$I_a = (1 - n - m)I_1, \tag{8.6j}$$

in units of W (m^2 leaf)$^{-1}$, where the physiologically obvious constraint

$$1 - n - m > 0 \tag{8.6k}$$

applies.

I_1 and I_a can be integrated over the leaf area index L to give the total irradiance incident on, or absorbed by, the leaves within the canopy. First, integrating I_1 over the total leaf area index L, we obtain the total irradiance incident on the leaves within the canopy as

$$J_l = \frac{I_0}{1-m}\left(1 - e^{-kL}\right). \tag{8.7a}$$

The irradiance intercepted by the canopy as a whole can be derived directly from eqn (8.1a) by taking the difference between the irradiance at the top and the bottom of the canopy as

$$J_c = I_0(1 - e^{-kL}). \tag{8.7b}$$

This differs from eqn (8.7a) in that with $m \neq 0$ radiation can be intercepted, transmitted, intercepted again, and so on. However, if $m = 0$ the expressions are the same. The important distinction is that eqn (8.7a) relates to the irradiance incident on individual leaves within the canopy whereas (8.7b) is concerned with the irradiance intercepted by the canopy as a whole. The total irradiance J_a absorbed by the canopy is obtained by integrating I_a over the leaf area index L, giving

$$J_a = \left(\frac{1-n-m}{1-m}\right) I_0(1 - e^{-kL}). \tag{8.7c}$$

This expression involves both the reflection and transmission coefficients. The expression for J_a is of direct physiological relevance as it defines the solar energy absorbed by the canopy which can be used in determining the canopy transpiration rate (see Chapter 14). However, J_l and J_c are of no particular significance; they have just been considered here to indicate the distinction between the various components.

We can also calculate the reflected and transmitted components of irradiance. The irradiance reflected by the canopy as a whole is given by

$$J_r = n\int_0^L I_l \, dl = \left(\frac{n}{1-m}\right) I_0(1 - e^{-kL}), \tag{8.7d}$$

where it should be noted that in deriving this expression it is assumed that all reflected light is reflected away from the canopy and is not subsequently intercepted; also, no account is taken of any irradiance which may be reflected by the ground. The transmitted irradiance cannot be obtained by integrating the factor mI_l through the canopy since this would include irradiance which was transmitted, intercepted, and transmitted again and so on (although this integral does have a physical interpretation (see Exercise 8.3)). This would lead to the same type of problem associated with eqns (8.7a) and (8.7b). The transmitted irradiance is simply the level of irradiance at the bottom of the canopy:

$$J_t = I_0 e^{-kL}, \tag{8.7e}$$

and it can be seen that eqns (8.7c)–(8.7e) sum to give

$$J_a + J_r + J_t = I_0 \tag{8.7f}$$

as required.

A final point to note is that at the top of the canopy, where $l \to 0$, eqn (8.6i) is

$$I_1 = \frac{k}{1-m} I_0, \tag{8.7g}$$

which, for $k/(1-m) > 1$, gives the physically impossible result that the irradiance incident on the leaf surface at the top of the canopy is greater than the unobstructed irradiance. This is due to the assumption that the parameter ξ in eqn (8.6a) is constant through the depth of the canopy. Clearly, at the top of canopy $\xi = 1$ for all leaf distributions and is not greater than unity. Extinction coefficients are seldom observed to be greater than unity, and so this problem is unlikely to occur in practice.

8.2.2 *Mixtures*

Now consider a binary mixture, and assume that the leaves of the components are homogeneously distributed in the horizontal plane. This means that there is no clumping of one species relative to the other, but does not preclude different vertical leaf distributions where one species may dominate the upper region of the sward. The following equations have been derived by Ross, Henzell, and Ross (1972).

Let subscript i denote either component of the sward, and consider an increment in leaf area index comprising $\mathrm{d}l_1$ and $\mathrm{d}l_2$ from each component. Analogous to the theory for monocultures, the irradiance incident on the leaves of each component is

$$I_{1,i} = \xi_i \zeta_i I. \tag{8.8a}$$

Again assuming that the scattered irradiance moves upwards and the transmitted irradiance moves downwards, the irradiance lost from I by being intercepted by $\mathrm{d}l_i$ is

$$(1 - m_i) I_{1,i}\, \mathrm{d}l_i = k_i I\, \mathrm{d}l_i, \tag{8.8b}$$

where the definition of k_i is analogous to eqn (8.6f). Hence

$$\frac{\partial I}{\partial l_i} = -k_i I. \tag{8.8c}$$

Now, if z denotes depth within the sward, then from the calculus the full derivative of I with respect to z is

$$\frac{\mathrm{d}I}{\mathrm{d}z} = \frac{\partial I}{\partial l_1}\frac{\mathrm{d}l_1}{\mathrm{d}z} + \frac{\partial I}{\partial l_2}\frac{\mathrm{d}l_2}{\mathrm{d}z}, \tag{8.8d}$$

which, combined with eqn (8.8c), gives

$$\frac{\mathrm{d}I}{\mathrm{d}z} = -I\left(k_1 \frac{\mathrm{d}l_1}{\mathrm{d}z} + k_2 \frac{\mathrm{d}l_2}{\mathrm{d}z}\right), \tag{8.8e}$$

and this can be integrated to obtain

$$I = I_0 \exp[-(k_1 l_1 + k_2 l_2)]. \tag{8.8f}$$

This can be used directly in eqn (8.8a) for the irradiance on the leaf surface:

$$I_{1,i} = \frac{k_i}{1 - m_i} I_0 \exp[-k_1 l_1 + k_2 l_2)] = \frac{k_i}{1 - m_i} I. \tag{8.8g}$$

This equation is analogous to eqn (8.6i) for monocultures. It incorporates the individual extinction coefficient k_i which relates to the local leaf distribution as well as the factor $\exp[-(k_1 l_1 + k_2 l_2)]$ which defines the light attenuation at that point in the canopy. It is clear, therefore, that I_1 depends on the relative distributions of each component through the canopy.

Now consider the total irradiance absorbed by the canopy, which is given by

$$J_a = J_{a,1} + J_{a,2}, \tag{8.9a}$$

where

$$J_{a,i} = (1 - n_i - m_i) \int_0^{L_i} I_{1,i} \, dl_i, \tag{8.9b}$$

and

$$L = L_1 + L_2, \tag{8.9c}$$

defines the total leaf area index of the canopy. It is convenient to write the integral (8.9b) as

$$J_{a,i} = (1 - n_i - m_i) \int_0^{L} I_{1,i} \frac{dl_i}{dl} \, dl, \tag{8.9d}$$

where the cumulative leaf area index of the canopy is

$$l = l_1 + l_2. \tag{8.9e}$$

This integral can only be evaluated if the relative leaf distribution through the depth of the canopy of the components is known. Most reports in the literature are for grass-legume mixtures. For experimental swards the legume is generally observed to dominate the upper region of the sward, as reported, for example, by Stern and Donald (1962) for subterranean clover and Wimmera ryegrass, and by Davidson, Robson, and Denis (1982) for white clover and perennial ryegrass. However, for mixed grass–clover swards which have adapted to either continuous grazing or frequent cutting, the relative leaf distribution is approximately homogeneous; this has been observed for continuously grazed swards (perennial ryegrass and white clover) by Johnson, Parsons, and Ludlow (1989), and for frequently cut swards (*Setaria*, a tropical C_4 grass, and *Desmodium*, a tropical legume) by Wilson and Ludlow (1983). It follows, therefore, that a realistic and useful special case to consider is where the relative vertical leaf distribution is homogeneous, i.e.

$$l_i = \frac{L_i}{L} l. \tag{8.9f}$$

Substituting this in eqn (8.8f) for I gives

$$I = I_0 \exp(-k_e l), \tag{8.9g}$$

where

$$k_e = \frac{k_1 l_1 + k_2 l_2}{l} = \frac{k_1 L_1 + k_2 L_2}{L} \tag{8.9h}$$

is defined as the *effective* extinction coefficient for the canopy. The irradiance incident on the leaves within the canopy now becomes

$$I_{1,i} = \frac{k_i}{1 - m_i} I_0 \exp(-k_e l), \tag{8.9i}$$

and the irradiance absorbed by each component is

$$\begin{aligned} J_{a,i} &= k_i I_0 \frac{L_i}{L}\left(\frac{1 - n_i - m_i}{1 - m_i}\right) \int_0^L \exp(-k_e l)\,dl \\ &= I_0 \left(\frac{1 - n_i - m_i}{1 - m_i}\right) \frac{k_i L_i}{k_e L}[1 - \exp(-k_e L)], \end{aligned} \tag{8.9j}$$

so that the total irradiance absorbed by the canopy is

$$J_a = I_0 \left[\frac{1 - \exp(-k_e L)}{k_e L}\right]\left[\left(\frac{1 - n_1 - m_1}{1 - m_1}\right) k_1 L_1 + \left(\frac{1 - n_2 - m_2}{1 - m_2}\right) k_2 L_2\right]. \tag{8.9k}$$

The reflected irradiance can be derived, analogous to the monoculture, as

$$J_r = I_0 \left[\frac{1 - \exp(-k_e L)}{k_e L}\right]\left[\left(\frac{n_1}{1 - m_1}\right) k_1 L_1 + \left(\frac{n_2}{1 - m_2}\right) k_2 L_2\right], \tag{8.9l}$$

and the transmitted irradiance is

$$J_t = I_0[1 - \exp(-k_e L)]. \tag{8.9m}$$

Again, it can be seen that the necessary physical constraint

$$J_a + J_r + J_t = I_0, \tag{8.9n}$$

is true.

8.3 Row crops

The problem of describing light interception and attenuation by a row crop is more difficult than that for a canopy because of the irregularity of the crop. Most of the work in this area has been for tree stands, although many annual crops such as sunflower also fall into this category. Models in this area are often extremely complex and, while they may help in understanding the underlying principles of light interception by row crops, they often require detailed information about the geometry and spatial distribution of leaves, as well as details about solar angle and cloud cover. This information is not often available for crop modelling, and it is important to develop simpler models which can be used in crop growth models. The following treatment is based on the work of Jackson and Palmer (1979); in part it was developed in joint work with David Hand and John Warren Wilson, to whom we are indebted.

Consider a discontinuous canopy and let the light transmission to the ground be made up of two components. The first is light incident upon the ground which would have been transmitted even if the canopy were non-transmitting so that no light passed through the canopy. The second is that light which has passed through the canopy. Denoting these components by T_b, and T_c respectively, the total light T transmitted by the canopy is

$$T = T_b + T_c. \tag{8.10a}$$

Note that the irradiance level in the gaps for a non-transmitting canopy will be T_b, as defined, but this will not be the case for a transmitting canopy because some of the transmitted light will fall in the gaps. The total irradiance J_c (W (m^2 ground)$^{-1}$) intercepted by the canopy is, by definition,

$$J_c = I_0(1 - T) = I_0[1 - (T_b + T_c)], \tag{8.10b}$$

where I_0 (W (m^2 ground)$^{-1}$) is, as before, the irradiance above the canopy. The problem we are faced with is to define T_b and T_c.

First consider T_b, and let the height of the crop be h (m) and the gap between the rows g (m). The height is measured in the z direction, the row lies along the y direction, and the gap is measured in the x direction, as illustrated in Fig. 8.2. To calculate the irradiance on the ground in the gap, the brightness function is

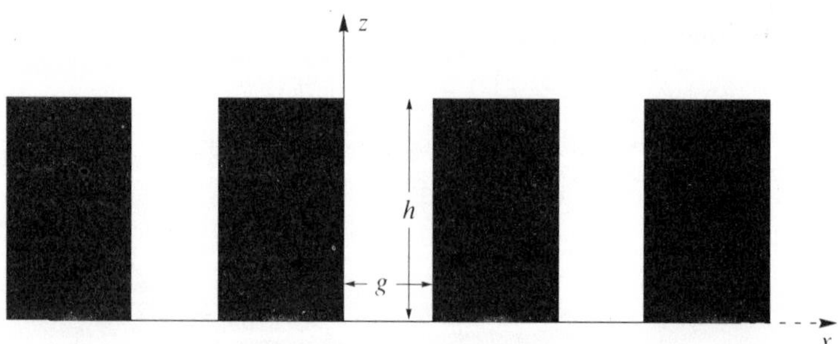

Fig. 8.2. Coordinate system for the row crop model. z denotes height, and the gaps between the rows lie in the x direction with the rows extending in the y direction. The height of the rows is h, and the gap between them is g. Four rows are shown.

integrated over the appropriate portion of the sky (cf. Section 8.1.1). We make the simplifying assumption that the height of the crop is considerably greater than the gap between the rows, so that the illuminating area can be taken as being restricted to that between the rows at the top of the canopy. The area element of the surface dS is a rectangle with sides g and $(h/\cos\theta)\,d\theta$, so that

$$dS = g\frac{h}{\cos\theta}\,d\theta. \tag{8.11a}$$

The solid angle subtended by this element of area is, by definition,

$$d\Omega = \frac{dS}{(h/\cos\theta)^2} = \frac{g}{h}\cos\theta\,d\theta. \tag{8.11b}$$

I_b is given by eqn (8.3c), with $\boldsymbol{r}\cdot\boldsymbol{n} = \cos\theta$, so that

$$I_b = \int B\cos\theta\,d\Omega \tag{8.11c}$$

which, using eqn (8.11b) for $d\Omega$, becomes

$$I_b = \frac{g}{h}\int_{-\pi/2}^{\pi/2} B\cos^2\theta\,d\theta. \tag{8.11d}$$

This defines I_b, and T_b is then given by

$$T_b = I_b/I_0. \tag{8.11e}$$

For example, if eqn (8.4a) is used for B, this becomes

$$T_b = \frac{g}{h}\frac{3}{\pi(\mu+2)}\int_{-\pi/2}^{\pi/2}[\mu + (1-\mu)\cos\theta]\cos^2\theta\,d\theta, \tag{8.11f}$$

which is (Exercise 8.4)

$$T_b = \frac{g}{h}\frac{3}{\pi(\mu+2)}\left[\frac{4}{3} + \mu\left(\frac{\pi}{2} - \frac{4}{3}\right)\right]. \tag{8.11g}$$

For a uniformly overcast sky ($\mu = 1$) $T_b = 0.5g/h$, and for a standard overcast sky ($\mu = 1/3$) $T_b = 0.58g/h$. Even in an extreme case, say $\mu = 1/10$, $T_b = 0.62g/h$. As a useful approximation, it may be reasonable to assume

$$0.5g/h < T_b < 0.6g/h \tag{8.11h}$$

and to assign a particular value depending on some fairly simple estimate of sky conditions. For example, on a clear bright day the value $0.6g/h$ might be used, whereas for a cloudy overcast day $0.5g/h$ might be more appropriate.

Now consider the transmission of the component of light T_c which is intercepted by the canopy. For a non-transmitting canopy, the total transmission is T_b, so that the remainder, $1 - T_b$, is intercepted by the canopy. Thus, we can write

$$T_c = (1 - T_b) \times \text{light attenuation by canopy}. \tag{8.12a}$$

Owing to the general inhomogeneity of row crops, the light attenuation is difficult to describe mathematically. However, Jackson and Palmer (1979) adapted the theory for uniform canopies (discussed in the previous section) in a simple but effective way. To do this, define the total leaf area index L as the total leaf area per total ground area, including row spaces. The effective leaf area index is defined, analogous to T_c, as

$$L' = \frac{L}{1 - T_b}, \tag{8.12b}$$

and the light attenuation is taken to be $\exp(-kL')$, so that

$$T_c = (1 - T_b)\exp(-kL'). \tag{8.12c}$$

k is the extinction coefficient within the canopy; for example, Jackson and Palmer (1979) estimated k to be 0.6 for apple trees.

Equations (8.12b), (8.12c), and (8.10b) define the light transmission though the canopy, and therefore the total irradiance intercepted. However, in calculations of photosynthesis, it is necessary to know the irradiance incident on the leaves within the canopy. The simplest approach is to extend the treatment for uniform canopies to row crops. Define the irradiance incident on the leaf surface at cumulative leaf area index l as

$$I_l = \frac{k}{1 - m} I_0 \exp(-kl'), \tag{8.13a}$$

where l' is defined (equivalent to L') as

$$l' = \frac{l}{1 - T_b}. \tag{8.13b}$$

The total irradiance absorbed by the canopy is

$$J_a = (1 - n - m)\int_0^L I_l \, dl, \tag{8.13c}$$

which, using eqns (8.12b), (8.13a), and (8.13b), is

$$J_a = I_0\left(\frac{1 - n - m}{1 - m}\right)(1 - T_b)[1 - \exp(-kL')]. \tag{8.13d}$$

Similarly, the reflected irradiance is

$$J_r = I_0\left(\frac{n}{1 - m}\right)(1 - T_b)[1 - \exp(-kL')], \tag{8.13e}$$

and the transmitted irradiance is

$$J_t = I_0[T_b + (1 - T_b)\exp(-kL')], \tag{8.13f}$$

where the first term is the component of irradiance which is not intercepted by the canopy and the second is the intercepted and transmitted component. Again it can be seen that $J_a + J_r + J_t = I_0$ as required for physical realism.

Finally, note that the analysis is consistent with that for uniform canopies in that as $g \to 0$, $T_b \to 0$ and $L' \to L$, so that the rows merge into a uniform canopy with the equations for light interception and attenuation being the same as those derived in the previous section.

8.4 Isolated plants

The isolated plant presents more difficulty than uniform canopies or row crops. The reason for this is that when considering canopies we were able to reduce the dimensions of the problem to a single coordinate, and for row crops it was still possible to develop this analysis in a fairly straightforward manner. However, no equivalent treatment has yet been developed for the isolated plant although Jackson and Palmer's (1979) method is applicable.

A rigorous description of the problem involves setting up a system of integro-differential equations to describe the radiation field at all points in space. A simpler semi-empirical approach, which can be used for plants of known structure, has been described by Charles-Edwards and Thornley (1973).

In Fig. 8.3 an example is shown for a plant whose structure is a triangle of revolution; other shapes could be used, such as an ellipsoid. Let dI_P be the irradiance per unit horizontal area at a point P within the plant, and assume that the Monsi–Saeki equation can be used to describe the attenuation along the pathlength s which the radiation passes to reach P and that the leaf area density is constant and equal to F (F is the leaf area per unit volume of space).

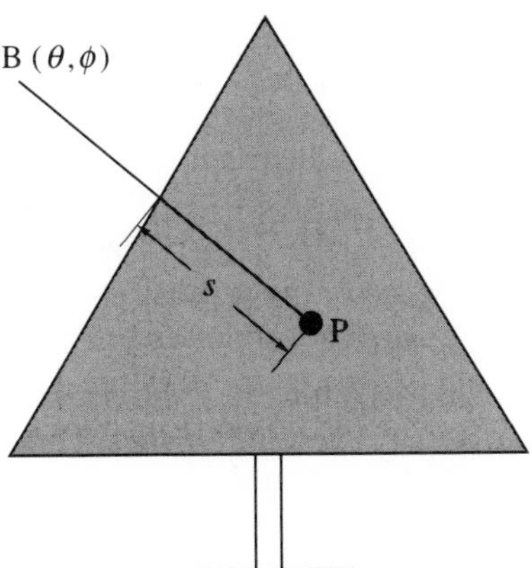

Fig. 8.3. Light interception by a single plant. The shaded area indicates the plant. Radiation from an element of sky, defined by the angles (θ, ϕ) and with brightness function $B(\theta, \phi)$ traverses a path of length s in order to reach the point P. Contributions from the whole sky (all θ and ϕ above the horizon) are summed to give the total irradiance reaching P.

Under these assumptions it follows that

$$dI_P = B(\theta, \phi)e^{-kFs}\cos\theta\sin\theta\,d\theta\,d\phi, \tag{8.14a}$$

which is equivalent to eqn (8.3e) with the light attenuation described by the factor e^{-kFs}, where Fs is the effective leaf area index traversed over the pathlength s and k is the extinction coefficient which is assumed constant. The pathlength s varies with the angles θ and ϕ. I_P is given by the integral

$$I_P = \int_0^{\pi/2}\int_0^{2\pi} B(\theta, \phi)e^{-kFs}\cos\theta\sin\theta\,d\theta\,d\phi. \tag{8.14b}$$

In practice, this integral must be evaluated numerically, even for simple plant shapes such as the triangular one shown in Fig. 8.3. However, computer programs can be written to calculate I_P (Charles-Edwards and Thorpe 1976, Palmer 1977, Whitfield 1980, Whitfield and Connor 1980).

Equation (8.14b) defines the irradiance incident on the horizontal plane within the plant. The irradiance I_l (W (m^2 leaf)$^{-1}$) on the leaf is defined by eqn (8.6i), and the total light intercepted by the plant is obtained by integrating I_l over the total leaf area index of the plant.

Exercises

8.1. Derive eqn (8.4c) for I_0, when the brightness function B is given by eqn (8.4a).

8.2. Derive the expression for I, corresponding to eqn (8.5g), for ζ given by

$$\zeta = (\zeta_m - \zeta_0)\frac{l}{l + K} + \zeta_0, \tag{E8.2a}$$

so that $\zeta(l = 0) = \zeta_0$ and $\zeta \to \zeta_m$ for large l. Show that the resulting expression for I_0 is consistent with eqn (8.5e) when $K \to 0$ (in which case $\zeta \to \zeta_m$ = constant).

8.3. The irradiance transmitted by a leaf within a monoculture canopy is mI_l. Integrate this over the total leaf area index L of the canopy. Consider the case $L \to \infty$ and give an interpretation of the result. In doing so, use the binomial expansion

$$(1 - m)^{-1} = (1 + m + m^2 + m^3 + \cdots). \tag{E8.3a}$$

8.4. Derive eqn (8.11g) for T_b from eqn (8.11f).

9
Leaf photosynthesis

9.1 Introduction

Photosynthesis is the primary process in crop growth and production, providing both energy and carbon for plant and crop processes. Consequently, one of the most important problems facing the plant scientist is the accurate prediction of photosynthesis in relation to the environment and the state of the crop.

Three of the various levels at which photosynthesis can be studied are the cell level, the leaf level, and the crop level. Models at these levels can be linked, since the summation of photosynthesis of individual cells leads to leaf photosynthesis, and the leaf photosynthetic contributions can in turn be summed to give crop photosynthesis. For crop growth models and many agronomic studies, an accurate quantitative description of crop photosynthesis is required. Models of crop photosynthesis are discussed in the next chapter. However, such models are usually based on some representation of the photosynthetic characteristics of the leaves, combined with the light environment which is determined by the light-intercepting properties of the canopy, as discussed in the last chapter. This chapter is concerned with various approaches to modelling leaf photosynthesis.

First, some biochemical and cellular models of leaf photosynthesis are discussed. Some of these use 'simplified' biochemistry, where only the main biochemical features are represented; others are much more detailed. In our view, none of the models of leaf photosynthesis which attempt to represent known biochemistry with a high degree of realism lead naturally to functions for leaf photosynthetic response suitable for connecting the leaf and crop levels of description. While there is undoubtedly much to be gained from complex models which represent detailed biochemistry, it is questionable whether there is value in applying simplifications that strain credibility to such models in order to obtain expressions for the leaf-crop modelling problem. Since the theme of this book is not specifically in detailed biochemistry, our account of this area will be limited, but gives an overview so that the simpler models of Section 9.3 (one of which leads to the valuable non-rectangular hyperbola photosynthetic response function) can be placed in context.

9.1.1 *Photosynthesis summarized*

The overall result of photosynthesis can be summarized by the reaction

$$CO_2 + H_2O \xrightarrow{\text{light}} \{CH_2O\} + O_2. \tag{9.1a}$$

$\{CH_2O\}$ denotes a carbohydrate unit; glucose ($C_6H_{12}O_6$) contains six carbohydrate units.

We consider two important groups of plants, known as C_3 and C_4, depending upon the physiology and biochemistry of photosynthesis. (An introductory account of photosynthesis is given by Jones (1983, Chapter 7), and more detail can be found in Edwards and Walker (1983).) All plants can carry out the process often referred to as 'dark' respiration (it occurs in the dark as well as in the light) which uses the biochemical processes of glycolysis, the citric acid cycle, and oxidative phosphorylation (see Chapters 11 and 12). This can be summarized by

$$\{CH_2O\} + O_2 \rightarrow CO_2 + H_2O. \tag{9.1b}$$

In the C_3 group of plants a light-stimulated form of respiration, known as photorespiration, can also take place in the light, and we can write this as

$$\{CH_2O\} + O_2 \xrightarrow{\text{light}} CO_2 + H_2O. \tag{9.1c}$$

In C_3 plants, photorespiration as in (9.1c) competes with photosynthesis as in (9.1a) to reduce the efficiency of carbon fixation. In C_4 plants, the employment of a specialized anatomy known as 'Kranz' or 'wreath' anatomy, together with a spatial separation of an initial carboxylation (phosphoenolpyruvate is the CO_2 acceptor) from a decarboxylation of a C_4 acid and an immediate recarboxylation (ribulose bisphosphate is the CO_2 acceptor as in C_3 plants), effectively prevents photorespiration (9.1c) from competing with photosynthesis (9.1a) (see Section 9.2.4 and Fig. 9.4).

Photosynthesis can be viewed as comprising several linked processes: (a) the supply of CO_2 from air to the carboxylation sites and the removal of O_2 as required by (9.1a)–(9.1c); (b) the absorption of light energy and the generation of chemical energy (ATP) and reducing power (which we write as NADPH); (c) the harnessing of this ATP and NADPH to drive the biochemical reactions and transport process associated with the reduction of CO_2 to sugars.

Note that NADPH is involved in the transfer of 2H, according to

$$NADP^+ + 2H \rightleftharpoons NADPH + H^+. \tag{9.1d}$$

Reactions such as (9.2a) are stoichiometric in H when (9.1d) is applied.

In the next section we discuss some biochemically based models of photosynthesis in order to clarify some of these problems from the modeling viewpoint. However, none of the models permits the derivation of analytical solutions, and space does not allow us to present any of the models in complete detail.

9.2 Biochemical models of leaf photosynthesis

9.2.1 *A simplifed scheme for the biochemistry of photosynthesis and photorespiration in C_3 plants*

Figure 9.1, the diagram depicting this scheme, shows that ribulose bisphosphate can be either carboxylated (accepting CO_2 and H_2O) or oxidized (accepting O_2);

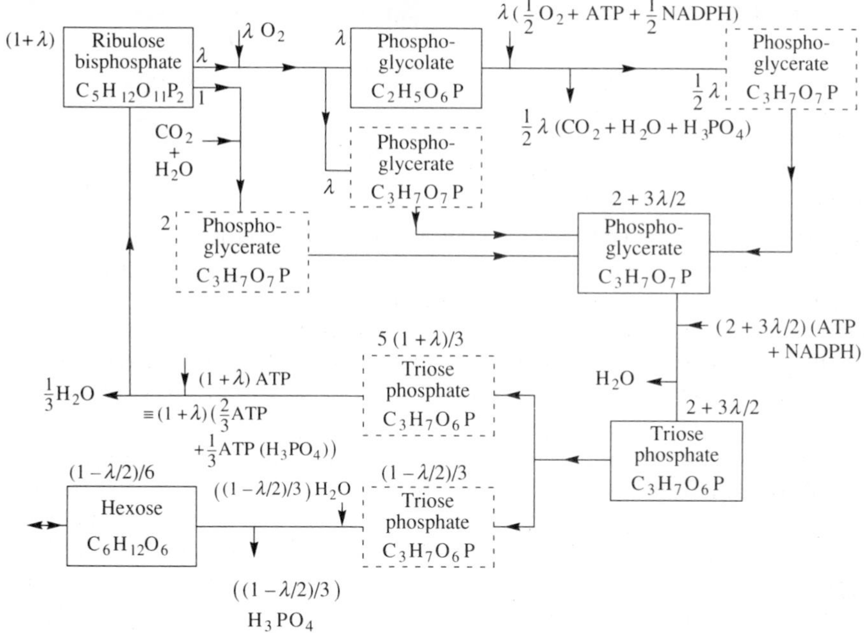

Fig. 9.1. A simplified scheme for the biochemistry of photosynthesis and photorespiration in C_3 plants. The five boxes enclosed by continuous lines are state variables of the system; the boxes enclosed by broken lines are drawn for clarity. The stoichiometry is complete except where ATP and NADPH are involved; NADPH transfers 2H. The flows shown are appropriate to a steady state in which λ molecules of ribulose bisphosphate are oxygenated for each molecule of ribulose bisphosphate that is carboxylated.

these are assumed to occur at a relative rate of $1:\lambda$. The subsequent series of reactions regenerates ribulose bisphosphate; the partitioning of triose phosphate between ribulose bisphosphate resynthesis and hexose production/utilization is such as just to regenerate the ribulose bisphosphate consumed in the initial carboxylation/oxidation. Since one-sixth of a hexose is equivalent to one carbohydrate unit $\{CH_2O\}$, the net result of carboxylation can be found by adding up the terms on the relevant pathway (the terms not involving λ) to give

$$CO_2 + 3ATP + 2NADPH \rightarrow \{CH_2O\} + H_2O. \quad (9.2a)$$

If a molecule of ribulose bisphosphate is oxidized by adding O_2, then adding up the terms involving λ gives

$$\tfrac{1}{2}\{CH_2O\} + \tfrac{3}{2}O_2 + \tfrac{7}{2}ATP + 2NADPH \rightarrow \tfrac{1}{2}CO_2 + \tfrac{5}{2}H_2O. \quad (9.2b)$$

For each λ O_2 reacting with ribulose bisphophate, the $\{CH_2O\}$ production by (9.2a) is reduced by λ/2. 3λ/2 O_2 are consumed in the scheme shown in Fig. 9.1. (λ/2 O_2 is involved at the phosphoglycolate to phosphoglycerate step; note that one O_2 is generated in the production of two NADPH so that the overall

$CO_2:O_2$ ratio is unity.) Adding (9.2b) multiplied by λ to (9.2a) gives

$$\left(1-\frac{\lambda}{2}\right)CO_2+\frac{3\lambda}{2}O_2+\left(3+\frac{7\lambda}{2}\right)ATP+(2+2\lambda)\,NADPH$$
$$\rightarrow\left(1-\frac{\lambda}{2}\right)\{CH_2O\}+\left(1+\frac{5\lambda}{2}\right)H_2O. \tag{9.2c}$$

The parameter λ, which denotes the relative rate of oxygenation to carboxylation of ribulose bisphosphate, is strongly influenced by the relative concentrations of O_2 and CO_2 in the chloroplast and, as already mentioned, C_4 plants can increase the local concentration of CO_2 relative to that of O_2 and thus keep λ small. Clearly, the larger is λ, the less CO_2 is converted to $\{CH_2O\}$ for a given amount of ATP and NADPH.

Assuming $\lambda = 0$, so that is no photorespiration and (9.2a) applies, we can estimate the efficiency of the dark reactions of photosynthesis as follows. The heat of combusion per $\{CH_2O\}$ is

$$\{CH_2O\}+O_2\rightarrow CO_2+H_2O,\quad \Delta H(\{CH_2O\})=467\ \text{MJ (kg mol)}^{-1}, \tag{9.2d}$$

where we have taken one-sixth of the value for glucose (Table 12.1, p. 290). The following free-energy changes are assumed:

$$ATP\rightarrow ADP,\quad \Delta F=-45\ \text{MJ (kg mol)}^{-1} \tag{9.2e}$$

$$NADPH+\tfrac{1}{2}O_2\rightarrow H_2O,\quad \Delta F=-218\ \text{MJ (kg mol)}^{-1}. \tag{9.2f}$$

Thus, 3ATP + 2NADPH as in (9.2a) delivers

$$3\times45+2\times218=571\ \text{MJ (kg mol)}^{-1}. \tag{9.2g}$$

The stored energy is 467 MJ $(\text{kg mol})^{-1}$ (9.2d), leading to

$$\text{efficiency (dark reaction, no photorespiration)}=467/571=0.82. \tag{9.2h}$$

We now indicate how the scheme of Fig. 9.1 can be used to construct a simple model of the dark reactions of photosynthesis and photorespiration in C_3 plants. Using square brackets to denote concentrations, the subscript i to denote internal (of concentrations of CO_2 and O_2, rather than in air), RuBP for ribulose bisphosphate, and an obvious notation for the rate parameters, and assuming simple mass action kinetics (rather than enzyme kinetic expressions such as the Michaelis–Menten equation), we obtain the rates of the reactions in Fig. 9.1 as follows:

$$k_{CO_2}[\text{RuBP}][CO_2]_i \tag{9.3a}$$

$$k_{O_2}[\text{RuBP}][O_2]_i \tag{9.3b}$$

$$k_{\text{glycol}}[\text{P-glycolate}][O_2]_i[\text{ATP}][\text{NADPH}] \tag{9.3c}$$

$$k_{\text{glycer}}[\text{P-glycerate}][\text{ATP}][\text{NADPH}] \tag{9.3d}$$

$$k_{\text{triose-P, RuBP}}[\text{triose-P}][\text{ATP}] \tag{9.3e}$$

$$k_{\text{triose-P, hexose}}[\text{triose-P}] \tag{9.3f}$$

$$k_{\text{hexose, transport}}([\text{hexose}] - \text{constant}). \tag{9.3g}$$

Using the stoichiometry shown in Fig. 9.1, we can construct the following differential equations:

$$\frac{d[\text{RuBP}]}{dt} = -[\text{RuBP}](k_{CO_2}[CO_2]_i + k_{O_2}[O_2]_i + \tfrac{3}{5}k_{\text{triose-P, RuBP}}[\text{triose-P}][\text{ATP}] \tag{9.3h}$$

$$\frac{d[\text{P-glycolate}]}{dt} = k_{O_2}[\text{RuBP}][O_2]_i - k_{\text{glycol}}[\text{P-glycolate}] \times [O_2]_i[\text{ATP}][\text{NADPH}] \tag{9.3i}$$

$$\frac{d[\text{P-glycerate}]}{dt} = 2k_{CO_2}[\text{RuBP}][CO_2]_i + \tfrac{1}{2}k_{\text{glycol}}[\text{P-glycolate}][O_2]_i[\text{ATP}][\text{NADPH}] + k_{O_2}[\text{RuBP}][O_2] \tag{9.3j}$$

$$\frac{d[\text{triose-P}]}{dt} = k_{\text{glycer}}[\text{P-glycerate}][\text{ATP}][\text{NADPH}] - k_{\text{triose-P, RuBP}}[\text{triose-P}][\text{ATP}]; \tag{9.3k}$$

$$\frac{d[\text{hexose}]}{dt} = \tfrac{1}{2}k_{\text{triose-P, hexose}}[\text{triose-P}] - k_{\text{hexose, transport}}([\text{hexose}] - \text{constant}). \tag{9.3l}$$

To specify the model so that the equations can be solved requires some further assumptions about (i) $[CO_2]_i$ and $[O_2]_i$ (e.g. the fluxes of CO_2 and O_2 are determined by diffusion from and to the air; see eqn (9.12l)) and (ii) regeneration of ATP and NADPH by the light reactions (e.g. it can be assumed that either or both of ATP and NADPH are important, and an equation such as eqn (9.7a) ($X + I_1 \rightarrow X^*$ where I_1 is the leaf irradiance) can be used to regenerate ATP and/or NADPH from ADP and/or NADP; a possible auxiliary assumption is that ATP + ADP = constant and NADPH + NADP = constant). The differential equations above can then be supplemented by equations for the rates of change of $[CO_2]_i$, $[O_2]_i$, [ATP], and [NADPH] (Exercise 9.1).

The gross photosynthetic rate P_g is given by

$$P_g = k_{CO_2}[\text{RuBP}][CO_2]_i, \tag{9.3m}$$

and the rate of photorespiration R_l by

$$R_l = \tfrac{1}{2}k_{glycol}[\text{P-glycolate}][O_2]_i[\text{ATP}][\text{NADPH}]. \tag{9.3n}$$

The model can then be solved numerically to give, for instance, the dependence of the rates of gross photosynthesis, photorespiration, and sugar export (the last term of eqn (9.3l)) on the leaf irradiance.

9.2.2 *Hahn's model for the biochemistry of photosynthesis and photorespiration in C_3 plant*

This model, depicted in Fig. 9.2, is at the level of detailed biochemistry, and Hahn's (1987) paper is an excellent example of how a complex model can be described with clarity. His model has 33 state variables, with a differential equation for each one. Perhaps his weakest assumption concerns the use of mass action kinetics to give the reaction rates. ATP alone drives the reactions. In investigating the steady state, he makes use of a number of conservation rules: (i) adenosine in ATP and ADP; (ii) uridine in UTP and UDP; (iii) phosphate

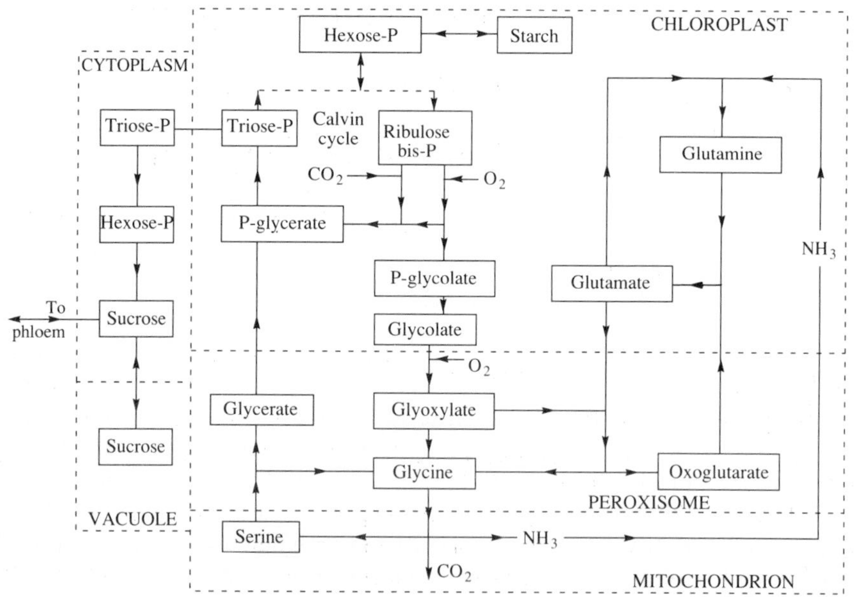

Fig. 9.2. A detailed scheme for the biochemistry of photosynthesis and photorespiration in C_3 plants (after Hahn 1987). The five spatial compartments of cytoplasm, vacuole, chloroplast, peroxisome, and mitochondrion are depicted. Not all the interconversions of the Calvin cycle between triose phosphate and ribulose bisphosphate, which Hahn (1987) took into account in his model, are shown (see Edwards and Walker, 1983, p. 108). P denotes phospho or phosphate. The involvement of ATP and NADPH in some of the reactions is not shown (cf. Fig. 9.1).

within the extended chloroplast; (iv) phosphate in the cytoplasm; (v) nitrogen in the extended chloroplast; (vi) glutamate, glutamine, and oxoglutarate. The rate constants are treated as phenomenological parameters, which are calculated by giving the state variables order-of-magnitude values. He examines the induction of photosynthesis by initially setting some state variables to zero, and then integrating the differential equations.

He finds that the steady state is indeed stable, and that the model itself behaves consistently and realistically. His assumptions about the biochemical pathways for glycerate and glycolate, and the use of a single pool for the chloroplast, peroxisome, and mitochondrion, are supported by the model's behaviour. He suggests, and this is important for our perspective as plant and crop modellers, that stoichiometry alone can be used to explain most of the responses of photosynthesis and photorespiration; details of reaction mechanisms and the use of complex expressions from enzyme kinetics may not be very important for our purposes, although, with different modelling objectives, these areas cannot always be ignored. This encourages our belief that the use of very simplified schemes (such as in Section 9.3) to construct appropriate models for leaf, plant, and crop investigations is legitimate and helpful.

9.2.3 *The light reactions of photosynthesis*

The biochemical reactions of photosynthesis require the provision of energy (ATP) and reducing power (NADPH) (e.g. (9.2a) and (9.2b)), and these are provided by the light reactions of photosynthesis which are driven by radiation in the visible region of the electromagnetic spectrum. The light reactions are not completely understood (Jones 1983, pp. 132–3, Edwards and Walker 1983, pp. 27–104), and Fig. 9.3 shows a highly simplified schematic representation of the process.

The scheme in Fig. 9.3 is for 'non-cyclic' photophosphorylation. Note that the ATP yield when an electron moves from point A on the diagram down to the pigment P700 is uncertain, and is considered to lie between 0.5 and 1. Multiplying the stoichiometry in Fig. 9.3 by 4 gives the overall result

$$8h\nu + 2H_2O + 2NADP^+ \rightarrow (2\text{–}4)ATP + 2(NADPH + H^+) + O_2. \quad (9.4a)$$

The effects of any charge separation are ignored. (It is possible that the proton H^+ produced at P690 is inside the thylakoid membrane, whereas the proton absorbed at B in NADPH production is taken from outside the membrane; the resulting proton concentration gradient could be used for further ATP production.)

'Cyclic' photophosphorylation refers to a process in which there is no other effect than the production of ATP; there is no oxidation of H_2O to O_2 and production of NADPH. For example, suppose that the electron at B in Fig. 9.3 is transferred to somewhere near A in the bridge part of the diagram and is able to return to P700, generating between 0.5 and 1 ATP while doing so. Therefore,

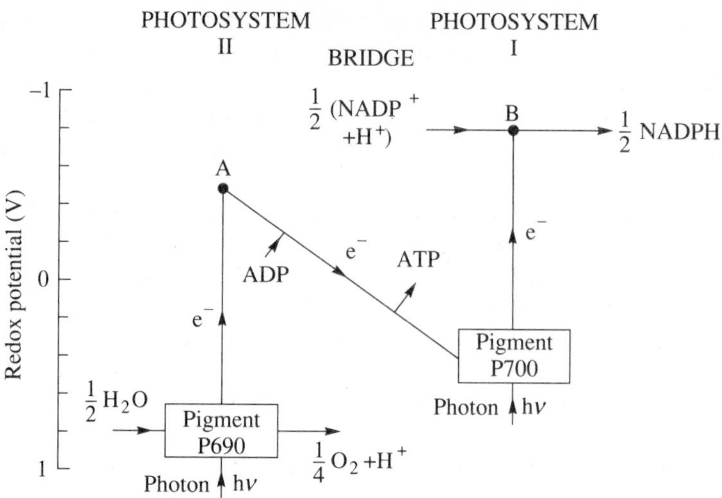

Fig. 9.3. Simple representation of the light reactions of photosynthesis: a single light quantum is absorbed by the pigment P690 of photosystem II; an electron (e^-) is removed from water (H_2O) and is moved to a state of higher energy (as electrons carry negative charge, this means its voltage becomes more negative); an electron transport bridge then carries the electron from point A to a lower energy state on the pigment P700 of photosystem I, and at the same time an uncertain number (from 0.5 to 1) of ATP are produced; a second light quantum is absorbed by pigment P700 of photosystem I, and the electron is moved to a higher energy state at B; at B, the electron is at a sufficiently negative voltage to reduce NADP to NADPH.

for eight photons absorbed by P700 which generate eight electrons which return to P700 via the bridge,

$$8hv \rightarrow (4\text{–}8)\text{ATP}. \tag{9.4b}$$

A similar overall effect can be obtained if the NADPH of (9.4a) is oxidized to water according to

$$\text{NADPH} + \tfrac{1}{2}\text{O}_2 \rightarrow \text{H}_2\text{O} + 2\text{ATP}. \tag{9.4c}$$

(Whether the ATP yield of NADPH oxidation is two or three ATP is arguable.) Adding two of (9.4c) to (9.4a) gives

$$8hv \rightarrow [4 + (2\text{–}4)]\text{ATP}. \tag{9.4d}$$

This last reaction, summarized in (9.4d), in which equal amounts of oxygen are evolved (9.4a) and taken up (9.4c), is sometimes referred to as 'pseudocyclic', to distinguish it from a 'cyclic' reaction such as (9.4b) in which oxygen is not involved.

There is another pseudocyclic process in which the Mehler reaction is involved. Suppose that the process shown in Fig. 9.3 at B, namely

$$\tfrac{1}{2}(NADP^+ + H^+) + e^- \rightarrow \tfrac{1}{2}NADPH \qquad (9.4e)$$

is replaced by the Mehler reaction

$$\tfrac{1}{2}O_2 + H^+ + e^- \rightarrow \tfrac{1}{2}H_2O_2 \rightarrow \tfrac{1}{2}H_2O + \tfrac{1}{4}O_2. \qquad (9.4f)$$

The last reaction producing oxygen occurs in the presence of the enzyme catalase. The overall result then becomes (cf. (9.4a))

$$8h\nu \rightarrow (2\text{–}4)ATP. \qquad (9.4g)$$

Note that cyclic photophosphorylation (9.4b) involves photosystem I only, whereas the two pseudocyclic processes (9.4d) and (9.4g)) involve both photosystems.

It is not known to what extent these various possibilities contribute under *in vivo* conditions. Therefore we restrict further discussion to developing (9.4a), which provides between two and four ATP per two NADPH. It is possible that the ATP:NADPH ratio is regulated, either directly or indirectly, by using interconversions as in (9.4c) so that the biochemical needs for ATP and NADPH are satisfied. (Note that the ATP:NADPH ratio for photosynthesis as in (9.2a) is 3:2, whereas that for photorespiration as in (9.2b) is 3.5:2.).

Next, we assume that the stoichiometry of the light reactions in (9.4a) is

$$8h\nu + 2H_2O + 2NADP^+ \rightarrow 3ATP + 2(NADPH + H^+) + O_2, \qquad (9.4h)$$

and estimate the efficiency of the light reactions of photosynthesis. For red light of wavelength 680 nm, the energy delivered is

$$176 \text{ MJ (kg mol)}^{-1} \text{ of 680 nm radiation.} \qquad (9.4i)$$

Taking the values used earlier in (9.2g), 3ATP + 2NADPH has a free energy of

$$571 \text{ MJ (kg mol)}^{-1}. \qquad (9.4j)$$

Therefore the efficiency of (9.4h) is

$$\text{efficiency (light reaction)} = \frac{571}{8 \times 176} = 0.40. \qquad (9.4k)$$

Combining (9.2h) with (9.4k) gives the overall efficiency of photosynthesis as

$$\text{efficiency (light and dark reactions)} = 0.4 \times 0.82 = 0.33. \qquad (9.4l)$$

9.2.4 *C_4 photosynthesis: outline of a simplified biochemical model*

In C_3 photosynthesis CO_2 and O_2 compete for ribulose bisphosphate (Fig. 9.2), and the result of oxidation, summarized in (9.2b), is a loss in energy (ATP), reducing power (NADPH), and sugar ($\{CH_2O\}$). In C_4 plants the concentration of CO_2 in the bundle sheath cells where the C_3 biochemistry occurs (for the C_3 mechanisms are still present in C_4 plants) is increased by using energy (ATP) to pump CO_2 into these cells. The higher local CO_2 concentration now

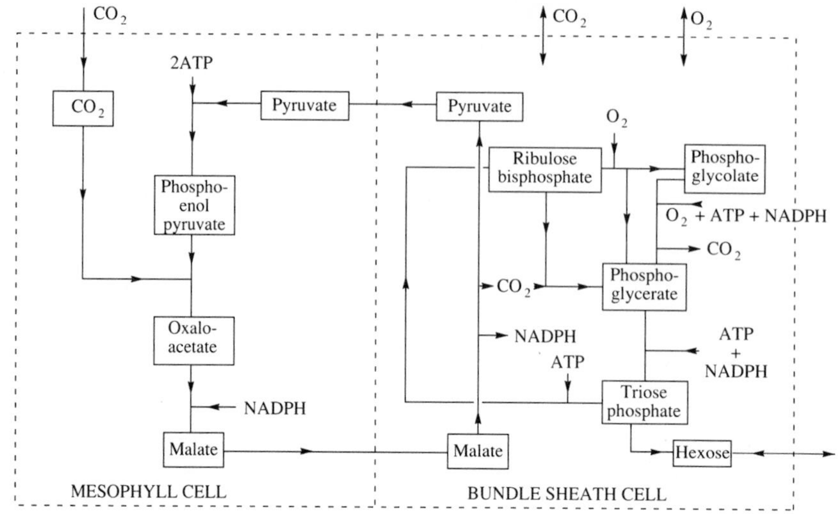

Fig. 9.4. C_4 photosynthesis: simplified scheme for the NADP–malic enzyme type of biochemistry which occurs in maize, sugarcane, and sorghum (Edwards and Walker, 1983, chapter 11). The C_4 reactions occur in the mesophyll cells and in the malate–pyruvate decarboxylation in the bundle sheath cells. The representation of the C_3 photosynthesis and photorespiratory pathways in the bundle sheath cells is taken from Fig. 9.1, simplified by omitting the boxes in the broken lines and details of stoichiometry.

competes more favourably with the O_2 concentration, so that oxidation of ribulose bisphosphate (and photorespiration) hardly occurs. All C_4 plants use phosphoenolpyruvate as the CO_2 acceptor in the mesophyll cells, but they differ in how the carbon is transported into the bundle sheath cells, and in how the CO_2 is then released. Figure 9.4 shows the scheme for the NADP–malic enzyme type of C_4 photosynthesis which occurs in the important crop plants maize, sugarcane, and sorghum.

A molecule of CO_2 is absorbed in the mesophyll cell and pyruvate is converted into malate at a cost of two ATP and one NADPH. The malate is transferred to a bundle sheath cell, where a molecule of CO_2 is released with one NADPH and pyruvate is regenerated. Pyruvate is transferred back to the mesophyll cell, and the CO_2 molecule released is able to enter the usual pathways of C_3 photosynthesis. If the NADPH absorbed in the mesophyll cell is equivalent to the NADPH produced in the bundle sheath cell, the overall cost of the CO_2 transfer is

$$CO_2(\text{mesophyll}) + 2\text{ATP} \rightarrow CO_2(\text{bundle sheath}). \qquad (9.5a)$$

Assuming diffusion and simple mass action kinetics (cf. (9.3g) and Fig. 9.1)), the rates of the reactions in the C_4 loop in Fig. 9.4 can be written as follows:

$$\frac{[CO_2(\text{air})] - [CO_2(\text{mesophyll})]}{r_{cm}} \tag{9.5b}$$

$$k_{CO_2m}[CO_2(\text{mesophyll})][\text{phosphoenolpyruvate}] \tag{9.5c}$$

$$k_{oxa}[\text{oxaloacetate}][\text{NADPH}] \tag{9.5d}$$

$$\frac{[\text{malate(mesophyll)}] - [\text{malate(bundle sheath)}]}{r_{malate}} \tag{9.5e}$$

$$k_{malate}[\text{malate(bundle sheath)}] \tag{9.5f}$$

$$\frac{[\text{pyruvate(bundle sheath)}] - [\text{pyruvate(mesophyll)}]}{r_{pyr}} \tag{9.5g}$$

$$k_{pyr}[\text{pyruvate(mesophyll)}][\text{ATP}]. \tag{9.5h}$$

The k denote rate constants and the r denote diffusion resistances, subscripted appropriately. If the ATP and NADPH pools are assumed to be common for the mesophyll and bundle sheath cells, there are seven extra state variables to be considered (compared with those required for C_3 photosynthesis and photorespiration). For example, two of the additional differential equations are

$$\frac{d[\text{oxaloacetate}]}{dt} = k_{CO_2m}[CO_2(\text{mesophyll})][\text{phosphoenolpyruvate}] - k_{oxa}[\text{oxaloacetate}][\text{NADPH}] \tag{9.5i}$$

$$\frac{d[CO_2(\text{bundle sheath})]}{dt} = \frac{[CO_2(\text{air})] - [CO_2(\text{bundle sheath})]}{r_{cbs}} + k_{malate}[\text{malate(bundle sheath)}] - k_{CO_2bs}[\text{RuBP}][CO_2(\text{bundle sheath})]. \tag{9.5j}$$

The remaining equations are easily constructed to give a soluble mathematical model for C_4 photosynthesis, which includes C_3 photosynthesis and photorespiration as a submodel.

9.3 Leaf photosynthesis models for plant and crop modelling

The models presented in this section are based on highly aggregated biochemistry, with the main criteria being that the resulting equations should be reasonably accurate at the leaf level and sufficiently tractable mathematically and computationally for use as submodels in possibly much larger models of plant and crop growth. Biochemical and physiological realism, while still desirable, are of secondary importance, although the more these are present, the more plausible it is to give the model parameters a biological interpretation.

Let the leaf irradiance be I_l (W m^{-2} of photosynthetically active radiation (PAR)) and the CO_2 concentration in the air be C_a (kg CO_2 m^{-3}; see Chapter 2, p. 49). We wish to derive an expression for the net photosynthetic rate P_n per unit area of leaf [kg CO_2 (m^2 leaf)$^{-1}$s^{-1}] as a function of I_l and C_a:

$$P_n(I_l, C_a). \tag{9.6a}$$

Because there is usually a CO_2 concentration gradient driving diffusion between the air outside the leaf (at concentration C_a) and the reaction site within the leaf (where the CO_2 concentration is C_i (i denotes internal)), and there is an internal production of CO_2 from respiration, it is first necessary to describe the internal photosynthetic rate P in terms of the irradiance I_l and the internal CO_2 concentration C_i, i.e.

$$P(I_l, C_i), \tag{9.6b}$$

and then to relate C_i to the CO_2 concentration C_a in the air.

9.3.1 *Photosynthesis in relation to internal leaf CO_2 concentration C_i*

The analysis here derives from Thornley (1974). Consider a homogeneous leaf of thickness h (m), and assume that this is uniformly irradiated with irradiance I_l and that the internal CO_2 concentration at the photosynthetic sites is C_i, which is also assumed to be uniform. It is supposed that the light energy in I_l reacts with some molecular species X to produce an activated form X* according to

$$X + I_l \xrightarrow{k_1} X^*, \tag{9.7a}$$

where k_1 is a rate constant. The activated form X* then reacts with CO_2 within the leaf to give carbohydrate $\{CH_2O\}$ and regenerate X, according to

$$X^* + CO_2 \xrightarrow{k_2} X + \{CH_2O\}, \tag{9.7b}$$

where k_2 is a second rate constant. These two reactions can perhaps be viewed as a simplified version of (9.4h) for the light reactions of photosynthesis in which ATP and NADPH are produced, and then used to drive the CO_2 reduction cycle (Fig. 9.1 and eqn (9.2a)). The rate of formation of X* is (from 9.7(a) and 9.7(b))

$$\frac{dX^*}{dt} = k_1 I_l X - k_2 X^* C_i.$$

Assuming that

$$X_o = X + X^* \tag{9.7c}$$

is constant and is the total concentration of X and X^* present, then substituting for X gives

$$\frac{dX^*}{dt} = k_1 I_l (X_o - X^*) - k_2 X^* C_i. \tag{9.7d}$$

From (9.7b), the rate of photosynthesis (that is, formation of carbohydrate or utilization of CO_2) is

$$P = hk_2X^*C_i. \tag{9.7e}$$

In the steady state $dX^*/dt = 0$, and therefore from (9.7d)

$$X^* = \frac{k_1 I_l X_o}{k_1 I_l + k_2 C_i}. \tag{9.7f}$$

Substituting for X^* in (9.7e) gives

$$P = \frac{hk_1 k_2 X_o C_i I_l}{k_1 I_l + k_2 C_i}. \tag{9.7g}$$

Multiplying the top and bottom of (9.7g) by hX_o and writing

$$\alpha = hk_1 X_o \quad \text{and} \quad r_x = (hk_2 X_o)^{-1} \tag{9.7h}$$

gives

$$P = \frac{\alpha I_l C_i / r_x}{\alpha I_l + C_i / r_x}. \tag{9.7i}$$

In (9.7h), α is known as the photochemical efficiency (kg CO_2 J^{-1}) and r_x as the carboxylation resistance (s m^{-1}).

Equation (9.7i) describes a rectangular hyperbola for both the $P:I_l$ and $P:C_i$ relationships, as illustrated in Figs 9.5(a) and 9.5(b). The effect of changing irradiance and the transition between different steady state rates of photosynthesis can be analysed through eqns (9.7e) (differentiated with respect to time) and (9.7d); this is discussed by Thornley (1974, 1976, Chapter 5). Here we restrict our attention to the steady state solution (9.7i).

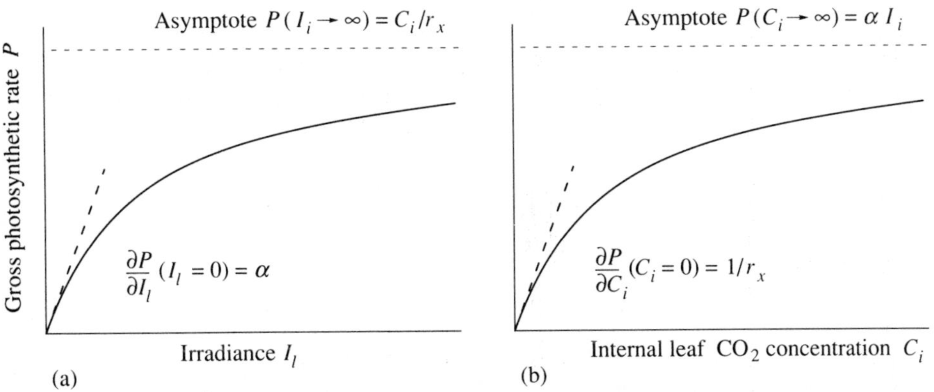

Fig. 9.5. Rectangular hyperbola describing single leaf gross photosynthesis P as given by eqn (9.7i): (a) as a function of the irradiance I_l of the leaf surface; (b) as a function of the internal leaf CO_2 concentration C_i.

9.3.2 *Photosynthesis in relation to the external CO_2 concentration C_a*

Equation (9.7i) describes the rate of gross photosynthesis as a function of the CO_2 concentration C_i within the leaf. However, C_i will be influenced by the external CO_2 concentration C_a, by the rate P of CO_2 fixation within the leaf, and also by any other sources of CO_2 within the leaf (such as respiration). We now make the following assumptions: (i) CO_2 moves between the sites of photosynthesis and the external environment by diffusion; (ii) respiration occurs at a constant rate R_d (kg CO_2 m^{-2} s^{-1}) (the so-called 'dark' respiration; see (9.1b)).

The net photosynthetic rate P_n (kg CO_2 m^{-2} s^{-1}) is defined as the difference between the gross photosynthetic rate P and the dark respiration rate R_d, where

$$P_n = P - R_d. \qquad (9.8a)$$

From assumption (i),

$$P_n = \frac{C_a - C_i}{r_d}, \qquad (9.8b)$$

where C_a is the CO_2 in the external environment and r_d is a diffusion resistance (m^{-1} s). Now use (9.8a) to substitute for P and (9.8b) to substitute for C_i in (9.7i), giving (Exercise 9.2)

$$0 = P_n^2 r_d - P_n[\alpha I_1(r_x + r_d) + C_a - R_d r_d] + \alpha I_1 C_a - R_d(\alpha I_1 r_x + C_a) \qquad (9.9a)$$

This describes both the $P_n : I_1$ and $P_n : C_a$ relationships as non-rectangular hyperbolas (the non-rectangular hyperbola is discussed on p. 56).

Equation (9.9a) is quadratic in P_n and therefore has two roots, only one of which is biologically acceptable. If we put $I_1 = 0$, so that dark respiration alone occurs, (9.9a) reduces to

$$0 = P_n^2 r_d - P_n(C_a - R_d r_d) - R_d C_a, \qquad (9.9b)$$

with solutions

$$P_n = \frac{(C_a - R_d r_d) \pm (C_a + R_d r_d)}{2r_d} = \frac{C_a}{r_d} \quad \text{or} \quad -R_d. \qquad (9.9c)$$

Clearly the negative root is the required solution (in general, the positive root corresponds to $C_i \leqslant 0$, which is not physiological).

The initial slope and asymptote with respect to irradiance I_1 can be shown to be (Exercise 9.2)

$$\frac{\partial P_n}{\partial I_1}(I_1 = 0) = \alpha \quad \text{and} \quad \lim_{I_1 \to \infty} P_n = \frac{C_a - R_d r_x}{r_x + r_d}. \qquad (9.9d)$$

If P_n is considered as a function of C_a, the asymptote is

$$\lim_{C_a \to \infty} P_n = \alpha I_1 - R_d. \qquad (9.9e)$$

The evaluation of $\partial P_n/\partial C_a(C_a = 0)$ is not as straightforward as that of $\partial P_n/\partial I_l(I_l = 0)$ of (9.9d); this is because $C_a = 0$ does not imply that $C_i = 0$, because dark respiration R_d is a source of CO_2 within the leaf; hence P (9.7i) may be non-zero when $C_a = 0$, owing to refixation of respired CO_2. However, it can be shown that

$$P_n(C_a = 0) = \frac{1}{2}\left[\alpha I_l\left(1 + \frac{r_x}{r_d}\right) - R_d\right]\left\{1 - \left[1 + \frac{4\alpha I_l R_d r_x/r_d}{\alpha I_l(1 + r_x/r_d) - R_d}\right]\right\} \quad (9.9f)$$

and

$$\frac{\partial P_n}{\partial C_a}(C_a = 0) = \frac{\alpha I_l - R_d - P_n(C_a = 0)}{\alpha I_l(r_x + r_d) - R_d r_d - 2r_d P_n(C_a = 0)}. \quad (9.9g)$$

These expressions are derived in Exercise 9.2.

Equation (9.9a) defines net photosynthesis as a function of the environment and leaf properties, and net photosynthesis is often the appropriate and required quantity. However, because the physiological roles of photosynthesis and respiration are different, it is sometimes useful to consider gross photosynthesis instead. Combining eqns (9.8a) and (9.9a) to eliminate P_n leads to the equation (Exercise 9.3)

$$P^2 r_d - P[\alpha I_l(r_d + r_x) + R_d r_d + C_a] + \alpha I_l(R_d r_d + C_a) = 0. \quad (9.10a)$$

This expression, like that for net photosynthesis, is a non-rectangular hyperbola for both $P{:}I_l$ and $P{:}C_a$ (as before, only the lower root of the quadratic is physiologically admissible (9.9c)).

Next, consider the characteristics of (9.10a) when C_a is constant. The initial slope and the asymptote with respect to I_l are (Exercise 9.3)

$$\frac{dP}{dI_l}(I_l = 0) = \alpha \quad (9.10b)$$

and

$$\lim_{I_l \to \infty} P(I_l) = \frac{R_d r_d + C_a}{r_d + r_x}. \quad (9.10c)$$

The (maximum) gross photosynthetic rate at saturating irradiance is denoted P_m, with (eqn (9.10c))

$$P_m = \frac{R_d r_d + C_a}{r_d + r_x}. \quad (9.10d)$$

For convenience, we define a parameter θ by

$$\theta = \frac{r_d}{r_d + r_x}. \quad (9.10e)$$

With eqns (9.10d) and (9.10e), eqn (9.10a) for P simplifies to

$$\theta P^2 - (\alpha I_l + P_m)P + \alpha I_l P_m = 0. \tag{9.10f}$$

With $\theta = 0$, eqn (9.10f) reduces to a rectangular hyperbola:

$$P = \frac{\alpha I_l P_m}{\alpha I_l + P_m}. \tag{9.10g}$$

With $\theta = 1$, a limiting response with two intersecting straight lines is obtained:

$$P = \begin{cases} \alpha I_l & I_l \leqslant P_m/\alpha \\ P_m & I_l > P_m/\alpha \end{cases}. \tag{9.10h}$$

The general solution for P, with $0 < \theta < 1$, is given by the lower root of (9.10f):

$$P = \frac{1}{2\theta}\{\alpha I_l + P_m - [(\alpha I_l + P_m)^2 - 4\theta\alpha I_l P_m]^{1/2}\}. \tag{9.10i}$$

Taking the limit $\theta \to 0$ and with $\theta = 1$, (9.10i) reduces to (9.10g) and (9.10h) respectively (Exercise 9.3); intermediate values of θ generate responses lying between these two extremes, as illustrated in Fig. 9.6 for θ values of 0, 0.5, 0.9, 0.95, and 1.

The special cases where $\theta = 0$ or 1 deserve brief mention. From (9.10e), $\theta = 0$ implies $r_d = 0$, and the diffusion resistance between the environment and the

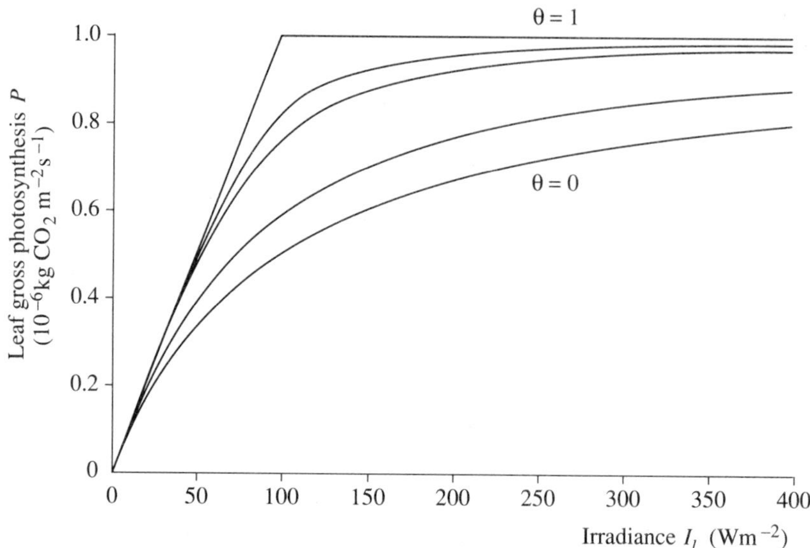

Fig. 9.6. Non-rectangular hyperbola describing single-leaf gross photosynthesis as a function of irradiance I_l for different values of the parameter θ (eqn (9.10i)). $\theta = 0$ (rectangular hyperbola, eqn (9.10g)) and $\theta = 1$ (limiting factor response, eqn (9.10h)) are indicated, and the intermediate curves are for θ values of 0.5, 0.9, and 0.95. Parameters are $P_m = 10^{-6}$ kg CO_2 m^{-2} s^{-1} and $\alpha = 10^{-8}$ kg CO_2 J^{-1}.

sites of photosynthesis is zero. $\theta = 1$ is obtained when $r_x = 0$, in which case the gross photosynthesis rate is, from eqns (9.7i) or (9.10h), $P = \alpha I_l$. Thus, for $\theta = 1$, the gross photosynthetic rate at the reaction sites increases linearly as irradiance increases until photosynthesis is limited by the diffusion of CO_2 from the air. Both these limiting cases are physiologically unlikely. For $\theta = 1$, the discontinuity in the derivative of $P(I_l)$ at $I_l = P_m/\alpha$ can cause analytical problems. The case of $\theta = 0$, however, produces a simplified expression for $P(I_l)$ (eqn (9.10g))–a rectangular hyperbola–which is useful for studying general concepts of single-leaf photosynthesis and developing analytical techniques when considering canopy photosynthesis.

The non-rectangular hyperbola derived here (eqn (9.10i)) for the response of leaf gross photosynthesis to irradiance is based on a very simplified picture of photosynthesis. Nevertheless, it leads to a straightforward and versatile expression for leaf photosynthesis where the parameters α, θ, and P_m have some underlying physiological meaning. Moreover, viewed as an empirical model, eqn (9.10i) is able to give an excellent fit to a wide range of leaf photosynthesis data (e.g. Marshall and Biscoe 1980). It is thus reasonable to regard the non-rectangular hyperbola as a valuable curve for studying leaf and crop photosynthesis.

9.3.3 *Effect of temperature*

To study the effects of temperature on photosynthesis through analytical development of the models described in Sections 9.3.1 and 9.3.2 is to overstretch the underlying concepts. It is preferable to observe empirically how the three parameters of eqn (9.10i)–θ, α, and P_m–are influenced by temperature (see Section 5.8.1, p. 139).

In experiments on leaves of white clover and perennial ryegrass, Woledge and Dennis (1982) observed considerable variation with temperature in the rate of photosynthesis at saturating irradiance P_m. For example, from their Fig. 6, P_m increases from 0.4×10^{-6} to 1.2×10^{-6} kg CO_2 m^{-2} s^{-1} as the temperature is increased from 5 to 25 °C. Over this temperature range, their data suggest a linear relationship between P_m and temperature T (°C):

$$P_m = \begin{cases} 0 & T < T^* \\ P_m(20)\left(\dfrac{T - T^*}{20 - T^*}\right) & T \geqslant T^* \end{cases}. \tag{9.11a}$$

P_m (20) is the value of P_m at the reference temperature of 20 °C; T^* can be interpreted as the temperature at which photosynthetic activity ceases, but, more important, it should be chosen to give a good fit to data over a typical temperature range, say 5–25 °C. In eqn (9.11a), P_m increases unboundedly with increasing temperature T; however, this expression is to be regarded as an approximation over a limited range to the more general response illustrated in Fig. 5.1 (p. 121) where there is an optimum temperature followed by a decreasing

P_m with further increases in T. For many purposes, a simple linear expression for $P_m(T)$ may suffice (e.g. eqn (8a) of Johnson and Thornley, 1985).

The effect of temperature on the photochemical efficiency α is much less than it is on P_m. For example, Ku and Edwards (1978) observed a decrease of 8 per cent in α for wheat when the temperature was increased from to 15 to 25 °C, and similar data have been reported by other workers. We are not aware of published data on the response of θ to temperature. However, the tentative physiological interpretation of θ in eqn (9.10e) suggests that there is unlikely to be a significant effect.

Summarizing, the temperature response of leaf photosynthesis is dominated by the temperature response of P_m; it can be assumed that α and θ do not vary with temperature. This is the basis of our analysis of the temperature dependence of canopy photosynthesis (Chapter 10, p. 252).

9.3.4 *Leaf photosynthesis and photorespiration in C_3 plants*

The simple leaf photosynthesis model of Section 9.3.1 can be extended to take account of photorespiration. Supplementing the two equations (9.7a) and (9.7b) we now have three reactions summarizing photosynthesis:

$$\mathrm{X} + I_l \xrightarrow{k_1} \mathrm{X}^* \tag{9.12a}$$

$$\mathrm{X}^* + \mathrm{CO_2} \xrightarrow{k_2} \mathrm{X} + \{\mathrm{CH_2O}\} \tag{9.12b}$$

$$\mathrm{X}^* + \mathrm{O_2} \xrightarrow{k_3} \mathrm{X} + \mathrm{CO_2}. \tag{9.12c}$$

Here the ks are rate constants. X* is the activated form of X. $\{CH_2O\}$ denotes carbohydrate. Define X_o as the total concentration of X and X* present, so that

$$X + X^* = X_o. \tag{9.12d}$$

Let C_i and O_i be the internal CO_2 and O_2 concentrations at the sites of photosynthesis; then, using (9.12d), the differential equation for X^* is

$$\frac{\mathrm{d}X^*}{\mathrm{d}t} = k_1 I_l(X_o - X^*) - X^*(k_2 C_i + k_3 O_i). \tag{9.12e}$$

In the steady state $\mathrm{d}X^*/\mathrm{d}t = 0$, and therefore

$$X^* = \frac{k_1 I_l X_o}{k_1 I_l + k_2 C_i + k_3 O_i}. \tag{9.12f}$$

The photosynthetic rate P (this is now photosynthesis minus photorespiration) is

$$P = h(k_2 X^* C_i - k_3 X^* O_i)$$

$$= \frac{h X_o k_1 I_l (k_2 C_i - k_3 O_i)}{k_1 I_l + k_2 C_i + k_3 O_i}. \tag{9.12g}$$

Multiplying the top and bottom of (9.12g) by hX_o and substituting

$$\alpha = hk_1 X_o, \quad r_x = (hk_2 X_o)^{-1}, \quad r_p = (hk_3 X_o)^{-1} \tag{9.12h}$$

gives

$$P = \frac{\alpha I_l (C_i/r_x - O_i/r_p)}{\alpha I_l + C_i/r_x + O_i/r_p} = P_g - R_l. \tag{9.12i}$$

(Note that, while the units of r_x are m^{-1} s, those of r_p are m^{-1} s kg O_2 (kg CO_2)$^{-1}$.) P_g and R_l are the rates of gross photosynthesis and photorespiration (kg CO_2 m^{-2} s^{-1}) and are given by

$$P_g = \frac{\alpha I_l C_i/r_x}{\alpha I_l + C_i/r_x + O_i/r_p}$$
$$R_l = \frac{\alpha I_l O_i/r_p}{\alpha I_l + C_i/r_x + O_i/r_p}. \tag{9.12j}$$

Equation (9.12i) can be compared with (9.7i).

Next, we wish to relate the photosynthetic rate to the CO_2 and O_2 concentrations in air, C_a and O_a. Continuing with the approach used in Section 9.3.1, we make the following assumptions.

1. There is a constant component of respiration R_d so that the net photosynthetic rate P_n is given by

$$P_n = P - R_d. \tag{9.12k}$$

2. CO_2 and O_2 move by diffusion between the sites of photosynthesis and photorespiration with diffusion constants r_{dc} and r_{do}, so that

$$P_n = \frac{C_a - C_i}{r_{dc}} \qquad P_n = \frac{O_i - O_a}{r_{do}}. \tag{9.12l}$$

(Note that the units of r_{do} in (9.12l) are s m^{-1} kg O_2 (kg CO_2)$^{-1}$.) Substituting in (9.12i) for P with (9.12k), and for C_i and O_i with (9.12l), and multiplying out gives

$$(P_n + R_d)\left(\alpha I_l + \frac{C_a - P_n r_{dc}}{r_x} + \frac{O_a + P_n r_{do}}{r_p}\right) = \alpha I_l\left(\frac{C_a - P_n r_{dc}}{r_x} - \frac{O_a + P_n r_{do}}{r_p}\right).$$

Multiplying both sides by $r_x r_p$ leads to

$$0 = -(P_n + R_d)[r_x r_p \alpha I_l + r_p(C_a - P_n r_{dc}) + r_x(O_a + P_n r_{do})$$
$$+ \alpha I_l[r_p(C_a - P_n r_{dc}) - r_x(O_a + P_n r_{do})]. \tag{9.12m}$$

Collecting terms, therefore, we have

$$O = P_n^2(r_p r_{dc} - r_x r_{do})$$
$$- P_n[r_p C_a + r_x O_a + \alpha I_l(r_x r_p + r_p r_{dc} + r_x r_{do}) - R_d(r_p r_{dc} - r_x r_{do})]$$
$$+ \alpha I_l(r_p C_a - r_x O_a) - R_d(r_x r_p \alpha I_l + r_p C_a + r_x O_a). \tag{9.12n}$$

This five-parameter quadratic equation describes the $P_n : I_l$, $P_n : C_a$, $P_n : O_a$ relationships. It can be shown to be consistent with eqn (9.9a) for the simpler model without photorespiration by allowing $r_p \to \infty$ (retain only the terms involving r_p, and then divide through by r_p). Note that $I_l = 0$ and $P_n = -R_d$ is a solution to (9.12n). To solve (9.12n), define coefficients a_2, a_1, and a_0:

$$a_2 = r_p r_{dc} - r_x r_{do}$$

$$a_1 = -[r_p C_a + r_x O_a + \alpha I_l(r_x r_p + r_p r_{dc} + r_x r_{do}) - R_d(r_p r_{dc} - r_x r_{do})]$$

$$a_0 = \alpha I_l(r_p C_a - r_x O_a) - R_d(r_x r_p \alpha I_l + r_p C_a + r_x O_a). \tag{9.12o}$$

The desired solution of the quadratic equation

$$0 = a_2 P_n^2 + a_1 P_n + a_0$$

is

$$P_n = \frac{-a_1 - (a_1^2 - 4a_2 a_0)^{1/2}}{2a_2}. \tag{9.12p}$$

To find the underlying rates of gross photosynthesis P_g and photorespiration R_l eqns (9.12l) are used to give C_i and O_i, and these values are substituted into eqns (9.12j) to give P_g and R_l. A range of responses is shown in Fig. 9.7.

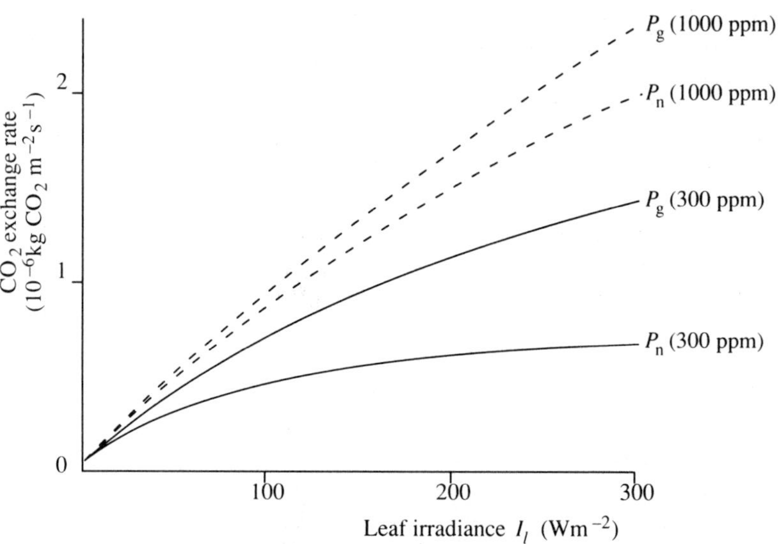

Fig. 9.7. Response to irradiance I_l of gross photosynthesis P_g and net photosynthesis P_n for a simple C_3 leaf model. These are calculated using eqn (9.12n) for P_n, and eqns (9.12j) for P_g. The solid lines are for $C_a = 0.549 \times 10^{-3}$ kg CO_2 m^{-3} (300 ppm at 20 °C), and the broken lines are for $C_a = 1.83 \times 10^{-3}$ kg CO_2 m^{-3} (1000 ppm at 20 °C). The other parameters are as follows: $O_a = 0.2661$ kg O_2 m^{-3} (20 per cent O_2 at 20 °C), $\alpha = 10^{-8}$ kg CO_2 J^{-1}, $r_{dc} = 500$ m^{-1} s, $r_{do} = 500$ m^{-1} s kg O_2 (kg CO_2)$^{-1}$, $r_p = 100\,000$ m^{-1} s kg O_2 (kg CO_2)$^{-1}$, $r_x = 50$ m^{-1} s, and $R_d = 0$ kg CO_2 m^{-2} s^{-1}. Because the dark respiration R_d is set to zero, the difference between the P_g and P_n curves is the photorespiration R_l.

It can be shown that the initial slope of the irradiance response curve is given by (Exercise 9.4)

$$\frac{\partial P_n}{\partial I_l}(I_l = 0) = \alpha \frac{r_p C_a - r_x O_a + R_d(r_p r_{dc} + r_x r_{do})}{r_p C_a + r_x O_a + R_d(r_p r_{dc} - r_x r_{do})}. \tag{9.13a}$$

Note that for $r_p \to \infty$ (no photorespiration), (9.13a) $\to \alpha$, as in eqns (9.9d). Interestingly, the effect of dark respiration R_d is to increase the apparent α; this is because the internal CO_2 concentration is raised, so that carboxylation competes better with oxygenation; in principle, it is possible for (9.13a) to give initial slope values above α for some (perhaps unrealistic) parameter values.

The asymptote of the irradiance response curve ($I_l \to \infty$) can be shown to be (in eqn (9.12n) retain only the terms in I_l and solve for P_n)

$$\lim_{I_l \to \infty} P_n = \frac{r_p C_a - r_x O_a - R_d r_x r_p}{r_x r_p + r_p r_{dc} + r_x r_{do}}. \tag{9.13b}$$

Again, in the limit $r_p \to \infty$ (no photorespiration) eqn (9.13b) agrees with eqns (9.9d) for the simpler model.

9.3.5 *C_4 photosynthesis: scheme with uncoupled mesophyll*

In this section and the next, two simple schemes are described to take account of C_4 characteristics (Fig. 9.4). These are both extensions of the model discussed in the last section for C_3 photosynthesis and photorespiration. In the first scheme (with uncoupled mesophyll) the energy supply available to the mesophyll for CO_2 pumping into the bundle sheath is assumed to be independent of the energy supply to the bundle sheath for C_3 photosynthesis and photorespiration. In the second scheme (with coupled mesophyll) a common supply of energy to both mesophyll and bundle sheath is assumed, so that CO_2 pumping occurs at the expense of energy which would otherwise have been available to the bundle sheath. The two schemes are presented for three reasons: first, to illustrate an approach to the construction of C_4 models which allows a variety of responses intermediate between C_3 and C_4 to be obtained; second, it is interesting to see how a seemingly innocuous assumption (independent or common energy availability for mesophyll and bundle sheath) can give rise to different mathematics and behaviour; third, it seems possible, if not likely, that there is in reality partial energetic coupling between mesophyll and bundle sheath, in which case some combination of the two models may be required.

The following scheme is assumed for the mesophyll:

$$\mathrm{Y} + I_l \xrightarrow{k_{1m}} \mathrm{Y}^* \tag{9.14a}$$

$$\mathrm{Y}^* + CO_2(\text{mesophyll}) \xrightarrow{k_4} CO_2(\text{bundle sheath}) + \mathrm{Y} \tag{9.14b}$$

$$CO_2(\text{air}) \xleftrightarrow{r_{dcm}} CO_2(\text{mesophyll}) \tag{9.14c}$$

where k_{1m} and k_4 are rate constants, r_{dcm} is a CO_2 diffusion constant, and I_l is the leaf irradiance. Y* denotes the activated form of Y, with

$$Y + Y^* = Y_o, \tag{9.14d}$$

where Y_o is a constant. The analysis proceeds similarly to that of Section 9.3.1 (eqns (9.7a)–(9.9a)). In the steady state

$$k_{1m}(Y_o - Y^*)I_l = k_4 Y^* C_m,$$

giving

$$\frac{Y^*}{Y_o} = \frac{k_{1m}I_l}{k_{1m}I_l + k_4 C_m} = \frac{\alpha_m I_l}{\alpha_m I_l + C_m/r_4}, \tag{9.14e}$$

where

$$\alpha_m = hk_{1m}Y_o \quad \text{and} \quad r_4 = \frac{1}{hk_4 Y_o}.$$

h is the leaf thickness (m). Let F_4 (kg CO_2 m^{-2} s^{-1}) denote the CO_2 flux from the mesophyll to the bundle sheath; then

$$F_4 = \frac{C_a - C_m}{r_{dcm}}$$

$$F_4 = hk_4 Y^* C_m = \frac{\alpha_m I_l C_m/r_4}{\alpha_m I_l + C_m/r_4}, \tag{9.14f}$$

where eqn (9.14e) and the definition of r_4 have been used. Eliminating C_m between eqns (9.14f) gives (cf. eqn (9.9a) with $R_d = 0$)

$$F_4{}^2 r_{dcm} - F_4[\alpha_m I_l(r_4 + r_{dcm}) + C_a] + \alpha_m I_l C_a = 0. \tag{9.14g}$$

It is easily shown that (cf. eqns (9.9d) and (9.9f))

$$F_4(I_l = 0) = 0$$

$$\frac{\partial F_4}{\partial I_l}(I_l = 0) = \alpha_m \tag{9.14h}$$

$$F_4(I_l \to \infty) = \frac{C_a}{r_4 + r_{dcm}}.$$

If we define the coefficients of the quadratic (eqn (9.14g)) by

$$a_2 = r_{dcm}$$

$$a_1 = -[\alpha_m I_l(r_4 + r_{dcm}) + C_a] \tag{9.14i}$$

$$a_0 = \alpha_m I_l C_a,$$

then the CO_2 flux into the bundle sheath F_4 is given by

$$F_4 = \frac{-a_1 - (a_1{}^2 - 4a_2 a_0)^{1/2}}{2a_2}. \tag{9.14j}$$

The analysis of C_3 photosynthesis and photorespiration is similar to that of Section 9.3.4 (p. 230), modified for the CO_2 flux of F_4 from the mesophyll which is injected into the bundle sheath. Thus the processes are

$$\mathrm{X} + I_l \xrightarrow{k_1} \mathrm{X}^* \tag{9.15a}$$

$$\mathrm{X}^* + \mathrm{CO_2(bundle\ sheath)} \xrightarrow{k_2} \{\mathrm{CH_2O}\} + \mathrm{X} \tag{9.15b}$$

$$\mathrm{X}^* + \mathrm{O_2(bundle\ sheath)} \xrightarrow{k_3} \mathrm{CO_2} + \mathrm{X} \tag{9.15c}$$

$$\mathrm{CO_2(air)} \overset{r_{\mathrm{dcbs}}}{\longleftrightarrow} \mathrm{CO_2(bundle\ sheath)}, \tag{9.15d}$$

where k_1, k_2, and k_3 are rate constants, and r_{dcbs} is the diffusion constant between air and the bundle sheath. X* is the activated form of X, with

$$X^* + X = X_o \tag{9.15e}$$

where X_o is constant. Following the earlier analysis (eqns (9.12e)–(9.12j)), for the steady state

$$\frac{X^*}{X_o} = \frac{\alpha I_l}{\alpha I_l + C_{\mathrm{bs}}/r_x + O_{\mathrm{bs}}/r_p} \tag{9.15f}$$

$$P = P_g - R_l = \frac{\alpha I_l(C_{\mathrm{bs}}/r_x - O_{\mathrm{bs}}/r_p)}{\alpha I_l + C_{\mathrm{bs}}/r_x + O_{\mathrm{bs}}/r_p}, \tag{9.15g}$$

where

$$\alpha = hk_1 X_o \qquad r_x = \frac{1}{hk_2 X_o} \qquad r_p = \frac{1}{hk_3 X_o},$$

and C_{bs} and O_{bs} denote the CO_2 and O_2 concentrations in the bundle sheath. From (9.15g), the gross photosynthetic rate P_g, and the photorespiration rate R_l are given by

$$\begin{aligned} P_g &= \frac{\alpha I_l C_{\mathrm{bs}}/r_x}{\alpha I_l + C_{\mathrm{bs}}/r_x + O_{\mathrm{bs}}/r_p} \\ R_l &= \frac{\alpha I_l O_{\mathrm{bs}}/r_p}{\alpha I_l + C_{\mathrm{bs}}/r_x + O_{\mathrm{bs}}/r_p}. \end{aligned} \tag{9.15h}$$

The net photosynthetic rate P_n and the photosynthetic rate P of eqn (9.15g) are related by (cf. eqn (9.12k))

$$P = P_n + R_d, \tag{9.15i}$$

where R_d is the dark respiration rate. The diffusion equations, analogous to eqns (9.12l), are

$$\begin{aligned} P_n &= \frac{C_a - C_{\mathrm{bs}}}{r_{\mathrm{dcbs}}} + F_4 \\ P_n &= \frac{O_{\mathrm{bs}} - O_a}{r_{\mathrm{do}}} \end{aligned} \tag{9.15j}$$

leading to

$$C_{bs} = C_a + \frac{F_4 - P_n}{r_{dcbs}}. \tag{9.15k}$$

$$O_{bs} = O_a + P_n r_{do}$$

Substituting eqns (9.15i)–(9.15k) in eqn (9.15g) and simplifying leads to (cf. eqns (9.12o) and (9.12p)) the quadratic equation

$$0 = b_2 P_n^2 + b_1 P_n + b_0, \tag{9.15l}$$

where

$$\begin{aligned} b_2 &= r_p r_{dcbs} - r_x r_{do} \\ b_1 &= -[r_p(C_a + F_4 r_{dcbs}) + r_x O_a + \alpha I_l(r_x r_p + r_p r_{dcbs} + r_x r_{do}) \\ &\quad - R_d(r_p r_{dcbs} - r_x r_{do})] \\ b_0 &= \alpha I_l[r_p(C_a + F_4 r_{dcbs}) - r_x O_a] \\ &\quad - R_d[r_x r_p \alpha I_l + r_p(C_a + F_4 r_{dcbs}) + r_x O_a], \end{aligned} \tag{9.15m}$$

with the solution

$$P_n = \frac{-b_1 - (b_1^2 - 4b_2 b_0)^{1/2}}{2b_2}. \tag{9.15n}$$

It can be shown (by retaining only the terms in I_l in (9.15m) and putting these into (9.15n)) that

$$P_n(I_l \to \infty) = \frac{r_p[C_a + F_4(I_l \to \infty) r_{dcbs}] - r_x O_a - r_x r_p R_d}{r_x r_p + r_{dcbs} r_p + r_{do} r_x}, \tag{9.15o}$$

where (eqns (9.14h))

$$F_4(I_l \to \infty) = \frac{C_a}{r_4 + r_{dcm}}.$$

This is consistent with eqn (9.13b) for C_3 photosynthesis and photorespiration. It can be verified that

$$I_l = 0, \quad F_4 = 0, \quad P_n = -R_d$$

is a solution of eqns (9.15l) and (9.15m). It can also be shown that the initial slope $\partial P_n/\partial I_l(I_l = 0)$ of the light response curve is given by

$$\frac{\partial P_n}{\partial I_l}(I_l = 0) = \alpha \frac{r_p C_a - r_x O_a + R_d(r_p r_{dcbs} + r_x r_{do})}{r_p C_a + r_x O_a + R_d(r_p r_{dcbs} - r_x r_{do})}. \tag{9.15p}$$

Note that this is essentially unchanged from eqn (9.13a), and the C_4 process makes no contribution to the initial slope. However, it can be seen that the slightly sigmoidal response of the net photosynthetic rate P_n to irradiance I_l from

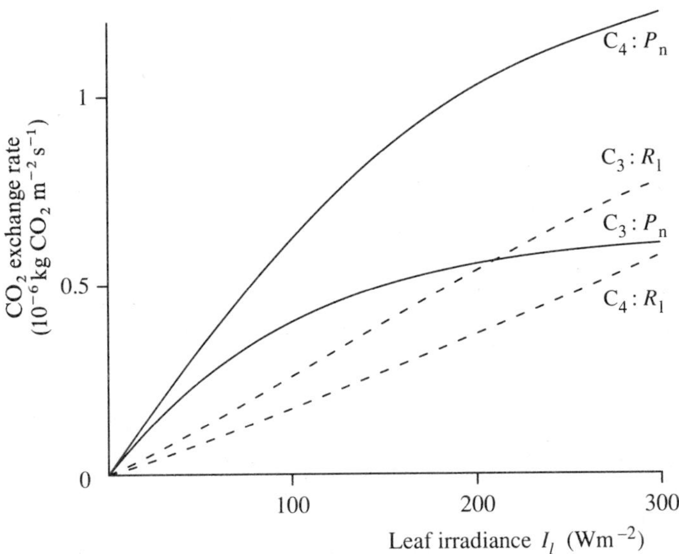

Fig. 9.8. The response to the irradiance I_l of net photosynthesis P_n and photorespiration R_l for a C_4 model with uncoupled mesophyll (responses denoted by C_4:) is compared with the same model in the absence of the CO_2 pumping process (denoted by C_3:) The C_4 results are calculated from eqn (9.15n) for P_n and from eqns (9.15h) for R_l. The solid lines denote P_n and the broken lines denote R_l. The parameters are $C_a = 0.549 \times 10^{-3}$ kg CO_2 m^{-3} (300 ppm at 20 °C), $O_a = 0.2661$ kg O_2 m^{-3} (20 per cent O_2 at 20 °C), $\alpha = 10^{-8}$ kg CO_2 J^{-1}, $r_{dcbs} = 500$ m^{-1} s, $r_{dcm} = 500$ m^{-1} s, $r_{do} = 500$ m^{-1} s kg O_2 (kg $CO_2)^{-1}$, $r_p =$ 100 000 m^{-1} s kg O_2 (kg $CO_2)^{-1}$, $r_x = 50$ m^{-1} s, $r_4 = 50$ m^{-1} s, $R_d = 0$ kg CO_2 m^{-2} s^{-1}. The C_3 calculations and parameters are as in Fig. 9.7. The asymptotes are as follows: C_4 (eqn (9.15o)), $P_n(I_l \rightarrow \infty) = 1.66 \times 10^{-6}$ kg CO_2 m^{-2} s^{-1}; C_3 (eqn (9.13b)), $P_n(I_l \rightarrow \infty) =$ 0.76×10^{-6} kg CO_2 m^{-2} s^{-1}.

this model gives rise to an apparent initial slope that is higher than in C_3 photosynthesis by some 20 per cent (Fig. 9.8).

In Fig. 9.8 the C_4 model (with uncoupled mesophyll) is compared with the corresponding C_3 model in respect of the responses of net photosynthetic rate and photorespiration to irradiance. The results appear reasonable.

9.3.6 *C_4 photosynthesis: scheme with coupled mesophyll*

As discussed at the beginning of the last section (p. 233), in this section we analyse a C_4 scheme (Fig. 9.4) where a common energy supply to both mesophyll and bundle sheath is assumed, so that CO_2 pumping occurs at the expense of energy which would otherwise have been available to the bundle sheath. The following scheme is assumed (cf. eqns (9.12a)–(9.12d)); the Y and X of the uncoupled C_4 scheme become the same (eqns (9.14a), (9.14b), and (9.15a)–(9.15c)):

$$\mathrm{X} + I_1 \xrightarrow{k_1} \mathrm{X}^* \tag{9.16a}$$

$$\mathrm{X}^* + \mathrm{CO_2(bundle\ sheath)} \xrightarrow{k_2} \mathrm{X} + \{\mathrm{CH_2O}\} + \mathrm{O_2(bundle\ sheath)} \tag{9.16b}$$

$$\mathrm{X}^* + \mathrm{O_2(bundle\ sheath)} \xrightarrow{k_3} \mathrm{X} + \mathrm{CO_2(bundle\ sheath)} \tag{9.16c}$$

$$\mathrm{X}^* + \mathrm{CO_2(mesophyll)} \xrightarrow{k_4} \mathrm{CO_2(bundle\ sheath)} + \mathrm{X}. \tag{9.16d}$$

Here the k_i, $i = 1$–4, are rate constants. The reactions (9.16b) and (9.16c) are non-stoichiometric, as they are intended to show the factors that will appear in the rate equations. X* is the activated form of X and $\{CH_2O\}$ denotes carbohydrate. Define X_o as the total concentration of X and X* present, so that

$$X + X^* = X_o. \tag{9.16e}$$

If we denote the CO_2 concentrations in air, mesophyll and bundle sheath by C_a, C_m, and C_{bs} (kg CO_2 m^{-3}) respectively, and the O_2 concentrations in the air and bundle sheath by O_a and O_{bs} (kg O_2 m^{-3}), then the following diffusive processes are assumed:

$$\text{air–mesophyll} \quad C_a \xleftrightarrow{r_{dcm}} C_m \tag{9.16f}$$

$$\text{air–bundle sheath} \quad C_a \xleftrightarrow{r_{dcbs}} C_{bs} \tag{9.16g}$$

$$\text{air–bundle sheath} \quad O_a \xleftrightarrow{r_{do}} O_{bs}. \tag{9.16h}$$

r_{dcm}, r_{dcbx}, and r_{do} are diffusion resistances (m^{-1} s). It is assumed that O_2 is only released and consumed in the bundle sheath, and there is no CO_2 diffusion between mesophyll and bundle sheath.

The differential equation for X^* (using (9.16e)) is

$$\frac{dX^*}{dt} = k_1 I_1(X_o - X^*) - X^*(k_2 C_{bs} + k_3 O_{bs} + k_4 C_m). \tag{9.16i}$$

In the steady state

$$X^* = \frac{k_1 I_1 X_o}{k_1 I_1 + k_2 C_{bs} + k_3 C_m}. \tag{9.16j}$$

The photosynthetic rate P (photosynthesis less photorespiration) is

$$\begin{aligned} P &= h(k_2 C_{bs} - k_3 O_{bs})X^* \\ &= \frac{hk_1 X_o I_1(k_2 C_{bs} - k_3 O_{bs})}{k_1 I_1 + k_2 C_{bs} + k_3 O_{bs} + k_4 C_m}. \end{aligned} \tag{9.16k}$$

Multiplying the top and bottom of (9.16k) by hX_o and substituting

$$\begin{aligned} \alpha = hk_1 X_o \qquad & r_x = (hk_2 X_o)^{-1} \\ r_p = (hk_3 X_o)^{-1} \qquad & r_4 = (hk_4 X_o)^{-1} \end{aligned} \tag{9.16l}$$

gives

$$P = \frac{\alpha I_l (C_{bs}/r_x - O_{bs}/r_p)}{\alpha I_l + C_{bs}/r_x + O_{bs}/r_p + C_m/r_4} = P_g - R_l. \tag{9.16m}$$

P_g and R_l (kg CO_2 m^{-2} s^{-1}) are the rates of gross photosynthesis and photorespiration, and are given by

$$P_g = \frac{\alpha I_l C_{bs}/r_x}{\alpha I_l + C_{bs}/r_x + O_{bs}/r_p + C_m/r_4}$$
$$R_l = \frac{\alpha I_l O_{bs}/r_p}{\alpha I_l + C_{bs}/r_x + O_{bs}/r_p + C_m/r_4}. \tag{9.16n}$$

These equations can be compared with eqns (9.12j).

Next we wish to relate the photosynthetic rate to the CO_2 and O_2 concentrations C_a and O_a in air. It is assumed that there is a constant component of dark respiration R_d so that the net photosynthetic rate P_n is given by

$$P_n = P - R_d, \tag{9.16o}$$

where P_n can be equated to the net CO_2 flux into the leaf or the net O_2 flux out:

$$P_n = \frac{C_a - C_m}{r_{dcm}} + \frac{C_a - C_{bs}}{r_{dcbs}} \tag{9.16p}$$

$$P_n = \frac{O_{bs} - O_a}{r_{do}}. \tag{9.16q}$$

An additional equation in needed to solve the steady state problem, namely that the rate of CO_2 diffusion into the mesophyll equals the rate of transfer of CO_2 from the mesophyll to the bundle sheath (9.16d), giving

$$\frac{C_a - C_m}{r_{dcm}} = \frac{\alpha I_l C_m/r_4}{\alpha I_l + C_{bs}/r_x + O_{bs}/r_p + C_m/r_4}. \tag{9.16r}$$

There are now five equations (9.16m), (9.16o), (9.16p), (9.16q), and (9.16r) for the five unknowns P, C_m, C_{bs}, O_{bs}, and P_n. In fact, the procedure we use is to eliminate P (9.16o), O_{bs} (9.16q), and C_{bs} (9.16p) in (9.16m) and (9.16r), and then to eliminate C_m between the two resulting equations, leading to a cubic equation in P_n (Exercise 9.5):

$$y = u_3 P_n^3 + u_2 P_n^2 + u_1 P_n + u_0, \tag{9.16s}$$

where

$$u_3 = -ab + ae - da^2 + h$$
$$u_2 = b^2 + ac - be + abd + ade + 2hd - g$$
$$u_1 = -2bc + ce - acd - bde + hd^2 - 2dg \tag{9.16t}$$
$$u_0 = c^2 + cde - gd^2$$

The coefficients in (9.16t) are given by

$$a\left(\frac{1}{r_4} - \frac{r_{\rm dcbs}}{r_{\rm dcm} r_{\rm x}}\right) = \frac{r_{\rm dcbs}}{r_{\rm x}} - \frac{r_{\rm do}}{r_{\rm p}}$$

$$b\left(\frac{1}{r_4} - \frac{r_{\rm dcbs}}{r_{\rm dcm} r_{\rm x}}\right) = \alpha I_{\rm l}\left(\frac{r_{\rm dcbs}}{r_{\rm x}} + \frac{r_{\rm do}}{r_{\rm p}} + 1\right) + \frac{C_{\rm a}(1 + r_{\rm dcbs}/r_{\rm dcm})}{r_{\rm x}} + \frac{O_{\rm a}}{r_{\rm p}} - R_{\rm d}\left(\frac{r_{\rm dcbs}}{r_{\rm x}} - \frac{r_{\rm do}}{r_{\rm p}}\right)$$

$$c\left(\frac{1}{r_4} - \frac{r_{\rm dcbs}}{r_{\rm dcm} r_{\rm x}}\right) = \alpha I_{\rm l}\left[\frac{C_{\rm a}(1 + r_{\rm dcbs}/r_{\rm dcm})}{r_{\rm x}} - \frac{O_{\rm a}}{r_{\rm p}} - R_{\rm d}\right] - R_{\rm d}\left[\frac{C_{\rm a}(1 + r_{\rm dcbs}/r_{\rm dcm})}{r_{\rm x}} + \frac{O_{\rm a}}{r_{\rm p}}\right]$$

$$d\left(\frac{1}{r_4} - \frac{r_{\rm dcbs}}{r_{\rm dcm} r_{\rm x}}\right) = \alpha I_{\rm l}\frac{r_{\rm dcbs}}{r_{\rm dcm} r_{\rm x}} + R_{\rm d}\left(\frac{1}{r_4} - \frac{r_{\rm dcbs}}{r_{\rm dcm} r_{\rm x}}\right) \qquad (9.16u)$$

$$e\left(\frac{1}{r_4} - \frac{r_{\rm dcbs}}{r_{\rm dcm} r_{\rm x}}\right) = \alpha I_{\rm l}\left(1 + \frac{r_{\rm dcm}}{r_4}\right) + \frac{O_{\rm a}}{r_{\rm p}} + C_{\rm a}\left(\frac{1}{r_{\rm x}} - \frac{1}{r_4} + \frac{2 r_{\rm dcbs}}{r_{\rm dcm} r_{\rm x}}\right)$$

$$f\left(\frac{1}{r_4} - \frac{r_{\rm dcbs}}{r_{\rm dcm} r_{\rm x}}\right) = \frac{r_{\rm dcbs}}{r_{\rm x}} - \frac{r_{\rm do}}{r_{\rm p}}$$

$$g\left(\frac{1}{r_4} - \frac{r_{\rm dcbs}}{r_{\rm dcm} r_{\rm x}}\right) = C_{\rm a}\left[\alpha I_{\rm l} + \frac{C_{\rm a}(1 + r_{\rm dcbs}/r_{\rm dcm})}{r_{\rm x}} + \frac{O_{\rm a}}{r_{\rm p}}\right]$$

$$h\left(\frac{1}{r_4} - \frac{r_{\rm dcbs}}{r_{\rm dcm} r_{\rm x}}\right) = C_{\rm a}\left(\frac{r_{\rm dcbs}}{r_{\rm x}} - \frac{r_{\rm do}}{r_{\rm p}}\right).$$

It can be shown that the photosynthetic rate at light saturation is given by (Exercise 9.5)

$$\lim_{I_{\rm l}\to\infty} P_{\rm n} = \left\{\left[\frac{C_{\rm a}(1 + r_{\rm dcbs}/r_{\rm dcm})}{r_{\rm x}} - \frac{O_{\rm a}}{r_{\rm p}} - R_{\rm d}\right]\left(1 + \frac{r_{\rm dcm}}{r_4}\right) - C_{\rm a}\frac{r_{\rm dcbs}}{r_{\rm dcm} r_{\rm x}}\right\} \times \left[\left(\frac{r_{\rm dcbs}}{r_{\rm x}} + \frac{r_{\rm do}}{r_{\rm p}} + 1\right)\left(1 + \frac{r_{\rm dcm}}{r_4}\right)\right]^{-1}. \qquad (9.16v)$$

This expression agrees exactly with eqn (9.15o) for the C_4 uncoupled mesophyll scheme. Also, in the limit $r_4 \to \infty$ (no C_4 photosynthesis), eqn (9.16v) reduces to eqn (9.13b) for C_3 photosynthesis and photorespiration.

Obtaining the initial slope $\partial P_{\rm n}/\partial I_{\rm l}$ ($I_{\rm l} = 0$) of the light response curve involves much algebra. Solution 9.5 indicates how this might be carried out—it is possibly a good exercise for a class.

Solutions to eqn (9.16s) for net photosynthetic rate $P_{\rm n}$ can be obtained by using the formulae for a cubic equation, or alternatively [see eqns (S9.5o)–(S9.5q),

p. 613]. After finding P_n numerically, the CO_2 concentration C_m in the mesophyll can be obtained from (eqn (S9.5i), p. 612)

$$C_m = \frac{aP_n^2 - bP_n + c}{P_n + d}. \tag{9.16w}$$

Then the internal CO_2 and O_2 concentrations C_{bs} and O_{bs} in the bundle sheath are calculated using eqns (9.16p) and (9.16q). The rates of gross photosynthesis and photorespiration P_g and R_l are give by eqns (9.16n), and the two components of the net photosynthetic rate P_n which enter via the mesophyll or by direct diffusion into the bundle sheath are given by eqn (9.16p). It can also be useful to calculate the fraction X^*/X_o of X_o which is in the light-activated state (9.16a); from eqns (9.16j) and (9.16l) this is given by

$$\frac{X^*}{X_o} = \frac{\alpha I_l}{\alpha I_l + C_{bs}/r_x + O_{bs}/r_p + C_m/r_4}. \tag{9.16x}$$

Some responses of this C_4 model are shown in Fig. 9.9, where they are compared with the C_3 model. Figure 9.9 for the coupled mesophyll scheme can

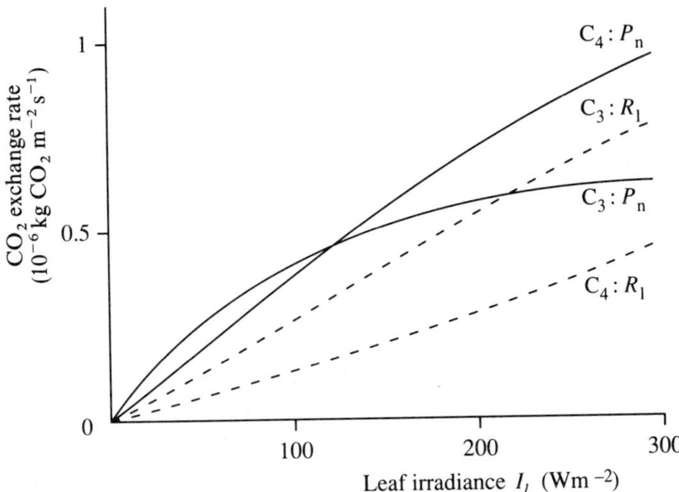

Fig. 9.9. The response to the irradiance I_l of net photosynthesis P_n and photorespiration R_l for a C_4 model with coupled mesophyll (responses denoted by C_4:) is compared with the same model in the absence of the CO_2 pumping process (denoted by C_3:). The C_4 results are calculated by solving eqn (9.16s) for P_n and eqns (9.16n) for R_l. The solid lines denote P_n and the broken lines denote R_l. The parameters are $C_a = 0.549 \times 10^{-3}$ kg CO_2 m^{-3} (300 ppm at 20 °C), $O_a = 0.2661$ kg O_2 m^{-3} (20 per cent O_2 at 20 °C), $\alpha = 10^{-8}$ kg CO_2 J^{-1}, $r_{dcbs} = 500$ m^{-1} s, $r_{dcm} = 500$ m^{-1} s, $r_{do} = 500$ m^{-1} s kg O_2 (kg CO_2)$^{-1}$, $r_p = 100000$ m^{-1} s kg O_2 (kg CO_2)$^{-1}$, $r_x = 50$ m^{-1} s, $r_4 = 50$ m^{-1} s, $R_d = 0$ kg CO_2 m^{-2} s^{-1}. The C_3 calculations and parameters are as in Fig. 9.7. The asymptotes are as follows: C_4 (eqn (9.16v)), $P_n(I_l \to \infty) = 1.66 \times 10^{-6}$ kg CO_2 m^{-2} s^{-1}; C_3 (eqn (9.13b)), $P_n(I_l \to \infty) = 0.76 \times 10^{-6}$ kg CO_2 m^{-2} s^{-1}.

be compared with Fig. 9.8 for the uncoupled mesophyll scheme: although the asymptote is the same in both figures, the approach to the asymptote is much slower for the coupled mesophyll scheme.

Exercises

9.1. For the scheme in Fig. 9.1, assume diffusion resistances r_{dc} and r_{do} for the movement of CO_2 and O_2 between the air (where the CO_2 and O_2 concentrations are $[CO_2]_a$ and $[O_2]_a$) and the metabolic sites (where the CO_2 and O_2 concentrations are $[CO_2]_i$ and $[O_2]_i$), and assume that ATP and NADPH are generated according to

$$[ADP] + I_l \rightarrow [ATP]$$
$$[ADP] + [ATP] = [AP]_o \tag{E9.1a}$$
$$[NADP] + I_l + H_2O \rightarrow [NADPH] + 1/2\, O_2,$$
$$[NADP] + [NADPH] = [NP]_o, \tag{E9.1b}$$

where I_l is the leaf irradiance, $[AP]_o$ and $[NP]_o$ are constants, and the rate constants are k_{ATP} and k_{NADPH}.

Write down, with reference to Fig. 9.1 and eqns (9.3h)–(9.3l), the differential equations for the rates of change of $[CO_2]_i$, $[O_2]_i$, $[ATP]$, and $[NADPH]$.

9.2. Derive eqn (9.9a) for the rate of single-leaf net photosynthesis P_n. Derive also the initial slope, asymptote, and light compensation point for $P_n(I_l)$ (eqns (9.9d) and (9.9e)). For the CO_2 response of photosynthesis $P_n(C_a)$, derive the asymptote (eqn (9.9e)); show that the initial slope of $P_n(C_a)$ in terms of $P_n(C_a = 0)$ is given by eqn (9.9g). Demonstrate that $P_n(C_a = 0)$ is given by eqn (9.9f). Explore and discuss $P_n(C_a = 0)$ and $\partial P_n/\partial C_a(C_a = 0)$ for the cases (a) $r_x/r_d \rightarrow 0$ and $\alpha I_l \gg R_d$, and (b) $r_x/r_d \gg 1$ and $\alpha I_l \gg R_d$.

9.3. Derive eqn (9.10a) for P, and show that the initial slope and asymptote of $P(I_l)$ are given by eqns (9.10b) and (9.10c). Obtain eqn (9.10i) and show that the limits $\theta \rightarrow 0$ and $\theta = 1$ lead to eqns (9.10g) and (9.10h).

9.4. For the C_3 photosynthesis and photorespiration model of Section 9.3.4, derive eqn (9.13a) for the initial slope of the light response curve from eqn (9.12n).

9.5. For the C_4 photosynthesis and photorespiration model of Section 9.3.6, derive eqn (9.16s) relating the ambient CO_2 and O_2 concentrations C_a and O_a to the net photosynthetic rate P_n; do this by combining eqns (9.16m) and (9.16o)–(9.16r) so as to eliminate P, C_{bs}, O_{bs}, and C_m. Use the Newton–Raphson method to obtain solutions numerically. Derive an expression for the light-saturated rate of photosynthesis $P_n(I_l \rightarrow \infty)$, and show that it reduces to eqn (9.13b), the corresponding expression for C_3 photosynthesis and photorespiration, in the limit $r_4 \rightarrow \infty$. A challenging problem algebraically is to derive an equation for the initial slope of the light response curve $dP_n/dI_l(I_l = 0)$.

10
Canopy photosynthesis

10.1 Introduction

In the previous two chapters we have considered light interception by crop canopies and single-leaf photosynthesis respectively: here some of these ideas are combined to develop models of canopy photosynthesis for monocultures, mixtures, and row crops. These are areas where accurate models are crucial in the development of crop growth models. Much of the work in this chapter derives from Johnson and Thornley (1984) and Johnson, Parsons, and Ludlow (1989).

Our emphasis is on deriving analytical expressions for the instantaneous and daily canopy photosynthetic rates. Analytical progress requires making assumptions concerning the variation of leaf photosynthetic characteristics through the depth of the canopy; for example, variation in the light-saturated photosynthetic rate in response to growth irradiance or leaf age may be important. In some cases, these effects may be allowed for within an analytical framework, but in other situations this is no longer possible without making unrealistic assumptions. In these cases, the basic approach presented below should be used and numerical techniques applied to evaluate the integrals leading to the canopy photosynthetic rates. One such situation may be where the leaves in the upper region of a canopy are suffering from some degree of water stress owing to high radiation levels whereas the leaves lower down may be relatively unstressed, so that the leaves within the canopy may have quite different photosynthetic characteristics. Another reason why it may sometimes be desirable not to use the analytical models presented below is that these are based on the non-rectangular hyperbola model for leaf photosynthesis, whereas an alternative leaf model may be required, for instance when investigating canopy photosynthetic response to CO_2 and O_2 levels. With these caveats, the following models are widely applicable in many crop modelling situations.

10.1.1 *Light interception and attenuation*

The light attenuation through canopies is described by Beer's law, as discussed in Chapter 8. For monocultures this is (eqn (8.1a))

$$I(l) = I_0 e^{-kl}, \tag{10.1a}$$

where I_0 and $I(l)$ W (m^2 ground)$^{-1}$ of photosynthetically active radiation (PAR) are the irradiances per unit horizontal area above and within the canopy respectively at cumulative leaf area index l and k is the extinction coefficient. The irradiance incident on the surface of a leaf within the canopy at depth l is (eqn (8.6i))

$$I_l(l) = \frac{k}{1-m} I(l) \text{ W (m}^2\text{ leaf)}^{-1}, \tag{10.1b}$$

where m is the leaf transmission coefficient. The developments of this approach to include mixed canopies and row crops, which were discussed in Chapter 8, are also used.

10.1.2 *Single-leaf photosynthesis*

Throughout the analysis, the term photosynthesis is taken to be single-leaf gross photosynthesis minus photorespiration, and is defined by eqn (9.10i):

$$P_l(I_l) = \frac{1}{2\theta}\{\alpha I_l + P_m - [(\alpha I_l + P_m)^2 - 4\theta\alpha I_l P_m]^{1/2}\}, \tag{10.2a}$$

with units kg CO_2 (m^2 leaf)$^{-1}$ s^{-1}. α is the photochemical efficiency, with units kg CO_2 J^{-1}, P_m is the asymptotic value of P_l at saturating irradiance, and θ is a dimensionless parameter ($0 \leqslant \theta \leqslant 1$). (Note that the notation used here is slightly different to that used in Chapter 9 in that the subscript l is used to denote single leaf photosynthesis.) For $\theta = 0$ eqn (10.2a) reduces to the rectangular hyperbola

$$P_l = \frac{\alpha I_l P_m}{\alpha I_l + P_m}, \tag{10.2b}$$

and with $\theta = 1$ it simplifies to the Blackman limiting response given by

$$P_l = \begin{cases} \alpha I_l & I_l \leqslant P_m/\alpha \\ P_m & I_l > P_m/\alpha. \end{cases} \tag{10.2c}$$

Intermediate values of θ produce curves lying between these two extremes, as illustrated in Chapter 9, Fig. 9.6, for θ taking the values (0, 0.5, 0.9, 0.95, 1). The discontinuity in the slope of this curve at $I_l = P_m/\alpha$ for $\theta = 1$ causes problems in the analysis for canopy photosynthesis, and since there is little evidence that it has any physiological meaning (see Section 9.3.2), the constraint

$$0 \leqslant \theta < 1 \tag{10.2d}$$

is adopted. The analysis could be developed to include $\theta = 1$ but there seems little point in doing so.

In Fig. 10.1 leaf photosynthesis data taken from Ludlow and Wilson (1971) are shown, along with the fitted non-rectangular hyperbolae, for Siratro (*Phaseolus atropurpureus*), a tropical legume, and green panic (*Panicum maximum*), a tropical C_4 grass. It can be seen that the non-rectangular hyperbola gives a

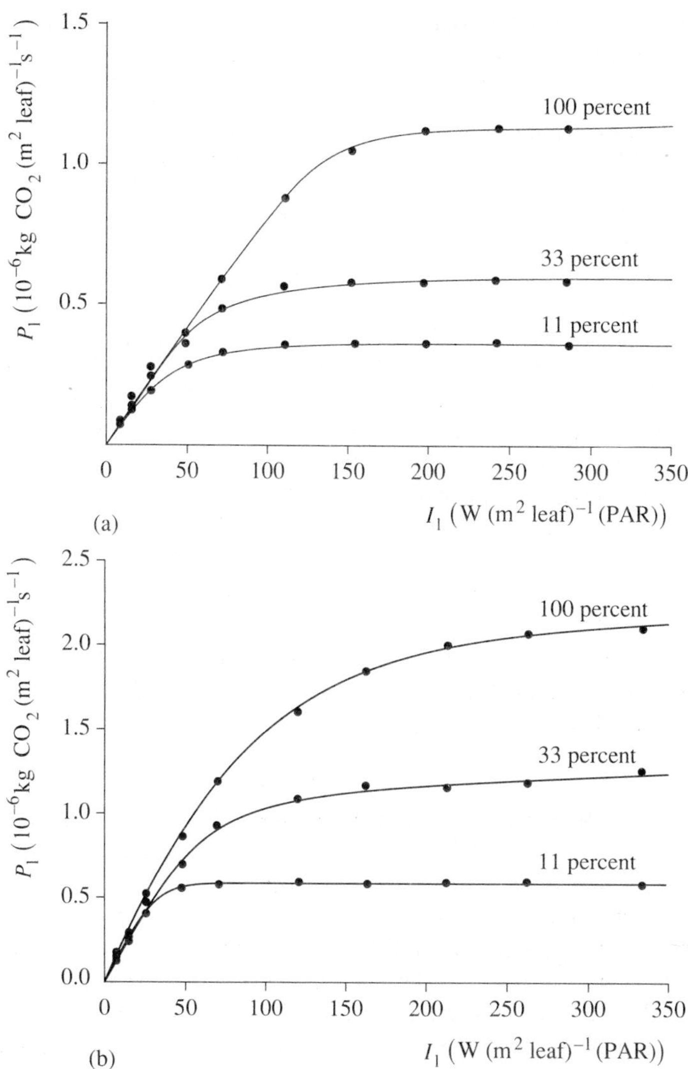

Fig. 10.1. Non-rectangular hyperbola fitted to single-leaf photosynthesis data for (a) Siratro (*Phaseolus atropurpureus*), a tropical legume, and (b) green panic (*Panicum maximum*), a tropical C_4 grass. The data are taken from Ludlow and Wilson (1971). The leaves were grown in 100 per cent, 33 per cent or 11 per cent daylight as indicated. Parameter values are presented in Table 10.1.

Table 10.1. Fitted non-rectangular hyperbola parameter values for the rate of single-leaf photosynthesis for Siratro and green panic

	Percentage growth irradiance	P_m (mg CO_2 m^{-2} s^{-1})	α (mg CO_2 J^{-1})	θ
Siratro	100	1.16	0.0086	0.98
	33	0.61	0.0092	0.90
	11	0.37	0.0080	0.89
Panic	100	2.32	0.020	0.81
	33	1.30	0.019	0.78
	11	0.60	0.017	0.96

very good fit to the data. The fitted parameter values (derived by minimizing the residual sum of squares with respect to P_m, α, and θ) are presented in Table 10.1.

The effect of temperature on leaf photosynthesis was discussed in Section 9.3.3, where it was argued that the main influence is on P_m, as described by eqn (9.11a):

$$P_m = P_m(20)\left(\frac{T - T^*}{20 - T^*}\right), \tag{10.2e}$$

where T (°C) is temperature, T^* is a critical temperature below which P_l is zero, and $P_m(20)$ is the value of P_m at the reference temperature of 20 °C. Throughout the following analysis we shall consider

$$T \geqslant T^*. \tag{10.2f}$$

For temperatures below T^*, P_l, and hence canopy photosynthesis, are zero.

Effect of growth irradiance As well as being influenced by the current environment, the rate of leaf photosynthesis is also influenced by the growth conditions, as is immediately apparent from Fig. 10.1. Of particular importance is the irradiance environment in which the leaf grows, as this may vary greatly through the depth of the canopy. Comparing the parameter values in Table 10.1 for leaves grown in full and 33 per cent daylight shows that α and θ are relatively unaffected by the growth irradiance, whereas there is a marked effect on P_m. The values of α and θ for 11 per cent daylight are affected to some degree, but much less so than P_m. However, since the leaves saturate at lower levels and irradiances, the fitted values for α and θ are less reliable. It is therefore reasonable to assume that the only significant effect of the growth irradiance is on P_m, and that α and θ are constant throughout the depth of the canopy. The fitted P_m values from Table 10.1 are illustrated in Fig. 10.2 in relation to their relative growth environment.

This feature that P_m depends upon the irradiance in which the leaves have been grown has been established for many years (e.g. Woledge 1971). To facilitate analytical progress in calculating the rate of canopy photosynthesis, Ludlow and Charles-Edwards (1980), based on the work of Acock *et al.* (1978), assumed

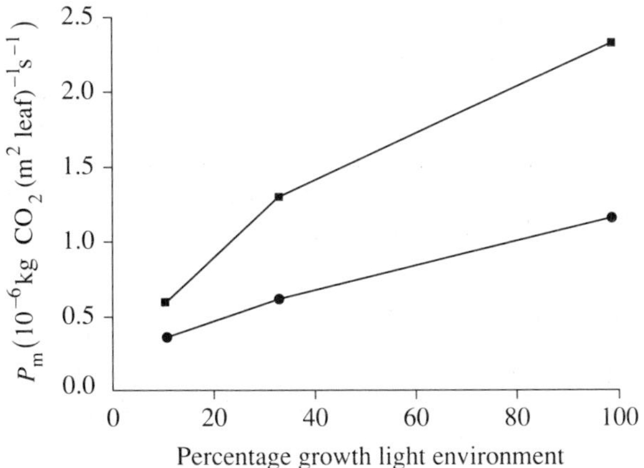

Fig. 10.2. Light-saturated rate of single-leaf photosynthesis P_m in relation to percentage growth light environment for Siratro (●) and green panic (■), corresponding to Fig. 10.1.

$$P_m = P_m{}^0 e^{-kl}, \tag{10.3a}$$

where $P_m{}^0$ is the value of P_m for those leaves at the top of the canopy. This equation implies a linear relationship between P_m and the irradiance in which the leaf was grown which passes through the origin. Acock *et al.* (1978) measured the leaf photosynthetic rate at three different levels within a tomato canopy and observed a decline in P_m. However, rather than supporting eqn (10.3a), their data indicated a similar trend to that in Fig. 10.2 (Acock *et al.* 1978, Table 3). An alternative approach, as used by Johnson and Thornley (1984) and suggested by Acock *et al.* (1978), is to define P_m appropriate to those leaves, or parts of leaves, at the top of the canopy, while recognizing that P_m does vary. The basis for this assumption is that leaves grown in a low irradiance have virtually the same rate of photosynthesis *at that irradiance* as do leaves grown in a much higher irradiance, indicating that the leaves maintain sufficient photosynthetic machinery to function efficiently in their growth environment. Further support for this hypothesis comes from the data of Prioul, Brangeon, and Reyss (1980a,b). It is clear that both these approaches have limitations, and that the data actually suggest an intermediate response. The rationale for adopting them is that they allow analytical progress in calculating the rate of canopy photosynthesis (see below).

A more realistic expression for the relation between P_m and the growth irradiance is

$$P_m = P_m{}^0\left[1 - \lambda\left(1 - \frac{I}{I_0}\right)\right]. \tag{10.3b}$$

When $\lambda = 0$, P_m is constant, and when $\lambda = 1$, P_m is given by eqn (10.3a). This equation is adopted on the basis of the data illustrated in Fig. 10.2 and Acock *et al.* (1978). It will break down as I approaches zero, but for most of the leaves in the canopy it is a reasonable approximation. Since it is clearly more appropriate to have P_m accurately defined in the upper region of the canopy, λ is evaluated from the values for P_m for Siratro and green panic leaves grown in full and 33 per cent daylight only; any error for leaves lower down the canopy will be negligible. The values are 0.71 and 0.70 respectively; the similarity between these numbers is probably just coincidence.

10.2 Monocultures

We will first consider monocultures; this provides a basis for looking at mixtures and row crops. We begin by looking at the instantaneous rate of canopy photosynthesis and then examine the daily rate.

10.2.1 *Instantaneous canopy photosynthesis*

The instantaneous rate of canopy photosynthesis is, in general, given by

$$P = \int_0^L P_l(I_l)\,\mathrm{d}l, \tag{10.4a}$$

with units kg CO_2 (m^2 ground)$^{-1}$ s^{-1}, where l is the cumulative leaf area index and L is the total leaf area index.

The two situations where P_m is constant through the depth of the canopy ($\lambda = 0$ in eqn (10.3b)) and where P_m is proportional to the level of irradiance within the canopy ($\lambda = 1$ in eqn (10.3b)) are considered separately. The accuracy of each approximation is then considered, and a simple modification is then presented which gives an improvement on both approaches.

P_m constant The instantaneous rate of photosynthesis of a monoculture requires combining eqns (10.1a), (10.1b), and (10.2a) in (10.4a). Differentiating eqn (10.1b) and using (10.1a) gives

$$\mathrm{d}l = -\frac{\mathrm{d}I_l}{kI_l}, \tag{10.5a}$$

so that eqn (10.4a) becomes

$$P = -\frac{1}{k}\int_{I_l(l=0)}^{I_l(l=L)} P_l(I_l)\frac{\mathrm{d}I_l}{I_l}. \tag{10.5b}$$

This is a cumbersome integral to evaluate. Fortunately, the solution is available in tables of standard integrals (e.g. Gradshteyn and Ryzhik 1980, pp. 81–4), and P is given by

$$P = \frac{1}{2\theta k}\{F[I_l(l=0)] - F[I_l(l=L)]\}, \tag{10.5c}$$

where $F(x)$ is defined by

$$F(x) = \alpha x - [(\alpha x)^2 + 2P_m(1-2\theta)\alpha x + P_m{}^2]^{1/2}$$
$$- P_m \ln\left\{\frac{[(\alpha x)^2 + 2P_m(1-2\theta)\alpha x + P_m{}^2]^{1/2} + \alpha x + P_m(1-2\theta)}{[(\alpha x)^2 + 2P_m(1-2\theta)\alpha x + P_m{}^2]^{1/2} + (1-2\theta)\alpha x + P_m}\right\}$$
$$+ 2\theta P_m \ln\{[(\alpha x)^2 + 2P_m(1-2\theta)\alpha x + P_m{}^2]^{1/2} + \alpha x + P_m(1-2\theta)\}. \qquad (10.5d)$$

This definition of $F(x)$ differs slightly from that presented by Johnson and Thornley (1984), although the two expressions for canopy photosynthetic rate are identical. The main difference is that two constant terms, specifically $P_m \ln P_m$ and $2\theta P_m \ln 2$, have been omitted. This is quite permissible since P involves the difference of $F(x)$ evaluated for two values of x, so that any constant terms will disappear.

For the special case $\theta = 0$ (rectangular hyperbola), it is readily shown (Exercise 10.1) by taking the limit $\theta \to 0$ that

$$P = \frac{P_m}{k}\ln\left[\frac{\alpha k I_0 + P_m(1-m)}{\alpha k I_0 e^{-kL} + P_m(1-m)}\right], \qquad (10.5e)$$

which is the same as that derived by Thornley (1976).

The initial slope and asymptote of $P(I_0)$ can be shown to be (Exercise 10.1)

$$\frac{dP}{dI_0}(I_0 = 0) = \frac{\alpha}{1-m}(1 - e^{-kL}) \qquad (10.5f)$$

and

$$\lim_{I_0 \to \infty} P = P_m L. \qquad (10.5g)$$

Both these results are physiologically realistic (Exercises 10.1 and 10.2).

P_m proportional to the level of irradiance within the canopy Now consider the case where P_m is defined by eqn (10.3a). In this case, with eqns (10.1a) and (10.1b) for light attenuation and eqn (10.2a) (the non-rectangular hyperbola) for the rate of single-leaf photosynthesis, eqn (10.4a) becomes

$$P = P_l[I_l(l = 0)]\int_0^L e^{-kl}\,dl, \qquad (10.6a)$$

which is simply

$$P = P_l[I_l(l = 0)]\frac{(1 - e^{-kL})}{k}, \qquad (10.6b)$$

where P_l and I_l are given by eqns (10.2a) and (10.1b). Equation (10.6b) is algebraically simpler than eqn (10.5c), although both are analytical expressions for the rate of canopy photosynthesis in terms of the same set of parameters.

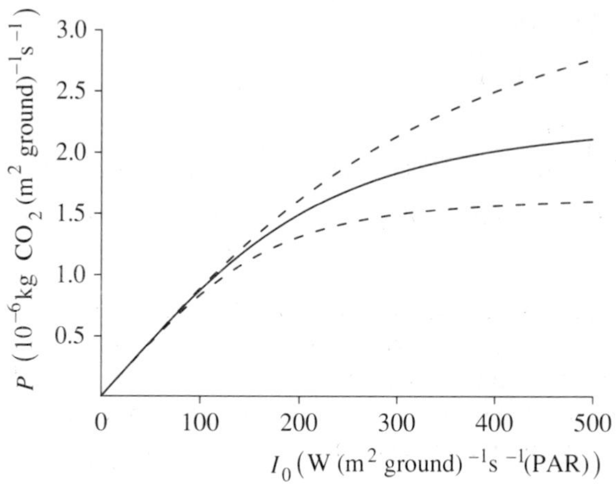

Fig. 10.3. Numerical solution for the rate of canopy photosynthesis, incorporating the variation in P_m in response to the growth environment (solid line). The upper broken line is eqn (10.5c) and the lower one is eqn (10.6b). The parameter values are $P_m = 10^{-6}$ kg CO_2 (m^2 leaf)$^{-1}$ s^{-1}, $\alpha = 0.01 \times 10^{-6}$ kg CO_2 J^{-1}, $\theta = 0.85$, $\lambda = 0.7$, $m = 0.1$, $L = 4$, and $k = 0.5$.

Comparison of the two approaches In order to assess the accuracy of these two expressions for canopy photosynthesis (eqns (10.5c) and (10.6b)) they are compared with the rate of canopy photosynthesis evaluated numerically using eqn (10.2a) in eqn (10.4a), with P_m defined by eqn (10.3b). The results are presented in Fig. 10.3 for typical parameter values. It can be seen that all the curves give a similar response at low irradiance levels. However, as the irradiance increases, eqn (10.5c) overestimates the rate of canopy photosynthesis, while eqn (10.6b) underestimates it. This result is not surprising, and the general trends of Fig. 10.3 apply over a wide range of parameter values

The problem can be overcome quite satisfactorily by ascribing a modified value to P_m:

$$P_m = P_m{}^0\left[1 - \frac{\lambda}{2}\left(1 - e^{-kL}\right)\right]. \tag{10.7a}$$

When L is small, so that $1 - e^{-kL} \simeq 0$, P_m approaches the value for those leaves at the top of the canopy, since all leaves will be growing in high light. Conversely, when L is large and the canopy is dense, $1 - e^{-kL} \simeq 1$, so that $P_m \simeq P_m{}^0(1 - \lambda/2)$. Comparing this with eqn (10.3b) it follows that P_m corresponds to those leaves growing in 50 per cent daylight, which occurs at

$$l = \frac{\ln 2}{k} = \frac{0.69}{k}, \tag{10.7b}$$

and will generally be close to the top of the canopy. For example, if $k = 0.5$, then

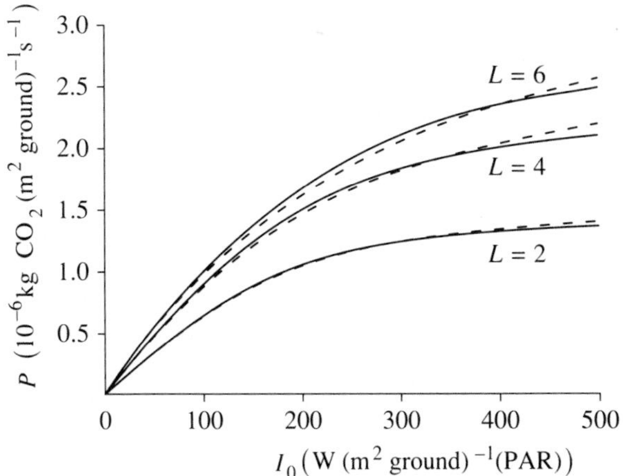

Fig. 10.4. Numerical solution for the rate of canopy photosynthesis, incorporating the variation in P_m in response to the growth environment (solid lines). The broken line is eqn (10.5c), but with P_m given by eqn (10.7a). Parameter values are as in Fig. 10.3, but with $L = 2, 4, 6$ as indicated.

the corresponding value of l is 1.4. Equation (10.5c) with P_m defined by eqn (10.7a) is illustrated in Fig. 10.4, where it is compared with the numerical solution for $L = 2, 4, 6$; in each case the two models give a very similar response. This is typical of the response we observed over a wide range of all the parameter values. There is no direct theoretical basis for eqn (10.7a). However, it is a physiologically plausible approximation, and seems to incorporate the variation in P_m in a manner which works well in practice, as indicated by Fig. 10.4.

Equations (10.5c) and (10.5d) with P_m given by eqn (10.7a) are therefore taken to define the instantaneous rate of canopy photosynthesis. These expressions involve the irradiance I_0 incident on the canopy, the leaf photosynthetic parameter P_m, α, and θ, and the canopy architecture through the extinction coefficient k. Temperature is also included through the dependence of P_m on temperature as given by eqn (10.2e) and discussed in Section 9.3.3.

10.2.2 *Daily canopy photosynthesis*

Write the instantaneous rate of canopy photosynthesis as

$$P = P(I_0, T). \tag{10.8a}$$

I_0 and T will vary throughout the day and, for given diurnal patterns, the daily rate of photosynthesis P_d kg CO_2 (m^2 ground)$^{-1}$ day^{-1} is given by

$$P_d = \int_0^h P \, dt, \tag{10.8b}$$

where t(s) is time and h (s day^{-1}) is the length of the photoperiod.

It is clear that, even for the simple expression for P where the rectangular hyperbola is used to describe the rate of single-leaf photosynthesis (eqn (10.5e)), there is little chance of making analytical progress for arbitrarily defined $I_0(t)$ and $T(t)$. However, using carefully chosen plausible functions for $I_0(t)$ and ignoring variation in $T(t)$, Charles-Edwards and Acock (1977) and Monteith (1981) have derived expressions for P_d using the rectangular hyperbola (eqn (10.2b)) and the Blackman response curve (eqn (10.2c)) respectively for the rate of single-leaf photosynthesis. (It should be noted that Monteith adopted a binomial light attenuation model rather than the Monsi–Saeki model (eqn (10.1a)), but the two approaches are similar.) Such analytical progress is not possible with general expressions for $I_0(t)$ and $T(t)$ or where the non-rectangular hyperbola is used.

To overcome this problem, Thornley's (1976, p. 89) approach to considering photosynthesis in the presence of a variable irradiance environment is adopted and extended to include temperature variation (Johnson and Thornley, 1984). The technique involves expanding $P(I_0, T)$ as a Taylor series about the mean values $\bar{I}_0$ and $\bar{T}$ of I_0 and T up to and including the second-derivative terms. This is

$$P(I_0, T) = P + (I_0 - \bar{I}_0)\frac{\partial P}{\partial I_0} + (T - \bar{T})\frac{\partial P}{\partial T} + \frac{1}{2}(I_0 - \bar{I}_0)^2\frac{\partial^2 P}{\partial I_0{}^2} + (I_0 - \bar{I}_0)(T - \bar{T})\frac{\partial^2 P}{\partial I_0 \partial T} + \frac{1}{2}(T - \bar{T})^2\frac{\partial^2 P}{\partial T^2}, \tag{10.8c}$$

where P and all the derivatives on the right-hand side are evaluated at $I_0 = \bar{I}_0$ and $T = \bar{T}$. Substituting in eqn (10.8b) and integrating gives

$$P_d = hP + \frac{h}{2}\sigma_{I_0}{}^2\frac{\partial^2 P}{\partial I_0{}^2} + h\sigma_{I_0 T}\frac{\partial^2 P}{\partial I_0 \partial T} + \frac{h}{2}\sigma_T{}^2\frac{\partial^2 P}{\partial T^2}; \tag{10.8d}$$

again all terms on the right-hand side are evaluated at $(\bar{I}_0, \bar{T})$. $\sigma_{I_0}{}^2$ and $\sigma_T{}^2$ are the variances of I_0 and T respectively and $\sigma_{I_0 T}$ is their covariance:

$$\sigma_{I_0}{}^2 = \frac{1}{h}\int_0^h (I_0 - \bar{I}_0)^2\,\mathrm{d}t \tag{10.8e}$$

$$\sigma_T{}^2 = \frac{1}{h}\int_0^h (T - \bar{T})^2\,\mathrm{d}t \tag{10.8f}$$

$$\sigma_{I_0 T} = \frac{1}{h}\int_0^h (I_0 - \bar{I}_0)(T - \bar{T})\,\mathrm{d}t. \tag{10.8g}$$

In integrating eqn (10.8d) the first derivatives vanish since, by definition,

$$\int_0^h (I_0 - \bar{I}_0)\,\mathrm{d}t = \int_0^h (T - \bar{T})\,\mathrm{d}t = 0. \tag{10.8h}$$

It is more convenient to express P_d in terms of the coefficients of variation of I_0 and T, rather than the variances. These are defined by

$$v_{I_0} = \frac{\sigma_{I_0}}{\bar{I}_0} \qquad v_T = \frac{\sigma_T}{\bar{T}}, \tag{10.8i}$$

and the covariance is then

$$\sigma_{I_0,T} = \rho\sigma_{I_0}\sigma_T = \rho\bar{I}_0\bar{T}v_{I_0}v_T, \tag{10.8j}$$

where ρ is the correlation coefficient between I_0 and T. Equation (10.8d) now becomes

$$P_d = hP + \frac{h}{2}\bar{I}_0^2 v_{I_0}^2\frac{\partial^2 P}{\partial I_0^2} + h\rho\bar{I}_0\bar{T}v_{I_0}v_T\frac{\partial^2 P}{\partial I_0 \partial T} + \frac{h}{2}\bar{T}^2 v_T^2\frac{\partial^2 P}{\partial T^2}, \tag{10.8k}$$

where, as before, all terms on the right-hand side are evaluated at $(\bar{I}_0, \bar{T})$.

The second-derivative terms must now be evaluated in order to estimate P_d. Rather than working with the expression for P as given by eqn (10.5c) with (10.5d), it is more convenient to work with the integral for P (eqn (10.8b) with (10.2a)) which, using (10.5a), can be written

$$-2\theta kP = \int_{I_{1,0}}^{I_{1,L}} \{(\alpha I_1 + P_m - [(\alpha I_1 + P_m)^2 - 4\theta\alpha I_1 P_m]^{1/2}\}\frac{dI_1}{I_1}, \tag{10.9a}$$

where

$$I_{1,0} = I_1(l = 0) \qquad I_{1,L} = I_1(l = L), \tag{10.9b}$$

with I_1 defined by eqn (10.1b). To calculate the I_0 derivatives we use the following standard theorem of calculus: if

$$\phi(u) = \int_{x_0(u)}^{x_1(u)} \psi(\xi)\, d\xi, \tag{10.9c}$$

then

$$\frac{d\phi}{du} = \psi[x_1(u)]x_1'(u) - \psi[x_0(u)]x_0'(u), \tag{10.9d}$$

where the prime denotes the derivative with respect to u. There are certain restrictions on the form of $\psi(\xi)$ for this theorem to apply, although these do not concern us here: the interested reader should consult any standard calculus text. It therefore follows from eqn (10.9a) that

$$2\theta k\frac{\partial P}{\partial I_0} = \frac{\alpha}{I_0}(I_{1,0} - I_{1,L}) - \frac{1}{I_0}\{[(\alpha I_{1,0} + P_m)^2 - 4\theta\alpha I_{1,0}P_m]^{1/2} - [(\alpha I_{1,L} + P_m)^2 - 4\theta\alpha I_{1,L}P_m]^{1/2}\}, \tag{10.9e}$$

from which we can derive

$$\frac{\partial^2 P}{\partial I_0{}^2} = \frac{1}{I_0{}^2}[f(I_{1,0}, P_m) - f(I_{1,L}, P_m)], \tag{10.9f}$$

where

$$f(I_1, P_m) = \frac{P_m}{2\theta k}\left\{\frac{(1-2\theta)\alpha I_1 + P_m}{[(\alpha I_1 + P_m)^2 - 4\theta\alpha I_1 P_m]^{1/2}}\right\}. \tag{10.9g}$$

To calculate $\partial^2 P/\partial T^2$, first consider $\partial^2 P/\partial P_m{}^2$. Differentiating eqn (10.9a) under the integral sign gives

$$-2\theta k\frac{\partial P}{\partial P_m} = \int_{I_{1,0}}^{I_{1,L}}\left\{1 - \frac{(1-2\theta)\alpha I_1 + P_m}{[(\alpha I_1 + P_m)^2 - 4\theta\alpha I_1 P_m]^{1/2}}\right\}\frac{dI_1}{I_1}, \tag{10.10a}$$

from which, after differentiating again and carrying out some analysis, we obtain

$$\frac{\partial^2 P}{\partial P_m{}^2} = \frac{2(1-\theta)}{k}\int_{I_{1,0}}^{I_{1,L}}\left\{\frac{\alpha I_1}{[(\alpha I_1 + P_m)^2 - 4\theta\alpha I_1 P_m]^{1/2}}\right\}\frac{dI_1}{I_1}. \tag{10.10b}$$

It follows from Gradshteyn and Ryzhic (1980, p. 83) that

$$\frac{\partial^2 P}{\partial P_m{}^2} = \frac{1}{P_m{}^2}[f(I_{1,0}, P_m) - f(I_{1,L}, P_m)], \tag{10.10c}$$

where $f(I_1, P_m)$ is given by eqn (10.9g).

$\partial^2 P/\partial I_0 \partial P_m$ is evaluated directly from (10.9e) as

$$\frac{\partial^2 P}{\partial I_0 \partial P_m} = -\frac{1}{I_0 P_m}[f(I_{1,0}, P_m) - f(I_{1,L}, P_m)]. \tag{10.10d}$$

Finally, to obtain the derivatives with respect to T rather than P_m, note that from eqn (10.2e)

$$\frac{\partial}{\partial T} = \frac{P_m(20)}{20 - T^*}\frac{\partial}{\partial P_m} \tag{10.10e}$$

and

$$\frac{\partial^2}{\partial T^2} = \left[\frac{P_m(20)}{20 - T^*}\right]^2\frac{\partial^2}{\partial P_m{}^2} \tag{10.10f}$$

which can be rewritten as

$$\frac{\partial}{\partial T} = \frac{P_m}{T - T^*}\frac{\partial}{\partial P_m} \tag{10.10g}$$

and

$$\frac{\partial^2}{\partial T^2} = \left(\frac{P_m}{T - T^*}\right)^2\frac{\partial^2}{\partial P_m{}^2}, \tag{10.10h}$$

so that the required derivatives are

$$\frac{\partial^2 P}{\partial I_0{}^2} = \frac{1}{I_0{}^2}[f(I_{1,0}, P_m) - f(I_{1,L}, P_m)] \tag{10.10i}$$

$$\frac{\partial^2 P}{\partial I_0 \partial T} = -\frac{1}{I_0(T - T^*)}[f(I_{1,0}, P_m) - f(I_{1,L}, P_m)] \tag{10.10j}$$

$$\frac{\partial^2 P}{\partial T^2} = \frac{1}{T - T^*}[f(I_{1,0}, P_m) - f(I_{1,L}, P_m)]. \tag{10.10k}$$

It is left to the reader to show that (Exercise 10.3)

$$f(I_{1,0}, P_m) - f(I_{1,L}, P_m) \leqslant 0, \tag{10.11a}$$

(where the equality holds when $L = 0$) so that the variance correction terms are negative and the covariance term is positive. Note that for $\theta = 0$ (so that the leaf photosynthetic rate is described by the rectangular hyperbola) $f(I_1, P_m)$ is indeterminate, but it can be shown (Exercise 10.3) that as $\theta \to 0$

$$f(I_{1,0}, P_m) - f(I_{1,L}, P_m) \to -\frac{P_m}{k}\left[\left(\frac{\alpha I_{1,0}}{\alpha I_{1,0} + P_m}\right)^2 - \left(\frac{\alpha I_{1,L}}{\alpha I_{1,L} + P_m}\right)^2\right]. \tag{10.11b}$$

When eqns (10.10i)–(10.10k) are substituted in eqn (10.8k), P_d is obtained as

$$P_d = hP(\bar{I}_0, \bar{T}) - \frac{h}{2}[f(\bar{I}_{1,0}, \bar{P}_m) - f(\bar{I}_{1,L}, \bar{P}_m)]g(\bar{T}), \tag{10.12a}$$

where $\bar{P}_m = P_m(\bar{T})$, $\bar{I}_1 = I_1(\bar{I}_0)$, and $g(\bar{T})$ is defined by

$$g(\bar{T}) = \left[v_{I_0} - v_T\left(\frac{\bar{T}}{\bar{T} - T^*}\right)\right]^2 + 2v_{I_0}v_T(1 - \rho)\left(\frac{\bar{T}}{\bar{T} - T^*}\right). \tag{10.12b}$$

Equation (10.12a) defines the daily rate of photosynthesis in terms of the single-leaf photosynthesis parameters, the canopy architecture, the mean daily irradiance, the mean daily temperature, the irradiance and temperature coefficients of variation and their correlation coefficient, and the day length.

Assessment of the accuracy of the approximation technique In order to obtain some indication of the accuracy of this approximation technique for calculating the daily rate of photosynthesis, consider the daily irradiance and temperature distributions given by

$$I_0 = \bar{I}_0 \frac{\pi}{2} \sin\left(\frac{\pi t}{h}\right) \qquad 0 \leqslant t \leqslant h \tag{10.13a}$$

$$T = \tau_1 + \tau_2 \sin\left[\frac{\pi}{h}(t - \phi)\right] \qquad 0 \leqslant t \leqslant h, \tag{10.13b}$$

where

$$\tau_1 = \frac{\bar{T} - (2T_m/\pi)\cos(\pi\phi/h)}{1 - (2/\pi)\cos(\pi\phi/h)} \tag{10.13c}$$

$$\tau_2 = \frac{T_m - \bar{T}}{1 - (2/\pi)\cos(\pi\phi/h)}, \tag{10.13d}$$

and T_m is the maximum daily temperature occurring at time

$$t = \frac{h}{2} + \phi. \tag{10.13e}$$

The equation for I_0 is that proposed by Monteith (1965) for a cloudless day, or a day with constant cloud cover, and where I_0 takes its maximum value at mid-day. The maximum temperature T_m is shifted by ϕ s from mid-day, and typically $\phi = 10\,800$ s $= 3$ h. These expressions for I_0 and T are illustrated in Fig. 10.5 for the environmental conditions given in Table 10.2 (which correspond to good summer conditions in southern Britain). It can be shown (Exercise 10.4) that ν_{I_0}, ν_T, and ρ are given by

$$\nu_{I_0} = \left(\frac{\pi^2}{8} - 1\right)^{1/2} \tag{10.13f}$$

$$\nu_T = \frac{(T_m/\bar{T} - 1)[\pi^2/8 - \cos^2(\pi\phi/h)]^{1/2}}{\pi/2 - \cos(\pi\phi/h)} \tag{10.13g}$$

$$\rho = \left[\frac{\pi^2/8 - 1}{\pi^2/8 - \cos^2(\pi\phi/h)}\right]^{1/2} \cos\left(\frac{\pi\phi}{h}\right). \tag{10.13h}$$

A numerical estimate of the daily photosynthetic rate, denoted by $P_{d,num}$, was

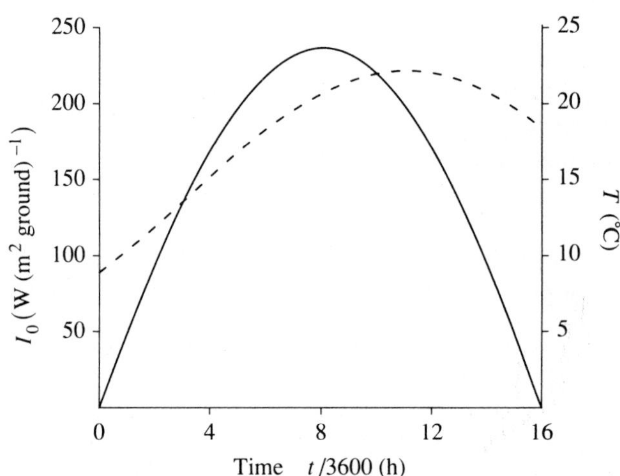

Fig. 10.5. Diurnal patterns of irradiance (solid line) and temperature (broken line) as given by eqns (10.13a)–(10.13d) and Table 10.2.

Table 10.2. Parameter values used to calculate daily canopy photosynthesis

(a) *Environmental parameters*

$\bar{I}_0$ ($W\,m^{-2}$)	$\bar{T}$ (°C)	T_m (°C)	h (s day^{-1})	ϕ (s)
150	18	22	57 600	10 800

(b) *Leaf photosynthetic and canopy architectural parameters*

$P_m(20)$ (kg $CO_2\,m^{-2}\,s^{-1}$)	α (kg $CO_2\,J^{-1}$)	T^* (°C)	k	m
10^{-6}	10^{-8}	−5	0.5	0.1

calculated using eqns (10.13a) and (10.13b) with (10.13c), (10.13d), and the parameters given in Table 10.2. The results obtained using the approximation technique of eqn (10.12a) for P_d were compared with $P_{d,num}$ by considering the percentage errors

$$\varepsilon_1 = 100 \times \frac{|P_{d,num} - P_{d,1}|}{P_{d,num}}, \tag{10.13i}$$

where

$$P_{d,1} = hP(\bar{I}_0, \bar{T}) \tag{10.13j}$$

and

$$\varepsilon = 100 \times \frac{|P_{d,num} - P_d|}{P_{d,num}}. \tag{10.13k}$$

ε_1 is the percentage error which results from using the mean values only of I_0 and T (i.e. only the first term on the right-hand side of (10.12a) is used), whereas ε is the percentage error when the approximation procedure of eqn (10.12a) is adopted. These error estimates are presented in Table 10.3 for $\theta = (0, 0.5, 0.95)$ and $L = (1, 3, 5, 10)$, along with the numerical values of $P_{d,num}$, P_d, and $P_{d,1}$. (The error estimates may not all be obtained directly from Table 10.3 since the figures have been rounded after being calculated.) It is apparent that the first term $P_{d,1}$ in the expression for P_d provides a good approximation to $P_{d,num}$; the largest relative error is $\varepsilon_1 = 5.9$ per cent which occurs for $L = 1$ and $\theta = 0.95$. Using the next-order terms in eqn (10.12a), the errors are seen to decrease in all cases, and the largest is now $\varepsilon = 1.3$ per cent (again for $L = 1$ and $\theta = 0.95$). Evidently, the reasonably accurate results obtained with the leading-order terms are a consequence of the correction terms being of second order. When the correction terms are included, the results indicate that the approximation procedure provides an extremely good estimate to the numerical value of the daily photosynthetic rate.

Table 10.3. Daily gross photosynthetic integral and error estimates

L	$P_{\mathrm{d,num}}$ (kg CO_2 m^{-2} day^{-1})	P_{d} (kg CO_2 m^{-2} day^{-1})	$P_{\mathrm{d,1}}$ (kg CO_2 m^{-2} day^{-1})	ε (per cent)	ε_1 (per cent)
(a) $\theta = 0$					
1	0.0208	0.0210	0.0219	1.1	5.5
3	0.0467	0.0470	0.0488	0.8	4.6
5	0.0584	0.0587	0.0607	0.6	4.1
10	0.0653	0.0656	0.0677	0.6	3.7
(b) $\theta = 0.5$					
1	0.0241	0.0243	0.0255	0.7	5.8
3	0.0533	0.0535	0.0557	0.4	4.5
5	0.0657	0.0660	0.0683	0.3	3.9
10	0.0728	0.0730	0.0754	0.3	3.6
(c) $\theta = 0.95$					
1	0.0324	0.0320	0.0343	1.3	5.9
3	0.0677	0.0673	0.0700	0.6	3.4
5	0.0811	0.0807	0.0834	0.5	2.9
10	0.0883	0.0879	0.0907	0.5	2.6

This technique for calculating the daily canopy photosynthetic rate requires the mean daily values of irradiance and temperature as well as their variances and covariance as environmental inputs. Meteorological stations generally measure irradiance and temperature (as well as other parameters) at frequent intervals. It may be of considerable value if such data are routinely analysed to provide the inputs for models such as the one presented here, since a few parameters may provide an acceptably accurate description of the diurnal pattern of weather variation.

10.3 Mixtures

We now turn our attention to mixed canopies. For convenience, we restrict attention to a binary mixture of grass and clover, although the development to more components is straightforward. The procedure is to derive the analogous equations to (10.5c) for the instantaneous rate of photosynthesis of each component, and then use expressions of the form (10.7a) for their P_{m} values. The treatment for the daily canopy photosynthetic rate is then derived in a similar manner as for the monoculture.

10.3.1 *Instantaneous canopy photosynthesis of the components*

The rate of canopy photosynthesis of either component is

$$P_i = \int_0^{L_i} P_{1,i}(I_{1,i})\,\mathrm{d}l_i, \qquad (10.14a)$$

in units of kg CO_2 (m^2 ground)$^{-1}$ s^{-1}, where i = g or c refers to either the grass or clover component. L_g and L_c are the total leaf area indices of the grass and clover respectively. The biological interpretation of eqn (10.14a) is that the photosynthetic rate of the individual leaves of the ith component $P_{l,i}$ is summed over the total leaf area index of that component as it varies according to the irradiance incident $I_{l,i}$ on those leaves. The irradiance incident on the leaves for either component is given by eqn (8.8g):

$$I_{l,i} = \left(\frac{k_i}{1 - m_i}\right) I_0 \exp[-(k_g l_g + k_c l_c)] = \left(\frac{k_i}{1 - m_i}\right) I, \qquad (10.14b)$$

which will depend on both l_g and l_c. It is convenient to write eqn (10.14a) as

$$P_i = \int_0^L P_{l,i}(I_{l,i}) \frac{dl_i}{dl} dl \qquad (10.14c)$$

where

$$l = l_g + l_c \qquad L = L_d + L_c. \qquad (10.14d)$$

Equation (10.14c) defines the rate of photosynthesis of either component of the canopy. In order to evaluate this integral dl_i/dl must be defined, which requires defining the relative distribution of each component through the depth of the sward. For arbitrary leaf distributions this integral will not, in general, be analytically soluble. However, provided that the leaf distribution is known, the solution can be evaluated numerically.

According to our investigations, there are two realistic situations for which eqn (10.14c) can be integrated analytically, regardless of whether the non-rectangular hyperbola (eqn (10.2a)) or the simpler rectangular hyperbola (eqn (10.2b)) is used. The first is for crops which do not intersect (e.g. trees and grass) in which case the analysis is a simple extension of that for monocultures and is not considered further. The second is when l_g and l_c are homogeneously distributed throughout the canopy which, as demonstrated by Johnson, Parsons, and Ludlow (1989) and discussed in Chapter 8, applies to continuously grazed or frequently cut grass–legume swards and therefore has an important practical application. In this case the light attenuation and interception by the canopy is described by (see eqns (8.9f)–(8.9i))

$$l_i = \frac{L_i}{L} l \qquad (10.15a)$$

and

$$I_{l,i} = \frac{k_i}{1 - m_i} I_0 \exp(-k_e l), \qquad (10.15b)$$

where the effective extinction coefficient k_e is given by

$$k_e = \frac{k_g l_g + k_c l_c}{l} = \frac{k_g L_g + k_c L_c}{L} \qquad (10.15c)$$

Thus

$$\frac{\mathrm{d}l_i}{\mathrm{d}l} = \frac{L_i}{L}, \tag{10.15d}$$

and hence eqn (10.14c) can be written

$$P_i = -\frac{1}{k_e}\frac{L_i}{L}\int_{I_{1,i}(l=0)}^{I_{1,i}(l=L)} P_{1,i}(I_{1,i})\frac{\mathrm{d}I_{1,i}}{I_{1,i}}. \tag{10.15e}$$

Note that the extinction coefficient factor appearing outside the integral is k_e and not k_i. P_i is readily evaluated as a direct extension of the monoculture theory to give

$$P_i = \frac{L_i}{L}\frac{1}{2\theta_i k_e}\{F[I_{1,i}(l=0)] - F[I_{1,i}(l=L)]\}, \tag{10.15f}$$

where $F(x)$ is defined by eqn (10.5d). Equation (10.15f) defines the canopy photosynthetic rate of either component. As for the monoculture, $P_{m,g}$ and $P_{m,c}$ are defined to incorporate the effect of variation due to the growth light environment so that, analogous to eqn (10.7a),

$$P_{m,i} = P_{m,i}{}^0\left\{1 - \frac{\lambda}{2}[1 - \exp(-k_e L)]\right\}. \tag{10.15g}$$

It is worth noting that, in terms of the total canopy photosynthetic rate

$$P = P_g + P_c, \tag{10.15h}$$

there is no simple expression relating P_g or P_c to P. (This applies to the full range of θ values, $0 \leqslant \theta < 1$.)

To illustrate the response of the photosynthetic rate of the mixture to irradiance, the following parameter values are used:

$$k_g = 0.5 \qquad k_c = 0.8$$

and

$$P_m{}^0 = 0.75 \times 10^{-6}\,\mathrm{kg\,CO_2\,(m^2\,leaf)^{-1}\,s^{-1}} \qquad \alpha = 10^{-8}\,\mathrm{kg\,CO_2\,J^{-1}} \qquad \theta = 0.65, \tag{10.16a}$$

which are typical for grass and clover. Grass and clover have similar photosynthetic characteristics (Woledge and Dennis, 1982), and so their leaf photosynthetic parameters are taken to be identical. The essential feature to note is that $k_c > k_g$ which is a consequence of the more prostrate nature of the clover leaves. In Fig. 10.6(a) the total canopy photosynthetic rate P (eqn (10.15h)) is shown for $L = 4$ with $(L_g, L_c) = (2, 2), (1, 3), (3, 1)$. It can be seen that there is little difference between P for the various grass–clover contributions. The corresponding fractional contribution by the clover

$$f_c = P_c/P \tag{10.16b}$$

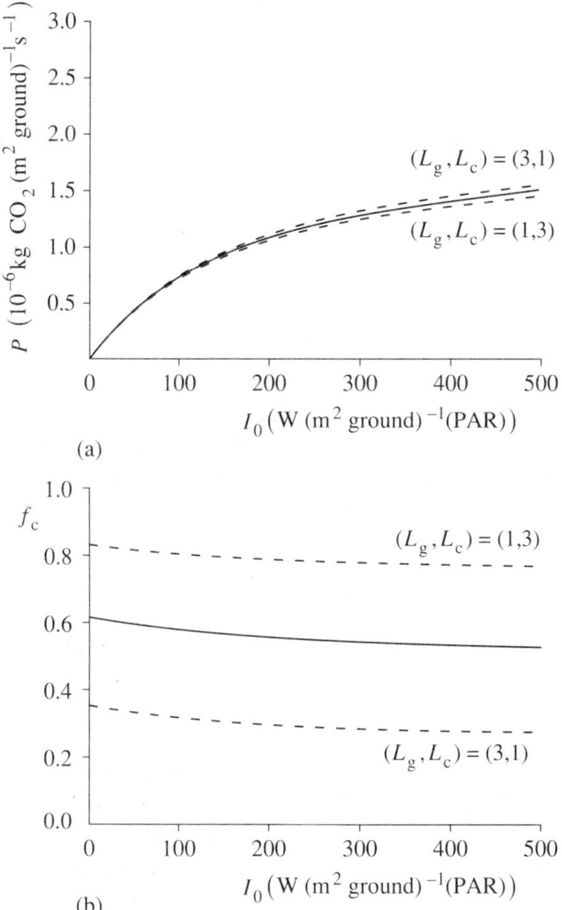

Fig. 10.6. (a) Total canopy photosynthetic rate P for the mixed canopy, given by eqn (10.15h) with (10.15f); (b) the corresponding contribution f_c made by the clover (eqn (10.16b)). The solid lines are for $(L_g, L_c) = (2, 2)$, and the broken lines are for $(L_g, L_c) = (1, 3), (3, 1)$ as indicated.

is shown in Fig. 10.6b, and it is apparent that this is always greater than the fraction L_c/L. This is due to the parameter values for k_g and k_c.

10.3.2 *Daily canopy photosynthesis of the components*

We can now extend the monoculture theory for the daily canopy photosynthetic rate to mixtures. Care must be taken, since eqn (10.15f) for the instantaneous rate of photosynthesis of the mixture includes k_i and k_e, whereas eqn (10.5c) for the monoculture has only a single k parameter. The procedure is to define an adjusted leaf area index parameter L_i^* as

$$L_i^* = \frac{k_e}{k_i} L \tag{10.17a}$$

so that eqn (10.15f) can be written

$$P_i = \frac{L_i}{L} \frac{1}{2\theta_i k_e} \left\{ F\left[\left(\frac{k_i}{1 - m_i} \right) I_0 \right] - F\left[\left(\frac{k_i}{1 - m_i} \right) I_0 \exp(-k_i L_i^*) \right] \right\} \tag{10.17b}$$

Thus

$$P_i = \frac{k_i L_i}{k_e L} P_{mono}(k_i, m_i, L_i^*), \tag{10.17c}$$

where $P_{mono}(k_i, m_i, L_i^*)$ is the photosynthetic rate of a monoculture as given by eqn (10.5c) evaluated for extinction coefficient k_i, transmission coefficient m_i, and leaf area index L_i^*. The daily canopy photosynthetic rate of the ith component of the mixture is then obtained directly as

$$P_{d,i} = \frac{k_i L_i}{k_e L} P_{d,mono}(k_i, L_i^*), \tag{10.17d}$$

where $P_{d,mono}$ is defined analogously to P_{mono}. (Exercise 10.5)

10.4 Row crops

In Chapter 8 we gave a very simplified account of light interception and attenuation in row crops. This involved calculating the parameter T_b for the proportion of the irradiance which is transmitted to the ground under the conditions that no light passes through the canopy (T_b is not to be confused with temperature) and then defining the irradiance incident on the leaves of the canopy as (eqn (8.13a))

$$I_1 = \left(\frac{k}{1 - m} \right) I_0 \exp(-kl'), \tag{10.18a}$$

where

$$l' = \frac{l}{1 - T_b}. \tag{10.18b}$$

The theory for monocultures, for both instantaneous and daily canopy photosynthetic rates, can therefore be applied directly to row crops, first by calculating T_b (see Section 8.3) and then by using

$$L' = \frac{L}{1 - T_b} \tag{10.18c}$$

for the leaf area index.

10.5 Discussion

We have presented analytical models for the rate of photosynthesis in monoculture, mixture, and row crops. These models are based on using the non-rectangular hyperbola to describe the rate of leaf photosynthesis in response to incident irradiance and Beer's law for light attenuation through the canopy. To calculate the daily rate of canopy photosynthesis, the analysis has incorporated the effects of temperature by assuming that P_{m} is linearly related to temperature. This is likely to be quite realistic in many situations. There will, of course, be situations where the level of approximation at the leaf photosynthesis level, or the description of light interception, is oversimplified. For example, we have not included the possibility of diurnal variation in the stomatal conductance (in response to factors such as leaf water potential or carbohydrate content) which may affect the rate of leaf photosynthesis. In such cases it will be necessary use the general concepts developed here and evaluate the integrals—either through the depth of the canopy or throughout the day—numerically.

Exercises

10.1. (a) Show that in the limit $\theta \to 0$, the rate of photosynthesis P for the monoculture canopy, as defined by eqns (10.5c) and (10.5d), reduces to eqn (10.5e).

(b) Derive eqns (10.5f) and (10.5g) for the initial slope and asymptote of the canopy photosynthetic rate P for the monoculture in response to irradiance. Explain why these expressions are physiologically realistic.

10.2. Using eqn (10.5e) for the rate of canopy photosynthesis, with $P_{\mathrm{m}} = 10^{-6}$ kg CO_2 m^{-2} s^{-1}, $\alpha = 10^{-8}$ kg CO_2 J^{-1} and $m = 0.1$, plot $P(I_0)$ for $L = (2, 8)$ and $k = (0.2, 0.5, 0.8)$. Comment on the features of these curves.

10.3. (a) Show that the inequality (10.11a) is true. (*Hint*: consider the derivative $\partial f/\partial I_1$.)

(b) Derive the limit given by eqn (10.11b).

10.4. Derive v_{I_0}, v_T, and ρ (eqns (10.13f)–(10.13h)) for I_0 and T defined by eqns (10.13a)–(10.13d).

10.5. Give a physiological interpretation of the parameters k_{e} (eqn (10.15c)), L_i^* (eqn (10.17a)), and the factor $k_i L_i / k_{\mathrm{e}} L$ in eqn (10.17c).

11
Whole-plant respiration and growth energetics

11.1 Introduction

Respiratory activity and the production of CO_2 are closely related to the growth and yield of plants and crops. The two conservation laws, of matter and of energy, provide an irreducible framework within which phenomenological or mechanistic considerations of the problem must fit. Indeed, application of these laws to plant growth allows some useful general relationships to be constructed. The continuity equation, denoting the conservation of the various species of matter, can be expressed for any or all of the chemical constituents of a plant—carbon, oxygen, nitrogen, phosphorus, etc.—but because carbon plays a dominant role in plant growth, it is usual to focus on carbon as the most important component in plant metabolism.

Respiration, i.e. the evolution of CO_2 and the consumption of O_2, has long been a subject for investigation in animal studies, and a terminology has developed in conjunction with this work. This terminology reflects the fact that animals tend to spend a period of their lives in vigorous growth, with the major portion of their life-span being passed in a mature and approximately steady state condition. Thus respiratory activity is frequently considered to result from these two processes: growth and maintenance of the status quo, although it is unclear what this division means at the biochemical level.

From this viewpoint, Pirt (1965) demonstrated experimentally that the respiration rate R (kg CO_2 day^{-1}) of some bacterial populations could be separated theoretically into two components, one proportional to the growth rate dW/dt, where W is dry mass (kg) and t is time (day), and the second proportional to the dry mass W. This gives

$$R = a\frac{\mathrm{d}W}{\mathrm{d}t} + bW, \tag{11.1a}$$

where a (kg CO_2 (kg dry mass)$^{-1}$) and b (kg CO_2 (kg dry mass)$^{-1}$ day^{-1}) are constants. A simple regression of the respiration rate R on the growth rate $\mathrm{d}W/\mathrm{d}t$ and on dry mass W gives the constants a and b, which are often interpreted as being concerned with the underlying processes of growth and maintenance.

Relationships such as eqn (11.1a) have been shown to be valid in the plant world (McCree 1970), and there have been various attempts to elucidate, at levels ranging from detailed biochemistry to that of the whole plant, what this all means and what it implies for crop growth and yield (the subject has been reviewed recently by Johnson (1987)). In the present chapter these matters are considered at the whole-plant level, or at the level of hightly aggregated biochemistry. In the next chapter we focus on the biochemical and chemical approaches to plant growth and respiration.

11.2 Phenomenology of respiration

McCree (1970) described some experiments on white clover plants which were concerned with the relationships between respiration rate R (kg CO_2 day^{-1} plant^{-1}), gross photosynthesis P (kg CO_2 day^{-1} plant^{-1}) during the light period, and plant dry mass W (kg CO_2 plant^{-1}). It is convenient to express plant mass W in terms of CO_2 equivalents. Both photosynthesis P and plant mass W were varied, and as a result the respiration rate R also varied. McCree found that the following equation described his data:

$$R = kP + cW, \tag{11.2a}$$

where k and c are constants. For white clover

$$k = 0.25 \qquad c = 0.015 \text{ day}^{-1}. \tag{11.2b}$$

These values imply that the carbon lost through respiration amounts to 25 per cent of photosynthesis and 1.5 per cent of dry mass per day. This equation is similar to eqn (11.1a) for bacterial growth, and analogously divides respiration into two components which can tentatively be identified with growth and maintenance. The physiological ideas behind eqn (11.2a) are first that there is a loss in material when converting the immediate products of photosynthesis into plant material, and second that some basal metabolism is required to maintain the current status of the plant. It is found that the coefficient k varies considerably with the type of plant tissue, as would be expected on simple energetic and biochemical considerations, and that the coefficient c associated with maintenance is also very variable, depending on tissue, organ, and plant age. Many of the approaches to respiration can be regarded as attempts to gain some understanding of the k and c coefficients of eqn (11.2a).

11.3 Substrate balance analysis of respiration

Considerations of substrate balance formed the basis of Pirt's (1965) analysis of bacterial growth and respiration; this approach was extended to plants, principally by Thornley (1970, 1971, 1976), and an account of this is given below.

It is assumed that a plant or a part of a plant is supplied with a quantity of substrate Δs during a time interval Δt. The substrate is measured in carbon units, and it is convenient to think of the time interval Δt as being equal to 1 day. It is also assumed that the substrate supplied is completely utilized by the end of the time interval. It is considered that the substrate can only be used in two ways: for growth of the plant or for its maintenance. Thus we write

$$\Delta s = \Delta s_g + \Delta s_m, \tag{11.3a}$$

where Δs_g is used for growth and Δs_m is used for maintenance. Whilst all the substrate in the maintenance term Δs_m is respired, only a part of the substrate in the growth term Δs_g is respired and the remainder is converted into an increment Δw in plant material, where w is dry mass (in carbon units). If Δs_r is the part of the growth substrate Δs_g which is respired, then conservation of material gives

$$\Delta s_g = \Delta w + \Delta s_r. \tag{11.3b}$$

Based on this equation, a conversion efficiency or yield Y_g for the growth process alone (in the absence of maintenance) can be defined as

$$Y_g = \frac{\Delta w}{\Delta s_g} = \frac{\Delta w}{\Delta w + \Delta s_r}. \tag{11.3c}$$

Note that Y_g will depend upon the type of plant material that is under consideration (see Table S12.1, p. 625).

A second conversion efficiency or yield Y can also be defined; this includes both growth and maintenance processes, and is therefore given by

$$Y = \frac{\Delta w}{\Delta s} = \frac{\Delta w}{\Delta w + \Delta s_r + \Delta s_m}. \tag{11.3d}$$

It is clear that the overall conversion efficiency Y of eqn (11.3d) must always be less than the growth conversion efficiency Y_g of eqn (11.3c). The relationship between Y and Y_g can be obtained by inverting eqn (11.3d) and using eqn (11.3c) to substitute for $(\Delta w + \Delta s_r)/\Delta w$, giving

$$\frac{1}{Y} = \frac{1}{Y_g} + \frac{\Delta s_m}{\Delta w}. \tag{11.3e}$$

It is convenient to define a maintenance coefficient m (per day) by

$$m = \frac{1}{w}\frac{\Delta s_m}{\Delta t}. \tag{11.3f}$$

Therefore, substituting into eqn (11.3e),

$$\frac{1}{Y} = \frac{1}{Y_g} + mw\frac{\Delta t}{\Delta w}. \tag{11.3g}$$

The specific growth rate μ of the plant is defined by

$$\mu = \frac{1}{w}\frac{\Delta w}{\Delta t}. \tag{11.3h}$$

Substituting eqn (11.3h) into eqn (11.3g) gives

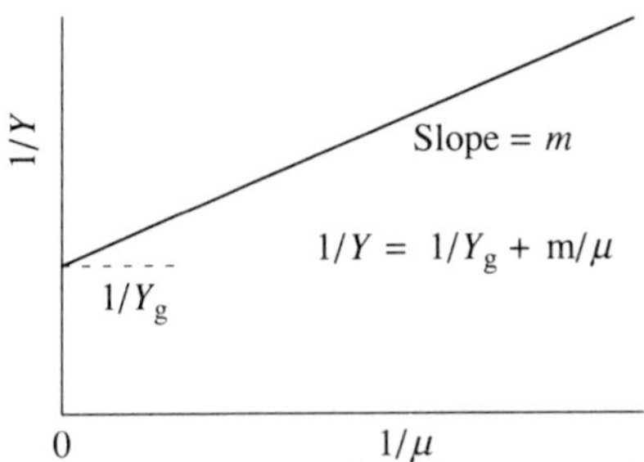

Fig. 11.1. Relationship between the conversion efficiency Y and the specific growth rate μ, illustrating eqn (11.3i). Y_g is the growth conversion efficiency (eqn (11.3c)) and m is the maintenance coefficient (eqn (11.3f)). The straight line describes the situation where Y_g and m are constant, where $1/Y_g$ is given by the intercept and m by the slope as shown.

$$\frac{1}{Y} = \frac{1}{Y_g} + \frac{m}{\mu}. \tag{11.3i}$$

If it is possible to vary the specific growth rate μ experimentally (for instance by varying substrate supply or photosynthetic rate) and the conversion efficiency Y can be determined, then the plot of $1/Y$ again $1/\mu$ shown in Fig. 11.1 tests the assumption that the growth conversion efficiency Y_g and the maintenance coefficient μ are constant. Given that the data lie on a straight line, Y_g and m are easily calculated. Note that for fast rates of growth, maintenance costs become less important and the conversion efficiency Y approaches its maximum value of Y_g; for slower specific growth rates, the conversion efficiency falls progressively ($1/Y$ increases) as the maintenance costs become relatively more important.

If there is a component of the maintenance requirement that is proportional to the specific growth rate μ, then this is indistinguishable from a reduced growth conversion efficiency Y_g. This can be seen by substituting

$$m = m_o + a\mu \tag{11.3j}$$

into eqn (11.3i); m_o and a are constants.

The respiration rate R (kg CO_2 day^{-1}) is given by

$$R\Delta t = \Delta s_m + \Delta s_r, \tag{11.3k}$$

since these are the components of eqns (11.3a) and (11.3b) that are actually respired. Therefore, combining eqns (11.3a) and (11.3b) and substituting into eqn (11.3k), gives

$$\Delta s = \Delta w + R\Delta t. \tag{11.3l}$$

From eqn (11.3c)

$$\Delta w = \Delta s_r \left[\frac{Y_g}{(1 - Y_g)} \right]. \tag{11.3m}$$

Eliminating Δs_m between eqns (11.3f) and (11.3k) gives

$$R\Delta t = mw\Delta t + \Delta s_r \tag{11.3n}$$

and then using eqn (11.3m) to eliminate Δs_r gives

$$R\Delta t = mw\Delta t + \left(\frac{1 - Y_g}{Y_g}\right)\Delta w. \tag{11.3o}$$

This equation gives the respiration occurring during the time interval Δt in terms of the net dry mass increment Δw. Substituting for Δw with eqn (11.3l) gives an alternative equation in terms of the gross supply of substrate Δs, i.e.

$$R\Delta t = Y_g mw\Delta t + (1 - Y_g)\Delta s. \tag{11.3p}$$

This equation is equivalent to eqn (11.2a), which was used by McCree (1970) for describing respiration phenomenologically.

Defining a specific respiration rate r by

$$r = R/w \tag{11.3q}$$

and dividing eqn (11.3o) through by $w\Delta t$, we therefore obtain (with eqn (11.3h))

$$r = \left(\frac{1 - Y_g}{Y_g}\right)\frac{1}{w}\frac{\Delta w}{\Delta t} + m = \mu\left(\frac{1 - Y_g}{Y_g}\right) + m. \tag{11.3r}$$

If the time course of dry mass w and the specific respiration rate r have been experimentally determined and are summarized in two empirical equations

$$w = w(t) \qquad r = r(t), \tag{11.3s}$$

then these equations can be substituted into eqn (11.3r) to test the applicability and assumptions of the above analysis; if eqns (11.3r) and (11.3s) are consistent with each other, then estimates of the parameters Y_g and m can be obtained (see the cotton boll analysis below).

11.3.1 *Two applications of substrate balance analysis*

White clover The results of McCree's (1970) investigation of white clover growth are described in eqns (11.2a) and (11.2b). We interpret these in terms of eqn (11.3p), giving

$$1 - Y_g = 0.25 \qquad Y_g m = 0.015. \tag{11.4a}$$

This leads to values for the two parameters of

$$Y_g = 0.75 \qquad m = 0.02\ \text{day}^{-1}. \tag{11.4b}$$

The interpretation of the experimental regression coefficients is changed subtly from that given after eqn (11.2b): now the maintenance loss is 2.0 per cent of the dry mass per day. The difference arises from the basis of the respiration equation in net or in gross increments, as seen by comparing eqns (11.3o) and (11.3p).

Cotton bolls Thornley and Hesketh (1972) applied eqn (11.3r) to measurements of growth and respiration of cotton bolls as a function of time. Their analysis makes use of the growth function method (Chapter 3) and gives parameter estimates that apply over a time period.

An exponential cubic equation is sometimes found to give a reasonable fit to dry mass : time data:

$$\ln w = a_0 + a_1 t + a_2 t^2 + a_3 t^3 \tag{11.4c}$$

where t is time (day), and a_0, a_1, a_2, and a_3 are constants. Differentiation of this equation gives

$$\frac{1}{w}\frac{\Delta w}{\Delta t} = a_1 + 2a_2 t + 3a_3 t^2 \tag{11.4d}$$

for the specific growth rate (eqn (11.3h)). Substitution of eqn (11.4d) into eqn (11.3r) leads to

$$r = \left[m + \left(\frac{1 - Y_g}{Y_g}\right)a_1\right] + \left(\frac{1 - Y_g}{Y_g}\right)2a_2 t + \left(\frac{1 - Y_g}{Y_g}\right)3a_3 t^2 \tag{11.4e}$$

for the specific respiration rate, which is a quadratic equation of the form

$$r = b_0 + b_1 t + b_2 t^2, \tag{11.4f}$$

where b_0, b_1, and b_2 are constants given by

$$b_0 = m + \left(\frac{1 - Y_g}{Y_g}\right)a_1 \tag{11.4g}$$

$$b_1 = 2\left(\frac{1 - Y_g}{Y_g}\right)a_2 \tag{11.4h}$$

$$b_2 = 3\left(\frac{1 - Y_g}{Y_g}\right)a_3. \tag{11.4i}$$

Thornley and Hesketh (1972) followed the procedure of fitting the exponential cubic of eqn (11.4c) to the $w : t$ data, and obtained four coefficients $a_0, \ldots, a_3$; an independent fit of a quadratic equation to the respiration versus time data $r : t$ gives the three coefficients b_0, b_1 and b_2. The coefficients are related to each other by three eqns (11.4g)–(11.4i) with only two parameters Y_g and m. Thus the problem is overdetermined.

Fitting the $w : t$ data with eqn (11.4c) and the $r : t$ data with eqn (11.4f) gave

$$\begin{aligned} &a_0 = -1.786 \pm 0.081 \qquad a_1 = 0.2721 \pm 0.0204 \\ &a_2 = -0.0077 \pm 0.0013 \qquad a_3 = -0.000079 \pm 0.000022 \\ &b_0 = 0.1004 \pm 0.0072 \qquad b_1 = -0.0053 \pm 0.0018 \\ &b_2 = 0.00007 \pm 0.000012. \end{aligned} \tag{11.4j}$$

Inserting the values of b_0, b_1, a_1, and a_2 into eqns (11.4g) and (11.4h) and solving for Y_g and m gives

$$Y_g = 0.74 \pm 0.10 \qquad m = 0.006 \pm 0.10 \text{ day}^{-1}. \tag{11.4k}$$

The errors are approximate. These values satisfy eqn (11.4i) within the error. They are not inconsistent with the white clover data (eqn (11.4b)), nor with other theoretical and experimental estimates of these parameters (eqn (12.23f), p. 353; Section 12.4, p. 364). Note for later use that

$$\frac{1 - Y_g}{Y_g} = 0.35. \tag{11.4l}$$

11.4 Recycling model of growth and maintenance respiration

Thornley (1977, 1982) proposed an approach to the analysis of respiration in which both the respiration components, growth and maintenance, are produced by the same synthetic process; it is assumed that a component of the structural dry mass of the plant is degraded and can be resynthesized, giving rise to a component of respiration that can be interpreted as maintenance. This model is also discussed by Barnes and Hole (1978) and Loehle (1982).

The model is depicted in Fig. 11.2, in which the variables and parameters are also defined. It has three state variables: the storage dry mass w_s, a degradable component of structural mass w_d, and a non-degradable component of structural mass w_n. The total dry mass w is given by

$$w = w_s + w_d + w_n. \tag{11.5a}$$

It is convenient to measure all the components of dry mass and all fluxes, including the photosynthetic and respiratory fluxes, in units of kg carbon or kg carbon day^{-1}.

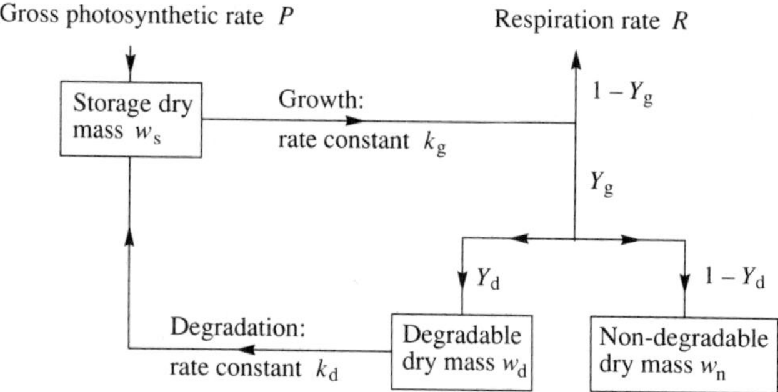

Fig. 11.2. Recycling model of growth and maintenance respiration. Plant dry mass w is considered in terms of the three components shown. Y_g and Y_d are conversion factors, with $0 \leqslant Y_g, Y_d \leqslant 1$.

First-order kinetics is assumed for the two processes of growth and degradation with rate constants k_g and k_d (day^{-1}). As above (cf. eqn (11.3c)), Y_g is the conversion efficiency or yield of the growth process, and Y_d gives the fraction of the newly synthesized material that is degradable; the remaining fraction $1 - Y_d$ is non-degradable. From Fig. 11.2, the differential equations for the state variables are

$$\frac{\mathrm{d}w_s}{\mathrm{d}t} = P - k_g w_s + k_d w_d \tag{11.5b}$$

$$\frac{\mathrm{d}w_d}{\mathrm{d}t} = Y_d Y_g k_g w_s - k_d w_d \tag{11.5c}$$

$$\frac{\mathrm{d}w_n}{\mathrm{d}t} = (1 - Y_d) Y_g k_g w_s. \tag{11.5d}$$

Adding these equations gives

$$\frac{\mathrm{d}w}{\mathrm{d}t} = P - k_g(1 - Y_g)w_s, \tag{11.5e}$$

and the respiration rate R is

$$R = P - \frac{\mathrm{d}w}{\mathrm{d}t} = k_g(1 - Y_g)w_s, \tag{11.5f}$$

as can be seen from Fig. 11.2.

A useful alternative equation for the respiration rate R can be derived. From eqns (11.5a), (11.5c), and (11.5d)

$$\frac{\mathrm{d}w}{\mathrm{d}t} - \frac{\mathrm{d}w_s}{\mathrm{d}t} = \frac{\mathrm{d}w_d}{\mathrm{d}t} + \frac{\mathrm{d}w_n}{\mathrm{d}t} = Y_g k_g w_s - k_d w_d. \tag{11.5g}$$

Substituting for $k_g w_s$ in eqn (11.5f) gives

$$R = \frac{1 - Y_g}{Y_g}\left(\frac{\mathrm{d}w}{\mathrm{d}t} - \frac{\mathrm{d}w_s}{\mathrm{d}t} + k_d w_d\right). \tag{11.5h}$$

The composition of the plant can be defined by the ratio f_s of storage dry mass to total dry mass, and the ratio f_d of degradable structure to total structure with

$$f_s = w_s/w \qquad f_d = w_d/w. \tag{11.5i}$$

With a constant plant composition, the fractions f_s and f_d are constant, so that

$$\frac{\mathrm{d}w_s}{\mathrm{d}t} = f_s \frac{\mathrm{d}w}{\mathrm{d}t}. \tag{11.5j}$$

Therefore, for the special case of a constant plant composition, eqn (11.5h) for the respiration rate R becomes

$$R = \frac{1 - Y_g}{Y_g}\left[(1 - f_s)\frac{\mathrm{d}w}{\mathrm{d}t} + f_d k_d w\right]. \tag{11.5k}$$

11.4.1 *Comparison with substrate balance analysis*

To facilitate comparison of this approach with the simpler substrate balance analysis (Section 11.3), eqn (11.3o) is written

$$R = \frac{1 - Y_g}{Y_g}\frac{\Delta w}{\Delta t} + mw. \tag{11.6a}$$

Comparison of the coefficients of dw/dt and w in eqns (11.5k) and (11.6a) gives a substantially altered interpretation of respiration parameters, and plant composition now enters into consideration.

The effects of this can be seen by reinterpreting the Y_g and m values of eqns (11.4k) and (11.4l) obtained using the substrate balance analysis method for cotton boll respiration. Assume that the storage and degradable structure fractions are each 10 per cent of the dry mass, with

$$f_s = 0.1 \qquad f_d = 0.1. \tag{11.6b}$$

From eqns (11.4l) and (11.5k)

$$0.35 = \left(\frac{1 - Y_g}{Y_g}\right)(1 - f_s) = \left(\frac{1 - Y_g}{Y_g}\right) \times 0.9. \tag{11.6c}$$

giving

$$Y_g = 0.72. \tag{11.6d}$$

Also, from eqns (11.4k) and (11.5k),

$$\frac{1 - Y_g}{Y_g} f_d k_d = 0.006, \tag{11.6e}$$

which with eqns (11.6b) and (11.6d) gives

$$k_d = 0.15 \text{ day}^{-1}. \tag{11.6f}$$

The parameter values of eqns (11.6d) and (11.6f) give a rather different view of respiration than those of eqns (11.4k); these are based on the level of biochemical aggregation assumed in eqns (11.5a), which might provide a useful level of description intermediate between the plant/organ level and that of detailed biochemistry.

11.4.2 *The linearized model*

An advantage of a state-variable approach to respiration, as outlined in Fig. 11.2 and eqns (11.5b)–(11.5d), as opposed to the substrate balance analysis (Section 11.3), is that it allows the consideration of more possibilities, including transients with changing plant composition, and exponential growth. In this section we examine the behaviour of the model for the special case of balanced exponential growth. These solutions are independent of the initial values of the state variables, and this allows the basic characteristics of the model to be explored in a simpler manner.

For balanced exponential growth solutions to exist at all, it is required that all the terms in eqns (11.5b)–(11.5d) scale linearly with the state variables. Thus it must be assumed that gross photosynthesis P is proportional to one of the components of dry mass (or a linear combination of these). It seems to be both reasonable physiologically and simple mathematically to assume that P is proportional to the degradable component of dry mass w_d, giving

$$P = kw_d. \tag{11.7a}$$

There are several techniques for solving eqns (11.5b)–(11.5d), and we outline one of these. As the equations are linear, we look for a solution for w_s of the form

$$w_s = Ae^{\lambda t}, \tag{11.7b}$$

where A and λ are constants. Substituting eqns (10.7a) and (10.7b) into eqn (11.5b) leads to

$$w_d = \frac{\lambda + k_g}{k + k_d} Ae^{\lambda t}. \tag{11.7c}$$

Substituting eqns (11.7b) and (11.7c) into eqn (11.5c) gives

$$Ae^{\lambda t}[(\lambda + k_g)(\lambda + k_d) - Y_d Y_g k_g(k + k_d)] = 0. \tag{11.7d}$$

We are not interested in the trivial cases where $Ae^{\lambda t} = 0$, so that

$$(\lambda + k_g)(\lambda + k_d) - Y_d Y_g k_g(k + k_d) = 0. \tag{11.7e}$$

This quadratic equation gives two solutions for λ:

$$\lambda_1 = \tfrac{1}{2}\{-(k_g + k_d) + [(k_g - k_d)^2 + 4Y_g Y_d k_g(k + k_d)]^{1/2}\} \tag{11.7f}$$

$$\lambda_2 = \tfrac{1}{2}\{-(k_g + k_d) - [(k_g - k_d)^2 + 4Y_g Y_d k_g(k + k_d)]^{1/2}\}. \tag{11.7g}$$

These solutions will only be equal if $k_g = k_d$ and at least one of k_g, Y_g, and Y_d is zero. This will not usually be the case, and the general solution for w_s will be a linear combination of the solutions, so that (cf. eqn (11.7b))

$$w_s = A_1 \exp(\lambda_1 t) + A_2 \exp(\lambda_2 t). \tag{11.7h}$$

Substituting each term of this equation into eqn (11.7c) gives

$$w_d = \frac{1}{k + k_d}[(\lambda_1 + k_g)A_1 \exp(\lambda_1 t) + (\lambda_2 + k_g)A_2 \exp(\lambda_2 t)]. \tag{11.7i}$$

The constants A_1 and A_2 are calculated from the initial values of $w_s(0)$ and $w_d(0)$. Setting $t = 0$ in eqns (11.7h) and (11.7i) and solving the two resulting equations for A_1 and A_2 leads to

$$A_1 = \frac{-w_s(0)(\lambda_2 + k_g) + w_d(0)(k + k_d)}{\lambda_1 - \lambda_2} \quad (11.7j)$$

$$A_2 = \frac{-w_s(0)(\lambda_1 + k_g) + w_d(0)(k + k_d)}{\lambda_2 - \lambda_1}. \quad (11.7k)$$

Finally, the non-degradable component is found by integrating eqn (11.5d):

$$w_n = w_n(0) + (1 - Y_d) Y_g k_g \int_0^t w_s \, dt, \quad (11.7l)$$

which combined with eqn (11.7h) gives

$$w_n = w_n(0) + (1 - Y_d) Y_g k_g \left\{ \frac{A_1[\exp(\lambda_1 t) - 1]}{\lambda_1} + \frac{A_2[\exp(\lambda_2 t) - 1]}{\lambda_2} \right\}. \quad (11.7m)$$

Equations (11.7h), (11.7i), and (11.7m) for w_s, w_d, and w_n with A_1 and A_2 given by eqns (11.7j) and (11.7k) provide a general solution to the linearized problem, i.e where eqn (11.7a) for photosynthesis is assumed to hold..

For future use, we note that eqns (11.7f) and (11.7g) lead to two equations for the sum and product of the roots of the quadratic equation (11.7e), i.e

$$\lambda_1 + \lambda_2 = -(k_g + k_d) \quad (11.7n)$$

and

$$\lambda_1 \lambda_2 = k_g k_d - Y_g Y_d k_g (k + k_d). \quad (11.7o)$$

Balanced exponential growth Balanced exponential growth occurs if one of the roots in eqns (11.7f) and (11.7g) is positive, so that in eqns (11.7h), (11.7i), and (11.7m) for the dry mass components, there is a term that increases at a constant exponential rate. The λ_2 root of eqn (11.7g) is always negative, and in eqns (11.7h), (11.7i), and (11.7m) this represents a transient part of the solution that depends on initial values and dies away.

The λ_1 root of eqn (11.7g) is the larger of the two roots, and is positive if

$$Y_g Y_d k_g (k + k_g) > k_g k_d,$$

which leads to

$$k > \left(\frac{1 - Y_g Y_d}{Y_g Y_d} \right) k_d. \quad (11.7p)$$

This equation gives the lower limit on the photosynthetic rate required for growth. In Fig. 11.3(a) the ratio k/k_d of the photosynthetic rate to the degradation rate is plotted against the product $Y_g Y_d$, which gives the conversion yield of degradable dry mass from storage dry mass (Fig. 11.2). The physiological meaning of eqn (11.7p) can be seen more easily by writing the equality (when zero growth is just sustained)

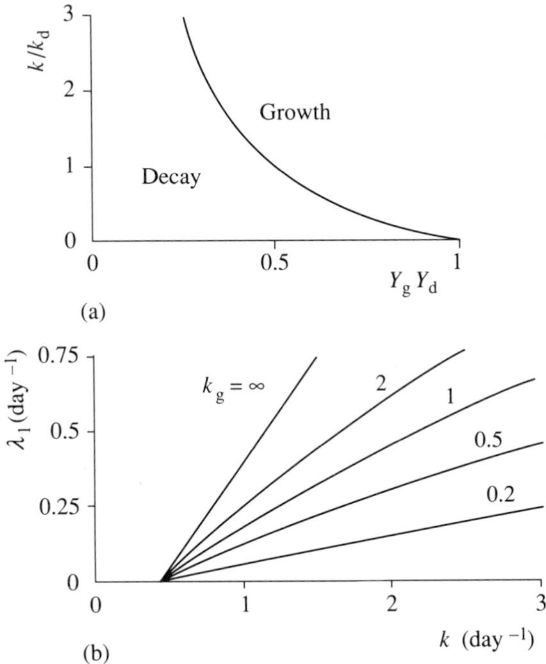

Fig. 11.3. Balanced exponential growth. (a) The minimum value of the photosynthetic rate constant k for positive growth: the ratio of the minimum k to the degradation rate constant k_d is plotted against the product of the conversion factors Y_g and Y_d (Fig. 11.2) as given by eqn (11.7p). Growth solutions only exist above the curve. (b) The specific growth rate (from eqn (11.7f)) is plotted against the photosynthetic rate constant k for different values of the growth rate constant k_g as shown. Other parameters are $Y_g Y_d = 0.7$ and $k_d = 1$ day^{-1}.

$$kY_g Y_d + k_d Y_g Y_d = k_d. \tag{11.7q}$$

In unit time k_d units of degradable dry mass degraded are just replaced by $kY_g Y_d$ dry mass units from photosynthesis plus $k_d Y_g Y_d$ dry mass units that are resynthesized from the extra storage material (k_d units) arising from the original degradation (Fig. 11.2). Note that the limits for growth in eqn (11.7p) only depend on the degradation rate constant k_d and not on the growth rate constant k_g.

Figure 11.3(b) shows the dependence of the specific growth rate λ_1 from eqn (11.7f) on the photosynthetic rate constant k for various values of the growth rate constant k_g. Although the general response to photosynthesis is curvilinear, for either high values of k_g or low values of k_g the response is linear. This is of some practical importance because low or high temperatures might cause such limiting responses to be approached. For $k_g \to \infty$

$$\lambda_1 = kY_g Y_d - k_d(1 - Y_g Y_d). \tag{11.7r}$$

For $k_g \to 0$

$$\lambda_1 = k_g\left[Y_g Y_d\left(1 + \frac{k}{k_d}\right) - 1\right]. \tag{11.7s}$$

Note that for both eqn (11.7r) and eqn (11.7s) $\lambda_1 = 0$ when eqn (11.7p) is satisfied. In Fig. 11.3(b), since it is assumed that $k_d = 1$ day^{-1}, effectively λ_1/k_d is plotted against k/k_d.

11.4.3 $^{14}CO_2$ *distribution following a pulse label*

Consider a photosynthetic pulse of $^{14}CO_2$. For a short pulse it can be assumed that at $t = 0$ the label is entirely in the storage pool. Provided that the system remains linear (eqn (11.7a)), the fate of the label will be independent of any subsequent photosynthesis and growth. Since the injection of the label ceases at $t = 0$, it is legitimate to take

$$k = 0 \tag{11.8a}$$

when solving eqns (11.7f) and (11.7g) for λ_1 and λ_2. The initial values to be used in eqns (11.7j) and (11.7k) are

$$w_s(0) = 1 \qquad w_d(0) = w_n(0) = 0. \tag{11.8b}$$

The constants A_1 and A_2 become

$$A_1 = -\left(\frac{\lambda_2 + k_g}{\lambda_1 - \lambda_2}\right) \qquad A_2 = -\left(\frac{\lambda_1 + k_g}{\lambda_2 - \lambda_1}\right). \tag{11.8c}$$

The distribution of $^{14}CO_2$ is obtained from eqns (11.7h), (11.7i), and (11.7m). The respiration rate R is given by eqns (11.5f), (11.7h), and (11.8c), and is

$$R(t) = \frac{k_g(1 - Y_g)}{\lambda_1 - \lambda_2}[(\lambda_2 + k_g)\exp(\lambda_1 t) - (\lambda_1 + k_g)\exp(\lambda_2 t)]. \tag{11.8d}$$

The two exponents λ_1 and λ_2 are both negative (as can be seen readily from eqns (11.7f) and (11.7g) with $k = 0$ and Y_g, $Y_d < 1$), giving a respiration rate that decays according to two exponential terms.

Denote the amount of label remaining in the plant at time t by $L(t)$. This can be calculated in two ways: by summing the label in the three compartments with eqns (11.5a), (11.7h), (11.7i), (11.7m), and (11.8c), or alternatively by integrating the respiration rate of eqn (11.8d) to give the label lost by respiration. Using this second method

$$L(t) = 1 - \int_0^t R(t')\,dt', \tag{11.8e}$$

where t' is a dummy time variable. With eqn (11.8d), this gives

$$L(t) = 1 + \frac{k_g(1 - Y_g)(\lambda_1 + \lambda_2 + k_g)}{\lambda_1 \lambda_2}$$

$$+ \frac{k_g(1 - Y_g)}{\lambda_1 - \lambda_2}\left[\left(\frac{\lambda_2 + k_g}{\lambda_1}\right)\exp(\lambda_1 t) - \left(\frac{\lambda_1 + k_g}{\lambda_2}\right)\exp(\lambda_2 t)\right]. \quad (11.8f)$$

Because λ_1 and λ_2 are both negative ($k = 0$), the exponential terms approach zero and the limiting amount of label remaining in the system is

$$\lim_{t\to\infty} L(t) = 1 + \frac{k_g(1 - Y_g)(\lambda_1 + \lambda_2 + k_g)}{\lambda_1 \lambda_2} \quad (11.8g)$$

By using eqns (11.7n) and (11.7o) with $k = 0$, λ_1 and λ_2 can be eliminated to give

$$\lim_{t\to\infty} L(t) = \frac{Y_g(1 - Y_d)}{1 - Y_g Y_d}. \quad (11.8h)$$

Equation (11.8f) defines the label remaining in the plant as a two-exponential decay, which approaches the finite value given by eqn (11.8h) when all the label left in the plant is in the non-degradable component. Interestingly, if eqn (11.8h) is expanded as an infinite series

$$\lim_{t\to\infty} L(t) = Y_g(1 - Y_d)[1 + Y_g Y_d + (Y_g Y_d)^2 + (Y_g Y_d)^3 + \cdots], \quad (11.8i)$$

it can be seen with reference to Fig. 10.2 how the material accumulates in the non-degradable compartment as a result of successive cycles of the labelled material.

In an experiment of this type, Ryle, Cobby, and Powell (1976) fitted an equation of the form

$$L(t) = (1 - b - c) + b\exp(\lambda_1 t) + c\exp(\lambda_2 t) \quad (11.8j)$$

where b and c are constants, and the equation is written so that

$$L(t = 0) = 1. \quad (11.8k)$$

By fitting eqn (11.8j) to the experimental data, estimates of b, c, λ_1, and λ_2 are obtained. By using the present model, these values can be interpreted in terms of the parameters k_g, k_d, Y_g, and Y_d by comparing eqn (11.8j) with eqn (11.8f); we describe how this is accomplished below.

Equating the coefficients of the exponential terms in eqns (11.8j) and (11.8f) gives

$$\frac{b}{c} = -\left(\frac{\lambda_2 + k_g}{\lambda_1}\right)\left(\frac{\lambda_2}{\lambda_1 + k_g}\right), \quad (11.8l)$$

which is equivalent to

$$k_g = -\left(\frac{b\lambda_1{}^2 + c\lambda_2{}^2}{b\lambda_1 + c\lambda_2}\right). \quad (11.8m)$$

This equation enables k_g to be calculated. Equation (11.7n) can be written as

$$k_d = -(\lambda_1 + \lambda_2 + k_g), \tag{11.8n}$$

thus allowing k_d to be obtained. Equating the constant terms in eqns (11.8f) and (11.8j), therefore, we have

$$1 + \frac{k_g(1 - Y_g)(\lambda_1 + \lambda_2 + k_g)}{\lambda_1 \lambda_2} = 1 - b - c,$$

which can be written (using also eqn (11.8n))

$$Y_g = 1 - \frac{(b + c)\lambda_1 \lambda_2}{k_g k_d}. \tag{11.8o}$$

Finally, from eqn (11.7o),

$$Y_d = \frac{1}{Y_g}\left(1 - \frac{\lambda_1 \lambda_2}{k_g k_d}\right). \tag{11.8p}$$

From the four experimentally estimated parameters b, c, λ_1, and λ_2 of eqn (11.8j), eqns (11.8m)–(11.8p) allow the four parameters of the model k_g, k_d, Y_g, and Y_d (Fig. 10.2) to be obtained. This method was applied by Thornley (1977), whose results are in error, and by Barnes and Hole (1978) to the data of Ryle *et al.* (1976) for barley and maize; the estimated parameters and the deduced model parameters are given in Table 11.1. The model provides a possible interpretation of the

Table 11.1. Experimental parameters (eqn (11.8j)) and derived model parameters (Fig. 11.2 and eqn (11.8f)) for respiratory losses of $^{14}CO_2$ in barley and maize

	Barley	Maize
Experimental parameters†		
λ_1 (day^{-1})	−0.20	−0.18
λ_2 (day^{-1})	−2.10	−2.10
$1 - b - c$	0.35	0.44
b/c	1.28	0.77
Model parameters‡		
k_g (day^{-1})	1.89	1.98
k_d (day^{-1})	0.41	0.30
Y_g	0.64	0.65
Y_d	0.71	0.56

† From Ryle *et al.* 1976, Table 1.
‡ Deduced using eqns (11.8m)–(11.8p).

data at the level of aggregated biochemistry. There is scope for developing the model further: for example, the assumption of linear kinetics could be relaxed; also other pools might be introduced into the scheme—McCree (1982) includes a starch pool which is filled from photosynthesis during the light period and empties into a soluble storage pool during the dark period.

11.5 Root respiration

In analysis so far we have concentrated on the growth and maintenance categories of respiration. If there are other significant respiratory costs, then this simple view may break down or at least be incomplete. This is likely to be the case in the root system of plants, where there may be energy costs of nutrient uptake, reduction, and maintaining ion gradients.

Observations of root respiration which have been analysed using the simple growth and maintenance dichotomy of Section 11.3 indicate that the growth efficiency Y_g in roots is substantially lower than that in shoots, suggesting that root synthesis requires more energy than shoot synthesis. This conflicts with the biochemical information given in the next chapter, since roots and shoots differ little in composition. Here, we develop the substrate balance approach so as to incorporate nitrate uptake (Johnson 1983) as well as plant structure synthesis.

11.5.1 *Analysis*

The principal assumption is that the growth efficiency Y_g is constant throughout the plant. Clearly, if the protein content varies substantially over the plant there will be some effect on Y_g; however, this is unlikely to be large. For example, from Table 12.8 and eqn (12.23d) (Chapter 12), for a plant comprising carbohydrates and nitrogenous compounds only with nitrate supplied, the effect of increasing the nitrogenous fraction from 10 to 30 per cent—a three-fold increase—is to reduce Y_g from 0.87 to 0.76, about a 12 per cent change. It is therefore a useful and reasonable approximation to consider a constant Y_g, although this could be relaxed. Only the effect of nitrate uptake is evaluated (this constitutes about 90 per cent of the total anion uptake (Veen 1981)), although it is straightforward to include the uptake of other minerals.

The rate at which nitrogen is required for the growth of a plant of structural dry mass W_g (kg total structural dry mass) is

$$f_N \frac{dW_g}{dt} \text{ kg nitrogen day}^{-1}, \tag{11.9a}$$

where f_N (kg nitrogen (kg structural dry mass)$^{-1}$) is the fractional nitrogen content of plant structure. It is assumed that no nitrogen is associated with the storage component of the plant. The respiratory cost associated with this nitrogen uptake flux is

$$\alpha f_{\mathrm{N}} \frac{\mathrm{d}W_{\mathrm{g}}}{\mathrm{d}t} \text{ kg carbon respired day}^{-1}, \tag{11.9b}$$

where α (kg carbon (kg nitrogen)$^{-1}$) is the carbon respired per unit of nitrogen taken up.

Note that in eqn (11.5k) for the respiration rate R the dry mass w has units of kilograms of carbon, rather than the units of kilograms of total dry mass used with W_{g} in eqns (11.9a) and (11.9b). To convert from one to the other, we write

$$w_{\mathrm{g}} = f_{\mathrm{C}} W_{\mathrm{g}}, \tag{11.9c}$$

where the fraction f_{C} of carbon in dry mass has units of kg carbon (kg total structural dry mass)$^{-1}$. Substituting for W_{g}, the respiratory cost (11.9b) becomes

$$\alpha \frac{f_{\mathrm{N}}}{f_{\mathrm{C}}} \frac{\mathrm{d}w_{\mathrm{g}}}{\mathrm{d}t}. \tag{11.9d}$$

The nitrate uptake costs of (11.9d) can be considered within the framework of the recycling model of respiration (Section 11.4 and Fig. 11.2). The structural dry mass w_{g} is defined by (cf. eqn (11.5a))

$$w_{\mathrm{g}} = 1 - w_{\mathrm{s}} = w_{\mathrm{d}} + w_{\mathrm{n}}, \tag{11.9e}$$

and with eqns (11.5i), therefore,

$$w_{\mathrm{g}} = w(1 - f_{\mathrm{s}}). \tag{11.9f}$$

Substitution of eqn (11.9f) into eqn (11.9d) gives

$$\alpha \frac{f_{\mathrm{N}}}{f_{\mathrm{C}}} (1 - f_{\mathrm{s}}) \frac{\mathrm{d}w}{\mathrm{d}t}. \tag{11.9g}$$

This expression describes an additional respiratory flux associated with growth which can be added to the equations for growth and maintenance respiration derived earlier. Equation (11.5k) gives the respiration flux assuming a constant storage fraction f_{s}; this was derived for the whole plant, but it can be applied to the organs of the plant separately. In eqn (11.5k) we simplify by writing the maintenance coefficient m (day^{-1}) as (cf. eqns (11.5k) and (11.6a))

$$m = \left(\frac{1 - Y_{\mathrm{g}}}{Y_{\mathrm{g}}} \right) f_{\mathrm{d}} k_{\mathrm{d}}, \tag{11.9h}$$

so that eqn (11.5k) becomes

$$R = \left(\frac{1 - Y_{\mathrm{g}}}{Y_{\mathrm{g}}} \right) (1 - f_{\mathrm{s}}) \frac{\mathrm{d}w}{\mathrm{d}t} + mw. \tag{11.9i}$$

Equations (11.9i) and (11.9g) can be applied to the whole plant, the shoot, and the root to give

$$R_p = w_p r_p = \left(\frac{1 - Y_g}{Y_g} + \alpha\frac{f_N}{f_C}\right)(1 - f_s)\frac{dw_p}{dt} + m_p w_p \tag{11.9j}$$

$$R_{sh} = w_{sh} r_{sh} = \left(\frac{1 - Y_g}{Y_g}\right)(1 - f_s)\frac{dw_{sh}}{dt} + m_{sh} w_{sh} \tag{11.9k}$$

$$R_r = w_r r_r = \left(\frac{1 - Y_g}{Y_g}\right)(1 - f_s)\frac{dw_r}{dt} + \alpha(1 - f_s)\frac{f_N}{f_C}\frac{dw_p}{dt} + m_r w_r, \tag{11.9l}$$

where the subscripts p, sh, and r denote the plant, shoot, and root respectively. The shoot, root, and plant dry masses are related by

$$w_p = w_{sh} + w_r. \tag{11.9m}$$

The r_i, i = p, sh, r, are specific respiration rates (day^{-1}).

By fitting shoot and whole-plant respiration data to the equations

$$r_i = g_i \frac{1}{w_i}\frac{dw_i}{dt} + m_i \qquad i = \text{p, sh}, \tag{11.9n}$$

where the g_i are the coefficients of the terms in eqns (11.9j) and (11.9k) given by

$$g_p = \left(\frac{1 - Y_g}{Y_g} + \alpha\frac{f_N}{f_C}\right)(1 - f_s)$$
$$g_{sh} = \left(\frac{1 - Y_g}{Y_g}\right)(1 - f_s), \tag{11.9o}$$

we obtain

$$Y_g = \frac{1 - f_s}{1 - f_s - g_{sh}}$$
$$\alpha = \frac{f_C(g_p - g_{sh})}{f_N(1 - f_s)}. \tag{11.9p}$$

Szaniawski and Kielkiewicz (1982) obtained the values

$$g_p = 0.33 \qquad g_{sh} = 0.28 \tag{11.9q}$$

for young sunflower plants. Inserting these values into eqn (11.9p) with

$$f_s = 0.1 \qquad f_C = 0.4 \qquad f_N = 0.025 \tag{11.9r}$$

yields

$$Y_g = 0.76 \qquad \alpha = 0.89 \text{ kg carbon (kg nitrogen)}^{-1}. \tag{11.9s}$$

This model provides a possible interpretation for the differences between shoot and root respiration rates by incorporating the energetics of nitrate ion uptake into whole-plant dynamics in a simple manner. If the assumption that the growth efficiency Y_g is constant throughout the plant is relaxed, then eqn (11.9l) is also needed to evaluate the results.

The maintenance respiration coefficient in roots is consistently observed to be greater than that in shoots (e.g. Hansen and Jensen 1977, Szaniawski and Kielkiewicz 1982). If the degradable fraction f_d of dry mass is similar in shoots and roots, then this difference is not readily explained without additional assumptions such as a higher degradation rate in the root or ion leakage (Section 12.4.4, p. 366). It is also possible that the shoot tissue uses light energy directly rather than via carbon substrate; this could lead to differing growth efficiencies Y_g in shoot and root, and, through eqn (11.5k) or eqn (11.9i), to differing maintenance coefficients m_i in eqns (11.9j)–(11.9l).

11.6 Temperature effects on respiration

The models of respiration described above emphasize the processes of growth and maintenance respiration, and the costs associated with nutrient uptake. These processes involve essentially three parameters: the growth efficiency Y_g (eqn (11.3c) and Fig. 11.2); the degradation constant k_d (Fig. 11.2) or the maintenance coefficient m (eqn (11.3f)); the respiratory cost parameter α for nitrate uptake (eqn (11.9d)). The growth efficiency is dimensionless, and can be regarded as a stoichiometric coefficient or alternatively as determined by the ratio of different rate constants. Provided that temperature does not influence the relative use of different metabolic pathways, possibly leading to a different composition for plant material, Y_g is expected to be independent of temperature. Similarly, the cost parameter α for nitrate uptake can be regarded as a stoichiometric parameter, and if the relative use of alternative pathways is assumed not to change with temperature, then α should not change with temperature. The possible temperature independence of Y_g and α does not preclude the expected marked dependence of growth and nitrate uptake on temperature. In contrast, the maintenance coefficient m and the degradation coefficient k_d are rate constants with dimensions of time^{-1}; they are therefore expected to be temperature dependent (Chapter 5).

In their experiments to investigate the effects of root temperature on respiration, referred to in the last section, Szaniawski and Kielkiewicz (1982) fit the relationship between specific respiration rate and specific growth rate as a straight line (eqns (11.3r) and (11.9n)). Interpreting their results as already described indicates that both the growth efficiency Y_g and the nitrate uptake cost parameter α are independent of temperature (the coefficients g_i did not change). However, the maintenance coefficients m_i respond strongly to temperature: increasing root temperature from 10 to 30 °C caused the maintenance coefficient in the root m_r to increase from 0.018 to 0.075 day^{-1}, and that in the shoot m_{sh} to increase from 0.014 to 0.02 day^{-1}. The substantial increase in m_r suggests a strong direct effect of temperature; the response in the shoot maintenance coefficient m_{sh} to the change in root temperature may be partly due to a response of plant composition to temperature, as in eqn (11.9h).

In the experiments just discussed, respiration was measured at the growth temperature of plants grown in constant conditions. Many authors (e.g. Treharne and Nelson 1975, Woledge and Dennis 1982) have observed short-term changes in respiration in response to temperature changes. Such responses are typically changes in plant-specific growth rate and growth respiration (usually indicating a constant growth efficiency Y_g) and maintenance respiration. Since such experiments involve adaptive responses, with possibly changes in substrate levels and plant composition, it may be inappropriate or misleading to draw firm conclusions about growth and maintenance respiration.

Another frequently applied technique for measuring maintenance respiration and its temperature response is to place plants in darkness. Eventually, the respiration rate often approaches a constant level which is assumed to represent maintenance costs (e.g. McCree 1974). However, Breeze and Elston (1983) have shown that the photosynthetic capacity of such plants is severely impaired, casting doubt on the assumption that plants treated in this way are maintaining themselves.

The interpretation of short-term temperature responses of respiration is hazardous; steady state experiments where transient effects are avoided and plant composition is measured may provide more useful results.

11.7 Discussion

Several models of respiration in plants have been discussed. A central feature of these models is that the principal component of respiration arises from the new synthesis of plant material from substrate. This component is identified as growth respiration, whereas respiration that occurs from the resynthesis of degraded material is identified as maintenance respiration. Other processes, such as maintaining concentration gradients against leakage (Section 12.4.4, p. 366) or the operation of metabolic cycles (Section 12.2.11, p. 349) which cannot be shut down, will also contribute to maintenance respiration.

There is a reasonable consensus concerning the representation of growth respiration, which is taken as the efficiency of conversion of substrate to structure. This has a sound biochemical underpinning (Chapter 12) which, while it can no doubt be developed to provide easier application, supplies theoretical estimates of organ- and plant-level parameters, allowing for the varying composition of plant material.

Maintenance respiration is an area of great uncertainty, reviewed by Amthor (1984), and discussed from a biochemical standpoint in the next chapter. The concept that maintenance results from the resynthesis of degraded material and the maintenance of concentration gradients is largely accepted; the extent to which these occur and the factors affecting them are largely obscure. The operation of metabolic cycles may also be important. It may be too simplistic to relate part of the maintenance respiration to protein content alone: there is evidence that different groups of enzymes are degraded at different rates; also, if enzyme breakdown is partially related to their activity, which may be synthetic, this component is more properly regarded as a cost on the growth process. Degradation in developing or older tissues, especially in nitrogen-deficient plants, may be a mechanism for remobilizing nitrogen for use at the same location or elsewhere, perhaps for synthesizing different types of proteins.

Other possibly relevant items include the costs of transport of ions, sugars, or other compounds, the importance of the sites of nitrate reduction in relation to the sites of nitrate utilization, and wasteful respiration (related to metabolic cycles). Some of these topics are discussed more fully in the next chapter, but their potential importance to the way in which we represent respiration at the plant and crop levels of organization should not be underestimated.

While the substrate balance and recycling models of respiration provide a valuable framework for experimental work and crop growth modelling, it is essential that the assumptions on which they are based are not forgotten. However, by growing plants under steady state conditions and measuring quantities such as the composition of plant material, useful parameters such as growth efficiencies and degradation rates can be estimated. The definition and understanding of the principal components of respiration provides a firm basis for quantifying the role and contribution of respiration in determining crop yield and productivity.

Exercises

11.1. The equation

$$\frac{\mathrm{d}w}{\mathrm{d}t} = P - R \tag{E11.1a}$$

relates the growth rate $\mathrm{d}w/\mathrm{d}t$ to the gross photosynthetic rate P and the respiration rate R (kg carbon day^{-1}). Use eqn (11.3o) (or otherwise) to derive the equation

$$\frac{\mathrm{d}w}{\mathrm{d}t} = Y_g(P - mw) \tag{E11.1b}$$

Assume that the photosynthetic rate responds linearly to plant size according to

$$P = \alpha I \lambda f_{\mathrm{leaf}} w, \tag{E11.1c}$$

where α is a photosynthetic efficiency (0.3×10^{-8} kg carbon J^{-1}), I is the mean light flux density incident on the leaves (2 MJ m^{-2} day^{-1}), λ is a specific leaf area (50 m^2 leaf (kg carbon)$^{-1}$), and f_{leaf} is the fraction of plant carbon in leaf tissue (0.5).

Derive an expression for the plant-specific growth rate $(1/w)(\mathrm{d}w/\mathrm{d}t)$, and calculate a numerical estimate of this, assuming that the growth efficiency Y_g is 0.7 and the maintenance coefficient m is 0.01 day^{-1}.

11.2. The growth in dry mass w of many plants and crops can be approximated by the logistic growth equation (Section 3.4, p. 78), which in differential form is defined by (eqn (3.4d))

$$\frac{\mathrm{d}w}{\mathrm{d}t} = \mu_0 w\left(1 - \frac{w}{w_f}\right) \tag{E11.2a}$$

where μ_0 (day^{-1}) and w_f (kg carbon) are parameters (the specific growth rate extrapolated back to zero mass and the final dry mass respectively). Use eqn (11.3r) for the specific respiration rate r, which is based on the substrate balance approach to respiration, to derive an expression for the $r:w$ relationship. Make a sketch of the equation. Derive an analogous $r:w$ equation based on the Gompertz growth equation (eqn (3.5l, p. 82)).

11.3. The recycling model of respiration (Fig. 11.2) has been extended by McCree (1982). Figure E11.1 shows a similar modification to that of McCree (1982) applied to Fig. 11.2, which has a starch pool in addition to the more mobile soluble sugar pool. Write down the four differential equations corresponding to eqns (11.5b)–(11.5d). Define the total dry mass w by

$$w = w_{st} + w_{sol} + w_d + w_n, \tag{E11.3a}$$

and the storage dry mass w_S by

$$w_S = w_{st} + w_{sol}. \tag{E11.3b}$$

Then, making any changes needed, follow through the derivation given in eqns (11.5e)–(11.5k) for the respiration rate R.

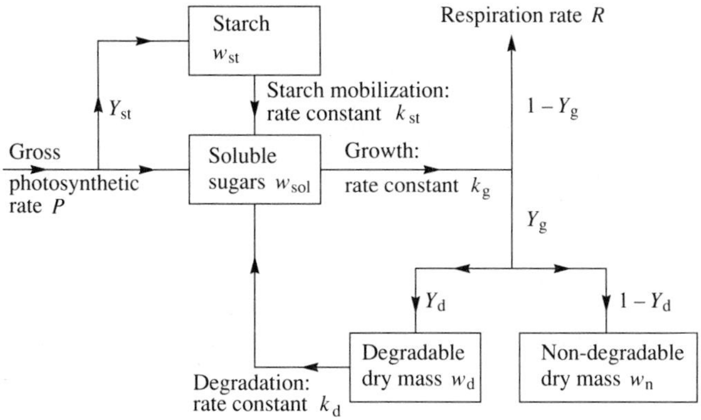

Fig. E11.1. Extended recycling model of growth and maintenance respiration (cf. Fig. 11.2, p. 270) modified for starch synthesis and mobilization along the lines suggested by McCree (1982). Y_{st} is the fraction of gross photosynthesis placed in the starch compartment, which is assumed to break down linearly with rate constant k_{st}.

11.4. As in Section 11.4.2, linearize the four differential eqns (S11.3a)–(S11.3d) derived in the last exercise by assuming $P = kw_d$ (cf. eqn (11.7a)). Construct the equations that characterize balanced exponential growth solutions where all four state variables are increasing according to the factor $\exp(\lambda t)$; derive the cubic equation for the specific growth rate λ and show that, in the limit $k_{st} \to \infty$, this reduces to eqn (11.7e).

11.5. Consider experiments in which plants are maintained in balanced exponential growth under constant light and temperature, and then placed in the dark, with their respiration rate being monitored. Analyse this situation using the recycling model of Section 11.4.

12
Biochemical and chemical approaches to plant growth

12.1 Introduction

In the last chapter we dealt principally with phenomenological approaches to the problem of plant respiration. Such approaches are widely used in crop models (Chapter 16), their mechanistic content is limited but none the less valuable, and their parameters are usually determined empirically (by direct appeal to observational data) or, less certainly, by investigation at the underlying levels of chemistry and biochemistry. The latter methods are the basis of the present chapter.

The biochemical approach to plant growth and respiration has been pioneered by Penning de Vries (1975a, b) and Penning de Vries *et al.* (1974); it uses biochemical knowledge about reaction pathways, stoichiometries, and energetics to predict the efficiencies of synthesizing plant compounds from substrates, including the costs of maintaining all aspects of plant structure. Although complex, the method can in principle be rigorous, noting that, as applied by Penning de Vries, the approach is that of a static rather than a fully dynamic model (cf. Gill, Thornley, Black, Oldham, and Beever 1984). As stressed by Beevers (1970), such approaches seem essential if we are aiming at more than limited empirical relations for individual species.

Owing to the complexity of the biochemical approach, attempts have been made to find simpler methods based on elemental composition and heats of combustion (McDermitt and Loomis 1981a,b, Williams, Percival, Merino, and Mooney 1987). These methods are approximate rather than rigorous; they are still in the early stages of evaluation, but it seems likely that they will be able to provide, more quickly and easily than resorting to detailed biochemistry, estimates of growth parameters for ecological or crop physiological objectives which are acceptably accurate.

12.2 Biochemistry of plant growth

Our analysis of this topic focuses on glucose as the starting point and principal substrate for plant processes. Glucose fills three important roles: it can be oxidized to provide energy (ATP) for driving the plant processes of synthesis and maintenance; it can be oxidized to provide reducing power (NADH) for changing the reduction level of compounds; lastly, it provides the basic building blocks that are used in the construction of new plant material. While plant growth and plant maintenance have many elements in common, we shall assume that they can be separated, and in this section we discuss growth, which is mostly concerned with the synthesis of new materials. The analysis of maintenance costs is more difficult and speculative, and is the topic of a later section.

For the cases considered, we present details of the pathway assumed. There is more scope for alternative pathway assumptions than is often realized, and it is important that the basis of any particular calculation is made clear.

12.2.1 *Some general considerations and assumptions*

In this section, the principal assumptions on which the calculations are based are set out. This is so that the tables can be used directly, with some idea of their possible or probable inadequacies. In Section 12.2.2 there is, amongst other things, a more detailed account of the production of energy and reducing power, which is summarized under point 8 immediately below.

1. Where the pathways of synthesis appear to be well established for plants, then those pathways are assumed to apply. Where there are alternative pathways, it is usually assumed that the most efficient pathway is followed. Most of the biochemical pathway information is derived from microbial or animal studies, and is often unsubstantiated for plant processes. It should be noted that the relative efficiencies of the different routes may depend quite strongly on the metabolic state of the cells where the synthesis is occurring. Moreover, the most energy efficient pathway may not always occur with sufficient rapidity for other purposes (the most thermodynamically efficient reactions often occur at vanishingly small rates), and a less efficient process releasing more energy will frequently proceed more rapidly. This is a difficult area where definitive answers are not always available.
2. Our principal biochemical sources are Mahler and Cordes (1966), Dagley and Nicholson (1970), Lehninger (1975), Schulz (1978), and Michal (1982). The style in which the material is presented owes much to Schulz (1978), and is aimed at providing sufficient detail so that the approach can be easily modified.
3. It is assumed (but see Section 12.2.7) that nitrogen is derived from ammonia, and that sulphur is derived from H_2S.
4. All nucleotide triphospates are assumed to be interconvertible, and are expressed as ATP. It is also assumed that the reaction

$$\mathrm{ATP} \rightarrow \mathrm{AMP} + \mathrm{PP_i}, \tag{12.1a}$$

is equivalent to two

$$\mathrm{ATP} \rightarrow \mathrm{ADP} + \mathrm{P}_i \tag{12.1b}$$

reactions. Mahler and Cordes (1966, p. 201) report that the standard free energies of hydrolysis of (12.1a) and (12.1b) are 31.8 MJ (kg mol)$^{-1}$ and 30.9 MJ (kg mol)$^{-1}$ respectively, and the free energy of hydrolysis of PP_i to $2P_i$ is 19.2 MJ (kg mol)$^{-1}$ (Lehninger, 1975, p. 411). Thus this assumption, which equates 31.8 + 19.2 with 2 × 30.9 is not very accurate, but this is not a significant factor in the calculations.

5. It is assumed that the cytosol and plastids are equivalent with respect to ATP and NADH (cf. Schulz (1978) who assumes that one ATP is required to transport NADH from the cytosol into a mitochondrion).
6. The reducing agents NADH, NADPH, and $FADH_2$, are accounted for separately, but then they are all assumed to be equivalent to NADH (Schulz (1978) assumes that NADH + ATP is equivalent to NADPH; note also that the oxidation of $FADH_2$ in the respiratory chain yields 2ATP rather than the 3ATP for NADH oxidation (Lehninger, 1975, p. 495)). To simplify notation, we write the reaction

$$AH_2 + NAD^+ \rightarrow A + NADH + H^+$$

as

$$AH_2 \rightarrow A + NADH,$$

remembering that NADH, NADPH, and $FADH_2$ always transfer two hydrogen atoms.
7. Any costs of transporting small-molecule metabolities such as acetyl-CoA etc. across membranes are ignored.
8. After applying 4 and 6, ATP and NADH requirements are converted into required glucose equivalents according to (cf. (12.1f) and (12.1m))

$$\text{glucose equivalents} = ATP/36 + NADH/12. \qquad (12.1c)$$

9. The ATP and NADH requirements of the stoichiometric equations are converted into a molar oxygen requirement and a molar CO_2 production according to (based on (12.1f) and (12.1m))

$$O_2 \text{ required} = ATP/6 \qquad (12.1d)$$

and

$$CO_2 \text{ produced} = ATP/6 + NADH/2. \qquad (12.1e)$$

10. Water is often omitted from the biochemical equations, although it is usually put back into the overall stoichiometric equation.

12.2.2 *The production of coenzymes*

Energy and reducing power Energy is represented by the 'energy-rich' compound ATP which can give up energy by a coupled hydrolysis:

$$ATP \rightarrow ADP + 77.3 \text{ MJ (kg mol)}^{-1}$$

where inorganic phosphate (P_i) and water have been omitted. Energy is produced by glucose oxidation using part of the glycolytic pathway (extra-mitochondrial) with oxidative phosphorylation in a mitochondrion (Lehninger, 1975, pp. 422, 445, 517):

$$\text{glucose} \rightarrow 2\ \text{pyruvate} + 2\text{ATP} + 2\text{NADH}$$

$$2\ \text{pyruvate} \rightarrow 2\ \text{acetyl-CoA} + 2\text{CO}_2 + 2\text{NADH}$$

$$2\ \text{acetyl-CoA} \rightarrow 4\text{CO}_2 + 8\text{NADH}$$

$$12\text{NADH} + 6\text{O}_2 \rightarrow 36\text{ATP}$$

$$\text{glucose} + 6\text{O}_2 \rightarrow 6\text{CO}_2 + 6\text{H}_2\text{O} + 38\text{ATP}.$$

Note that in these conversion equations, we omit to give explicitly the alternative form of the coenzymes, e.g. ADP, H-CoA, and NAD^+, and also H_2O and P_i. Note also that there is uncertainty as to how the electrons from the glucose–pyruvate conversion are transferred into the mitochondrion, and concerning the ATP yield from these electrons which may be 2ATP instead of 3ATP as above (Lehninger, 1975, pp. 517, 533 ff.). A more conservative equation for ATP production is therefore

$$\text{glucose} + 6\text{O}_2 \rightarrow 6\text{CO}_2 + 6\text{H}_2\text{O} + 36\text{ATP}. \tag{12.1f}$$

We shall use (12.1f) in preference to the preceding equation, and this is the basis of (12.1c) above. In actuality, glycolytic $NADH_2$ can usually be used directly in the cytosol so that the difference in ATP yield may not be of overall significance (Loomis, personal communication).

The heat of combustion of glucose is 2803 MJ $(\text{kg mol})^{-1}$ (Table 12.1), and the energy released when ATP is hydrolized under standard conditions is 30.5 MJ $(\text{kg mol})^{-1}$ (this is highly dependent upon conditions (Lehninger, 1975, p. 402)). Thus the efficiency of respiration is $36 \times 30.5/2803 = 0.39$. (Blaxter (1962, p. 263) suggests that the energy of the pyrophosphate bond in ATP under physiological conditions (in animals) is about 48 MJ $(\text{kg mol})^{-1}$, giving an efficiency of $36 \times 48/2803 = 0.62$.) A consequence of the use of an overall average as in (12.1c) is that occasionally biochemical efficiencies higher than unity will be obtained (e.g. (12.7h)); these are an artefact of the method. The calculated biochemical efficiencies need to be viewed cautiously since the heats of combustion used are not always appropriate to physiological conditions. This may also apply where heats of combustion have been calculated from heats of formation; also, where the *in vivo* reactions are far from equilibrium, the efficiency may be considerably reduced.

Reducing power, represented principally by the coenzymes NADH or NADPH, is produced in two ways. First the production of NADH by means of part of the glycolytic pathway and the citric acid cycle can be written (Lehninger, 1975, pp. 422, 445)

Table 12.1. Chemical compounds, abbreviations, relative molecular masses (RMM), heats of combustion ΔH_c (where available), and chemical formulae

Substance	Symbol	RMM	ΔH_c[†]	Formula	Substance	Symbol	RMM	ΔH_c[†]	Formula
Monosacchandes and disaccharides; polysaccharides									
Ribose		150	2339[‡]	$C_5H_{10}O_5$	Deoxyribose		134		$C_5H_{10}O_4$
Xylose		150	2339	$C_5H_{10}O_5$	Arabinose		150	2339[‡]	$C_5H_{10}O_5$
Fructose		180	2812	$C_6H_{12}O_6$	Mannose		180	2803[‡]	$C_6H_{12}O_6$
Galactose		180	2804	$C_6H_{12}O_6$	Glucose		180	2803	$C_6H_{12}O_6$
Sucrose		342	5641	$C_{12}H_{22}O_{11}$					
Araban		132	2339[‡]	$C_5H_8O_4$	Xylan		132	2339[‡]	$C_5H_8O_4$
Cellulose		162	2796[§]	$C_6H_{10}O_5$	Fructan		162	2796[§]	$C_6H_{10}O_5$
Starch		162	2796	$C_6H_{10}O_5$					
Hemicellulose		1044[¶]	18180[‖]						
Sugar alcohols									
Glycerol		92	1661	$C_3H_8O_3$	Inositol		180	2770	$C_6H_{12}O_6$
Sugar acids and polymers									
Galacturonic acid		194		$C_6H_{10}O_7$	Glucuronic acid		194		$C_6H_{10}O_7$
Galacturonan (poly-galacturonic acid)		176		$C_6H_8O_6$					
Phenylpropane alcohols (components of lignin)									
p-hydroxycinnamyl alcohol		150		$C_9H_{10}O_2$	Coniferyl alcohol		180		$C_{10}H_{12}O_3$
Sinapyl alcohol		210		$C_{11}H_{14}O_4$					
(For the units in lignin, add oxygen (O, RMM = 16) to the above, giving RMMs of 166, 196, and 226)									
Organic acids									
Ascorbate		176		$C_6H_8O_6$	Cis-aconitate		174		$C_6H_6O_6$
Citramalate		148		$C_5H_8O_5$	Citrate		192	1961	$C_6H_8O_7$
Formate		46	255	CH_2O_2	Fumarate		116	1335	$C_4H_4O_4$
Glyoxylate		74		$C_2H_2O_3$	α-ketoglutarate		146		$C_5H_6O_5$

Malate		134	1328	$C_4H_6O_5$	Malonate		104	861	$C_3H_4O_4$
Oxalate		90	246	$C_2H_2O_4$	Oxaloacetate		132		$C_4H_4O_5$
Pyruvate		88		$C_3H_4O_3$	Shikimate		174		$C_7H_{10}O_5$
Succinate		118	1491	$C_4H_6O_4$	Tartarate		150	1149	$C_4H_6O_6$
Fatty acids									
Linoleic acid		280		$C_{18}H_{32}O_2$	Linolenic acid		278		$C_{18}H_{30}O_2$
Oleic acid		282	11119	$C_{18}H_{34}O_2$	Palmitic acid		256	9978	$C_{16}H_{32}O_2$
Stearic acid		284	11281	$C_{18}H_{36}O_2$					
Triglycerides of fatty acids									
Linoleic acid		878		$C_{57}H_{98}O_6$	Linolenic acid		872		$C_{57}H_{92}O_6$
Oleic acid		884		$C_{57}H_{104}O_6$	Palmitic acid		806		$C_{51}H_{98}O_6$
Stearic acid		890		$C_{57}H_{110}O_6$					
Deoxyribonucleotides									
Deoxyadenylic acid	dAMP	331		$C_{10}H_{14}N_5O_6P$	Deoxyguanylic acid	dGMP	347		$C_{10}H_{14}N_5O_7P$
Deoxycytidylic acid	dCMP	307		$C_9H_{14}N_3O_7P$	Deoxythymidylic acid	dTMP	322		$C_{10}H_{15}N_2O_8P$
Mean (1307/4)		327							
(For the dXMP residues in DNA subtract 18 from the RMMs)									
Ribonucleotides									
Adenylic acid	AMP	347		$C_{10}H_{14}N_5O_7P$	Guanylic acid	GMP	363		$C_{10}H_{14}N_5O_8P$
Cytidylic acid	CMP	323		$C_9H_{14}N_3O_8P$	Uridylic acid	UMP	324		$C_9H_{13}N_2O_9P$
mean (1357/4)		339							
(For the XMP residues in RNA subtract 18 from the RMMs)									
Amino acids									
Alanine	Ala	89	1622	$C_3H_7NO_2$	Arginine	Arg	174		$C_6H_{10}N_4O_2$
Asparagine	Asn	132		$C_4H_8N_2O_3$	Aspartic acid	Asp	133		$C_4H_7NO_4$
Cysteine	Cys	121		$C_3H_7NO_2S$	Cystine	Cyi	240		$C_6H_{12}N_2O_4S_2$
Glutamic acid	Glu	147		$C_5H_9NO_4$	Glutamine	Gln	146		$C_5H_{10}N_2O_3$
Glycine	Gly	75	973	$C_2H_5NO_2$	Histidine	His	155		$C_6H_9N_3O_2$

Table 12.1 *(continued)*

Substance	Symbol	RMM	ΔH_c †	Formula	Substance	Symbol	RMM	ΔH_c †	Formula
Iso-leucine	Ile	131		$C_6H_{13}NO_2$	Leucine	Leu	131	3580	$C_6H_{13}NO_2$
Lysine	Lys	146		$C_6H_{14}N_2O_2$	Methionine	Met	149		$C_5H_{11}NO_2S$
Phenylalanine	Phe	165	4650	$C_9H_{11}NO_2$	Proline	Pro	115		$C_5H_9NO_2$
Serine	Ser	105		$C_3H_7NO_3$	Threonine	Thr	119		$C_4H_9NO_3$
Tryptophan	Trp	204		$C_{11}H_{12}N_2O_2$	Tyrosine	Tyr	181	4478	$C_9H_{11}NO_3$
Valine	Val	117		$C_5H_{11}NO_2$	Mean (2735/20, no Cyi)		137		
Sum (no Cyi)		2735		$C_{107}H_{197}O_{49}N_{29}S_2$					
Sum of residues (no Cyi)		2375		$C_{107}H_{157}O_{29}N_{29}S_2$					
Others									
Ammonia		17		NH_3	Carbon dioxide			44	CO_2
Hydrogen sulphide		34		H_2S	Oxygen			32	O_2
Phosphorus		31		P	Sulphur			32	S
Water		18		H_2O					

† In MJ (kg mol)$^{-1}$. The products of combustion are liquid H_2O and gaseous CO_2 and nitrogen. The values given were obtained by multiplying the values in Weast (1980, p. D-283) by 4.184 J cal^{-1}.

‡ Estimated as xylose or glucose.

§ Estimated as starch.

¶ 4 glucose units + 3 xylose units − $7H_2O$.

‖ Estimated as 3 xylose + 4 glucose − 7 (glucose − starch) = 3 × 2339 + 4 × 2803 − 7 (2803 − 2796).

$$\text{glucose} \rightarrow 2\ \text{pyruvate} + 2\text{ATP} + 2\text{NADH} \tag{12.1g}$$

$$2\ \text{pyruvate} \rightarrow 2\ \text{acetyl-CoA} + 2\text{CO}_2 + 2\text{NADH} \tag{12.1h}$$

$$2\ \text{acetyl-CoA} \rightarrow 4\text{CO}_2 + 8\text{NADH} \tag{12.1i}$$

$$\boxed{\text{glucose} + 6\text{H}_2\text{O} \rightarrow 6\text{CO}_2 + 2\text{ATP} + 12\text{NADH}.} \tag{12.1j}$$

The pentose phosphate pathway provides an alternative route for glucose oxidation (Lehninger, 1975, pp. 423, 471). This pathway is generally more active in tissues that have a high demand for reducing power, e.g. those involved in fatty acid synthesis; it appears to be more important in animals than in plants. The biochemistry can be summarized by

$$\text{glucose} + \text{ATP} \rightarrow \text{glucose 6-phosphate} \tag{12.1k}$$

$$\text{glucose 6-phosphate} \rightarrow 6\text{CO}_2 + 12\text{NADPH}$$

$$\boxed{\text{glucose} + 6\text{H}_2\text{O} + \text{ATP} \rightarrow 6\text{CO}_2 + 12\text{NADPH}.} \tag{12.1l}$$

For the overall production of reducing power, we shall assume that (see (12.1c))

$$\text{glucose} + 6\text{H}_2\text{O} \rightarrow 6\text{CO}_2 + 12\text{NADH}. \tag{12.1m}$$

See Exercise 12.1.

Acetyl coenzyme-A Adding (12.1g) and (12.1h) gives a useful equation for the formation of acetyl coenzyme-A:

$$\text{glucose} \rightarrow 2\ \text{acetyl-CoA} + 2\text{CO}_2 + 2\text{ATP} + 4\text{NADH}. \tag{12.1n}$$

Tetrahydrofolic acid (FH_4) Tetrahydrofolic acid takes part in an important class of reactions involving the transfer of one-carbon fragments. This involvement is quite widespread in metabolism, and we need to be able to assess the glucose cost of these contributions. We give a simplified account of this topic (Mahler and Cordes 1966, p. 352; Lehninger 1975, p. 345).

Methyl tetrahydrofolate ($CH_3{\cdot}FH_3$) can be used to transfer a methyl or CH_3 group, i.e. to replace H by CH_3, as in methyl tetrahydrofolate + homocysteine $\rightleftharpoons$ tetrahydrofolate + methionine, which can be written more explicitly as

$$\begin{aligned} &\text{CH}_3{\cdot}\text{FH}_3 + \text{HS}{\cdot}\text{CH}_2{\cdot}\text{CH}_2{\cdot}\text{CHNH}_2{\cdot}\text{COOH} \\ &\qquad \rightleftharpoons \text{FH}_4 + \text{CH}_3{\cdot}\text{S}{\cdot}\text{CH}_2{\cdot}\text{CHNH}_2{\cdot}\text{COOH}. \end{aligned} \tag{12.2a}$$

Methylene tetrahydrofolate, which we write as $\cdot CH_2{\cdot}FH_2\cdot$ to indicate that the CH_2 group is linked cyclically into the FH_4 molecule at the expense of two H atoms, can be employed to replace an H by CH_2OH, aided by an H_2O, as in

methylene tetrahydrofolate + glycine + $H_2O \rightleftharpoons$ tetrahydrofolate + serine,

i.e.

$$\cdot CH_2 \cdot FH_2 \cdot + H \cdot CHNH_2 \cdot COOH + H_2O$$
$$\rightleftharpoons FH_4 + CH_2OH \cdot CHNH_2 \cdot COOH. \quad (12.2b)$$

Methenyl tetrahydrofolate is written as $:CH \cdot (FH_2)^+$ to indicate that the three disposable bonds of the CH group are to the FH_4 molecule at the expense of two H atoms and that it has lost an electron, leaving it positively charged. Methenyl tetrahydrofolate can be used to replace an H by a CHO group, aided by an OH^-. An illustration is from purine (inosinic acid) biosynthesis (Mahler and Cordes, 1966, p. 722; see also (12.15d) and above):

methenyl tetrahydrofolate + phospho-ribosylglycinamide

→ tetrahydrofolate + phospho-ribosylformylglycinamide,

which we can write, representing only the relevant $\cdot NH_2$ group of the phospho-ribosylglycinamide molecule, as

$$:CH \cdot (FH_2)^+ + \cdot NH_2 + OH^- \rightarrow FH_4 + \cdot NH \cdot CHO. \quad (12.2c)$$

Formyl tetrahydrofolate has two forms, prefixed with N^{10} or N^5, depending on which nitrogen atom in the tetrahydrofolate molecule the CHO group is affixed to. Here we consider the N^{10} form alone. We denote formyl tetrahydrofolate by $CHO \cdot FH_3$; similarly to methenyl tetrahydrofolate, it can be used to replace a hydrogen atom by CHO, as in Mahler and Cordes (1966, p. 722) and the reaction immediately above (12.15d):

formyl tetrahydrofolate + phospho-ribosylaminoimidazol-carboxamide

→ tetrahydrofolate + inosine monophosphate.

We can write thus, representing only the relevant $\cdot NH_2$ group, as

$$CHO \cdot FH_3 + \cdot NH_2 \rightarrow FH_4 + \cdot NH \cdot CHO. \quad (12.2d)$$

To calculate the glucose cost of producing these one-carbon donors, we assume that they are all produced from formate according to (Mahler and Cordes, 1966, p. 353)

tetrahydrofolate + formate + ATP → formyl tetrahydrofolate + H_2O (12.2e)

formyl tetrahydrofolate → methenyl tetrahydrofolate + OH^- (12.2f)

methenyl tetrahydrofolate + NADPH

→ methylene tetrahydrofolate + $NADP^+$ (12.2g)

methylene tetrahydrofolate + NADH → methyl tetrahydrofolate. (12.2h)

We assume that formate is generated from glucose via acetyl CoA, the citric acid cycle, glyoxalate, and oxalate (Michal 1982), giving

$$0.5 \text{ glucose} \rightarrow \text{acetyl-CoA} + CO_2 + \text{ATP} + 2\text{NADH}$$

$$\text{acetyl-CoA} + \text{oxaloacetate} \rightarrow \text{citrate}$$

$$\text{citrate} \rightarrow \text{succinate} + \text{glyoxylate}$$

$$\text{succinate} \rightarrow \text{fumarate} + \text{FADH}_2$$

$$\text{fumarate} \rightarrow \text{oxaloacetate} + \text{NADH}$$

$$\text{glyoxylate} + 0.5O_2 \rightarrow \text{oxalate}$$

$$\text{oxalate} \rightarrow \text{formate} + CO_2$$

$$\boxed{0.5 \text{ glucose} + 0.5O_2 + 2H_2O \rightarrow \text{formate} + 2CO_2 + 3\text{NADH} + \text{FADH}_2 + \text{ATP}.} \quad (12.2i)$$

Adding (12.2e) and (12.2i) gives

$$0.5 \text{ glucose} + \text{tetrahydrofolate} + 0.5O_2 + H_2O \rightarrow \text{formyl tetrahydrofolate} + 2CO_2 + 3\text{NADH} + \text{FADH}_2. \quad (12.2j)$$

Combining this equation with (12.2f) leads to

$$0.5 \text{ glucose} + \text{tetrahydrofolate} + 0.5O_2 + H_2O \rightarrow \text{methenyl tetrahydrofolate} + 2CO_2 + OH^- + 3\text{NADH} + \text{FADH}_2. \quad (12.2k)$$

With (12.2g), therefore,

$$0.5 \text{ glucose} + \text{tetrahydrofolate} + 0.5O_2 + H_2O + \text{NADPH} \rightarrow \text{methylene tetrahydrofolate} + 2CO_2 + 3\text{NADH} + \text{FADH}_2. \quad (12.2l)$$

Finally, adding (12.2h) to (12.2k) gives

$$0.5 \text{ glucose} + \text{tetrahydrofolate} + 0.5O_2 + H_2O + \text{NADPH} \rightarrow \text{methyl tetrahydrofolate} + 2CO_2 + 2\text{NADH} + \text{FADH}_2. \quad (12.2m)$$

These four reactions give an estimate of the glucose cost of the one-carbon fragments provided by the various forms of tetrahydrolate.

12.2.3 *The production of some useful compounds and simple carbohydrates*

In this section we describe the synthesis from glucose of a number of compounds, some of which can be viewed as 'final' products but many of which are convenient starting points for other syntheses. The results are summarized in Table 12.2.

Table 12.2. Stoichiometric coefficients (mol (mol product)$^{-1}$), glucose requirements (kg glucose (kg product)$^{-1}$), biochemical efficiencies, CO_2 production (kg CO_2 (kg product)$^{-1}$), and oxygen requirements (kg O_2 (kg product)$^{-1}$) for the formation of some carbohydrates, polysaccharides, sugar alcohols, and pectins from glucose

Product	Equation	Glucose	CO_2	O_2	ATP	NADH	NADPH	$FADH_2$	Glucose requirement[†]	Biochemical efficiency[‡]	CO_2 production[§]	O_2 requirement[¶]
Ribose	(12.3h)	−1	1		−1		2		1.033	0.969	0.049	0.036
Galactose	(12.4d)	−1							1.000	1.000	0	0
Mannose	(12.4e)	−1							1.000	1.000	0	0
Fructose	(12.4f)	−1							1.000	1.000	0	0
Sucrose	(12.4g)	−2			−3				1.096	0.966	0.064	0.047
Cellulose, starch	(12.5a)	−1			−2				1.173	0.945	0.091	0.066
Fructans	(12.5e)	−1			−3				1.204	0.921	0.136	0.099
Hemicellulose												
Xylan	(12.5k)	−1	1		−2	2			1.212	0.939	0.111	0.081
Glucose–xylose polymer	(12.5n)	−7	3		−14		6		1.188	0.942	0.098	0.072
Pectins												
Araban	(12.6e)	−1	1		−2	2			1.212	0.939	0.111	0.081
Galacturonan	(12.6b)	−1			−2	2			0.909		−0.167	0.061

Lignins											
p-Hydroxy-cinnamyl alcohol	(12.6p)	−1.5			−2	1	−3		1.867		0.353	0.064
Coniferyl alcohol	(12.6q)	−2	2	−0.5	−2	4	−4	1	1.811		0.412	0.136
Sinapyl alcohol	(12.6r)	−2.5	4	−1	−2	7	−5	2	1.770		0.454	0.189
Glycerol	(12.3b)	−0.5			−1	−1			1.196	0.970	0.319	0.058
Inositol	(12.4k)	−1			−1				1.028	0.962	0.041	0.030

Some relevant data (relative molecular masses and heats of combustion) are given in Table 12.1. The second column gives the equation reference where details of the assumed pathway can be found. A negative coefficient indicates a requirement, and that item appears on the left-hand side of the overall equation; a positive coefficient indicates a product appearing on the right-hand side of the equation. The reduced forms NADH, NADPH, and $FADH_2$ are each equivalent to 2H. Water is omitted.

† Calculated by applying point 8, p. 288, and (12.1c) to adjust the molar glucose requirement of the third column.

‡ Calculated by dividing the heat of combustion of the product, where available, by the heat of combustion of the glucose required to synthesize that product.

§ See (12.1e).

¶ See (12.1d).

Glycerol phosphate and glycerol According to Lehninger (1975, pp. 423–6) and Mahler and Cordes (1966, p. 433) glycerol phosphate and glycerol are synthesized as follows:

glucose + ATP → glucose 6-phosphate

glucose 6-phosphate → fructose 6-phosphate

fructose 6-phosphate + ATP → fructose 1,6-diphosphate

fructose 1,6-diphosphate → 2 dihydroxy-acetone phosphate

2 dihydroxy-acetone phosphate + 2NADH → 2 glycerol 3-phosphate

glucose + 2ATP + 2NADH → 2 glycerol 3-phosphate.	(12.3a)

To complete the conversion to glycerol, we add

2 glycerol 3-phosphate → 2 glycerol

to give

$$\text{glucose} + 2\text{ATP} + 2\text{NADH} \rightarrow 2\ \text{glycerol}. \tag{12.3b}$$

The overall stoichiometry is given by applying eqn (12.1c) to give

$$(1 + 1/18 + 1/6)\ \text{glucose} \rightarrow 2\ \text{glycerol}. \tag{12.3c}$$

Per unit mass of product, with the relative molecular masses from Table 12.1, therefore

$$1.196\ \text{kg glucose} \rightarrow 1\ \text{kg glycerol}. \tag{12.3d}$$

The biochemical efficiency is obtained by applying the heats of combustion from Table 12.1 to eqn (12.3c):

$$\text{efficiency} = \frac{2 \times 1661}{(1 + 1/18 + 1/6) \times 2803} = 0.970. \tag{12.3e}$$

Ribose and ribose 5-phosphate From Mahler and Cordes (1966, pp. 454–5), Dagley and Nicholson (1970, p. 141), and Michal (1982)

5 glucose + 5ATP → 5 glucose 6-phosphate

5 glucose 6-phosphate → 6 ribose 5-phosphate

5 glucose + 5ATP → 6 ribose 5-phosphate	(12.3f)

and with

$$6 \text{ ribose 5-phosphate} \rightarrow 6 \text{ ribose},$$

therefore

$$5 \text{ glucose} + 5\text{ATP} \rightarrow 6 \text{ ribose}. \tag{12.3g}$$

An alternative pathway to ribose from glucose is (Mahler and Cordes 1966, p. 452, Dagley and Nicholson 1970, p. 141)

glucose + ATP → glucose-phosphate

glucose-phosphate → phospho-gluconate + NADPH

phospho-gluconate → ribulose-phosphate + CO_2 + NADPH

ribulose-phosphate → ribose

$$\boxed{\text{glucose} + \text{ATP} \rightarrow \text{ribose} + CO_2 + 2\text{NADPH}.} \tag{12.3h}$$

From (12.1c) with (12.3h) the overall stoichiometry is

$$(1 + 1/36 - 2/12) \text{ glucose} \rightarrow 1 \text{ ribose}. \tag{12.3i}$$

Therefore

$$1.033 \text{ kg glucose} \rightarrow 1 \text{ kg ribose}. \tag{12.3j}$$

The efficiency is given by

$$\text{efficiency} = \frac{2339}{(1 - 5/36) \times 2803} = 0.969. \tag{12.3k}$$

Do Exercise 12.2.

Erythrose-phosphate From Lehninger (1975, p. 422), Mahler and Cordes (1966, pp. 454–5) and Michal (1982)

0.5 glucose + ATP → 0.5 glucose-phosphate

0.5 glucose-phosphate → 0.5 fructose-phosphate

0.5 fructose-phosphate + 0.5ATP → 0.5 fructose-diphosphate

0.5 fructose-diphosphate → glyceraldehyde-phosphate

$$\boxed{0.5 \text{ glucose} + \text{ATP} \rightarrow \text{glyceraldehyde-phosphate}.} \tag{12.4a}$$

Also

glucose + ATP → fructose-phosphate

fructose-phosphate + glyceraldehyde-phosphate

→ erythrose-phosphate + xylulose-phosphate

xylulose-phosphate → ribulose-phosphate − [glucose + ATP

→ ribulose-phosphate + CO_2 + 2NADPH]

glyceraldehyde-phosphate + CO_2 + 2NADPH

→ erythrose-phosphate. (12.4b)

Adding (12.4a) and (12.4b)

0.5 glucose + CO_2 + ATP + 2NADPH → erythrose-phosphate. (12.4c)

Galactose and mannose From Dagley and Nicholson (1970, pp. 137, 138),

glucose → galactose (12.4d)

glucose → mannose. (12.4e)

The stoichiometric coefficients, the glucose requirements (kg glucose per kg product), and the biochemical efficiencies are all unity (cf. Penning de Vries *et al.* 1974, Williams *et al.* 1987).

Fructose and sucrose From Michal (1982) and Lehninger (1975, pp. 643–4), assuming that UTP is equivalent to ATP,

glucose → fructose. (12.4f)

Again, as with galactose and mannose above, the stoichiometric coefficient, the glucose requirement, and the biochemical efficiency are all unity.

Adding (12.4f) to

glucose + ATP → glucose-6-phosphate

glucose-6-phosphate → glucose-1-phosphate

fructose + ATP → fructose-6-phosphate

UTP + glucose-1-phosphate → UDP-glucose + PP_i

$PP_i \rightarrow 2P_i$,

UDP-glucose + fructose-6-phosphate → sucrose-6-phosphate + UDP

sucrose-6-phosphate → sucrose + P_i

UDP + ATP → UTP + ADP

$$2\ \text{glucose} + 3\text{ATP} \rightarrow \text{sucrose}. \tag{12.4g}$$

Using (12.1c), the stoichiometric equation is

$$(2 + 3/36)\ \text{glucose} \rightarrow \text{sucrose}. \tag{12.4h}$$

The glucose requirement is therefore

$$1.096\ \text{kg glucose} \rightarrow 1\ \text{kg sucrose}. \tag{12.4i}$$

The biochemical efficiency is

$$\text{efficiency} = \frac{5641}{(2 + 1/12) \times 2803} = 0.966. \tag{12.4j}$$

Inositol This sugar alcohol is formed by (Michal 1982)

glucose + ATP → glucose-6-phosphate

glucose-6-phosphate → inositol phosphate

inositol phosphate → inositol.

$$\text{glucose} + \text{ATP} \rightarrow \text{inositol}. \tag{12.4k}$$

The stoichiometric equation is

$$(1 + 1/36)\ \text{glucose} \rightarrow \text{inositol}, \tag{12.4l}$$

giving

$$1.028\ \text{kg glucose} \rightarrow 1\ \text{kg inositol}. \tag{12.4m}$$

The biochemical efficiency of inositol synthesis is

$$\text{efficiency} = \frac{2770}{(1 + 1/36) \times 2803} = 0.962. \tag{12.4n}$$

Phytin is used as a storage compound in seeds, for storing phosphate, calcium, magnesium, and potassium. It consists of the calcium, magnesium, or potassium salts of phytic acid, inositol hexaphosphate. We assume that inositol hexaphosphate is synthesized according to

$$\text{glucose} + 6\text{ATP} \rightarrow \text{inositol hexaphosphate}. \tag{12.4o}$$

If inositol hexaphosphate can combine with six equivalents of metal ion, then approximately 1/36 mol glucose is respired to store 1 mol of potassium, or 1/2 mol of calcium or magnesium.

12.2.4 *The formation of polysaccharides and lignins*

Starch and cellulose These are composed of glucose monomers. Starch can have two components: α-amylose, which is unbranched, and amylopectin, which is highly branched. We assume that all these are produced according to (Lehninger 1975, pp. 422, 645–6):

$$\text{glucose} + \text{ATP} \rightarrow \text{glucose-6-phosphate}$$

$$\text{glucose-6-phosphate} \rightarrow \text{glucose-1-phosphate}$$

$$\text{glucose-1-phosphate} + \text{XTP} \rightarrow \text{XDP-glucose} + \text{PP}_\text{i}$$

$$\text{XDP-glucose} + (\text{starch})_n \rightarrow (\text{starch})_{n+1} + \text{XDP}$$

$$\text{PP}_\text{i} \rightarrow 2\text{P}_\text{i}.$$

$$\text{XDP} + \text{ATP} \rightarrow \text{XTP} + \text{ADP}$$

$$\boxed{\text{glucose} + 2\text{ATP} + (\text{starch})_n \rightarrow (\text{starch})_{n+1}.} \tag{12.5a}$$

XTP is a ribonucleoside triphosphate which, as XDP-glucose, can act as a glucose donor; this may be GTP, UTP, or ATP, depending upon whether the glucose is being donated to starch or to cellulose (Lehninger 1975, pp. 646, 649).

With (12.1c), the stoichiometry of starch and cellulose synthesis is, per glucose moiety,

$$(1 + 2/36)\ \text{glucose} \rightarrow \text{glucose in starch or cellulose.} \tag{12.5b}$$

The glucose requirement is therefore

$$1.173\ \text{kg glucose} \rightarrow 1\ \text{kg starch or cellulose.} \tag{12.5c}$$

The biochemical efficiency is

$$\text{efficiency} = \frac{2796}{(1 + 1/18) \times 2803} = 0.945. \tag{12.5d}$$

Fructans Fructans are composed of fructose units, apart from a terminal sucrose group (Goodwin and Mercer 1972, pp. 169–70). The mechanism of fructan biosynthesis is not resolved. We assume that (with (12.4g))

$$2\ \text{glucose} + 3\text{ATP} \rightarrow \text{sucrose}$$

$$\text{sucrose} + (\text{fructan})_n \rightarrow (\text{fructan})_{n+1} + \text{glucose}$$

$$\boxed{\text{glucose} + 3\text{ATP} + (\text{fructan})_n \rightarrow (\text{fructan})_{n+1}.} \tag{12.5e}$$

With (12.1c), the stoichiometric equation is

$$(1 + 3/36) \text{ glucose} \rightarrow \text{fructose in fructan}. \tag{12.5f}$$

The glucose requirement is therefore

$$1.204 \text{ kg glucose} \rightarrow 1 \text{ kg fructan}. \tag{12.5g}$$

The biochemical efficiency is

$$\text{efficiency} = \frac{2796}{(1 + 1/12) \times 2803} = 0.921. \tag{12.5h}$$

Hemicelluloses The hemicelluloses are not closely related to cellulose. They are a very diverse group of compounds, and often different plants have different hemicelluloses. Some are polymers of pentoses, especially D-xylose, but they can also be polymers of arabinose. Hexoses can be involved with sidechains of other sugars (Lehninger 1975, p. 267, Goodwin and Mercer 1972, pp. 159; 170–1, Bauer, Talmadge, Keegstra, and Albersheim 1973).

For the synthesis of xylose and xylans, we assume the following (Goodwin and Mercer 1972, pp. 170–1):

$$\text{glucose} + \text{ATP} \rightarrow \text{glucose-6-phosphate}$$

$$\text{glucose-6-phosphate} \rightarrow \text{glucose-1-phosphate}$$

$$\text{UDP} + \text{ATP} \rightarrow \text{UTP} + \text{ADP}$$

$$\text{glucose-1-phosphate} + \text{UTP} \rightarrow \text{UDP-glucose} + \text{PP}_i$$

$$\text{PP}_i \rightarrow 2\text{P}_i$$

$$\text{UDP-glucose} \rightarrow \text{UDP-glucuronate} + 2\text{NADH}$$

$$\boxed{\text{glucose} + 2\text{ATP} + \text{UDP} \rightarrow \text{UDP-glucuronate} + 2\text{NADH}.} \tag{12.5i}$$

Adding (12.5i) to the two equations

$$\text{UDP-glucuronate} \rightarrow \text{UDP-xylose} + CO_2$$

$$\text{UDP-xylose} + (\text{xylan})_n \rightarrow (\text{xylan})_{n+1} + \text{UDP}$$

leads to

$$\text{glucose} + 2\text{ATP} + (\text{xylan})_n \rightarrow (\text{xylan})_{n+1} + CO_2 + 2\text{NADH}. \tag{12.5j}$$

With (12.1c), the stoichiometric equation is

$$(1 + 2/36 - 2/12) \text{ glucose} \rightarrow \text{xylose in xylan}. \tag{12.5k}$$

The glucose requirement is

$$1.212 \text{ kg glucose} \rightarrow 1 \text{ kg xylan}. \tag{12.5l}$$

The biochemical efficiency is

$$\text{efficiency} = \frac{2339}{(1 + 1/18 - 2/12) \times 2803} = 0.939. \tag{12.5m}$$

Also, we make an estimate on the basis of the work of Bauer *et al.* (1973), whose results on sycamore hemicellulose indicate a heptasaccharide unit consisting of four glucose and three xylose residues. With eqns (12.5a) and (12.5j), therefore,

$$7 \text{ glucose} + 14\text{ATP} \rightarrow 1 \text{ heptasaccharide unit} + 3CO_2 + 6\text{NADPH}. \tag{12.5n}$$

Using (12.1c), the stoichiometric equation is

$$(7 + 14/36 - 6/12) \text{ glucose} \rightarrow 1 \text{ heptasaccharide unit}. \tag{12.5o}$$

This leads to a glucose requirement of (Table 12.2)

$$1.188 \text{ kg glucose} \rightarrow 1 \text{ kg hemicellulose}. \tag{12.5p}$$

The biochemical efficiency is

$$\text{efficiency} = \frac{18\,180}{(7 + 7/18 - 1/2) \times 2803} = 0.942. \tag{12.5q}$$

Pectins Making use of (12.5i), we assume that galacturonan is synthesized according to (Goodwin and Mercer 1972, p. 171)

$$\text{glucose} + 2\text{ATP} + \text{UDP} \rightarrow \text{UDP-glucuronate} + 2\text{NADH} \tag{12.6a}$$

$$\text{UDP-glucuronate} \rightarrow \text{UDP-galacturonate}$$

$$\text{UDP-galacturonate} + (\text{galacturonan})_n \rightarrow (\text{galacturonan})_{n+1} + \text{UDP}.$$

$$\boxed{\text{glucose} + 2\text{ATP} + (\text{galacturonan})_n \rightarrow (\text{galacturonan})_{n+1} + 2\text{NADH}.} \tag{12.6b}$$

With (12.1c) this gives a stoichiometric equation of

$$(1 + 2/36 - 2/12) \text{ glucose} \rightarrow 1 \text{ galacturonate unit}. \tag{12.6c}$$

The glucose requirement is

$$0.909 \text{ kg glucose} \rightarrow 1 \text{ kg galacturonan}. \tag{12.6d}$$

The biochemical efficiency could not be estimated owing to the unavailability of the heat of combustion.

In fact, some of the carboxyl groups of the galacturonic acid residues occur as methyl esters, with the methyl groups coming from methionine. In Exercise 12.3, the stoichiometry and glucose requirement of a fully methylated galacturonan are calculated.

Arabans are polymers of arabinose. With the scheme proposed by Goodwin and Mercer (1972, p. 172), the formation of arabans is similar to that of xylans, so that, from (12.5j)–(12.5m),

$$\text{glucose} + 2\text{ATP} + (\text{araban})_n \rightarrow (\text{araban})_{n+1} + CO_2 + 2\text{NADH}. \quad (12.6e)$$

With (11.1c), the stoichiometric equation is

$$(1 + 2/36 - 2/12)\ \text{glucose} \rightarrow \text{arabinose in araban}. \quad (12.6f)$$

The glucose requirement is

$$1.212\ \text{kg glucose} \rightarrow 1\ \text{kg araban}. \quad (12.6g)$$

The biochemical efficiency is

$$\text{efficiency} = \frac{2339}{(1 + 1/18 - 2/12) \times 2803} = 0.939. \quad (12.6h)$$

Lignins Neither the structure of lignin nor its synthesis is fully understood (Goodwin and Mercer 1972, pp. 62–70). The main units of the polymers are phenylpropane blocks (C_6—C_3) of three types: *p*-hydroxycinnamyl alcohol (coumaryl alcohol), coniferyl alcohol, and sinapyl alcohol. We assume that these are all synthesized from *p*-hydroxycinnamate, which is formed from phenylalanine (12.19g) according to

$$1.5\ \text{glucose} + NH_3 + 3\text{NADPH} + 2\text{ATP} \rightarrow \text{phenylalanine} + \text{NADH}$$

$$\text{phenylalanine} \rightarrow \text{cinnamate} + NH_3$$

$$\text{cinnamate} + H_2O \rightarrow p\text{-hydroxycinnamate} + \text{NADH}$$

$$1.5\ \text{glucose} + 3\text{NADPH} + 2\text{ATP} \rightarrow p\text{-hydroxycinnamate} + 2\text{NADH}. \quad (12.6i)$$

p-hydroxycinnamyl alcohol is obtained by reducing the COOH group to CH_2OH by means of (it is assumed)

$$p\text{-hydroxycinnamate} + 2\text{NADH} \rightarrow p\text{-hydroxycinnamyl alcohol} + H_2O. \quad (12.6j)$$

Summing (12.6i) (12.6j) gives

$$1.5\ \text{glucose} + 3\text{NADPH} + 2\text{ATP} \rightarrow p\text{-hydroxycinnamyl alcohol} + 7H_2O. \quad (12.6k)$$

Coniferyl alcohol and sinapyl alcohol differ from *p*-hydroxycinnamyl alcohol in having one methoxy (OCH_3) and two methoxy groups respectively in place of the H atoms of the aromatic ring. Designating a member of the aromatic ring of *p*-hydroxycinnamyl alcohol by ·CH:, it is assumed that this occurs according to (with (12.2m))

$$\cdot CH: + H_2O \rightarrow \cdot COH: + NADH$$

$$\cdot COH: + \text{methionine} \rightarrow \cdot COCH_3: + \text{homocysteine}$$

$$\text{homocysteine} + \text{methyl tetrahydrofolate} \rightarrow \text{methionine} + \text{tetrahydrofolate}$$

$$0.5\ \text{glucose} + \text{tetrahydrofolate} + 0.5O_2 + NADPH \rightarrow \text{methyl tetrahydrofolate} + 2CO_2 + 2NADH + FADH_2$$

$$\boxed{\cdot CH: + 0.5\ \text{glucose} + 0.5O_2 + NADPH + H_2O \rightarrow \cdot COCH_3: + 2CO_2 + 3NADH + FADH_2.} \tag{12.6l}$$

Although it appears that methoxylation occurs before the COOH group is reduced to a CH_2OH group (Goodwin and Mercer 1972, p. 67), we can represent the stoichiometry by adding (12.6l) once or twice to (12.6k) to give

$$2\ \text{glucose} + 0.5O_2 + 4NADPH + 2ATP \rightarrow \text{coniferyl alcohol} + 2CO_2 + 6H_2O + 3NADH + FADH_2 \tag{12.6m}$$

and

$$2.5\ \text{glucose} + O_2 + 5NADPH + 2ATP \rightarrow \text{sinapyl alcohol} + 4CO_2 + 5H_2O + 6NADH + 2FADH_2. \tag{12.6n}$$

In contrast with the formation of polysaccharides, the polymerization does not occur by elimination of water but by addition of oxygen (Goodwin and Mercer 1972, p. 69). We assume that

$$\text{alcohol} + (\text{lignin})_n + H_2O \rightarrow (\text{lignin})_{n+1} + NADH, \tag{12.6o}$$

where alcohol indicates one of the three alcohols discussed above. Per moiety in lignin therefore, from (12.5k), (12.6m), and (12.6n),

$$1.5\ \text{glucose} + 3NADPH + 2ATP + [\text{lignin}\ (p\text{-hydroxycinnamyl alcohol})]_n \rightarrow [\text{lignin}\ (p\text{-hydroxycinnamyl alcohol})]_{n+1} + NADH + 6H_2O \tag{12.6p}$$

$$2\ \text{glucose} + 0.5O_2 + 4NADPH + 2ATP + [\text{lignin coniferyl alcohol})]_n \rightarrow [\text{lignin (coniferyl alcohol)}]_{n+1} + 2CO_2 + 5H_2O + 4NADH + FADH_2 \tag{12.6q}$$

and

$$2.5 \text{ glucose} + O_2 + 5\text{NADPH} + 2\text{ATP} + [\text{lignin (sinapyl alcohol)}]_n$$
$$\rightarrow [\text{lignin (sinapyl alcohol)}]_{n+1} + 4CO_2 + 4H_2O + 7\text{NADH} + 2\text{FADH}_2. \quad (12.6r)$$

The stoichiometric equations are (with (12.1c), and (12.6p)–(12.6r))

$$(1.5 + 2/12 + 2/36) \text{ glucose} \rightarrow \text{lignin (p-hydroxycinnamyl alcohol)} \quad (12.6s)$$

$$(2 - 1/12 + 2/36) \text{ glucose} \rightarrow \text{lignin (coniferyl alcohol)} \quad (12.6t)$$

$$(2.5 - 4/12 + 2/36) \text{ glucose} \rightarrow \text{lignin (sinapyl alcohol).} \quad (12.6u)$$

The glucose requirements are

$$1.867 \text{ kg glucose} \rightarrow 1 \text{ kg lignin } (p\text{-hydroxycinnamyl alcohol}) \quad (12.6v)$$

$$1.811 \text{ kg glucose} \rightarrow 1 \text{ kg lignin (coniferyl alcohol)} \quad (12.6w)$$

$$1.770 \text{ kg glucose} \rightarrow 1 \text{ kg lignin (sinapyl alcohol).} \quad (12.6x)$$

12.2.5 *The formation of organic acids*

A notable feature of many plant tissues is their ability to accumulate significant amounts of free organic acids (Goodwin and Mercer 1972, p. 140). The results are summarized in Table 12.3.

Pyruvate From (12.1g)

$$\text{glucose} \rightarrow 2 \text{ pyruvate} + 2\text{ATP} + 2\text{NADH}. \quad (12.7a)$$

With (12.1c), the stoichiometry of pyruvate formation is

$$(1 - 2/36 - 2/12) \text{ glucose} \rightarrow 2 \text{ pyruvate.} \quad (12.7b)$$

The glucose requirement is therefore

$$0.795 \text{ kg glucose} \rightarrow 1 \text{ kg pyruvate.} \quad (12.7c)$$

Malate We assume that in plants the reaction

$$\text{pyruvate} + CO_2 + \text{NADPH} \rightarrow \text{malate} \quad (12.7d)$$

is the means by which the citric acid cycle components are replenished (Goodwin and Mercer 1972, p. 141). Combining (12.7a) and (12.7d) gives

$$0.5 \text{ glucose} + CO_2 + \text{NADPH} \rightarrow \text{malate} + \text{ATP} + \text{NADH}. \quad (12.7e)$$

Using our accounting procedure, the CO_2 is 'free'; the use of (12.1f) to generate CO_2 also provides six ATP per CO_2, which, with (12.1c), spares one-sixth of a glucose. The stoichiometric equation is

$$(0.5 - 1/36) \text{ glucose} \rightarrow \text{malate,} \quad (12.7f)$$

Table 12.3. Stoichiometric coefficients (mol (mol product)$^{-1}$), glucose requirements (kg glucose (kg product)$^{-1}$), biochemical efficiencies, CO_2 production (kg CO_2 (kg product)$^{-1}$), and oxygen requirements (kg O_2 (kg product)$^{-1}$) for the formation of some organic acids

Product	Equation	Glucose	CO_2	O_2	ATP	NADH	NADPH	$FADH_2$	Glucose Requirement[†]	Biochemical efficiency[‡]	CO_2 production[§]	O_2 requirement[¶]
Ascorbate	(12.10d)	−1		−0.5	−2	2	−1		0.994		−0.042	0.152
Cis-aconitate	(12.7q)	−1			2	4	−1		0.718		−0.464	−0.061
Citramalate	(12.10l)	−1	1		1	3			0.878		−0.198	−0.036
Citrate	(12.7o)	−1			2	4	−1		0.651	1.007	−0.420	−0.056
Formate	(12.9j)	−0.5	2	−0.5	1	4			0.543	0.655	−0.159	0.232
Fumarate	(12.8l)	−1	2		2	7	−1		0.690	1.072	−0.506	−0.092
Glyoxylate	(12.9a)	−0.5	1		1	3			0.338		−0.694	−0.072
α-ketoglutarate	(12.8b)	−1	1		2	5	−1		0.753		−0.402	−0.073
Malate	(12.7e)	−0.5	−1		1	1	−1		0.634	1.003	−0.383	−0.040
Malonate	(12.8t)	−0.5			1	2			0.529	1.005	−0.494	−0.051
Oxalate	(12.9e)	−0.5	1	−0.5	1	3		1	0.278	0.632	−0.570	0.119
Oxaloacetate	(12.7l)	−0.5	−1		1	2	−1		0.530		−0.556	−0.040
Pyruvate	(12.7c)	−0.5			1	1			0.795		−0.333	−0.061
Shikimate	(12.10a)	−1	−1		1	1	−3		1.236		0.042	0.031
Succinate	(12.8f)	−1	2		3	6	−1		0.763	1.064	−0.373	−0.136
Tartarate	(12.10h)	−0.5	−1	−2	−3	−1	−1	−1	1.000	0.492	0.293	0.533

Some relevant data (relative molecular masses and heats of combustion) are given in Table 12.1. The second column gives the equation reference where details of the assumed pathway can be found. A negative stoichiometric coefficient indicates a requirement, and that item appears on the left-hand side of the overall equation; a positive coefficient indicates a product appearing on the right side of the equation. The reduced forms NADH, NADPH and $FADH_2$ are each equivalent to 2H. Water is omitted.

[†] Calculated by applying point 8, p. 288, and (12.1c) to adjust the molar glucose requirement of the third column.

[‡] Calculated by dividing the heat of combustion of the product, where available, by the heat of combustion of the glucose required to synthesize that product.

[§] See (12.1e)

[¶] See (12.1d)

giving a glucose requirement of

$$0.634 \text{ kg glucose} \rightarrow 1 \text{ kg malate}. \tag{12.7g}$$

The biochemical efficiency is

$$\text{efficiency} = \frac{1328}{(0.5 - 1/36) \times 2803} = 1.003. \tag{12.7h}$$

In view of this efficiency of greater than unity, note the comment after (12.1c). See also Exercise 12.4.

Oxaloacetate There are two possibilities here. First, from Lehninger (1975, p. 464),

$$\text{pyruvate} + CO_2 + \text{ATP} \rightarrow \text{oxaloacetate}. \tag{12.7i}$$

Combining (12.7a) and (12.7i), therefore, we obtain

$$0.5 \text{ glucose} + CO_2 \rightarrow \text{oxaloacetate} + \text{NADH}. \tag{12.7j}$$

An alternative is to use the citric acid cycle reaction (Lehninger 1975, p. 446)

$$\text{malate} \rightarrow \text{oxaloacetate} + \text{NADH}, \tag{12.7k}$$

and with (12.7e)

$$0.5 \text{ glucose} + CO_2 + \text{NADPH} \rightarrow \text{oxaloacetate} + \text{ATP} + 2\text{NADH}. \tag{12.7l}$$

Note that, with the assumption that NADH and NADPH are equivalent, (12.7l) is more efficient than (12.7j) by one ATP; with Schulz's (1978) assumption that NADH + ATP is equivalent to NADPH, the two pathways are exactly equivalent. Following our assumption in Section 12.2.1, point 6 (p. 288) for a consistent treatment, we take (12.7l).

The stoichiometric equation is

$$(0.5 - 1/12 - 1/36) \text{ glucose} \rightarrow \text{oxaloacetate}, \tag{12.7m}$$

giving a glucose requirement of

$$0.530 \text{ kg glucose} \rightarrow 1 \text{ kg oxaloacetate}. \tag{12.7n}$$

Citrate, isocitrate With (12.1n), (12.7l), and Lehninger (1975, p. 446)

$$0.5 \text{ glucose} \rightarrow \text{acetyl-CoA} + CO_2 + \text{ATP} + 2\text{NADH}$$

$$0.5 \text{ glucose} + CO_2 + \text{NADPH} \rightarrow \text{oxaloacetate} + 2\text{NADH} + \text{ATP}$$

$$\text{acetyl-CoA} + \text{oxaloacetate} \rightarrow \text{citrate}.$$

$$\boxed{\text{glucose} + H_2O + \text{NADPH} \rightarrow \text{citrate} + 4\text{NADH} + 2\text{ATP}.} \tag{12.7o}$$

With (12.1c), the stoichiometry is

$$(1 - 3/12 - 2/36) \text{ glucose} \rightarrow \text{citrate}, \tag{12.7p}$$

giving a glucose requirement of

$$0.651 \text{ kg glucose} \rightarrow 1 \text{ kg citrate}. \tag{12.7q}$$

The biochemical efficiency is

$$\text{efficiency} = \frac{1961}{(1 - 3/12 - 2/36) \times 2803} = 1.007. \tag{12.7r}$$

Cis-aconitate This is formed from citrate by loss of water:

$$\text{citrate} \rightarrow cis\text{-aconitate} + H_2O \tag{12.7s}$$

Adding this to (12.7o) gives

$$\text{glucose} + \text{NADPH} \rightarrow cis\text{-aconitate} + 4\text{NADH} + 2\text{ATP}. \tag{12.7t}$$

The stoichiometry is unchanged at

$$(1 - 3/12 - 2/36) \text{ glucose} \rightarrow cis\text{-aconitate}, \tag{12.7u}$$

giving a glucose requirement of

$$0.718 \text{ kg glucose} \rightarrow 1 \text{ kg } cis\text{-aconitate}. \tag{12.7v}$$

α-ketoglutarate Adding (12.7o) to (Lehninger 1975, p. 446)

$$\text{citrate} \rightarrow \alpha\text{-ketoglutarate} + \text{NADH} + CO_2 \tag{12.8a}$$

gives

$$\text{glucose} + H_2O + \text{NADPH} \rightarrow \alpha\text{-ketoglutarate} + CO_2 + 5\text{NADH} + 2\text{ATP}. \tag{12.8b}$$

The stoichiometry is

$$(1 - 4/12 - 2/36) \text{ glucose} \rightarrow \alpha\text{-ketoglutarate}, \tag{12.8c}$$

giving a glucose requirement of

$$0.753 \text{ kg glucose} \rightarrow \alpha\text{-ketoglutarate}. \tag{12.8d}$$

Succinate Succinate is obtained from α-ketoglutarate by means of (Lehninger 1975, p. 446):

$$\alpha\text{-ketoglutarate} \rightarrow \text{succinate} + CO_2 + \text{NADH} + \text{GTP}, \tag{12.8e}$$

which, when added to (12.8b), gives

$$\text{glucose} + H_2O + \text{NADPH} \rightarrow \text{succinate} + 2CO_2 + 6\text{NADH} + 2\text{ATP} + \text{GTP}. \tag{12.8f}$$

The stoichiometry is (assuming that GTP and ATP are equivalent)

$$(1 - 5/12 - 3/36)\ \text{glucose} \rightarrow \text{succinate}, \tag{12.8g}$$

and the glucose requirement is

$$0.763\ \text{kg glucose} \rightarrow 1\ \text{kg succinate}. \tag{12.8h}$$

The biochemical efficiency is

$$\text{efficiency} = \frac{1491}{(1 - 5/12 - 3/36) \times 2803} = 1.064. \tag{12.8i}$$

Fumarate This is formed by means of (Lehninger 1975, p. 459)

$$\text{succinate} \rightarrow \text{fumarate} + \text{FADH}_2, \tag{12.8j}$$

and adding (12.8f) gives

$$\text{glucose} + \text{H}_2\text{O} + \text{NADPH} \rightarrow \text{fumarate} + 2\text{CO}_2 + 6\text{NADH} + \text{FADH}_2 + 2\text{ATP} + \text{GTP}. \tag{12.8k}$$

Because oxidative phosphorylation of $FADH_2$ produces only two ATP (compared with three ATP per NADH (Lehninger 1975, p. 495)) and to be consistent with our use of (12.1c) for calculating stoichiometries, we assume here that $FADH_2$ + GTP is equivalent to NADH, so that (12.8k) becomes

$$\text{glucose} + \text{H}_2\text{O} + \text{NADPH} \rightarrow \text{fumarate} + 2\text{CO}_2 + 7\text{NADH} + 2\text{ATP}. \tag{12.8l}$$

The stoichiometry is

$$(1 - 6/12 - 2/36)\ \text{glucose} \rightarrow \text{fumarate}, \tag{12.8m}$$

giving

$$0.690\ \text{kg glucose} \rightarrow 1\ \text{kg fumarate}. \tag{12.8n}$$

The biochemical efficiency is

$$\text{efficiency} = \frac{1335}{(1 - 6/12 - 2/36) \times 2803} = 1.072. \tag{12.8o}$$

Malonate There are two pathways for malonate formation, and it is not known which is the more important. Malonate can be formed from oxaloacetate by means of an oxidative decarboxylation (Goodwin and Mercer 1972, p. 142):

$$\text{oxaloacetate} + 0.5\text{O}_2 \rightarrow \text{malonate} + \text{CO}_2. \tag{12.8p}$$

Adding (12.7l) and (12.8p), therefore,

$$0.5\ \text{glucose} + \text{NADPH} + 0.5\text{O}_2 \rightarrow \text{malonate} + \text{ATP} + 2\text{NADH}. \tag{12.8q}$$

The other pathway is involved in fatty acid metabolism, and combines (12.1n) and (12.11a):

$$0.5 \text{ glucose} \rightarrow \text{acetyl-CoA} + CO_2 + 2\text{NADH} + \text{ATP} \quad (12.8r)$$

$$\text{acetyl-CoA} + CO_2 + \text{ATP} \rightarrow \text{malonyl-CoA} \quad (12.8s)$$

and

$$\text{malonyl-CoA} \rightarrow \text{malonate}$$

to give

$$0.5 \text{ glucose} + H_2O \rightarrow \text{malonate} + 2\text{NADH} + \text{ATP}. \quad (12.8t)$$

We calculate the stoichiometry from (12.8t):

$$(0.5 - 2/12 - 1/36) \text{ glucose} \rightarrow \text{malonate}. \quad (12.8u)$$

Therefore

$$0.529 \text{ kg glucose} \rightarrow 1 \text{ kg malonate}; \quad (12.8v)$$

The biochemical efficiency is

$$\text{efficiency} = \frac{861}{(0.5 - 2/12 - 1/36) \times 2803} = 1.005. \quad (12.8w)$$

Glyoxylate Formation of glyoxylate occurs according to ((12.8r) and Lehninger (1975, pp. 446, 466))

$$0.5 \text{ glucose} \rightarrow \text{acetyl-CoA} + CO_2 + 2\text{NADH} + \text{ATP}$$

$$\text{succinate} \rightarrow \text{oxaloacetate} + \text{NADH} + FADH_2$$

$$\text{oxaloacetate} + \text{acetyl-CoA} \rightarrow \text{citrate}$$

$$\text{citrate} \rightarrow \text{isocitrate}$$

$$\text{isocitrate} \rightarrow \text{succinate} + \text{glyoxylate}$$

$$\boxed{\begin{aligned} &0.5 \text{ glucose} + 2H_2O \\ &\qquad \rightarrow \text{glyoxylate} + CO_2 + 3\text{NADH} + FADH_2 + \text{ATP}. \end{aligned}} \quad (12.9a)$$

The stoichiometry is

$$(0.5 - 4/12 - 1/36) \text{ glucose} \rightarrow \text{glyoxylate} \quad (12.9b)$$

and

$$0.338 \text{ kg glucose} \rightarrow 1 \text{ kg glyoxalate}. \quad (12.9c)$$

Oxalate We assume (after Goodwin and Mercer 1972, p. 143) that

$$\text{glyoxylate} + 0.5O_2 \rightarrow \text{oxalate}, \quad (12.9d)$$

which, with (12.9a), leads to

$$0.5 \text{ glucose} + 0.5O_2 + 2H_2O \rightarrow \text{oxalate} + CO_2 + 3NADH + FADH_2 + ATP. \quad (12.9e)$$

The stoichiometry and efficiency are

$$(0.5 - 4/12 - 1/36) \text{ glucose} \rightarrow \text{oxalate} \quad (12.9f)$$

$$0.278 \text{ kg glucose} \rightarrow 1 \text{ kg oxalate} \quad (12.9g)$$

and

$$\text{efficiency} = \frac{246}{(0.5 - 4/12 - 1/36) \times 2803} = 0.632. \quad (12.9h)$$

Formic acid We assume that this is formed by decarboxylation of oxalate:

$$\text{oxalate} \rightarrow \text{formate} + CO_2. \quad (12.9i)$$

Adding this to (11.9e), therefore, gives

$$0.5 \text{ glucose} + 0.5O_2 + 2H_2O \rightarrow \text{oxalate} + 2CO_2 + 4NADH + ATP. \quad (12.9j)$$

The stoichiometry and efficiency are

$$(0.5 - 4/12 - 1/36) \text{ glucose} \rightarrow \text{formate} \quad (12.9k)$$

$$0.543 \text{ kg glucose} \rightarrow 1 \text{ kg formate} \quad (12.9l)$$

and

$$\text{efficiency} = \frac{255}{(0.5 - 4/12 - 1/36) \times 2803} = 0.655. \quad (12.9m)$$

Shikimic acid This aromatic acid is found in lucerne. It is synthesized according to ((12.4c) and Lehninger, pp. 422, 694, 708)

$$0.5 \text{ glucose} + CO_2 + ATP + 2NADPH \rightarrow \text{erythrose-phosphate},$$

$$0.5 \text{ glucose} \rightarrow \text{phosphoenolpyruvate} + NADH$$

$$\text{phosphoenolpyruvate} + \text{erythrose-phosphate} \rightarrow \text{dehydroshikimate}$$

$$\text{dehydroshikimate} + NADPH \rightarrow \text{shikimate}$$

$$\boxed{\text{glucose} + CO_2 + ATP + 3NADPH \rightarrow \text{shikimate} + NADH + 3H_2O.} \quad (12.10a)$$

With (12.1c), the stoichiometry of formation is

$$(1 + 1/36 + 2/12) \text{ glucose} \rightarrow \text{shikimate}, \quad (12.10b)$$

giving

$$1.236 \text{ kg glucose} \rightarrow 1 \text{ kg shikimate.} \qquad (12.10c)$$

Ascorbate Following Lehninger (1975, pp. 641–3) (see also Goodwin and Mercer 1972, pp. 160–1),

glucose + ATP → glucose-6-phosphate

glucose-6-phosphate → glucose-1-phosphate

glucose-1-phosphate + UTP → UDP-glucose + PP_i

$PP_i \rightarrow 2P_i$

UDP-glucose → UDP-glucuronate + 2NADH

UDP-glucuronate → glucuronate + UDP

UDP + ATP → UTP + ADP

glucuronate + NADPH → gulonate

gulonate → gulonolactone + H_2O

gulonolactone + O_2 → ascorbate + H_2O_2

$H_2O_2 \rightarrow H_2O + 0.5O_2$

$$\boxed{\text{glucose} + 0.5O_2 + \text{NADPH} + 2\text{ATP} \rightarrow \text{ascorbate} + 2\text{NADH} + H_2O.} \qquad (12.10d)$$

With (12.1c), therefore,

$$(1 + 2/36 - 1/12) \text{ glucose} \rightarrow \text{ascorbate} \qquad (12.10e)$$

$$0.994 \text{ kg glucose} \rightarrow 1 \text{ kg ascorbate.} \qquad (12.10f)$$

Tartrate This is a common plant acid. Its synthesis is uncertain. As suggested by Luckner (1984, pp. 124–5), we assume that ascorbic acid is cleaved oxidatively to give tartrate and oxalate:

$$\text{ascorbate} + 2O_2 \rightarrow \text{tartrate} + \text{oxalate.} \qquad (12.10g)$$

Adding (12.10d) and (12.10g) and subtracting (12.9e) gives

$$0.5 \text{ glucose} + CO_2 + 2O_2 + \text{NADH} + \text{NADPH} + FADH_2 + 3\text{ATP} \rightarrow \text{tartrate} + 3H_2O. \qquad (12.10h)$$

The stoichiometry is

$$(0.5 + 3/12 + 3/36)\ \text{glucose} \rightarrow \text{tartrate} \quad (12.10i)$$

$$1.000\ \text{kg glucose} \rightarrow 1\ \text{kg tartrate.} \quad (12.10j)$$

The biochemical efficiency is

$$\text{efficiency} = \frac{1149}{(0.5 + 3/12 + 3/36) \times 2803} = 0.492. \quad (12.10k)$$

Citramalate This acid is found in lucerne. Following Michal (1982), it is formed from glutamate (12.18k):

glucose + NH_3 → glutamate + CO_2 + 3NADH + ATP

glutamate → methyl aspartate

methyl aspartate → mesaconate + NH_3

mesaconate + H_2O → citramalate

glucose + H_2O → citramalate + CO_2 + 3NADH + ATP.	(12.10l)

The stoichiometry is

$$(1 - 3/12 - 1/36)\ \text{glucose} \rightarrow \text{citramalate} \quad (12.10m)$$

$$0.878\ \text{kg glucose} \rightarrow 1\ \text{kg citramalate.} \quad (12.10n)$$

12.2.6 *The formation of fatty acids and triglycerides*

The results obtained in this section are summarized in Table 12.4.

Saturated fatty acids: palmitic and stearic acids The first step in fatty acid synthesis (Harwood 1975, Lehninger 1975, pp. 660–7, Stumpf 1977) is to add CO_2 to acetyl-CoA ($CH_3{\cdot}CO{\cdot}CoA$) to give malonyl-CoA ($COOH{\cdot}CH_2{\cdot}CO{\cdot}CoA$) according to

$$\text{acetyl-CoA} + CO_2 + \text{ATP} \rightarrow \text{malonyl-CoA.} \quad (12.11a)$$

For each turn of the fatty acid synthesis cycle, a malonyl-CoA adds two carbon units to a growing fatty acid chain, which is attached to coenzyme A, and releases a CO_2. Thus the overall stoichiometry for the synthesis of palmitate acid, a saturated fatty acid with 16 carbon atoms ($CH_3{\cdot}(CH_2)_{14}{\cdot}COOH$), is

$$\text{acetyl-CoA} + 7\ \text{malonyl-CoA} + 14\text{NADPH} \rightarrow \text{palmitate} + 7CO_2. \quad (12.11b)$$

Eight coenzyme A molecules and water are released on the right of the reaction but, following our convention, we do not show them. Combining (12.11a) and

Table 12.4. Stoichiometric coefficients (mol (mol of product)$^{-1}$), glucose requirements (kg glucose (kg product)$^{-1}$) biochemical efficiencies, CO_2 production (kg CO_2 (kg product)$^{-1}$), and oxygen requirements (kg O_2 (kg product)$^{-1}$) for the formation of some long-chain fatty acids and triacyl glycerides from glucose

Product	Equation	n_C	n_{db}	Glucose	CO_2	O_2	ATP	NADH	NADPH	Glucose requirement[†]	Biochemical efficiency[‡]	CO_2 production[§]	O_2 requirement[¶]
Palmitate	(12.11e)	16	0	−4	8		1	16	−14	2.676	0.935	1.174	−0.021
Stearate	(12.11f)	18	0	−4.5	9		1	18	−16	2.729	0.935	1.214	−0.019
Oleate	(12.12d)	18	1	−4.5	9	−1	1	18	−17	2.801	0.904	1.300	0.095
Linoleate	(12.12e)	18	2	−4.5	9	−2	1	18	−18	2.875		1.388	0.210
Linolenate	(12.12f)	18	3	−4.5	9	−3	1	18	−19	2.950		1.477	0.326
Triacyl glycerides of													
palmitate	(12.13k)	16	0	−12.5	24		−4	47	−42	2.723		1.210	0.026
stearate	(12.13k)	18	0	−14	27		−4	53	−48	2.770		1.244	0.024
oleate	(12.13e)	18	1	−14	27	−3	−4	53	−51	2.839		1.327	0.133
linoleate	(12.13k)	18	2	−14	27	−6	−4	53	−54	2.910		1.412	0.243
linolenate	(12.13k)	18	3	−14	27	−9	−4	53	−57	2.982		1.497	0.355

Some relevant data (relative molecular masses and heats of combustion) are given in Table 12.1. The second column gives the equation reference where details of the assumed pathway can be found or evaluated. n_C and n_{db} are the numbers of C atoms and double bonds in the fatty acids. A negative stoichiometric coefficient indicates a requirement, and that item appears on the left-hand side of the overall equation; a positive coefficient indicates a product appearing on the right-hand side of the equation. The reduced forms NADH and NADPH are each equivalent to two H atoms. Water is omitted.

[†] Calculated by applying point 8, p. 288, and (12.1c) to adjust the molar glucose requirement of the fifth column.

[‡] Calculated by dividing the heat of combustion of the product, where available, by the heat of combustion of the glucose required to synthesize that product.

[§] See (12.1e).

[¶] See (12.1d).

(12.11b), therefore, we obtain

$$8 \text{ acetyl-CoA} + 14\text{NADPH} + 7\text{ATP} \rightarrow \text{palmitate.} \qquad (12.11c)$$

Similarly, stearic acid, a saturated long-chain fatty acid with 18 carbon atoms ($CH_3 \cdot (CH_2)_{16} \cdot COOH$) is produced according to

$$9 \text{ acetyl-CoA} + 16\text{NADPH} + 8\text{ATP} \rightarrow \text{stearate.} \qquad (12.11d)$$

The conversions in (12.11c) and (12.11d) can be combined with (12.1n), i.e.

$$0.5 \text{ glucose} \rightarrow \text{acetyl-CoA} + CO_2 + 2\text{NADH} + CO_2,$$

to give

$$4 \text{ glucose} + 14\text{NADPH} \rightarrow \text{palmitate} + 8CO_2 + 6H_2O + 16\text{NADH} + \text{ATP} \qquad (12.11e)$$

and

$$4.5 \text{ glucose} + 16\text{NADPH} \rightarrow \text{stearate} + 9CO_2 + 7H_2O + 18\text{NADH} + \text{ATP.} \qquad (12.11f)$$

Using (12.1c), the stoichiometry is given by

$$(4 - 2/12 - 1/36) \text{ glucose} \rightarrow \text{palmitate} \qquad (12.11g)$$

$$(4.5 - 2/12 - 1/36) \text{ glucose} \rightarrow \text{stearate.} \qquad (12.11h)$$

These lead to

$$2.676 \text{ kg glucose} \rightarrow 1 \text{ kg palmitate} \qquad (12.11i)$$

$$2.729 \text{ kg glucose} \rightarrow 1 \text{ kg stearate.} \qquad (12.11j)$$

The biochemical efficiencies are

$$\text{efficiency (palmitate)} = \frac{9978}{(4 - 2/12 - 1/36) \times 2803} = 0.935 \qquad (12.11k)$$

$$\text{efficiency (stearate)} = \frac{11281}{(4.5 - 2/12 - 1/36) \times 2803} = 0.935. \qquad (12.11l)$$

Unsaturated fatty acids: oleic, linoleic, and linolenic acids The biosynthetic pathways for unsaturated fatty acids are uncertain (Mahler and Cordes 1966, p. 630). From Harwood (1975, p. 62), we assume that stearic acid is reduced to oleic acid according to

$$CH_3 \cdot (CH_2)_{16} \cdot COOH + \text{NADPH} + O_2 \rightarrow CH_3 \cdot (CH_2)_7 \cdot CH{:}CH \cdot (CH_2)_7 \cdot COOH + 2H_2O. \qquad (12.12a)$$

Note that: denotes a C=C double bond, and that NADPH effectively donates two H atoms to the reductive process.

Oleic acid is reduced to linoleic acid by means of

$$CH_3{\cdot}(CH_2)_7{\cdot}CH{:}CH{\cdot}(CH_2)_7{\cdot}COOH + NADPH + O_2$$
$$\rightarrow CH_3{\cdot}(CH_2)_4{\cdot}CH{:}CH{\cdot}CH_2{\cdot}CH{:}CH{\cdot}(CH_2)_7{\cdot}COOH + 2H_2O. \quad (12.12b)$$

Linoleic acid is reduced to linolenic acid by means of

$$CH_3{\cdot}(CH_2)_4{\cdot}CH{:}CH{\cdot}CH_2CH{\cdot}(CH_2)_7{\cdot}COOH + NADPH + O_2$$
$$\rightarrow CH_3{\cdot}CH_2{\cdot}CH{:}CH{\cdot}CH_2{\cdot}CH{:}CH{\cdot}CH_2{\cdot}CH{:}CH{\cdot}(CH_2)_7{\cdot}COOH + 2H_2O. \quad (12.12c)$$

Combining these equations with (12.11f) gives the stoichiometry of the production of the unsaturated fatty acids from glucose as

$$4.5 \text{ glucose} + 17\text{NADPH} + O_2 \rightarrow \text{oleate} + 9CO_2 + 9H_2O + \text{ATP} + 18\text{NADH} \quad (12.12d)$$

$$4.5 \text{ glucose} + 18\text{NADPH} + 2O_2 \rightarrow \text{linoleate} + 9CO_2 + 11H_2O + \text{ATP} + 18\text{NADH} \quad (12.12e)$$

$$4.5 \text{ glucose} + 19\text{NADPH} + 3O_2 \rightarrow \text{linolenate} + 9CO_2 + 13H_2O + \text{ATP} + 18\text{NADH}. \quad (12.12f)$$

With (12.1c), the stoichiometric expressions are

$$(4.5 - 1/12 - 1/36) \text{ glucose} \rightarrow \text{oleate} \quad (12.12g)$$

$$(4.5 - 1/36) \text{ glucose} \rightarrow \text{linoleate} \quad (12.12h)$$

$$(4.5 + 1/12 - 1/36) \text{ glucose} \rightarrow \text{linolenate}. \quad (12.12i)$$

These give

$$2.801 \text{ kg glucose} \rightarrow 1 \text{ kg oleate} \quad (12.12j)$$

$$2.875 \text{ kg glucose} \rightarrow 1 \text{ kg linoleate} \quad (12.12k)$$

$$2.950 \text{ kg glucose} \rightarrow 1 \text{ kg linolenate}. \quad (12.12l)$$

The only biochemical efficiency that we can evaluate is

$$\text{efficiency (oleate)} = \frac{11\,119}{(4.5 - 1/12 - 1/36) \times 2803} = 0.904. \quad (12.12m)$$

Synthesis of triglycerides: glycerol trioleate The precursors for triglyceride synthesis are glycerol 3-phosphate and the fatty acyl-CoA molecules. From (12.3a), glycerol 3-phosphate is obtained by

$$\text{glucose} + 2\text{ATP} + 2\text{NADH} \rightarrow 2 \text{ glycerol 3-phosphate.} \qquad (12.13a)$$

We assume that the free fatty acids are released from the fatty acid synthetase system and that, before triglyceride synthesis can take place, the fatty acid must be rejoined to acetyl-CoA according to Schulz (1978, p. 248) for, say, palmitic acid,

$$\text{palmitic acid} + 2\text{ATP} \rightarrow \text{palmityl-CoA.} \qquad (12.13b)$$

For the general fatty acid ROH, this equation becomes

$$\text{ROH} + 2\text{ATP} \rightarrow \text{R-CoA.} \qquad (12.13c)$$

For general fatty acyl–CoA molecules, R_1–CoA, R_2–CoA, and R_3–CoA (the free fatty acids are R_1OH etc.), the overall conversion is according to

$$\text{glycerol 3-phosphate} + R_1\text{–CoA} + R_2\text{–CoA} + R_3\text{–CoA}$$
$$\rightarrow CH_2OR_1{\cdot}CHOR_2{\cdot}CH_2OR_3. \qquad (12.13d)$$

To obtain the synthetic requirements for glycerol trioleate, we add three of (12.12d), half of (12.13a), three of (12.13c), and one of (12.13d) to give

$$14 \text{ glucose} + 51\text{NADPH} + 3O_2 + 4\text{ATP}$$
$$\rightarrow \text{glycerol trioleate} + 27CO_2 + 30H_2O + 53\text{NADH}. \qquad (12.13e)$$

The stoichiometric equations are

$$(14 - 2/12 + 4/36) \text{ glucose} \rightarrow \text{glycerol trioleate} \qquad (12.13f)$$

$$2.839 \text{ kg glucose} \rightarrow 1 \text{ kg glycerol trioleate.} \qquad (12.13g)$$

Some results for triglyceride synthesis are summarized in Table 12.4.

General expressions for fatty acid and triglyceride synthesis We denote a general fatty acid with n_C C atoms (n_C even only) and n_{db} double bonds by $\{n_C, n_{db}\}$. The fatty acid $\{n_C, n_{db}\}$ contains n_C C atoms, $2(n_C - n_{db})$ H atoms, and two O atoms. Generalizing from (12.11e) and (12.11f), we have

$$\frac{n_C}{4} \text{glucose} + (n_C - 2)\text{NADPH}$$
$$\rightarrow \{n_C, 0\} + \frac{n_C}{2} CO_2 + \frac{n_C - 4}{2} H_2O + n_C\text{NADH} + \text{ATP}. \qquad (12.13h)$$

From (12.12a)–(12.12c)

$$\{n_C, 0\} + n_{db}\text{NADPH} + n_{db}O_2 \rightarrow \{n_C, n_{db}\} + 2n_{db}H_2O. \qquad (12.13i)$$

Adding (12.13h) and (12.13i) gives

$$\frac{n_C}{4}\text{glucose} + (n_C + n_{db} - 2)\text{NADPH} + n_{db}O_2$$

$$\rightarrow \{n_C, n_{db}\} + \frac{n_C}{2}CO_2 + \left(2n_{db} + \frac{n_C - 4}{2}\right)H_2O$$

$$+ n_C\text{NADH} + \text{ATP}. \qquad (12.13j)$$

For a triglyceride with $R_1 = R_2 = R_3$ (see (12.13c)), we add half (12.13a), three (12.13c), one (12.13d), and three (12.13j) to give

$$\frac{2 + 3n_C}{4}\text{ glucose} + 3n_{db}O_2 + 4\text{ATP} + 3(n_C + n_{db} - 2)\text{NADPH}$$

$$\rightarrow \text{glycerol tri-}\{n_C, n_{db}\} + \frac{3n_C}{2}CO_2 + 3\left(2n_{db} + \frac{n_C - 2}{2}\right)H_2O$$

$$+ (3n_C - 1)\text{NADH}. \qquad (12.13k)$$

For a triglyceride with different fatty acids simply substitute

$$3n_C = n_C(1) + n_C(2) + n_C(3) \qquad (12.13l)$$

and

$$3n_{db} = n_{db}(1) + n_{db}(2) + n_{db}(3). \qquad (12.13m)$$

in eqn (12.13k). See Exercise 12.5.

12.2.7 *The source of nitrogen and sulphur for plant growth*

Some important classes of plant compounds, i.e. amino acids, proteins, and nucleic acids, contain nitrogen. Sulphur is less importance, as it occurs mostly only in the amino acids, cysteine, and methionine. Before the efficiency of formation of these substances is considered, it is necessary to digress briefly and consider how nitrogen and sulphur become available to the plant.

When the synthesis of nitrogen- and sulphur-containing compounds is considered, it will be assumed that the starting points are ammonia (NH_3) and hydrogen sulphide (H_2S). We now describe how these are obtained from the raw materials available to the plant. We discuss only chemical costs, and not any transport costs; the latter are more uncertain but may not be negligible.

Nitrate reduction Nitrate in the soil, which arises from various sources, both manmade and natural, is often the most important source of plant nitrogen. Following Mahler and Cordes (1966, p. 281), we assume that (remembering that each NADH transfers two H atoms)

$$HNO_3 + 4\text{NADH} \rightarrow NH_3 + 3H_2O. \qquad (12.14a)$$

With (12.1c), the glucose requirement for nitrate reduction is

$$(4/12) \text{ glucose per } NH_3, \quad (12.14b)$$

giving a value of

$$4.29 \text{ kg glucose (kg ammonia nitrogen)}^{-1}. \quad (12.14c)$$

Nitrogen fixation Following Simpson (1987, eqn (2)), it is assumed that dinitrogen fixation occurs according to

$$0.5N_2 + 2NADH + 8ATP \rightarrow NH_3 + 0.5H_2. \quad (12.14d)$$

Although some organisms can recover part of the energy in the hydrogen as reducing equivalents or ATP, this seems unlikely to make a substantial contribution to the energetics and is therefore ignored. (At most, it appears that recycling all the hydrogen may reduce the ATP requirement in (12.14d) by one ATP, giving a reduction in the glucose requirement from 14/36 to 13/36). The glucose requirement of (12.14d) is

$$(2/12 + 8/36) \text{ glucose per } NH_3, \quad (12.14e)$$

giving

$$5.00 \text{ kg glucose (kg ammonia nitrogen)}^{-1}. \quad (12.14f)$$

Efficiencies of nitrate reduction and nitrogen fixation The heats of combustion relative to the standard combustion products of nitrogen gas, liquid water, and CO_2 can be calculated from the energies of formation (Weast 1980, pp. D-67–D-73, D-289; there is uncertainty in these calculations because it is not known which value of the several tabulated values is most appropriate) (see Exercise 12.6); they are

$$\Delta H_c(NH_3) = 383 \text{ MJ (kg mol)}^{-1} \quad (12.14g)$$

$$\Delta H_c(HNO_3) = -30 \text{ MJ (kg mol)}^{-1}. \quad (12.14h)$$

The biochemical efficiencies of the reductive conversions can then be calculated as (12.14b) and (12.14e)

$$\text{efficiency } (NO_3^- \rightarrow NH_3) = \frac{383 + 30}{(4/12) \times 2803} = 0.44 \quad (12.14i)$$

$$\text{efficiency } (0.5N_2 \rightarrow NH_3) = \frac{383}{(2/12 + 8/36) \times 2803} = 0.35. \quad (12.14j)$$

Not only is nitrate reduction more efficient in energetic terms, but, as seen in (12.14c) and (12.14f), the glucose requirement per unit of ammonia nitrogen produced is less.

Corrections to stoichiometric results due to a change in the nitrogen source The source of nitrogen can be NO_3^-, N_2, or NH_3. First write a simplified stoichiometric equation as

$$g \text{ glucose} + a\text{ATP} + h\text{NADH} + n\text{NH}_3 \rightarrow 1 \text{ product}, \quad (12.14k)$$

defining the coefficients g, a, h, and n. For nitrate reduction, applying (2.14a) to this gives

$$g \text{ glucose} + a\text{ATP} + (h + 4n)\text{NADH} + n\text{HNO}_3 \rightarrow 1 \text{ product}, \quad (12.14l)$$

and, with (12.1c), the increase in the glucose requirement is

$$\frac{(4n/12) \times 180}{\text{RMM(product)}} = \frac{60n}{\text{RMM}} \text{ kg glucose (kg product)}^{-1}. \quad (12.14m)$$

where RMM is the relative molecular mass.

For nitrogen fixation, (12.14d) and (12.14k) give

$$g \text{ glucose} + (a + 8n)\text{ATP} + (h + 2n)\text{NADH} + (n/2)\text{N}_2 \rightarrow 1 \text{ product}, \quad (12.14n)$$

and, with (12.1c), the increase in the glucose requirement is

$$\frac{(8n/36 + 2n/12) \times 180}{\text{RMM(product)}} = \frac{70n}{\text{RMM}} \text{ kg glucose (kg product)}^{-1}. \quad (12.14o)$$

Equations (12.14m) and (12.14o), or their equivalents corresponding to different underlying assumptions, enable the glucose requirements for different nitrogen sources to be calculated easily.

Sulphate reduction to H_2S Following Michal (1982) and Wilson and Reuveny (1976), it is assumed that (replacing one ($\text{ATP} \rightarrow \text{AMP} + \text{PP}_i$ and $\text{PP}_i \rightarrow 2\text{P}_i$) reaction by two ($\text{ATP} \rightarrow \text{ADP} + \text{P}_i$) reactions)

$$\text{H}_2\text{SO}_4 + 4\text{NADPH} + 3\text{ATP} \rightarrow \text{H}_2\text{S} + 4\text{H}_2\text{O}. \quad (12.14p)$$

Assume the simplified stoichiometric equation

$$g \text{ glucose} + a\text{ATP} + h\text{NADH} + h'\text{NADPH} + s\text{H}_2\text{S} \rightarrow 1 \text{ product}, \quad (12.14q)$$

defining the coefficients g, a, h, h', and s. Combining (12.14p) and (12.14q) gives

$$g \text{ glucose} + (a + 3s)\text{ATP} + h\text{NADH} + (h' + 4s)\text{NADPH} + s\text{H}_2\text{SO}_4 \rightarrow 1 \text{ product}. \quad (12.14r)$$

If we assume that NADPH and NADH are equivalent and apply (12.1c), the glucose requirement is increased by

$$\frac{(3s/36 + 4s/12) \times 180}{\text{RMM(product)}} = \frac{75s}{\text{RMM}} \text{ kg glucose (kg product)}^{-1}. \quad (12.14s)$$

12.2.8 *The formation of purines, pyrimidines, and nucleic acids*

Some of the results obtained in this section are summarized in Table 12.5.

Inosine monophosphate (IMP) IMP is an important purine intermediate in the synthetic pathways of adenosine and guanosine monophosphates. From Dagley

Table 12.5. Stoichiometric coefficients (mol (mol product)$^{-1}$), glucose requirements (kg glucose (kg product)$^{-1}$), CO_2 production (kg CO_2 (kg product)$^{-1}$), and oxygen requirements (kg O_2 (kg product)$^{-1}$) for the formation of some nucleoside monophosphates, nucleoside monophosphate residues in RNA, and deoxynucleoside monophosphate residues in DNA from glucose and ammonia

Product	Equation	Glucose	NH_3 or NO_3^-	O_2	CO_2	ATP	NADH		NADPH	$FADH_2$	Glucose requirement[†]		CO_2 production[‡]		O_2 requirement[¶]
							NH_3	NO_3^-			NH_3	NO_3	NH_3	NO_3^-	
AMP	(12.15h)	−2.5	−5	−1	5	−11	8	−12	3	2	0.893	1.758	0.042	1.310	0.261
GMP	(12.15k)	−2.5	−5	−1	5	−12	9	−11	3	2	0.826	1.653	0.000	1.212	0.264
UMP	(12.16b)	−1.5	−2			−6	1	−7	2		0.787	1.157	−0.068	0.475	0.099
CMP	(12.16f)	−1.5	−3			−7	1	−11	2		0.805	1.362	−0.045	0.772	0.116
Nucleoside monophosphate residues in RNA															
AMP	(12.16l)	−2.5	−5	−1	5	−13	8	−12	3	2	0.973	1.884	0.089	1.427	0.308
GMP	(12.16m)	−2.5	−5	−1	5	−14	9	−11	3	2	0.899	1.768	0.043	1.318	0.309
UMP	(12.16n)	−1.5	−2			−8	1	−7	2		0.866	1.258	−0.024	0.551	0.139
CMP	(12.16o)	−1.5	−3			−9	1	−11	2		0.885	1.475	0.000	0.866	0.157
Sum		−8	−15	−2	10	−44	19	−41	10	4	0.907	1.607	0.029	1.056	0.232
Deoxynucleoside monophosphate residues in DNA															
dAMP	(12.17a)	−2.5	−5	−1	5	−13	8	−12	2	2	1.070	2.029	0.164	1.570	0.324
dGMP	(12.17b)	−2.5	−5	−1	5	−14	9	−11	2	2	0.988	1.900	0.111	1.449	0.324
dTMP	(12.17d)	−1.67	−2			−9	1	−7			1.086	1.480	0.145	0.724	0.158
dCMP	(12.17c)	−1.5	−3			−9	1	−11	1		0.986	1.609	0.076	0.990	0.166
Sum		−8.17	−15	−2	10	−45	19	−41	5	4	1.032	1.761	0.125	1.194	0.246

Some relevant data (relative molecular masses) are given in Table 12.1. The second column gives the equation reference where details of the assumed pathway can be found or evaluated. A negative stoichiometric coefficient indicates a requirement, and that item appears on the left-hand side of the overall equation; a positive coefficient indicates a product appearing on the right-hand side of the equation. The reduced forms NADH, NADPH, and $FADH_2$ are each equivalent to two H atoms. Water is omitted. The nitrogen source can be ammonia or nitrate; these are related by (12.14a), (12.14l), and (12.14m), and the effects of the nitrogen source on NADH production and the glucose requirement are given.

[†] Calculated by applying point 8, p. 288, and (12.1c) to adjust due molar glucose requirement of the third column.

[‡] See (12.1e)

[¶] See (12.1d)

and Nicholson (1970, p. 160–2)

ribose + ATP → phospho-ribose

phospho-ribose + ATP → phospho-ribosylpyrophosphate + AMP

phospho-ribosylpyrophosphate + glutamine

→ phospho-ribosylamine + pyrophosphate + glutamate

phospho-ribosylamine + glycine + ATP

→ phospho-ribosylglycineamide,

$$\text{ribose} + 3\text{ATP} + \text{glutamine} + \text{glycine} \rightarrow \text{phospho-ribosylglycineamide} + \text{glutamate} + 2\text{ADP} + \text{AMP} + \text{PP}_i + \text{P}_i. \quad (12.15a)$$

In the final equation we give the forms ADP and AMP explicitly because of the occurrence of pyrophosphate PP_i. Making use of the assumed equivalence of (12.1a) and (12.1b), we replace one ATP → AMP + PP_i by two ATP → 2ADP + $2P_i$, so that (12.15a) becomes

$$\text{ribose} + 4\text{ATP} + \text{glutamine} + \text{glycine} \rightarrow \text{phospho-ribosylglycineamide} + \text{glutamate}. \quad (12.15b)$$

Making use of

$$\text{glutamate} + NH_3 + \text{ATP} \rightarrow \text{glutamine},$$

reaction (12.18m) for glycine biosynthesis, i.e.

$$0.5\ \text{glucose} + NH_3 \rightarrow \text{glycine} + CO_2 + 2\text{NADH} + \text{NADPH},$$

and (12.3h) for ribose synthesis

$$\text{glucose} + \text{ATP} \rightarrow \text{ribose} + CO_2 + 2\text{NADPH},$$

we add these to (12.15b) to give

$$1.5\ \text{glucose} + 2NH_3 + 6\text{ATP} \rightarrow \text{phospho-ribosylglycineamide} + 2CO_2 + 2\text{NADH} + 3\text{NADPH}. \quad (12.15c)$$

Continuing, we have

phospho-ribosylglycineamide + methenyl-tetrahydrofolate

→ phospho-ribosylformylglycineamide

phospho-ribosylformylglycineamide + glutamine + ATP

→ phospho-ribosylformylglycineamidine

phospho-ribosylformylglycineamidine + ATP

→ phospho-ribosylaminoimidazole,

phospho-ribosylaminoimidazole + CO_2

→ phospho-ribosylaminoimidazole-carboxylate

phospho-ribosylaminoimidazole-carboxylate + aspartate + ATP

→ phospho-ribosylaminoimidazole-carboxamide + fumarate

phospho-ribosylaminoimidazole-carboxamide

+ formyl-tetrahydrofolate → inosine monophosphate.

phospho-ribosylglycineamide + methenyl-tetrahydrofolate

+ glutamine + 3ATP + CO_2 + aspartate

+ formyl-tetrahydrofolate

→ inosine monophosphate + glutamate + fumarate. (12.15d)

Further simplification of this requires various relations: from (12.2k) and (12.2j)

0.5 glucose + tetrahydrofolate + $0.5O_2$

→ methenyl tetrahydrofolate + $2CO_2$ + $FADH_2$ + 3NADH

0.5 glucose + tetrahydrofolate + $0.5O_2$

→ formyl tetrahydrofolate + $2CO_2$ + $FADH_2$ + 3NADH

glutamate + NH_3 + ATP → glutamine

fumarate + NH_3 → aspartate.

Adding these four equations to (12.15d) and simplifying, therefore, we have

phospho-ribosylglycineamide + glucose + $2NH_3$ + O_2 + 4ATP

→ inosine monophosphate + $3CO_2$ + $2FADH_2$ + 6NADH. (12.15e)

Finally, adding (12.15c) to (12.15e) gives

2.5 glucose + O_2 + $4NH_3$ + 10ATP → inosine monophosphate

+ $5CO_2$ + 3NADPH + $2FADH_2$ + 8NADH. (12.15f)

Adenosine monophosphate (AMP)

IMP + aspartate + GTP → AMP + fumarate + GDP + P_i

ATP + GDP → GTP + ADP

fumarate + NH_3 → aspartate.

$$\boxed{IMP + ATP + NH_3 \rightarrow AMP + ADP + P_i.} \quad (12.15g)$$

Here we have given the full equations (apart from water) to make it clear that the AMP molecule is being synthesized *de novo* and is not a product of the energy source ATP. Combining (12.15f) and (12.15g) gives

$$\begin{aligned} &2.5 \text{ glucose} + O_2 + 5NH_3 + 11ATP \\ &\rightarrow AMP + 5CO_2 + 3NADPH + 2FADH_2 + 8NADH + 11ADP + 10P_i. \end{aligned} \quad (12.15h)$$

The glucose requirement is (with (12.1c))

$$(2.5 + 11/36 - 13/12) \text{ glucose} \rightarrow AMP \quad (12.15i)$$

$$0.893 \text{ kg glucose} \rightarrow 1 \text{ kg AMP}. \quad (12.15j)$$

See Exercise 12.7.

Guanosine monophosphate (*GMP*)

$$IMP + NH_3 + ATP \rightarrow GMP + NADH + AMP + PP_i,$$

which, with (12.1a), (12.1b), and (12.15f) gives

$$\begin{aligned} &2.5 \text{ glucose} + O_2 + 5NH_3 + 12ATP \\ &\qquad \rightarrow GMP + 5CO_2 + 3NADPH + 2FADH_2 + 9NADH. \end{aligned} \quad (12.15k)$$

With (12.1c), the glucose requirement is

$$(2.5 + 12/36 - 14/12) \text{ glucose} \rightarrow GMP \quad (12.15l)$$

$$0.826 \text{ kg glucose} \rightarrow 1 \text{ kg GMP}. \quad (12.15m)$$

Orotic acid The synthesis of the pyrimidine phosphates (uridine, thymidine, and cytidine phosphates) is simpler than that of the purines, with a key intermediate being orotic acid. It is achieved by means of ((12.18h) and Lehninger (1975, pp. 580, 736))

$$2ATP + CO_2 + NH_3 \rightarrow \text{carbamoyl phosphate}$$

$$0.5 \text{ glucose} + NH_3 + CO_2 \rightarrow \text{aspartate}$$

$$\text{carbamoyl phosphate} + \text{aspartate} \rightarrow \text{orotic acid} + NADH.$$

$$\boxed{0.5 \text{ glucose} + 2CO_2 + 2NH_3 + 2ATP \rightarrow \text{orotic acid} + NADH.} \quad (12.16a)$$

Uridine monophosphate (UMP) This is synthesized according to

$$\text{glucose} + \text{ATP} \rightarrow \text{ribose} + CO_2 + 2\text{NADPH}$$

$$\text{ribose} + 3\text{ATP} \rightarrow \text{phospho-ribosylpyrophosphate} + 3\text{ADP}$$

$$\text{orotic acid} + \text{phospho-ribosylpyrophosphate} \rightarrow \text{UMP} + CO_2.$$

$$0.5\ \text{glucose} + 2CO_2 + 2NH_3 + 2\text{ATP} \rightarrow \text{orotic acid} + \text{NADH}$$

$$\boxed{1.5\ \text{glucose} + 2NH_3 + 6\text{ATP} \rightarrow \text{UMP} + 2\text{NADPH} + \text{NADH}.} \quad (12.16b)$$

The first of these equations is from (12.3h), the second is from the first two conversions of this section under the inosine monophosphate subheading, assuming that the reaction $\text{ATP} \rightarrow \text{AMP} + PP_i$ can be replaced by $2\text{ATP} \rightarrow 2\text{ADP} + 2P_i$, the third is from Mahler and Cordes (1966, p. 727), and the last is from (12.16a).

The stoichiometric equations are, using (12.1c),

$$(1.5 + 6/36 - 3/12)\ \text{glucose} \rightarrow \text{UMP} \quad (12.16c)$$

$$0.787\ \text{kg glucose} \rightarrow 1\ \text{kg UMP}. \quad (12.16d)$$

Cytidine triphosphate (CTP) and cytidine monophosphate (CMP) The amination of UTP is the only known route for the synthesis of cytidine nucleotides (Mahler and Cordes, 1966, pp. 727–8), with

$$1.5\ \text{glucose} + 2NH_3 + 6\text{ATP} \rightarrow \text{UMP} + 2\text{NADPH} + \text{NADH}$$

$$\text{UMP} + 2\text{ATP} \rightarrow \text{UTP}$$

$$\text{UTP} + NH_3 + \text{ATP} \rightarrow \text{CTP}$$

$$\boxed{1.5\ \text{glucose} + 3NH_3 + 9\text{ATP} \rightarrow \text{CTP} + 2\text{NADPH} + \text{NADH}.} \quad (12.16e)$$

Assuming that CTP is equivalent to CMP + 2ATP, (12.16e) becomes

$$1.5\ \text{glucose} + 3NH_3 + 7\text{ATP} \rightarrow \text{CMP} + 2\text{NADPH} + \text{NADH}. \quad (12.16f)$$

With (12.1c), the stoichiometric equations are

$$(1.5 + 7/36 - 3/12)\ \text{glucose} \rightarrow \text{CMP} \quad (12.16g)$$

$$0.805\ \text{kg glucose} \rightarrow 1\ \text{kg CMP}. \quad (12.16h)$$

RNA synthesis For both RNA and DNA synthesis, each XMP unit is attached to a growing chain of *n* XMP units by means of (Lehninger 1975, pp. 895, 916)

$$XTP + (XMP)_n \rightarrow (XMP)_{n+1} + PP_i, \tag{12.16i}$$

and we also assume that the reaction

$$PP_i \rightarrow 2P_i \tag{12.16j}$$

takes place, so that the cost of this process is two ATP → ADP reactions.

The nucleoside monophosphates AMP, GMP, UMP, and CMP, which are the components of RNA, are converted to nucleoside triphosphates by means of

$$XMP + 2ATP \rightarrow XTP. \tag{12.16k}$$

Therefore, from (12.15h), (12.15k), (12.16b), and (12.16e),

$$\begin{aligned} &2.5\ \text{glucose} + O_2 + 5NH_3 + 13ATP \\ &\quad \rightarrow ATP + 5CO_2 + 3NADPH + 2FADH_2 + 8NADH \\ &\quad\quad + 13ADP + 12P_i \end{aligned} \tag{12.16l}$$

$$\begin{aligned} &2.5\ \text{glucose} + O_2 + 5NH_3 + 14ATP \\ &\quad \rightarrow GTP + 5CO_2 + 3NADPH + 2FADH_2 + 9NADH \end{aligned} \tag{12.16m}$$

$$1.5\ \text{glucose} + 2NH_3 + 8ATP \rightarrow UTP + 2NADPH + NADH \tag{12.16n}$$

$$1.5\ \text{glucose} + 3NH_3 + 9ATP \rightarrow CTP + 2NADPH + NADH. \tag{12.16o}$$

These four equations give the stoichiometry of synthesis per nucleoside monophosphate residue in RNA (XMP less H_2O). With (12.1c)

$$(2.5 + 14/36 - 13/12)\ \text{glucose} \rightarrow \text{AMP residue} \tag{12.16p}$$

$$(2.5 + 14/36 - 14/12)\ \text{glucose} \rightarrow \text{GMP residue} \tag{12.16q}$$

$$(1.5 + 8/36 - 3/12)\ \text{glucose} \rightarrow \text{UMP residue} \tag{12.16r}$$

$$(1.5 + 9/36 - 3/12)\ \text{glucose} \rightarrow \text{CMP residue.} \tag{12.16s}$$

Using the relative molecular masses in Table 12.1 less 18 (to allow for the H_2O that is eliminated in (12.16i)), therefore,

$$0.973\ \text{kg glucose} \rightarrow 1\ \text{kg AMP in RNA} \tag{12.16t}$$

$$0.899\ \text{kg glucose} \rightarrow 1\ \text{kg GMP in RNA} \tag{12.16u}$$

$$0.866\ \text{kg glucose} \rightarrow 1\ \text{kg UMP in RNA} \tag{12.16v}$$

$$0.885\ \text{kg glucose} \rightarrow 1\ \text{kg CMP in RNA.} \tag{12.16w}$$

The glucose requirement for forming RNA consisting of equal molar proportions of the four bases is also given in Table 12.5. A useful figure is the increased glucose requirement resulting from higher ATP costs; for each ATP per base of extra cost, the value of 0.907 would be increased by

$$\frac{1}{36} \times \frac{180}{339 - 18} = 0.016 \text{ kg glucose (kg RNA)}^{-1}. \qquad (12.16x)$$

Deoxynucleoside triphosphates (dXTP) In the cases of ATP, GTP, and CTP, ribose is reduced to deoxyribose by means of NADPH (Lehninger 1975, p. 738):

$$\text{XTP} + \text{NADPH} \rightarrow \text{dXTP},$$

giving (from (12.16l), (12.16m), and (12.16o))

$$2.5 \text{ glucose} + O_2 + 5NH_3 + 13ATP$$
$$\rightarrow dATP + 5CO_2 + 2FADH_2 + 2NADPH + 8NADH \qquad (12.17a)$$

$$2.5 \text{ glucose} + O_2 + 5NH_3 + 14ATP$$
$$\rightarrow dGTP + 5CO_2 + 2FADH_2 + 2NADPH + 9NADH \qquad (12.17b)$$

$$1.5 \text{ glucose} + 3NH_3 + 9ATP \rightarrow dCTP + NADPH + NADH. \qquad (12.17c)$$

DNA contains the pyrimidine thymine instead of the uracil present in RNA. The pathway for the synthesis of deoxythymidylic acid (dTMP) is (Mahler and Cordes 1966, p. 730, Lehninger 1975, p. 739)

$$CDP \rightarrow dCDP \rightarrow dUMP \rightarrow dTMP.$$

Simplifying, we can write (using (12.16e))

$$1.5 \text{ glucose} + 3NH_3 + 9ATP \rightarrow CTP + 2NADPH + NADH$$

$$CTP + ADP \rightarrow CDP + ATP$$

$$CDP + NADPH \rightarrow dCDP$$

$$dCDP + ADP \rightarrow dUMP + ATP + NH_3$$

$$dUMP + NADPH + \{CH_2O\} \rightarrow dTMP$$

$$(1/6) \text{ glucose} \rightarrow \{CH_2O\}$$

$$dTMP + 2ATP \rightarrow dTTP$$

$$\boxed{(1 + 2/3) \text{ glucose} + 2NH_3 + 9ATP \rightarrow dTTP + NADH.} \qquad (12.17d)$$

DNA synthesis This is similar to RNA synthesis, as given by (12.16i) and (12.16j). Thus, per deoxynucleoside monophosphate in DNA, (12.1c) and 12.17(a)–(12.17d) lead immediately to the stoichiometric relations

$$(2.5 + 13/36 - 12/12)\ \text{glucose} \rightarrow \text{dAMP residue} \quad (12.17e)$$

$$(2.5 + 14/36 - 13/12)\ \text{glucose} \rightarrow \text{dGMP residue} \quad (12.17f)$$

$$(1 + 2/3 + 9/36 - 1/12)\ \text{glucose} \rightarrow \text{dTMP residue} \quad (12.17g)$$

$$(1.5 + 9/36 - 2/12)\ \text{glucose} \rightarrow \text{dCMP residue.} \quad (12.17h)$$

Using the relative molecular masses in Table 12.1 less 18 (for the H_2O eliminated in (12.16i)), therefore, we have

$$1.070\ \text{kg glucose} \rightarrow 1\ \text{kg dAMP in DNA} \quad (12.17i)$$

$$0.988\ \text{kg glucose} \rightarrow 1\ \text{kg dGMP in DNA} \quad (12.17j)$$

$$1.086\ \text{kg glucose} \rightarrow 1\ \text{kg dTMP in DNA} \quad (12.17k)$$

$$0.986\ \text{kg glucose} \rightarrow 1\ \text{kg dCMP in DNA.} \quad (12.17l)$$

The glucose requirement for the formation of DNA with equal molar proportions of the bases is given in Table 12.5. It should be noted that, if the costs per base are increased by 1 ATP, this increases the value of 1.032 by

$$(1/36) \times 180/(327 - 18) = 0.016\ \text{kg glucose (kg equimolar DNA)}^{-1}. \quad (12.17m)$$

General expressions for RNA and DNA synthesis We denote a single-stranded RNA molecule which contains n_A AMP residues, n_G GMP residues, n_C CMP residues and n_U UMP residues as RNA$[n_A, n_G, n_C, n_U]$. Then with (12.16i) and (12.16j)

$$n_A\text{ATP} + n_G\text{GTP} + n_C\text{CTP} + n_U\text{UTP}$$
$$\rightarrow \text{RNA}[n_A, n_G, n_C, n_U] + 2(n_A + n_G + n_C + n_U)\text{P}_i. \quad (12.17n)$$

Therefore, with eqns (12.16l)–(12.16o) we obtain

$$\frac{5n_A + 5n_G + 3n_C + 3n_U}{2}\ \text{glucose} + (n_A + n_G)\text{O}_2$$
$$+ (5n_A + 5n_G + 3n_C + 2n_U)\text{NH}_3$$
$$+ (13n_A + 14n_G + 9n_C + 8n_U)\text{ATP}$$
$$\rightarrow \text{RNA}[n_A, n_G, n_C, n_U] + 5(n_A + n_G)\text{CO}_2$$
$$+ (3n_A + 3n_G + 2n_C + 2n_U)\text{NADPH}$$
$$+ 2(n_A + n_G)\text{FADH}_2 + (8n_A + 9n_G + n_C + n_U)\text{NADH}. \quad (12.17o)$$

For a double-stranded DNA molecule, the number of dAMP residues is equal to the number of dTMP residues, n_A say, and the number of dGMP residues equals the number of dCMP residues, n_G say. Therefore, denoting DNA by DNA$[n_A, n_G]$, with (12.16i) and (12.16j) we obtain

$$n_A\text{dATP} + n_G\text{dGTP} + n_G\text{dCTP} + n_A\text{dTTP} \rightarrow \text{DNA}[n_A, n_G] + 4(n_A + n_G)\text{P}_i. \quad (12.17\text{p})$$

Making use of eqns (12.17a)–(12.17d), multiplying (12.17a) + (12.17d) by n_A, and adding $n_G \times [(12.17b) + (12.17c)]$, it follows that

$$[(4 + 1/6)n_A + 4n_G] \text{ glucose} + (n_A + n_G)\text{O}_2 + (7n_A + 8n_G)\text{NH}_3 + (22n_A + 23n_G)\text{ATP} \rightarrow \text{DNA}[n_A, n_G] + 5(n_A + n_G)\text{CO}_2 + 2(n_A + n_G)\text{FADH}_2 + (2n_A + 3n_G)\text{NADPH} + (9n_A + 10n_G)\text{NADH}. \quad (12.17\text{q})$$

Nitrate as the nitrogen source instead of ammonia If nitrate is used as the source of nitrogen instead of ammonia, as assumed, then with (12.14a), the NADH requirement is increased by four per nitrogen required. The stoichiometric coefficients change from those obtained, and the glucose requirements (applying (12.1c) and (12.14m) are increased. These results are listed in Table 12.5 (see Exercises 12.7 and 12.9). If nitrogen fixation were used to supply the nitrogen required, then (12.14d), (12.14n), and (12.14o) enable the appropriate stoichiometry and glucose requirement to be evaluated.

Hydrolysis and resynthesis of RNA Protein synthesis occurs by translation of single-stranded messenger RNA (mRNA), which is unstable. Hydrolysis of mRNA gives nucleoside monophosphates (XMP) and, before resynthesis of the RNA can occur this must be converted to the triphosphate, XTP (see above) which requires two ATP. Thus, resynthesis of a three-subunit codon, coding for a single amino acid, requires six ATP. If the lifetime of the mRNA is such that the synthesis of n_{pp} polypetide chains occurs (on average) before the RNA is hydrolysed, then, per amino acid incorporated into protein, the ATP cost of maintaining the RNA in existence is

$$(6/n_{pp}) \text{ ATP per amino acid residue.} \quad (12.17\text{r})$$

12.2.9 *The formation of amino acids and proteins*

In this section we give less detail than hitherto. In each case the stoichiometric equation is given (e.g. (12.18a)). However, the glucose requirements and efficiencies (where these can be calculated) are given only in Table 12.6, except for the case of alanine where the calculation is done in full; these are obtained by applying (12.1c) and, for example, following the method of (12.3b)–(12.3e). The effects of different sources of nitrogen and sulphur on the stoichiometry can be obtained as outlined in Section 12.2.7 (p. 320).

Table 12.6. Stoichiometric coefficients (mol (mol product)$^{-1}$), glucose requirements (kg glucose (kg product)$^{-1}$), biochemical efficiencies, CO_2 production (kg CO_2 (kg product)$^{-1}$), and oxygen requirements (kg O_2 (kg product)$^{-1}$) for the formation of some amino acids and amino acid residues in protein

Product	Equation	Glucose	NH_3 or NO_3^-	H_2S	O_2	CO_2	ATP		NADH		NADPH	$FADH_2$
							Amino acid	Residue	NH_3	NO_3^-		
Ala	(12.18a)	−0.5	−1				1	−3	0	−4		
Arg	(12.18f)	−0.5	−4			−3	−6	−10	−4	−20	−1	
Asn	(12.18g)	−0.5	−2			−1	−1	−5	0	−8		
Asp	(12.18h)	−0.5	−1			−1	0	−4	0	−4		
Cys	(12.18i)	−0.5	−1	−1			0	−4	1	−3		
Cyi	(12.18j)	−1	−2	−2			0	−4	3	−5		
Gln	(12.18l)	−1	−2			1	0	−4	3	−5		
Glu	(12.18k)	−1	−1			1	1	−3	3	−1		
Gly	(12.18m)	−0.5	−1			1	0	−4	2	−2	1	
His	(12.18q)	−1.5	−3		−0.5	3	−8	−12	6	−6	0	1
Ile	(12.19a)	−1	−1				−1	−5	−2	−6	−1	
Leu	(12.19b)	−1.5	−1			3	3	−1	3	−1		
Lys	(12.19c)	−1	−2				−1	−5	−1	−9	−1	
Met	(12.19d)	−1	−1	−1	−0.5	1	−3	−7	1	−3	−2	1
Phe	(12.19g)	−1.5	−1				−2	−6	1	−3	−3	
Pro	(12.19h)	−1	−1			1	0	−4	2	−2	−1	
Ser	(12.19i)	−0.5	−1				0	−4	1	−3		
Thr	(12.19k)	−0.5	−1			−1	−2	−6	−1	−5	−1	
Trp	(12.19n)	−2	−2			1	−7	−11	4	−4	−3	
Tyr	(12.19o)	−1.5	−1				−2	−6	2	−2	−3	
Val	(12.19p)	−1	−1			1	2	−2	0	−4		
Sum‖		−19	−29	−2	−1	7	−26	−106	21	−95	−15	2
Restricted sum††		−4.5	−12			−3	−5	−33	3	−45	−1	

Product	Glucose requirement[†]				Efficiency[‡]	CO_2 production[§]				O_2 requirement[¶]	
	Amino acid		Residue			Amino acid		Residue		Amino acid	Residue
Ala	NH_3	NO_3^-	NH_3	NO_3^-		NH_3	NO_3^-	NH_3	NO_3^-		
Ala	0.955	1.629	1.479	2.324	0.936	−0.082	0.906	0.310	1.549	−0.060	0.225
Arg	1.121	2.500	1.378	2.917		0.126	2.149	0.329	2.585	0.184	0.342
Asn	0.720	1.629	1.009	2.061		−0.278	1.056	−0.064	1.480	0.040	0.234
Asp	0.677	1.128	0.957	1.478		−0.331	0.331	−0.128	0.638	0.000	0.186
Cys	0.620	1.116	0.922	1.505		−0.182	0.545	0.071	0.926	0.000	0.207
Cyi											
Gln	0.925	1.747	1.211	2.148		−0.151	1.055	0.057	1.432	0.000	0.167
Glu	0.884	1.293	1.163	1.628		−0.200	0.399	0.000	0.682	−0.036	0.124
Gly	0.600	1.400	1.140	2.193	0.842	−0.293	0.880	0.129	1.673	0.000	0.374
His	1.323	2.484	1.642	2.956		0.237	1.940	0.482	2.409	0.378	0.584
Ile	1.756	2.214	2.212	2.743		0.560	1.232	0.909	1.687	0.041	0.236
Leu	1.603	2.061	2.035	2.566	0.978	0.336	1.008	0.649	1.428	−0.122	0.047
Lys	1.473	2.295	1.836	2.773		0.352	1.557	0.630	2.005	0.037	0.208
Met	1.309	1.711	1.641	2.099		0.443	1.034	0.728	1.399	0.215	0.407
Phe	1.879	2.242	2.245	2.653	0.884	0.356	0.889	0.599	1.197	0.065	0.218
Pro	1.435	1.957	1.907	2.526		0.191	0.957	0.529	1.436	0.000	0.220
Ser	0.714	1.286	1.092	1.782		−0.210	0.629	0.084	1.096	0.000	0.245
Thr	1.092	1.597	1.485	2.079		0.123	0.863	0.436	1.307	0.090	0.317
Trp	1.863	2.451	2.151	2.796		0.359	1.222	0.552	1.498	0.183	0.315
Tyr	1.630	1.961	1.933	2.301	0.891	0.203	0.689	0.405	0.945	0.059	0.196
Val	1.453	1.966	1.919	2.525		0.251	1.003	0.593	1.481	−0.091	0.108
Sum[‖]	1.254	1.890	1.613	2.345		0.118	1.051	0.383	1.457	0.062	0.252
Restricted sum[††]	0.869	1.647	1.181	2.081		−0.150	0.990	0.083	1.403	0.029	0.220

Table 12.6 (*continued*)

Some relevant data (relative molecular masses) are given in Table 12.1. The second column gives the equation reference where details of the assumed pathway can be found or evaluated. A negative stoichiometric coefficient indicates a requirement, and that item appears on the left-hand side of the overall equation; a positive coefficient indicates a product appearing on the right-hand side of the equation. The reduced forms NADH, NADPH, and $FADH_2$ are each equivalent to two H atoms. Water is omitted. The nitrogen source can be ammonia or nitrate; these are related by (12.14a), (12.14l), and (12.14m), and the effects of the nitrogen source on NADH production and the glucose requirement are given. Under ATP, the amino acid column denotes the coefficient for formation of the amino acid, whereas residue denotes the ATP coefficient for formation of the amino acid residue in protein (12.20a). To compute the efficiency, the heats of combustion in Table 12.1 for the amino acids are reduced by 383 MJ (kg mol)$^{-1}$ for each NH_3 required (12.14g); the efficiency is for the free amino acid formed from NH_3. A detailed example is given after (12.18a) for alanine.

[†] Calculated by applying point 8, p. 288, and (12.1c) to adjust the molar glucose requirement of the third column (upper part).

[‡] Calculated by dividing the heat of combustion of the product, where available, by the heat of combustion of the glucose required to synthesize that product.

[§] See (12.1e).

[¶] See (12.1d).

[‖] Equal molar proportions of 20 amino acids (not Cyi).

[††] Equal molar proportions of Ala, Arg, Ser, Asn, Asp, Gln, Glu (see Table 12.8).

Alanine Making use of (12.1g) and Lehninger (1975, pp. 694, 696), we have

0.5 glucose → pyruvate + NADH + ATP,

α-ketoglutarate + NH_3 + NADH → glutamate

pyruvate + glutamate → alanine + α-ketoglutarate

$$0.5 \text{ glucose} + NH_3 \rightarrow \text{alanine} + H_2O + \text{ATP}. \tag{12.18a}$$

Note that this does not agree with Schulz (1978, p. 245), who finds

$$0.5 \text{ glucose} + NH_3 + \text{ATP} \rightarrow \text{alanine} + H_2O.$$

With (12.1c), (12.18a) leads to a glucose requirement of

$$(0.5 - 1/36) \text{ glucose} \rightarrow \text{alanine},$$

giving

$$0.955 \text{ kg glucose} \rightarrow 1 \text{ kg alanine}.$$

The biochemical efficiency of this is (the heat of combustion of 1622 MJ (kg mol)$^{-1}$ given in Table 12.1 is reduced by 383 MJ (kg mol)$^{-1}$ for each NH_3 required; see eqn (12.14g) and above)

$$\frac{1622 - 383}{(0.5 - 1/36) \times 2803} = 0.936.$$

Schulz's stoichiometric equation leads to 1.067 kg glucose for the glucose requirement and an efficiency of

$$\frac{1622 - 383}{(0.5 + 1/36) \times 2803} = 0.838.$$

Continuing to use (12.18a), if nitrate replaces NH_3, then there is an extra cost of four NADH (12.14a), giving a glucose requirement of

$$(0.5 - 1/36 + 4/12) \text{ glucose} \rightarrow \text{alanine}$$

and

$$1.629 \text{ kg glucose} \rightarrow 1 \text{ kg alanine}.$$

For the alanine residue in protein, there is an extra cost of four ATP (12.20a), giving, for formation of the alanine residue from NH_3,

$$(0.5 + 3/36) \text{ glucose} \rightarrow \text{alanine residue in protein}$$

and (using 89 − 18 for the relative molecular mass of the alanine residue)

$$1.479 \text{ kg glucose} \rightarrow 1 \text{ kg alanine residue}.$$

For nitrate as the nitrogen source, this becomes (12.14b)

$$(0.5 + 3/36 + 4/12)\ \text{glucose} \rightarrow \text{alanine residue}$$

and

$$2.324\ \text{kg glucose} \rightarrow 1\ \text{kg alanine residue.}$$

Arginine The first reaction of the sequence involves acetyl-CoA and glutamate, and leads to the synthesis of ornithine. Using (12.18k) and Lehninger (1975, pp. 694, 705) we obtain

glucose + NH_3 → glutamate + CO_2 + 3 NADH + ATP

acetylglutamate + ATP → acetylglutamylphosphate

acetylglutamylphosphate + NADPH → acetylglutamicsemialdehyde

acetylglutamicsemialdehyde + glutamate

→ acetylornithine + α-ketoglutarate

α-ketoglutarate + NH_3 + NADH → glutamate

acetylornithine + glutamate → ornithine + acetylglutamate

$$\boxed{\text{glucose} + 2NH_3 + \text{NADPH} \rightarrow \text{ornithine} + CO_2 + 2H_2O + 2\text{NADH.}} \quad (12.18b)$$

Arginine synthesis occurs from ornithine via (Lehninger 1975, p. 706; (12.1a))

ornithine + carbamoyl-phosphate → citrulline

citrulline + aspartate + ATP → argininosuccinate + AMP + PP_i

argininosuccinate → arginine + fumarate

$$\boxed{\text{ornithine} + \text{carbamoyl-phosphate} + \text{aspartate} + 2\text{ATP} \rightarrow \text{arginine} + \text{fumarate.}} \quad (12.18c)$$

Carbamoyl-phosphate is synthesized according to (Lehninger 1975, pp. 695, 735)

glutamine + CO_2 + H_2O + 2ATP

→ carbamoyl-phosphate + glutamate

glutamate + NH_3 + ATP → glutamine

giving

$$CO_2 + NH_3 + 3\text{ATP} \rightarrow \text{carbamoyl-phosphate.} \quad (12.18d)$$

It is assumed that fumarate is spared according to ((12.1g) and Lehninger 1975, pp. 446, 464)

$$\text{glucose} \rightarrow 2\ \text{pyruvate} + 2\text{ATP} + 2\text{NADH}$$

$$\text{pyruvate} + CO_2 + H_2O + \text{ATP} \rightarrow \text{oxaloacetate}$$

$$\text{pyruvate} \rightarrow \text{acetyl-CoA} + CO_2 + \text{NADH}$$

$$\text{oxaloacetate} + \text{acetyl-CoA} + 2H_2O \rightarrow \text{fumarate} + 2CO_2 + 3\text{NADH},$$

$$\boxed{\text{glucose} + 2H_2O \rightarrow \text{fumarate} + 2CO_2 + 6\text{NADH} + \text{ATP}.} \quad (12.18e)$$

The overall equation for arginine synthesis is obtained by adding (12.18b), (12.18c), (12.18d), and (12.18h), and subtracting (12.18e), to give

$$0.5\ \text{glucose} + 4NH_3 + 3CO_2 + 4\text{NADH} + \text{NADPH} + 6\text{ATP} \rightarrow \text{arginine} + 7H_2O \quad (12.18f)$$

See Exercise 12.10.

Asparagine Asparagine is synthesized from aspartate by means of (Lehninger 1975, p. 696)

$$\text{aspartate} + NH_3 + \text{ATP} \rightarrow \text{asparagine}.$$

Combining this with (12.18h) gives the overall reaction

$$0.5\ \text{glucose} + CO_2 + 2NH_3 + \text{ATP} \rightarrow \text{asparagine} + 2H_2O. \quad (12.18g)$$

Aspartic acid With (12.1g) and Lehninger (1975, pp. 464, 694, 696) we obtain

$$0.5\ \text{glucose} \rightarrow \text{pyruvate} + \text{NADH} + \text{ATP}$$

$$\text{pyruvate} + CO_2 + \text{ATP} \rightarrow \text{oxaloacetate}$$

$$\text{glutamate} + \text{oxaloacetate} \rightarrow \text{aspartate} + \alpha\text{-ketoglutarate}$$

$$\alpha\text{-ketoglutarate} + NH_3 + \text{NADH} \rightarrow \text{glutamate}$$

$$\boxed{0.5\ \text{glucose} + CO_2 + NH_3 \rightarrow \text{aspartate} + H_2O.} \quad (12.18h)$$

Cysteine We assume that cysteine is synthesized via serine (12.19i) according to (Lehninger 1975, p. 422, Mahler and Cordes 1966, pp. 543, 675, 676)

$$0.5\ \text{glucose} + NH_3 \rightarrow \text{serine} + \text{NADH}$$

$$\text{serine} + H_2S \rightarrow \text{cysteine}$$

$$0.5\ \text{glucose} + NH_3 + H_2S \rightarrow \text{cysteine} + H_2O + \text{NADH}. \tag{12.18i}$$

It is assumed that transamination is effected by means of glutamate. See Exercise 12.8 for the stoichiometry with sulphate as the sulphur source.

Cystine Cystine provides the disulphide bridges of proteins; these are formed by oxidation following the incorporation of cysteine into the polypeptide chain. We assume (cf. Lehninger 1975, p. 568) that

$$2\ \text{cysteine} \rightarrow \text{cystine} + \text{NADH}$$

$$\text{glucose} + 2NH_3 + 2H_2S \rightarrow 2\ \text{cysteine} + 2\text{NADH}$$

$$\text{glucose} + 2NH_3 + 2H_2S \rightarrow \text{cystine} + 2H_2O + 3\text{NADH}. \tag{12.18j}$$

Glutamic acid With (12.1g), Mahler and Cordes (1966, p. 544), and Lehninger (1975, pp. 445, 446, 694) we obtain

$$\text{glucose} \rightarrow 2\ \text{pyruvate} + 2\text{NADH} + 2\text{ATP}$$

$$\text{pyruvate} + CO_2 + \text{ATP} \rightarrow \text{oxaloacetate}$$

$$\text{pyruvate} \rightarrow \text{acetyl-CoA} + CO_2 + \text{NADH}$$

$$\text{oxaloacetate} + \text{acetyl-CoA} \rightarrow \text{citrate}$$

$$\text{citrate} \rightarrow \alpha\text{-ketoglutarate} + CO_2 + \text{NADH}$$

$$\alpha\text{-ketoglutarate} + NH_3 + \text{NADH} \rightarrow \text{glutamate}$$

$$\text{glucose} + NH_3 \rightarrow \text{glutamate} + CO_2 + 3\text{NADH} + \text{ATP}. \tag{12.18k}$$

Glutamine Using (12.18k) and Lehninger (1975, p. 695) we obtain

$$\text{glucose} + NH_3 \rightarrow \text{glutamate} + CO_2 + 3\text{NADH} + \text{ATP}$$

$$\text{glutamate} + NH_3 + \text{ATP} \rightarrow \text{glutamine}$$

$$\text{glucose} + 2NH_3 \rightarrow \text{glutamine} + CO_2 + H_2O + 3\text{NADH}. \tag{12.18l}$$

Glycine We assumed that glycine is synthesized via serine (12.19i), with the scheme (Mahler and Cordes 1966, pp. 280, 352, 353, 675) (see (12.2f) and (12.2g)):

0.5 glucose + NH_3 → serine + NADH

serine + tetrahydrofolate → glycine + methylene-tetrahydrofolate

methylene-tetrahydrofolate → formyl-tetrahydrofolate + NADPH

formyl-tetrahydrofolate → tetrahydrofolate + formate

formate → CO_2 + NADH

$$0.5\ \text{glucose} + NH_3 + H_2O \rightarrow \text{glycine} + CO_2 + 2\text{NADH} + \text{NADPH}. \tag{12.18m}$$

Histidine It is convenient to break this down into several parts. With (12.3h), the first two reactions of Section 12.2.8, p. 324 (replacing ATP → AMP by 2ATP → 2ADP + $2P_i$), Lehninger (1975, p. 707), and noting that, unusually, in the fourth reaction ATP is actually a substrate that is used up and is regenerated in the sixth reaction derived from (12.15h) for AMP biosynthesis, we obtain

glucose + ATP → ribose + CO_2 + 2NADH

ribose + ATP → phospho-ribose

phospho-ribose + 2ATP → phospho-ribosylpyrophosphate

phospho-ribosylpyrophosphate + ATP
→ phosphoribosyl-AMP + $2PP_i$

$2PP_i \rightarrow 4P_i$

2.5 glucose + O_2 + $5NH_3$ + 13ATP
→ ATP + $5CO_2$ + 3NADPH + $2FADH_2$ + 8NADH
+ 13ADP + $13P_i$

phosphoribosyl-AMP
→ phosphoribulosyl-formimino-amino-(phosphoribosyl)-imidazole-carboxamide

phosphoribulosyl-formimino-amino-(phosphoribosyl)-imidazole-carboxamide + glutamine
→ imidazoleglycerolphosphate
+ phospho-ribosylaminoimidazole-carboxamide + glutamate

glutamate + NH_3 + ATP → glutamine

$$3.5\ \text{glucose} + 18\text{ATP} + O_2 + 6NH_3 \rightarrow \text{imidazoleglycerolphosphate} + 6CO_2 + 10\text{NADH} + 3\text{NADPH} + 2FADH_2 + \text{phospho-ribosylaminoimidazole-carboxamide}. \quad (12.18n)$$

The last named compound of this reaction is a byproduct of histidine biosynthesis that occurs in the biosynthetic pathway of the purines (Section 12.2.8, p. 325). We assume that its production spares the use of substrates according to the stoichiometry obtained in purine biosynthesis. Summing the first five of the six reactions before (12.15d), p. 325 (or subtracting the last reaction before (12.15d) from (12.15d) and rearranging) gives

$$\text{phospho-ribosylglycineamide} + \text{methenyl-tetrahydrofolate} + \text{glutamine} + 3\text{ATP} + CO_2 + \text{aspartate} \rightarrow \text{phospho-ribosylaminoimidazole-carboxamide} + \text{glutamate} + \text{fumarate}.$$

This equation is added to (12.15c) and the first, third, and fourth reactions after (12.15d), i.e.

$$1.5\ \text{glucose} + 2NH_3 + 6\text{ATP} \rightarrow \text{phospho-ribosylglycineamide} + 2CO_2 + 2\text{NADH} + 3\text{NADPH}$$

$$0.5\ \text{glucose} + \text{tetrahydrolate} + 0.5O_2 \rightarrow \text{methenyl-tetrahydrofolate} + 2CO_2 + FADH_2 + 3\text{NADH}$$

$$\text{glutamate} + NH_3 + \text{ATP} \rightarrow \text{glutamine}$$

$$\text{fumarate} + NH_3 \rightarrow \text{aspartate}$$

$$2\ \text{glucose} + 4NH_3 + 0.5O_2 + 10\text{ATP} \rightarrow \text{phospho-ribosylaminoimidazole-carboxamide} + 3CO_2 + 3\text{NADPH} + FADH_2 + 5\text{NADH}. \quad (12.18o)$$

The remaining part of the histidine biosynthetic pathway from imidazoleglycerolphosphate is

$$\text{imidazoleglycerolphosphate} \rightarrow \text{imidazoleacetolphosphate}$$

$$\text{imidazoleacetolphosphate} + \text{glutamate} \rightarrow \text{histidinol} + P_i + \alpha\text{-ketoglutarate}$$

$$\alpha\text{-ketoglutarate} + NH_3 + \text{NADH} \rightarrow \text{glutamate}$$

histidinol → histidinal + NADH

histidinal → histidine + NADH

imidazoleglycerolphosphate + NH_3 → histidine + NADH.	(12.18p)

Finally, adding (12.18n) to (12.18p) and subtracting (12.18o) gives

$$1.5 \text{ glucose} + 3NH_3 + 8ATP + 0.5O_2 \rightarrow \text{histidine} + 3CO_2 + 2H_2O + 6NADH + FADH_2. \quad (12.18q)$$

Isoleucine This is synthesized from threonine, (12.19k), and Lehninger (1975, pp. 694, 704):

0.5 glucose + CO_2 + NH_3 + 2ATP + NADPH + NADH → threonine

threonine → α-ketobutyrate + NH_3

α-ketobutyrate + pyruvate → acetohydroxybutyrate + CO_2

0.5 glucose → pyruvate + ATP + NADH

acetohydroxybutyrate + NADH → dihydroxymethylvalerate

dihydroxymethylvalerate + glutamate → isoleucine + α-ketoglutarate

α-ketoglutarate + NH_3 + NADH → glutamate

glucose + NH_3 + ATP + NADPH + 2NADH → isoleucine + $4H_2O$.	(12.19a)

Leucine The first four steps are as for valine synthesis (12.19p) and sum to

glucose → α-ketoisovalerate + CO_2 + 2ATP + NADH.

They are followed by (Lehninger (1975, pp. 694, 705) and (12.1n))

α-ketoisovalerate + acetyl-CoA → isopropylmalate

0.5 glucose → acetyl-CoA + CO_2 + ATP + 2NADH

isopropylmalate → α-ketoisocaproate + NADH + CO_2

α-ketoisocaproate + glutamate → leucine + α-ketoglutarate

α-ketoglutarate + NH_3 + NADH → glutamate

$$1.5 \text{ glucose} + NH_3 \rightarrow \text{leucine} + 3CO_2 + H_2O + 3ATP + 3NADH. \quad (12.19b)$$

Lysine There are two major pathways for lysine synthesis, and the diaminopimelic route is common in higher plants (Lehninger, pp. 458, 694, 701–3). This begins with asparate (12.18h), and makes use of (12.1g):

0.5 glucose + CO_2 + NH_3 → aspartate

aspartate + ATP → aspartate phosphate

aspartate phosphate + NADH → aspartate semialdehyde + P_i

aspartate semialdehyde + pyruvate → dihydrodipicolinate

0.5 glucose → pyruvate + ATP + NADH

dihydrodipicolinate + NADPH → piperideine dicarboxylate

piperideine dicarboxylate + succinyl-CoA + glutamate
→ succinyl-diaminopimelate + α-ketoglutarate + H-CoA

α-ketoglutarate + NH_3 + NADH → glutamate

succinyl-diaminopimelate → diaminopimelate + succinate

succinate + GTP + H-CoA → succinyl-CoA + GDP

ATP + GDP → ADP + GTP

diaminopimelate → lysine + CO_2

$$\text{glucose} + 2NH_3 + ATP + NADPH + NADH \rightarrow \text{lysine} + 4H_2O \quad (12.19c)$$

Methionine This is generated from homoserine (12.19j) (Lehninger 1975, p. 701), making use also of Lehninger (1975, p. 458), (12.18i), and (12.2m) for tetrahydrofolate:

0.5 glucose + CO_2 + NH_3 + ATP + NADPH + NADH
→ homoserine

homoserine + succinyl-CoA → succinyl-homoserine + H-CoA

succinyl-homoserine + cysteine → cystathionine + succinate

succinate + H-CoA + GTP → succinyl-CoA + GDP

GDP + ATP → GTP + ADP

0.5 glucose + NH_3 + H_2S → cysteine + NADH

cystathionine → homocysteine + NH_3 + pyruvate

− [0.5 glucose → pyruvate + ATP + NADH]

homocysteine + methyl-tetrahydrofolate

→ methionine + tetrahydrofolate

0.5 glucose + tetrahydrofolate + NADPH + $0.5O_2$

→ methyltetrahydrofolate + $2CO_2$ + 2NADH + $FADH_2$

$$\boxed{\begin{aligned}&\text{glucose} + NH_3 + H_2S + 0.5O_2 + 3\text{ATP} + 2\text{NADPH}\\&\quad\rightarrow \text{methionine} + CO_2 + 3H_2O + \text{NADH} + FADH_2.\end{aligned}} \quad (12.19d)$$

Phenylalanine Using Lehninger (1975, pp. 422, 694, 708) and (12.4c)

0.5 glucose → phosphoenolpyruvate + NADH

0.5 glucose + CO_2 + ATP + 2NADPH → erythrose-phosphate

phosphoenolpyruvate + erythrose-phosphate

→ dehydroshikimic acid

dehydroshikimic acid + NADPH → shikimic acid

shikimic acid + ATP → shikimate acid phosphate

shikimate acid phosphate + phosphoenolpyruvate

→ enolpyruvyl shikimic acid phosphate

0.5 glucose → phosphoenolpyruvate + NADH

enolpyruvyl shikimic acid phosphate → chorismic acid

$$\boxed{\begin{aligned}&1.5\ \text{glucose} + CO_2 + 2\text{ATP} + 3\text{NADPH}\\&\quad\rightarrow \text{chorismic acid} + 2\text{NADH}.\end{aligned}} \quad (12.19e)$$

Proceeding down the phenylalanine pathway, we have

chorismic acid → prephrenic acid

prephrenic acid → phenylpyruvate + CO_2 + H_2O

phenylpyruvate + glutamate → phenylalanine + α-ketoglutarate

α-ketoglutarate + NH_3 + NADH → glutamate

$$\boxed{\text{chorismic acid} + NH_3 + \text{NADH} \rightarrow \text{phenylalanine} + CO_2 + 2H_2O.} \quad (12.19f)$$

Adding (12.19e) and (12.19f) gives

$$1.5 \text{ glucose} + NH_3 + 2\text{ATP} + 3\text{NADPH} \rightarrow \text{phenylalanine} + \text{NADH} + 7H_2O. \quad (12.19g)$$

Proline This is synthesized from glutamate (12.18k) (Lehninger 1975, p. 695)

glucose + NH_3 → glutamate + CO_2 + 3NADH + ATP

glutamate + ATP + NADH → glutamate semialdehyde

glutamate semialdehyde → pyrroline carboxylate + H_2O

pyrroline carboxylate + NADPH → proline

$$\boxed{\text{glucose} + NH_3 + \text{NADPH} \rightarrow \text{proline} + CO_2 + 2H_2O + 2\text{NADH}.} \quad (12.19h)$$

Serine From Lehninger (1975, p. 422) and Mahler and Cordes (1966, pp. 543, 675), we have

0.5 glucose → phosphoglycerate + NADH

phosphoglycerate → phosphohydroxypyruvate + NADH

phosphohydroxypyruvate + glutamate → phosphoserine + α-ketoglutarate

α-ketoglutarate + NH_3 + NADH → glutamate

phosphoserine → serine

$$\boxed{0.5 \text{ glucose} + NH_3 \rightarrow \text{serine} + \text{NADH}.} \quad (12.19i)$$

Threonine This is obtained from aspartic acid (12.18h) via homoserine (Lehninger 1975, p. 700)

$$0.5 \text{ glucose} + CO_2 + NH_3 \rightarrow \text{aspartate}$$

$$\text{aspartate} + \text{ATP} \rightarrow \text{aspartylphosphate}$$

$$\text{aspartylphosphate} + \text{NADPH} \rightarrow \text{aspartic semialdehyde}$$

$$\text{aspartic semialdehyde} + \text{NADH} \rightarrow \text{homoserine}$$

$$0.5 \text{ glucose} + CO_2 + NH_3 + \text{ATP} + \text{NADPH} + \text{NADH} \rightarrow \text{homoserine}. \quad (12.19j)$$

Homoserine synthesis is followed by

$$\text{homoserine} + \text{ATP} \rightarrow \text{homoserine phosphate}$$

$$\text{homoserine phosphate} \rightarrow \text{threonine} + P_i$$

to give (adding these two reactions to (12.19j))

$$0.5 \text{ glucose} + CO_2 + NH_3 + 2\text{ATP} + \text{NADPH} + \text{NADH} \rightarrow \text{threonine} + 2H_2O. \quad (12.19k)$$

Tryptophan The first step is to synthesis anthranilate from chorismate ((12.19e) and Lehninger (1975, pp. 695, 708):

$$1.5 \text{ glucose} + CO_2 + 2\text{ATP} + 3\text{NADPH} \rightarrow \text{chorismate} + 2\text{NADH}$$

$$\text{chorismate} + \text{glutamine} \rightarrow \text{anthranilate} + \text{glutamate} + \text{pyruvate}$$

$$\text{glutamate} + NH_3 + \text{ATP} \rightarrow \text{glutamine}$$

$$-[0.5 \text{ glucose} \rightarrow \text{pyruvate} + \text{ATP} + \text{NADH}]$$

$$\text{glucose} + CO_2 + NH_3 + 4\text{ATP} + 3\text{NADPH} \rightarrow \text{anthranilate} + 6H_2O + \text{NADH}. \quad (12.19l)$$

Summing the first three reactions listed under histidine, and with Lehninger (1975, pp. 422, 709) and (12.19i), we have

$$\text{glucose} + 4\text{ATP} \rightarrow \text{phospho-ribosylpyrophosphate} + CO_2 + 2\text{NADH}$$

$$\text{phospho-ribosylpyrophosphate} + \text{anthranilate} \rightarrow \text{phospho-ribosyl anthranilate} + PP_i$$

$PP_i \rightarrow 2P_i$

phospho-ribosyl anthranilate → indoleglycerolphosphate + CO_2

indoleglycerolphosphate + serine

→ tryptophan + glyceraldehydephosphate

0.5 glucose + NH_3 → serine + NADH

− [0.5 glucose + ATP → glyceraldehydephosphate]

glucose + anthranilate + NH_3 + 3ATP

→ tryptophan + $2CO_2$ + $2H_2O$ + 3NADH. (12.19m)

Finally, adding (12.19l) and (12.19m) gives

2 glucose + $2NH_3$ + 7ATP + 3NADPH

→ tryptophan + CO_2 + $8H_2O$ + 4NADH. (12.19n)

Tyrosine From (12.19e) and Lehninger (1975, pp. 694, 708) we have

1.5 glucose + CO_2 + 2ATP + 3NADPH → chorismate + 2NADH

chorismate → prephrenate

prephrenate → hydroxyphenylpyruvate + CO_2 + NADH

α-ketoglutarate + NH_3 + NADH → glutamate

hydroxyphenylpyruvate + glutamate → tyrosine + α-ketoglutarate

1.5 glucose + NH_3 + 2ATP + 3NADPH

→ tyrosine + $6H_2O$ + 2NADH. (12.19o)

Valine This is synthesized from pyruvate ((12.1g) and Lehninger (1975, pp. 694, 704):

glucose → 2 pyruvate + 2ATP + 2NADH

pyruvate + pyruvate → acetolactate + CO_2

acetolactate + NADH → dihydroxyisovalerate

dihydroxyisovalerate → α-ketoisovalerate

α-ketoisovalerate + glutamate → valine + α-ketoglutarate

α-ketoglutarate + NH_3 + NADH → glutamate

$$\boxed{\text{glucose} + NH_3 \rightarrow \text{valine} + CO_2 + 2H_2O + 2\text{ATP}.} \quad (12.19p)$$

Protein synthesis It is assumed that the energy cost of protein synthesis is two ATP per amino acid for the formation of amino acyl transfer RNA (tRNA), one ATP for elongation of the peptide chain by one amino acid residue, and one ATP for shifting the peptide chain from the A site to the P site on the ribosome and also moving the ribosome to the next mRNA codon (Schulz 1978, p. 244, Lehninger 1975, pp. 933–9). Thus a total of four ATP is required per peptide bond. Any possible additional costs arising from the initiation of the peptide chain are ignored, as are costs arising from post-translational covalent modification of the polypeptide chain, such as phosphorylation, methylation, and hydroxylation, or orientational and conformational costs. The cysteine–cystine oxidation occurs post-translation, and this is allowed for in (12.18j). For a polypeptide chain of n_{aa} amino acid residues, we can write

$$n_{aa} \text{ amino acids} + 4n_{aa}\text{ATP} \rightarrow \text{polypeptide chain of length } n_{aa}. \quad (12.20a)$$

Another cost of protein synthesis arises from the unstable nature of the mRNA, and the necessity to resynthesize the RNA. As outlined in (12.17r) and above (p. 331), if n_{pp} is the average number of polypeptide chains that is synthesized on a given mRNA molecule before that molecule is hydrolysed, then the ATP cost is

$$6/n_{pp} \text{ ATP per amino acid residue.} \quad (12.20b)$$

Protein synthesis may not occur at the sites where the amino acids are synthesized; transport across membranes, requiring ATP, may be necessary. Schulz (1978, p. 245) mentions the use of a transport factor for amino acids in his calculations, although it is not clear what value he uses. We denote the amino acid transport cost by $c_{aa,T}$, such that the ATP cost is

$$c_{aa,T} \text{ ATP per amino acid.} \quad (12.20c)$$

Summing (12.20a)–(12.20c) gives the non-specific ATP cost of protein synthesis as

$$(4 + c_{aa,T} + 6/n_{pp}) \text{ ATP per amino acid.} \quad (12.20d)$$

The bottom row of Table 12.6 gives the stoichiometry of formation of a polypeptide chain consisting of equal molar proportions of the 20 amino acid (ignoring cystine). No costs for transport (12.20c) or mRNA hydrolysis (12.20b) are included. For each ATP per amino acid of extra cost included, the glucose requirement value of

$$1.613 \text{ kg glucose per kg average amino acid residue in protein} \quad (12.20e)$$

is increased by (using (12.1c))

$$\frac{1}{36} \times \frac{180}{137 - 18} = 0.042. \quad (12.20f)$$

12.2.10 *Mineral content*

The details of ion uptake by plants and ion transport within the plant are not fully understood. We consider the problem semiquantitatively, making several assumptions.

Table 12.7 gives some measurements of mineral content of a maize plant. An average relative molecular mass RMM(average) for plant minerals can be defined by

$$\text{mass of minerals per kg}$$
$$= \text{RMM(average)} \times \text{total mineral concentration (kg mol kg}^{-1}). \quad (12.21a)$$

This gives (using Table 12.7)

$$(5.63 - 0.93) \times 10^{-2} = \text{RMM(average)} \times 2052 \times 10^{-6}, \quad (12.21b)$$

and therefore

$$\text{RMM(average)} = 23. \quad (12.21c)$$

Table 12.7. Mineral content of a *Zea mays* plant

Element	Relative molecular mass (kg (kg mol)$^{-1}$)	Fraction of total dry mass ($\times$ 100)	Concentration (kg mol (kg total dry mass)$^{-1}$ $\times 10^6$)
Nitrogen	14	1.46	1043
Silicon	28	1.17	418
Potassium	39	0.92	240
Calcium	40	0.23	57
Phosphorus	31	0.20	65
Magnesium	24	0.18	75
Sulphur	32	0.17	53
Chlorine	35.5	0.14	39
Aluminium	27	0.11	41
Iron	56	0.08	14
Manganese	55	0.04	7
Undetermined		0.93	
Total		5.63	2052

After Sutcliffe (1962, p. 3) and based on an ash analysis by Miller (1938).

If nitrogen is omitted from the calculation, then RMM(average) = 32. Since the undetermined components are likely to have higher relative molecular masses, we assume that

$$\text{RMM(average)} = 40. \tag{12.21d}$$

It is assumed that the ATP cost of ion uptake and transport is

$$1 \text{ ATP per ion.} \tag{12.21e}$$

1 kg of minerals is, with (12.21d), 1/40 kg mol of ions, which with (12.21e) and (12.1c) to transform the ATP cost into a glucose cost, gives a glucose requirement of

$$\frac{1}{40} \times \frac{180}{36} \text{ kg glucose} \rightarrow 1 \text{ kg minerals} \tag{12.21f}$$

and therefore

$$0.125 \text{ kg glucose} \rightarrow 1 \text{ kg minerals.} \tag{12.21g}$$

The CO_2 production associated with this is (with (12.1e))

$$\begin{aligned} CO_2 \text{ production} &= \frac{44 \times 1/6}{40} \\ &= 0.183 \text{ kg } CO_2 \text{ (kg minerals)}^{-1} \end{aligned} \tag{12.21h}$$

and the oxygen requirement is (with (12.1d))

$$\begin{aligned} O_2 \text{ requirement} &= \frac{32 \times 1/6}{40} \\ &= 0.133 \text{ kg } O_2 \text{ (kg minerals)}^{-1}. \end{aligned} \tag{12.21i}$$

12.2.11 *Metabolic cycles*

Metabolic cycles (sometimes called substrate cycles or futile cycles) are cyclical systems of reactions whose effect is to dissipate or degrade energy, but at the same time they may provide sensitive regulative possibilities. Some review material is given by Newsholme and Start (1973), Stein and Blum (1978), and Hue (1982). Rabkin and Blum (1985), who investigated intermediary metabolism in rat liver hepatocytes experimentally and by means of models, considered five such cycles. Mostly, the problem has been considered in the context of mammalian systems, and there is little that is directly relevant to plant metabolism (Preiss and Kosuge 1976). Two examples of metabolic cycles are as follows.

Fructose 6-phosphate–fructose 1,6-biphosphate These reactions occur in glycolysis:

fructose 6-phosphate + ATP → fructose 1,6-biphosphate + ADP

fructose 1,6-biphosphate → fructose 6-phosphate + P_i

$$\boxed{ATP \rightarrow ADP + P_i.} \tag{12.22a}$$

Phosphoenolpyruvate–pyruvate–oxaloacetate

phosphoenolpyruvate + ADP → pyruvate + ATP

pyruvate + CO_2 + ATP → oxaloacetate + ADP + P_i

oxaloacetate + ATP → phosphoenolpyruvate + CO_2 + ADP

$$\boxed{ATP \rightarrow ADP + P_i.} \tag{12.22b}$$

Other possible cycles involve the pairs glucose–glucose 6-phosphate, glycogen–glucose 6-phosphate, and acetate–acetyl-CoA. Leaky membranes, across which a concentration gradient must be maintained, also operate as a type of a dissipative cycle. The net effect of both (12.22a) and (12.22b) is simply to hydrolyse ATP, with a consequent loss in available energy. Such cycles provide a means of generating heat which is used by some insects and mammals.

However, the rationale for metabolic cycles in plants, if they are significant, may be to provide a way of getting rid of surplus carbohydrate and also for the regulatory possibilities. For example, if the two reactions shown above (12.22a) are working at a high level in opposite directions, then a small change in either reaction may cause a large net flux in one direction or the other (e.g. Stein and Blum 1978).

In our consideration of plant growth efficiency, metabolic cycles are ignored although their possible relevance should not be forgotten.

12.2.12 *Growth efficiency of plant material*

Plant composition varies widely, with both type of plant, part of plant considered, and age of plant. For example, considering the effects of stage of maturity in pasture plants, Osbourn (1980, p. 73) shows that increasing maturity changes plant composition as follows: protein, 33 to 7 per cent; lipid, 10 to 3 per cent; sugars, 10 to 25 per cent; minerals, 12 to 5 per cent; hemicellulose, 14 to 23 per cent; cellulose, 18 to 30 per cent; lignin, 3 to 7 per cent (see Exercise 12.11 and Table E12.1). In the solution to Exercise 12.11 (p. 368) it is clear that the costs of the direct formation of mature plant tissue would be less than those of young tissue. However, mature tissue is formed from the ageing of younger tissues, and the reduction in free energy that occurs during this process is likely to be mostly lost to the plant; it is therefore the costs of formation of young tissues that are required for the initial considerations of plant growth efficiency, and tissue ageing and senescence will lead to lower overall efficiencies.

Long (1968, pp. 937–1060) has an extensive section on the chemical composition of plant and microbiological tissues, which reveals the complexity of the problem. In order to demonstrate how estimates of growth efficiency can be obtained, a simplified plant composition is assumed in Table 12.8. Nucleic acids are omitted entirely from consideration, since it appears that they are usually only present in concentrations of 0.2–0.7 per cent of dry matter (Long 1968, p. 996). From the bottom row of Table 12.8, it is seen that, for 1 kg of plant material, 1.316 kg glucose are required and 0.218 kg CO_2 are respired. If nitrate is the nitrogen source instead of ammonia, these values are increased to 1.463 kg glucose and 0.435 kg CO_2.

Next, we consider how the growth efficiency or yield parameter Y_g (defined in eqn (11.3c), p. 266) is related to the present approach. The definition of Y_g is

$$Y_g = \frac{\Delta w}{\Delta w + \Delta s_r}, \tag{12.23a}$$

where Δw is an increment in plant dry matter (in units of carbon) and Δs_r is the carbon respired when Δw is synthesized. Thus the denominator of eqn (12.23a) is the total carbon initially available to the plant for the synthetic process (maintenance is ignored). If we denote the glucose requirement (kg glucose per kg plant dry matter) by

$$R_{\mathrm{DM,glc}} \tag{12.23b}$$

and the CO_2 production (kg CO_2 per kg plant dry matter) by

$$P_{\mathrm{DM,CO_2}} \tag{12.23c}$$

then, accounting for the carbon atoms, these quantities are related to Y_g by (72/180 = 12/30 is the fraction of glucose by weight that is carbon)

$$Y_g = \frac{(72/180)R_{\mathrm{DM,glc}} - (12/44)P_{\mathrm{DM,CO_2}}}{(72/180)R_{\mathrm{DM,glc}}},$$

which becomes

$$Y_g = 1 - \frac{30}{44}\frac{P_{\mathrm{DM,CO_2}}}{R_{\mathrm{DM,glc}}}. \tag{12.23d}$$

Applying (12.23d) to the average plant material in Table 12.8, using the coefficients in the bottom row, gives

$$Y_g = 0.89\ (0.80) \tag{12.23e}$$

where the value in parentheses is for nitrate rather than ammonia as the nitrogen source. Now, applying eqn (12.23d) to the grass coefficients in Table S12.1

Table 12.8. Glucose requirements, CO_2 production and oxygen requirements for the formation of 1 kg of average plant material

Main category	Fraction	Subcategory	Subfractions	Glucose requirement	CO_2 production	O_2 requirement
Carbohydrates	0.55	Cellulose	0.35	0.226	0.018	0.013
		Hemicellulose (xylan)	0.25	0.167	0.015	0.011
		Starch	0.25	0.161	0.013	0.009
		Pectin (araban)	0.05	0.033	0.003	0.002
		Sucrose	0.10	0.060	0.004	0.003
		Subtotal	1.00	0.647	0.053	0.038
Nitrogenous compounds[†]	0.20	Amino acids[‡]	0.10	0.017 (0.033)	−0.003 (0.020)	0.001
		Proteins[§]	0.90	0.290 (0.422)	0.069 (0.263)	0.046
		Subtotal	1.00	0.307 (0.455)	0.066 (0.283)	0.047
Lignin	0.10	Coniferyl alcohol	1.00	0.181	0.045	0.020
Organic acids[¶]	0.05	Malate	0.50	0.016	−0.010	−0.001
		Citrate	0.50	0.016	−0.011	−0.001
		Subtotal	1.00	0.032	−0.021	−0.002
Lipids	0.05	Glycerol trioleate	1.00	0.142	0.066	0.007
Ash[‖]	0.05		1.00	0.006	0.009	0.007
Total	1.00			1.315 (1.463)	0.218 (0.435)	0.117

These are all expressed in kilograms per kilogram of total plant material. The fractions and subfractions are all expressed on a mass basis. The glucose and oxygen requirements, and CO_2 production are all in kilograms per kilogram of total dry matter. These are obtained from Tables 12.2, 12.3, 12.4, and 12.6.

[†] From NH_3 and H_2S; values from nitrate are in parentheses.

[‡] See Long (1968, p. 991) and Table 11.6, bottom row. Assumed free amino acid composition is equal molar proportions of Ala, Arg, Ser, Asn, Asp, Gln, and Glu.

[§] Equal molar amounts of the 20 amino acids (not cystine).

[¶] Long (1968, pp. 958–75) discusses plant non-volatile carboxylic acids.

[‖] The assumed average RMM is 40; see (12.21d)–(12.21i).

p. 625), we have

$$Y_g = 0.84\ (0.71) \tag{12.23f}$$

for young vegetative grass, and

$$Y_g = 0.90\ (0.77) \tag{12.23g}$$

for mature flowering grass.

Equations (12.23e)–(12.23g) provide biochemically based estimates of the yield parameter Y_g. However, the use of such values in models of crop or plant growth (p. 464) may be complicated by consideration of the changes in tissue composition that occur as a result of the processes of maturation, ageing, and senescence, which are discussed next.

12.2.13 *Maturation, ageing, and senescence of plant tissue*

As is apparent from Table E12.1 (p. 368), there are substantial changes in tissue composition with age, and the qualitative changes shown here for pasture grasses occur in other plants. Biochemical pathway analysis allows the derivation of a yield parameter Y_g, as above in eqns (12.23e)–(12.23g), but it is apparent that a simple consideration of the costs of a direct synthesis of the components of plant material from glucose ignores the effects of subsequent transformations on respiration and efficiency.

It seems reasonable that the yield parameter used in a crop model should correspond to that for the synthesis of new tissue. It is then less clear how the ongoing compositional changes should be represented. It could be supposed that the loss in protein and minerals occurs via the export of amino acids and ions from the ageing tissue, presumably with some associated energy cost. It might be assumed that any lipid lost (Table E12.1, p. 368) is totally oxidized to ATP, and this ATP is used to meet the costs of export and also of increases in lignin, cellulose, and hemicellulose contents. In photosynthetic tissue, other contributions may also come into the accounting. It does seem that there is a need for a plant tissue life-cycle model, perhaps along the lines of the leaf expansion and growth model of Lainson and Thornley (1982) combined with the aggregated biochemical approach applied by Gill *et al.* (1984) to ruminant metabolism.

12.3 Elemental composition, growth requirements, and heats of combustion

The pathway analysis approach to plant growth efficiency described in the preceding section is rigorous within the limitations of a static framework (it may be unsuitable for consideration of, for example, multiple pathways used with varying weightings and metabolic cycles operating at changing intensities). However, it is complex and tedious to apply and owing to the limitations of biochemical knowledge and the numerous assumptions that are necessary, it is only approximate. Thus, in both crop physiology and ecology, the development of simpler methods has received attention (McDermitt and Loomis 1981a,b, Williams *et al.* 1987). The methods we describe in this section revolve around a number of relationships that exist between (i) elemental composition of plant material (or a component thereof), (ii) the glucose requirement to satisfy the needs of this material for carbon atoms, for reducing/oxidizing power, and for energy (to drive the reactions and/or allow for thermodynamic inefficiencies), and (iii) the heat of combustion of the product material.

12.3.1 *Elemental composition, stoichiometry, and energy*

Requirement for carbon skeletons and reducing power Consider a tissue or compound with the formula

$$C_cH_hO_xN_nS_s \tag{12.24a}$$

where c, h, x, n, and s are constants ($\geqslant 0$) that express the composition. It is assumed that this is synthesized from glucose, ammonia, sulphuric acid, and reducing power [H], according to

$$(c/6)C_6H_{12}O_6 + nNH_3 + sH_2SO_4 + r[H]$$

$$\rightarrow C_cH_hO_xN_nS_s + wH_2O. \tag{12.24b}$$

The constants r (the reducing equivalents required) and w (for water) are constants which may be positive or negative (in which case the terms go to the other side of the equation). Note that free oxygen (O_2) does not appear in (12.24b), but since $2[H] + 0.5O_2 = H_2O$ it is trivial to rewrite it to give the oxidizing equivalents required. If r is positive, the product is, relative to the substrates shown, in a more reduced state, and vice versa. Equating the numbers of hydrogen and oxygen atoms on each side of (12.24b) gives

$$\text{H: } 2c + 3n + 2s + r = h + 2w \tag{12.24c}$$

and

$$\text{O: } c + 4s = x + w. \tag{12.24d}$$

Eliminating w leads to

$$r = h - 2x - 3n + 6s. \tag{12.24e}$$

Thus the number r of reducing equivalents required is obtained directly from the coefficients of the elemental formula (12.24a). Note that the coefficients in eqn (12.24e) can be regarded as related to the oxidation state of the substrate atoms, e.g. H^+, O^{2-}, N^{3-}, and S^{6+}. It is easily verified that, for different nitrogen and sulphur substrates, eqn (12.24e) can be written (Exercise 12.12)

$$r(NH_3, H_2SO_4) = h - 2x - 3n + 6s \tag{12.24f}$$

$$r(N_2, H_2SO_4) = h - 2x + 6s \tag{12.24g}$$

$$r(HNO_3, H_2SO_4) = h - 2x + 5n + 6s \qquad (12.24h)$$

$$r(NH_3, H_2S) = h - 2x - 3n - 2s \qquad (12.24i)$$

$$r(N_2, H_2S) = h - 2x - 2s \qquad (12.24j)$$

$$r(HNO_3, H_2S) = h - 2x + 5n - 2s. \qquad (12.24k)$$

The number w of water molecules produced is obtained from, for instance, (12.24d), (S12.12b), and (S12.12e) (p. 625). Again, we see that the coefficients of h, x, n, and s in eqns (12.24f)–(12.24k) are the oxidation numbers (related to the number of electrons donated when the element is in the combined state, i.e. the valence) of the elements as provided, for example $+1$ for hydrogen, -2 for oxygen, -3 for nitrogen in NH_3, 0 for nitrogen in N_2, $+5$ for nitrogen in HNO_3, $+6$ for sulphur in H_2SO_4, and -2 for sulphur in H_2S. Carbon is regarded as having a variable oxidation number between 4 (CO_2) and -4 (CH_4). In any compound the sum of the oxidation numbers is zero.

The carbon requirement of (12.24a) for stoichiometry alone is satisfied by

$$(c/6) \text{ glucose}. \qquad (12.24l)$$

The requirement for reducing equivalents $r[H]$ in eqn (12.24b) could be met in various ways: hydrogen directly available from the environment, or as a by-product of photosynthesis or nitrogen fixation. We assume, with reference to oxidation of glucose by glycolysis and the citric acid cycle and (12.1m) (p. 293), that

$$C_6H_{12}O_6 + 6H_2O \rightarrow 6CO_2 + 12NADH, \qquad (12.24m)$$

remembering that 2[H] is equivalent to NADH. Adding

$(r/24) \times$ (12.24m) to (12.24b) leads to

$$(c/6 + r/24)C_6H_{12}O_6 + nNH_3 + sH_2SO_4$$

$$\rightarrow C_cH_hO_xN_nS_s + (r/4)CO_2 + (w - r/4)H_2O. \qquad (12.24n)$$

We now have, in eqn (12.24n), an equation which gives the glucose requirement both for carbon skeletons and to satisfy the needs of the product for reducing equivalents.

See Exercise 12.13, which extends (12.24a), (12.24b), and (12.24e) to phosphorus.

Energy requirements of synthesis We first consider the energy produced and applied through the production of reducing equivalents, as in (12.24m) and (12.24n) and the efficiency of this process; then we consider any additional energy requirement which may be met by the provision of ATP.

It should be noted that (12.24n) incorporates (12.24m), and (12.24m), which provides reducing equivalents, is exothermic. We can estimate the energy evolved when glucose is oxidized as in (12.24m) by means of

$$C_6H_{12}O_6 + 6O_2 \rightarrow 6CO_2 + 6H_2O \qquad \Delta H_c = 2803 \text{ MJ (kg mol)}^{-1} \qquad (12.24o)$$

where ΔH_c is the heat of combustion, and

$$NADH + 0.5O_2 \rightarrow H_2O \qquad \Delta G = -218 \text{ MJ (kg mol)}^{-1}. \qquad (12.24p)$$

ΔG is the free-energy change (a negative value means that heat is evolved).

The value of -218 MJ (kg mol)$^{-1}$ is estimated from

$$\Delta G = -nF\,\Delta E, \qquad (12.24q)$$

where $n = 2$ is the number of electrons transferred in the oxidation–reduction reaction, F is the Faraday (96 MJ (V kg equivalent)$^{-1}$ and ΔE (0.815 − (−0.32) V) is the difference between the standard reduction potentials (Mahler and Cordes 1966, pp. 207, 210).

Combining (12.24o) and (12.24p) gives

$$C_6H_{12}O_6 + 6H_2O \rightarrow 6CO_2 + 12NADH$$

$$\Delta G = -(2803 - 12 \times 218) = -187 \text{ MJ (kg mol)}^{-1} \qquad (12.24r)$$

The energetic efficiency of NADH production from glucose is $12 \times 218/2803 = 0.93$, and the heat production associated with the residual inefficiency will drive the reaction to the right.

When glucose is oxidized to give reducing power (12.24m), this occurs with high efficiency and most of the free energy of glucose is captured in NADH. When reducing power is applied as in (12.24b) to give product formation, much of the free energy of NADH is captured in the product, which is more highly reduced. This can be seen in the strong correlation between heat of combustion and the oxido-reductive state of a compound; McDermitt and Loomis (1981a, Fig. 1) discuss the linear relationship existing between the glucose requirement as given by $(c/6 + r/24)$ of (12.24n) and the heat of combustion (see also below). However, there is the possibility that to drive the product formation represented in (12.24n) requires the provision of extra energy that is not linked to reducing power.

If this additional energy is represented by ATP, (12.24n) becomes

$$(c/6 + r/24)C_6H_{12}O_6 + nNH_3 + sH_2SO_4 + aATP$$

$$\rightarrow C_cH_hO_xN_nS_s + (r/4)CO_2 + (w - r/4)H_2O \qquad (12.24s)$$

where a is the number of moles of ATP needed. The ADP + P_i appearing on the right-hand side of this equation are not shown. The fate of the energy applied may be to be captured in the product, to be harnessed to drive the reactions at an acceptable rate (ending up as heat), or simply to be wasted as heat with little useful result. For ATP production from glucose, we assume (12.1f) (p. 289), i.e.

$$C_6H_{12}O_6 + 6O_2 \rightarrow 6CO_2 + 6H_2O + 36ATP. \qquad (12.24t)$$

Adding $(a/36) \times$ (12.24t) to (12.24s) gives

$$(c/6 + r/24 + a/36)C_6H_{12}O_6 + (a/6)O_2 + nNH_3 + sH_2SO_4$$

$$\rightarrow C_cH_hO_xN_nS_s + (r/4 + a/6)CO_2 + (w - r/4 + a/6)H_2O \qquad (12.24u)$$

This equation summarizes the reactions that result in the formation of the product. The similarity of the terms for the glucose requirement, CO_2 production, and oxygen requirement with equations (12.1c)–(12.1e) (p. 288) can be noted. While stoichiometry alone determines the coefficients c (for carbon skeletons) and r (for reducing equivalents), only detailed biochemistry can provide the value of a (ATP required), and also supply the numbers 24 and 36 ((12.24m) and (12.24t)) which specify how efficiently the biochemical machinery can produce reducing power and energy from glucose.

12.3.2 *Glucose requirement from elemental composition compared with those from pathway analysis*

In the last section, we discussed how the glucose requirement for the synthesis of the compound (12.24a) can be broken down into the three components of requirement for carbon skeletons (c in (12.24b)), for reducing power produced from glucose (r of (12.24n)), and for energy also produced from glucose (a of (12.24u)). The oxido-reductive state of biochemical compounds varies greatly, from fatty acids and lipids, which are highly reduced, to organic acids, some of which are highly oxidized. In this section we calculate, for a range of compounds, the molar glucose requirement for carbon skeletons and reducing equivalents alone from (12.24s) and (12.24i), assuming that the nitrogen and sulphur substrates are NH_3 and H_2S. These can be compared with the biochemical pathway estimates obtained previously. These are listed in Table 12.9. See Exercise 12.14.

The carbon requirement c_{stoich}, and the reducing requirement r_{stoich}, for stoichiometry are obtained from eqns (12.24b) and (12.24i). The glucose requirement $R_{glc,stoich}$ from stoichiometry is calculated by means of (see (12.24s))

$$R_{glc,stoich} = c_{stoich}/6 + r_{stoich}/24. \qquad (12.25a)$$

The corresponding glucose requirement $R_{glc,pathway}$ from biochemical pathway analysis is given by (see (12.1c))

$$R_{glc,pathway} = c_{pathway}/6 + r_{pathway}/24 + a_{pathway}/36 \qquad (12.25b)$$

where $a_{pathway}$ is the ATP requirement of the reaction. Note that the carbon required from glucose might not agree for stoichiometry and pathway analysis, since CO_2 might be evolved (ribose) or taken up (tartrate). The same is true of the reducing equivalents required, owing to the possible involvement of CO_2 and/or oxygen. With an obvious notation and referring to the coefficients of Tables 12.2 etc., these requirements obey ($c_{pathway} = -6 \times$ coefficient in the glucose column)

$$c_{stoich} = c_{pathway} - CO_2, \qquad (12.25c)$$

and

$$r_{stoich} = r_{pathway} + 4CO_2 + 4O_2. \qquad (12.25d)$$

Combining eqns (11.25a)–(11.25d) gives

Table 12.9. Glucose requirement (mol glucose (mol product)$^{-1}$) from stoichiometry compared with pathway analysis

Substance	Glucose requirement (stoichiometry)			Glucose requirement (pathway analysis)				Percentage difference
	c	r	$R_{glc,stoich}$	c	r	a	$R_{glc,pathway}$	
Carbohydrates, polysaccharides, and sugar alcohols (Table 12.2)								
Ribose	5	0	0.833	6	−4	1	0.861	−3.2
Sucrose	12	0	2.000	12	0	3	2.083	−4.0
Cellulose, starch	6	0	1.000	6	0	2	1.056	−5.3
Fructans	6	0	1.000	6	0	3	1.083	−7.7
Hemicellulose								
Xylan	5	0	0.833	6	−4	2	0.889	−6.3
Glucose–xylose polymer	39	0	6.500	42	−12	14	6.889	−5.6
Pectins								
Araban	5	0	0.833	6	−4	2	0.889	−6.3
Galacturonan	6	0	0.833	6	−4	2	0.889	−6.3
Lignin								
Coniferyl	10	4	1.833	12	−2	2	1.972	−7.0
Glycerol	3	2	0.583	3	2	1	0.611	−4.5
Inositol	6	0	1.000	6	0	1	1.028	−2.7
Organic acids (Table 12.3)								
Ascorbate	6	−4	0.833	6	−2	2	0.972	−14.3
Cis-aconitate	6	−6	0.750	6	−6	−2	0.694	8.0
Citramalate	5	−2	0.750	6	−6	−1	0.722	3.8
Citrate	6	−6	0.750	6	−6	−2	0.694	8.0
Formate	1	−2	0.083	3	−8	−1	0.139	−40.0
Fumarate	4	−4	0.500	6	−12	−2	0.444	12.5
Glyoxylate	2	−4	0.167	3	−8	−1	0.139	20.0
α-keto glutarate	5	−4	0.667	6	−8	−2	0.611	9.1
Malate	4	−4	0.500	3	0	−1	0.472	5.9
Malonate	3	−4	0.333	3	−4	−1	0.306	9.1
Oxalate	2	−6	0.083	3	−8	−1	0.139	−40.0
Oxaloacetate	4	−6	0.417	3	−2	−1	0.389	7.1
Pyruvate	3	−2	0.417	3	−2	−1	0.389	7.1
Shikimate	7	0	1.167	6	4	−1	1.194	2.4
Succinate	4	−2	0.583	6	−10	−3	0.500	16.7
Tartrate	4	−6	0.417	3	6	3	0.833	−50.0
Long-chain fatty acids (Table 12.4)								
Palmitate	16	28	3.833	24	−4	−1	3.806	0.7
Stearate	18	32	4.333	27	−4	−1	4.306	0.6
Oleate	18	30	4.250	27	−2	−1	4.389	−3.2
Linoleate	18	28	4.167	27	0	−1	4.472	−6.8
Linolenate	18	26	4.083	27	2	−1	4.556	−10.4

Table 12.9 (*continued*)

Substance	Glucose requirement (stoichiometry)			Glucose requirement (pathway analysis)				Percentage difference
	c	r	$R_{glc,stoich}$	c	r	a	$R_{glc,pathway}$	
Triacyl glycerides of long-chain fatty acids (Table 12.4)								
Palmitate	51	86	12.083	75	−10	4	12.194	−0.9
Stearate	57	98	13.583	84	−10	4	13.694	−0.8
Oleate	57	92	13.333	84	−4	4	13.944	−4.4
Linoleate	57	86	13.083	84	2	4	14.194	−7.8
Linolenate	57	80	12.833	84	8	4	14.444	−11.2
Nucleoside monophosphates from NH_3 (Table 12.5)								
AMP	10	−10	1.250	15	−26	11	1.722	−27.4
GMP	10	−12	1.167	15	−28	12	1.667	−30.0
UMP	9	−6	1.250	9	−6	6	1.417	−11.8
CMP	9	−6	1.250	9	−6	7	1.444	−13.5
Nucleoside monophosphate residues in RNA from NH_3 (Table 12.5)								
AMP	10	−10	1.250	15	−26	13	1.778	−29.7
GMP	10	−12	1.167	15	−28	14	1.722	−32.3
UMP	9	−6	1.250	9	−6	8	1.472	−15.1
CMP	9	−6	1.250	9	−6	9	1.500	−16.7
Deoxynucleoside monophosphate residues in DNA from NH_3 (Table 12.5)								
dAMP	10	−8	1.333	15	−24	13	1.861	−28.4
dGMP	10	−10	1.250	15	−26	14	1.806	−30.8
dTMP	10	−2	1.583	10	−2	9	1.834	−13.7
dCMP	9	−4	1.333	9	−4	9	1.583	−15.8
Amino acids from NH_3 and H_2S (Table 12.6)								
Alanine	3	0	0.500	3	0	−1	0.472	5.9
Arginine	6	−2	0.917	3	10	6	1.083	−15.4
Asparagine	4	−4	0.500	3	0	1	0.528	−5.3
Aspartate	4	−4	0.500	3	0	0	0.500	0.0
Cysteine	3	−2	0.417	3	−2	0	0.417	0.0
Glutamine	5	−2	0.750	6	−6	0	0.750	0.0
Glutamate	5	−2	0.750	6	−6	−1	0.722	3.8
Glycine	2	−2	0.250	3	−6	0	0.250	0.0
Histidine	6	−4	0.833	9	−14	8	1.139	−26.8
Isoleucine	6	6	1.250	6	6	1	1.278	−2.2
Leucine	6	6	1.250	9	−6	−3	1.167	7.1
Lysine	6	4	1.167	6	4	1	1.194	−2.3
Methionine	5	2	0.917	6	0	3	1.083	−15.4
Phenylalanine	9	4	1.667	9	4	2	1.722	−3.2
Proline	5	2	0.917	6	−2	0	0.917	0.0
Serine	3	−2	0.417	3	−2	0	0.417	0.0

Table 12.9 (*continued*)

Substance	Glucose requirement (stoichiometry)			Glucose requirement (pathway analysis)				Percentage difference
	c	r	$R_{glc,stoich}$	c	r	a	$R_{glc,pathway}$	
Threonine	4	0	0.667	3	4	2	0.722	−7.7
Tryptophan	11	2	1.917	12	−2	7	2.111	−9.2
Tryrosine	9	2	1.583	9	2	2	1.639	−3.4
Valine	5	4	1.000	6	0	−2	0.944	5.9
Sum	107	8	18.167	114	−16	27	19.056	−4.7
Protein[†]	107	8	18.167	114	−16	106	21.278	−14.6

The stoichiometric requirement for glucose for carbon skeletons and reducing power is calculated from (12.24n) with (12.24i). Data from Tables 12.1–12.6 are used. The carbon requirement c and the reducing requirement r (stoichiometry) are defined in eqns (12.24b) and (12.24i); the glucose requirement from stoichiometry $R_{glc,stoich}$, is given by eqn (12.25a). Pathway analysis gives a carbon requirement which may be partly met directly from glucose (c) and partly from CO_2, r gives the reducing equivalents required ([H] is equivalent to 0.5NADH), and a is the ATP requirement. Equation (12.25b) (see (12.1c), p. 288) is applied to convert r and a into a total glucose requirement $R_{glc,pathway}$. Stoichiometry generally underestimates the glucose requirement, and so the percentage difference (last column, calculated as a percentage of the pathway value) is usually negative.
[†] Equal molar amounts of 20 amino acids.

$$R_{glc,pathway} = R_{glc,stoich} - O_2/6 + a_{pathway}/36. \tag{12.25e}$$

The results given in Table 12.9 can be applied to a range of plant materials, as in Table 12.8, to provide a glucose requirement from stoichiometry $R_{glc,stoich}$, and a glucose requirement from pathway analysis $R_{glc,pathway}$. McDermitt and Loomis (1981a, Table 5) performed this calculation for a range of seeds, and found approximately for the aggregated values that

$$R_{glc,pathway} = (1.14 \pm 0.01) R_{glc,stoich}. \tag{12.25f}$$

where ± 0.01 is the standard deviation. Given an elemental composition, the stoichiometric glucose requirement is easily obtained and then the application of eqn (12.25f) gives an estimate of the pathway analysis value.

12.3.3 *Stoichiometric glucose requirements and heats of combustion*

The heat of combustion of a compound is related to its stoichiometric glucose requirement. This is an empirical rather than a rigorous theoretical relationship. If eqn (12.25a) is used to give the stoichiometric glucose requirement $R_{glc,stoich}$, then multiplying this by the heat of combustion of glucose (2803 MJ (kg mol)$^{-1}$) gives an estimate of the heat of combustion of the product compound.

Heats of combustion are usually tabulated with nitrogen gas as the product of combustion for nitrogenous compounds. It is convenient, partly so that Table 12.9 can be used directly and partly because the biochemical pathway analysis begins with NH_3, to correct the heats of combustion of the nitrogenous compounds tabulated in Table 12.1 to NH_3 as a product of combustion by subtracting 383 MJ per kilogram mole of nitrogen in the compound from the values tabulated in Table 12.1 (see eqn (S12.6b), p. 622). Thus the heat of combustion of the compound

$$C_cH_hO_xN_n \tag{12.26a}$$

is corrected by means of

$$\Delta H_c(NH_3) = \Delta H_c(N_2) - 383n. \tag{12.26b}$$

This is compared with an estimate of the heat of combustion obtained from

$$\Delta H_c(NH_3, \text{estimate}) = 2803 R_{\text{glc, stoich}}, \tag{12.26c}$$

using values of $R_{\text{glc, stoich}}$ from Table 12.9. The result of applying eqns (12.26b) and (12.26c) are given in Table 12.10.

On examining the percentage differences in the last column of Table 12.10, it can be seen that the carbohydrates can be grouped together, and the fatty acids, amino acids, and organic acids (with the exceptions of formate, oxalate, and tartrate) form a second group. McDermitt and Loomis (1981b) give linear regressions of heat of combustion on the stoichiometric glucose requirement. In our units of MJ (kg mol)$^{-1}$ these are

$$\Delta H_c = 2824 R_{\text{glc, stoich}} - 6 \tag{12.26d}$$

for the carbohydrates and glycerol. For organic acids, fatty acids, amino acids, and proteins, they found that the heat of combustion with respect to nitrogen gas (in MJ (kg mol)$^{-1}$) is

$$\Delta H_c(N_2) = 2598 R_{\text{glc, stoich}}(N_2) + 42. \tag{12.26e}$$

From our results, which are not in complete agreement with those of McDermitt and Loomis (1981b), Table 12.10 shows that eqn (12.26c) gives a good estimate of the heat of combustion for the carbohydrates, whereas for other groups of compounds, eqn (12.26c) overestimates the heat of combustion by some 5–10 per cent.

Suppose that the heat of combustion ΔH of some plant material is known, either by direct measurement or otherwise. Then the glucose requirement can be defined by

$$R_{\text{glc}} = \frac{1}{\text{efficiency}} \frac{\Delta H_c}{2803}. \tag{12.26f}$$

This is a useful approach if the efficiency of synthesis is reasonably constant over a broad range of plant compositions. MacDermitt and Loomis (1981b) con-

Table 12.10. Heats of combustion (MJ (kg mol)$^{-1}$) estimated from the stoichiometric glucose requirement (from Table 12.9 and (12.26c)) compared with the measured value (Table 12.1) corrected to NH_3 as a product of combustion by eqn (12.26b)

Substance	$R_{glc,stoich}$	$\Delta H_c(NH_3$, estimate)	$\Delta H_c(NH_3)$	Percentage difference
Ribose	0.833	2335	2339	−0.2
Sucrose	2.000	5605	5641	−0.6
Cellulose, starch	1.000	2803	2796	0.3
Glycerol	0.583	1634	1661	−1.6
Inositol	1.000	2803	2770	1.2
Citrate	0.750	2102	1961	7.2
Formate	0.083	233	255	−8.8
Fumarate	0.500	1402	1335	5.0
Malate	0.500	1402	1328	5.6
Malonate	0.333	934	861	8.5
Oxalate	0.083	233	246	−5.3
Succinate	0.583	1634	1491	9.6
Tartarate	0.417	1169	1149	1.7
Palmitate	3.833	10744	9978	7.7
Stearate	4.333	12145	11281	7.7
Oleate	4.250	11913	11119	7.1
Alanine	0.500	1402	1239	13.2
Glycine	0.250	701	590	18.8
Leucine	1.250	3504	3207	9.2
Phenylalanine	1.667	4673	4277	9.2
Tyrosine	1.583	4437	4105	8.1

cluded that this was so, and that a value of

$$\text{efficiency} = 0.88 \tag{12.26g}$$

gives satisfactory results. Where possible, we have calculated, in Tables 12.2, 12.3, 12.4, and 12.6, our own values of the biochemical efficiency with which (12.26g) can be compared. There is considerable variation in the biochemical efficiencies of the least-cost pathways, and some of this variation may be reflected in the synthetic costs of plant material.

12.3.4 *Summary*

The methods described in this section provide some alternative approaches to the biochemical pathway method of estimating the synthetic costs of plant material.

Apart from mineral content (see Section 12.2.10, p. 348), these may provide approximate values of growth parameters of crop physiological and ecological interest more simply.

Given a stoichiometric formula as in (12.24a), the glucose requirement for synthesis can be estimated by means of eqns (12.24f), (12.24n), (12.25a), and (12.25f) for example. Alternatively, if the heat of combustion of the plant material has been measured or is known, eqns (12.26f) and (12.26g) can be used to estimate the glucose requirement.

12.4 The biochemistry of plant maintenance

As discussed in Chapter 11, the concept of plant maintenance with its associated respiratory costs has been much used at the organ, plant, and crop levels of description. Indeed, assuming that the growth : maintenance separation is meaningful, over a growing season the maintenance costs are about equal to the growth costs and so are important for crop yield considerations. Amthor (1984) has reviewed methods of estimating the maintenance coefficient m((day^{-1})) and also gives a table of measured values ranging from 0.01 to 0.1 day^{-1}. Penning de Vries (1975b) gives a valuable discussion of the topic from a biochemical point of view.

Metabolic cyles (e.g. Section 12.2.11, p. 349) inevitably degrade free energy, but they provide an essential means for adaption, maturation and senescence from the molecular level to the whole-plant level. It may be easier to define maintenance requirements theoretically, i.e. in terms of assumed biochemistry, than by attempting to specify an experimental procedure for estimating maintenance that is convincing and gives an interpretable result.

We define the maintenance requirement as comprising those energy inputs (converted to glucose equivalents) needed to maintain the functional and compositional status quo of the tissue. The growth : maintenance separation is of most value if the maintenance energy inputs are independent of tissue growth rate, which is unlikely. In this section we build on the analysis in Section 12.2 to obtain rough estimates of the costs of processes such as protein turnover, RNA turnover, and ion leakage across membrance.

12.4.1 *Protein turnover*

Protein synthesis and degradation always occur together, and they confer adaptability on the plant, allowing protein patterns to be altered and nitrogen to be remobilized. These processes generally apply to enzymatic proteins and not to structural proteins. The degradation of specific proteins usually requires ATP, whereas non-specific protein degradation does not require ATP. We now estimate the maintenance cost arising from non-specific protein degradation and resynthesis.

Assume that (Table E12.1, p. 368, Table 12.1, p. 291, and eqn (12.20a), p. 347; Penning de Vries 1975b, Table 1, gives protein turnover rates varying from 0 to 50 day^{-1}) the protein content is

$$f_p = 0.1 \text{ kg protein (kg dry matter)}^{-1}, \tag{12.27a}$$

the carbon content is

$$f_C = 0.4 \text{ kg C (kg dry matter)}^{-1}, \tag{12.27b}$$

the RMM of average amino acid residue is

$$M_{aa} = 120 \text{ kg (kg mol)}^{-1} \tag{12.27c}$$

the ATP cost of protein synthesis from amino acids is

$$c_{ATP,ps} = 4 \text{ ATP per peptide bond}, \tag{12.27d}$$

and the protein degradation rate is

$$k_{deg,p} = 1 \text{ day}^{-1}. \tag{12.27e}$$

The ATP cost per day is

$$\left(\frac{f_p}{M_{aa}}\right) k_{deg,p} c_{ATP,ps} \text{ kg mol ATP (kg dry matter)}^{-1} \text{ day}^{-1}. \tag{12.27f}$$

Since 36 ATP are equivalent to 1 $C_6H_{12}O_6$ (glucose) (12.1c), and 1 kg mol glucose is equivalent to 6 × 12 kg carbon, the carbon respired as CO_2 is

$$\frac{6 \times 12}{36}\left(\frac{f_p}{M_{aa}}\right) k_{deg,p} c_{ATP,ps} \text{ kg C respired (kg dry matter)}^{-1} \text{ day}^{-1} \tag{12.27g}$$

To put this in the usual units, division by the fractional carbon content f_C, of (12.27b) gives the maintenance coefficient from protein turnover m_{pt} as

$$m_{pt} = \frac{2 f_p k_{deg,p} c_{ATP,ps}}{M_{aa} f_C}. \tag{12.27h}$$

With the values in (12.27a)–(12.27e), therefore,

$$m_{pt} = 1/60 \approx 0.02 \text{ day}^{-1}. \tag{12.27i}$$

The value is typical of measured plant maintenance rates (p. 268).

12.4.2 *Nucleic acid turnover*

It appears that DNA turnover is negligible. Of the components of RNA, it seems that turnover of mRNA is the most significant. In so far as this is linked to the rate of protein synthesis, it can be allowed for by replacing the quantity $c_{ATP,ps}$ of eqn (12.27h) by the quantity given in (12.20d); this also includes any amino acid transport costs that may be incurred. If it is assumed (see (12.20d)) that the transport cost coefficient $c_{aa,T} = 1$ and that the number n_{pp} of polypeptide chains synthesized on a given mRNA molecule is 6, then (12.20d) and (12.27d) are increased from 4 to 6, and the maintenance coefficient from protein turnover, associated mRNA turnover, and additional amino acid transport costs becomes (cf. eqn (12.27i))

$$m_{pt} = 0.03 \text{ day}^{-1} \tag{12.27j}$$

There is also turnover of tRNA and rRNA, as well as the possibility of there being mRNA turnover which is not linked to protein synthesis. Define (Long 1968, pp. 995–7, Table 12.1, and (12.17r) and above) nucleic acid content

$$f_{NA} = 0.0005 \text{ kg nucleic acid (kg dry matter)}^{-1}, \tag{12.27k}$$

RMM of average nucleic acid residue

$$M_{NA} = 310 \text{ kg (kg mol)}^{-1}, \tag{12.27l}$$

ATP cost of nucleic acid synthesis from nucleosides

$$c_{ATP,NA} = 2 \text{ ATP per nucleoside residue}, \tag{12.27m}$$

and average nucleic acid degradation rate

$$k_{deg,NA} = 0.1 \text{ day}^{-1}. \tag{12.27n}$$

(12.27n) is conjecture. Following the derivation of eqn (12.27i), the maintenance coefficient m_{NA} from nuclei acid turnover is

$$m_{NA} = \frac{2 f_{NA} k_{deg,NA} c_{ATP,NA}}{M_{NA} f_C}. \tag{12.27o}$$

With (12.27b) and the values above, therefore,

$$m_{NA} = 1.5 \times 10^{-6} \text{ day}^{-1}. \tag{12.27p}$$

The contribution from nucleic acid turnover is at least three orders of magnitude smaller than that from protein turnover.

12.4.3 *Lipid turnover*

There is little relevant data (Penning de Vries 1975b) and we proceed by assumption. We define (Table E12.1, p. 368, and Table 12.1, p. 291) lipid content

$$f_{lipid} = 0.05 \text{ kg lipid (kg dry matter)}^{-1}, \tag{12.28a}$$

RMM of average lipid

$$M_{lipid} = 900 \text{ kg (kg mol)}^{-1}, \tag{12.28b}$$

ATP cost of lipid resynthesis from glycerol and free fatty acids

$$c_{ATP,lipid} = 7 \text{ ATP per triglyceride}, \tag{12.28c}$$

and average lipid degradation rate

$$k_{\mathrm{deg,lipid}} = 0.1\ \mathrm{day}^{-1}. \tag{12.28d}$$

The value of 7 ATP per triglyceride for the cost of triglyceride resynthesis is obtained by assuming that one ATP is required to give glycerol 3-phosphate from glycerol, and two ATP are required for each of the three fatty acids (12.13c). The value in (12.28d) is conjecture. Following the derivation of eqn (12.27i), the maintenance coefficient m_{lipid} from lipid turnover is

$$m_{\mathrm{lipid}} = \frac{2f_{\mathrm{lipid}}k_{\mathrm{deg,lipid}}c_{\mathrm{ATP,lipid}}}{M_{\mathrm{lipid}}f_{\mathrm{C}}}. \tag{12.28e}$$

With (12.27c) and the values above, therefore,

$$m_{\mathrm{lipid}} = 0.4 \times 10^{-3}\ \mathrm{day}^{-1}. \tag{12.28f}$$

12.4.4 *Ion cycling*

If ions leak through membranes and have to be pumped back using energy-consuming processes in order to maintain ionic concentrations, this makes a contribution to the tissue maintenance requirement. We define (Table E12.1 p. 368, (12.21e), and (12.21d)) mineral content

$$f_{\mathrm{minerals}} = 0.05\ \mathrm{kg\ minerals\ (kg\ dry\ matter)}^{-1}, \tag{12.28g}$$

average RMM of minerals

$$M_{\mathrm{minerals}} = 40\ \mathrm{kg\ (kg\ mol)}^{-1}, \tag{12.28h}$$

ATP cost of ion transport

$$c_{\mathrm{ATP,ion}} = 1\ \text{ATP per ion transported}, \tag{12.28i}$$

and average ion leakage rate

$$k_{\mathrm{leakage,ion}} = 0.1\ \mathrm{day}^{-1}. \tag{12.28j}$$

The last figure is pure conjecture. Following the derivation of eqn (12.27i),, the maintenance coefficient m_{ion} from ion leakage is

$$m_{\mathrm{ion}} = \frac{2f_{\mathrm{minerals}}k_{\mathrm{leakage,ion}}c_{\mathrm{ATP,ion}}}{M_{\mathrm{minerals}}f_{\mathrm{C}}}. \tag{12.28k}$$

With (12.27c) and the values above, therefore,

$$m_{\mathrm{ion}} = 0.007\ \mathrm{day}^{-1}. \tag{12.28l}$$

There may also be leakage of other ionic or non-ionic metabolites across membranes; these would also contribute to the tissue maintenance requirements.

Exercises

12.1. Calculate the number of equivalents of glucose required to produce 12 NADH or NADH using first (12.1j) with ATP equivalent to glucose/36, second (12.1j) ignoring the ATP yield of 2 molecules, and third via the pentose phosphate pathway (12.1l) with ATP equivalent to glucose/36.

12.2. Using (12.3g) rather than (12.3h) for ribose synthesis, find the stoichiometric equation (using (12.1c)), the glucose requirement, and the biochemical efficiency. Compare these with (12.3i)–(12.3k).

12.3. Calculate the stoichiometry and the glucose requirement for the synthesis of a fully methylated galacturonan. Start with (12.6b); assume that galacturonate is methylated by methionine giving methyl galacturonate and homocysteine, that methyl tetrahydrofolate reconverts homocysteine to methionine, and finally that (12.2m) regenerates methyl tetrahydrofolate from tetrahydrofolate.

12.4. Recalculate the stoichiometry and efficiency of malate synthesis (12.7e) assuming that NADH + ATP is equivalent to NADPH (as in Schulz 1978) and replacing eqns (12.7f)–(12.7h).

12.5. The stoichiometric coefficients for the synthesis of glycerol trioleate from glucose are given in (12.13e) and Table 12.4. Using the general equation (12.13k) for triglyceride synthesis, evaluate and check the stoichiometric coefficients for the other triglycerides given in Table 12.4.

12.6. Ammonia is combusted to H_2O (liquid) and nitrogen (gas) according to

$$NH_3 + \tfrac{3}{4}O_2 \rightarrow \tfrac{1}{2}N_2 + \tfrac{3}{2}H_2O. \qquad \text{(E12.6a)}$$

From Weast (1980, pp. D-67, D-71; see also p. D-289) the energies of formation from their elements of NH_3(gas) and H_2O(liquid) are -46.2 MJ (kg mol)$^{-1}$ and -285.9 MJ (kg mol)$^{-1}$ (the negative sign means that heat is evolved when the elements combine to form the compounds). Calculate the heat of combustion of NH_3(gas) in (E12.6a).

12.7. Equations (12.15h)–(12.15j) p. 326, define the stoichiometry and glucose requirement for AMP formation when ammonia is the nitrogen source. Make use of (12.14a), i.e.

$$HNO_3 + 4NADH \rightarrow NH_3 + 3H_2O, \qquad \text{(E12.7a)}$$

to modify (12.15h) and (12.15i), and recalculate the glucose requirement (see Table 12.5).

12.8. The stoichiometric equation for cysteine synthesis is (12.18i)

$$0.5\ \text{glucose} + NH_3 + H_2S \rightarrow \text{cysteine} + NADH. \qquad \text{(E12.8a)}$$

For sulphate reduction to H_2S, assume (12.14p), i.e. (ignoring H_2O)

$$H_2SO_4 + 4NADPH + 3ATP \rightarrow H_2S, \qquad \text{(E12.8b)}$$

modify (E12.8a) for a sulphate source of sulphur, and re-evaluate the four glucose requirement values of Table 12.6.

12.9. Evaluate, using the stoichiometric coefficients in Table 12.5 and (12.1d) and (12.1e), the coefficients for CO_2 produced and oxygen required (kilograms per kilogram of product formed) for AMP synthesis and RNA synthesis (equal molar amounts of the four bases) from ammonia and nitrate.

12.10. Evaluate, using the stoichiometric coefficients in Table 12.6 and (12.1d) and (12.1e), the coefficients for CO_2 produced and oxygen required (kilograms per kilogram of product formed) for the synthesis of arginine and the arginine residue in a polypeptide chain from ammonia and nitrate.

12.11. The composition of plant material can vary greatly with the stage of maturity of a crop. For example, Table E12.1 (based on Osbourn 1980, p. 73), shows the compositional range encountered in grasses. Assuming that the protein is equimolar in the 20 amino acids (Table 12.6), the lipid is glycerol trioleate (Table 12.4), the sugars are sucrose, the lignin is polyconiferyl alcohol (Table 12.2), and the cost of mineral uptake can be represented as in (12.21d) and (12.21e), calculate the glucose, CO_2 and oxygen requirements/production for 1 kg of dry matter.

Table E12.1. Composition of grass at different stages of maturity

Stage of maturity	Protein	Lipid	Sugars	Minerals	Hemi-cellulose	Cellulose	Lignin	Total
Young vegetative	33	10	10	12	14	18	3	100
Mature flowering	7	3	25	5	23	30	7	100

After Osbourn 1980, p. 73.

12.12. For a compound with the formula

$$C_cH_hO_xN_nS_s \tag{E12.12a}$$

prove, by writing down the stoichiometric equation for synthesis of this compound from glucose, HNO_3, H_2SO_4, and reducing equivalents [H], i.e.

$$(c/6)C_6H_{12}O_6 + nHNO_3 + sH_2SO_4 + r[H] \rightarrow C_cH_hO_xN_nS_s + wH_2O, \tag{E12.12b}$$

and equating the hydrogen and oxygen atoms on each side, that

$$r(HNO_3, H_2SO_4) = h - 2x + 5n + 6s. \tag{E12.12c}$$

Also show, by replacing the $nHNO_3$ term in (E12.12b) by $(n/2)N_2$, that

$$r(N_2, H_2SO_4) = h - 2x + 6s. \tag{E12.12d}$$

12.13. For a compound with the formula

$$C_cH_hO_xN_nS_sP_p \tag{E12.13a}$$

formed from substrates according to

$$(c/6)C_6H_{12}O_6 + nNH_3 + sH_2S + pH_3PO_4 + r[H] \rightarrow C_cH_hO_xN_nS_sP_p + wH_2O, \tag{E12.13b}$$

prove that

$$r = h - 2x - 3n - 2s + 5p. \tag{E12.13c}$$

12.14. The stoichiometric glucose requirements given in Table 12.9 for nitrogenous compounds (nucleic acids and amino acids) are calculated assuming that NH_3 is the nitrogen substrate (eqn (12.24i)) and are compared with the pathway analysis results from Tables 12.5 and 12.6.

If HNO_3 is the nitrogen substrate, from eqns (12.24k) and (E12.13c) the requirement for reducing equivalents is

$$r = h - 2x + 5n - 2s + 5p. \qquad \text{(E12.14a)}$$

Use Tables 12.5 and 12.6 to recalculate the coefficients for AMP and for protein (last row of Table 12.9).

13
Partitioning during vegetative growth

13.1 Introduction

During vegetative growth, the shoots and roots of plants have well-defined roles: the shoots assimilate carbon, and the roots acquire mineral nutrients and water. It is observed that if there is a reduction in, say, the supply of nitrogen then a greater proportion of plant growth is directed, or partitioned, to the roots. The teleonomic interpretation of this response is that the plant 'seeks' more nitrogen. Similar responses are observed for changes in the shoot environment (i.e. a reduction in irradiance results in greater partitioning to the shoots). It is apparent, therefore, that there is a continual interplay between the function of the shoots and roots, with the plant apparently attempting to balance its acquisition of resources. The interested reader should consult Brouwer's (1962) classic discussion for the experimental background to the subject. During reproductive growth the situation is quite different, as the plant's overriding objective is to produce seed (Chapter 7, Section 7.6, p. 193). In this chapter we concern ourselves solely with partitioning during vegetative growth. Furthermore, in our analysis root function is restricted to nitrogen uptake, which implies that other mineral nutrients and water are non-limiting (Chapter 15, Section 15.7.4, p. 451); a similar treatment is possible for any single limiting root factor, and a method for incorporating several root functions is outlined.

Partitioning is a subject which is particularly interesting from a modelling viewpoint, as it lends itself readily to the three modelling approaches discussed in Chapter 1—namely empirical, mechanistic, and teleonomic—and each approach is considered here. We argue that empirical models have little to offer, since once any attempt is made to incorporate realistic responses to variation in the shoot or root environment they are likely to be at least as complex as a teleonomic model, which has a sounder physiological base. Indeed, our treatment of partitioning centres around the teleonomic approach, as this provides a robust and simple model. Thornley's (1972a,b, 1976) mechanistic model of shoot : root partitioning is also discussed. It is more complex than the teleonomic model in that there are more parameters and they are less easily estimated. However, the mechanistic model is based solely on the principles of substrate transport around the plant and its utilization for structural growth. The model is not designed to behave in any preconceived manner, but is seen to display all the expected characteristics of shoot : root partitioning in response to variation to the environment.

13.1.1 *Basic simplifications*

Since our primary interest here is in partitioning, the other physiological processes of photosynthesis, respiration, and nutrient uptake are treated in a very simple manner. Other aspects such as structural degradation, senescence, and transpiration are ignored. All these features can be incorporated in a more general crop growth model once the basic principles of the partitioning models are understood (Chapter 16, Section 16.5, p. 464).

The unit of time throughout is the day, and so diurnal fluctuations in, say, the carbon substrate level within the plant are not considered. Since there is considerable variation in the sugar content of plants, care should be taken when relating the models to experimental data. It may be appropriate to measure plant substrate levels at the beginning and end of the photoperiod and take the mean value.

Dry matter variables are on a per plant basis, although the analysis is directly applicable to a crop canopy with the appropriate carbon and nitrogen inputs.

13.2 Empirical approach

Suppose that the net daily supply of carbon to the plant is ΔC kg C day^{-1}, which can be calculated using the methods of Chapter 10. If the dry masses of the shoot and root are denoted by W_{sh} and W_r respectively, the simplest empirical description of the partitioning of ΔC between the shoots and roots is

$$\Delta W_{sh} = \phi_{sh}\frac{\Delta C}{f_C} \qquad \Delta W_r = \phi_r\frac{\Delta C}{f_C}, \tag{13.1a}$$

where ΔW_{sh} and ΔW_r are the daily increments in shoot and root dry mass, f_C is the fractional carbon content of dry mass (kg C (kg dry mass)$^{-1}$), which is assumed to be the same for the shoot and root, and ϕ_{sh} and ϕ_r are the partitioning coefficients of growth between the shoot and root with

$$\phi_{sh} + \phi_r = 1. \tag{13.1b}$$

Clearly, if ϕ_{sh} and ϕ_r are assigned constant values, this approach has severe limitations. Most notably, there is no allowance for changes in either the environment or the plant's current structure. For example, following shoot defoliation of a grass crop, partitioning to the roots virtually ceases.

While this method can be generalized in several ways, for example by making ϕ_{sh} and ϕ_r functions of time or the environment, this does not provide a useful approach, nor is it consistent with the rate–state formalism discussed in Chapter 1. An alternative is to relate ϕ_{sh} and ϕ_r to some substrate level within the plant, and assign an empirical priority to the growth of a particular organ based on that substrate level. However, as will be seen in the next section, a more systematic approach at this level of complexity results in a simple and robust model.

It is not our intention to dismiss empirical partitioning models totally. For example, Robson (1973) observed that the ratio of leaf weight to stem weight is virtually constant over an entire growth period of experimentally simulated swards of perennial ryegrass, and A. J. Parsons (personal communication) has observed similar results in the field under a wide range of grazing managements. As a first approximation, therefore, it may be reasonable to define the partitioning of growth between leaf and stem in ryegrass swards as constant. However, for most crops partitioning between the shoots and roots is highly variable and depends on the environmental conditions (light and nutrient supply). We shall not pursue empirical models of shoot : root partitioning further (see France and Thornley (1984, pp. 128–9) and Thornley (1977, pp. 369–70) for some further discussion of empirical partitioning models), but will develop the teleonomic and mechanistic approaches.

13.3 Teleonomic approach

Partitioning readily lends itself to teleonomic modelling as some possible goals for the plant can be defined relatively easily. The mathematical description of such a proposed goal, associated with the work of White (1937), Brouwer (1962), and Davidson (1969), is that, during balanced exponential growth,

$$\text{shoot mass} \times \text{shoot specific activity} \propto \text{root mass} \times \text{root specific activity}. \tag{13.2a}$$

The shoot and root specific activities are the rates of photosynthesis and nitrogen uptake per unit of shoot or root mass respectively, and depend directly on the environmental conditions. This expression simply says that the absolute rate of photosynthesis is proportional to the absolute rate of nitrogen uptake and, as the environment varies, the plant adjusts its shoot and root masses accordingly so as to maintain this relationship. Equation (13.2a) is frequently referred to as the functional equilibrium between shoots and roots. The implication is that if, for example, the plant's supply of carbon is reduced then the response is to increase the shoot : root ratio. The teleonomic interpretation of this response is that the plant 'seeks' more carbon by partitioning a relatively greater proportion of growth to the shoot. Similarly, reducing the supply of nitrogen causes the shoot : root ratio to fall, so that the plant is now 'seeking' more nitrogen.

The functional balance could be used as a starting point for developing a partitioning model. However, this would require defining the partitioning response directly in terms of the environment—i.e. the shoot and root activities—and, as discussed in Chapter 1 (Section 1.5, p. 15), it is more appropriate to express the dynamics of the model in terms of the state variables rather than the environmental conditions directly. The environment will influence these state variables (e.g. the irradiance level will affect the carbon substrate concentration) and hence, in turn, plant growth.

Goal-seeking models of partitioning have been proposed by Reynolds and Thornley (1982) and Johnson (1985). The latter was an attempt to overcome some drawbacks associated with the former, the most significant of which is that the ratio C/N of the whole-plant substrate concentrations (where C and N denote plant carbon and nitrogen substrate concentrations) is constrained to settle to a fixed value as balanced exponential growth is attained, and this value is not influenced by variation in the external environment. (In balanced exponential growth all the state variables have the same specific growth rate; see eqn (13.5a) below.) However, experimental evidence indicates that this substrate ratio is affected considerably by the environment (e.g. Bhat, Nye, and Brereton 1979) which, intuitively, we would expect. A major advantage of the Reynolds and Thornley (1982) model is that it is possible to derive an optimization condition so that the plant's specific growth rate is maximized in balanced exponential growth. Unfortunately this requires defining partitioning in terms of the shoot and root specific activities, which is undesirable. In the analysis presented here this general approach is developed (Johnson and Thornley 1987, Mäkelä and Sievänen 1987) with the requirement that the specific growth rate is maximized, and (a) the C/N ratio is not constrained to settle to a fixed value in balanced exponential growth and (b) partitioning is defined solely in terms of the state variables of the system and not directly by the external environmental conditions.

13.3.1 *Description of the model*

The model is illustrated schematically in Fig. 13.1. The plant is divided into shoot and root structural dry mass components (kg structural dry mass), and carbon and nitrogen substrates (kg C or N). The total structural dry mass W_G is

$$W_G = W_{sh} + W_r \tag{13.3a}$$

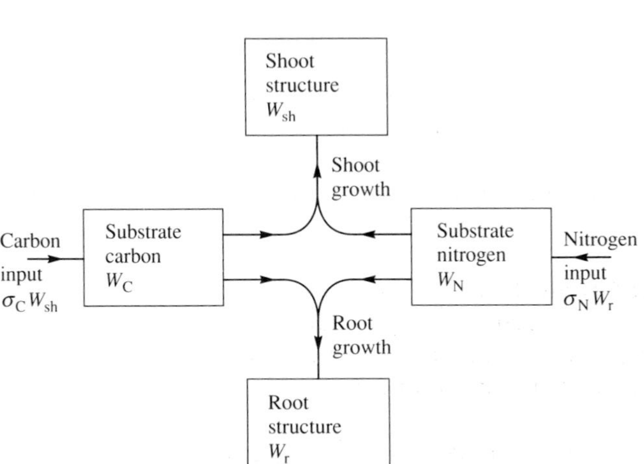

Fig. 13.1. Schematic representation of the teleonomic model.

and the total substrate dry weight W_S is

$$W_S = \frac{M_C}{12} W_C + \frac{M_N}{14} W_N, \qquad (13.3b)$$

where M_C and M_N are the molecular masses of the carbon and nitrogen substrates relative to ^{12}C and ^{14}N respectively. Equation (13.3b) implied that each substrate exists in only one form, and that there is no carbon associated with the substrate nitrogen. In the illustrations presented later, the substrates are taken to be sucrose and nitrate. The model could easily be generalized to include other sugars and amino acids, but with some increase in complexity. However, the form of the substrates does not affect the overall dynamics of the model, only the contribution of the substrates to the dry mass, and any errors are likely to be small. Combining eqns (13.3a) and (13.3b) gives the total plant dry mass W as

$$W = W_G + W_S. \qquad (13.3c)$$

The carbon and nitrogen substrate concentrations C and N are

$$C = W_C/W_G \qquad N = W_N/W_G, \qquad (13.3d)$$

with units of kilograms carbon or nitrogen per kilogram structural dry mass. The shoot and root fractions f_{sh} and f_r are given by

$$f_{sh} = W_{sh}/W_G \qquad f_r = W_r/W_G, \qquad (13.3e)$$

and the shoot : root ratio is

$$f_{sh}/f_r. \qquad (13.3f)$$

Strictly, this implies that the carbon and nitrogen substrates are uniformly distributed through the shoot and root structural dry matter. Although there may be concentration gradients of these substrates, as discussed in Section 13.4 when a mechanistic model of partitioning is presented which is based on substrate transport along concentration gradients, any errors in the shoot : root ratio based on eqns (13.3e) and (13.3f) are likely to be small.

Since the object of the present analysis is to examine partitioning, it is assumed that the net rates of carbon and nitrogen inputs are simply proportional to the shoot and root structural dry masses respectively, which are given by

$$\sigma_C W_{sh} \qquad \sigma_N W_r, \qquad (13.3g)$$

where σ_C and σ_N are the specific shoot and root activities respectively, with units kg C (or N) (kg structure)$^{-1}$ day^{-1} (Exercise 13.1). Once the behaviour of the partitioning model is understood, more general treatments of photosynthesis and nitrogen uptake can be incorporated.

Dynamic equations The shoot and root growth rates (kg structural dry mass day^{-1}) are assumed to be given by

$$\frac{dW_{sh}}{dt} = kCN\lambda_{sh}W_{sh} \tag{13.4a}$$

$$\frac{dW_r}{dt} = kCN\lambda_r W_r, \tag{13.4b}$$

where k is a growth rate coefficient and λ_{sh} and λ_r are dimensionless partitioning functions. In eqns (13.4a) and (13.4b) growth is taken to be directly proportional to the product of the substrate concentrations C and N. The following analysis is quite straightforward for the more general bi-substrate Michaelis–Menten equation (eqn (2.10m), p. 60), and is presented in brief below. However, as precise experimental information is lacking, eqns (13.4a) and (13.4b) are adopted for the main discussion. Since the absolute rate of growth is controlled by the magnitude of k, it is assumed that

$$\lambda_{sh} + \lambda_r = 1, \tag{13.4c}$$

which implies a reciprocal role for the shoot and root; that is, a greater emphasis on shoot growth necessarily involves a lesser emphasis on root growth and vice versa. Equation (13.4c) distinguishes between the plant growth characteristics—defined by k— and the partitioning response—defined by λ_{sh} and λ_r. It is convenient to define λ_{sh} and λ_r in terms of a partitioning function P by

$$\lambda_{sh} = \frac{P}{1+P} \qquad \lambda_r = \frac{1}{1+P}, \tag{13.4d}$$

where P can be a function of any, or all, of the internal variables of the system (W_{sh}, W_r, C, N, etc.). With this formalism, eqn (13.4c) is satisfied and, from eqns (13.4a), (13.4b), and (13.4d),

$$\frac{1}{W_{sh}}\frac{dW_{sh}}{dt} \Big/ \frac{1}{W_r}\frac{dW_r}{dt} = P, \tag{13.4e}$$

so that P is simply the ratio of the shoot and root specific growth rates. The essence of this partitioning model is in the definition of P, which is considered below.

First, we define the mass balance equations for the substrate pools:

$$\frac{dW_C}{dt} = \sigma_C W_{sh} - f_C\frac{dW_G}{dt} \tag{13.4f}$$

$$\frac{dW_N}{dt} = \sigma_N W_r - f_N\frac{dW_G}{dt}, \tag{13.4g}$$

where

$$\frac{dW_G}{dt} = \frac{dW_{sh}}{dt} + \frac{dW_r}{dt}. \tag{13.4h}$$

f_C and f_N are the fractional carbon and nitrogen contents of structure respectively, and are taken to be constant.

Partitioning We now derive the partitioning function P. In principle, we can define P in any way we choose, and then see how the model performs in comparison with our physiological understanding of partitioning. However, the objective in this teleonomic approach is to relate P to our perceived goals of the plant. To do so, consider the situation where the plant is in balanced exponential growth, in which case all the state variables have the same specific growth rate μ (day^{-1}), so that they satisfy the equation

$$\mu = \frac{1}{X}\frac{dX}{dt}. \tag{13.5a}$$

By looking at this special case, it is possible to simplify the mathematics and thereby develop the theory. This can be thought of as analogous to conducting experiments in controlled conditions, rather than under field conditions, to develop understanding. Furthermore, the mechanistic partitioning model described in Section 13.4 does have balanced exponential growth solutions, and it is reasonable to expect the plant to attain balanced exponential growth in uniform conditions during the early stages of growth. A further mathematical advantage is that these solutions are independent of our chosen initial conditions of the differential equations, thus preventing any unintended influence on the solutions by our choice of initial conditions.

From eqn (13.4e) with eqn (13.5a), it follows that

$$P = 1. \tag{13.5b}$$

This condition is necessary and sufficient for balanced exponential growth (i.e. if $P = 1$ then the plant is in balanced exponential growth, and if the plant is in balanced exponential growth then $P = 1$).

We could now define P such that eqn (13.5b) is satisfied and hence the model will attain balanced exponential growth. For example, Johnson (1985) defined

$$P = \xi\frac{NW_r}{CW_{sh}}, \tag{13.5c}$$

where ξ is a partitioning parameter. With eqn (13.5b) this gives

$$CW_{sh} = \xi NW_r \tag{13.5d}$$

in balanced exponential growth, which is similar to eqn (13.2a) (the functional equilibrium condition) but now involves the state variables rather than the environmental conditions. Equation (13.5d), while being physiologically reasonable, is still to some extent arbitrary.

To overcome this arbitrariness in the definition of P, we define the teleonomic goal of the plant and then derive P accordingly. We choose—and it is a choice—

that the plant partitions its growth between the shoot and root in such a manner as to maximize the specific growth rate μ in balanced exponential growth. Mathematically, this requires defining P such that, when the state variables satisfy eqn (13.5a) (balanced exponential growth),

$$\frac{d\mu}{df_{sh}} = 0, \tag{13.5e}$$

and μ is maximized rather than minimized.

In balanced exponential growth $P = 1$ (eqn (13.5b)) so that, from eqns (13.4d),

$$\lambda_{sh} = \lambda_r = \tfrac{1}{2}, \tag{13.5f}$$

and hence from either of eqns (13.4a) or (13.4b) combined with eqn (13.5a)

$$\mu = \tfrac{1}{2}kCN, \tag{13.5g}$$

and from eqns (13.4f) and (13.4g)

$$\mu C = \sigma_C f_{sh} - \mu f_C \tag{13.5h}$$

$$\mu N = \sigma_N f_r - \mu f_N. \tag{13.5i}$$

It can now be shown from eqns (13.5g)–(13.5i) (Exercise 13.2) that

$$\left(\frac{f_C}{C} + \frac{f_N}{N} + 3\right)\frac{d\mu}{df_{sh}} = \frac{\sigma_C}{C} - \frac{\sigma_N}{N}, \tag{13.5j}$$

so that a necessary and sufficient condition for $d\mu/df_{sh} = 0$ is

$$\sigma_C/C = \sigma_N/N. \tag{13.5k}$$

It is readily shown that this maximizes, rather than minimizes, μ (Exercise 13.2). Equation (13.5k) was derived by Reynolds and Thornley (1982) specifically for their model, and in the more general case by Mäkelä and Sievänen (1987); this is a crucial result in the teleonomic approach.

Now, eliminating μ from eqns (13.5h) and (13.5i), we obtain

$$(N + f_N)\sigma_C f_{sh} = (C + f_C)\sigma_N f_r \tag{13.5l}$$

so that, using eqn (13.5k), $d\mu/df_{sh} = 0$ requires

$$f_r \frac{N}{N + f_N} = f_{sh}\frac{C}{C + f_C}. \tag{13.5m}$$

Thus we can define a general partitioning function P by

$$P = \frac{f_r}{f_{sh}}\frac{N/(N + f_N)}{C/(C + f_C)}. \tag{13.5n}$$

Note that there are no adjustable parameters in (13.5n): f_r, f_{sh}, N, and C are variables ((13.3d) and (13.3e)); f_N and f_C define the composition of plant structure.

For balanced exponential growth $P = 1$ (eqn (13.5b)), so that eqn (13.5m) is satisfied and hence growth is maximized, i.e. $d\mu/df_{sh} = 0$.

Equation (13.5n), with the constraint $P = 1$, is consistent with the perceived teleonomic goals of the plant of optimum growth in response to variation in the substrate concentrations C and N, i.e. the response to a shortage in either substrate is to partition a relatively greater proportion of growth in order to acquire that substrate. If, for example, nitrogen is 'non-limiting', which can be interpreted as

$$\frac{N}{N + f_N} \to 1. \tag{13.5o}$$

then the shoot : root ratio is dominated by variation in C.

For the general situation, of which balanced exponential growth is a special case, eqn (13.4e) holds, i.e. P is the ratio of the shoot and root specific growth rates. The basis for generalizing from this specific case is that the partitioning is defined in a physiologically reasonable manner that is also experimentally testable (or, in the Popperian sense, falsifiable). Of course, the definition of the partitioning function is not unique. P could be replaced by some other function $g(P)$ such that

$$g(1) = 1. \tag{13.6a}$$

This would not affect the balanced exponential growth solution, which is independent of the choice of $g(P)$ (provided that eqn (13.6a) holds), but only the dynamic solution prior to balanced exponential growth. (Note that for some $g(P)$, e.g. $g(P) = 1/P$, the balanced exponential growth solution is unstable.) One possibility is to define

$$g(P) = P^q, \tag{13.6b}$$

where q is a parameter giving greater control characteristics over the solution. However, in the absence of evidence to the contrary, we shall retain the simplest realistic form for $g(P)$:

$$g(P) = P. \tag{13.6c}$$

This defines the model completely in terms of the environmental conditions σ_C and σ_N, the plant structural characteristics f_C and f_N, and the growth coefficient k. Clearly, the main effect of k will be to influence the internal substrate levels C and N; the actual growth and yield will depend on the supply of substrate. It is difficult to imagine defining the dynamics of partitioning in terms of a smaller parameter set.

However, if the bi-substrate Michaelis–Menten equation is used in eqns (13.4a) and (13.4b), so that the expression kCN in eqns (13.4a) and (13.4b) is replaced by the more general expression (p. 60)

$$F(C, N) = \frac{kCN}{1 + C/K_C + N/K_N + CN/K_{CN}}, \tag{13.7a}$$

where K_C, K_N, and K_{CN} are constants, then the required constraint to maximize μ in balanced exponential growth becomes

$$\frac{\sigma_C}{C(1 + C/K_C)} = \frac{\sigma_N}{N(1 + N/K_N)}, \tag{13.7b}$$

and the partitioning function can be shown to be (Exercise 13.3)

$$P = \frac{f_r}{f_{sh}} \frac{N(1 + N/K_N)/(N + f_N)}{C(1 + C/K_C)/(C + f_C)}. \tag{13.7c}$$

The remainder of the analysis is unchanged. Note that, in the limits K_C, K_N, $K_{CN} \to \infty$, eqns (13.7b) and (13.7c) are consistent with eqns (13.5k) and (13.5n).

Finally, if the root and shoot have different growth coefficients k_{sh} and k_r, then the partitioning coefficients become (Exercise 13.4)

$$\lambda_{sh} = \frac{k_r P}{k_{sh} + k_r P} \qquad \lambda_r = \frac{k_{sh}}{k_{sh} + k_r P}. \tag{13.7d}$$

In the following illustrations, we shall use the model in its basic optimized form, with the structural growth rates given by eqns (13.4a) and (13.4b) (rather than using eqn (13.7a)) and with $k_{sh} = k_r = k$, so that the partitioning function is defined by eqn (13.5n) and λ_{sh} and λ_r are defined by eqns (13.4d).

13.3.2 *Illustrations*

We now examine the behaviour of the model, which requires assigning values to the parameters k, f_C, and f_N. Experimentally, k can be estimated from eqn (13.5g), and for the present purposes is taken to be

$$k = 100 \text{ day}^{-1}[\text{C}]^{-1}[\text{N}]^{-1}. \tag{13.8a}$$

The fractional carbon and nitrogen contents of structural dry mass are readily measured, and the values

$$\begin{aligned} f_C &= 0.45 \text{ kg C (kg structure)}^{-1} \\ f_N &= 0.03 \text{ kg N (kg structure)}^{-1} \end{aligned} \tag{13.8b}$$

are used.

The differential equations describing the model (eqns (13.4a), (13.4b), (13.4f), and (13.4g) with eqns (13.4d), (13.4h), and (13.5n)) must be solved subject to prescribed initial conditions. Balanced exponential growth is eventually attained regardless of the choice of initial conditions, although they influence the transient solution prior to balanced exponential growth.

Balanced exponential growth solutions We first examine the balanced exponential growth solutions. This is particularly useful since these solutions are independent of the chosen initial conditions, thus allowing us to investigate the behaviour of the model uninfluenced by the choice of initial conditions. Once this has been done, and the model's performance is clearly understood, it can then be used more generally in the knowledge that its underlying features and limitations have been assessed.

In the following illustrations, to simulate changes in shoot and root environments, the effects of various values of specific shoot and root activities σ_C and σ_N on the balanced exponential growth solutions are examined, where either

$$\sigma_C = 0.15 \quad \text{and } \sigma_N \text{ varies,} \tag{13.8c}$$

or

$$\sigma_C \text{ varies} \quad \text{and } \sigma_N = 0.05. \tag{13.8d}$$

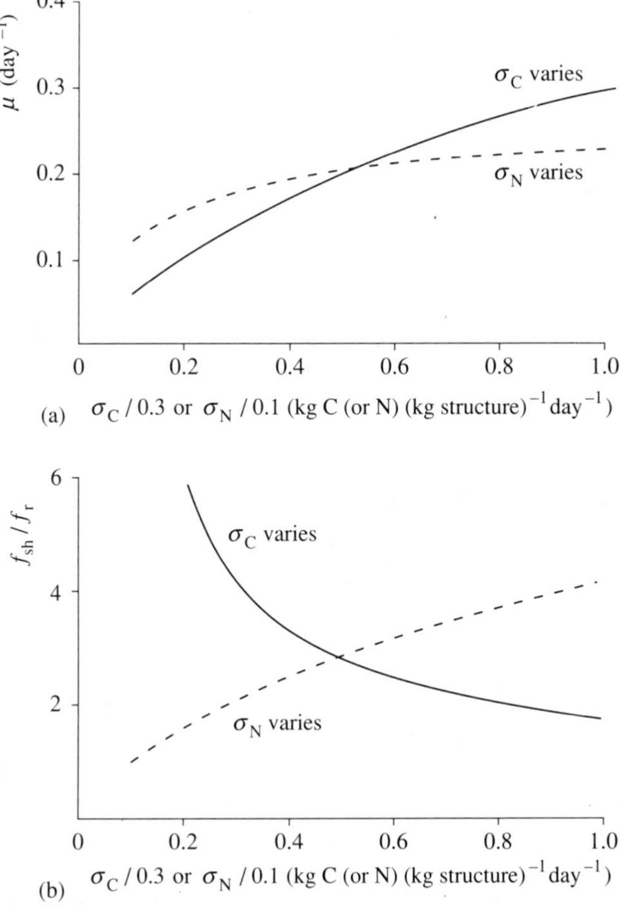

Fig. 13.2. Dependence of the balanced exponential growth solutions of the teleonomic model for (a) the specific growth rate, μ, and (b) the shoot : root ratio f_{sh}/f_r on variation in either $\sigma_C/$ 0.3 or $\sigma_N/0.1$ (kg C (or N) (kg structure)$^{-1}$ day^{-1}). See text for details.

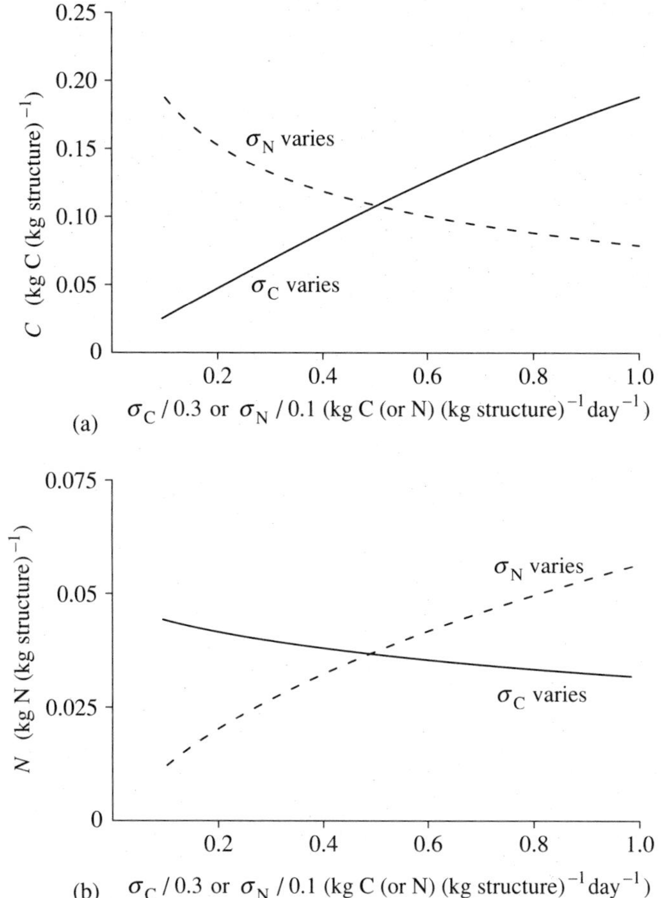

Fig. 13.3. Variation in (a) the carbon substrate concentration C and (b) the nitrogen substrate concentration N corresponding to Fig. 13.2.

Recall that σ_C and σ_N have units of kg C (or N) (kg structure)$^{-1}$ day^{-1}. In Fig. 13.2, μ and f_{sh}/f_r are shown. The responses are as expected: increasing either σ_C or σ_N causes μ to increase; f_{sh}/f_r increases with increased σ_N and decreases with increased σ_C. The substrate levels indicate the carbon and nitrogen status of the plant. In Fig. 13.3, C and N (eqns (13.3d)) are illustrated. Again these follow an expected pattern, with each rising as the supply of the relevant substrate rises.

Although the substrate concentrations are central to the plant's carbon and nitrogen status, frequently only the whole-plant carbon or nitrogen fractions are measured. These are given by

$$F_C = \frac{W_C + f_C W_G}{W} \qquad F_N = \frac{W_N + f_N W_G}{W}, \tag{13.8e}$$

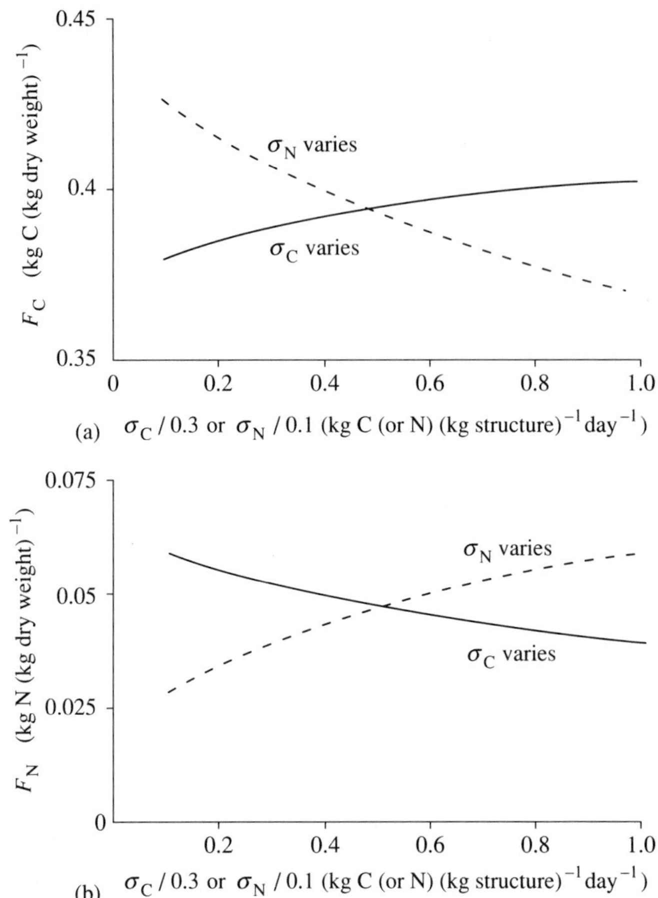

Fig. 13.4. Variation in (a) the whole-plant carbon fraction F_C and (b) the whole-plant nitrogen fraction F_N corresponding to Fig. 13.2. See eqns (13.8e).

and are presented in Fig. 13.4. Note that the individual fractions of carbon and nitrogen are influenced simultaneously by C and N. Thus, if (for example) the plant's carbon substrate level is increased and the nitrogen substrate level remains constant, both F_C and F_N will be affected. Consequently, while these fractions have important practical value, a fuller understanding of the plant's carbon and nitrogen status is obtained when the substrate concentrations are known.

Dynamic solutions We now turn out attention to the dynamic solution following a change in either the shoot or root specific activity. Consider the situation

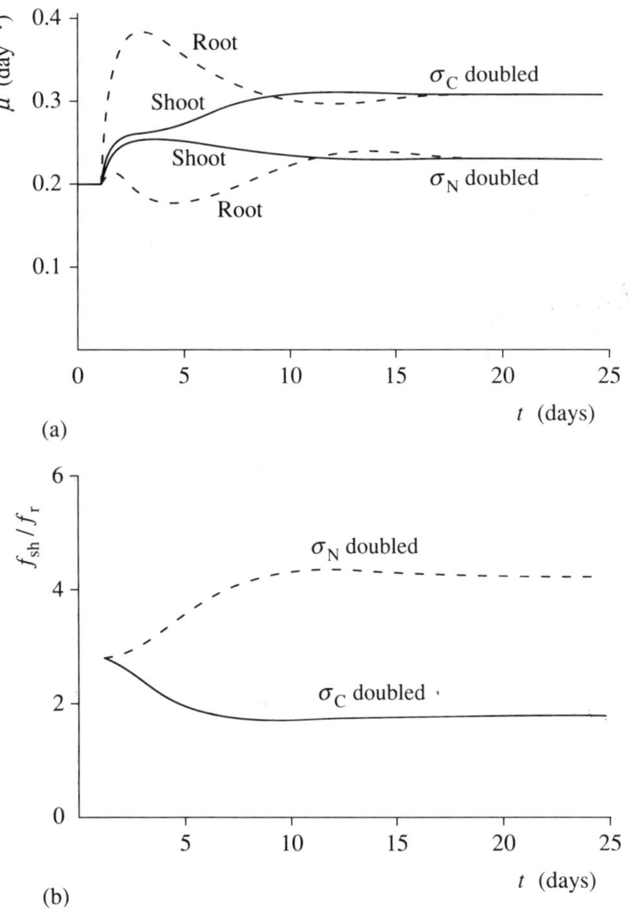

Fig. 13.5. Effect of an increase (a doubling) in specific shoot activity σ_C or specific root activity σ_N on (a) the shoot and root specific growth rates μ_{sh} and μ_r and (b) the shoot : root ratio f_{sh}/f_r. Prior to these changes, which occur at $t = 1$, the plant was in balanced exponential growth. See text for details.

where the plant has attained balanced exponential growth with

$$\sigma_C = 0.15 \qquad \sigma_N = 0.05, \tag{13.8f}$$

following which either σ_C or σ_N is doubled. These changes can be interpreted as increasing the light or nitrogen supply respectively. In Fig. 13.5 μ_{sh}, μ_r and f_{sh}/f_r are illustrated, and it can be seen that the effects of these environmental changes are as expected: increasing either the shoot or root specific activity increases the specific growth rate μ; the shoot : root ratio f_{sh}/f_r increases with increased nitrogen supply or decreases with increased carbon supply. In Fig. 13.6 the plant substrate concentrations C and N (eqns (13.3d)) are illustrated. Following a

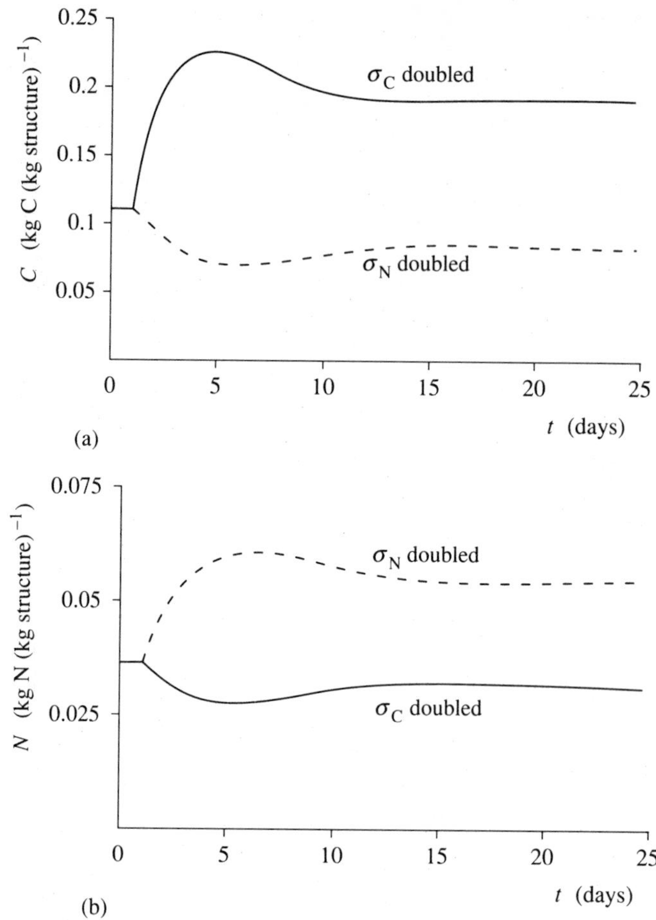

Fig. 13.6. Variation in (a) the carbon substrate concentration C and (b) the nitrogen substrate concentration N corresponding to Fig. 13.5 with a doubling in shoot or root specific activities.

sudden increase in the supply of either substrate, there is an initial marked increase in that substrate and a corresponding decrease in the other. As the plant adapts to its new environment (by adjusting its shoot : root ratio), the substrate levels approach the new equilibrium values as discussed above.

Shoot defoliation Consider the effect of severe shoot defoliation on the subsequent partitioning and plant growth. This is achieved by removing 95 per cent of the shoot structural dry weight and the corresponding fractions of the substrate components (assuming uniform distributions), again for plants in balanced exponential growth with σ_C and σ_N given by eqn (13.8f). The shoot : root

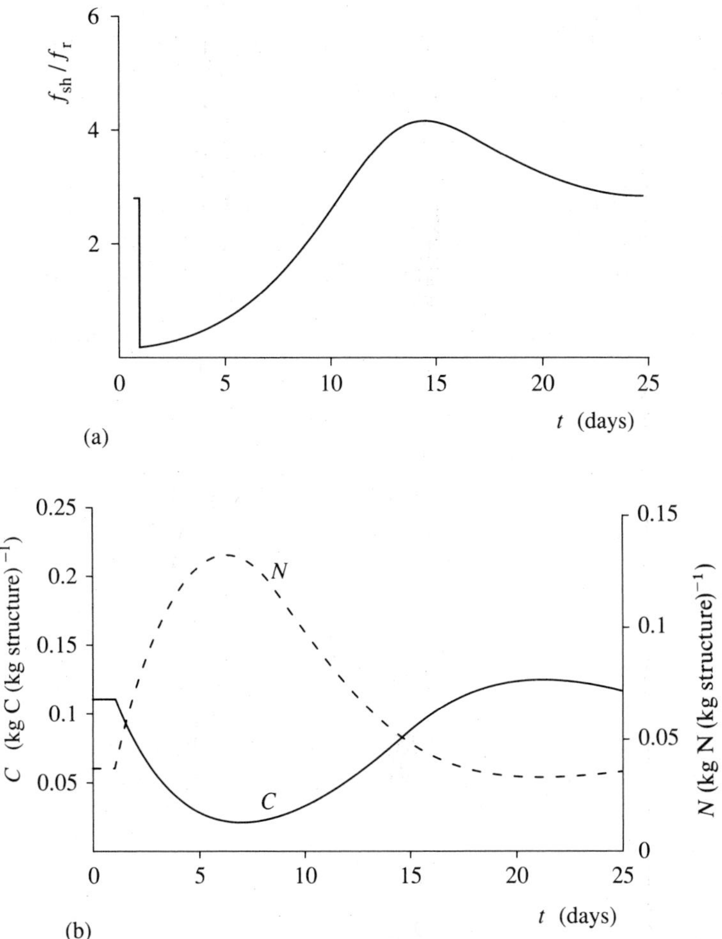

Fig. 13.7. Effect of 95 per cent shoot defoliation on (a) the shoot : root ratio f_{sh}/f_r, and (b) the carbon substrate concentration C (solid line) and the nitrogen substrate concentration N (broken line). Prior to defoliation, which occurs at $t = 1$, the plant was in balanced exponential growth. See text for details.

ratio f_{sh}/f_r and substrate concentrations following this defoliation are presented in Fig. 13.7. f_{sh}/f_r increases, with a slight oscillation, to its original value. There is an initial marked rise in the nitrogen substrate concentration and a fall in the carbon substrate concentration, following which they both return to their original values. These features are as expected: following shoot defoliation, the plant has a severely reduced supply of carbon and so a relatively excessive supply of nitrogen. The plant's response is to partition a greater proportion of growth to the shoot, thus reverting to the original balance prior to defoliation.

13.3.3 *Nutrient uptake influenced by substrate levels*

There is experimental evidence that nutrient uptake is affected by the substrate levels. Clement, Hopper, Jones, and Leafe (1978) have observed that the rate of uptake of nitrogen in perennial ryegrass fluctuates diurnally in response to photosynthesis. They also observed that nitrogen uptake fell by 90 per cent or more following shoot defoliation. There may be considerable respiratory costs involved in nitrogen uptake (Sections 11.5, p. 279, 12.2.7, p. 320, and 12.2.10, p. 348)), and the data of Clement *et al.* suggest that its rate depends on the level of carbon substrate (i.e. the energy available) in the plant. Furthermore, nitrogen uptake is an active process (Section 4.4, p. 102), i.e. the nitrogen is transported

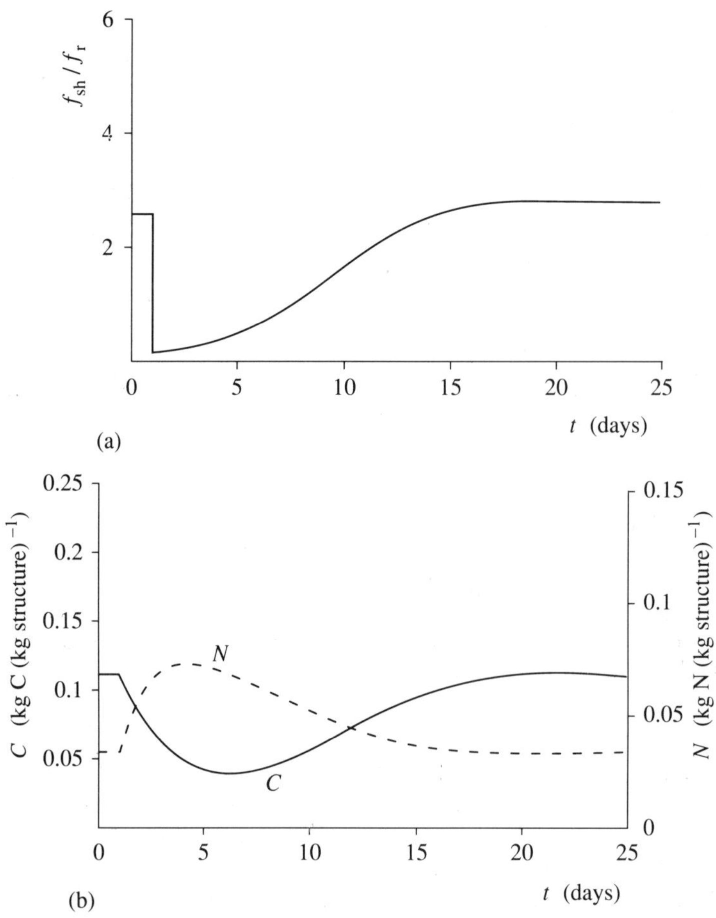

Fig. 13.8. Repeat of the illustrations in Fig. 13.7 for 95 per cent shoot defoliation, but with the specific root activity σ_N related to the substrate levels in the plant by eqn (13.9a). f_{sh}/f_r denotes the shoot:root ratio, and C and N the carbon and nitrogen substrate concentrations.

into the plant up an ion gradient, so that it is likely that it will be influenced by demand. This is discussed by Clement *et al.* and has been observed by White (1973) in relation to phosphorus uptake in lucerne. Consequently, the behaviour of the model following shoot defoliation, where a high level of N is attained (Fig. 13.7(b)) may be physiologically unrealistic.

These features suggest that nitrogen uptake is reduced either when carbon substrate is limiting, or by high levels of nitrogen substrate within the plant. One simple method of incorporating such responses is to relate σ_N to C and N by

$$\sigma_N = \frac{\tilde{\sigma}_N}{1 + (K_C/C)(1 + N/K_N)}, \tag{13.9a}$$

where $\tilde{\sigma}_N$, K_C, and K_N are constants with the same dimensions as σ_N, C, and N respectively. This is a standard relationship in enzyme kinetics for enzyme–substrate reactions with a fully competitive inhibitor, and is discussed in Section 2.3.5, p. 60. Carbon is the substrate and nitrogen is the inhibitor (note that K_C and K_N are not the same as those in eqn (13.7a)).

In Fig. 13.8, the illustrations shown in Fig. 13.7 are repeated and compared with the solution obtained when eqn (13.9a) is used with the parameter values

$$\tilde{\sigma}_N = 0.1 \qquad K_C = 0.05 \qquad K_N = 0.02. \tag{13.9b}$$

The responses are considerably dampened compared with those in Fig. 13.7. In particular, the nitrogen substrate level N rises to a much lesser extent. In Fig. 13.9, the corresponding time course of σ_N, as given by eqn (13.9a), is presented, and it can be seen that this falls by a considerable amount following defoliation, after which it gradually returns to its original value as the plant recovers.

Note that the analysis of eqns (13.5j) and (13.5k) in which the specific growth rate μ is maximized does not hold with eqn (13.9a) for σ_N. However, our basic

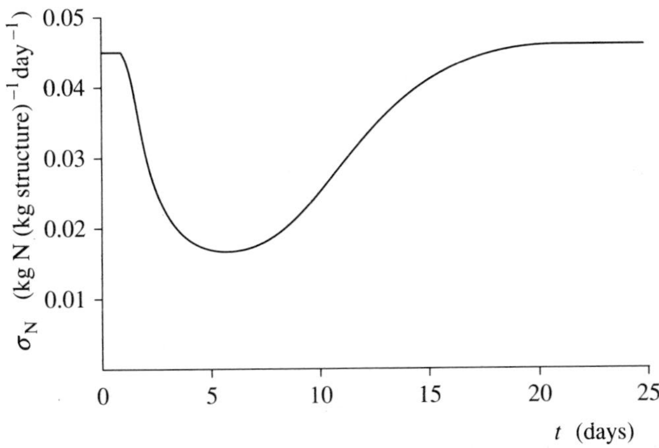

Fig. 13.9. Response of the specific root activity σ_N to 95 per cent shoot defoliation, corresponding to Fig. 13.8.

approach has been to derive the partitioning function under the simplest physiological assumptions, and to generalize from this basis where it seems necessary.

13.3.4 *Assessment of the functional balance*

The functional balance (eqn (13.2a)) relating the shoot and root activities, can be expressed in the above notation as

$$\sigma_C f_{sh} \propto \sigma_N f_r. \tag{13.10a}$$

Comparing this with eqn (13.5l), this is only true if

$$\eta = \frac{C + f_C}{N + f_N} = \text{constant}. \tag{13.10b}$$

This equation states that the total carbon and total nitrogen contents (cf. eqns (13.8e)) are proportional to each other: the total carbon : nitrogen ratio is constant. This requirement is independent of any hypothesis regarding the dynamics of partitioning, and is based solely on the principle of mass balance. Physiologically, eqn (13.10b) is unlikely to be satisfied since an increase in C is often accompanied by a decrease in N and vice versa. We can use the optimized teleonomic model described here to examine the response of η to changes in σ_C and σ_N for the balanced exponential growth solution presented above. In doing so, we should remember that the principal assumption influencing these solutions is that the plant maximizes its specific growth rate μ; the partition function P (and hence λ_{sh} and λ_r) only affects the dynamic solution prior to balanced exponential growth. In Fig. 13.10 η is shown corresponding to the illustrations

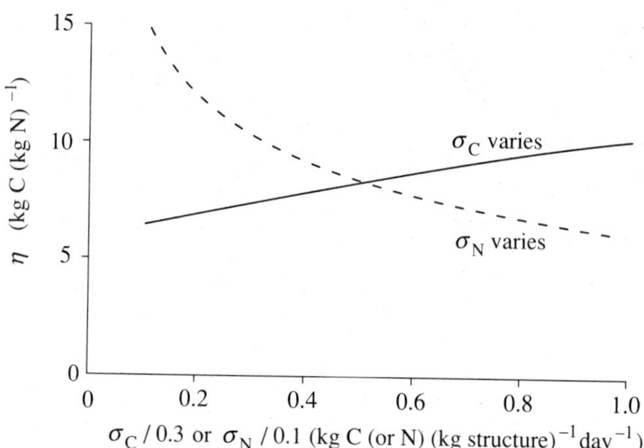

Fig. 13.10. Variation in the total carbon : total nitrogen ratio η (eqn (13.10b)) with changes in σ_C or σ_N, corresponding to Figs 13.2–13.4, to assess the functional balance hypothesis (eqn (13.2a)) for which η should be constant. See text for discussion.

in Figs 13.3 and 13.4. It is obvious that η is not constant, indicating that the functional balance equation, while being qualitatively reasonable, is not appropriate for detailed quantitative study.

13.4 Transport-resistance mechanistic model

The mechanistic approach to modelling partitioning which is now discussed is based on the work of Thornley (1972a,b, 1976). In this analysis, Thornley formalized the ideas expressed by White (1937). Basically, the method involves the two processes of substrate transport around the plant and chemical conversion of substrate to structure at the site of utilization. The model does not incorporate explicitly any of the 'goals' which the plant can be regarded as having, but its behaviour is consistent with these observations.

13.4.1 *Description and behaviour of the model*

The model, which is illustrated schematically in Fig. 13.11, differs from the model just described in that the substrates in the shoot and root are considered separately, rather than using whole-plant averages (the model uses local variables entirely). Thus, extending the previous notation (Figs 13.1 and 13.11), we have

$$W_{\mathrm{C}} = W_{\mathrm{C,sh}} + W_{\mathrm{C,r}} \tag{13.11a}$$

$$W_{\mathrm{N}} = W_{\mathrm{N,sh}} + W_{\mathrm{N,r}}, \tag{13.11b}$$

and the carbon and nitrogen substrate concentrations in the shoot and root are

$$C_{\mathrm{sh}} = \frac{W_{\mathrm{C,sh}}}{W_{\mathrm{sh}}} \qquad N_{\mathrm{sh}} = \frac{W_{\mathrm{N,sh}}}{W_{\mathrm{sh}}} \tag{13.11c}$$

$$C_{\mathrm{r}} = \frac{W_{\mathrm{C,r}}}{W_{\mathrm{r}}} \qquad N_{\mathrm{r}} = \frac{W_{\mathrm{N,r}}}{W_{\mathrm{r}}}. \tag{13.11d}$$

The shoot and root fractions of dry mass are now

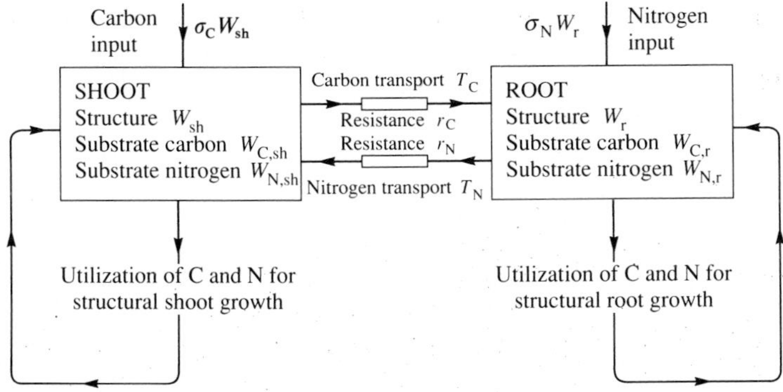

Fig. 13.11. Schematic representation of the mechanistic transport-resistance model.

$$f_{sh} = \frac{W_{sh} + (M_C/12)W_{C,sh} + (M_N/14)W_{N,sh}}{W} \qquad (13.11e)$$

$$f_r = \frac{W_r + (M_C/12)W_{C,r} + (M_N/14)W_{N,r}}{W}, \qquad (13.11f)$$

where W is the total mass (root and shoot, structure and storage). Again the shoot : root ratio is

$$f_{sh}/f_r. \qquad (13.11g)$$

The carbon and nitrogen inputs to the system are as in the previous model, i.e. $\sigma_C W_{sh}$ and $\sigma_N W_r$ respectively (eqn (13.3g)). The modification relating σ_N to the substrate levels in the plant (eqn (13.9a)) is not used here but could readily be incorporated.

The substrate transport fluxes T_C and T_N of carbon and nitrogen respectively between the shoot and root are taken to be driven by the concentration gradients, and can be expressed as

$$T_C = \frac{\beta(C_{sh} - C_r)}{r_C} \qquad T_N = \frac{\beta(N_r - N_{sh})}{r_N}, \qquad (13.11h)$$

with units kg C (or N) day^{-1}. r_C and r_N are resistances to movement of carbon and nitrogen, and β is a scaling factor to allow for the dependence of these resistances on plant size. The role of β in these expressions can be understood by considering the situation where all plant parts are growing at the same constant specific growth rate, i.e. the plant is in balanced exponential growth. In these circumstances the substrate concentrations will remain constant. However, as the shoot and root structural dry masses increase, a greater supply of substrate is required to maintain the specific growth rate, which would not be satisfied with $\beta = 1$. Indeed, such balanced exponential growth is only possible with β proportional to one of the state variables or to a linear combination of them. It is assumed that

$$\beta = W_G = W_{sh} + W_r, \qquad (13.11i)$$

where W_G is the total plant structural dry mass. This implies that resistance to transport is inversely proportional to plant size, which seems physiologically reasonable. Note that the units of r_C and r_N depend on those of β and, from eqn (13.11i), they are simply days.

Dynamic equations The shoot and root growth rates are assumed to be given by

$$\frac{dW_{sh}}{dt} = kC_{sh}N_{sh}W_{sh} \qquad (13.12a)$$

$$\frac{dW_r}{dt} = kC_rN_rW_r \qquad (13.12b)$$

in units of kg structural dry mass day^{-1}, analogous to eqns (13.4a) and (13.4b).

The equations for the substrate pools are

$$\frac{dW_{C,sh}}{dt} = \sigma_C W_{sh} - T_C - f_C \frac{dW_{sh}}{dt} \tag{13.12c}$$

$$\frac{dW_{N,sh}}{dt} = T_N - f_N \frac{dW_{sh}}{dt} \tag{13.12d}$$

$$\frac{dW_{C,r}}{dt} = T_C - f_C \frac{dW_r}{dt} \tag{13.12e}$$

$$\frac{dW_{N,r}}{dt} = \sigma_N W_r - T_N - f_N \frac{dW_r}{dt}. \tag{13.12f}$$

Equations (13.12a)–(13.12f), together with eqns (13.11c), (13.11d), (13.11h), and (13.11i) define the problem mathematically. Note that eqns (13.12c) and (13.12e) and eqns (13.12d) and (13.12f) sum to give the dynamic equations for the substrate pools in the previous models, i.e. eqn (13.4f) and eqn (13.4g) respectively. See Exercise 13.6.

Illustrations The differential equations describing the model must be solved subject to prescribed initial conditions. Again, balanced exponential growth is eventually attained (with eqn (13.11i) for β) regardless of the choice of initial conditions, although they will influence the transient solution (stable balanced exponential growth may not be a possible solution for all parameter values). The behaviour of the model is very similar to the teleonomic model, and so there is little value in reproducing all the previous illustrations. Nevertheless, a brief look at some of the solutions is worthwhile.

Parameter values of

$$r_C = r_N = 0.5 \text{ day} \tag{13.13a}$$

$$k = 50 \text{ day}^{-1}[\text{C}]^{-1}[\text{N}]^{-1} \tag{13.13b}$$

are adopted for the resistances and the growth coefficient. The value used for k is half that used previously (eqn (13.8a)) to allow for the λ_{sh} and λ_r factors (eqns (13.4a) and (13.4b)). The fractional carbon and nitrogen contents of structural dry weight f_C and f_N are as before (eqn (13.8b)), i.e. 0.45 and 0.03 respectively.

The balanced exponential growth solutions of μ and f_{sh}/f_r for σ_C and σ_N varying over the same range as used previously (Figs 13.2–13.4 and eqns (13.8c) and (13.8d)) are shown in Fig. 13.12. The responses are very similar to those presented in Fig. 13.2. Since the present model is defined in terms of shoot and root substrate concentrations specifically, rather than whole-plant concentrations, these are presented in Fig. 13.13. The patterns are similar to those in Fig. 13.3, but with concentration gradients between the shoot and root.

Finally, turning to the dynamic solutions, Fig. 13.14 shows μ_{sh}, μ_r, and f_{sh}/f_r for the changes in either σ_C or σ_N corresponding to Fig. 13.5 and, again, the

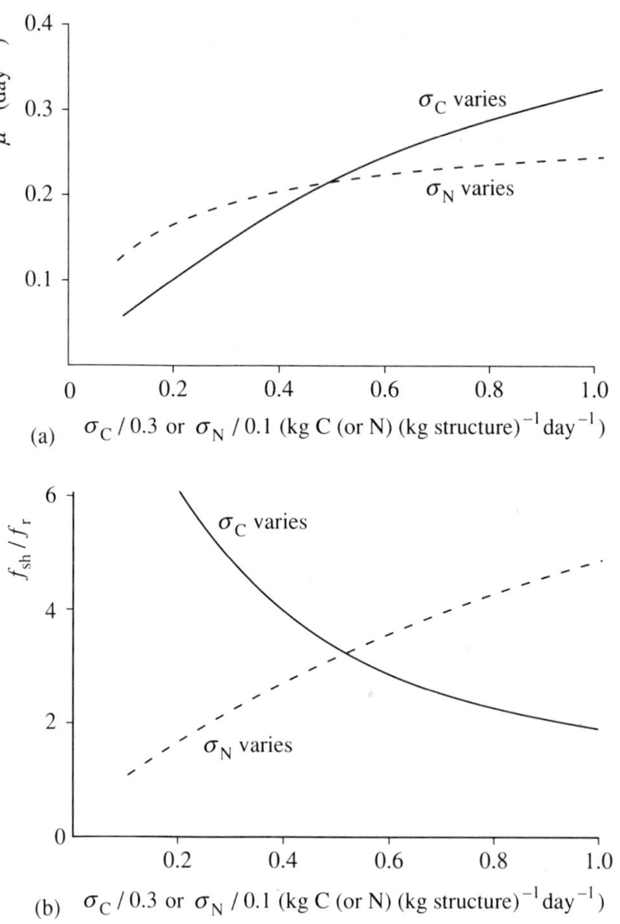

Fig. 13.12. Dependence of the balanced exponential growth solutions of the transport-resistance model for (a) the specific growth rate μ and (b) the shoot:root ratio f_{sh}/f_r on variation in either specific shoot activity σ_C or specific root activity σ_N. See text for details.

responses are quite similar although not identical. This may be due in part to the choice of parameter values as well as the different structures of the models.

13.5 Plant growth involving several mineral nutrients

The models described above involve only one root function—the acquisition of nitrogen. However, the roots are the source of all the mineral nutrients as well as water, possibly including essential compounds such as growth factors. We shall now outline the extensions of the models to incorporate a second root function. The theory can then be generalized to as many nutrients or other functions as required.

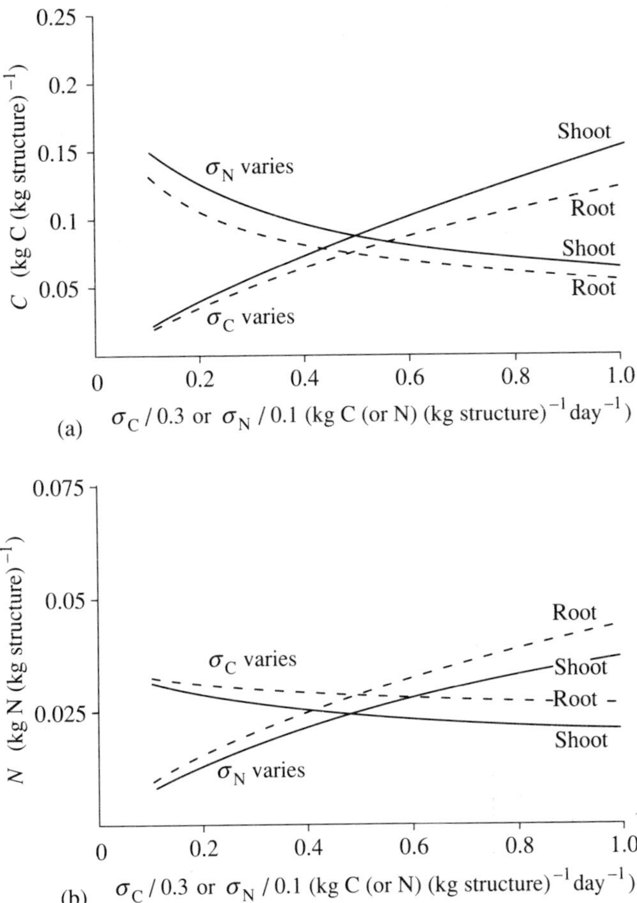

Fig. 13.13. Variation in (a) the shoot and root carbon substrate concentrations C_{sh} and C_r and (b) the shoot and root nitrogen substrate concentrations N_{sh} and N_r corresponding to Fig. 13.12.

Consider an element or compound E, in addition to nitrogen, with uptake or synthesis (analogous to eqn (13.3g))

$$\sigma_E W_r. \tag{13.14a}$$

The remainder of the notation involving E is an obvious extension of that for carbon and nitrogen. We consider each model separately.

Teleonomic model The shoot and root growth rates are now (by assumption)

$$\frac{dW_{sh}}{dt} = kCNE\lambda_{sh} W_{sh} \tag{13.15a}$$

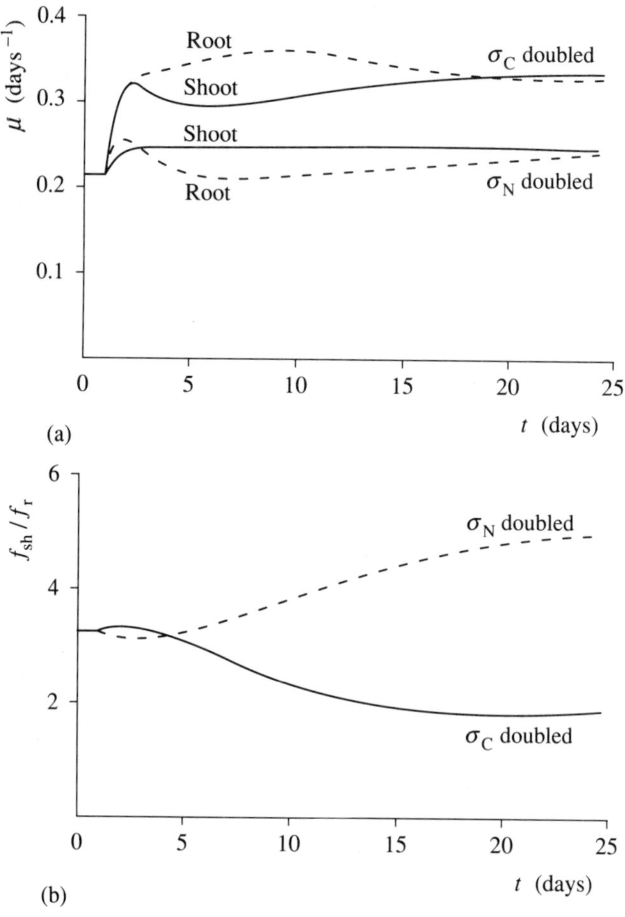

Fig. 13.14. Effect of an increase in specific shoot activity σ_C or specific root activity σ_N on: (a) the shoot and root specific growth rates μ_{sh} and μ_r and (b) the shoot : root ratio f_{sh}/f_r. Prior to these changes, which occur at $t = 1$, the plant was in balanced exponential growth. See text for details.

$$\frac{dW_r}{dt} = kCNE\lambda_r W_r, \tag{13.15b}$$

where, as before, λ_{sh} and λ_r are defined in terms of the partitioning function P as

$$\lambda_{sh} = \frac{P}{1+P} \qquad \lambda_r = \frac{1}{1+P}. \tag{13.15c}$$

The extra substrate equation is

$$\frac{dW_E}{dt} = \sigma_E W_r - f_E \frac{dW_G}{dt}. \tag{13.15d}$$

In balanced exponential growth, again

$$P = 1 \tag{13.16a}$$

and

$$\lambda_{sh} = \lambda_r = \tfrac{1}{2}, \tag{13.16b}$$

and eqn (13.15d) becomes

$$\mu E = \sigma_E f_r - \mu f_E. \tag{13.16c}$$

Generalizing the previous analysis, a necessary and sufficient condition for μ being maximized is (Exercise 13.5)

$$\frac{\sigma_C}{C} = \frac{\sigma_N}{N} + \frac{\sigma_E}{E}, \tag{13.16d}$$

and the partitioning function can now be derived as (Exercise 13.5)

$$P = \frac{f_r}{f_{sh}} \frac{[(N + f_N)/N + (E + f_E)/E]^{-1}}{C/(C + f_C)}. \tag{13.16e}$$

Equation (13.14a), and (13.15a)–(13.15d) with the partitioning function given by eqn (13.16e) define the problem to include the extra substrate E, and the analysis can be extended to any number of root functions in an obvious manner (cf. eqns (13.5n) and (13.16e)).

If E is regarded as 'non-limiting', relative to the nitrogen status of the plant, this can be interpreted as

$$\frac{E + f_E}{E} \ll \frac{N + f_N}{N}, \tag{13.17a}$$

in which case P reduces to

$$P \approx \frac{f_r}{f_{sh}} \frac{N/(N + f_N)}{C/(C + f_C)}, \tag{13.17b}$$

which is the partitioning function derived previously (eqn (13.5n)).

Mechanistic transport-resistance model Let eqns (13.12a) and (13.12b) for the structural growth rates be extended to

$$\frac{dW_{sh}}{dt} = kC_{sh}N_{sh}E_{sh}W_{sh} \tag{13.18a}$$

$$\frac{dW_r}{dt} = kC_r N_r E_r W_r. \tag{13.18b}$$

Again, a more general Michaelis–Menten equation could be used (cf. eqn (2.10m), p. 60, generalized to three substrates).

Transport of E between the root and shoot is given by

$$T_{\mathrm{E}} = \frac{\beta(E_{\mathrm{r}} - E_{\mathrm{sh}})}{r_{\mathrm{E}}}, \tag{13.18c}$$

and hence the extra substrate balance equations for E in the shoot and root are

$$\frac{\mathrm{d}W_{\mathrm{E,sh}}}{\mathrm{d}t} = T_{\mathrm{E}} - f_{\mathrm{E}}\frac{\mathrm{d}W_{\mathrm{sh}}}{\mathrm{d}t}, \tag{13.18d}$$

$$\frac{\mathrm{d}W_{\mathrm{E,r}}}{\mathrm{d}t} = \sigma_{\mathrm{E}} W_{\mathrm{r}} - T_{\mathrm{E}} - f_{\mathrm{E}}\frac{\mathrm{d}W_{\mathrm{r}}}{\mathrm{d}t}. \tag{13.18e}$$

Equations (13.18a)–(13.18e) define the extension of the model to include the extra substrate E, and, as for the teleonomic model, this analysis can be generalized in an obvious manner to include more root or shoot functions.

13.6 Discussion

Our account of shoot : root partitioning has focused on a teleonomic approach, based on the hypothesis that the plant attempts to maximize its specific growth rate by adjusting the shoot : root ratio and the levels of required substrates. We have also described the transport-resistance mechanistic model. Both models display the same general features of partitioning in response to varying shoot and root specific activities, and shoot defoliation. As mentioned in Section 13.1, the other physiological aspects involved in plant growth have been, to a large extent, ignored. When incorporating senescence and structural degradation, the shoot and root growth rate equations for $\mathrm{d}W_{\mathrm{sh}}/\mathrm{d}t$ and $\mathrm{d}W_{\mathrm{r}}/\mathrm{d}t$ should be interpreted as the gross rates of production of structure and not the net growth rates. If a more complete description of the other physiological processes involved is included in the model, the analysis deriving the optimized partitioning function P (eqn (13.5n)) in the teleonomic model (such that $P = 1$ maximizes the specific growth rate in balanced exponential growth) does not hold. However, the analysis provides a possible rationale for defining partitioning, and the approach is used in a crop growth model in Chapter 16, Section 16.5 (p. 464), where fertilizer responses are derived.

The value of the teleonomic approach is perhaps best seen in such embracing crop growth models in which partitioning is only one of several sub-models. While being simple and displaying the required characteristics, it is, nevertheless, an expression of our interpretation of how the plant responds to its environment. It may therefore be inappropriate to draw conclusions as to the underlying mechanisms involved in partitioning from this model. However, experiments which use this model as a working hypothesis could have great value in providing an underlying theoretical framework for a quantitative description of crop growth.

The mechanistic model, in contrast, provides a sound basis for studying the underlying principles of partitioning. No views of the plant's 'goals' are incorporated. The information which has been gained from experimental work in this area is essentially qualitative; that is, it extends no further than the general trends and patterns discussed above. We believe that it is timely to incorporate this type of mechanistic model in experimental programmes in order to make significant advances in understanding the problem. One disadvantage of the model is the need to estimate the resistances to substrate transport, which may prove difficult. Nevertheless, since these resistances seem to be central to the understanding of partitioning, the problem of estimating them should be addressed. In doing so, it is likely that insights into other unforeseen features of partitioning will be gained.

A parallel approach to studying partitioning with either of these models as the theoretical basis will benefit both our immediate needs of having a practical description for use in crop growth studies, and the more long-term objective of increasing our basic understanding of the problem.

Exercises

13.1. Show that the specific shoot activity $\sigma_C = 0.15$ kg C (kg structure)$^{-1}$ day^{-1} is physiologically reasonable by deriving a corresponding photosynthetic rate (kg CO_2 m^{-2} (leaf) s^{-1}) over a 16 h day. Assume that the specific leaf area is 20 m^2 (kg structure)$^{-1}$.

13.2. Derive eqn (13.5j) for $d\mu/df_{sh}$ in the teleonomic model, and show that eqn (13.5k) is a necessary and sufficient condition for μ to be maximized.

13.3. Replacing the factor kCN in the structural growth equations for the teleonomic model (eqns (13.4a) and (13.4b)) by the more general bi-substrate Michaelis–Menten equation

$$F(C, N) = \frac{kCN}{1 + C/K_C + N/K_N + CN/K_{CN}}, \quad \text{(E13.3a)}$$

derive the constraint (eqn (13.7b))

$$\frac{\sigma_C}{C(1 + C/K_C)} = \frac{\sigma_N}{N(1 + N/K_N)} \quad \text{(E13.3b)}$$

required for μ to be maximized in balanced exponential growth and show that the partitioning function now becomes (eqn (13.7c))

$$P = \frac{f_r}{f_{sh}} \frac{N(1 + N/K_N)/(N + f_N)}{C(1 + C/K_C)/(C + f_C)}. \quad \text{(E13.3c)}$$

13.4. If the shoot and root have different growth coefficients k, so that the structural growth equations for the teleonomic model (cf. eqns (13.4a) and (13.4b)) become

$$\frac{dW_{sh}}{dt} = k_{sh} CN \lambda_{sh} W_{sh} \quad \text{(E13.4a)}$$

$$\frac{dW_r}{dt} = k_r CN \lambda_r W_r, \quad \text{(E13.4b)}$$

show that eqns (13.4d) for λ_{sh} and λ_r can be replaced by (eqns (13.7d))

$$\lambda_{sh} = \frac{k_r P}{k_{sh} + k_r P} \qquad \lambda_r = \frac{k_{sh}}{k_{sh} + k_r P}. \tag{E13.4c}$$

13.5. Derive the constraint (eqn (13.16d)) for maximizing μ in balanced exponential growth with two mineral nutrients in the teleonomic model:

$$\frac{\sigma_C}{C} = \frac{\sigma_N}{N} + \frac{\sigma_E}{E}. \tag{E13.5a}$$

Show that the partitioning function P is now generalized to

$$P = \frac{f_r}{f_{sh}} \frac{\{(N + f_N)/N + (E + f_E)/E\}^{-1}}{C/(C + f_C)}, \tag{E13.5b}$$

with

$$P = 1 \tag{E13.5c}$$

in balanced exponential growth.

13.6. The mechanistic model using transport resistances given in Section 13.4 has seven parameters σ_C, σ_N, f_C, f_N, r_C, r_N, and k (assuming β is given by eqn (13.11i)). It is difficult to explore the behaviour of a model with so many parameters, and it is usually efficient to reduce the effective number of parameters as much as possible by transforming the variables. Transform the six differential equations of the model (eqns (13.12a)–(13.12f)), after substituting with eqns (13.11c), (13.11d), (13.11h), and (13.11i), to new time and substrate variables defined by

$$t = \frac{\tau}{k f_C f_N} \tag{E13.6a}$$

$$W_{C,sh} = f_C w_{C,sh} \qquad W_{C,r} = f_C w_{C,r} \tag{E13.6b}$$

$$W_{N,sh} = f_N w_{N,sh} \qquad W_{N,r} = f_N w_{N,r}. \tag{E13.6c}$$

Show that there are now effectively four parameters given by

$$\alpha_C = \frac{\sigma_C}{k f_C^2 f_N} \qquad \alpha_N = \frac{\sigma_N}{k f_N^2 f_C}$$
$$\gamma_C = \frac{1}{k f_C f_N r_C} \qquad \gamma_N = \frac{1}{k f_C f_N r_N}. \tag{E13.6d}$$

14
Transpiration by a crop canopy

14.1 Introduction

In many parts of the world, water stress poses the most severe limitation to crop growth. Often, this is due to low levels of rainfall, although areas with similar rainfall patterns may have markedly different losses of water through evaporation and transpiration as a result of other environmental conditions. For example, many coastal regions of Australia have levels of rainfall which are roughly the same as those in the United Kingdom but, owing to the higher solar radiation and temperature, water stress is generally more severe in Australia. Efficient use of water and planning of irrigation scheduling requires a sound understanding of crop water use in relation to the environment. This involves the physics of heat and mass (water vapour) transfer from the crop to the atmosphere, and how these processes are influenced by radiation, temperature, vapour density (or pressure) in the atmosphere, wind speed, and canopy height.

The theory of crop transpiration has received considerable attention during the last few decades, with major contributions being by Penman (1948) and Monteith (1965), resulting in the development of the celebrated Penman–Monteith equation. This approach, which is based on sound physical principles, is seen as the interface between crop physiology studies and environmental physics. There is much to be gained by plant and crop physiologists and modellers from having a good grasp of the advances in the understanding of crop water use made by physicists in recent years. For example, well-watered crops of similar leaf area index and stomatal conductance, but of different heights, may transpire at different rates. This type of response would be difficult to explain without understanding the underlying processes involved.

In this chapter we present the basic theory of crop transpiration as influenced by the environment, based on the Penman–Monteith equation. The theory provides a sound basis for modelling crop water use. Essentially, the approach is to combine the equation for heat transfer between the crop and the bulk air, which results from their temperature difference, with that for evaporation in relation to the vapour concentration difference between the evaporating surface and the bulk air. Both processes involve the crop temperature, but this can be eliminated by combining the two equations. However, it is equally possible to eliminate evaporation and so derive an expression for the crop temperature. Therefore, as a corollary, we also consider the temperature of the canopy in relation to the environmental conditions, crop structure, and water status. As well as providing insight into the response of canopy temperature to environmental conditions, this also allows us to assess the accuracy of the common assumptions used in the derivation and application of the Penman–Monteith equation.

Before considering crop transpiration, it is first necessary to discuss briefly the radiation balance at the crop surface, as this provides the energy for transpiration, and to consider turbulent boundary layer wind flow over the crop.

14.1.1 *Radiation balance at the crop surface*

If J_a and R_L are the solar radiation absorbed and net long-wave radiation transmitted by the canopy, with units W (m^2 ground)$^{-1}$, then the net radiation R_N at the surface of the canopy is given by

$$R_N = J_a - R_L. \tag{14.1a}$$

In Chapter 8 we considered the light relations within the canopy and derived J_a as (eqn (8.7c), p. 204)

$$J_a = \left(\frac{1 - n - m}{1 - m}\right) I_0(1 - e^{-kL}), \tag{14.1b}$$

where L is the leaf area index of the canopy, k is the extinction coefficient, and n and m are the leaf reflection and transmission coefficients respectively. Note that in eqn (14.1b) I_0 refers to total solar radiation and not just the photosynthetically active component; in this case n and m are typically of order 0.2–0.3 (Jones 1983). The factor $(1 - n - m)/(1 - m)$ is often written simply as $1 - r$ where r is a canopy reflection coefficient. r is measured by taking the ratio of the reflected to incident solar radiation although, as the absorbed fraction varies, r will depend on reflection from the canopy and the soil. This may give misleading results if the canopy and soil have different reflectivities. Even for a fully light-intercepting canopy (so that there is no reflection from the soil) using eqn (14.1b) (with $1 - e^{-kL} \approx 1$) has a sounder theoretical base than using a canopy reflection coefficient.

It is instructive to consider typical values of I_0 and R_L. I_0 is obviously strongly dependent on solar elevation and cloud cover, and on a clear summer's day in the United Kingdom is likely to reach values in the region of 800 W (m^2 ground)$^{-1}$; in other parts of the world it may exceed 1000 W (m^2 ground)$^{-1}$. R_L is the net effect of long-wave radiation emitted by the canopy as a function of its temperature, and that emitted from the atmosphere, mainly the CO_2 and water vapour. With no cloud cover, R_L may be around 100 W (m^2 ground)$^{-1}$, falling to less than 10 W (m^2 ground)$^{-1}$ with full cloud cover. Both I_0 and R_L will be subject to considerable variation due mainly to cloud cover, and so it is essential that meteorological records include both short- and long-wave radiation data: short-wave radiation affects photosynthesis (through the visible, photosynthetically active, component), and total radiation drives transpiration. For a more detailed discussion of atmospheric radiation see, for example, Monteith (1973).

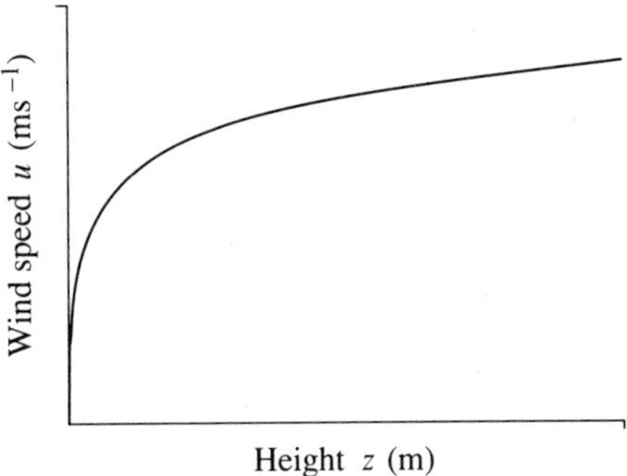

Fig. 14.1. Typical wind speed profile over an open surface (eqn (14.2a)).

14.1.2 *Boundary layer flow*

As air flows over a surface, it experiences drag by that surface. The standard assumption for such air flow is that there is no slip at the interface of the surface and the air, so that at the interface the velocity of the air is zero. Indeed, this boundary condition is almost universal in fluid dynamics, although there are exceptions (for example, ice sheet and glacier flow). It therefore follows that in the region between the free air stream and the surface there will be a wind speed gradient with the wind speed falling to zero, as illustrated in Fig. 14.1, and in this region there will be high shear stresses. This region is termed the boundary layer, and is usually arbitrarily defined as extending from the surface to the height at which the wind speed is 99 per cent of that of the free air stream.

In the boundary layer, the wind speed $u(z)$ (m s^{-1}), where z (*m*) is vertical height, is well represented by an equation of the form

$$u(z) = \frac{u^*}{\kappa} \ln\left(\frac{z + \zeta_m}{\zeta_m}\right) \tag{14.2a}$$

where u^* is the friction velocity (m s^{-1}), related to the shear stress in the boundary layer, κ is von Karman's constant and is generally taken to be 0.4 (Campbell 1977, p. 38), and ζ_m is the momentum roughness parameter (m). These parameters are discussed in more detail in Section 14.3, where the theoretical basis for eqn (14.2a) and the more general form for wind flow over a crop (eqn (14.2d) below) are presented. In the meantime, note that u^* represents the magnitude of the wind speed and ζ_m the vertical profile. This equation is only likely to be valid over relatively rough surfaces and z should be regarded as the mean height above the surface. Of course, $u(z)$ as defined by eqn (14.2a) continues to increase as z increases, so that it never approaches a constant value. However,

$$\frac{1}{u}\frac{\mathrm{d}u}{\mathrm{d}z} = \frac{1}{(z + \zeta_m)\ln[(z + \zeta_m)/\zeta_m]}, \tag{14.2b}$$

and ζ_m is typically an order of magnitude smaller than the roughness elements of the surface (Thom 1975), so that, other than very close to the surface at $z = 0$, the relative wind speed gradient will be small. For example, at $z = 2$ m with $\zeta_m = 0.005$ m (0.5cm),

$$\frac{1}{u}\frac{\mathrm{d}u}{\mathrm{d}z} = 0.08, \tag{14.2c}$$

so that $u(z)$ is increasing at less than 10 per cent m^{-1}.

When the wind blows over a crop, the general form of eqn (14.2a) still applies, but the profile is displaced vertically, so that now

$$u(z) = \frac{u^*}{\kappa}\ln\left(\frac{z + \zeta_m - d}{\zeta_m}\right), \tag{14.2d}$$

where d is the zero-plane displacement with

$$u(d) = 0. \tag{14.2e}$$

Equation (14.2d) is only valid for $z > h$, where h (m) is the crop height. However, the zero-plane displacement d has been empirically determined as

$$d = 0.64h \tag{14.2f}$$

(Campbell 1977, p. 38), so that the plane $z = d$ where $u(d) = 0$ is interpreted as the apparent sink of momentum.

Equation (14.2d) is the form proposed by Campbell (1977), although several other authors (e.g. Monteith 1973, Thom 1975) describe $u(z)$ by

$$u(z) = \frac{u^*}{\kappa}\ln\left(\frac{z - d}{\zeta_m}\right), \tag{14.2g}$$

so that now the zero-plane displacement is $\zeta_m + d$, with

$$u(\zeta_m + d) = 0. \tag{14.2h}$$

Since, as mentioned above, ζ_m is generally an order of magnitude smaller than the roughness elements of the surface, and hence d, eqns (14.2d) and (14.2g) are likely to give equally good fits to data. We adopt Campbell's (1977) notation (eqn (14.2d)) as this defines ζ_m and d in the clearest manner.

The above equations for the wind profile, and therefore the definition of the boundary layer itself, are unlikely to hold in the region where the wind passes from one type of surface to another. There will be a transition from the effects of each surface, and the height at which the influence of one surface on the boundary layer of another is apparent increases with distance from the interface. Consequently, it is important that measurements are taken below this height. The distance that the wind has blown over the uniform surface is termed the fetch, and it is generally assumed that the wind speed is virtually independent of surrounding surfaces to heights given by $0.01 \times$ fetch (Campbell 1977). Thus measurements made at 2 m above a crop are likely to be independent of edge effects provided that they are taken about 200 m downwind of the edge of the crop.

For very low wind speeds in the bulk air stream, the gradient of wind speed in the boundary layer will be small and so the flow will be laminar, which means that there is no vertical air motion. As the wind speed in the air stream increases, the wind speed gradient also increases (since u^* is related to the wind speed (see Section 14.3)) and the flow becomes turbulent, in which case eddies are created and there is vertical transport of air. These eddies are essentially swirls of air, which can be seen in smoke plumes, and are continually being created and dissipated in turbulent air. Whether the flow is turbulent or laminar depends on the wind speed and the roughness of the surface (Section 18.6.3, p. 531). Jones (1983) has calculated that, for a real leaf only 1 cm wide, the flow will be turbulent for wind speeds greater than 0.76 m s^{-1}.

In practice, therefore, most atmospheric air flow is turbulent and, as Monteith (1973) emphasizes, turbulent mixing is essential for creating a microclimate which is capable of sustaining life. Without this process, there would be enormous diurnal temperature fluctuations which plants could not endure. Furthermore, there would be large gradients of CO_2, water vapour, and oxygen, so that animals and humans would not be able to survive either (Monteith, 1973). As well as turbulent transport, transport in the boundary layer is also driven by buoyancy, which results from temperature gradients. In stable air, where vertical temperature gradients are typically less than 1 °C m^{-1} (Monteith 1973), buoyancy effects are small (Section 4.5.4, p. 110).

In the following analysis of conductance of momentum, heat, and water vapour, we restrict attention to pure turbulent flow with no buoyancy effects. These conditions are widely applicable, although more comprehensive treatments have been presented (e.g. Monteith 1973, Campbell 1977).

14.2 The Penman–Monteith equation

We now derive the Penman–Monteith equation for crop transpiration, and subsequently use it to examine the influence of the environment on transpiration. The analysis is restricted to the steady state, so that the flux density of water vapour in the boundary layer is independent of height. This means that rapid fluctuations in the environmental conditions (such as net radiation and wind speed) are ignored, and only the mean values over some time period are used. Wind speed is likely to be the environmental quantity most subject to rapid fluctuations, and Monteith (1973) recommends averaging data over a period of time which is substantially longer than the period of fluctuations, usually between 10 and 30 min. The other important assumption is that the canopy is considered to consist of only one layer, and therefore the temperatures at the evaporating surface and the region of heat exchange between the canopy and the air are equal. This assumption is generally thought not to lead to serious error, particularly for reasonably dense canopies where there is relatively little heat flux from the soil. Throughout the analysis, we use units of kg water $(m^2 \text{ ground})^{-1} s^{-1}$ for transpiration rates, although these are readily converted to the other common units of millimetres of water per unit of time (Exercise 14.1).

14.2.1 *Derivation*

The energy balance of the crop involves the following components:

1. the transpiration rate E (kg water $(m^2 \text{ ground})^{-1} s^{-1}$);
2. heat transfer H between the canopy and the air (W $(m^2 \text{ ground})^{-1}$), which is commonly referred to as the sensible heat flux;
3. soil heat flux G into the ground (W $(m^2 \text{ ground})^{-1}$);
4. heat storage by the canopy;
5. metabolic processes of photosynthesis and respiration.

Of these, the last two are generally negligible and are usually ignored. For a further discussion of these components, see Jones (1983).

The energy available to the crop is

$$\phi_N = R_N - G, \tag{14.3a}$$

where the soil heat flux is positive during much of the day and may range from 2 per cent of R_N for dense canopies with complete ground cover to values greater than 30 per cent of R_N in sparse canopies (Jones 1983). Thus, neglecting components 4 and 5 above, the energy balance for the crop can be written

$$\phi_N = H + \lambda E, \tag{14.3b}$$

where λ is the latent heat of vaporization of water (J kg^{-1}) and is given in Table 14.3 below for a range of temperatures.

The transpiration rate across the boundary layer is given by Fick's law of diffusion, and the sensible heat flux by Fourier's law of heat transfer. However, the latter can be written in terms of the diffusion equation and so, adopting the diffusion equation in its integrated form (eqn (4.8d), p. 95), we can write

$$H = \rho c_p (T_l - T_a) g_H \tag{14.3c}$$

and

$$E = (\rho_{vl} - \rho_{va}) g_W, \tag{14.3d}$$

where ρ is the density of dry air (kg m^{-3}), c_p is the specific heat capacity of dry air (J kg^{-1} K^{-1}), T_l and T_a are the leaf and bulk air temperatures respectively (K), g_H is the boundary layer conductance for heat (m s^{-1}), ρ_{vl} and ρ_{va} are the vapour densities at the evaporating surfaces within the leaves and in the bulk air respectively (kg water m^{-3}), and g_W is the boundary layer conductance for water vapour (m s^{-1}). Note that vapour density is often referred to as absolute humidity, and the ratio of vapour density to saturated vapour density (see eqn (14.3g) below) is the relative humidity.

The conductances g_H and g_W in relation to crop structure and the environmental conditions will be considered in Section 14.3. However, one important feature to note here is that these conductances are not constant, but are functions of height. This is apparent from eqns (14.3c) and (14.3d), which can be written as

$$g_H = \frac{H}{\rho c_p(T_l - T_a)} \tag{14.3e}$$

and

$$g_W = \frac{E}{\rho_{vl} - \rho_{va}}. \tag{14.3f}$$

Now, for steady state (which we are considering here) both H and E are independent of height, but the air temperature T_a and the vapour density ρ_{va} will vary with height (just as the wind speed profile is a function of height). Consequently, both g_H and g_W will vary with height, and therefore it is essential that these conductances, as well as all other varying quantities, are defined at a given reference height—this point is considered further in Section 14.3.

In using the density and specific heat capacity of dry air in eqn (14.3c), the heat content of the actual air (dry air plus water vapour) is assumed to be equal to that of dry air alone. In Table 14.1 the ratio of the saturation vapour density of water to the density of dry air is shown for a range of temperatures, and it is clear that the contribution of the water to the heat content of the air will be small, so that errors associated with this assumption are likely to be unimportant.

The object of the following analysis is to express transpiration in terms of readily measurable environmental quantities, which requires eliminating the leaf temperature. It is necessary to assume that the vapour at the evaporating surface is saturated. At a water surface, the relative humidity h_r is given by (Exercise 14.2)

$$h_r = \frac{\rho_v}{\rho_v'} = \exp\left(\frac{\psi M}{RT}\right), \tag{14.3g}$$

where ρ_v and ρ_v' (kg water m^{-3}) are the actual and saturated vapour densities respectively, R is the gas constant (8314 J (kg mol)$^{-1}$ K^{-1}), T is the absolute temperature at the evaporating surface (K), M is the relative molecular mass of water (18 kg (kg mol)$^{-1}$), and ψ is the water potential (J kg^{-1}). (The terms relative molecular mass, kilogram mole, and water potential, and their units, are discussed in Chapter 2.) The water potential in leaves is subject to considerable

Table 14.1. Ratio of saturation vapour density ρ_{va}' (kg water m^{-3}) to the density ρ of dry air (kg air m^{-3}) as a function of temperature T

T (°C)	0	5	10	15	20	25	30
ρ_v'/ρ	0.004	0.005	0.008	0.010	0.014	0.020	0.026

Data for calculation taken from Table 14.3 below.

Table 14.2. Relative humidities h_r at the evaporating surface within the leaf for various leaf water potentials ψ (J kg^{-1}) as defined by eqn (14.3g)

ψ	-200	-500	-1000	-2000	-4000
h_r	0.999	0.996	0.993	0.985	0.971

$T = 293$ K (20 °C).

variation, depending on the water status of the crop and the evaporative demand. This is discussed in Chapter 15, but for the present purpose it is sufficient to quote typical values: under normal growing conditions, the plant leaves might have water potentials of -1000 to -2000 J kg^{-1} during the day, and -200 to -300 J kg^{-1} at night. If the water potential falls to around -3000 to -4000 J kg^{-1} then the plant is likely to be stressed to such an extent that it wilts and dies. We should emphasize that these values are subject to considerable variation; for example, some plants may wilt at water potentials higher than -2000 J kg^{-1}. In Table 14.2, the relative humidity is presented for this range of values of the leaf water potential as defined by eqn (14.3g) (leaf temperature has a negligible effect on h_r and is taken to be 20 °C); it is clear that for all the water potentials in this range the relative humidity is virtually unity, so that the assumption that the air at the evaporating surface is saturated is reasonable.

Thus we can write

$$\rho_{\mathrm{vl}} = \rho_{\mathrm{v}}'(T_{\mathrm{l}}), \qquad (14.3\mathrm{h})$$

where $\rho_{\mathrm{v}}'(T_{\mathrm{l}})$ denotes the saturation vapour density as a function of leaf temperature T_{l}. Some values of $\rho_{\mathrm{v}}'(T_{\mathrm{l}})$ are presented in Table 14.3 below. Equation (14.3d) now becomes

$$E = [\rho_{\mathrm{v}}'(T_{\mathrm{l}}) - \rho_{\mathrm{va}}]g_{\mathrm{W}}. \qquad (14.3\mathrm{i})$$

We are now in a position to eliminate T_{l} from eqns (14.3c) and (14.3i). To do this, we take the first two terms in the Taylor series expansion (eqn (1.7e), p. 20) for $\rho_{\mathrm{v}}'(T_{\mathrm{l}})$ about T_{a}:

$$\rho_{\mathrm{v}}'(T_{\mathrm{l}}) = \rho_{\mathrm{v}}'(T_{\mathrm{a}}) + s(T_{\mathrm{l}} - T_{\mathrm{a}}), \qquad (14.4\mathrm{a})$$

where

$$s = \frac{\mathrm{d}\rho_{\mathrm{v}}'}{\mathrm{d}T}(T = T_{\mathrm{a}}) \qquad (14.4\mathrm{b})$$

is the slope of the saturation vapour density with respect to temperature at temperature T_{a}. Sometimes the parameter s is defined at some temperature between T_{l} and T_{a}, but this is of little value since the object of the exercise is to eliminate the leaf temperature T_{l}. Unless the leaf and air temperatures differ by a large margin, eqn (14.4a) will provide a good approximation for $\rho_{\mathrm{v}}'(T_{\mathrm{l}})$. Using eqn (14.4a), eqn (14.3i) becomes

$$E = [s(T_{\mathrm{l}} - T_{\mathrm{a}}) + \Delta\rho_{\mathrm{va}}]g_{\mathrm{W}}, \qquad (14.4\mathrm{c})$$

where

$$\Delta\rho_{va} = \rho_v'(T_a) - \rho_{va} \tag{14.4d}$$

is the atmospheric vapour density deficit (kg water m^{-3}). Substituting for $T_l - T_a$ from eqn (14.3c), eqn (14.4c) becomes

$$E = \left(\frac{sH}{\rho c_p g_H} + \Delta\rho_{va}\right) g_W. \tag{14.4e}$$

We now eliminate H using the energy balance equation (eqn (14.3b)) to obtain (after some algebra)

$$E = \frac{s\phi_N + \lambda\gamma g_H \Delta\rho_{va}}{\lambda(s + \gamma g_H/g_W)}, \tag{14.4f}$$

where

$$\gamma = \frac{\rho c_p}{\lambda} \tag{14.4g}$$

is commonly termed the psychrometric constant. This can be slightly misleading, since γ is not actually constant because of the dependence of ρ, c_p and λ on temperature and atmospheric pressure. It is perhaps more appropriate, therefore, to use the term psychrometric parameter rather than psychrometric constant. Some values of γ for different temperatures at standard pressure are presented in Table 14.3 below.

Now, the water vapour transfer pathway includes movement out of the substomatal cavities from the evaporating surfaces to the leaf surfaces and then from the leaf surfaces across the boundary layer to the bulk air stream. We can therefore partition the conductance for water vapour into substomatal and boundary layer components, with conductances g_c and g_a respectively, where

$$\frac{1}{g_W} = \frac{1}{g_a} + \frac{1}{g_c}. \tag{14.4h}$$

Here the boundary layer conductance g_a of the canopy is the sum of the conductances of the water vapour pathway from the leaf surfaces over 1 m^2 of ground to the reference height above the canopy. An expression for g_a is derived in Section 14.3, but it is sufficient at this stage to use typical values where necessary and to state the following:

1. g_a increases with crop height;
2. g_a increases with wind speed;
3. g_a decreases with height above the canopy.

This third point is a consequence of eqn (14.3f), and its is worth emphasizing once again that all atmospheric measurements must be made at a fixed reference height. The canopy conductance g_c is the corresponding sum from the water surfaces within the canopy to the leaf surfaces. Since the pathways for water transfer from the evaporating surface within the leaves to the leaf surfaces through the stomata are in parallel, it follows that

$$g_c = \sum_i (g_{c,i}{}^{ad} + g_{c,i}{}^{ab})L_i, \qquad (14.4i)$$

where $g_{c,i}{}^{ad}$ and $g_{c,i}{}^{ab}$ are the adaxial (upper surface) and abaxial (lower surface) stomatal conductances of leaf area index components L_i.

For turbulent flow, transport of both heat and water vapour is dominated by eddy diffusion so that the processes are the same and (Jones 1983)

$$g_H = g_a. \qquad (14.4j)$$

Hence, combining eqns (14.4f), (14.4h), and (14.4j), the transpiration rate from the canopy is given by

$$E = \frac{s\phi_N + \lambda\gamma g_a \Delta\rho_{va}}{\lambda[s + \gamma(1 + g_a/g_c)]}. \qquad (14.4k)$$

This is the Penman–Monteith equation and describes the transpiration rate as a function of the energy ϕ_N available for evaporation, the vapour density deficit $\Delta\rho_{va}$, the canopy and boundary layer conductances g_c and g_a, and the physical parameters s, λ, and γ. All these parameters, apart from the conductances, are readily measurable or available from standard tables. The physical parameters required for the Penman–Monteith equation are presented in Table 14.3 for a range of temperatures. Temperature will affect E through its influence on the

Table 14.3. Temperature-dependent physical quantities required for the Penman–Monteith equation (eqn (14.4 k))

T (°C)	ρ (kg m^{-3})	λ (MJ kg^{-1})	γ ($10^{-3}\,\text{kg m}^{-3}\,\text{K}^{-1}$)	ρ'_v ($10^{-3}\,\text{kg m}^{-3}$)	s ($10^{-3}\,\text{kg m}^{-3}\,\text{K}^{-1}$)
0	1.29	2.50	0.521	4.9	0.33
5	1.27	2.49	0.515	6.8	0.45
10	1.25	2.48	0.509	9.4	0.60
15	1.23	2.47	0.503	12.8	0.78
20	1.20	2.45	0.495	17.3	1.01
25	1.18	2.44	0.488	23.1	1.30
30	1.16	2.43	0.482	30.4	1.65
35	1.15	2.42	0.480	39.7	2.07
40	1.13	2.41	0.474	51.2	2.57

After Campbell (1977, Tables A1, A3) and Jones (1983, Appendix 3). ρ is the density of dry air, λ is the latent heat of vaporization of water, ρ'_v is the saturation vapour density of water, $s = d\rho'_v/dT$, and the psychrometric parameter γ is calculated from eqn (14.4g) using c_p ($T = 20\,°\text{C}$) $= 1010\,\text{J kg}^{-1}\,\text{K}^{-1}$ (c_p is the specific heat capacity of water, and is assumed constant over the temperature range considered). Values correspond to standard pressure, and should be adjusted for different pressures.

saturation vapour density of the air $\rho_v'(T_a)$. (Note that these parameter values correspond to standard pressure, and must be adjusted for deviations from this.) The canopy conductance g_c will depend on the water status of the crop through its influence on the stomatal conductance, as discussed in Chapter 15 (eqns (15.21d) (15.22a), and (15.22b); an expression for the boundary layer conductance g_a is derived in Section 14.3.

Before considering the behaviour of eqn (14.4k), we should point out that vapour pressure rather than vapour density is frequently used in the treatment of evaporation and transpiration (e.g. Monteith 1973, Jones 1983). It can be shown (Exercise 14.3) that vapour density and pressure are related by

$$\rho_v = \rho\varepsilon \frac{p_v}{P - p_v}, \tag{14.5a}$$

where P is the total atmospheric pressure (dry air plus water vapour) (Pa), p_v is the vapour pressure (or partial vapour pressure) (Pa), ρ, as defined above, is the density of dry air (kg m^{-3}), and ε is the ratio of the relative molecular mass of water to the relative molecular mass of dry air ($\varepsilon = 0.622$ (Exercise 14.3)). In practice

$$p_v/P \ll 1, \tag{14.5b}$$

so that eqn (14.5a) becomes

$$\rho_v = \frac{\rho\varepsilon p_v}{P}. \tag{14.5c}$$

When this expression is used, the Penman–Monteith equation (eqn(14.4k)) becomes

$$E = \frac{\tilde{s}\phi_N + c_p\rho g_a \Delta p_v}{\lambda[\tilde{s} + \tilde{\gamma}(1 + g_a/g_c)]}, \tag{14.5d}$$

where

$$\Delta p_v = p_v' - p_v \tag{14.5e}$$

is the atmospheric vapour pressure deficit and p_v' is the saturation vapour pressure. The parameters $\tilde{s}$ and $\tilde{\gamma}$ are equivalent to s and γ in eqn (14.4k), and are defined by (cf. eqns (14.4b) and (14.4g))

$$\tilde{s} = \frac{dp_v'}{dT}(T = T_a) \tag{14.5f}$$

and

$$\tilde{\gamma} = \frac{c_p P}{\gamma\varepsilon} \tag{14.5g}$$

respectively, with units of Pa K^{-1}.

Equations (14.4k) and (14.5d) are therefore equivalent expressions for the crop transpiration rate E, subject to the approximation (14.5b) for eqn (14.5a). Perhaps the best argument for using vapour pressure rather than vapour density is that this gives the number of molecules per unit volume more directly, and is independent of relative molecular mass. Both expressions are subject to variation in response to atmospheric pressure and temperature. We prefer to use vapour density, which is consistent with Campbell (1977), as the basic equation for evaporation (eqn (14.3d)) is defined in terms of this quantity.

14.2.2 *Illustrations*

First, note that E can be separated into two components

$$E_e = \frac{s\phi_N}{\lambda[s + \gamma(1 + g_a/g_c)]} \tag{14.6a}$$

and

$$E_d = \frac{\gamma g_a \Delta\rho_{va}}{s + \gamma(1 + g_a/g_c)}, \tag{14.6b}$$

where E_e is energy driven and E_d is diffusion driven. Transpiration will be dominated by the energy term when

$$\lambda\gamma g_a \Delta\rho_{va} \ll s\phi_N, \tag{14.6c}$$

which will occur in humid still environments with low crops. Conversely, the diffusion term will dominate when

$$\lambda\gamma g_a \Delta\rho_{va} \gg s\phi_N, \tag{14.6d}$$

in which case conditions are likely to be windy with low incident radiation and the crop tall. Figures 14.2(a) and 14.2(b) show E in response to variation of either ϕ_N or $\Delta\rho_{va}$, with E_e and E_d indicated. It is apparent from these illustrations that, while the partitioning of transpiration into diffusion- and energy-driven components is valuable in helping understand the processes involved, it is unlikely that either of eqns (14.6a) or (14.6b) can be used with confidence in place of the full Penman–Monteith equation (eqn (14.4k)) for extended periods of crop growth.

The effect of varying the canopy and boundary layer conductances on transpiration for given radiation and humidity conditions is illustrated in Figs 14.3(a) and 14.3(b). It is clear that increasing the canopy conductance g_c causes E to increase. The effect of increasing the boundary layer conductance g_a is not so straightforward, since an increase in g_a causes E_e to fall and E_d to rise. As a result, at low values of g_c, E actually decreases as g_a increases, and at high values of g_c this trend is reversed. Indeed, it is apparent that there is a critical value of g_c for which the transpiration rate E is independent of variation in the boundary layer

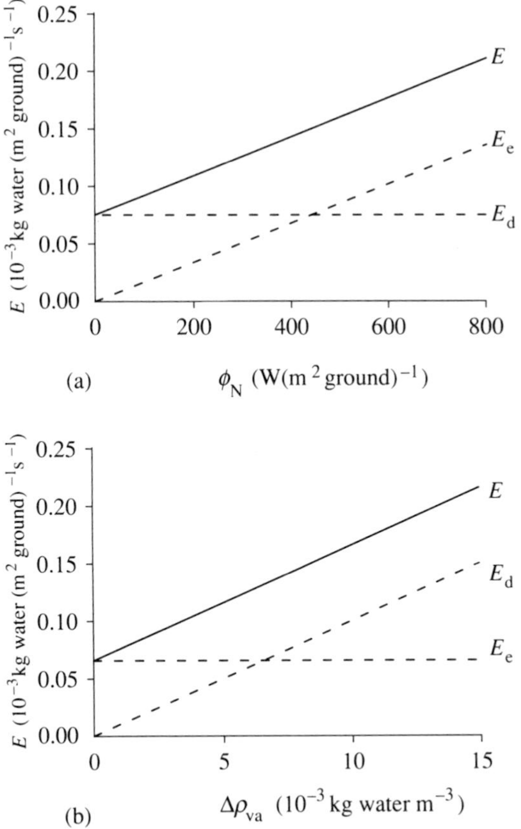

Fig. 14.2. Transpiration rate E (—), given by eqn (14.4k), illustrated as a function of (a) available energy ϕ_N with $\Delta\rho_{va} = 0.0075$ kg m^{-3} and (b) vapour density deficit $\Delta\rho_{va}$, with $\phi_N = 400$ W m^{-2}. The separate energy- and diffusion-driven components E_e and E_d (eqns (14.6a) and (14.6(b) respectively) are also shown (- - -). In both illustrations $g_a = 0.05$ m s^{-1} and $g_c = 0.025$ m s^{-1}. The other parameters are taken from Table 14.3 for $T = 20\,°$C. (Note that the value of $\Delta\rho_{va}$ used here corresponds to $h_r = 0.57$.)

conductance g_a, and this value is (Exercise 14.4)

$$g_c = \frac{s\phi_N}{\lambda\Delta\rho_{va}(s+\gamma)}, \tag{14.6e}$$

in which case E is given by

$$E = \frac{s\phi_N}{\lambda(s+\gamma)}. \tag{14.6f}$$

(For the parameters used in Fig. 14.3(b) the value of g_c satisfying eqn (14.6e) is 0.015 m s^{-1}, and is shown in the figure.) Consequently, we cannot talk of a

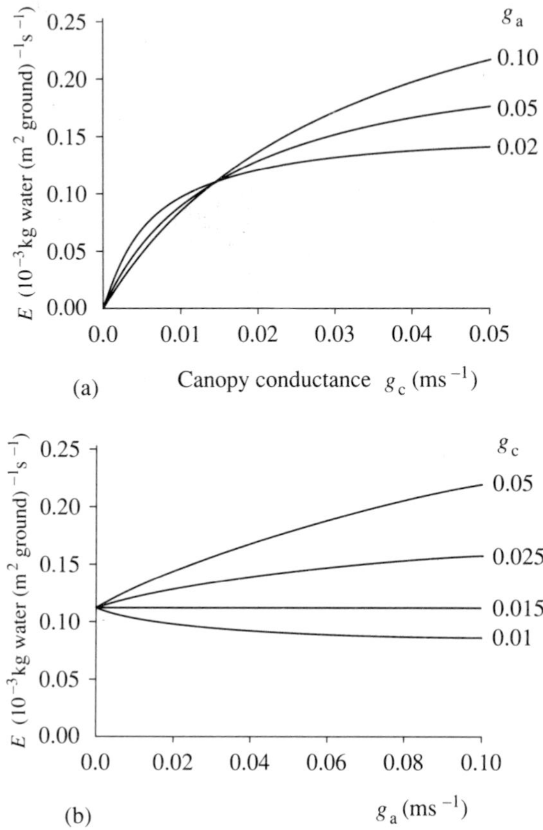

Fig. 14.3. Transpiration rate E, given by eqn (14.4k), illustrated as a function of (a) canopy conductance g_c with $g_a = (0.02, 0.05, 0.1)$ m s^{-1} as indicated and (b) boundary layer conductance g_a with $g_c = (0.01, 0.015, 0.025, 0.05)$ m s^{-1}. In both illustrations $\phi_N = 400$ W m^{-2} and $\Delta\rho_{va} = 0.0075$ kg m^{-3}. As for Fig. 14.2, the remaining parameters are taken from Table 14.3 for $T = 20\,°$C. Note that, in Fig. 14.3(b), with these parameter values, when $g_c = 0.015$ m s^{-1}, E is independent of variation in g_a (see text for details).

response of transpiration to any environmental factor in isolation, but knowledge of the complete environment is required in order to understand crop transpiration.

Another important feature to note from Fig. 14.3(a) is that, owing to the influence of crop structure on the boundary layer conductance g_a, tall crops may transpire at different rates than shorter crops of the same leaf area index in identical environmental conditions. (Recall that g_a increases with crop height.) In particular, note that for freely transpiring crops where g_c is large, the transpiration rate of the taller crop may be considerably greater than that of the shorter crop.

14.2.3 *Potential transpiration*

If the canopy conductance is so high that there is no resistance to water flow from the evaporating surface to the leaf surface compared with the resistance from this surface to the bulk air stream, in which case

$$g_c \gg g_a, \tag{14.7a}$$

the transpiration rate is given by (from eqn (14.4k))

$$E_p = \frac{s\phi_N + \lambda\gamma g_a \Delta\rho_{va}}{\lambda(s + \gamma)} \tag{14.7b}$$

and is termed the potential transpiration rate. Clearly, E_p gives an upper limit to E. Equation (14.7b) will be valid when the stomata are wide open and the boundary layer conductance is low, or for wet surfaces. Note that E_p is a monotonically increasing function of g_a and hence wind speed (since g_a increases with wind speed). Equation (14.7b) will apply to any evaporating surface that is wet and explains why, when wet, one feels colder in a breeze than in still air.

14.2.4 *Dew fall*

For dew to fall, it follows that the canopy (or ground) surface is wet. In this case eqn (14.7b) applies and dew will fall if E_p is negative, which occurs when

$$s\phi_N < -\lambda\gamma g_a \Delta\rho_{va}. \tag{14.7c}$$

This is likely to occur at night when solar radiation is zero and there is a net efflux of long-wave radiation.

14.3 Boundary layer conductance

We now derive an expression for the boundary layer conductance g_a in terms of the wind speed profile above the canopy and the canopy structure. The following analysis is for pure turbulent flow with no buoyancy effects (see Section 14.1.2), which is perhaps the simplest set of realistic conditions for which g_a can be derived, is widely applicable, and should provide the reader with a useful background with which to pursue the subject further. The equations for turbulent transfer are analogous to the one-dimensional diffusion equation (eqn (3.8d)) and are, for the flux density of momentum, heat, and water vapour (Campbell 1977, p. 37),

$$\tau = K_m \rho \frac{du}{dz} \tag{14.8a}$$

$$H = -K_H \rho c_p \frac{dT}{dz} \tag{14.8b}$$

$$E = -K_v \frac{d\rho_v}{dz} \tag{14.8c}$$

respectively, where K_m, K_H, and K_v are the eddy viscosity, eddy thermal diffusivity, and eddy vapour diffusivity, with units $m^2\ s^{-1}$ (as for diffusion coefficients), τ is the horizontal shear stress, or drag force, caused by the horizontal motion of the air, and z (m) is the vertical height above ground level. These equations for turbulent transfer are analogous to those for molecular transport, which are respectively Newtons's law of viscosity, Fick's law of diffusion (see Chapter 4), and Fourier's law of heat transfer. Equation (14.8a) is perhaps made clear by noting that momentum flux density has the same units as shear stress (kg m^{-1} s^{-2}). Strictly, the density of the air ρ should appear inside the derivative terms in eqns (14.8a) and (14.8b), but changes in ρ are assumed to be negligibly small within the boundary layer. (Note that, as in the derivation of the Penman–Monteith equation, we use the density and specific heat capacity of dry air.)

For a boundary layer in the steady state, it is assumed that τ, H and E do not vary with height, so that the parameters K_m, K_H, and K_v increase with height. Furthermore, we would expect τ, H, and E to increase with wind speed and the surface roughness features of the canopy. To incorporate the wind speed effect, it is convenient to define the 'friction velocity' u^* (m s^{-1}) by

$$u^* = (\tau/\rho)^{1/2}. \tag{14.8d}$$

This parameter, which is constant in the boundary layer (since τ is constant), characterizes the wind speed effects through its influence on the shear stress τ. Assuming that the K coefficients are directly proportional to u^* and increase linearly with height, we can now write (Campbell 1977, p. 38)

$$K_m = \kappa u^*(z + \zeta_m - d) \tag{14.8e}$$

$$K_H = \kappa u^*(z + \zeta_H - d) \tag{14.8f}$$

$$K_v = \kappa u^*(z + \zeta_v - d), \tag{14.8g}$$

where (as mentioned in Section 14.1.2) κ is known as von Karman's constant, with value empirically determined as 0.4 (Campbell 1977, p. 38), d is the zero-plane displacement, which is the apparent sink of momentum so that $u(d) = 0$, and ζ_m, ζ_H, and ζ_v are roughness parameters which indicate the effects of the surface on turbulent transfer. The derivation of these parameters will be discussed shortly. Equations (14.8e)–(14.8g), while being physically reasonable, may seem somewhat arbitrary, and this is discussed further below. However, it is first necessary to proceed and derive the profiles of wind speed, temperature, and water vapour within the boundary layer, and subsequently derive the boundary layer conductance.

Combining eqns (14.8a)–(14.8c) with eqns (14.8d)–(14.8g) and integrating from height $z = d$ to the reference height $z = Z$, where by definition

$$T(Z) = T_a \qquad \rho_v(Z) = \rho_{va} \qquad u(Z) = u_a, \tag{14.9a}$$

gives

$$u_a = \frac{u^*}{\kappa}\ln\left(\frac{Z + \zeta_m - d}{\zeta_m}\right) \tag{14.9b}$$

$$T_a = T(d) - \frac{H}{\rho c_p \kappa u^*}\ln\left(\frac{Z + \zeta_H - d}{\zeta_H}\right) \tag{14.9c}$$

$$\rho_{va} = \rho_v(d) - \frac{E}{\kappa u^*}\ln\left(\frac{Z + \zeta_v - d}{\zeta_v}\right). \tag{14.9d}$$

Equation (14.9b) was introduced in Section 14.1.2 (eqn (14.2d)), and it is clear that u^* relates to the magnitude of the wind speed and the logarithmic term gives the variation with height.

Now consider $T(d)$ and $\rho_v(d)$. As we have stated, the canopy is assumed to comprise a single layer, which implies that

$$T(d) = T_l. \tag{14.9e}$$

The definition of $\rho_v(d)$ is not so straightforward, because vapour transport involves the two processes of diffusion from the evaporating surfaces within the leaves (which are saturated at density $\rho_v'(T_l)$) to the external leaf surfaces and then turbulent transport across the boundary layer to the bulk air stream. The conductances for these two processes are the canopy and boundary layer conductances g_c and g_a respectively, and, since eqn (14.9d) has been derived for turbulent transport only, $\rho_v(d)$ therefore corresponds to the vapour density at the external leaf surfaces.

We can now evaluate g_a. First, since we are considering steady state flow,

$$E = [\rho_v'(T_l) - \rho_v(d)]g_c = [\rho_v(d) - \rho_{va}]g_a, \tag{14.9f}$$

and hence

$$g_a = \frac{E}{\rho_v(d) - \rho_{va}}. \tag{14.9g}$$

Combining this with eqn (14.9d) gives

$$g_a = \frac{\kappa u^*}{\ln[(Z + \zeta_v - d)/\zeta_v]}, \tag{14.9h}$$

which, using eqn (14.9b) to eliminate u^*, becomes

$$g_a = \frac{\kappa^2 u_a}{\ln[(Z + \zeta_v - d)/\zeta_v]\ln[(Z + \zeta_m - d)/\zeta_m]}. \tag{14.9i}$$

This expression for g_a has been derived using the equation for vapour transport (eqn (14.9g)). We can also derive g_a from the heat flux equation (eqn (14.3c)) which, recalling that $g_H = g_a$ for turbulent flow (eqn (14.4j)), can be written

$$g_a = \frac{H}{\rho c_p(T_l - T_a)} \tag{14.9j}$$

and on substitution into eqn (14.9c) yields

$$g_a = \frac{\kappa u^*}{\ln[(Z + \zeta_H - d)/\zeta_H]}, \tag{14.9k}$$

so that now, using eqn (14.9b) to eliminate u^*, g_a becomes

$$g_a = \frac{\kappa^2 u_a}{\ln[(Z + \zeta_H - d)/\zeta_H]\ln[(Z + \zeta_m - d)/\zeta_m]}. \tag{14.9l}$$

We now have two expressions for g_a—eqns (14.9i) and (14.9l)—which involve the roughness parameters for vapour and heat exchange respectively, and from which it follows that

$$\zeta_H = \zeta_v = \zeta \qquad \text{say}, \tag{14.9m}$$

and hence the boundary layer conductance can be written

$$g_a = \frac{\kappa^2 u_a}{\ln[(Z + \zeta - d)/\zeta]\ln[(Z + \zeta_m - d)/\zeta_m]}. \tag{14.9n}$$

The parameters κ, ζ, ζ_m, and d are determined empirically, and Campbell (1977, pp. 38–9) quotes the values

$$\kappa = 0.4 \qquad \zeta = 0.026\text{h} \qquad \zeta_m = 0.13h \qquad d = 0.64h \tag{14.9o}$$

where, as defined above, h (m) is the crop height. (This expression for d was given in Section 14.1.2) Note that u^* does not need to be defined in order to define g_a, but just the wind speed u_a (at the reference height $z = Z$).

The wind speed profile u_a (eqn (14.9b)) and the canopy conductance g_a (eqn (14.9n)) obtained using these relationships, with the values

$$u^* = 0.3 \text{ m s}^{-1} \qquad h = (0.5, 1) \text{ m}, \tag{14.10a}$$

are shown in Fig. 14.4(a) and Fig. 14.4(b) as functions of the reference height $z = Z$. Since these quantities vary with the reference height, we emphasize once again that it is essential that all measurements be made at a fixed reference height, and we cannot talk of, for example, a boundary layer conductance without defining the reference height.

The analysis presented in this chapter is for steady state air flow, so that E is constant throughout the boundary layer. Therefore when g_a is substituted in the Penman–Monteith equation (eqn (14.4k)), the right-hand side of that equation must be independent of the reference height $z = Z$. This is achieved by variation of $\Delta\rho_{va}$ with height. In the simple case where $E = E_p$ (eqn (14.7b)), it follows that

$$g_a \Delta\rho_{va} = \text{constant}. \tag{14.10b}$$

For non-potential evaporation ($E \neq E_p$) the vertical variation of $\Delta\rho_{va}$ will not be so straightforward, since the canopy conductance will affect the vapour density at the canopy surface and hence the vapour density deficit. However, provided that the wind speed and vapour density deficit are measured at a fixed reference height, the analysis is still valid.

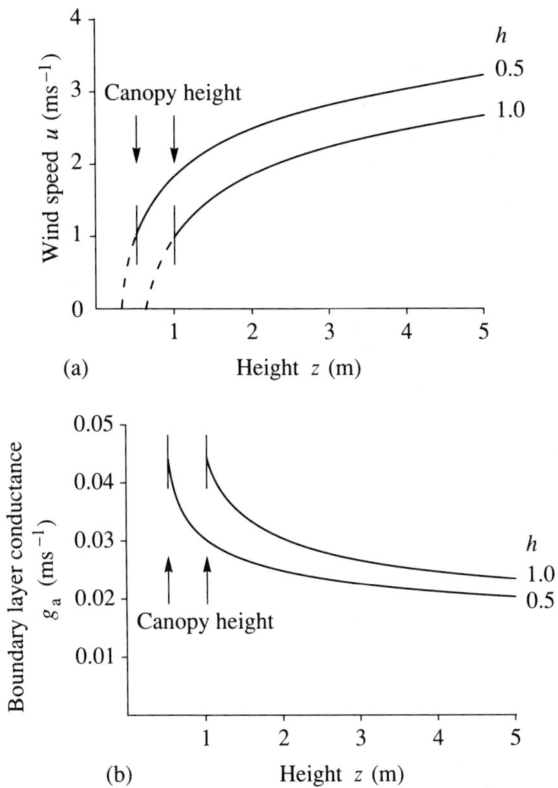

Fig. 14.4. (a) Wind speed profile u_a, given by eqn (14.9b), and (b) boundary layer conductance g_a, given by eqn (14.9n), as functions of the reference height above the canopy. The parameter values are defined by eqns (14.9o) and (14.10a). The broken line in (a) indicates the extrapolated wind speed profile below $Z = h$.

The above analysis is based on the assumptions defining the K coefficients by eqns (14.8e)–(14.8g). The rationale for adopting the assumption for K_m is that it leads to eqn (14.9b) for the wind speed profile u_a above the canopy which generally gives a good fit to data. K_H and K_v are then defined analogously. There is no other sound basis for these relationships. Nevertheless, they are physically reasonable, and do lead to realistic expressions for u_a, T_a, and ρ_{va}. It would be reasonable, therefore, to regard the above analysis as a plausibility argument for defining these profiles.

14.4 Canopy temperature

In the analysis so far we have used the equations for heat and water vapour transfer (eqns (14.3c) and (14.3d)) to derive an expression for the crop transpiration rate, which is independent of the canopy temperature (i.e. eqn (14.4k), the

Penman–Monteith equation). It is equally possible to eliminate the transpiration and obtain an expression for the canopy temperature.

14.4.1 *Derivation*

Recall eqn (14.3c) for heat transfer between the canopy and the air with $g_H = g_a$ (eqn (14.4j)):

$$H = \rho c_p (T_l - T_a) g_a, \tag{14.11a}$$

eqn (14.4c) for the transpiration rate

$$E = [s(T_l - T_a) + \Delta\rho_{va}] g_W, \tag{14.11b}$$

and the energy balance equation (eqn (14.3b)):

$$\phi_N = H + \lambda E. \tag{14.11c}$$

Eliminating the transpiration rate E from eqns (14.11b) and (14.11c) and then using eqn (14.11c) to substitute for H gives, after some algebra,

$$T_l = T_a + \frac{\phi_N - \lambda g_W \Delta\rho_{va}}{\lambda(s g_W + \gamma g_a)} \tag{14.11d}$$

for the canopy temperature. (Note that $\lambda\gamma = \rho c_p$ (eqn 14.4g).) As before, the conductance g_W for water vapour can be separated into the stomatal and boundary layer components g_c and g_a (where g_c is the canopy conductance and g_a the boundary layer conductance), so that eqn (14.4h) for g_W in terms of g_c and g_a applies, and eqn (14.11d) becomes

$$T_l - T_a = \frac{\phi_N(1/g_a + 1/g_c) - \lambda\Delta\rho_{va}}{\lambda[s + \gamma(1 + g_a/g_c)]}. \tag{14.11e}$$

This expression for the difference between the canopy and air temperature is analogous to the Penman–Monteith equation for canopy transpiration, and shows that $T_l - T_a$ increases linearly with ϕ_N and decreases linearly with $\Delta\rho_{va}$.

14.4.2 *Illustrations*

The following illustrations of $T_l - T_a$ as a function of ϕ_N, $\Delta\rho_{va}$, g_c, and g_a correspond to Figs 14.2 and 14.3 for the canopy transpiration rate E as a function of these variables. In Figs 14.5(a) and 14.5(b) the temperature difference $T_l - T_a$, given by eqn (14.11e), is shown in response to variation in ϕ_N and $\Delta\rho_{va}$ respectively. In both cases, there is a fairly small difference between the canopy and air temperatures. The response to variation in the conductances g_c and g_a is presented in Figs 14.6(a) and 14.6(b), and it can be seen that $T_l - T_a$ decreases with an increase in either conductance.

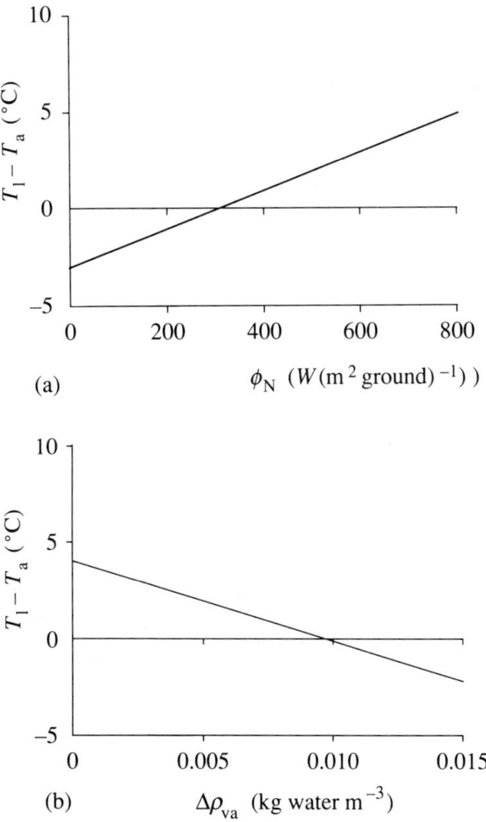

Fig. 14.5. Temperature difference $T_l - T_a$ between the crop and the air given by eqn (14.11e), illustrated as a function of (a) the available energy ϕ_N and (b) the vapour density deficit $\Delta\rho_{va}$. Parameter values are the same as those in Fig. 14.2.

These illustrations provide a useful means of assessing the validity of the assumptions used in the present analysis. First, recall that we are only considering turbulent flow with no buoyancy effects. This assumption will break down in very still conditions where there are large temperature gradients between the canopy and the air. From Figs 14.5 and 14.6 it is clear that larger temperature gradients will occur when either of the conductances g_a or g_c is very small. This is to be expected. If the canopy conductance is small, then there is little evaporation and consequently the incident radiation is used mainly for heating the crop. Equally, if the boundary layer conductance is small, turbulent transfer of heat across the boundary layer is small which also results in heating the crop. Under these conditions, buoyancy becomes important and the present theory will be inaccurate. The other point to note is the assumption relating the saturation vapour density at the canopy temperature to the air temperature (eqn (14.4a))

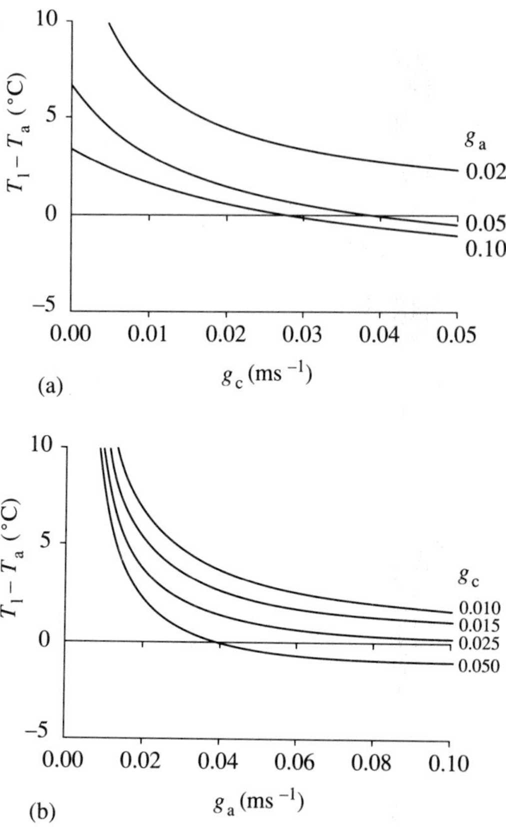

Fig. 14.6. Temperature difference $T_l - T_a$ between the crop and the air, given by eqn (14.11e), illustrated as a function of (a) the canopy conductance g_c and (b) the boundary layer conductance g_a. Parameter values are the same as those in Fig. 14.3.

which relies on a small temperature difference between the canopy and the air. Thus, as well as the effects of buoyancy not being included for high temperature gradients associated with low canopy and boundary layer conductances, the treatment of turbulent transport also becomes inaccurate under these conditions owing to the breakdown of eqn (14.4a).

Nevertheless, we can conclude that, in general terms, the analysis is likely to be valid for other than very still conditions—resulting in a low boundary layer conductance g_a—or where the crop is quite severely water stressed—in which case the canopy conductance g_c will be reduced.

14.5 Discussion

We have derived expressions for the instantaneous crop transpiration rate and the canopy temperature in terms of the net radiation, vapour density deficit, wind speed, and crop height. The analysis centres around the Penman–Monteith equation which is soundly based in physics, and the inputs required to apply the models do not present insuperable difficulties. Probably the main assumption is that the crop consists of a single layer, but it is generally thought that this does not lead to serious error.

We have used vapour density rather than vapour pressure as a measure of atmospheric vapour content in the analysis. Many authors use vapour pressure, although there are exceptions (e.g. Campbell 1977). We prefer to use density as the basic equation for vapour flux density from the crop across the boundary layer (eqn (14.3d)) is defined in terms of this quantity. Since, as we have demonstrated, the two approaches are equivalent, there seems little point in converting from vapour density to vapour pressure.

The analysis centres around the approximation of eqn (14.4a) relating the saturation vapour density at the leaf temperature to that at air temperature. As the difference between the leaf and air temperatures increases, so this approximation will become increasingly inaccurate. Indeed, the percentage error is

$$\frac{\rho_v'(T_l) - [\rho_v'(T_a) + s(T_l - T_a)]}{\rho_v'(T_l)} \times 100 \text{ per cent.} \qquad (14.12a)$$

This can be calcualted from Table 14.3. For example, if $T_l = 25\,°C$ and $T_a = 20\,°C$ then the error is 3 per cent; for a more extreme situation where $T_l = 30\,°C$ and $T_a = 40\,°C$ the error is still only 8 per cent. Consequently, in many situations this approximation is likely to be sufficiently accurate. However, large temperature gradients are likely to affect the analysis through their effect on the flux of momentum, heat, and water vapour across the boundary layer, since the analysis is only for turbulent transport. If there are large temperature gradients between the crop and the bulk air, then buoyancy effects will be significant and will have to be included in the analysis (for a discussion of buoyancy see Campbell (1977)).

One frequently cited limitation of the Penman–Monteith equation is that it fails to account adequately for advection effects in hot arid regions where the crop is surrounded by relatively unvegetated areas which are at considerably higher temperatures than the crop. (This is generally referred to as the oasis effect.) However, provided that measurements of the environmental variables are confined to the boundary layer of the crop (see Section 14.1.2), advection is properly treated in the analysis. It is more likely that, for such regions, either buoyancy effects must be included or the canopy conductance is not properly treated. Even for well-watered crops, if the plants cannot maintain the transpiration stream, they may experience water stress during the hotter parts of the day and the canopy conductance may fall. In such cases it is unlikely that any useful correlation between the actual and potential transpiration rates can be established.

The analysis is for the instantaneous rate of transpiration, although in crop models the daily transpiration is required (Section 15.5.1, p. 440). The most satisfactory method for calculating the daily transpiration is to integrate throughout the light period with frequent environmental data (e.g. averaged over 15 min intervals). If such extensive data are not available, or this calculation is impractical, then it may be necessary to use some approximation technique. While various simplications to the theory have been proposed, our view is that any such approach should derive from the Penman–Monteith equation owing to its rigorous basis.

Exercises

14.1. Evaporation and transpiration are frequently given in units of millimetres of water per unit time. The density of water at 4 °C is 1000 kg m^{-3}. Using this value, convert the units kg water (m^2 ground)$^{-1}$ s^{-1} to mm water s^{-1}, mm water $hour^{-1}$, and mm water day^{-1}. At 20 °C the density of water is, more accurately, 998.2 kg m^{-3}; estimate the error from using the rounded value given above.

14.2. The gas laws state that, for a perfect gas,

$$p = \frac{\rho}{M} RT, \tag{E14.2a}$$

where the symbols are defined as follows: p, pressure (Pa); R, gas constant 8314 J (kg mol)$^{-1}$ K^{-1}; T, temperature (K); ρ, density (kg m^{-3}); M, relative molecular mass (kg (kg mol)$^{-1}$). For an adiabatic system (no heat transfer), the first law of thermodynamics (conservation of energy) states that

$$dU + p\,dV = 0, \tag{E14.2b}$$

where dU (J) is the change in internal energy and dV(m^3) is the change in volume. Now, if the volume of gas is V(m^3) and there are n kg mol present, the density can be written

$$\rho = \frac{Mn}{V} \tag{E14.2c}$$

so that eqn (E14.2a) becomes

$$pV = nRT. \tag{E14.2d}$$

Defining the water potential ψ (J kg^{-1}) by

$$\psi = \frac{\text{energy of water}}{\text{mass of water}} \tag{E14.2e}$$

and assuming that

$$\frac{\rho_v}{\rho_v'} = \frac{p_v}{p_v'}, \tag{E14.2f}$$

where ρ_v and p_v are vapour density and pressure respectively and the prime denotes saturation, derive eqn (14.3g), i.e.

$$h_r = \frac{\rho_v}{\rho_v'} = \exp\left(\frac{\psi M}{RT}\right). \tag{E14.2g}$$

To do this, define the water potential of saturated air as zero. The assumption in eqn (E14.2f) is considered in the next exercise.

14.3. For a mixture of perfect gases, Dalton's law of partial pressures extends eqn (E14.2a) to

$$P = \sum_i p_i = \sum_i \frac{\rho_i}{M_i} RT, \tag{E14.3a}$$

where P (Pa) is the pressure of the mixture and subscript i denotes the constituents. At normal temperature and pressure, 1 kg mol of any perfect gas occupies a volume of 22.4136 m^3, with concentration

$$\frac{1}{22.4136} = 0.0446158 \text{ kg mol m}^{-3}. \tag{E14.3b}$$

The density $\rho_{i,\text{pure}}$ of a pure gas is therefore

$$\rho_{i,\text{pure}} = M \times 0.0446158 \text{ kg gas m}^{-3}. \tag{E14.3c}$$

In a mixture of gases, it follows that the density of the ith constituent is

$$\rho_i = \rho_{i,\text{pure}} \times \text{fractional content of the } i\text{th constituent}. \tag{E14.3d}$$

The relative molecular masses and fractional contents of the main constituents of dry air (from Monteith 1973, p. 6) are given in Table E14.1.

(a) Calculate the density of dry air and, applying eqn (E14.3c) to air, evaluate its effective relative molecular mass. Show that

$$\varepsilon = \frac{M_{\text{v}}}{M_{\text{a}}} = 0.622 \tag{E14.3e}$$

where subscripts a and v denote air and water vapour respectively.

(b) Using the gas laws stated above, derive eqn (14.5a), i.e.

$$\rho_{\text{v}} = \rho\varepsilon\frac{p_{\text{v}}}{P - p_{\text{v}}}. \tag{E14.3f}$$

(Recall that ρ is the density of dry air.)

(c) Using eqn (E14.3f) estimate the percentage error in using the approximation

$$\frac{\rho_{\text{v}}}{\rho'_{\text{v}}} = \frac{p_{\text{v}}}{p'_{\text{v}}} \tag{E14.3g}$$

and calculate this error at temperatures 10, 20, and 30 °C.

Table E14.1. Composition of dry air

Gas	Nitrogen	Oxygen	Argon	CO_2
M	28.01	32.00	38.98	44.01
f	0.7809	0.2095	0.0093	0.0003

From Monteith 1973, p. 6
M is the relative molecular mass (kg (kg mol)$^{-1}$) and f is the fractional content.

14.4. Defining the rate of transpiration from a crop by the Penman–Monteith equation (eqn (14.4k)) show, by differentiating this equation with respect to g_{a}, that when

$$g_{\text{c}} = \frac{s\phi_{\text{N}}}{\lambda\,\Delta\rho_{\text{va}}(s + \gamma)} \tag{E14.4a}$$

E is independent of g_{a} and that

$$E = \frac{s\phi_{\text{N}}}{\lambda(s + \gamma)}. \tag{E14.4b}$$

15
Crop water relations

15.1 Introduction

In this chapter, the elements needed for modelling crop water use as shown in Fig. 15.1 are discussed. For modelling crop growth, the effects of water status on physiological processes such as photosynthesis, leaf area expansion, root : shoot partitioning, growth, respiration, the uptake of nitrogen and other nutrients, and perhaps nitrogen fixation are required; we discuss some of the possibilities. Ludlow (1987) gives a valuable critical introduction to the topic; Taylor, Jordan, and Sinclair (1983) give a more detailed account.

The model in Fig. 15.1 is highly simplified, showing only the main components required in a crop water use model. We know that considerable water potential gradients can exist in the plant and in the soil; for some purposes it may be necessary to represent these by treating the components of the plant, root, stem, and leaves as separate compartments, and using a series of compartments to represent the different soil horizons.

Let Q_{pl} and Q_s be the quantities of water in the plant and soil in units of kg H_2O (m^2 ground)$^{-1}$. Rainfall R_w and drainage D_w are defined in the usual units of m day^{-1}, and to convert these to units of kg H_2O (m^2 ground)$^{-1}$ day^{-1} they are multiplied by the density of water ρ_w (kg H_2O m^{-3}) to give

$$\rho_w R_w \text{ and } \rho_w D_w \text{ kg H}_2\text{O (m}^2\text{ ground)}^{-1}\text{ day}^{-1}. \tag{15.1a}$$

The flux density of water from the plant to the atmosphere is denoted by $F_{pl,atm}$, and that from the soil into the plant by $F_{s,pl}$, both in units of kg H_2O (m^2 ground)$^{-1}$ day^{-1}. The dynamics of the system are given by two differential equations for the state variables Q_{pl} and Q_s, i.e.

$$\frac{dQ_{pl}}{dt} = F_{s,pl} - F_{pl,atm} \tag{15.1b}$$

$$\frac{dQ_s}{dt} = \rho_w R_w - \rho_w D_w - F_{s,pl}. \tag{15.1c}$$

Much of this chapter is concerned with the calculation of the variables of Fig. 15.1 (θ_{pl}, ψ_{pl}, θ_s, and ψ_s) from the (extensive) state variables (W_G, W_S, Q_{pl}, and Q_s), and thence the two flux densities $F_{s,pl}$ and $F_{pl,atm}$ on the right-hand side of these equations can be obtained.

15.2 The soil water characteristic

The soil water characteristic is concerned with the relationship between the soil relative water content θ_s (m^3 water (m^3 soil)$^{-1}$) and the soil water potential ψ_s (J

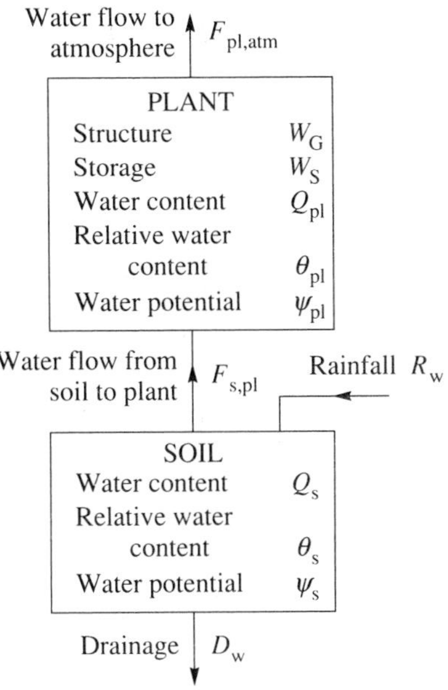

Fig. 15.1. A simple model for crop water use. The state variables of the system are W_G, W_S, Q_{pl}, and Q_s; the other quantities shown are derived (θ_{pl}, ψ_{pl}, θ_s, and ψ_s) or given (R_w).

$(\text{kg } H_2O)^{-1}$). (Water potential and its units are discussed in Section 2.2.6, p. 50.) θ_s is related to the quantity of water per unit area Q_s (kg H_2O (m^2 ground)$^{-1}$) of Fig. 15.1 and (15.1c) by means of

$$\theta_s = \frac{Q_s}{\rho_w d_s}, \tag{15.2a}$$

where ρ_w (kg H_2O m^{-3}) is the density of water d_s (m) is a soil depth parameter required to define an effective soil depth in the simple model of Fig. 15.1.

It is widely observed (e.g Gregson, Hector, and McGowan, 1987) that the equation

$$\psi_s = \psi_{s,max} \frac{1}{(\theta_s/\theta_{s,max})^b} \tag{15.2b}$$

gives a good fit to $\psi_s : \theta_s$ data. Here, b is a positive dimensionless constant, usually lying in the range $2 < b < 18$; $\theta_{s,max}$ is the maximum value of the soil water content θ_s observed at 'field capacity' when free run-off has occurred, which takes place down to a soil water potential value $\psi_{s,max}$ of about -10 J (kg water)$^{-1}$. (The point ($\theta_{s,max}, \psi_{s,max}$) is not well defined experimentally.) This equation is illustrated in Fig. 15.2, where a typical range of b values is used. Free-draining sandy soils have low values of b, whereas heavy clay soils have high values.

Equation (15.2b) contains three parameters: $\psi_{s,max}$, $\theta_{s,max}$, and b. It can be written as

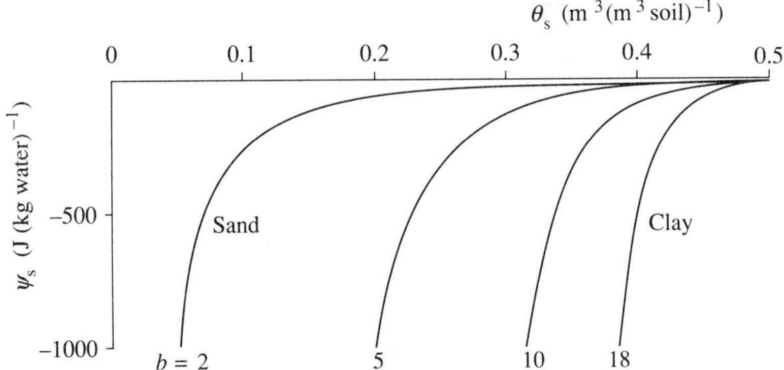

Fig. 15.2. Soil water characteristic. Equation (15.2b) relating the soil water potential ψ_s to the soil relative water content θ_s is drawn for $\psi_{s,max} = -10$ J (kg water)$^{-1}$, $\theta_{s,max} = 0.5$ m^3 water (m^3 soil)$^{-1}$, and various values of b as indicated.

$$\ln \psi_s = \ln(\psi_{s,max}\, \theta^b_{s,max}) - b \ln \theta_s = a - b \ln \theta_s. \tag{15.2c}$$

Two measurements of (θ_s, ψ_s) are required to determine the values of a and b.

In order to simplify the data required for defining the soil water characteristic, Gregson *et al.* (1987), after considering a wide range of data, concluded that the parameters a and b in (15.2c) are not independent, and that accurate predictions (within a few percent) can be made using a single adjustable parameter with (the equation after eqn (4) in Gregson *et al.* (1987) has been written in the units used here)

$$\psi_s = -0.375\,(0.557)^b \frac{1}{\theta_s^{\,b}}. \tag{15.2d}$$

The determination of a single (θ_s, ψ_s) point is now sufficient to define b in (15.2d).

15.2.1 *A pore-size distribution model for the soil water characteristic*

The water potential inside a concave meniscus of radius r (m) is, relative to a free water surface (Exercise 15.1),

$$\psi = -\frac{2T_w}{\rho_w r} \tag{15.3a}$$

where T_w is the surface tension of water (0.073 N m^{-1} at 20 °C) and ρ_w is the density of water (998 kg m^{-3} at 20 °C). Equation (15.3a) also assumes a zero angle of contact between the wall of the capillary and the meniscus. Thus

$$\psi = -\frac{0.146 \times 10^{-3}}{r} \text{ J (kg water)}^{-1} \quad \text{at } 20\,°\text{C}. \tag{15.3b}$$

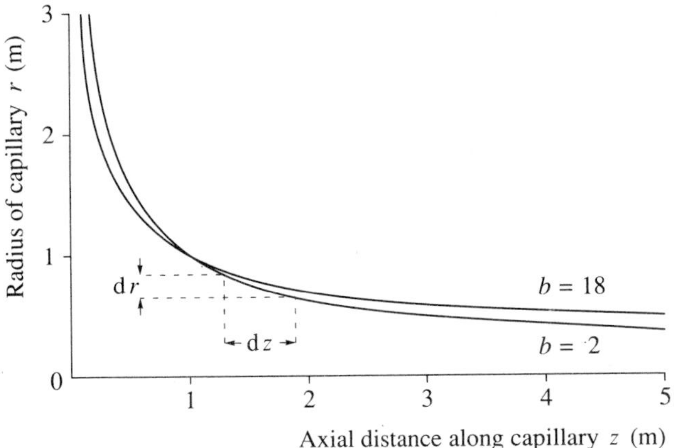

Fig. 15.3. Model for the shape of soil pores, as given by eqn (15.3e). Parameter values are $r_o = 1$ m and $z_0 = 1$ m (the scale is arbitrary). Two values of q are illustrated: $q = 2/3$ and $q = 18/35$. With eqn (15.3n), these correspond to $b = 2$ and $b = 18$ respectively.

Assume that there are n identical capillaries per cubic meter of soil (n does not vary with water content); these have a circular cross-section with radius r (m), and r varies with axial distance z (m) according to

$$\frac{r}{r_0} = \left(\frac{z}{z_0}\right)^{-q} \tag{15.3c}$$

where q is a constant (dimensionless) and the equation passes through the point (r_0, z_0). This is plotted in Fig. 15.3 for two values of q ($q = 2/3$ and $q = 18/35$) which, with this model, span the range of b found in nature (see (15.2b), $2 < b < 18$, and (15.3n) below). It is assumed that $q > 1/2$ (without this assumption parts of the analysis are slightly different).

It is useful to derive a pore volume distribution function

$$f(r)\,\mathrm{d}r \tag{15.3d}$$

with units of m^3 (m^3 soil)$^{-1}$; $f(r)\,\mathrm{d}r$ is the volume of pore with radius between r and $r + \mathrm{d}r$ per unit volume of soil. To relate this to pore shape (15.3c), we invert (15.3c) to give

$$\frac{z}{z_0} = \left(\frac{r}{r_0}\right)^{-1/q}. \tag{15.3e}$$

Differentiating gives

$$\mathrm{d}z = \frac{z_0}{r_0}\left(-\frac{1}{q}\right)\left(\frac{r}{r_0}\right)^{-1-1/q}\mathrm{d}r. \tag{15.3f}$$

A hypothetical dx and dz are illustrated in Fig. 15.3. For a positive dr, dz is negative. The pore-size distribution function f(r) dr is then given by

$$f(r)\,\mathrm{d}r = -n\pi r^2\,\mathrm{d}z = \frac{n\pi r_0 z_0}{q}\left(\frac{r}{r_0}\right)^{1-1/q}\mathrm{d}r. \qquad (15.3g)$$

As far as the relative water content θ_s is concerned, it makes no difference whether we regard unit volume of soil as containing n identical pores (as above) or just a single pore of much greater length (e.g. write (15.3c) as $r/r_0 = (Z/Z_0)^{-q}$ where $Z = nz$ and $Z_0 = nz_0$)), but this does affect hydraulic conductivity which depends on how the pores are connected.

Assume also that all the water in the soil is contained in these capillaries, which fill with water until (15.3a) is satisfied. We now deduce an equation for the soil moisture characteristic, relating the soil water potential ψ_s (J (kg water)$^{-1}$) to the soil relative water content θ_s (m^3 water (m^3 soil)$^{-1}$).

The volume v (m^3) of water in a capillary filled up as far as (r_s, z_s) is

$$v = \int_{z_s}^{\infty} \pi r^2\,\mathrm{d}z. \qquad (15.3h)$$

As the soil relative water content is $\theta_s = nv$, it can be shown by substituting (15.3c) in (15.3h) that (Exercise 15.2)

$$\theta_s = nv = \frac{n\pi z_0 r_0^{\,2}}{2q-1}\left(\frac{r_s}{r_0}\right)^{(2q-1)/q}. \qquad (15.3i)$$

By direct differentiation of (15.3i), or using $\mathrm{d}\theta_s = n\,\mathrm{d}v = -n\pi r^2\,\mathrm{d}z$ from (15.3i) and (15.3g), it follows that

$$\mathrm{d}\theta_s = f(r)\,\mathrm{d}r = \frac{n\pi r_0 z_0}{q}\left(\frac{r}{r_0}\right)^{1-1/q}\mathrm{d}r. \qquad (15.3j)$$

Thus the pore volume distribution function introduced in (15.3d) is equal to the change in the soil relative water content θ_s.

Solving equation (15.3i) for $1/r_s$ gives

$$\frac{1}{r_s} = \frac{1}{r_0}\left(\frac{n\pi z_0 r_0^{\,2}}{2q-1}\right)^{q/(2q-1)}\left(\frac{1}{\theta_s}\right)^{q/(2q-1)}. \qquad (15.3k)$$

Substituting this in (15.3a) gives

$$\psi_s = -\frac{2T_w}{\rho_w r_0}\left(\frac{n\pi z_0 r_0^{\,2}}{2q-1}\right)^{q/(2q-1)}\left(\frac{1}{\theta_s}\right)^{q/(2q-1)}. \qquad (15.3l)$$

This can be compared with the empirically observed soil water characteristic in (15.2b), i.e.

$$\psi_s = \psi_{s,\max}\frac{1}{(\theta_s/\theta_{s,\max})^b}. \qquad (15.3m)$$

Equating the coefficients of θ_s in the two expressions gives

$$b = \frac{q}{2q-1}. \tag{15.3n}$$

The wide range of b observed experimentally ($2 < b < 18$) is covered in this model by a very narrow range of q ($2/3 > q > 18/35$). Equating the coefficients of θ_s^{-b} in (15.3l) and (15.3m) and using (15.3n) gives

$$\psi_{s,\max}\,\theta^b_{s,\max} = -\frac{2T_w}{\rho_w r_0}\left(\frac{n\pi z_0 r_0^2}{2q-1}\right)^b. \tag{15.3o}$$

This relationship may suggest a tentative interpretation of soil water parameters.

It is also useful to write down the relationship between the maximum radius r_s of the water-filled pores and the soil water content θ_s. Combining eqns (15.3k) and (15.3n) gives

$$r_s = r_0[n\pi z_0 r_0^2(2b-1)]^{-b}\theta_s^{\,b}. \tag{15.3p}$$

15.3 Hydraulic conductivity

We introduce the concept of hydraulic conductivity by first considering Poiseuille's formula for streamline flow through a tube of circular cross-section with radius r (m) and length Δx (m) across which a pressure difference ΔP (kg m^{-1} s^{-2} or Pa) is applied:

$$q = -\frac{\pi r^4}{8\eta}\frac{\Delta P}{\Delta x}, \tag{15.4a}$$

where q (m^3 s^{-1}) is the volume flow rate and η (kg m^{-1} s^{-1}) is the viscosity. The negative sign reflects the fact that the flow is down the pressure gradient. This equation can be written in terms of a mass flux density F (kg water m^{-2} s^{-1}):

$$F = \frac{\rho_w q}{\pi r^2} = -\frac{\rho_w r^2}{8\eta}\frac{\Delta P}{\Delta x}, \tag{15.4b}$$

where ρ_w is the density of water (kg water m^{-3}). Finally, we write pressure P in terms of the pressure component of the water potential ψ_p (J (kg water)$^{-1}$) using

$$P = \rho_w \psi_p. \tag{15.4c}$$

Substituting in (15.4b) gives

$$F = -\frac{\rho_w^2 r^2}{8\eta}\frac{\Delta\psi_p}{\Delta x}. \tag{15.4d}$$

The hydraulic conductivity K (kg m^{-3} s) is defined by

$$K = \frac{\rho_w^2 r^2}{8\eta}; \tag{15.4e}$$

then the flux density (eqn (15.4d)) becomes

$$F = -K\frac{\Delta\psi_p}{\Delta x}. \tag{15.4f}$$

Sometimes the hydraulic conductivity is given in units of, for instance, m s^{-1}. The volume flux of water F_V ($m^3\ m^{-2}\ s^{-1}$) is related to the mass flux F (kg m^{-2} s^{-1}) by

$$F_V = F/\rho_w. \tag{15.4g}$$

The driving force can be expressed as a head of water of height h (m). With

$$\psi_p = \frac{P}{\rho_w} = \frac{\rho_w gh}{\rho_w} = gh, \tag{15.4h}$$

where $g = 9.81$ m s^{-2} is the acceleration due to gravity, eqn (15.4f) can be written

$$F_V = -\left(\frac{Kg}{\rho_w}\right)\frac{\Delta h}{\Delta x}. \tag{15.4i}$$

In this representation, the 'hydraulic conductivity' Kg/ρ_w, has units of m s^{-1}; this differs by a factor of about 100 from our definition given in (15.4e) and (15.4f).

15.3.1 *Hydraulic conductivity of the soil*

It has been observed that an equation similar in form to (15.2b) which relates soil relative water content θ_s (m^3 water (m^3 soil)$^{-1}$) to soil water potential θ_s (J (kg water)$^{-1}$) can be used to relate θ_s to soil hydraulic conductivity K_s (kg m^{-3} s):

$$K_s = K_{s,max}\left(\frac{\theta_s}{\theta_{s,max}}\right)^c \tag{15.5a}$$

where $K_{s,max}$ is the maximum value of the soil hydraulic conductivity obtained when θ_s has its maximum value of $\theta_{s,max}$ and c is a positive dimensionless constant. Typically, $K_{s,max}$ can take values ranging from 10^{-4} kg m^{-3} s for a clay soil to 10^{-3} kg m^{-3} s for a sandy soil. The range of c is large: from about 7 for a sandy soil to values as high as 39 for some clay soils (see (15.5b) below, and Fig. 15.2 with (15.2b)). Equation (15.5a) is drawn in Fig. 15.4; comparison with Fig. 15.2, where the effect of the soil water content on the soil water potential is given, shows the marked effect of the soil relative water content on the soil hydraulic conductivity.

It has been found (e.g. Campbell 1974) that satisfactory prediction of both the soil water potential ψ_s and the soil hydraulic conductivity K_s from the soil water content θ_s can be obtained if c of (15.5a) is related to the parameter b of (15.2b) by

$$c = 2b + 3. \tag{15.5b}$$

There have been several discussions of the interpretation of this relationship (e.g. Childs 1969, pp. 184–9, Campbell 1985, Chapter 6), which are roughly as follows.

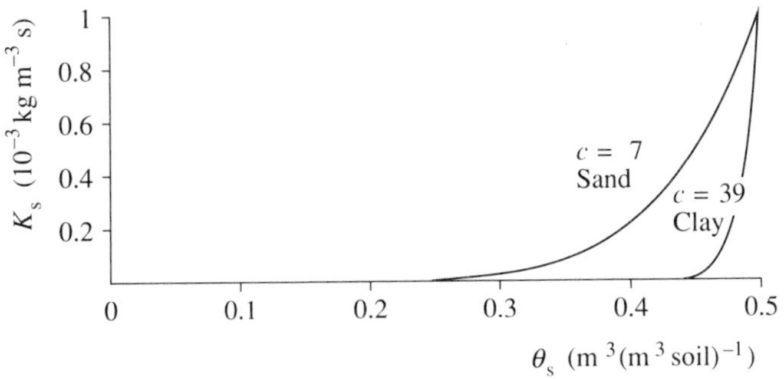

Fig. 15.4. Soil hydraulic conductivity characteristic. Equation (15.5a) relating the soil hydraulic conductivity K_s, to the soil relative water content θ_s, is drawn for $K_{s,max} = 0.001$ kg m^{-3} s, $\theta_{s,max} = 0.5$ m^3 water (m^3 soil)$^{-1}$, and two values of c as indicated.

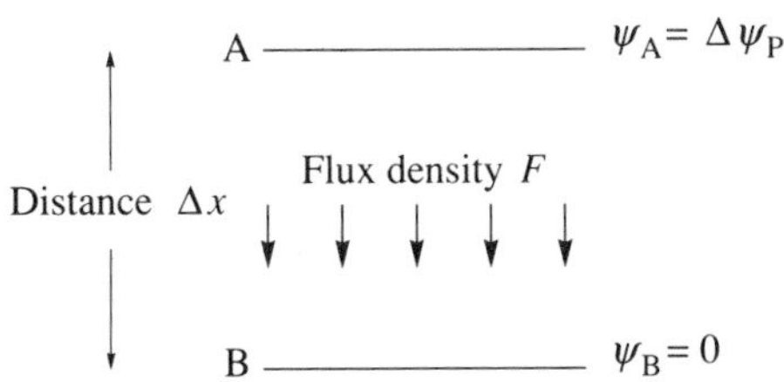

Fig. 15.5. Water transport through the soil. A and B are planes of unit area separated by distance Δx (m). Between A and B is a water potential difference $\Delta\psi_p$ (J (kg water)$^{-1}$, the pressure component of the water potential). The flux density between A and B is F (kg water m^{-2} s^{-1}).

With reference to Fig. 15.5, A and B represent two planes in the soil which are distance Δx apart; they are each of unit area. As in eqn (15.4f), a pressure potential $\Delta\psi_p$ is applied between A and B, and the flux density F is measured. If we locate ourselves randomly on the plane at A, then the soil relative water content θ_s (m^3 water (m^3 soil)$^{-1}$) gives the probability of encountering water. The fraction of unit area at B over which there is water is thus also θ_s. Imagine that the water-covered area at A is in the form of n_p discs, each of area πr^2 (the units of n_p are m^{-2}). Then the total water-covered area is $n_p \pi r^2$, and therefore

$$\theta_s = n_p \pi r^2. \tag{15.5c}$$

Equations (15.4e) and (15.4f) give the flux density for a single pore with reference to the cross-sectional area of the pore alone. To obtain the flux density through unit area with n_p pores each of area πr^2, the flux density for a single pore must be multiplied by $n_p \pi r^2$, i.e. θ_s. Thus the soil hydraulic conductivity K_s, becomes (cf. (15.4e))

$$K_s = \theta_s \frac{\rho_w{}^2 r^2}{8\eta}. \tag{15.5d}$$

Equation (15.5d) for the hydraulic conductivity assumes that the pores run continuously and directly from A and B, and thus the distribution of water on plane A is exactly the same as that on plane B. The next stages in the argument may be thought somewhat unconvincing. If it is assumed that the distributions of water on planes A and B are completely uncorrelated, then the probability of a filament of water running from A to B is θ_s^2, rather than θ_s, so that (15.5d) is replaced by

$$K_s = \theta_s^2 \frac{\rho_w^2 r^2}{8\eta}. \tag{15.5e}$$

Next, as the soil water content θ_s declines, the length of the path between planes A and B increases because a more tortuous route is necessarily followed; Δx of Fig. 15.5 is replaced by $\Delta x/\theta_s$. Since, in (15.4f), Δx appears in the denominator, this multiplies the apparent hydraulic conductivity by a further factor of θ_s, so that (15.5e) becomes

$$K_s = \theta_s^3 \frac{\rho_w^2 r^2}{8\eta}. \tag{15.5f}$$

Finally, it is assumed that only the pores with the maximum radius r_s contribute. Substituting for $r = r_s$ from (15.3p), therefore,

$$K_s = \frac{\rho_w^2 r_0^2}{8\eta[n\pi z_0 r_0^2(2b-1)]^b} \theta_s^{2b+3}. \tag{15.5g}$$

15.4 The flow of water between the soil and the plant

In this section we examine the main determinants of the flux of water from the soil into the plant. In order to construct some equations with the appropriate characteristics, the simple root model of Fig. 15.6 is considered.

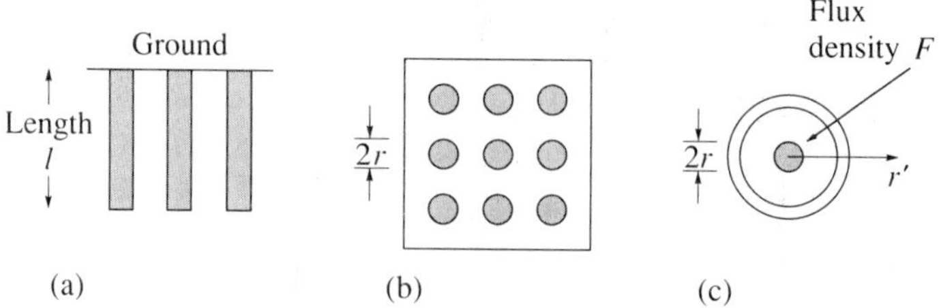

Fig. 15.6. Simple root model showing n roots (shaded) of length l and radius r per unit ground area (branching is ignored): (a) vertical section; (b) horizontal section; (c) radial movement of water, flux density F, into a single root of radius r (r' is a radial variable; two concentric cylinders of radius r' and $r' + dr'$ are drawn around the root; F is constant round a cylinder).

15.4.1 *Flux of water from the root surface into the plant*

It is assumed that there are n uniform unbranched roots of length l (m) and radius r (m) per unit ground area. Equation (15.4d) gives the flux density for a single tube. To obtain the flux density per unit of ground area, (15.4d) must be multiplied by πr^2 to give the flux through a single tube, and then by n (for the number of tubes per unit ground area); also we put $\Delta x = -l/2$ as an approximation since water enters along the root through the walls, and write $\Delta\psi_p = \psi_{rs} - \psi_{pl}$, where ψ_{rs} (J kg^{-1}) is the water potential at the root surface and ψ_{pl} (J kg^{-1}) is the plant water potential. This gives

$$F = \frac{n\rho_w^2 \pi r^4}{8\eta} \frac{\psi_{rs} - \psi_{pl}}{l/2}. \tag{15.6a}$$

Therefore, denoting the root dry mass by W_r (kg dry matter (m^2 ground)$^{-1}$) and assuming that the roots are hollow tubes of wall thickness s (kg dry matter (m^2 wall)$^{-1}$) we have

$$W_r = ns2\pi rl \quad \text{or} \quad r = \frac{W_r}{ns2\pi l}. \tag{15.6b}$$

Equation (15.6a) is written as

$$F = \frac{\psi_{rs} - \psi_{pl}}{r_{rs,pl}} \tag{15.6c}$$

where

$$r_{rs,pl} = \frac{4\eta l}{\rho_w^2 n\pi r^4}. \tag{15.6d}$$

$r_{rs,pl}$ (kg^{-1} m^4 s^{-1}) is the resistance between the root surface and the plant. Equation (15.6d) suggests how the resistance from the root surface to the plant may vary with the number of roots per unit area n, the root length l, and the root radius r. These quantities are not independent, but are related to each other and to the root dry matter W_r through (15.6b). For instance, as the root dry matter W_r increases, we might expect the four root parameters n, s, r, and l to increase also.

15.4.2 *Flux of water from the bulk soil to the root surface*

First, we write down the usual flux density equation for movement through the soil (cf. (15.4f))

$$F = -K_s \frac{d\psi}{dr'}, \tag{15.7a}$$

where F (kg m^{-2} s^{-1}) is the flux density, K_s is the hydraulic conductivity of the soil (kg m^{-3} s), ψ is the water potential (J (kg water)$^{-1}$), and r' (m) is the cylindrical radial variable shown in Fig. 15.6(c). Then the flux ϕ (kg s^{-1}) into a single root is given by

$$\phi = -2\pi r' l F = 2\pi l K_s r' \frac{d\psi}{dr'}. \tag{15.7b}$$

This applies to a steady state, when the total flux into any cylinder (Fig. 15.6(c)) is the same; i.e. ϕ is constant and independent of r'. We write (15.7b) as

$$\phi \int_{r_{bs}}^{r} \frac{dr'}{r'} = \int_{\psi_s}^{\psi_{rs}} 2\pi l K_s \, d\psi, \tag{15.7c}$$

where ψ_{rs} (J kg^{-1}) is the water potential at the root surface where the radius is r (m) and ψ_s (J kg^{-1}) is the bulk soil water potential which is at some imagined radius of r_{bs} (m). Integrating gives

$$\phi \ln\left(\frac{r_{bs}}{r}\right) = 2\pi l K_s(\psi_s - \psi_{rs}). \tag{15.7d}$$

We assume that the radius r_{bs} from which water movement occurs is given by

$$r_{bs} = 1/n^{1/2}, \tag{15.7e}$$

where n denotes the number of roots per unit ground area ((m^2 ground)$^{-1}$) (Fig. 15.6b). Substituting (15.7e) in (15.7d) gives

$$\phi \ln\left(\frac{1}{rn^{1/2}}\right) = 2\pi l K(\psi_s - \psi_{rs}). \tag{15.7f}$$

For n roots, the flux density per unit ground area F is

$$F = n\phi = \frac{2\pi l K_s n(\psi_s - \psi_{rs})}{\ln[1/(rn^{1/2})]}. \tag{15.7g}$$

Corresponding to (15.6d), we write this as

$$F = \frac{\psi_s - \psi_{rs}}{r_{s,rs}} \quad \text{with} \quad r_{s,rs} = \frac{\ln[1/(rn^{1/2})]}{2\pi l K_s n}, \tag{15.7h}$$

where $r_{s,rs}$ (kg^{-1} m^4 s^{-1}) is the resistance between the bulk soil and the root surface. This resistance depends on the root geometry, as exemplified by the parameters n (number of roots per unit ground area), l (root length), and r (root radius), as well as on the soil hydraulic conductivity K_s.

15.4.3 *Root structure and the resistance to water movement from the soil to the plant*

First we combine eqns (15.6d) and (15.7h), eliminating the water potential ψ_{rs} at the root surface and neglecting any storage of water at the root surface, to give the flux density $F_{s,pl}$ (kg water (m^2 ground)$^{-1}$ s^{-1}) of water from the soil to the plant as

$$F_{s,pl} = \frac{\psi_s - \psi_{pl}}{r_{s,pl}}, \tag{15.8a}$$

where the resistance $r_{s,pl}$ (kg^{-1} m^4 s^{-1}) between the soil and the plant is given by

$$r_{s,pl} = r_{s,rs} + r_{rs,pl}, \tag{15.8b}$$

which, with eqns (15.6d) and (15.7h), gives

$$r_{s,pl} = \frac{4\eta l}{\rho_w{}^2 n\pi r^4} + \frac{\ln[1/(rn^{1/2})]}{2\pi l K_s n}. \tag{15.8c}$$

In order to put this expression into a useful form for crop modelling, it is necessary to make some assumptions about how the roots grow. We outline two out of many possible approaches.

(i) Root growth is accompanied by an increase in rooting density n alone; wall thickness s, root length l and radius r are constant. From (15.6b)

$$n = \frac{W_r}{2\pi rsl}. \tag{15.9a}$$

Substituting in (15.8c) gives

$$r_{s,pl} = \frac{8\eta sl^2}{\rho_w{}^2 r^3 W_r} + \frac{rs\ln\{[2\pi ls/(rW_r)]^{1/2}\}}{K_s W_r}. \tag{15.9b}$$

This expression gives contributions to the soil–plant resistance that decrease as the root dry matter W_r increases. It is interesting to put some actual values into (15.9b). We assume

$$l = 0.2\ \text{m} \qquad r = 10^{-3}\ \text{m} \qquad s = 0.02\ \text{kg dry matter m}^{-2}, \tag{15.9c}$$

which gives (using (15.9a))

$$n = 1000\ (\text{m}^2\ \text{ground})^{-1} \qquad \text{corresponds to}$$
$$W_r = 0.025\ \text{kg root dry matter (m}^2\ \text{ground})^{-1}.$$

In addition, it is assumed that

$$\rho_w = 998\ \text{kg m}^{-3} \qquad \eta = 1.002 \times 10^{-3}\ \text{kg m}^{-1}\ \text{s}^{-1}. \tag{15.9d}$$

These values apply to a temperature of 20 °C. The soil hydraulic conductivity K_s is taken to be (Fig. 15.4)

$$K_s = 10^{-3}\ \text{kg m}^{-3}\ \text{s}. \tag{15.9e}$$

Substituting (15.9c)–(15.9e) into (15.9b) leads to

$$r_{s,pl} = \frac{0.0064}{W_r} + \frac{0.02}{W_r}\ln\left(\frac{11.2}{W_r^{1/2}}\right). \tag{15.9f}$$

With $W_r = 0.1$ kg root dry matter (m^2 ground)$^{-1}$, this gives

$$r_{s,pl} = 0.06 + 0.71 = 0.77 \text{ kg}^{-1} \text{ m}^4 \text{ s}^{-1} \tag{15.9g}$$

for the root surface–plant, bulk soil–root surface, and total soil–plant resistances. With a water potential difference of 1 J kg^{-1} ($\equiv$ 1000 Pa $\equiv$ 0.01 bar), this gives a flow rate of (using (15.8a)) 1.3 kg water (m^2 ground)$^{-1}$ s^{-1}.

(ii) Root growth is accompanied by increases in root density n and root length l, but the root wall density s and radius r are constant. Assume that the root density n and the root length l are proportional to the square root of root dry matter W_r, with

$$n = aW_r^{1/2} \quad \text{and} \quad l = bW_r^{1/2}, \tag{15.10a}$$

where a (kg$^{-1/2}$ m^{-1}) and b (kg$^{-1/2}$ m^2) are constants. Putting these in (15.9a) gives

$$ab2\pi rs = 1, \tag{15.10b}$$

which a and b must satisfy. Substituting (15.10a) into (15.8c) leads to

$$r_{s,pl} = \frac{4\eta b}{\rho_w^2 \pi r^4 a} + \frac{\ln[1/(ra^{1/2}W_r^{1/4})]}{2\pi K_s ab W_r}. \tag{15.10c}$$

Considering numerical values, suppose that when $W_r = 0.25$ kg root dry matter (m^2 ground)$^{-1}$, $n = 10^4$ (m^2 ground)$^{-1}$, and $l = 0.5$ m. From (15.10a), therefore,

$$a = 2 \times 10^4 \text{ kg}^{-1/2} \text{ m}^{-1} \qquad b = 1 \text{ kg}^{-1/2} \text{ m}^2. \tag{15.10d}$$

Assuming that the root radius is constant at

$$r = 10^{-3} \text{ m}; \tag{15.10e}$$

then (15.10b) and (15.10d) give

$$s = 0.008 \text{ kg dry matter m}^{-2}. \tag{15.10f}$$

Equations (15.10e) and (15.10f) can be compared with (15.9c). With the values given in (15.9d) and (15.9e), eqn (15.10c) for the resistance between the soil and the plant becomes

$$r_{s,pl} = 0.064 + \frac{0.008}{W_r}\ln\left(\frac{7.07}{W_r^{1/4}}\right). \tag{15.10g}$$

For a root dry mass of $W_r = 0.1$ kg root dry matter (m^2 ground)$^{-1}$,

$$r_{s,pl} = 0.064 + 0.203 = 0.27 \text{ kg}^{-1} \text{ m}^4 \text{ s}^{-1}. \tag{15.10h}$$

This can be compared with (15.9g). Both (15.9g) and (15.10h) give unreasonably low values of resistance.

15.4.4 *A simplified approach*

It is not yet possible to treat water uptake by plants with a satisfactory mechanistic analysis. In a recent review, Taylor and Klepper (1978) say 'We suppose that we again may have created a feeling of hopelessness about modeling water uptake by roots in field situations. We do not agree that the situation is hopeless. It is complex and very difficult; ...'. We resort therefore to empiricism, guided by the results that have been obtained using simple models, namely in eqns (15.8c), (15.9b), and (15.10c).

First, we introduce a root density function ρ_r (kg root dry matter m^{-3}) in order that the root has some adaptive capacity. The root depth d_r (m) is defined by

$$d_r = \frac{W_r}{\rho_r}, \tag{15.11a}$$

where W_r (kg root dry matter m^{-2}) is the root dry mass. We can think of root depth d_r as being closely related to the length l of the roots in Fig. 15.6 and eqn (15.8c). We write down the following equation for the soil–plant resistance $r_{s,pl}$ (kg^{-1} m^4 s^{-1}) which is analogous to (15.8b) with two terms:

$$r_{s,pl} = \frac{a}{\rho_r}\left(\frac{W_r + K_r}{W_r}\right) + \frac{b\rho_r}{K_s W_r}, \tag{15.11b}$$

where K_s (kg m^{-3} s) is the soil hydraulic conductivity, and a (m s^{-1}), b (m^2), and K_r (kg dry matter m^{-2}) are constants. This equation is not valid for very low values of root dry matter W_r, when in any case water loss from the ground is likely to be dominated by direct evaporation from the soil rather than movement through the plant. In (15.11b) a more extensive root system (i.e. for a given root mass W_r, a system with a lower root density ρ_r and a larger root depth d_r) has a higher root surface–plant resistance (the first term in (15.11b); the pathways are longer) and a lower bulk soil–root surface resistance (the second term in (15.11b); there are more pathways in parallel). It should be noted that the soil–plant resistance $r_{s,pl}$ in (15.11b) is not dependent on the plant relative water content θ_{pl} (see (15.15d) below). It may be thought that both terms in (15.11b) should be proportional to some power of $1/\theta_{pl}$: as the plant is dehydrated, θ_{pl} falls below unity, and presumably the pathways of water transport become more restricted and also the root surface and the root radius decrease.

In order to use (15.11b) in a plant model, some additional equations are needed. It is assumed that root growth rate $\mathrm{d}W_r/dt$ is given by

$$\frac{\mathrm{d}W_r}{\mathrm{d}t} = G_r - S_r, \tag{15.11c}$$

where G_r (kg root dry matter m^{-2} day^{-1}) is the rate of production of new root material, S_r (kg root dry matter m^{-2} day^{-1}) is the rate of root senescence, and t (days) is the time variable. New root material has a root density of $\rho_{r,new}$

(kg dry matter m^{-3}) which is influenced by the soil water content θ_s ((m^3 water) (m^3 soil)$^{-1}$), according to (for instance)

$$\rho_{r,new} = \rho_{r,max} - c(\theta_{s,max} - \theta_s); \tag{15.11d}$$

$\rho_{r,max}$ (kg dry matter m^{-3}) is the maximum value of root density, obtained when the soil is at field capacity with $\theta_s = \theta_{s,max}$ (see eqn (15.2b) *et seq*., and c is a constant (kg dry matter (m^3 water)$^{-1}$). Finally, the root density ρ_r is updated by means of the differential equation

$$\frac{d\rho_r}{dt} = \frac{G_r}{W_r}(\rho_{r,new} - \rho_r); \tag{15.11e}$$

this equation gives an asymptotic approach of ρ_r to $\rho_{r,new}$ at a rate of G_r/W_r.

The flux density $F_{s,pl}$ (kg water (m^2 ground)$^{-1}$ day^{-1}) of water from the soil into the plant is given by

$$F_{s,pl} = \frac{86\,400(\psi_s - \psi_{pl})}{r_{s,pl}}. \tag{15.11f}$$

Numerical considerations To find numerical values for the parameters of the above equations, we base our considerations on ryegrass growing under UK conditions. A volume flux density of 0.002 m^3 water (m^2 ground)$^{-1}$ day^{-1} is equivalent to (multiplying by 1000 kg m^{-3} for the density of water)

$$F_{s,pl} = 2 \text{ kg } (m^2 \text{ ground})^{-1} \text{ day}^{-1}. \tag{15.12a}$$

If a water potential gradient of 20 J (kg water)$^{-1}$ is assumed, (15.11f) leads to

$$r_{s,pl} = 0.864 \times 10^6 \text{ kg}^{-1} \text{ m}^4 \text{ s}^{-1}. \tag{15.12b}$$

For grassland, with typically a root dry mass W_r of 1 kg dry matter m^{-2} and a root depth d_r of 0.2 m, this gives

$$\rho_r = 5 \text{ kg dry matter } (m^3 \text{ soil})^{-1}. \tag{15.12c}$$

Assume that the two terms in (15.11b) are equal, and that each is 0.432×10^6 kg^{-1} m^4 s^{-1} (half of (15.12b)). Taking $W_r = K_r = 1$ kg dry matter m^{-2} and with (15.12c), we have

$$a = 1.08 \times 10^6 \text{ m s}^{-1}. \tag{15.12d}$$

In the second term of (15.11b), take $K_s = 10^{-3}$ kg m^{-3} s to give

$$b = 86.4 \text{ m}^2. \tag{15.12e}$$

Finally, in eqn (15.11d), assume that $\rho_{r,max} = 5$ kg dry matter m^{-3}, $\theta_{s,max} = 0.5$ m^3 water (m^3 soil)$^{-1}$, and $\rho_{r,new} = 2.5$ kg dry matter m^{-3} with $\theta_s = 0$ m^3 water (m^3 soil)$^{-1}$, which gives

$$c = 5 \text{ kg dry matter } (m^3 \text{ water})^{-1}. \tag{15.12f}$$

15.5 The flow of water between the plant and the atmosphere

Referring to Fig. 15.1, the second flux that is required to define the model is that between the plant and the atmosphere, $F_{\mathrm{pl,atm}}$ (kg water (m^2 ground)$^{-1}$ day^{-1}). Chapter 14 gives a detailed account of crop transpiration and the Penman–Monteith equation (14.4k), on the basis of which we proceed and which takes the form

$$E = \frac{s\phi_{\mathrm{N}} + \lambda\gamma g_{\mathrm{a}}\,\Delta\rho_{\mathrm{va}}}{\lambda[s + \gamma(1 + g_{\mathrm{a}}/g_{\mathrm{c}})]}. \tag{15.13a}$$

The symbols are as follows. E is the transpiration rate (kg water (m^2 ground)$^{-1}$ s^{-1}), and

$$s = \frac{\mathrm{d}\rho'_{\mathrm{v}}}{\mathrm{d}T}(T = T_{\mathrm{a}}) = s(T_{\mathrm{a}}) \text{ kg water m}^{-3}\text{ K}^{-1} \tag{15.13b}$$

is the slope of the saturation vapour pressure density : temperature response at air temperature T_{a} where $s(T_{\mathrm{a}})$ denotes that s is a function of temperature ((14.4b) and Table 14.3).

$$\phi_{\mathrm{N}} = R_{\mathrm{N}} - G \text{ W (m}^2\text{ ground)}^{-1}. \tag{15.13c}$$

is the energy available to the crop (eqn (14.3a)) with net radiation R_{N} (14.1a) and soil heat flux G; for closed canopies G is small, and we assume that

$$\phi_{\mathrm{N}} = R_{\mathrm{N}}. \tag{15.13d}$$

λ (J (kg water)$^{-1}$) is the latent heat of evaporation of water (Table 14.3); we assume that λ varies with air temperature, $\lambda(T_{\mathrm{a}})$, as given in Table 14.3. γ (kg m^{-3} K^{-1}) is the psychrometric parameter, which also varies with air temperature, $\gamma(T_{\mathrm{a}})$ (Table 14.3).

$$\Delta\rho_{\mathrm{va}} = \rho'_{\mathrm{v}}(T_{\mathrm{a}}) - \rho_{\mathrm{va}} \text{ (kg water m}^{-3}) \tag{15.13e}$$

is the atmospheric vapour density deficit ((14.4d) and Table 14.3), and ρ_{va} (kg water m^{-3}) is the actual vapour density in the atmosphere. g_{a} (m s^{-1}) is the boundary layer conductance, with (eqn (14.9n,o))

$$g_{\mathrm{a}} = \frac{\kappa^2 u_{\mathrm{a}}}{\ln[(Z + \zeta - d)/\zeta]\ln[Z + \zeta_{\mathrm{m}} - d)/\zeta_{\mathrm{m}}]}, \tag{15.13f}$$

where

$$\kappa = 0.4 \qquad \zeta = 0.026h \qquad \zeta_{\mathrm{m}} = 0.13h \qquad d = 0.64h, \tag{15.13g}$$

h (m) is the crop height, and u_{a} (m s^{-1}) is the air speed at reference height Z (m) above the ground. Finally, g_{c} (m s^{-1}) is the canopy conductance, which we assume is given by (cf. (14.4i))

$$g_{\mathrm{c}} = L(g_{\mathrm{c,ad}} + g_{\mathrm{c,ab}}), \tag{15.13h}$$

where L (m^2 leaf (m^2 ground)$^{-1}$) denotes the canopy leaf area index, and $g_{c,ad}$ and $g_{c,ab}$ (m s^{-1}) are average values of the stomatal conductances for the abaxial and adaxial leaf surfaces in the canopy.

Remembering that there are 86 400 s in 1 day, the daily transpiration rate $F_{pl,atm}$ is given by

$$F_{pl,atm} = 86\,400E = \frac{sJ}{\lambda[s + \gamma(1 + g_a/g_c)]} + \frac{86\,400\lambda\gamma g_a\,\Delta\rho_{va}}{\lambda[s + \gamma(1 + g_a/g_c)]}, \quad (15.13i)$$

where J (J day^{-1}) is the net radiation receipt on a particular day.

15.5.1 *Example of the calculation of a day's transpiration for a closed canopy*

Assume that the leaf area index L is 2, the height of the canopy h is 0.2 m, the adaxial and abaxial stomatal conductances are $g_{c,ad} = g_{c,ab} = 0.01$ m s^{-1}, the air temperature T_a is 20 °C, the atmospheric vapour density ρ_{va} is 8.65×10^{-3} kg water m^{-3} (50 per cent relative humidity), the day's net radiation J is 10^7 J m^{-2} (about 200 W m^{-2} for a 14 h day), and the average windspeed u_a measured at a reference height of $Z = 2$ m is 1 m s^{-1}.

From (15.13h), the canopy conductance $g_c = 0.04$ m s^{-1}. Equations (15.13f) (15.13g) give the boundary layer conductance $g_a = 0.0063$ m s^{-1}. At 20 °C (Table 14.3), $\lambda = 2.45 \times 10^{-6}$ J kg^{-1}, $\gamma = 0.495 \times 10^{-3}$ kg m^{-3} K^{-1}, and $s = 1.01 \times 10^{-3}$ kg m^{-3} K^{-1}. The denominator of (15.13i) is therefore 3.88×10^3. The term $\Delta\rho_{va}$ is (eqn (15.13e)) $(17.3 - 8.65) \times 10^{-3} = 8.65 \times 10^{-3}$ kg water m^{-3}. Thus the radiation-driven and vapour-density-gradient-driven terms in (15.13i) are 2.6 and 1.5 kg water (m^2 ground)$^{-1}$ day^{-1}.

It can be seen that for normal summer conditions in the UK, the radiation-driven and vapour-density-gradient-driven transpiration terms are comparable. For this calculation of transpiration, apart from the environmental data, the quantities that must be supplied by a crop model are leaf area index L, stomatal conductances $g_{c,ad}$ and $g_{c,ab}$, and crop height h.

15.6 Plant water relations

In this section, consideration is given to the relationships between variables such as the quantity of water in the plant, the relative water content of the plant, the components of plant water potential (pressure and osmotic), the structural and storage components of plant dry matter, and cell wall elasticity (discussions of this topic with Bob Grange were much appreciated).

15.6.1 *Some definitions*

Young's modulus Y and a cell rigidity modulus ε Given a rod of cross-sectional area a_0 (m^2) and length l_0 (m) subject to a force F (kg m s^{-2}) along its length, the increase Δl (m) in length resulting from the application of the force is defined by the equation

$$\frac{F}{a_0} = Y\frac{\Delta l}{l_0}, \tag{15.14a}$$

where Y (kg m^{-1} s^{-2} or Pa) is Young's modulus and the force per unit area F/a_0, is equivalent to a (negative) pressure (Pa). A higher value of Y means a more rigid material, with a smaller extension for a given applied force.

Consider a solid homogenous cube of volume V_0 subject to an external pressure ΔP (Pa). The bulk modulus K (Pa) is defined by

$$\Delta P = -K\frac{\Delta V}{V_0}. \tag{15.14b}$$

Usually the bulk modulus K and Young's modulus Y are about equal. (In fact, Y and K are related by $3K(1 - 2\nu) = Y$, where ν is the Poisson ratio (see (15.14e) and below)) and $0 < \nu < 0.5$.

We now define a rigidity modulus ε (Pa) for a plant cell by

$$\Delta P = \varepsilon\frac{\Delta V}{V}, \tag{15.14c}$$

where ΔP (Pa) is an increment in pressure within a plant cell and ΔV(m^3) is the corresponding increase in volume V (m^3). ε is often referred to as the 'bulk' modulus, although it is not a true bulk modulus, as in (15.14b). ε can be related to Young's modulus, the Poisson ratio, and the cell geometry as follows.

Consider a spherical cell of radius r_0 (m) with wall thickness h_0 (m). If the inside of the cell is subject to a positive pressure P (Pa), then the wall is subject to a tension T (force per unit length (kg s^{-2})) given by (cf. (E15.1b) with $h\rho g = P$))

$$P = \frac{2T}{r_0}. \tag{15.14d}$$

Imagine a small square element of cell wall, of unstretched dimensions $l_0 \times l_0 \times h_0$. The outward force F (kg m s^{-2}) acting on each side of the square is (to first order) $F = l_0 T$; the cross-sectional area is $a_0 = l_0 h_0$. Substituting directly in (15.14a) gives $\Delta l/l_0 = T/Yh_0$, but as there are two stretching forces acting at right angles in this case it is necessary to correct by the Poisson ratio ν (dimensionless) to give

$$\frac{\Delta l}{l_0} = \frac{(1 - \nu)T}{Yh_0}. \tag{15.14e}$$

(Poisson's ratio gives the ratio of the lateral contraction (e.g. $-\mathrm{d}x/x$) that occurs when a stress is applied in, say, the z direction to give a strain $\mathrm{d}z/z$ and the sides are free to move; if the volume change is negligible, then $\nu = 0.5$ because $\mathrm{d}z/z + \mathrm{d}x/x + \mathrm{d}y/y = 0$, $\mathrm{d}x/x = \mathrm{d}y/y$ (assumed), and therefore $\nu = (-\mathrm{d}x/x)/(\mathrm{d}z/\mathrm{d}z) =$

0.5; usually v is less than 0.5.) To first order

$$\frac{\Delta V}{V_0} = \frac{3\Delta l}{l_0}. \tag{15.14f}$$

Therefore (using (15.14d)–(15.14f))

$$\frac{\Delta V}{V_0} = \frac{3\Delta l}{l_0} = \frac{3(1-v)T}{Yh_0} = \frac{3(1-v)Pr_0}{2Yh_0}. \tag{15.14g}$$

Thus, comparing (15.14g) and (15.14c),

$$\varepsilon = \frac{2Yh_0}{3(1-v)r_0}. \tag{15.14h}$$

The apparent rigidity modulus of the cell depends on the wall thickness and the cell size, as well as on Young's modulus and the Poisson ratio for the cell wall material. Therefore, for a plant cell we use the equation

$$\frac{\Delta V}{V} = \frac{1}{\varepsilon}\Delta P, \tag{15.14i}$$

where ε is given by an equation such as (15.14h), depending on cell geometry and cell wall elastic properties. Assuming that ε is constant (the evidence is that, as the cell is stretched, the cell wall becomes more rigid so that ε increases), then integrating (15.14i) from $V = V_0$, $P = 0$ to $V = V_0 + \Delta V$, $P = P$, and approximating $\ln(V/V_0)$ by $\Delta V/V_0$, gives

$$\frac{\Delta V}{V_0} = \frac{1}{\varepsilon}P. \tag{15.14j}$$

This can be written as

$$P = \varepsilon\left(\frac{V}{V_0} - 1\right). \tag{15.14k}$$

Plant water potential and plant relative water content In plants the water potential ψ_{pl} (J kg^{-1}) comprises three components:

$$\psi_{pl} = \psi_P + \psi_O + \psi_m, \tag{15.15a}$$

where ψ_P denotes the pressure potential, ψ_O the osmotic potential due to solutes, and ψ_m the matric potential, all with units of J (kg water)$^{-1}$. We assume that, for plant water, the matric potential ψ_m can be neglected; this may not be true of apoplastic water (water held in the cell wall), but this is usually only 10–15 per cent of the total water, the rest being in the cytoplasm. Thus we approximate

$$\psi_{pl} = \psi_P + \psi_O. \tag{15.15b}$$

The hydrostatic pressure P (Pa) is related to the pressure potential ψ_P by

$$P = \rho_w \psi_P, \tag{15.15c}$$

where ρ_w (kg water m^{-3}) is the density of water.

The plant relative water content is denoted by θ_{pl}, defined by

$$\theta_{pl} = \frac{\text{actual quantity of water in the plant}}{\text{quantity of water in the plant at full turgor}}, \tag{15.15d}$$

where full turgor means that the plant is in equilibrium with water at zero water potential. θ_{pl} is truly dimensionless (cf. soil relative water content in (15.2a) and above); the quantity of water can be measured by volume or by mass. Experimentally, θ_{pl} is measured by drying and weighing, with

$$\theta_{pl} = \frac{\text{fresh weight } (\psi_{pl}) - \text{dry weight}}{\text{fresh weight } (\psi_{pl} = 0) - \text{dry weight}}. \tag{15.15e}$$

In our notation, we define plant relative water content by (Fig. 15.1)

$$\theta_{pl} = \frac{Q_{pl}(\psi_{pl})}{Q_{pl}(\psi_{pl} = 0)}. \tag{15.15f}$$

Structure and storage: the concentration of osmotica The plant dry matter W (kg (m^2 ground)$^{-1}$) is regarded as being divisible into a structural fraction W_G (kg (m^2 ground)$^{-1}$) and a storage fraction W_S (kg (m^2 ground)$^{-1}$), where

$$W = W_G + W_S. \tag{15.16a}$$

A fraction f_{cw} of the structural dry matter W_G is cell wall material, and therefore, assuming a spherical cell of unstretched radius r_0 (m) and wall thickness h_0 (m),

$$f_{cw} W_G = n_c 4\pi r_0{}^2 h_0 \rho_{cw}, \tag{15.16b}$$

where ρ_{cw} (kg m^{-3}) is the density of the cell wall material and n_c ((m^2 ground)$^{-1}$) is the number of cells per unit ground area. The unstretched cytoplasmic volume V_0 is

$$V_0 = n_c \tfrac{4}{3}\pi r_0{}^3. \tag{15.16c}$$

If we assume the ratio f_h of wall thickness to cell radius is given by

$$h_0 = f_h r_0, \tag{15.16d}$$

then

$$f_{cw} W_G = 3 V_0 f_h \rho_{cw}. \tag{15.16e}$$

With $V = Q_{pl}/\rho_w$, we have

$$\frac{V}{V_0} = \frac{3 f_h \rho_{cw} Q_{pl}}{\rho_w f_{cw} W_G}. \tag{15.16f}$$

Hence, from (15.14k),

$$\psi_P = \frac{\varepsilon(cQ_{pl}/W_G - 1)}{\rho_w} \quad \text{with } \psi_P \geqslant 0, \tag{15.16g}$$

where the constant c (kg structure (kg water)$^{-1}$) is given by

$$c = \frac{3f_h\rho_{cw}}{f_{cw}\rho_w}. \tag{15.16h}$$

Notwithstanding this model, we regard c in (15.16g) as an adjustable parameter.

A fraction f_O of the storage dry matter W_S is soluble (osmotically active) with a mean relative molecular mass of μ (kg (kg mol)$^{-1}$). The concentration of osmotica is $f_O W_S/\mu V$ (kg mol m^{-3}) (p. 48) and hence the plant osmotic potential (using the van't Hoff equation for osmotic pressure $\psi_P = P/\rho_w$ and $V\rho_w = Q_{pl}$) is

$$\psi_O = -\frac{RTf_O W_S}{\mu Q_{pl}}. \tag{15.16i}$$

R is the gas constant (8314 J (kg mol)$^{-1}$ K^{-1}) and T is the absolute temperature (K).

15.6.2 *Derivation of the plant water characteristic*

It is assumed that all the water is cytoplasmic and accessible to the osmotica, and that there is thermodynamic and mechanical equilibrium.

Combining (15.16g) and (15.16i) with (15.15b) gives a quadratic equation relating plant water content Q_{pl} to plant water potential ψ_{pl}:

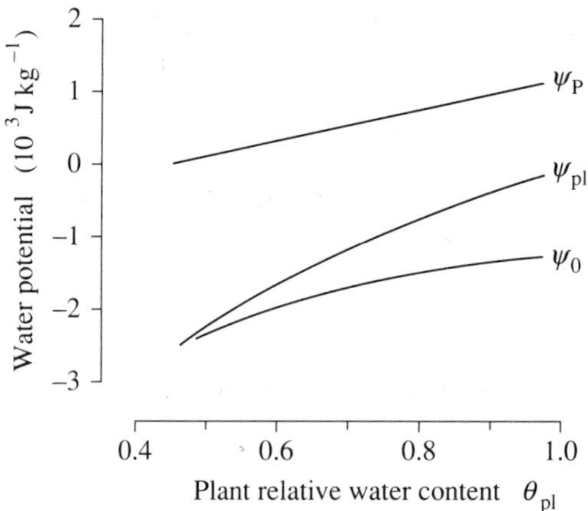

Fig. 15.7. Plant water characteristic. The dependence of plant water potential ψ_{pl} and its two components, the pressure potential ψ_P and the osmotic potential ψ_O, on the plant relative water content θ_{pl}, are shown. The results were obtained using (15.17a) and the procedure outlined after (15.17a), with $\varepsilon = 10^6$ Pa, $c = 0.15$ kg structure (kg water)$^{-1}$, $\rho_w = 1000$ kg m^{-3}, $W_G = 0.4$ kg structure (m^2 ground)$^{-1}$, $R = 8314$ J (kg mol)$^{-1}$ K^{-1}, $T = 293.15$ K, $f_O = 1$, $W_S = 0.0577$ kg storage (m^2 ground)$^{-1}$, and $\mu = 20$ kg (kg mol)$^{-1}$.

$$0 = \frac{\varepsilon c}{\rho_w W_G} Q_{pl}^2 - \left(\psi_{pl} + \frac{\varepsilon}{\rho_w}\right) Q_{pl} - \frac{RTf_0 W_S}{\mu}. \tag{15.17a}$$

Q_{pl} is given by the positive root of this quadratic, and can be obtained for various values of ψ_{pl}.

To obtain the plant water characteristic, a range of values of the state variable Q_{pl} is taken, ψ_P, ψ_O, and ψ_{pl} are obtained from (15.16g), (15.16i), and (15.15b), the quantity of water $Q_{pl}(\psi_{pl} = 0)$ in the plant at full turgor is found by solving (15.17a) with $\psi_{pl} = 0$, and substitution into (15.15f) then gives the plant relative water content θ_{pl}. Typical responses are shown in Fig. 15.7; it is seen that they approximate to measured plant water characteristics.

15.6.3 *Dependence of the plant water characteristic on plant status*

The plant relative water content θ_{pl} is calculated relative to the quantity of water in the plant at full turgor $Q_{pl}(\psi_{pl} = 0)$ from (15.15f). It is possible for this to vary even when the quantity of water in the plant is constant, owing to changing osmotic and cell wall factors. Thus θ_{pl} may be a rather labile variable, and not as useful as is sometimes assumed.

From (15.16g), it can be seen that wilting occurs when

$$Q_{pl} = W_G/c. \tag{15.18a}$$

However, increased osmotic effects (f_O, W_S larger; μ, ε, W_G smaller) cause $Q_{pl}(\psi_{pl} = 0)$ to increase (from (15.17a)). Thus wilting occurs at lower values of relative water content under such conditions. There are clearly several mechanisms by which osmoregulation can enable a plant to adapt to drought conditions.

Dainty (1972, p. 92, 1976, Tables 2.1, 2.2, 2.3) gives values for ε ranging from zero to 1000×10^5 Pa (1 bar $\equiv 10^5$ Pa), although typical values for leaves are 20×10^5 Pa measured at low values of turgor pressure P and 100×10^5 Pa measured at high P. The increases in ε (rigidity) as pressure increases and the cell is stretched may be due to changes in Young's modulus Y or the Poisson ratio (cf. (15.14h) for the spherical cell model); cell stretching itself (increasing radius r_0, decreasing wall thickness h_0) seems likely to lead to a reduction in ε.

Short-term adaptive changes (over a few days) might be achieved by adjustment of osmotic effects through W_S (the storage dry matter), f_O (fraction of W_S that is osmotically active), and μ. In the longer term (weeks and months: ontogenesis, ageing, adaption) changes in the structural dry matter W_G and in the fraction of this material in the cell wall f_{cw} may be more important (15.16h).

15.6.4 *Effects of a non-constant cell elasticity*

The analysis leading to the relations shown in Fig. 15.7 between the components of the plant water potential and the plant relative water content is only valid if the cell elasticity parameter ε is constant, and for small volume changes (cf. (15.14j)). However, as mentioned above, ε is not constant and we indicate here how the more general problem can be treated.

Equation (15.14j) is equivalent to

$$\frac{V - V_0}{V_0} = \frac{1}{\varepsilon}P \quad \text{or} \quad V = V_0 + \frac{V_0}{\varepsilon}P. \tag{15.19a}$$

We need the general equilibrium volume : pressure curve, denoted by

$$V = f(P), \tag{15.19b}$$

where $f(P)$ is some function of pressure P, to replace (15.19a). Tyree and Hammett (1972) have used an equation equivalent to

$$V = V_0 + kP^n, \tag{15.19c}$$

where k is a constant and n varies from 0.27 to 0.53 for some tree species. This is clearly an unacceptable function at low values of P where $\mathrm{d}V/\mathrm{d}P$ approaches infinity as $P \to 0$. A well-behaved $V = f(P)$ function can be constructed by assuming that the parameter ε increases as the cell is stretched according to

$$\varepsilon = \varepsilon_0 + \varepsilon_1\left(\frac{V - V_0}{V_0}\right), \tag{15.19d}$$

where ε_0 and ε_1 (Pa) are constants. With $\varepsilon_1 = \varepsilon_0$, a doubling in cell volume ($V = 2V_0$) doubles the cell rigidity. Substituting in (15.19a) and solving for $V - V_0$ gives

$$\frac{V - V_0}{V_0} = \frac{-\varepsilon_0 + (\varepsilon_0{}^2 + 4\varepsilon_1 P)^{1/2}}{2\varepsilon_1}. \tag{15.19e}$$

This is shown in Fig. 15.8. The curvilinearity of the response could be further increased by quadratic or higher terms in (15.19d), e.g. $\varepsilon_2[(V - V_0)/V_0]^2$.

In order to treat the general case (15.19b), substitute $\psi_P = P/\rho_w$, (15.16i), and $Q_{pl} = V\rho_w$ into (15.15b) to give

$$P = \rho_w\psi_{pl} + \frac{RTf_0 W_S}{\mu V}. \tag{15.19f}$$

Inverting (15.19b) to give

$$P = f'(V), \tag{15.19g}$$

where f' denotes the inverse function, and substituting in (15.19f) gives

$$f'(V) = \rho_w\psi_{pl} + \frac{RTf_0 W_S}{\mu V}. \tag{15.19h}$$

This equation is solved for V or Q_{pl}/ρ_w. The analysis can then continue from (15.17a) to give the plant water characteristic for the general case (Exercise 15.3).

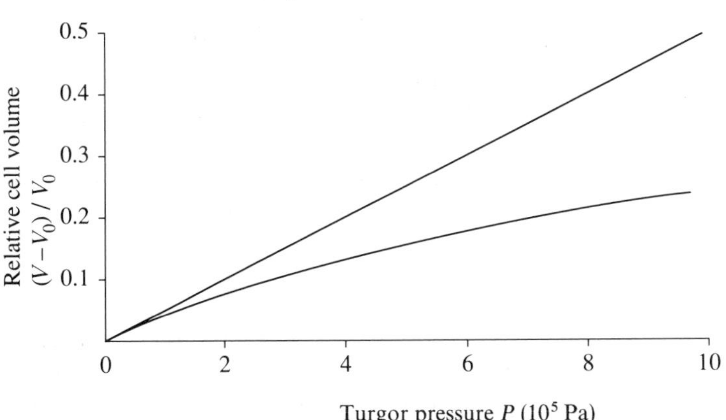

Fig. 15.8. Effects of non-constant cell elasticity. The straight line shows (15.19a) with $\varepsilon = 2 \times 10^6$ Pa. The curve depicts the effect of assuming that stretching increases the rigidity of the cell wall, as given by (15.19d) with $\varepsilon_0 = 2 \times 10^6$ Pa and $\varepsilon_1 = 8 \times 10^6$ Pa, and is calculated from (15.19e).

15.7 Physiological effects of plant water status

In this section we discuss the effects of plant water status on the rates of physiological processes so that the objective of modelling plant and crop growth in water-limited conditions can be achieved. Plant water status is a multidimensional quantity which can be described in several ways: the plant relative water content (θ_{pl} of (15.15d)), the plant water potential ψ_{pl} ((J kg water)$^{-1}$), and the components of plant water potential (pressure and osmotic potentials (15.15a)). For some purposes the chemical activity of water a_w may be important; this is defined by

$$a_w = \exp\left(\frac{\mu\psi_{pl}}{RT}\right), \qquad (15.20a)$$

where μ is the relative molecular mass of water (18 kg (kg mol)$^{-1}$), R is the gas constant, and T is the absolute temperature. Note that at 20 °C, $RT = 2.44 \times 10^6$ J (kg mol)$^{-1}$ and, with $\psi_{pl} = -1000$ J (kg water)$^{-1}$ ($\equiv -10$ bar), $a_w = 0.993$. For our purposes, we can write

$$1 - a_w = -\frac{\mu\psi_{pl}}{RT}. \qquad (15.20b)$$

15.7.1 *Photosynthesis*

The leaf photosynthesis response equation recommended for crop studies is the non-rectangular hyperbola of (9.10f) (p. 228), i.e.

$$\theta P^2 - (\alpha I_l + P_m)P + \alpha I_l P_m = 0, \tag{15.21a}$$

where P (kg CO_2 (m^2 leaf)$^{-1}$ s^{-1})) is the photosynthetic rate, I_l (J m^{-2} s^{-1}) is the light flux density incident on the leaf, α (kg CO_2 J^{-1}) is a biochemical efficiency parameter, P_m (kg CO_2 (m^2 leaf)$^{-1}$ s^{-1}) is a parameter giving the maximum value of P ($I_l \to \infty$), and θ is a dimensionless parameter. P_m and θ depend on a diffusion resistance r_d (s m^{-1}) according to (9.10d) and (9.10e):

$$P_m = \frac{R_d r_d + C_a}{r_d + r_x} \qquad \theta = \frac{r_d}{r_d + r_x}, \tag{15.21b}$$

where R_d (kg CO_2 (m^2 leaf)$^{-1}$ s^{-1}) is a dark respiration rate, r_x (s m^{-1}) is a biochemical resistance, and C_a (kg CO_2 m^{-3}) is the ambient CO_2 concentration. The diffusion resistance is often considered as comprising three components:

$$r_d = r_m + r_s + r_a, \tag{15.21c}$$

where r_m (s m^{-1}) is a mesophyll resistance, r_s (s m^{-1}) is a stomatal resistance, and r_a (s m^{-1}) is a CO_2 boundary layer resistance (see Chapter 14, p. 416).

The approach sometimes adopted is to assume that plant water status only affects the stomatal resistance r_s and ignore possible effects on the other parameters of (15.21a)–(15.21c). It seems that there is no unique relationship between stomatal resistance and plant water potential (Ludlow 1987), and that it might be better to relate stomatal resistance to plant relative water content θ_{pl}. For instance, the equation

$$r_s = r_{s,min} + r_{s,\theta}(1 - \theta_{pl}), \tag{15.21d}$$

with the minimum value of $r_s = r_{s,min} \approx 0$ s m^{-1} at $\theta_{pl} = 1$ and $r_{s,\theta} = 25\,000$ s m^{-1}, gives $r_s = 5000$ s m^{-1} at a relative water content θ_{pl} of 0.8, which is equivalent to fully closed stomata.

However, it may be thought that this approach (a) strains the model of photosynthesis too far and (b) ignores possibly potent effects of water activity ((15.20a) and (15.20b)) on enzyme activity and thence on the biochemical resistance r_x. We suggest that an equally effective and rather simpler approach may be to write the parameter P_m (maximum photosynthetic rate) of (15.21a) as

$$P_m = P_{m,max}[1 - c(1 - \theta_{pl})], \tag{15.21e}$$

where c is a dimensionless constant which reduces P_m below its maximum value of $P_{m,max}$ as θ_{pl} falls below unity; for example, with $c = 2$, a θ_{pl} of 0.5 gives a P_m of zero.

There is a view that stomatal aperture remains unaffected over a wide range of total leaf water potential, but then stomatal closure occurs rapidly once a critical leaf water potential is reached. Zur and Jones (1981, eqn (7)) assume that stomatal resistance is function of the pressure potential ψ_P below a certain value of ψ_P.

Boyer, Armond, and Sharp (1987, Fig. 5.4) observed that the quantum yield (equivalent to the parameter α in (15.21a)) is depressed to about 30 per cent of its maximal value in sunflower leaves as the leaf water potential ψ_l, is reduced. Their data show that the conversion efficiency α is reduced approximately linearly from a maximum value of α_{max} at full turgor, where $\psi_l = \psi_{l,0} = -500\ J\ kg^{-1}$, towards an asymptotic value of $\alpha_{min} \approx 0.3\alpha_{max}$ when $\psi_l \approx -3000\ J\ kg^{-1}$. This behaviour can be described by the equation

$$\alpha = \alpha_{max} - (\alpha_{max} - \alpha_{min})\frac{(\psi_l - \psi_{l,0})}{(\psi_l - \psi_{l,0}) + (\psi_{l,h} - \psi_{l,0})}, \qquad (15.21f)$$

where $\psi_{l,h}$ ($J\ kg^{-1}$) is the value of the leaf water potential ψ_l, where α is half-way between α_{max} and α_{min}. The data of Boyer *et al.* (1987) suggest that $\psi_{l,h} \approx 1300\ J\ kg^{-1}$.

If seems that there is a need for the measurement of photosynthetic responses of the leaf to light and CO_2, the fitting of models such as (15.21a) to the data, and examination of the dependence of the fitted parameters on the water status of the leaf for a range of conditions.

15.7.2 *Transpiration*

As calculated by the Penman–Monteith equation (15.13a), this depends on the canopy conductance g_c, which is assumed to be obtainable from the leaf area index, and the abaxial and the adaxial stomatal conductances (to water vapour) as in (15.13h). With the total stomatal conductance to CO_2 given by

$$g_{s,CO_2} = 1/r_s, \qquad (15.22a)$$

and, for instance, the CO_2 stomatal resistance described by (15.21d), the stomatal conductance for water vapour transfer is

$$g_{s,H_2O} = g_{s,CO_2}\left(\frac{\mu_{CO_2}}{\mu_{H_2O}}\right)^{1/2}, \qquad (15.22b)$$

where μ_{CO_2} and μ_{H_2O} are the relative molecular masses of CO_2 and H_2O (44 kg (kg mol^{-1}) and 18 kg (kg mol)$^{-1}$) respectively.

15.7.3 *Leaf area expansion*

There is evidence that a major effect of water stress on leaf area expansion is through cell enlargement, and cell division is little affected except under very severe stress (Hsiao and Bradford 1983). In crop growth models generally, the leaf area expansion submodel is one of the more crucial submodels, and the assumptions made here can greatly affect predictions of growth and yield. Even for unstressed leaves, there have been few mechanistic models of leaf area expansion, although Lainson and Thornley (1982) investigated cucumber leaf growth using a model which differentiated between the primary and secondary cell wall components and associated the primary component with leaf area and the secondary component with leaf dry matter per unit area. However, although this model suggests ways in which water stress and leaf expansion might be modelled mechanistically, such models are too complex for use in crop growth models, which is our objective here.

Some authors (e.g. Zur and Jones 1981, eqn (9)) adjust a relative leaf area expansion rate according to turgor pressure ψ_P, using empirical coefficients that can vary with the stage of growth and with species. Others use concepts such as a yield threshold for the pressure potential, the water potential difference between the leaf cells and their surroundings, a conductance parameter determining how easily water moves into the leaf cells, and a cell wall elasticity parameter (see Hsiao and Bradford 1983, eqns (5) and (6)). However, Ludlow (1987) suggests that there are no unique (and therefore useful) relationships between leaf area expansion rate and either the total leaf water potential or its pressure component. Barlow (1986) concludes that leaf expansion is more controlled by properties such as extensibility and yield threshold (which we feel is a dubious concept), rather than by water relations parameters such as turgor pressure and hydraulic conductivity.

A truly mechanistic model of leaf expansion would make use of local variables (local variables are variables that relate to the leaf directly, i.e. there is no 'action at a distance' as in, say, assuming that soil relative water content affects leaf area expansion rate); the model of Lainson and Thornley (1982) is mechanistic and deals with local variables only, and, although their model does not take account of leaf water status, it could be extended in this direction. As an example of a simpler and possibly more suitable technique, we outline an extension of the method used by Johnson and Thornley (1985) for leaf area expansion. Essentially, they used an equation for the production of new leaf area of the form

$$\frac{\mathrm{d}A}{\mathrm{d}t} = \eta_{\mathrm{m}}(1 - \zeta C) f_{\mathrm{lam}} f_{\mathrm{sh}} \frac{\mathrm{d}W_{\mathrm{G}}}{\mathrm{d}t}, \tag{15.23a}$$

where A (m^2) is the leaf area, t (days) is the time variable, η_{m} (m^2 leaf (kg structural dry matter)$^{-1}$) is a constant describing the structural specific leaf area, ζ (kg carbohydrate C (kg structural dry matter)$^{-1}$) is a constant that describes the effect of carbohydrate substrate C (kg carbohydrate C (kg structural dry matter)$^{-1}$) on leaf area expansion, f_{lam} is the fraction of new shoot structural dry matter that is incorporated into the lamina, f_{sh} is the fraction of new plant structural dry matter that is partitioned to the shoot, and $\mathrm{d}W_{\mathrm{G}}/\mathrm{d}t$ (kg structural dry matter day^{-1}) is the gross rate of synthesis of new structure for the whole plant.

When a plant is stressed for water, the leaf area expansion rate falls. Equation (15.23a) demonstrates that this can expected in any case as a secondary effect: enzyme activities and growth rates $\mathrm{d}W_{\mathrm{G}}/\mathrm{d}t$ are generally reduced as water activity falls; partitioning changes in favour of the root, with f_{sh} being reduced; partitioning to the lamina f_{lam} may fall, with other shoot components being favoured; carbohydrate levels C may rise due to reduced growth and maintained photosynthesis, giving a reduction in $\mathrm{d}A/\mathrm{d}t$. The question remaining is: are there, in addition, direct effects of plant or leaf water status on the parameter η_m? It seems that this question has not been definitively answered.

We suggest that a method of introducing a direct effect is to add a factor to (15.23a) to give

$$\frac{\mathrm{d}A}{\mathrm{d}t} = \eta_m[1 - c_A(1 - \theta_{pl})](1 - \zeta C) f_{lam} f_{sh} \frac{\mathrm{d}W_G}{\mathrm{d}t}. \qquad (15.23b)$$

where c_A is a constant (probably about unity) and θ_{pl} is the plant relative water content (15.15d) and (15.15e).

15.7.4 *Root : shoot partitioning*

The application of water stress alters the root : shoot ratio in favour of the root. There is no consensus as to how this response can best be represented, and here we indicate how the partitioning models of Chapter 13 can be modified to allow for the consequences of water stress.

The teleonomic model of (13.5n) can be extended in an intuitive way to give

$$P = \frac{\mathrm{d}W_{sh}/W_{sh}}{\mathrm{d}W_r/W_r} = \left[\frac{W_r N/(N + f_N)}{W_{sh} C/(C + f_C)}\right] \theta_{pl}{}^q, \qquad (15.24a)$$

where P is a dimensionless function determining the ratio of the proportional amounts of newly synthesized material added to the shoot (structural dry matter W_{sh}) and the root (structural dry matter W_r), N and C are the whole-plant nitrogen and carbon substrate concentrations, θ_p is the plant relative water content, and q is a constant. As θ_{pl} falls below unity, material is increasingly directed towards root synthesis, as also happens if the substrate nitrogen level N falls. To complete the model, an equation is needed to determine $\mathrm{d}\theta_{pl}/\mathrm{d}t$ (or $\mathrm{d}Q_{pl}/\mathrm{d}t$ of (15.1b), and (15.15e) is used to give θ_{pl}). It might be necessary to allow for the effects of reduced water content on growth by, for instance, multiplying the right-hand side of the growth equations (13.4a) and (13.4b) by θ_{pl} or some function of water activity (15.20a). See also Section 13.5 (p. 392); to some extent, water can be regarded as an extra mineral nutrient.

The mechanistic model of Section 13.4 (p. 389) shown in Fig. 13.11 has six state variables. To allow for the effects of water stress mechanistically at this level of detail, it is necessary to consider two more state variables: Q_{sh} and Q_r, the quantities of water present in the shoot and in the root. Their rates of change will be determined by flux equations like (15.1b). Equations relating the state variables to the dependent water relations quantities (components of water potential, relative water content) can be constructed as outlined in Section 15.6 (p. 440). Equations (13.12a)–(13.12f) can be modified by the relative water contents θ_{sh} and θ_r in the shoot and root: for instance by making quantities such as shoot and root growth rates ((13.12a) and (13.12b)), photosynthetic rate σ_C, nitrogen uptake rate σ_N, and the transport resistances (r_C and r_N of (13.11h)) depend to a greater or lesser extent on, say, the relative water content or water activity (15.20a).

15.7.5 *Respiration*

Referring to the recycling model of respiration (Section 11.4, p. 270; Fig. 11.2), we would expect the rate constants k_g and k_d to decrease as the water activity (15.20a) drops below unity. For a plant of constant composition, the growth efficiency Y_g and the maintenance coefficient (see (11.5k), p. 271) might be expected to be independent of and dependent on water status respectively.

15.7.6 *Enzymatic reactions*

The effects of pressure P and water activity a_w on the rates of enzyme-catalysed reactions have been largely ignored by biochemists and enzymologists, although these may be important in understanding plant responses to water stress. Microbiologists have considered the effects of water activity, and this can be an important factor in determining microbial growth rates. The effects of water activity can also be strongly influenced by pH. For instance, Sinell (1980) reports that at pH 7 *Clostridium botulinum* can grow down to $a_w = 0.94$, whereas at pH 5.3 growth stops at $a_w = 0.99$. Similar effects may be expected to occur in plants. It may be reasonable to assume that most enzyme systems function optimally at an a_w close to unity.

Exercises

15.1. For a liquid of density ρ (kg m^{-3}) and surface tension T (N m^{-1}) placed in a tube of circular cross-section with radius r (m), and assuming zero contact angle between the wall of the tube and the liquid surface, derive eqn (15.3a) for the water potential ψ (J (kg water)$^{-1}$) inside the meniscus relative to a free surface, namely

$$\psi = -\frac{2T}{\rho r}. \tag{E15.1a}$$

Use the formula for the rise in height h (m) in a capillary:

$$h\rho g = \frac{2T}{r} \tag{E15.1b}$$

where g (9.81 m s^{-2}) is the acceleration due to gravity.

15.2. Using eqn (15.3c) for pore shape, i.e.

$$\frac{r}{r_0} = \left(\frac{z}{z_0}\right)^{-q}, \tag{E15.2a}$$

evaluate the integral

$$\theta_s = nv = n\int_{z_s}^{\infty} \pi r^2\, dz \tag{E15.2b}$$

to give an expression for θ_s versus r_s (the pores are filled from $r = 0$ up to $r = r_s$, where (r_s, z_s) satisfies (E15.2a)). Assume that in (E15.2a), $q > 1/2$.

15.3. For a non-linearly stretching cell obeying (15.19a) and (15.19d) construct an equation equivalent to (15.17a) which will allow the plant water characteristics (as in Fig. 15.7) to be calculated.

16
Crop responses

16.1 Introduction

In this chapter, some simple but useful crop response functions are first presented: these are concerned with the relation between crop yield and planting density, plant competition, and static fertilizer responses. Essentially the approach used here is more empirical than mechanistic. In the last section we describe a vegetative crop growth model both as an elementary example of a crop growth simulator and also to demonstrate how such a model can be used to obtain fertilizer responses dynamically. This model, which is highly simplified, illustrates the point that, for particular objectives, it may be desirable to build the simplest model possible; such simplified models can yield more in terms of understanding (because of their greater transparency) than far more comprehensive crop simulators.

16.2 Crop yield and planting density

The relationship between crop yield Y (kg m^{-2}) and planting density ρ (number of plants (m^2 ground)$^{-1}$) is of considerable practical importance. We first survey empirical approaches, and then describe a simple mechanistic analysis (France and Thornley (1984, pp. 156–61) give references to background material).

16.2.1 *Empirical response equations*

Two types of $Y{:}\,\rho$ response have been observed (Fig. 16.1): the hyperbolic response, in which the yield Y approaches an asymptote, and the parabolic response, in which an optimum is reached followed by a subsequent decline (a parabola is often used to fit the data).

If it is assumed that the $Y{:}\,\rho$ equation necessarily passes through the origin ($\rho = 0$, $Y = 0$), a typical parabolic response equation is

$$Y = a\rho - b\rho^2, \tag{16.1a}$$

where a and b are constants, with $a > 0$ and $b > 0$. This equation is symmetric about the maximum, although this constraint can be changed by modifying (16.1a) to

$$Y = a\rho - b\rho^{1/2}, \tag{16.1b}$$

again with the constants $a > 0$ and $b > 0$. Both (16.1a) and (16.1b) behave unreasonably at high values of planting density ρ, as Y can become negative. An

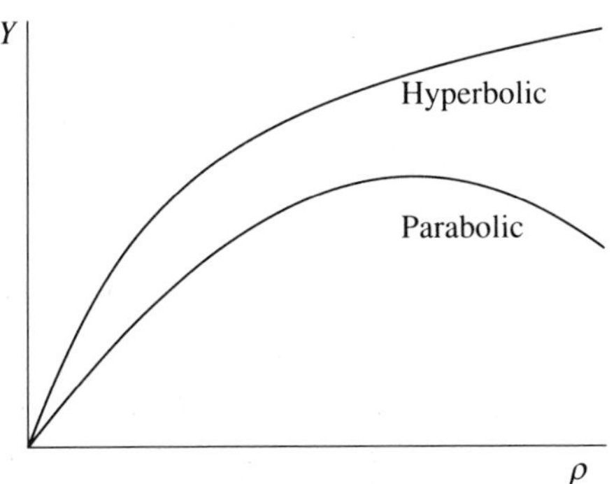

Fig. 16.1. Response of crop yield Y to planting density ρ illustrating the two main responses observed: the hyperbolic response approaches an asymptote; the parabolic response reaches a maximum followed by a decline.

equation that is better behaved but still shows a maximum is

$$Y = a\rho e^{-b\rho}, \tag{16.1c}$$

with the constants $a > 0$ and $b > 0$. (Exercise 16.1.)

Another equation that has been used is based on the allometric equation (eqn (3.7e), p. 88)

$$Y = a\rho^{b}, \tag{16.1d}$$

with the constants $a > 0$ and $0 < b < 1$. This equation has neither a maximum nor an asymptote.

Asymptotic functions are generally more appropriate to biological problems (Section 2.3, p. 51) than eqns (16.1a)–(16.1d). The simple rectangular hyperbola

$$Y = \frac{a\rho}{1 + b\rho}, \tag{16.2a}$$

where a and b are positive constants, has been much used. This familiar equation can be modified to give a maximum:

$$Y = \frac{a\rho}{1 + b\rho + c\rho^{2}}, \tag{16.2b}$$

where c is an additional constant. This equation can describe responses ranging from the asymptotic to the parabolic (Exercise 16.2).

16.2.2 *A simple mechanistic yield : density response equation*

An equation can usually be found from those given above that will successfully fit the data for a particular experimental situation. Attempts have been made to find a biological basis for these responses, and here we describe an approach using the established responses of plants to photosynthesis and nutrient uptake (Thornley 1983).

For a planting density of ρ (number of plants (m^2 ground)$^{-1}$), the ground area A (m^2) associated with each plant is

$$A = 1/\rho. \tag{16.3a}$$

A soil volume V (m^3) may also be associated with each plant, where

$$V = (1/\rho)^{3/2}. \tag{16.3b}$$

While (16.3a) and (16.3b) ignore many factors such as canopy extent and root density, activity, and depth, we regard them as giving a first approximation to a complex situation.

Assume that carbon and nitrogen are the substrates limiting plant growth, and let C (kg carbon plant^{-1}) and N (kg nitrogen plant^{-1}) denote the quantities of carbon and nitrogen substrates available to the plant for growth. We now assume that the plant dry mass w (kg plant^{-1}) depends upon C and N according to

$$w = w_m\left(\frac{1}{1 + K_C/C + K_N/N + K_{CN}/CN}\right), \tag{16.3c}$$

where K_C, K_N, and K_{CN} are constants and w_m is the maximum plant mass obtained when both carbon and nitrogen are present in large amounts. Equation (16.3c) can be tentatively justified by nothing that two-substrate reactions obey this equation (Section 2.3.4, p. 59), and fertilizer responses are often of this type (Section 16.4, p. 462).

Next assume that the quantities C and N of carbon and nitrogen available to each plant are related to the area A and volume V assigned to each plant in (16.3a) and (16.3b) according to

$$C = b_C A = b_C/\rho \qquad N = b_N V = b_N/\rho^{3/2}, \tag{16.3d}$$

where b_C and b_N are constants. b_C may depend on light receipt and b_N may depend on the soil nitrogen level. At low planting densities (16.3d) will break down, because the plants will not be able to exploit the available space fully; however, at low ρ and therefore high C and N (eqn (16.3d)), eqn (16.3c) gives a plant mass w that approaches its asymptotic value w_m, so that this failure is not very important.

Substituting (16.3d) into (16.3c), therefore, we have

$$w = w_m\frac{1}{1 + K_C\rho/b_C + K_N\rho^{3/2}/b_N + K_{CN}\rho^{5/2}/b_C b_N}. \tag{16.3e}$$

Defining simpler parameters by

$$g_1 = \frac{K_C}{b_C} \qquad g_2 = \frac{K_N}{b_N} \qquad g_3 = \frac{K_{CN}}{b_C b_N} \tag{16.3f}$$

gives

$$w = w_m\frac{1}{1 + g_1\rho + g_2\rho^{3/2} + g_3\rho^{5/2}}. \tag{16.3g}$$

Crop yield Y is given by

$$Y = \rho w = w_{\mathrm{m}} \frac{\rho}{1 + g_1 \rho + g_2 \rho^{3/2} + g_3 \rho^{5/2}}. \tag{16.3h}$$

Comparison of (16.3h) with the empirical (16.2b) shows that (16.3h) is able to describe a range of responses, including the hyperbolic and parabolic responses (Fig. 16.1). Equations (16.3g) and (16.3h) are drawn in Fig. 16.2 for a range of the parameters g_1, g_2, and g_3 to show the extremes of the responses that can be obtained (Exercise 16.3).

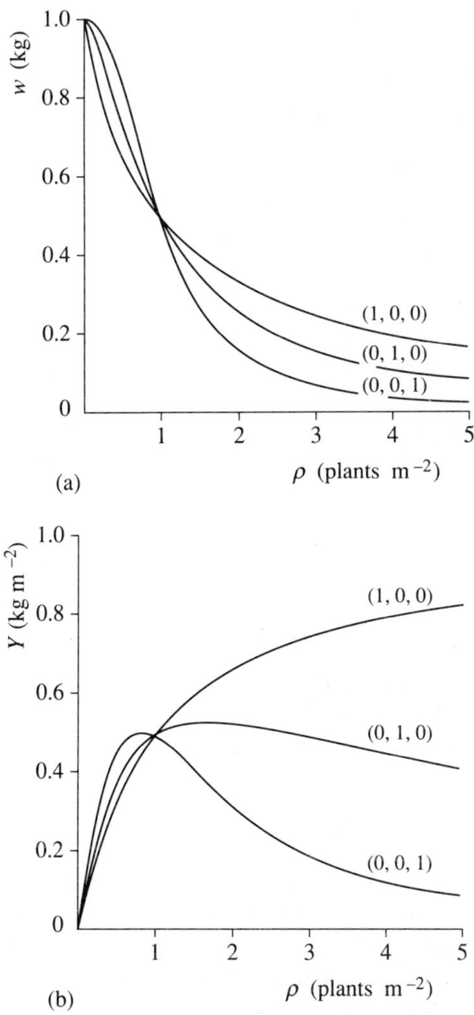

Fig. 16.2. Response of plant mass w and crop yield Y to plant density ρ according to (16.3g) and (16.3h) with $w_{\mathrm{m}} = 1$ and values of (g_1, g_2, g_3) as indicated.

Equation (16.3h) can be used to examine the effects of environment on the crop yield : planting density response by considering the influence of environment on the parameters g_1, g_2, and g_3. For example, it can be assumed that b_C and b_N of (16.3d) are given by

$$b_C = \alpha J \qquad b_N = \beta N_s, \tag{16.3i}$$

where α and β are constants, J is light energy receipt, and N_s is soil nitrogen. Substituting (16.3i) in (16.3f) gives

$$g_1 = \frac{K_C}{\alpha J} \qquad g_2 = \frac{K_N}{\beta N_s} \qquad g_3 = \frac{K_{CN}}{\alpha J \beta N_s}. \tag{16.3j}$$

Varying the growing conditions (J and N_s), with (16.3j), allows the effects of environment on the crop yield : planting density response (16.3h) to be assessed. The (1, 0, 0) curve of Fig. 16.2(b) represents a light-limited high-nutrient case; the (0, 1, 0) curve represents a nutrient-limited high-light case.

16.3 Plant competition

Competition of different plants in a closed canopy for light is considered in physiological terms in Chapter 10, Section 10.3 (p. 258). The competition between plants can take many forms (Harper 1977), and here we discuss the phenomenon empirically. While competition between different plant species can sometimes be interpreted, for instance, in terms of differing physiological features of the canopies or the root systems, it is also of some interest to consider competition between identical plants, as this is related to self-thinning. To this end, we use a simple plant growth function, and, making empirical assumptions about plant competition, we outline two possible approaches to the problem. In particular, we show that, depending on parameter values and initial plant dry mass values, competing plants may coexist in equilibrium or one may destroy the other.

16.3.1 *A simple plant growth function*

If the net growth of a single plant, of dry mass w, is regarded as being the result of growth less a degradation or maintenance term (Fig. 11.2, p. 270), we can write down an empirical growth rate function as

$$\frac{\mathrm{d}w}{\mathrm{d}t} = \frac{kw}{1 + gw^n} - mw, \tag{16.4a}$$

where t is the time variable, k (time^{-1}), g (mass$^{1/n}$), and n are constants in the growth term, and m (time^{-1}) is a degradation rate. For $k > m$ and $n > 0$, (16.4a) is sigmoidal and asymptotic (Fig. 16.3 and Exercise 16.4). Note that with $n = 1/3$, the growth term at large w is proportional to $w^{2/3}$, which may reflect surface-limited growth. Differing values of k, g, and n can represent different types of growth limitation. Note that growth equations such as the logistic, Gompertz, or Chanter equations (pp. 78–85) are not suitable for examining plant competi-

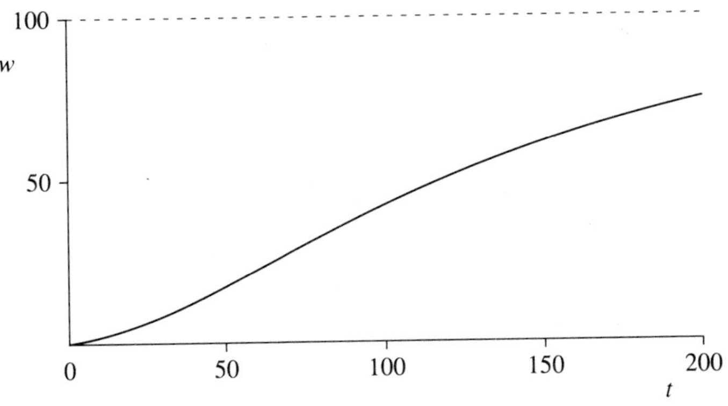

Fig. 16.3. Plant growth curve. Equation (16.4a) is drawn for $k = 0.11$, $m = 0.01$, $g = 0.1$, $n = 1$, and $w(t = 0) = 1$. The asymptote at $w = 100$ is shown by the broken line.

tion in situations in which death can occur in the less successful competitor; this is because these growth functions do not allow the growth rate to become negative.

16.3.2 *Two competing plants*

Consider two interacting plants of dry masses w_1 and w_2. The total dry mass w is given by

$$w = w_1 + w_2. \tag{16.5a}$$

Indeterminate plants occupying the same space For this case, taking the first term on the right of (16.4a), we assume that the total gross production of dry mass is given by

$$\frac{kw}{1 + gw^n}. \tag{16.5b}$$

This gross production is allocated to plants 1 and 2 according to the factors

$$\frac{w_1^{\,q}}{w_1^{\,q} + cw_2^{\,q}} \qquad \frac{cw_2^{\,q}}{w_1^{\,q} + cw_2^{\,q}}, \tag{16.5c}$$

where c and q are parameters that summarize the competitive situation between the plants. These two factors add up to unity. The two growth equations for the model now become

$$\frac{dw_1}{dt} = \frac{w_1^{\,q}}{w_1^{\,q} + cw_2^{\,q}} \frac{kw}{1 + gw^n} - mw_1 \tag{16.5d}$$

$$\frac{dw_2}{dt} = \frac{cw_2^{\,q}}{w_1^{\,q} + cw_2^{\,q}} \frac{kw}{1 + gw^n} - mw_2. \tag{16.5e}$$

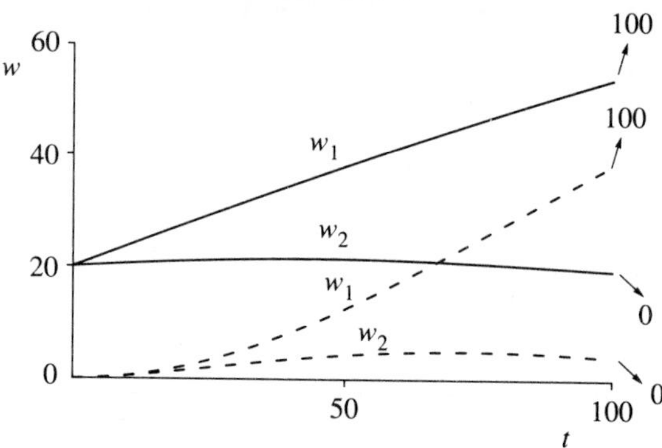

Fig. 16.4. Competition between two plants. Equations (15.5d) and (15.5e) with (16.5a) are drawn for $k = 0.11$, $m = 0.01$, $g = 0.1$, $n = 1$: in one case (solid lines) $c = 0.5$, $q = 1$, and $w_1(t = 0) = w_2(t = 0) = 20$; in the second case (broken lines) $c = 1$, $q = 2$, $w_1(t = 0) = 0.51$, and $w_2(t = 0) = 0.49$.

Adding these together gives (16.4a), and so the growth of the combined system is unchanged, as is the asymptote for $w_1 + w_2$. In this model the presence of the maintenance terms in (16.5d) and (16.5e) permits one plant to destroy the other, once it has a small advantage. With $c = 1$ (identical plants) and $q = 1$ (pro rata division of 'assimilate'), destructive competition does not occur: whatever the initial values, the $w_1 : w_2$ ratio is maintained as $w_1 + w_2$ increases up to its asymptotic value. With $c \neq 1$ (non-identical plants, e.g. grass and clover in a mixed sward) and $q = 1$, then for any initial values, with $c < 1$, w_1 increases to its asymptotic value and $w_2 \to 0$. With $c = 1$ and $q > 1$ (e.g. $q = 2$), any equilibrium is unstable and w_1 can destroy w_2. Some results are shown in Fig. 16.4.

Two determinate plants at a distance In many crops, the young small plants are not in competition with each other, and by the time that inter-plant competition does occur, the plants are sufficiently advanced along the growth curve that they do not annihilate each other but coexist. Equations (16.5d) and (16.5e) do not, as they stand, describe such behaviour, as they predict destructive competition at all plant masses. We describe a model more suited to this situation.

Two plants of dry masses w_1 and w_2 are assigned radii r_1 and r_2, where $r_1 \approx w_1^{2/3}$ and $r_2 \approx w_2^{2/3}$. If h is the distance between the two plants, the fraction of overlap θ between the two plants is approximated by $\theta = 1 - ah/(r_1 + r_2)$, with $0 \leqslant \theta \leqslant 1$; a is a constant. We summarize these considerations by writing the overlap fraction θ as

$$\theta = 1 - \frac{b}{w_1^{2/3} + w_2^{2/3}}, \tag{16.5f}$$

where b is a constant. If this calculation gives $\theta < 0$, then we write $\theta = 0$. Again, we use (16.4a) as a growth function, modifying the growth term. We replace (16.5d) and (16.5e) by

$$\frac{dw_1}{dt} = \frac{w_1}{w_1 + c\theta w_2} \frac{kw_1}{1 + gw_1{}^n} - mw_1 \tag{16.5g}$$

$$\frac{dw_2}{dt} = \frac{w_2}{w_2 + \theta w_1/c} \frac{kw_2}{1 + gw_2{}^n} - mw_2. \tag{16.5h}$$

Comparing these with (16.5d) and (16.5e), it can be seen that the growth terms are obtained from the individual plant masses; for separated plants this seems to be more appropriate than using a combined plant mass, and also the q parameter has been put equal to unity in order to simplify, giving a less aggressive competitive interaction. If there is no overlap ($\theta = 0$), then the two plants do not affect each other. If there is total overlap ($\theta = 1$), then with $c = 1$ the two plants affect each other equally. If $c < 1$, then w_1 has a greater effect on the growth of w_2 than vice versa. Solutions to this model are shown in Fig. 16.5. In Fig. 16.5(a) the plants are equivalent with $c = 1$; if the initial dry masses w_1 and w_2 at

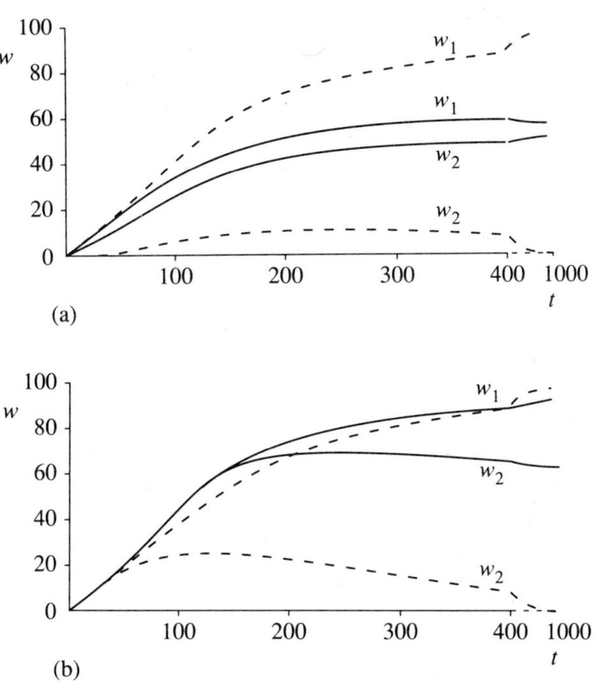

Fig. 16.5. Competition between two plants. Equations (16.5g) and (16.5h) are drawn for $k = 0.11$, $m = 0.01$, $g = 0.1$, and $n = 1$. In (a), $c = 1$ (equivalent plants) and $b = 10$: solid lines denote initial values of $w_1 = 1$ and $w_2 = 0.5$ at $t = 0$ (stable equilibrium); broken lines denote initial values of $w_1 = 1$ and $w_2 = 0.01$ at $t = 0$ (unstable, with the second plant destroyed). In (b), $c = 0.5$ (inequivalent plants), $w_1 = w_2 = 1$ at $t = 0$: solid lines denote $b = 30$ (widely separated weakly competing plants) giving a stable equilibrium; broken lines denote $b = 10$ (closer more strongly competing plants) giving instability, and the second plant is destroyed.

time $t = 0$ are similar, the plants coexist in a stable equilibrium; if they are substantially different, the larger plant destroys the smaller. In Fig. 16.5(b), the first plant is superior to the second, with $c = 0.5$; when the plants are close together ($b = 10$ in (16.5f)), the first plant destroys the second; when the plants are further part ($b = 30$), the plants can coexist in a stable equilibrium.

16.3.3 *Three-halves thinning law*

It has been widely observed (Yoda *et al.* 1963) that in a population of plants of the same species, the dry mass per plant w (kg plant^{-1}) is approximately related to the plant density ρ (number of plants (m^2 ground)$^{-1}$) by

$$w = c\rho^{-3/2}, \qquad (16.6a)$$

where c is a constant. A similar relationship applies to grass tillers. In (16.6a), the plant dry mass w does not represent the asymptote of a plant growth curve, as in (16.4a) which has the asymptote given by (S16.4a), but rather a point reached along a continuing growth curve when self-thinning has led to a plant density ρ.

The planting density of an agricultural crop is usually such that self-thinning does not occur, since this would be wasteful of seed; in many crops, by the time inter-plant competition occurs, the plants are in a stable competitive situation (Fig. 16.5(a)); for unlike plants (e.g. weeds), Fig. 16.5(b) shows that early competition may be more effective than late competition. In Section 16.2 (p. 454) we discussed the relationship between planting density ρ and plant dry mass w at harvest. Equation (16.3e) was derived to describe the $w : \rho$ response, and it is clear that this equation is similar to (16.6a) if the $\rho^{3/2}$ term dominates.

Equation (16.3e) with a dominant $\rho^{3/2}$ term is equivalent to the following rationalization of the three-halves law of (16.6a). As in (16.3a) and (16.3b) the area associated with each plant is $1/\rho$. Thus a volume of $\rho^{-3/2}$ may be associated with each plant. If plant mass w is proportional to this volume, then $w \propto \rho^{-3/2}$.

Pickard (1983), in an excellent paper, has summarized the status of the three-halves thinning law as follows: 'It seems unlikely ... that a mathematically and physiologically rigorous demonstration of the three-halves law will ever be found. On the other hand, it seems likely—as shown by a number of approximate treatments—that constraints on the plant force it into a biomass *vs.* number density relationship which approximates a three-halves law.' We concur with this view.

16.4 Static responses to fertilizer

These describe the crop yield Y (kg m^{-2}) as a function of the level of fertilizer applied X (kg of X m^{-2}) (see Fig. 1.1, p. 6). Two typical empirical response equations are

$$Y = a_0 + a_1 X + a_2 X^2 \qquad (16.7a)$$

$$Y = \frac{aX}{b + X}, \qquad (16.7b)$$

where a_0, a_1, a_2, a, and b are constants. While polynomials such as (16.7a) are statistically convenient, they are not well behaved biologically, and we do not consider such equations further. Equations such as (16.7b) are sometimes known as inverse polynomials, as they can be written

$$\frac{1}{Y} = \frac{1}{a} + \frac{b}{aX}, \tag{16.7c}$$

giving a linear form. Our discussion is based on (16.7c) or extensions to this equation.

16.4.1 *N, P, and K responses*

An equation that can be used to describe the response of yield Y (kg m^{-2}) to the three fertilizers N, P, and K (kg N, P, or K m^{-2}) is

$$\frac{1}{Y} = \frac{1}{Y_{\max}} + \frac{1}{b_N N} + \frac{1}{b_P P} + \frac{1}{b_K K}, \tag{16.8a}$$

where $Y_{\max}$ is the maximum yield (for N, P, $K \to \infty$), and b_N, b_P, and b_K are constants. The response to K given by (16.8a) is drawn in Fig. 16.6. This can be written as

$$Y = \frac{Y_{\max} b_N b_P b_K NPK}{b_N b_P b_K NPK + Y_{\max}(b_P b_K PK + b_N b_K NK + b_N b_P NP)}. \tag{16.8b}$$

Comparison of (16.8a) or (16.8b) with (2.7f) and (2.7i) (p. 52) shows that response to any single fertilizer (N say) is of the Michaelis–Menten type, with an initial slope that is independent of P and K, but an asymptote (Y as $N \to \infty$) that depends on P and K (Exercise 16.5).

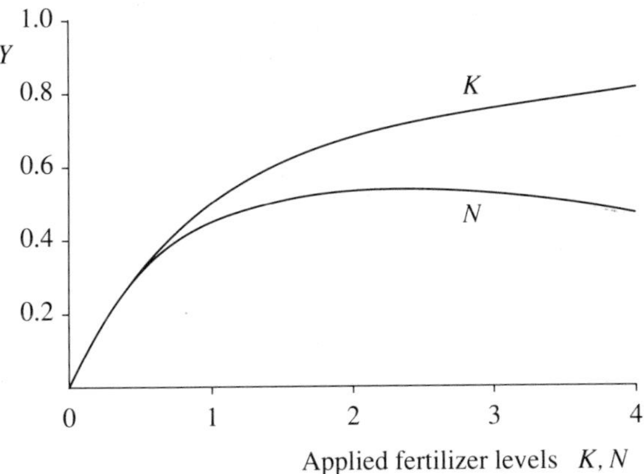

Fig. 16.6. Static response of crop yield Y to applied fertilizer. The K response is according to (16.8a), with $Y_{\max} = 1$, $N = P = \infty$ and $b_K = 1$. The N response is given by (16.8c) with $Y_{\max} = 1$, $P = K = \infty$, $a = 10$ and $b_N = 1$. The units are arbitrary.

It is sometimes observed that high nitrogen levels depress yield. A modification of (16.8a) which achieves this is

$$\frac{1}{Y} = \frac{1}{1 - N/a}\left(\frac{1}{Y_{\max}} + \frac{1}{b_{\mathrm{N}} N} + \frac{1}{b_{\mathrm{P}} P} + \frac{1}{b_{\mathrm{K}} K}\right), \tag{16.8c}$$

where a is a constant. The constant a causes the response to N of (16.8c) to reach a maximum and then decline, as shown in Fig. 16.6. Further discussion of the static approach to describing fertilizer responses is given by France and Thornley (1984, pp. 144–51).

16.5 Dynamic response to fertilizer using a crop model

In this section we describe how the simple crop model shown in Fig. 16.7 can be used to give a fertilizer response. In contrast with the approach described in the last section, using a dynamic model for fertilizer response calculations, although more complex, can allow naturally for environmental conditions during the growing season, and is a far more promising general method. The model discussed is a simplified and modified version of one presented by Johnson and Thornley (1985).

16.5.1 *The model*

The basic structure of the model is shown in Fig. 16.7, where the six state variables of the system are shown in the boxes. The total structural dry mass of the crop is W_{G} (kg structural dry mass m^{-2}), comprising shoot and root components:

$$W_{\mathrm{G}} = W_{\mathrm{sh}} + W_{\mathrm{r}}. \tag{16.9a}$$

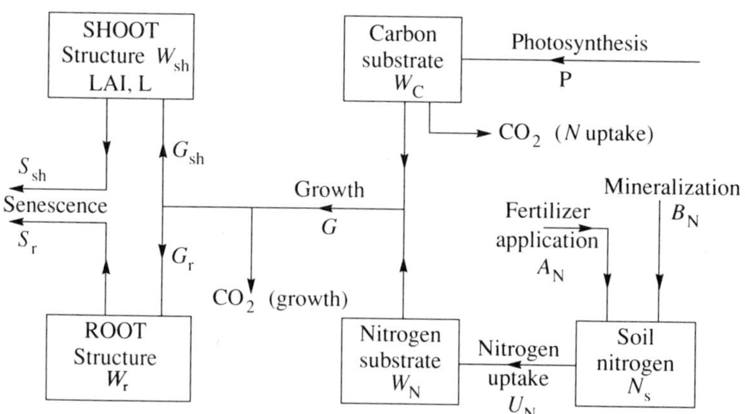

Fig. 16.7. Simple crop model for dynamic fertilizer responses. The six state variables are defined in the boxes. Two components of respiration are shown: that associated with growth, and that associated with nitrogen uptake.

The two plant substrates, carbon and nitrogen, with masses W_C and W_N (kg C (N) m^{-2}) are assumed to be uniformly distributed (or available) throughout the plant. The total storage dry mass W_S (kg storage dry mass m^{-2}) is given by

$$W_S = \frac{M_C}{12} W_C + \frac{M_N}{14} W_N, \tag{16.9b}$$

where M_C and M_N are the molecular masses of the carbon and nitrogen substrates relative to ^{12}C and ^{14}N. N carbon is associated with the nitrogen substrate; we take carbohydrate units (CH_2O) and nitrate (NO_3) as representing the substrates so that

$$M_C = 30 \text{ kg (kg mol)}^{-1} \qquad M_N = 62 \text{ kg (kg mol)}^{-1}. \tag{16.9c}$$

The carbon and nitrogen substrate concentrations C and N (kg C (N) (kg structural dry mass)$^{-1}$) are defined by

$$C = \frac{W_C}{W_G} \qquad N = \frac{W_N}{W_G}. \tag{16.9d}$$

The shoot and root structural fractions f_{sh} and f_r are given by

$$f_{sh} = \frac{W_{sh}}{W_G} \qquad f_r = \frac{W_r}{W_G}. \tag{16.9e}$$

The total shoot and root dry masses $W_{sh,t}$ and $W_{r,t}$ (kg dry mass m^{-2}) are obtained from

$$W_{sh,t} = W_{sh} + f_{sh} W_S \qquad W_{r,t} = W_r + f_r W_S. \tag{16.9f}$$

Light interception and photosynthesis As discussed in Chapter 10 (pp. 243–9), the Monsi–Saeki law for light attenuation in the canopy and a non-rectangular hyperbola for the leaf response to irradiance are used. Thus, given the five parameters

$$\begin{aligned} &\alpha = 10^{-8} \text{ kg } CO_2 \text{ J}^{-1} \qquad P_m = 10^{-6} \text{ kg } CO_2 \text{ m}^{-2} \text{ s}^{-1} \\ &\theta = 0.95 \qquad k = 0.5 \qquad m = 0.1, \end{aligned} \tag{16.10a}$$

and given a light flux density I_0 (J PAR m^{-2} s^{-1}) above the canopy, an instantaneous canopy photosynthetic rate P_c (kg CO_2 m^{-2} s^{-1}) can be calculated from (10.5c) (p. 248). We assume a constant light environment with daily radiation receipt J (J PAR m^{-2} day^{-1}), where

$$J = 5 \times 10^6 \text{ J PAR m}^{-2} \text{ day}^{-1}. \tag{16.10b}$$

For an assumed constant day length of 12 h with day length h (s day^{-1}) given by

$$h = 43\,200 \text{ s day}^{-1}, \tag{16.10c}$$

the average irradiance level above the canopy is obtained from

$$I_0 = J/h. \tag{16.10d}$$

The daily photosynthetic input P (kg C m^{-2} day^{-1}) is calculated from P_c (10.5c) with

$$P = 43\,200 \times \frac{12}{44} P_c. \tag{16.10e}$$

Nitrogen uptake and soil nitrogen The nitrogen uptake rate U_N (kg N m^{-2} day^{-1}) is given by

$$U_N = \frac{\sigma_N W_r N_s}{1 + (K_C/C)(1 + N/K_N)}, \tag{16.11a}$$

where the parameters are

$$\begin{aligned} \sigma_N &= 3000 \text{ kg dry soil (kg root structure)}^{-1} \text{ day}^{-1} \\ K_C &= 0.05 \text{ kg C (kg structure)}^{-1} \\ K_N &= 0.005 \text{ kg N (kg structure)}^{-1}. \end{aligned} \tag{16.11b}$$

The soil nitrogen concentration N_s has units of kg N (kg dry soil)$^{-1}$. A fertilizer application of A_N (kg N m^{-2}) increments N_s by an amount

$$\frac{A_N}{d_r \rho_s}, \tag{16.11c}$$

where d_r (m) is the root depth and ρ_s (kg dry soil m^{-3}) is the soil density. We take

$$d_r = 0.2 \text{ m} \qquad \rho_s = 1500 \text{ kg m}^{-3}. \tag{16.11d}$$

Note also that nitrogen utilization at a rate U_N gives depletion of the soil nitrogen at a rate

$$\frac{U_N}{d_r \rho_s} \tag{16.11e}$$

in units of kg N (kg dry soil)$^{-1}$ day^{-1}.

Mineralization at a rate of B_N (kg N (kg dry soil)$^{-1}$ day^{-1}) also provides a flux into the soil nitrogen compartment. As will be seen later (Fig. 16.8), the value ascribed to B_N has a considerable effect on the fertilizer response.

Growth and partitioning For growth and partitioning we employ the teleonomic model of Section 13.3 (p. 372). For the two growth functions G_{sh} and G_r (kg structure m^{-2} day^{-1}), (13.4a) and (13.4b) give

$$G_{sh} = \mu C N \lambda_{sh} W_{sh} \qquad G_r = \mu C N \lambda_r W_r, \tag{16.12a}$$

where the parameter μ is given by

$$\mu = 150 \text{ day}^{-1} \text{ (kg C (kg structure)}^{-1})^{-1} \text{ (kg N (kg structure)}^{-1})^{-1}. \tag{16.12b}$$

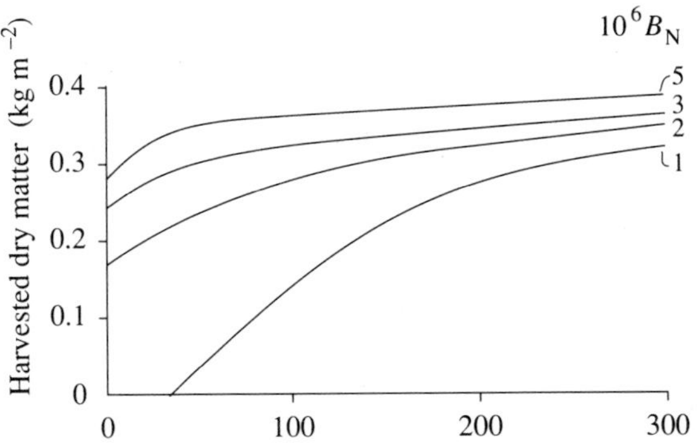

Fig. 16.8. Fertilizer responses calculated dynamically using the crop growth model of Fig. 16.7. The response is illustrated for four values of the mineralization parameter B_N (kg N (kg dry soil)$^{-1}$ day^{-1}).

The partitioning parameters λ_{sh} and λ_r are given by (from (13.4d) and (13.5n))

$$\lambda_{sh} = \frac{f_r N/(N + f_N)}{f_{sh} C/(C + f_C) + f_r N/(N + f_N)}$$
$$\lambda_r = \frac{f_{sh} C/(C + f_C)}{f_{sh} C/(C + f_C) + f_r N/(N + f_N)}, \tag{16.12c}$$

where the parameters f_C and f_N define the carbon and nitrogen contents of the plant structural dry matter and are given by

$$f_C = 0.45 \text{ kg C (kg structure)}^{-1}$$
$$f_N = 0.03 \text{ kg N (kg structure)}^{-1}. \tag{16.12d}$$

Leaf area index A fraction f_l of the shoot structural growth (16.12a) is used in lamina production and, with an incremental specific leaf area η (m^2 leaf (kg structure)$^{-1}$), this gives a rate of leaf area production of

$$\eta f_l G_{sh}. \tag{16.13a}$$

The parameter f_l is taken to be

$$f_l = 0.7. \tag{16.13b}$$

The incremental specific leaf area η is assumed to depend on the carbon substrate level C according to

$$\eta = \eta_m(1 - \zeta C), \tag{16.13c}$$

where the two parameters are given by

$$\eta_{m} = 25 \text{ m}^2 \text{ (kg structure)}^{-1}$$
$$\zeta = 2.5 \text{ (kg C (kg structure)}^{-1})^{-1}. \qquad (16.13d)$$

Equation (16.13c) gives the desired responses of specific leaf area to light and temperature (although temperature is not explicitly represented in this model).

Substrate utilization New structure is synthesized at a rate $G_{sh} + G_r$, with (16.12a). Carbon and nitrogen substrates are therefore required at the rates

$$f_C(G_{sh} + G_r) \qquad f_N(G_{sh} + G_r). \qquad (16.14a)$$

However, owing to the respiration of some of the carbon substrate (eqn (11.3c), p. 266), the rate of utilization of carbon substrate is

$$\frac{f_C(G_{sh} + G_r)}{Y}. \qquad (16.14b)$$

The parameter Y is given by

$$Y = 0.75. \qquad (16.14c)$$

Another component of carbon substrate utilization is associated with the nitrogen uptake rate U_N (16.11a), and is given by

$$\alpha_N U_N, \qquad (16.14d)$$

where the cost parameter for nitrogen uptake is

$$\alpha_N = 0.5 \text{ kg C (kg N)}^{-1}. \qquad (16.14e)$$

Senescence It is assumed that shoot and root senescence occur at rates given by

$$\gamma_{sh} W_{sh} \qquad \gamma_r W_r, \qquad (16.15a)$$

where the two rate constants have the values

$$\gamma_{sh} = 0.04 \text{ day}^{-1} \qquad \gamma_r = 0.01 \text{ day}^{-1}. \qquad (16.15b)$$

Similarly, the leaf area index is assumed to decline due to senescence at a rate of $\gamma_{sh} L$.

16.5.2 *Mathematical statement of the model*

The six state variables of the model are described by the following equations:

$$\frac{dW_C}{dt} = P - \frac{f_C}{Y}(G_{sh} + G_r) - \alpha_N U_N \qquad \text{with (16.10e), (16.14b), (16.14d);} \qquad (16.16a)$$

$$\frac{dW_N}{dt} = U_N - f_N(G_{sh} + G_r) \qquad \text{with (16.11a), (16.14a);} \qquad (16.16b)$$

$$\frac{dW_{sh}}{dt} = G_{sh} - \gamma_{sh} W_{sh} \qquad \text{with (16.12a), (16.15a);} \tag{16.16c}$$

$$\frac{dW_r}{dt} = G_r - \gamma_r W_r \qquad \text{with (16.12a), (16.15a);} \tag{16.16d}$$

$$\frac{dL}{dt} = \eta f_l G_{sh} - \gamma_{sh} L \qquad \text{with (16.13a), (16.15c);} \tag{16.16e}$$

$$\frac{dN_s}{dt} = B_N - \frac{U_N}{d_r \rho_s} \qquad \text{with (16.11f), (16.11g).} \tag{16.16f}$$

Initial values The initial values for the first five state variables above are

$$\begin{aligned} W_C &= 0.015 \text{ kg C m}^{-2} \qquad W_N = 0.004 \text{ kg N m}^{-2} \\ W_{sh} &= 0.2 \text{ kg structural dry mass m}^{-2} \\ W_r &= 0.2 \text{ kg structural dry mass m}^{-2} \\ L &= 0.8 \text{ m}^2 \text{ leaf (m}^2 \text{ ground)}^{-1}. \end{aligned} \tag{16.17a}$$

A fertilizer application at time $t = 0$ day of

$$100 \text{ kg N ha}^{-1} \tag{16.17b}$$

is equivalent, using (16.11c) and (16.11d), to an initial value of soil nitrogen N_s of

$$N_s = 33.33 \times 10^{-6} \text{ kg N (kg dry soil)}^{-1}. \tag{16.17c}$$

This initial value is varied from zero to 10^{-4} kg N (kg dry soil)$^{-1}$ in order to generate fertilizer response curves covering fertilizer application rates up to 300 kg N ha^{-1}.

16.5.3 *Simulated fertilized response*

A computer program was written using the modelling language ACSL to solve the mathematical equations described in the last section. The harvested dry matter was calculated from the growth in total shoot dry mass (16.9f) above the initial value over a 50 day period. A fourth-order integration routine with a fixed time step of 0.5 day was employed ((1.81), p. 23). The fertilizer responses generated are shown in Fig. 16.8. It can be seen that the responses given by the model are realistic, and that the dynamic approach to fertilizer application offers many possibilities that are excluded from the simple static method of Section 16.4.

Exercises

16.1. Show that the initial slope of (16.1a), (16.1b), and (16.1c) is $dY/d\rho(\rho = 0) = a$. Find the plant density ρ_{max} for maximum yield Y_{max}.

16.2. Find the planting density ρ_{max} for maximum yield Y_{max} for the modified hyperbolic equation (16.2b).

16.3. Derive a polynomial for the planting density ρ_{max} for maximum yield using the simple mechanistic response eqn (16.3h) for crop yield Y.

16.4. Derive an expression for the asymptote $w = w_f$ of the growth equation (16.4a). (*Hint:* equate (16.4a) to zero and solve for $w = w_f$.) Derive an equation for the dry mass w^* at the inflexion point.

16.5. Find the initial slope (it is easiest to calculate Y/N as $N \to 0$) and asymptote (Y as $N \to \infty$) of (16.8b).

17
Root growth

17.1 Introduction

In this chapter, our principal aim is to describe a model of root growth which simulates changes in the two most important characteristics of a root system: root mass and root distribution in the vertical direction. A wider objective is to develop a treatment of the root which is comparable with the well-established modelling approaches for handling above-ground quantities such as leaf area index, light interception (Chapter 8), and photosynthesis (Chapters 9 and 10). Such a treatment would greatly assist our attempts to simulate crop growth. Our discussion is largely based on papers by Brugge and Thornley (1985), and particularly Brugge (1985).

There are few models of root growth and extension: the problem is complex, and a knowledge of plant and soil characteristics and processes is needed which is still mostly missing. In Chapter 18, a variety of approaches to modelling branching processes are described. Branching models specifically of the root have been constructed (e.g. Lungley 1973), but these are mostly unrelated to plant status or soil conditions. Hackett and Rose (1972a,b) reported a branching model of the barley root; this was empirically based on observations of numbers, length, surface area, and volume of different root members during vegetative growth. Our approach here to modelling root growth and extension is to ignore explicit consideration of root morphology and branching characteristics, just as is done in many successful shoot growth models (Chapter 16, Section 16.5, p. 464).

Gerwitz and Page (1974) considered data on many plant root systems, including the effects of soil moisture, fertility, plant age, and mass. They concluded that a negative exponential function gave a satisfactory description of two-thirds of the data sets. Thus if W_r (kg root dry matter (m^2 ground)$^{-1}$) is the root mass and z (m) is the distance below the soil surface at $z = 0$, then the function

$$\lambda W_r e^{-\lambda z}, \tag{17.1a}$$

with units of kg m^{-3}, where λ (m^{-1}) is a constant, gives the root 'density' (per unit volume of soil) at depth z. Integrating (17.1a) with respect to z between $z = 0$ and z gives

$$W_r(1 - e^{-\lambda z}). \tag{17.1b}$$

Thus $1 - e^{-\lambda z}$ gives the fraction of the root mass that lies between the soil surface and depth z.

17.1.1 *Model of an exponential root distribution*

Assume that the way in which the root explores the soil approximates to a diffusion-like process with diffusion constant D ($\mathrm{m^2\ day^{-1}}$), and also that the root dry mass per unit volume of soil w_r ($\mathrm{kg\ m^{-3}}$) consumes itself in a process with rate constant m_r ($\mathrm{day^{-1}}$); the process may represent maintenance (Chapter 11) and senescence. If we assume horizontal homogeneity, the conservation equation is

$$\frac{\partial w_r}{\partial t} = D\frac{\partial^2 w_r}{\partial z^2} - m_r w_r. \tag{17.2a}$$

where z (m) denotes depth in the soil and t (days) denotes the time variable. In a steady state therefore

$$0 = D\frac{\mathrm{d}^2 w_r}{\mathrm{d}z^2} - m_r w_r. \tag{17.2b}$$

The required solution to (17.2b) is

$$w_r = w_0 \mathrm{e}^{-\lambda z} \quad \text{where } \lambda^2 = m_r/D \text{ and } w_r = w_0 \text{ at } z = 0. \tag{17.2c}$$

The total root dry matter W_r ($\mathrm{kg\ m^{-2}}$), is obtained by integrating w_r from $z = 0$ to $z = \infty$:

$$W_r = \int_0^\infty w_r\,\mathrm{d}z = \frac{w_0}{\lambda}. \tag{17.2d}$$

It is interesting that such simple assumptions about root growth give rise to realistic predictions for the vertical distribution of root dry matter. In Exercise 17.1, it is shown how an assumption that soil exploration is more like a facilitated diffusion process (Chapter 4, Section 4.3, p. 96) gives rise to a finite quadratic root distribution rather than an infinite negative exponential distribution.

17.2 A carbon–nitrogen model of distributed root growth

In Section 17.1.1 and Exercise 17.1 it is shown how quite realistic root distributions can be generated from simple assumptions. In many plant and crop models a useful level of detail is to consider the plant as divisible into structural and storage dry matter, to represent shoot : root partitioning, and to include carbon and nitrogen substrates since these are often the principal determinants of growth; this permits adaptation to the above-ground (carbon) and below-ground (nitrogen) environments, with varying shoot : root ratios and growth rates. In this section this approach is used to examine the distribution of root dry matter with depth in the soil. The model we describe is based on the work of Brugge (1985) and Brugge and Thornley (1985), where additional details can be found.

The root growth model is shown in Figs 17.1 and 17.2. The shoot model (Fig. 17.1) is highly simplified, since our objective here is to understand the spatial and

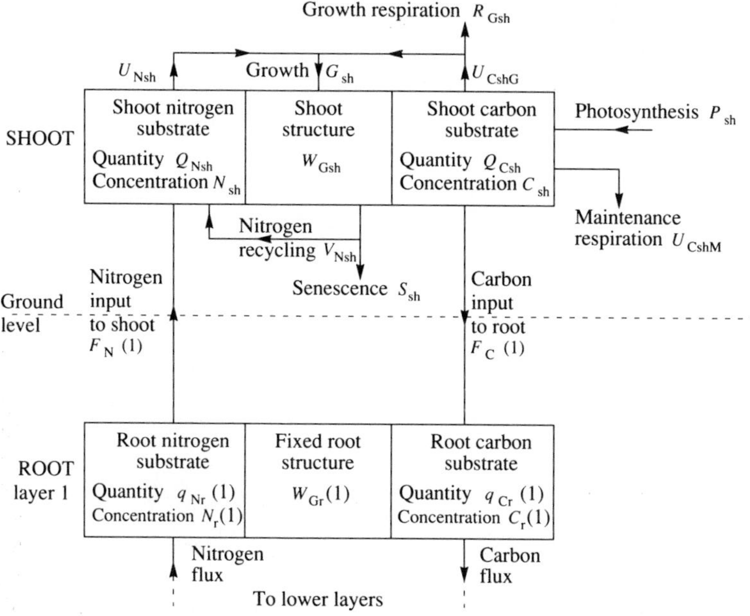

Fig. 17.1. Root growth model. The shoot is described by three state variables: Q_{Nsh}, Q_{Csh}, and W_{Gsh} (Table 17.1). The shoot interacts only with the top layer of the root system through the interchange of nitrogen and carbon substrates, as shown. Here, only the shoot processes and the shoot-to-root transfers are illustrated; see Fig. 17.2 for a full description of the root system.

temporal dynamics of root growth. Only vegetative growth is considered, the leaf area index is simply proportional to shoot structural dry matter W_{Gsh}, photosynthesis is seasonally dependent, and water and temperature are assumed to be non-limiting and are ignored. In the root (Fig. 17.2) nitrogen is absorbed from a constant soil nitrogen pool, with the soil nitrogen concentration N_s (kg nitrogen (m^3 soil)$^{-1}$) representing the availability of nitrate, ammonia, or dinitrogen (for legumes) to the root; root rub-off and substrate exudation are included, along with nitrogen recovery from senescing material. Horizontal homogeneity is assumed in all parts of the model. The principal symbols used are defined and their units given in Table 17.1.

17.2.1 *Basic definitions*

The below-ground part of the model consists of n_s layers of soil, each of depth Δz (m), extending downwards from the soil surface at $z = 0$ m. An integer index i addresses these layers, with $1 \leqslant i \leqslant n_s$. Any of these layers may contain root material, of which there are four types (Fig. 17.2): fixed root structural matter of mass $w_{fGr}(i)$ (kg structural C (m^3 soil)$^{-1}$), mobile root structural matter of mass

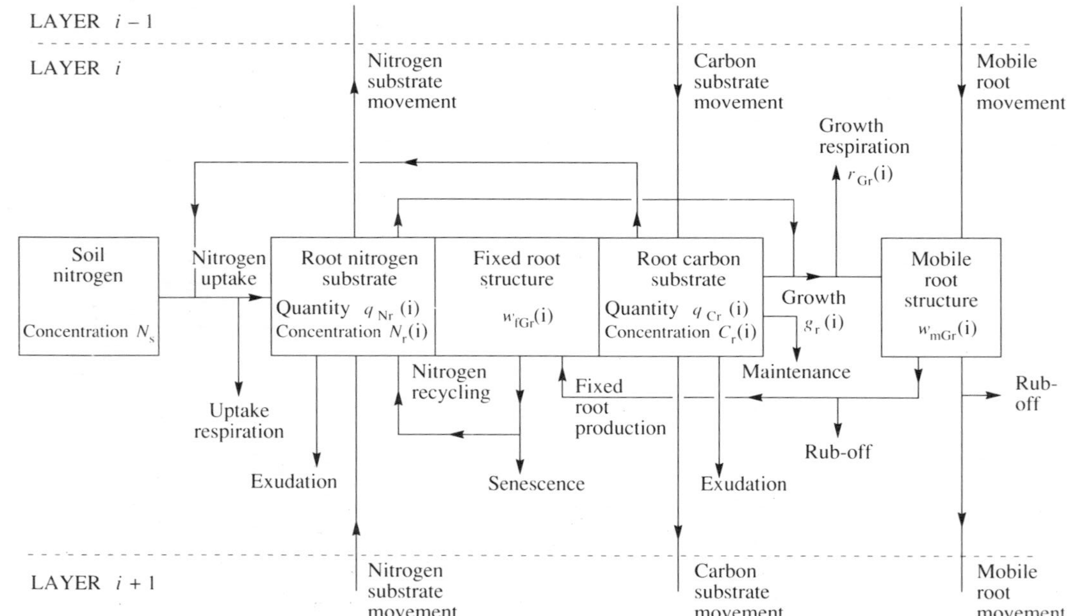

Fig. 17.2. Root growth model. There are n_s layers in the model, each corresponding to a depth Δz in the soil. The ith layer is described by four state variables: $q_{Nr}(i)$, $q_{Cr}(i)$, $w_{fGr}(i)$, and $w_{mGr}(i)$ (Table 17.1). The soil nitrogen concentration N_s is assumed to be constant.

Table 17.1. Principal symbols and their units (the equation where the symbol is first introduced is given)

Symbol	Definition	Units
Independent variables		
t	Time	days
z	Depth below ground	m
State variables		
($m^{-2} = (m^2 \text{ ground})^{-1}$; $m^{-3} = (m^3 \text{ soil})^{-1}$)		
Q_{Csh}	Quantity of C substrate in shoot (17.3d)	kg substrate C m^{-2}
Q_{Nsh}	Quantity of N substrate in shoot (17.3d)	kg substrate N m^{-2}
$q_{Cr}(i)$	Quantity of C substrate in root in ith soil layer (17.3c)	kg substrate C m^{-3}
$q_{Nr}(i)$	Quantity of N substrate in root in ith soil layer (17.3c)	kg substrate N m^{-3}
W_{Gsh}	Shoot structural dry matter (17.3d)	kg structural C m^{-2}
$w_{fGr}(i)$	Fixed structural root dry matter in ith soil layer (17.3a)	kg structural C m^{-3}
$w_{mGr}(i)$	Mobile structural root dry matter in ith soil layer (17.3a)	kg structural C m^{-3}
Other variables		
$C_r(i)$	C substrate concentration in root in ith soil layer (17.3c)	kg substrate C (kg structural C)$^{-1}$
C_{sh}	C substrate concentration in shoot (17.3d)	kg substrate C (kg structural C)$^{-1}$
$F_C(i)$, $F_N(i)$	Downward (upward) C (N) flux density into root in the ith soil layer ((17.7a), (17.7c), (17.11a), (17.11c))	kg substrate C (N) m^{-2} day^{-1}
$F_{mr}(i)$	Downward mobile root flux into the ith soil layer (17.10a)	kg structural C m^{-2} day^{-1}
$G_{mr}(i)$	Gain from movement of mobile root into ith soil layer (17.10d)	kg structural C m^{-3} day^{-1}
G_{sh}	Shoot structural growth rate (17.5a)	kg structural C m^{-2} day^{-1}
$g_r(i)$	Structural growth rate of root in ith soil layer (17.8a)	kg structural C m^{-3} day^{-1}
L	Leaf area index (17.4a) and (17.4b)	m^2 leaf (m^2 ground)$^{-1}$
$N_r(i)$	N substrate concentration in root in ith soil layer (17.3c)	kg substrate N (kg structural C)$^{-1}$
N_{sh}	N substrate concentration in shoot (17.3d)	kg substrate N (kg structural C)$^{-1}$
i	Index to the n_s soil layers (17.3a)	
P_{sh}	Daily photosynthetic rate (17.4a)	kg substrate C m^{-2} day^{-1}
R_{Gsh}	Shoot growth respiration rate (17.5c)	kg substrate C m^{-2} day^{-1}

Table 17.1 (*continued*)

Symbol	Definition	Units
$r_{Gr}(i)$	Root growth respiration rate in ith soil layer (17.8c)	kg substrate C m^{-2} day^{-1}
S_{sh}	Shoot senescence rate (17.6b)	kg structural C m^{-2} day^{-1}
$s_r(i)$	Root senescence rate in ith soil layer (17.12b)	kg structural C m^{-3} day^{-1}
U_{CshG}	C utilization rate for shoot growth (17.5d)	kg substrate C m^{-2} day^{-1}
U_{CshM}	C utilization for shoot maintenance (17.6a)	kg substrate C m^{-2} day^{-1}
U_{Nsh}	N utilization for shoot growth (17.5b)	kg substrate N m^{-2} day^{-1}
$u_{CrG}(i)$	Utilization rate of substrate C for root growth in ith soil layer (17.8d)	kg C substrate m^{-3} day^{-1}
$u_{CrM}(i)$	Utilization rate of substrate C for root maintenance in the ith soil layer (17.12a)	kg C substrate m^{-3} day^{-1}
$u_{Nr}(i)$	N utilization for root growth (17.8b)	kg substrate N m^{-2} day^{-1}
$\upsilon_{Nr}(i)$	Rate of N recycling from root senescence in ith soil layer (17.12c)	kg substrate N m^{-3} day^{-1}
V_{Nsh}	Rate of N recycling from senescence (17.6c)	kg substrate N m^{-2} day^{-1}
W_{Gr}	Total structural root day matter (17.3b)	kg structural C m^{-2}
$w_{Gr}(i)$	Total structural root dry matter in ith soil layer (17.3a)	kg structural C m^{-3}
$Z_C(i)$, $Z_N(i)$	Total substrate C (N) flux into root in layer i ((17.11b), (17.11d))	kg substrate C (N) m^{-3} day^{-1}
$\nu_{Nr}(i)$	Rate of formation of N substrate in root in ith soil layer (17.13a)	kg substrate N m^{-3} day^{-1}
Parameters and driving variables		
a	Leaf area constant ((17.4b), (17.4e))	m^2 leaf (kg structural $C)^{-1}$
c_N	Cost parameter for soil N uptake ((17.13c), (17.13d))	kg substrate C (kg substrate $N)^{-1}$
h	Day length (17.4d)	s day^{-1}
J	Daily light receipt (17.4c)	J m^{-2} day^{-1}
J_{Nv}	Soil N uptake parameter ((17.13a), (17.13d))	kg substrate N (kg structural $C)^{-1}$
K_{Cr}, K_{Nr}	Root growth constants ((17.8a), (17.8e)	kg substrate C (N) (kg structural $C)^{-1}$
K_{Csh}, K_{Nsh}	Shoot growth Michaelis–Menten constants ((17.5a), (17.5e))	kg substrate C (N) (kg structural $C)^{-1}$

Table 17.1 (*continued*)

Symbol	Definition	Units
K_{Cv}	Soil N uptake parameter ((17.13a), (17.13d))	kg substrate C (kg structural C)$^{-1}$
K_m	Maintenance constant ((17.6a), (17.6d))	kg substrate C (kg structural C)$^{-1}$
K_ϕ	N recycling constant ((17.6c), (17.6d))	kg substrate N (kg structural N)$^{-1}$
k	Extinction coefficient ((17.4a), (17.4e))	
k_{Nv}	Soil N uptake parameter ((17.13a), (17.13d))	m^3 (kg structural C)$^{-1}$ day^{-1}
k_r	Root growth constant ((17.8a), (17.8e))	day^{-1}
k_{sh}	Shoot growth constant ((17.5a), (17.5e))	day^{-1}
k_{s1}, k_{s2}	Senescence constants ((17.6b), (17.6d))	day^{-1}
k_{xC}, k_{xN}	Root exudation parameters for C and N substrates ((17.14a), (17.14b))	day^{-1}
m	Maintenance rate ((17.6a), (17.6d))	kg substrate C (kg structural C)$^{-1}$ day^{-1}
N_s	Soil N concentration ((17.13a), (17.13d))	kg N m^{-3}
n_s	Number of soil layers of thickness Δz (17.3a)	
P_m	Light-saturated photosynthetic rate ((17.4a), (17.4e))	kg CO_2 (m^2 leaf)$^{-1}$ s^{-1}
w_{Grmax}	maximum density of root structure ((17.8a), (17.8e))	kg structural C m^{-3}
w_{fGrmin}	root movement parameter ((17.10a), (17.10g))	kg structural C m^{-3}
Y	Conversion efficiency ((17.5c), (17.5e))	kg substrate C (kg structural C)$^{-1}$
α	Photosynthetic efficiency ((17.4a), (17.4e))	kg CO_2 J^{-1}
β	Mobile root to fixed root rate constant ((17.9a), (17.9d))	day^{-1}
γ_{mr}	Mobile root 'diffusion' constant ((17.10a), (17.10g))	m^2 day^{-1}
ε_r	Root rub-off fraction ((17.9b), (17.9d))	
ζ_C, ζ_N	C (N) substrate conductivity parameter between shoot and root ((17.7a)–(17.7d))	m day^{-1}
θ	CO_2 to C conversion factor ((17.4a), (17.4e))	kg C (kg CO_2)$^{-1}$
λ	N content of structural matter ((17.5b), (17.5e))	kg structural N (kg structural C)$^{-1}$
σ_{Cr}, σ_{Nr}	C (N) substrate transport parameters in root ((17.11a), (17.11c), (17.11e))	m^2 day^{-1}
τ	Leaf transmission coefficient ((17.4a), (17.4e))	
ϕ	N recycling constant ((17.6c), (17.6d))	kg substrate N (kg structural N)$^{-1}$
ω	C content of N substrate ((17.5d), (17.5e))	kg C (kg N substrate)$^{-1}$

$w_{mGr}(i)$ (kg structural C (m^3 soil)$^{-1}$), carbon substrate of mass, $q_{Cr}(i)$ (kg substrate C (m^3 soil)$^{-1}$), and nitrogen substrate of mass $q_{Nr}(i)$ (kg substrate N (m^3 soil)$^{-1}$). The division of root structural dry matter into spatially fixed and spatially mobile fractions (the latter is the recently synthesized structural material) allows the root to move into new soil horizons in much the same way as in the simple model in Section 17.1.1. It is suggested that this method is sufficiently accurate and far simpler than more rigorous treatments that take account of physical cell velocities along the root (e.g. Silk and Erickson 1979, Silk, Walker, and Labavitch 1984); our approximation assumes that the appearance of root dry matter in hitherto unexplored volumes of soil is due more to new or recent synthesis than to the movement of older root material.

The total structural dry matter $w_{Gr}(i)$ (kg structural C (m^3 soil)$^{-1}$) in the ith layer is

$$w_{Gr}(i) = w_{fGr}(i) + w_{mGr}(i) \qquad 1 \leqslant i \leqslant n_s. \tag{17.3a}$$

The total root structural mass W_{Gr} (kg structural C (m^2 ground)$^{-1}$) is given by

$$W_{Gr} = \sum_{i=1}^{n_s} w_{Gr}(i)\Delta z. \tag{17.3b}$$

The concentrations of root carbon and nitrogen substrates $C_r(i)$ and $N_r(i)$ (kg substrate C (N) (kg structural C)$^{-1}$) are

$$C_r(i) = \frac{q_{Cr}(i)}{w_{Gr}(i)} \qquad N_r(i) = \frac{q_{Nr}(i)}{w_{Gr}(i)}. \tag{17.3c}$$

The shoot (Fig. 17.1) has a structural dry mass W_{Gsh} (kg structural C (m^2 ground)$^{-1}$), which contains a quantities of substrate carbon and substrate nitrogen Q_{Csh} and Q_{Nsh} (kg substrate C (N) (m^2 ground)$^{-1}$); these are associated with shoot carbon and nitrogen substrate concentrations C_{sh} and N_{sh} (kg substrate C (N) (kg structural C)$^{-1}$), where

$$C_{sh} = \frac{Q_{Csh}}{W_{Gsh}} \qquad N_{sh} = \frac{Q_{Nsh}}{W_{Gsh}}. \tag{17.3d}$$

It is assumed that in both shoot and root the structural matter contains λ (kg structural N (kg structural C)$^{-1}$) mass units of nitrogen per mass unit of carbon, and also the nitrogen substrates in both shoot and root contain ω (kg C (kg substrate N)$^{-1}$) mass units of carbon per mass unit of substrate nitrogen.

17.2.2 *Processes and fluxes*

Photosynthesis and leaf area index The daily photosynthetic carbon substrate input to the shoot is P_{sh} (kg substrate C (m^2 ground)$^{-1}$ day^{-1}), where

$$P_{sh} = \frac{\theta P_m h}{k} \ln\left[\frac{\alpha kJ/h + (1-\tau)P_m}{(\alpha kJ/h)e^{-kL} + (1-\tau)P_m}\right]. \tag{17.4a}$$

This expression is derived from (10.5e), p. 249; θ (kg C (kg CO_2)$^{-1}$) is a CO_2 to carbon conversion factor, P_m (kg CO_2 (m^2 leaf)$^{-1}$ s^{-1}) is the light-saturated photosynthetic rate, h (s day^{-1}) is the day length, k is the extinction coefficient, α (kg CO_2 J^{-1}) is the photosynthetic efficiency, τ is the leaf transmission coefficient, J (J m^{-2} day^{-1} PAR) (PAR, photosynthetically active rate) is the daily light receipt, and L (m^2 leaf (m^2 ground)$^{-1}$) is the leaf area index. Equation (17.4a) is obtained from (10.5e) by replacing I_0 and m of (10.5e) by J/h and τ, and multiplying by θ and h.

It is assumed that the leaf area index L is proportional to the shoot structural dry mass W_{Gsh} with

$$L = aW_{Gsh}, \tag{17.4b}$$

where a (m^2 leaf (kg structural C)$^{-1}$) is a constant.

Equations (17.4a) and (17.4b) involve several approximations: the response of leaf gross photosynthesis to light flux density is a rectangular hyperbola; light-saturated gross photosynthetic rate is independent of leaf age and position in the canopy; the light level is constant throughout the day; the specific leaf area constant a does not change with plant status and growing conditions.

The daily light receipt J and the day length h are given by

$$J = \left\{0.7 + 8.2 \sin^2\left[\frac{\pi(t + 10)}{365}\right]\right\} \times 10^6 \text{ J PAR (m}^2 \text{ ground)}^{-1} \text{ day}^{-1} \tag{17.4c}$$

$$h = 28\,800\left\{1 + \sin^2\left[\frac{\pi(t + 10)}{365}\right]\right\} \text{s day}^{-1}. \tag{17.4d}$$

These expressions represent average conditions at the Grassland Research Institute, Hurley, UK; the time t (days) is measured from 1 January.

The parameters used in the calculations are

$$\begin{aligned} \theta &= 12/44 \text{ kg C (kg CO}_2)^{-1} \\ P_m &= 1.5 \times 10^{-6} \text{ kg CO}_2 \text{ m}^{-2} \text{ s}^{-1} \\ k &= 0.5 \\ \alpha &= 1.2 \times 10^{-8} \text{ kg CO}_2 \text{ J}^{-1} \\ \tau &= 0.1 \\ a &= 21 \text{ m}^2 \text{ leaf (kg structural C)}^{-1}. \end{aligned} \tag{17.4e}$$

The first five of these parameters are as used by Johnson and Thornley (1983); the last parameter a is as used by Johnson, Ameziane, and Thornley (1983).

Shoot growth New shoot structural material is synthesized at a rate of G_{sh} (kg structural C (m^2 soil)$^{-1}$ day^{-1}), where

$$G_{sh} = k_{sh} W_{Gsh} \frac{1}{1 + K_{Csh}/C_{sh} + K_{Nsh}/N_{sh}}. \tag{17.5a}$$

k_{sh} (day^{-1}) is the shoot growth rate parameter, and K_{Csh} and K_{Nsh} (kg substrate C (N) (kg structural C)$^{-1}$) are Michaelis–Menten constants.

The rate of nitrogen utilization due to this growth process is U_{Nsh} (kg substrate N (m^2 ground)$^{-1}$ day^{-1}), where

$$U_{Nsh} = \lambda G_{sh}. \tag{17.5b}$$

λ is defined after (17.3d). It is assumed that growth incurs respiration costs, represented by a conversion efficiency parameter Y (equivalent to the Y_g of Chapter 11, Section 11.3, p. 266), such that 1 unit of substrate carbon produces Y units of structural carbon and the rest is respired. Thus the growth respiration rate R_{Gsh} (kg substrate C (m^2 ground)$^{-1}$ day^{-1}) of the shoot is

$$R_{Gsh} = \frac{1 - Y}{Y} G_{sh}. \tag{17.5c}$$

Shoot carbon substrate is incorporated into new shoot structure at a rate of U_{CshG} (kg substrate C (m^2 ground)$^{-1}$ day^{-1}), with

$$U_{CshG} = G_{sh}(1 - \omega\lambda). \tag{17.5d}$$

The negative term in this expression arises because the nitrogen substrate brings some carbon with it; see the definition of ω after (17.3d).

The parameters have the values

$$\begin{aligned} k_{sh} &= 0.132 \text{ day}^{-1} \\ K_{Csh} &= 0.1 \text{ kg substrate C (kg structural C)}^{-1} \\ K_{Nsh} &= 0.01 \text{ kg substrate N (kg structural C)}^{-1} \\ \lambda &= 0.1 \text{ kg structural N (kg structural C)}^{-1} \\ Y &= 0.75 \\ \omega &= 2 \text{ kg C (kg substrate N)}^{-1}. \end{aligned} \tag{17.5e}$$

The rate constant k_{sh} has a value that gives realistic growth rates, the two Michaelis–Menten parameters are assigned values equivalent to the initial concentrations of the relevant substrates (this procedure works well provided that these initial concentrations are reasonable), λ is based on typical plant compositions, for Y see (11.4b) (p. 268), and ω is based on an approximate estimate of the likely nature of substrate nitrogen (see Table 12.1, p. 290).

Shoot maintenance and senescence Shoot maintenance costs are met directly from the shoot carbon substrate pool at a rate U_{CshM} (kg substrate C (m^2 ground)$^{-1}$ day^{-1}) with

$$U_{CshM} = mW_{Gsh}\frac{C_{sh}}{K_m + C_{sh}}, \tag{17.6a}$$

where m (kg substrate C (kg structural C)$^{-1}$ day^{-1}) is the maintenance coefficient and K_m (kg substrate C (kg structural C)$^{-1}$) is a parameter. Under conditions of abundant carbon substrate, the above expression tends to mW_{Gsh}.

The shoot senescence rate S_{sh} (kg structural C (m^2 ground)$^{-1}$ day^{-1}) is given by

$$S_{sh} = k_{s1}W_{Gsh} + k_{s2}W_{Gsh}\frac{K_m}{K_m + C_{sh}}. \tag{17.6b}$$

The first term on the right-hand side is the senescence rate under conditions of abundant carbon substrate, when the second term is negligible; k_{s1} (day^{-1}) is a senescence constant. From (17.6a), it can be shown that only a fraction $C_{sh}/(K_m + C_{sh})$ is fully maintained under non-abundant conditions of carbon substrate; an assumption that the non-maintained fraction decays at rate k_{s2} (day^{-1}) gives the second term in (17.6b).

The recycling of nitrogen from senescing shoot structural matter into the shoot substrate nitrogen pool is described by

$$V_{Nsh} = \frac{\phi}{1 + N_{sh}/K_\phi}\lambda S_{sh}, \tag{17.6c}$$

where V_{Nsh} (kg substrate N (m^2 ground)$^{-1}$ day^{-1}) is the nitrogen flux, ϕ (kg substrate N (kg structural N)$^{-1}$) is the maximum fraction recoverable, and K_ϕ (kg substrate N (kg structural C)$^{-1}$) is a constant. The denominator ensures that large amounts of nitrogen are recovered under conditions of low nitrogen status, and vice versa.

The parameters have the values

$$\begin{aligned} m &= 0.01 \text{ kg substrate C (kg structural C)}^{-1} \text{ day}^{-1} \\ K_m &= 0.002 \text{ kg substrate C (kg structural C)}^{-1} \\ k_{s1} &= 0.015 \text{ day}^{-1} \\ k_{s2} &= 0.1 \text{ day}^{-1} \\ \phi &= 0.6 \text{ kg substrate N (kg structural N)}^{-1} \\ K_\phi &= 0.03 \text{ kg substrate N (kg structural N)}^{-1}. \end{aligned} \tag{17.6d}$$

m is a typical plant maintenance rate ((10.4b), p. 268), K_m and K_ϕ are equivalent to the initial concentrations (see after (17.5e)), the rate constants k_{s1} and k_{s2} are set to plausible values, and ϕ is set to a reasonable value.

Carbon and nitrogen substrate fluxes between shoot and root The flux density $F_C(i = 1)$ (kg substrate C (m^2 ground)$^{-1}$ day^{-1}) of carbon substrate between the shoot and the first root layer (Fig. 17.1) is modelled by

$$F_C(1) = \zeta_C[C_{sh} - C_r(1)]\left[\frac{W_{Gsh}w_{Gr}(1)}{W_{Gsh} + w_{Gr}(1)\Delta z}\right], \tag{17.7a}$$

where ζ_C (m day^{-1}) is the carbon substrate conductivity parameter between shoot and root, with the value

$$\zeta_C = 0.126 \text{ m day}^{-1}. \tag{17.7b}$$

The behaviour and plausibility of (17.7a) is considered in Exercise 17.2.

Similarly, the upward nitrogen flux from the first root layer to the shoot is

$$F_N(1) = \zeta_N[N_r(1) - N_{sh}]\left[\frac{W_{Gsh} w_{Gr}(1)}{W_{Gsh} + w_{Gr}(1)\Delta z}\right], \tag{17.7c}$$

where ζ_N (m day^{-1}) is the nitrogen substrate conductivity parameter between shoot and root, with the value

$$\zeta_N = 0.0148 \text{ m day}^{-1}. \tag{17.7d}$$

This value gives reasonable rates of transport for acceptable values of the concentrations.

Root growth The rate of formation $g_r(i)$ (kg structural C (m^3 soil)$^{-1}$ day^{-1}) of mobile structure in layer i is

$$g_r(i) = k_r w_{Gr}(i)\frac{1}{1 + K_{Cr}/C_r(i) + K_{Nr}/N_r(i)}\left[1 - \frac{w_{Gr}(i)}{w_{Gr\,max}}\right] \qquad 1 \leqslant i \leqslant n_s, \tag{17.8a}$$

where k_r (day^{-1}) is a root growth rate parameter, $w_{Gr\,max}$ (kg structural carbon (m^3 soil)$^{-1}$) is the maximum density of structural root that the soil can accommodate, and K_{Cr} and K_{Nr} (kg substrate C (N) (kg structural C)$^{-1}$) are Michaelis–Menten constants.

Following the approach used for shoot growth (17.5a)–(17.5d), we obtain the rate $u_{Nr}(i)$ (kg substrate N (m^3 soil)$^{-1}$ day^{-1}) of nitrogen substrate utilization during root growth, the root growth respiration rate $r_{Gr}(i)$ (kg substrate C (m^3 soil)$^{-1}$ day^{-1}), and the rate $u_{CrG}(i)$ (kg substrate C (m^3 soil)$^{-1}$ day^{-1}) of utilization of carbon substrate into growth of new structure as

$$u_{Nr}(i) = \lambda g_r(i) \qquad 1 \leqslant i \leqslant n_s \tag{17.8b}$$

$$r_{Gr}(i) = \frac{1 - Y}{Y} g_r(i) \qquad 1 \leqslant i \leqslant n_s \tag{17.8c}$$

$$u_{CrG}(i) = g_r(i)(1 - \omega\lambda) \qquad 1 \leqslant i \leqslant n_s. \tag{17.8d}$$

The parameters of (17.8a) have the following values:

$$\begin{aligned} k_r &= 0.168 \text{ day}^{-1} \\ K_{Cr} &= 0.07 \text{ kg substrate C (kg structural C)}^{-1} \\ K_{Nr} &= 0.03 \text{ kg substrate N (kg structural C)}^{-1} \\ w_{Gr\,max} &= 5 \text{ kg structural C (m}^3 \text{ soil)}^{-1}. \end{aligned} \tag{17.8e}$$

The value of k_r gives reasonable growth rates and the K constants are equivalent to the initial concentrations; with soil possessing a mass of about 1200 kg m^{-3}, this value of $w_{Gr\,max}$ corresponds to a realistic value of root density.

Conversion of mobile root to fixed root It is assumed that this process occurs according to first-order kinetics with rate constant β (day^{-1}), so that the rate at which material is lost from the mobile pool (kg structural carbon (m^3 soil)$^{-1}$ day^{-1}) is

$$\beta w_{mGr}(i) \qquad 1 \leqslant i \leqslant n_s. \tag{17.9a}$$

This process is envisaged as simulating lateral root extension, and consequently there will be root rub-off due to friction between the soil particles and the mobile root. If a fraction ε_r is lost due to rub-off, then the fixed root pool ($w_{fGr}(i)$) increases at the rate

$$(1 - \varepsilon_r)\beta w_{mGr}(i) \qquad 1 \leqslant i \leqslant n_s. \tag{17.9b}$$

The rate at which material is lost in layer i is

$$\varepsilon_r \beta w_{mGr}(i) \qquad 1 \leqslant i \leqslant n_s. \tag{17.9c}$$

The parameters of (17.9a)–(17.9c) are

$$\beta = 0.75 \text{ day}^{-1} \qquad \varepsilon_r = 0.02. \tag{17.9d}$$

These values seem biologically reasonable, and give realistic results when used in the model.

Vertical movements of mobile root This provides the means by which the model simulates the vertical penetration of the root into uncolonized soil layers. The downward flux $F_{mr}(i)$ (kg structural C (m^2 ground)$^{-1}$ day^{-1}) of mobile root from layer $i - 1$ towards layer i is obtained from

$$F_{mr}(i = 1) = 0$$

$$F_{mr}(i) = \delta_1(i)\frac{\gamma_{mr}}{\Delta z}\{w_{mGr}(i-1) - w_{mGr}(i)\}\left[1 - \frac{w^*_{Gr}(i)}{w_{Gr\,max}}\right] \qquad 2 \leqslant i \leqslant n_s. \tag{17.10a}$$

$\delta_1(i)$ is defined by

$$\delta_1(i) = \begin{cases} 1 & \dfrac{w_{fGr}(i-1) + w_{fGr}(i)}{2} \geqslant w_{fGr\,min} \\[2ex] 0 & \dfrac{w_{fGr}(i-1) + w_{fGr}(i)}{2} < w_{fGr\,min} \end{cases}. \tag{17.10b}$$

This device (17.10b) ensures that movement between adjacent layers occurs only if there is sufficient root in the upper layer so that the root extends from the upper to the lower boundary of that layer, with $w_{\mathrm{fGr\,min}}$ (kg structural C (m^3 soil)$^{-1}$) defining this minimum.

γ_{mr} (m^2 day^{-1}) is the mobile root 'diffusion' coefficient. The rate of movement depends on the quantity of total structural carbon in the layer with the smaller mobile root pool $w^*_{\mathrm{Gr}}(i)$ (kg structural carbon (m^3 soil)$^{-1}$), where

$$w^*_{\mathrm{Gr}}(i) = \begin{cases} w_{\mathrm{Gr}}(i) & w_{\mathrm{mGr}}(i-1) \geqslant w_{\mathrm{mGr}}(i) \\ w_{\mathrm{Gr}}(i-1) & w_{\mathrm{mGr}}(i) > w_{\mathrm{mGr}}(i-1) \end{cases}. \tag{17.10c}$$

If $F_{\mathrm{mr}}(i)$ of (17.10a) is negative, then the flux is from layer i to layer $i-1$. If it is assumed that a fraction ε_{r} of the moving mobile root is lost through rub-off, then the gain $G_{\mathrm{mr}}(i)$ (kg structural C (m^3 soil)$^{-1}$ day^{-1}) in mobile root in layer i is

$$G_{\mathrm{mr}}(i) = \frac{1}{\Delta z}\{F_{\mathrm{mr}}(i)[1-\delta_2(i)\varepsilon_{\mathrm{r}}] - F_{\mathrm{mr}}(i+1)[1-(1-\delta_2(i+1))\varepsilon_{\mathrm{r}}]\}$$

$$1 \leqslant i \leqslant n_{\mathrm{s}} - 1 \tag{17.10d}$$

$$G_{\mathrm{mr}}(i) = \frac{1}{\Delta z}\{F_{\mathrm{mr}}(i)[1-\delta_2(i)\varepsilon_{\mathrm{r}}]\} \qquad i = n_{\mathrm{s}}.$$

Here

$$\begin{aligned} \delta_2(i) &= 1 \qquad F_{\mathrm{mr}}(i) > 0 \\ \delta_2(i) &= 0 \qquad F_{\mathrm{mr}}(i) \leqslant 0. \end{aligned} \tag{17.10e}$$

The $\delta_2(i)$ term ensures that the loss due to rub-off is only levied on the flux entering a compartment, and not on the flux leaving a compartment. Thus if $\delta_2(i+1) = 1$, the flux of material is from i to $i+1$ (the normal situation) and the rub-off factor ε_{r} does not affect the flux leaving compartment i.

It is useful to keep account of the material lost due to rub-off; in units of kg structural C (m^2 ground)$^{-1}$ day^{-1}, this is just the terms involving ε_{r} in (17.10d) with reversed sign. For layer i this is

$$\begin{aligned} &F_{\mathrm{mr}}(i)\delta_2(i)\varepsilon_{\mathrm{r}} - F_{\mathrm{mr}}(i+1)[1-\delta_2(i+1)]\varepsilon_{\mathrm{r}} \qquad 1 \leqslant i \leqslant n_{\mathrm{s}} - 1 \\ &F_{\mathrm{mr}}(i)\delta_2(i)\varepsilon_{\mathrm{r}} \qquad i = n_{\mathrm{s}}. \end{aligned} \tag{17.10f}$$

The losses are associated with the layer towards which the flux of material is directed.

The parameters of (17.10a) and (17.10b) are

$$\begin{aligned} \gamma_{\mathrm{mr}} &= 2.5 \times 10^{-4}\ \mathrm{m^2\ day^{-1}} \\ w_{\mathrm{fGr\,min}} &= 0.005\ \text{kg structural C (m}^3\text{ soil)}^{-1}. \end{aligned} \tag{17.10g}$$

These values give good results when used in the model, and seem biologically reasonable.

Movement of carbon and nitrogen substrates in the root It is assumed that carbon substrate moves in response to its concentration gradient, so that the downward flux density $F_C(i)$ (kg substrate C (m^2 ground)$^{-1}$ day^{-1}) between layers $i-1$ and i is

$$F_C(i) = \frac{\sigma_{Cr}}{\Delta z}[C_r(i-1) - C_r(i)]\left[\frac{w_{fGr}(i)w_{fGr}(i-1)}{w_{fGr}(i) + w_{fGr}(i-1)}\right] \qquad 2 \leqslant i \leqslant n_s. \quad (17.11a)$$

σ_{Cr} (m^2 day^{-1}) is a root carbon transport parameter. The second term in square brackets ensures that the flux density vanishes if either $w_{fGr}(i) \to 0$ or $w_{fGr}(i-1) \to 0$. $F_C(1)$ is defined in (17.7a). The rate $Z_C(i)$ (kg substrate C (m^3 soil)$^{-1}$ day^{-1}) at which carbon substrate moves into layer i is

$$Z_C(i) = \frac{F_C(i) - F_C(i+1)}{\Delta z} \qquad 1 \leqslant i \leqslant n_s - 1$$

$$Z_C(i) = \frac{F_C(i)}{\Delta z} \qquad i = n_s. \qquad (17.11b)$$

Similarly, the upward flux density $F_N(i)$ (kg substrate N (m^2 ground)$^{-1}$ day^{-1}) of nitrogen substrate layers i and $i-1$ is

$$F_N(i) = \frac{\sigma_{Nr}}{\Delta z}[N_r(i) - N_r(i-1)]\left[\frac{w_{fGr}(i)w_{fGr}(i-1)}{w_{fGr}(i) + w_{fGr}(i-1)}\right] \qquad 2 \leqslant i \leqslant n_s, \quad (17.11c)$$

where σ_{Nr} (m^2 day^{-1}) is a root nitrogen transport parameter and $F_N(1)$ is given by (17.7c). The rate $Z_N(i)$ (kg substrate N (m^3 soil)$^{-1}$ day^{-1}) at which nitrogen substrate moves into layer i is

$$Z_N(i) = \frac{F_N(i) - F_N(i+1)}{\Delta z} \qquad 1 \leqslant i \leqslant n_s - 1$$

$$Z_N(i) = \frac{F_N(i)}{\Delta z} \qquad i = n_s. \qquad (17.11d)$$

The carbon and nitrogen transport parameters have the values

$$\sigma_{Cr} = 0.1 \text{ m}^2 \text{ day}^{-1} \qquad \sigma_{Nr} = 0.1 \text{ m}^2 \text{ day}^{-1}. \qquad (17.11e)$$

Root maintenance and senescence These are represented in a similar manner to (17.6a)–(17.6c) for the shoot. Thus the rate $u_{CrM}(i)$ (kg C substrate (m^3 soil)$^{-1}$ day^{-1}) of utilization of carbon substrate from the carbon substrate pool in layer i is

$$u_{CrM}(i) = m w_{fGr}(i)\frac{C_r(i)}{K_m + C_r(i)} \qquad 1 \leqslant i \leqslant n_s, \qquad (17.12a)$$

where the parameters m and K_m are as before ((17.6a) and (17.6d)).

Root senescence $s_r(i)$ (kg structural C (m^3 soil)$^{-1}$ day^{-1}) in layer i is (cf. (17.6b))

$$s_r(i) = k_{s1} w_{fGr}(i) + k_{s2} w_{fGr}(i) \frac{K_m}{K_m + C_r(i)} \qquad 1 \leqslant i \leqslant n_s. \qquad (17.12b)$$

The recycling $v_{Nr}(i)$ (kg substrate N (m^3 soil)$^{-1}$ day^{-1}) of nitrogen from senescing root matter in layer i is (cf. (17.6c))

$$v_{Nr}(i) = \frac{\phi}{1 + N_r(i)/K_\phi} \lambda s_r(i) \qquad 1 \leqslant i \leqslant n_s. \qquad (17.12c)$$

The parameters are as in (17.6d).

Nitrogen substrate formation It is assumed that the soil nitrogen profile is constant. If the soil nitrogen derives from fertilizer applications, this may be a considerable simplification. However, the model can also be considered as representing legume growth, where the nitrogen substrate is provided by dinitrogen fixation. The model neglects water, which can strongly affect these and other processes. We therefore ignore the precise form of the soil nitrogen and mechanism of uptake, and assume that the rate $\nu_{Nr}(i)$ (kg substrate N (m^3 soil)$^{-1}$ day^{-1}) of uptake of soil nitrogen in layer i is given by

$$\nu_{Nr}(i) = k_{N\nu} N_s \left[\frac{C_r(i)}{K_{C\nu} + C_r(i)}\right] \left[\frac{1}{1 + N_r(i)/J_{N\nu}}\right] w_{fGr}(i) \qquad 1 \leqslant i \leqslant n_s, \quad (17.13a)$$

where $k_{N\nu}$ (m^3 (kg structural C)$^{-1}$ day^{-1}) is an uptake constant, N_s (kg N (m^3 soil)$^{-1}$) is the soil nitrogen concentration, $K_{C\nu}$ (kg substrate C (kg structural C)$^{-1}$) is a Michaelis–Menten constant, and $J_{N\nu}$ (kg substrate N (kg structural C)$^{-1}$) is an inhibition constant. The first term in square brackets ensures that the uptake rate is reduced if the carbon substrate concentration is low; the second term in square brackets means that less nitrogen is taken up under conditions of high nitrogen status in the root. (Note: (17.13a) describes non-competitive inhibition between carbon and nitrogen. This should be compared with (16.11a) (p. 466), where an expression for competitive inhibition is employed for nitrogen uptake; we are not aware of experimental work that resolves this issue. See Section 2.3 (p. 51) for a general discussion of the enzyme kinetics background, and also Section 4.4.2 (p. 102) for an alternative approach based on membrane transport.)

Carbon substrate is required to provide carbon skeletons for amino acids, and also to provide energy for the processes of uptake and reduction. The former process requires carbon substrate at the rate of

$$\omega \nu_{Nr}(i) \text{ kg C substrate (m}^3\text{ soil)}^{-1}\text{ day}^{-1} \qquad 1 \leqslant i \leqslant n_s \qquad (17.13b)$$

in layer i (for ω, see after (17.3d) and (17.5d)); the energy cost of nitrogen uptake is

$$c_N \nu_{Nr}(i) \text{ kg C substrate (m}^3\text{ soil)}^{-1}\text{ day}^{-1} \qquad 1 \leqslant i \leqslant n_s \qquad (17.13c)$$

where c_N (kg substrate C (kg substrate N)$^{-1}$) is the cost of nitrogen uptake and reduction. It is assumed that reduction occurs immediately after uptake and at the same location.

The parameters of nitrogen substrate formation have the following values:

$$k_{Nv} = 0.0018 \text{ m}^3 \text{ (kg structural C)}^{-1} \text{ day}^{-1}$$
$$N_s = 30 \text{ kg N (m}^3 \text{ soil)}^{-1}$$
$$K_{Cv} = 0.07 \text{ kg substrate C (kg structural C)}^{-1} \tag{17.13d}$$
$$J_{Nv} = 0.03 \text{ kg substrate N (kg structural C)}^{-1}$$
$$c_N = 0.55 \text{ kg substrate C (kg substrate N)}^{-1}.$$

Root exudates It is assumed that a fixed fraction of both the carbon and nitrogen substrates is lost per unit time by exudation, giving fluxes of

$$k_{xC} q_{Cr}(i) \qquad k_{xN} q_{Nr}(i) \qquad 1 \leqslant i \leqslant n_s \tag{17.14a}$$

in units of kg substrate C (N) (m^3 soil)$^{-1}$ day^{-1}. k_{xC} and k_{xN} are rate constants:

$$k_{xC} = 0.04 \text{ day}^{-1} \qquad k_{xN} = 0.04 \text{ day}^{-1}. \tag{17.14b}$$

17.2.3 *Dynamic equations of the model*

The differential equations for the three shoot state variables are

$$\frac{dW_{Gsh}}{dt} = G_{sh} - S_{sh} \qquad \text{with (17.5a), (17.6b)} \tag{17.15a}$$

$$\frac{dQ_{Csh}}{dt} = P_{sh} - R_{Gsh} - U_{CshG} - U_{CshM} - F_C(1)$$
$$\text{with (17.4a), (17.5c), (17.5d), (17.6a), (17.7a)} \tag{17.15b}$$

$$\frac{dQ_{Nsh}}{dt} = F_N(1) - U_{Nsh} + V_{Nsh} \qquad \text{with (17.7c), (17.5b), (17.6c).} \tag{17.15c}$$

The differential equations for the four root state variables are ($1 \leqslant i \leqslant n_s$)

$$\frac{dw_{fGr}(i)}{dt} = (1 - \varepsilon_r)\beta w_{mGr}(i) - s_r(i) \qquad \text{with (17.9b), (17.12b)} \tag{17.15d}$$

$$\frac{dw_{mGr}(i)}{dt} = g_r(i) - \beta w_{mGr}(i) + G_{mr}(i) \qquad \text{with (17.8a), (17.9a), (17.10d)} \tag{17.15e}$$

$$\frac{dq_{Cr}(i)}{dt} = -u_{CrG}(i) - r_{Gr}(i) + Z_C(i) - u_{CrM}(i) - (\omega + c_N)v_{Nr}(i) - k_{xC}q_{Cr}(i)$$
$$\text{with (17.8d), (17.8c), (17.11b), (17.12a), (17.13b), (17.13c), (17.14a)} \tag{17.15f}$$

$$\frac{dq_{Nr}(i)}{dt} = -u_{Nr}(i) + Z_N(i) + v_{Nr}(i) + \nu_{Nr}(i) - k_{xN}q_{Nr}(i)$$

with (17.8b), (17.11d), (17.12c), (17.13a), (17.14a). (17.15g)

The lower boundary condition is one of an impenetrable surface. The initial values of the state variables are as follows:

$$\begin{aligned} W_{Gsh} &= 0.05 \text{ kg structural C (m}^2\text{ ground)}^{-1} \\ Q_{Csh} &= 0.005 \text{ kg substrate C (m}^2\text{ ground)}^{-1} \\ Q_{Nsh} &= 0.0005 \text{ kg substrate N (m}^2\text{ ground)}^{-1} \\ w_{fGr}(1) &= 1 \text{ kg structural C (m}^3\text{ soil)}^{-1} \\ w_{fGr}(i) &= 0 \text{ kg structural C (m}^3\text{ soil)}^{-1} \qquad 2 \leqslant i \leqslant n_s \\ w_{mGr}(i) &= 0 \text{ kg structural C (m}^3\text{ soil)}^{-1} \qquad 1 \leqslant i \leqslant n_s \\ q_{Cr}(1) &= 0.07 \text{ kg substrate C (m}^3\text{ soil)}^{-1} \\ q_{Cr}(i) &= 0 \text{ kg substrate C (m}^3\text{ soil)}^{-1} \qquad 2 \leqslant i \leqslant n_s \\ q_{Nr}(1) &= 0.03 \text{ kg substrate N (m}^3\text{ soil)}^{-1} \\ q_{Nr}(i) &= 0 \text{ kg substrate N (m}^3\text{ soil)}^{-1} \qquad 2 \leqslant i \leqslant n_s. \end{aligned} \tag{17.15h}$$

Integration commenced at time $t = 91$ days (1 April) using the Euler integration (Chapter 1, Section 1.6.1, p. 19) and a time step of 0.001 day. The number n_s of soil layers and the vertical grid length Δz are given by

$$n_s = 12 \qquad \Delta z = 0.025 \text{ m}. \tag{17.15i}$$

These values are suitable for a shallow rooting crop such as grass. Additional details are given by Brugge (1985).

17.2.4 *Some simulations of the model*

The standard simulation makes use of the parameters and initial values specified in the last section. This is compared with various other situations, such as high and low soil nitrogen, where the soil nitrogen parameter is assigned values of 60 kg N (m^3 soil)$^{-1}$ and 15 kg N (m^3 soil)$^{-1}$ respectively (cf. the standard value of 30 in (17.13d)); also, the effects of defoliation are considered.

The time course of development of shoot and root structural dry matter is shown in Fig. 17.3. The roots attain a maximum about 50 days before the shoot maximum occurs, and the root : shoot ratio decreases with time, as noted by many authors (e.g. Brouwer 1962, Troughton 1978). The carbon and nitrogen substrate levels in the shoot (C_{sh} and N_{sh}) and in the first root compartment ($C_r(1)$ and $N_r(1)$) at various times during the growing season are given in Table 17.2. After the first month of simulated growth, the carbon substrate levels decrease

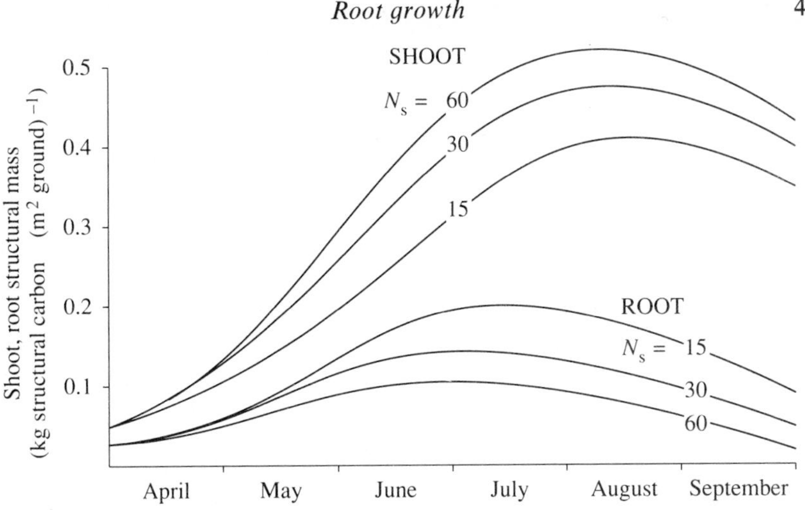

Fig. 17.3. Time course of shoot and root structural dry masses for three values of the soil nitrogen parameter N_s: standard, 30 kg N (m^3 soil)$^{-1}$; low, 15 kg N (m^3 soil)$^{-1}$; high, 60 kg N (m^3 soil)$^{-1}$.

Table 17.2. Carbon and nitrogen substrate concentrations in the shoot (C_{sh} and N_{sh}) and in the first root compartment ($C_r(1)$ and $N_r(1)$) on various dates in the growing season

Substrate	Location	Substrate concentration (kg substrate C (N) (kg structural C)$^{-1}$)				
		30 April	20 May	9 June	29 June	19 July
C_{sh}	Shoot	0.147	0.112	0.077	0.060	0.051
$C_r(1)$	Root	0.106	0.073	0.042	0.029	0.022
N_{sh}	Shoot	0.008	0.007	0.006	0.005	0.004
$N_r(1)$	Root	0.034	0.034	0.031	0.028	0.026

Growth begins on 1 April. Parameters and initial conditions are standard.

markedly throughout the plant, as has been observed in tomatoes (Cooper and Thornley 1976), soybean (Dunphy and Hanway 1976), and grasses (Waite and Boyd 1953). This arises because most processes utilizing carbon substrate are proportional to structural dry mass, whereas, owing to increased shading, there is no such proportionality between shoot structural dry mass and photosynthetic input.

The decrease in carbon substrate levels is most marked in the roots, and this is accompanied by a reduction in the nitrogen uptake rate in early June and a net root senescence from July onwards. Photosynthesis reaches a maximum at the end of June, and thereafter shoot growth rates decrease rapidly. The

Table 17.3. Root development for a high ($N_s = 60$ kg N (m^3 soil)$^{-1}$) and a low ($N_s = 15$ kg N (m^3 soil)$^{-1}$) soil nitrogen regime obtained using (17.3b)

Mid-layer depth (m)	Total structural root mass (kg structural C (m^3 soil)$^{-1}$)			
	30 April	30 May	9 July	27 Sept
High soil nitrogen status				
0.0125	1.53	2.07	1.78	0.54
0.0375	0.37	0.90	1.06	0.35
0.0625	0.14	0.44	0.61	0.21
0.0875	0.049	0.20	0.32	0.11
0.1125	0.015	0.083	0.15	0.053
0.1375	0.0015	0.031	0.066	0.024
0.1625		0.0096	0.026	0.0099
0.1875			0.0099	0.0040
0.2125				0.0015
Low soil nitrogen status				
0.0125	1.66	2.62	2.82	1.15
0.0375	0.42	1.31	2.03	0.92
0.0625	0.16	0.70	1.41	0.68
0.0875	0.056	0.34	0.88	0.47
0.1125	0.017	0.15	0.50	0.29
0.1375	0.0023	0.062	0.26	0.16
0.1625		0.021	0.12	0.084
0.1875		0.0041	0.049	0.040
0.2125			0.018	0.018
0.2375			0.0036	0.0072
0.2625				0.0025

detailed results suggest that root senescence rates reach a maximum in about mid-September, and then root growth rates begin to increase. This appears to agree with observations by Garwood (1967) that in grasses new adventitious roots are produced mainly in later winter and early spring; however, the autumn peak in growth rates that is sometimes observed is not reproduced by the model.

The effect of contrasting soil nitrogen regimes on root growth is shown in Table 17.3. Plants grown in soils with low nitrogen tend to produce heavier and deeper roots; there are many observations to support this prediction. Thus the plant can increase its nitrogen uptake, but at the expense of shoot development. A high soil nitrogen status results in the early attainment of shoot and root maximum structural masses, and under such conditions carbon substrate levels are reduced throughout the plant. whereas nitrogen substrate levels are increased.

The simulation of the depth profiles of the carbon and nitrogen substrate concentrations is interesting. With increasing depth the carbon substrate concentration falls off steadily, as would be expected. However, the nitrogen substrate concentration is virtually independent of depth. This arises because nitrogen is input at all levels in the root system, whereas carbon is only input into the top root horizon.

Brugge (1985) also discusses the effects of a 50 per cent shoot defoliation on shoot and root growth using his model.

17.2.5 *Discussion*

We have discussed this model in detail for two reasons. First, it demonstrates how partial differential equations and a spatial variable can be handled in a plant growth model. Second, it seems that the approach may be generally useful for modelling root growth and extension in a relatively simply manner at a level which corresponds to many successful treatments of the shoot system. The framework employed permits a simple parameterization of soil physical conditions, and extensions of the model to take account of water and nutrient gradients in the soil appear to be feasible.

Exercises

17.1. Assume that exploration of the soil is easier in regions where there is already some root; replace the diffusion constant D of (17.2a) by Dw_r, so that (D now has units of m^2 kg^{-1} day^{-1})

$$\frac{\partial w_r}{\partial t} = Dw_r \frac{\partial^2 w_r}{\partial z^2} - m_r w_r. \tag{E17.1a}$$

Solve the steady state problem, given that $w_r = w_0$ at $z = 0$, and that at some depth $z = z_0$, $w_r = dw_r/dz = 0$. Show that the root distribution function is a quadratic, and derive an equation for the total root dry mass W_r.

17.2. Write (17.7a) in the form

$$F_C(1) = \zeta_C \left[\frac{C_{sh} - C_r(1)}{\Delta z}\right]\left[\frac{W_{Gsh} w_{Gr}(1)\Delta z}{W_{Gsh} + w_{Gr}(1)\Delta z}\right] \tag{E17.2a}$$

and discuss what happens to the flux density $F_C(1)$ as the size of the plant is altered and as the layer thickness Δz is altered. Does the behaviour of this equation seem acceptable? Can you devise a reasonable alternative for the second term in square brackets?

Part III
Plant morphology

18
Branching

18.1 Introduction

The form and function of tree-like structures is relevant to both the above-ground and below-ground parts of plants. There is a wide range of naturally occurring branching structures, including river systems, the bronchial tree, and the pulmonary artery, in addition to the many examples from the plant world. Thus a good deal of attention has been devoted to these problems.

We shall only consider trees and not networks, although in leaves networks are clearly important to both structure and function. Mostly, we shall be concerned with paracladial trees, i.e. trees where if any branch is cut off, it has the same structural characteristics, apart from size, as the parent from which it is cut. Thus the branching processes that operate to generate the tree structure are assumed to function uniformly throughout the tree; they are not context dependent, i.e. they do not depend, for instance, upon the distance from sources of carbon or nitrogen substrates, or the uneven distribution of light within a canopy, or of nutrients within the soil.

The measurements that are taken on a tree structure are generally simple, but numerous. If we consider a branch as being the segment between two adjacent branch points, then it would be usual to number the branches so as to define their relative positions and sequence, and perhaps measure the length, mean diameter, and (less often) the angles of each branch, of which there may be many hundreds or thousands. The immediate problem is to summarize this mass of data, and various ordering systems have been devised for this purpose. The method devised by Horton (1945) for river systems, which is cumbersome to apply and requires a judgement as to which is the major daughter branch at each bifurcation, has been superseded by a modified method due to Strahler (1952). Strahler's method and a later method due to Shreve (1966) are illustrated in Fig. 18.1, and the rules are given in the figure caption. Neither of these methods is without problems. In Strahler's method, the addition of one more bifurcation can change the numbering markedly; also it is seen that a Strahler-enumerated component (e.g. with order number 2 or 3 in Fig. 18.1(a)) can encompass several branches as defined above. In Shreve's scheme several order numbers are entirely absent. Other methods have been constructed and applied by various workers (see MacDonald 1983); none of these is without defects, and thus no method has won general acceptance although Strahler's technique is the most widely used. The requirements of a good ordering scheme are as follows. The method must be unambiguous and easy to apply. The derived parameters should be capable

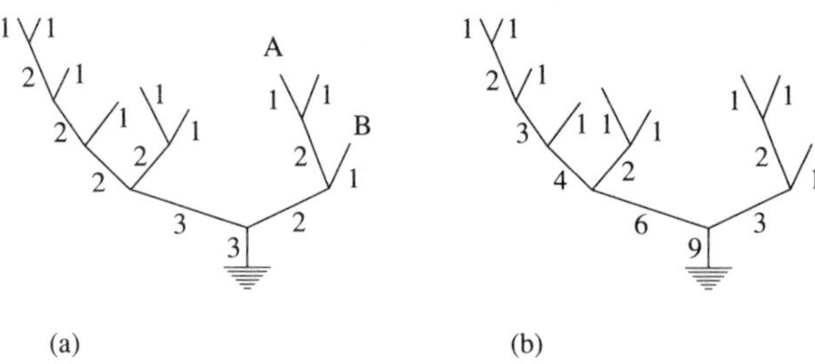

Fig. 18.1. (a) Strahler and (b) Shreve ordering of a branched structure. In (a) the rules are as follows: (i) all ends (except for the base of the tree) are assigned order number 1; (ii) if two branches of equal order meet, then the order of the parent branch is increased by 1; (iii) if two branches of unequal number meet, the parent branch is assigned the greater of the two daughter orders. In (b) the rules are as follows: (i) all free ends are assigned order 1; (ii) parents are assigned an order equal to the sum of the orders of the daughters, so that the order of a branch is equal to the number of free ends it supports.

of being visualized or interpreted in physiological or structural terms. It should be possible to relate the derived parameters to functional characteristics of the system.

There are essentially two approaches to the problem of describing a branching system: static methods give a snapshot description of the state of the system, and the ordering-system approach belongs to this category; dynamic or generative methods are concerned with the processes that give rise to the structure over time. Ideally the two should be complementary and convergent, although the dynamic approach will generally be more complex but more complete. Of the dynamic models, some consider a continuously extending system, within which, sometimes subject to some constraints such as a minimum age or diameter, bifurcation may occur at any point. The root system is often considered in this manner (e.g. Rose 1983). Other models are modular in nature, composed of what can be called morphological units. The above-ground parts of many plants and trees sometimes fit more naturally into this framework, with bifurcations being restricted to buds which appear at intervals along the stem.

Adding to the complexity of the picture, we can also categorize models of tree growth as empirical, mechanistic, or teleonomic (p. 5). To date most models have been empirical or teleonomic; mechanism still seems to be largely beyond our grasp, although this is a direction in which we must reach out. Necessarily, the teleonomic models are concerned with function and efficiency, and here it must be said that, even within the plant world alone, there appears to be a great diversity of possible functions. Some of these are light-gathering efficiency, soil exploration, nutrient and water absorption, seed dispersal, mechanical support, competitive ability, and the transport of various materials. In this chapter we shall explore several of these aspects, although it will be clear that there is no prospect at present of constructing a unified picture of this area.

Models which generate branched structures are described in Sections 18.2 and 18.3; these models serve to introduce terminology and methodology. Both sections can be omitted by readers who wish to proceed to the three remaining sections which discuss different aspects of branching.

18.2 Branching model with a delay in bud outgrowth

In this section we describe a simple rule-based branching model of a modular type using 'morphological units'. The objective of this type of modelling is to seek rules that can be applied repetitively and are able to generate some of the branched structures that are found naturally. The tree is constructed on a computer, where it is analysed using Strahler's method. The model is semi-mechanistic in that the principal assumption of the model, that of delayed bud outgrowth, can be given a natural physiological interpretation (this model derives from discussions many years ago with Alun Rees, to whom we are indebted).

18.2.1 *Generation of the tree structure*

It is assumed that, after a new bud has been formed, there is a delay before it is permitted to grow out. This could be envisaged as follows. If the apex is a source of inhibitor of bud outgrowth, then as a shoot continues to extend the apex becomes more distant from a given bud, the inhibitory effect of the apex becomes weaker, and eventually the outgrowth of the bud is no longer suppressed.

The morphological unit considered consists of a terminal bud (the apex) plus the first lateral bud below the apex (e.g. unit 4 in Fig. 18.2(a) with $n = 2$), or just below a lateral bud to just below the next lower lateral bud which may or may not have grown out (unit 2 in Fig. 18.2(a) with $n = 2$, or unit 2 in Fig. 18.2(c) with $n = 3$). Let t_1 be the time of formation of the first unit. Let the time interval Δt be the generation time or the cycle time of the system. It is assumed that each terminal bud (or apex) continues to grow and gives rise to one new unit after time Δt. A lateral bud can also grow out to give a unit with a terminal apex, but only if it is of age $a\,\Delta t$ where a is a constant. Once a lateral bud has begun to grow out, it continues to extend at each time step. Some of the branching patterns resulting from this scheme are shown in Fig. 18.2. (See Exercise 18.1 for a trifurcating system) Consider the situation at time t where

$$t = t_1 + n\,\Delta t, \tag{18.1a}$$

i.e. after n generations. There are potentially 2^n terminal morphological units (MUs), and

$$N_{\text{max}} = 2^{n+1} - 1 \quad \text{total MUs (terminal + non-terminal)}, \tag{18.1b}$$

where N_{max} is the maximum potential total number of units realized only if $a = 0$ as shown in Fig. 18.2(a). Thus in Fig. 18.2(a), with $n = 3$, there are eight terminal

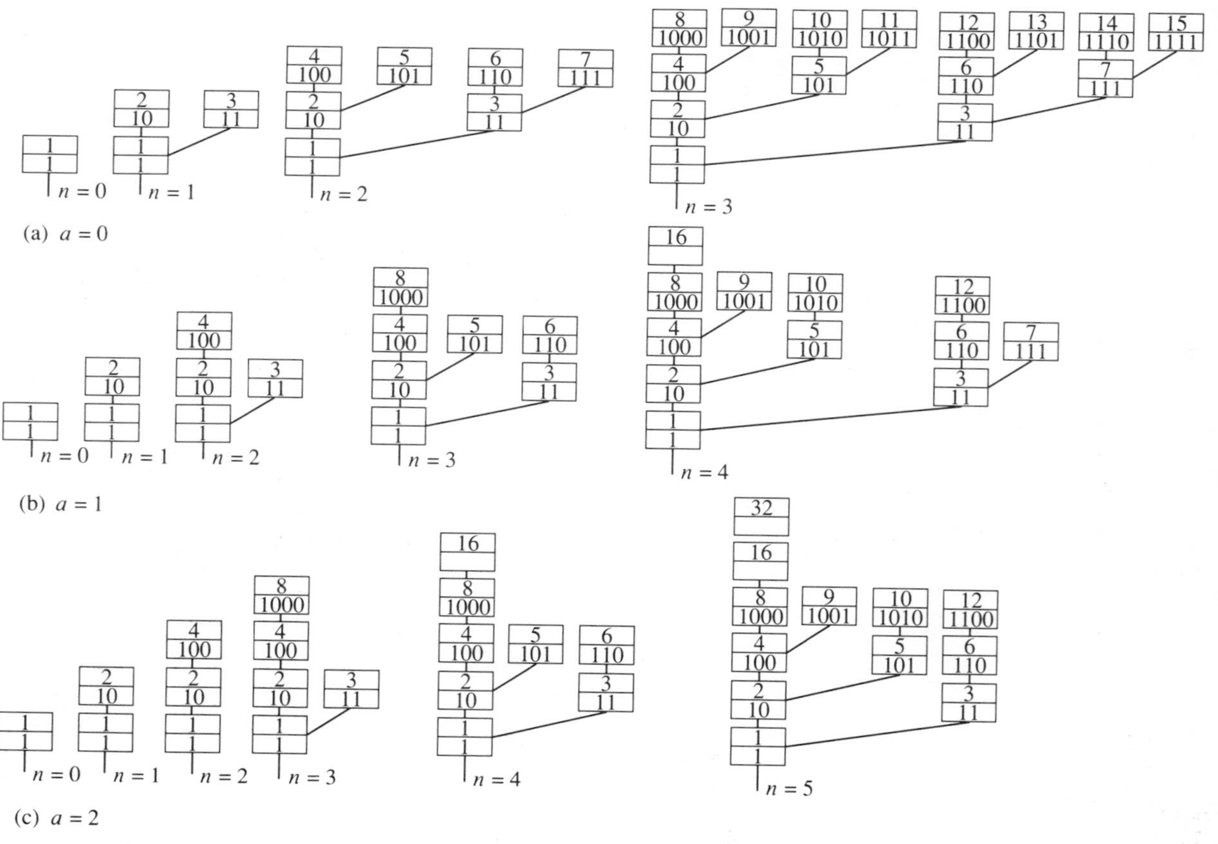

Fig. 18.2. Delayed bud outgrowth model. Each morphological unit contains one lateral bud which can grow out after a time $a\Delta t$ where a is a constant. Each morphological unit is assigned a decimal integer N which it is useful to represent by binary digits (18.1c). n is the generation number (18.1a).

buds and 15 units. As shown in Fig. 18.2, every unit is given a decimal number N. Consider the Nth unit, with $1 \leqslant N \leqslant 2^{n+1} - 1$. The decimal digit N can be represented by k binary digits:

$$N(i_1, i_2, \ldots, i_{k-1}, i_k), \tag{18.1c}$$

where k is given by

$$k = 1 + \text{integer part}(\log_2 N). \tag{18.1d}$$

For example, the seventh MU is represented by binary (111), as shown in Fig. 18.2. Also, in Fig. 18.2, we follow the convention that continuing to the next MU along the same axis adds a zero to the binary number, so that

$$N \rightarrow 2N; \tag{18.1e}$$

bifurcating to the right adds a 1 to the binary number, so that

$$N \rightarrow 2N + 1. \tag{18.1f}$$

Next, we consider the question: will the Nth unit exist at time $t = t_1 + n\,\Delta t$? Let t_N be the time of formation of the Nth unit, so that

$$t_N = t_1 + \Delta t\left(k - 1 + a \sum_{j=2}^{k} i_j\right). \tag{18.1g}$$

The operation of eqn (18.1g) is shown in Table 18.1, which should be compared

Table 18.1. Delayed bud outgrowth model: calculation of the time of outgrowth t_N of morphological unit N using eqn (18.1g) with $t_1 = 0$

N	Binary	k	$\sum_{j=2}^{k} i_j$	Time of outgrowth $t_N/\Delta t$		
				$a = 0$	$a = 1$	$a = 2$
1	1	1	0	0	0	0
2	10	2	0	1	1	1
3	11	2	1	1	2	3
4	100	3	0	2	2	2
5	101	3	1	2	3	4
7	111	3	2	2	4	6
8	1000	4	0	3	3	3
9	1001	4	1	3	4	5
15	1111	4	3	3	6	9

N is the decimal number of the unit (Fig. 18.2) which has the binary representation given in the second column; k is the number of digits in the binary representation (eqn (18.1c)) and is defined in eqn (18.1d); the sum in the fourth column is made from the left on the binary representation (18.1c), omitting the first binary digit which is always unity; in the last three columns, the constant a is the delay (in time units of Δt) applied to a lateral bud before it is allowed to grow out; the time of outgrowth t_N is expressed here in units of Δt.

with Fig. 18.2. Thus

$$\text{if } t_N \leqslant t, \quad \text{then } N \text{ exists}, \quad \text{and } e_N = 1; \qquad \text{otherwise } e_N = 0. \tag{18.1h}$$

e_N is used to denote the existence, or otherwise, of the morphological unit with number N.

Let τ_N denote the age of the Nth unit, given by

$$\begin{aligned}\tau_N &= t - t_N \\ &= n\,\Delta t - \left(k - 1 + a \sum_{j=2}^{k} i_j\right)\Delta t\end{aligned} \tag{18.1i}$$

from eqns (18.1a) and (18.1g). It is assumed that morphological unit N has length l_N and mean diameter d_N, which can be calculated from the length and diameter l_0 and d_0 at birth and the time τ_N that the unit has had to grow by means of a growth function such as the logistic or the Gompertz growth function (Chapter 3). For example, using the Gompertz growth equation for length l_N, we obtain

$$l_N = l_0 \exp\{k_0[1 - \exp(-k_1\tau_N]/k_1\} \tag{18.1j}$$

where k_0 and k_1 are parameters, and a similar equation for d_N. Note that the asymptotic value of length (and diameter) as time t approaches infinity is greater than its original value by a factor of $l_N(\tau_N = \infty)/l_0 = \exp(k_0/k_1)$; this is the maximum increase in length (and diameter) of a morphological unit that can occur using the Gompertz growth function.

In summary, each morphological unit N has the attributes of t_N (time of birth), e_N (denotes its existence), τ_N (age), l_N (length), and d_N (mean diameter). The 'daughter' morphological units of unit N are $2N$ and $2N + 1$. The above rules permit the generation of a branched structure which can be subjected to an analysis such as that provided by the Strahler ordering technique (Fig. 18.1).

18.2.2 *Application of Strahler's ordering rules*

We define the following quantities:

s_N, Strahler order of MU N (18.2a)

β_m, number of Strahler-ordered branches (not MUs) of Strahler order m (18.2b)

μ_m, number of MUs of Strahler order m (18.2c)

L_m, total length of all the branches (or MUs) of Strahler order m (18.2d)

D_m, sum of the diameters of all the MUs of Strahler order m. (18.2e)

β_m, μ_m, L_m, and D_m are initialized to zero for a range of m ($m = 1, 2, 3, \ldots$) sufficient to accommodate the structure to be described.

The algorithm to sort and count the morphological units according to the Strahler ordering rules is as follows.

N is successively assigned integer values from $N_{\max}$ down to 1, i.e.

$$N := N_{\max}, N_{\max} - 1, \ldots, 2, 1, \tag{18.2f}$$

where here and elsewhere the symbol := denotes 'is assigned the value of'. If morphological unit N does not exist, then N is simply decremented, so that

$$\text{if } e_N = 0, \text{ then decrement } N. \tag{18.2g}$$

If an MU is terminal (without daughters) it necessarily has a Strahler order of unity. Thus

$$\begin{aligned} &\text{if } 2N > N_{\max} \quad \text{or} \quad e_{2N} = e_{2N+1} = 0, \text{ then} \\ &s_N := 1, \quad \beta_1 := \beta_1 + 1, \quad \mu_1 := \mu_1 + 1, \\ &L_1 := L_1 + l_N, \quad D_1 := D_1 + d_N. \end{aligned} \tag{18.2h}$$

If the MU is not terminal but has a dormant bud, then its Strahler order is the same as that of its daughter ($2N$) and it belongs to the same Strahler-ordered branch. Hence

$$\begin{aligned} &\text{if } e_{2N} = 1 \quad \text{and} \quad e_{2N+1} = 0, \text{ then} \\ &s_N := s_{2N} \quad (\beta_{sN} \text{ is not incremented}), \\ &\mu_{sN} := \mu_{sN} + 1, \quad L_{sN} := L_{sN} + l_N, \\ &D_{sN} := D_N + d_N. \end{aligned} \tag{18.2i}$$

The two-letter subscript sN denotes the value of s_N. If the bud on the MU has grown out to give a lateral shoot, then MU N has two daughter MUs, numbered $2N$ and $2N + 1$; these either have equal or unequal Strahler order numbers, so that there are two cases to consider:

$$\begin{aligned} &\text{if } e_{2N} = e_{2N+1} = 1, \text{ then} \\ &\text{if } s_{2N} = s_{2N+1} \text{ then} \\ &\quad s_N := s_{2N} + 1, \quad \beta_{sN} := \beta_{sN} + 1, \\ &\quad \mu_{sN} := \mu_{sN} + 1, \quad L_{sN} := L_{sN} + l_N, \\ &\quad D_{sN} := D_{sN} + d_N; \end{aligned} \tag{18.2j}$$

$$\begin{aligned} &\text{if } s_{2N} \neq s_{2N+1} \text{ then} \\ &\quad s_N := s_{2N} \quad (\beta_{sN} \text{ is not incremented}), \\ &\quad \mu_{sN} := \mu_{sN} + 1, \quad L_{sN} := L_{sN} + l_N, \\ &\quad D_{sN} := D_{sN} + d_N. \end{aligned} \tag{18.2k}$$

Note that with the conventions shown in Fig. 18.2, $s_{2N} \geqslant s_{2N+1}$ always.

Finally, the mean length λ_m and the mean diameter δ_m of the branches of Strahler order m are calculated using

$$\lambda_m = L_m/\beta_m \qquad \delta_m = D_m/\mu_m. \tag{18.21}$$

The application of Shreve's ordering rules is considered in Exercise 18.2.

18.2.3 *Some simulated tree structures*

A computer program was written to carry out the operations described in the last section, and some results are now described.

Seven parameters are needed to define the structure of the tree. These are the generation time Δt, which is taken to be 1.0, the bud outgrowth delay parameter a, which is assigned values in the range 0–3.0, the time of formation t_1 of the first morphological unit, which is given the value 0.0, the length l_0 and diameter d_0 of a morphological unit when it is first formed, which are both taken to be 1.0, and the growth parameters k_0 and k_1 which are taken to be 1.0 and 0.43 respectively, giving a maximum growth factor of 10 (eqn (18.1j) *et seq.*). The tree structure can then be calculated for any given value of the time variable t, which, with Δt and t_1 given, determines the number of possible generations to be included (eqn (18.1a)). Three typical tree structures generated by the model are shown in Fig. 18.3; these are drawn with the MUs having the lengths generated

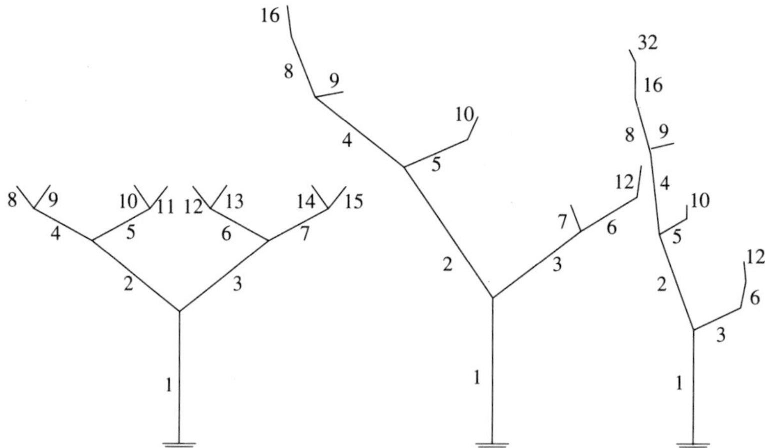

Fig. 18.3. Three tree structures generated by the delayed bud outgrowth model. The morphological units are numbered as discussed in eqns (18.1c)–(18.1f); the lengths shown are scaled according to eqn (18.1j) and the angles drawn are arbitrary. The parameters of the model are $\Delta t = 1$, $t_1 = 0$, $l_0 = 1$, $d_0 = 1$, $k_0 = 1$, $k_1 = 0.43$; in (a) $a = 0$ and $t = 3$, in (b) $a = 1$ and $t = 4$, and in (c) $a = 2$ and $t = 5$. Strahler analysis gives the following Strahler order (number of Strahler-ordered branches of that order) relationships: (a) 1 (8), 2 (4), 3 (2), 4 (1); (b) 1 (5), 2 (2), 3 (1); (c) 1 (4), 2 (1).

by the Gompertz growth equation (eqn (18.1j)). It can be seen that this simple algorithm is capable of generating a range of structures.

In order to apply the Strahler analysis large tree structures must be considered to avoid the 'lumpiness' in the analysis for small structures. The branching ratio R_B is defined by

$$R_B = \beta_m/\beta_{m+1}, \tag{18.3a}$$

where β_m is the number of Strahler-ordered branches of Strahler order m (18.2b). The branching ratio is perhaps the most important characteristic of a branching structure; it is the ratio between the number of branches in adjacent orders. In the three structures shown in Fig. 18.3 (a) has $R_B = 2$, and this increases in (b) and (c).

Table 18.2 contains a Strahler analysis of a large computer-generated tree structure. For this case, an average branching ratio can be calculated approximately by taking the eighth root of 1352 to give $R_B = 2.46$. As shown in Table 18.2, the branching ratio (the ratio of adjacent numbers in the second column, which is given in the third column) is effectively constant, and the present algorithm of delayed bud outgrowth does give rise to a characteristic branching ratio. Apart from the highest-order branch present ($m = 9$), the number of MUs per Strahler branch (fifth column) increases slowly with increasing order number; this ratio depends upon the delay parameter a (for $a = 0$, the ratio is unity), but

Table 18.2. Strahler analysis of a computer-generated tree structure using the delayed bud outgrowth model

m	β_m	β_m/β_{m+1}	μ_m	μ_m/β_m	Age$_m$	λ_m	δ_m
1	1352	2.44	1941	1.44	0.63	2.76	1.92
2	553	2.48	799	1.44	2.38	6.57	4.55
3	223	2.53	330	1.48	4.15	10.23	6.91
4	88	2.59	135	1.53	5.95	12.95	8.44
5	34	2.62	54	1.59	7.79	14.72	9.27
6	13	2.60	21	1.62	9.67	15.62	9.67
7	5	2.50	8	1.60	11.55	15.76	9.85
8	2	2.00	3	1.67	13.40	14.90	9.93
9	1	—	1	1.00	15.00	9.97	9.97

The parameters of the model are $\Delta t = 1$, $t_1 = 0$, $l_0 = 1$, $d_0 = 1$, $k_0 = 1$, $k_1 = 0.43$ (giving a maximum length and diameter of 10), $a = 0.8$ and $t = 15.1$; this gives 15 generations ($n = 15$ from eqn (18.1a)) and a maximum potential number of MUs of 65535 (N_{max} from eqn (18.1b)). The columns in the table are as follows: m is the Strahler order number; β_m is the number of branches of order m; the third column is the ratio of the branch numbers in successive orders, i.e. the branching ratio R_B of eqn (18.3a); μ_m is the number of MUs in the branches of order m; the ratio in the fifth column gives the number of MUs per Strahler branch; age$_m$ denotes the mean age of the branches in that order; λ_m and δ_m are the mean length and diameter from eqns (18.21).

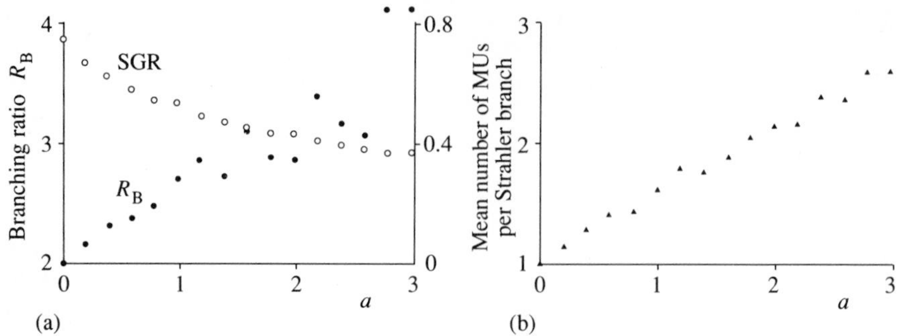

Fig. 18.4. Delayed bud outgrowth model; a is the delay in bud outgrowth. In (a) the branching ratio R_B and the specific growth rate SGR are shown. R_B is defined in eqn (18.3a), and was calculated approximately as described in the text; SGR was calculated from eqn (18.3d). In (b) the mean number of MUs per Strahler branch is shown. The parameters used for the results were $\Delta t = 1$, $t_1 = 0$, $l_0 = d_0 = 1$, $k_0 = 1$, $k_1 = 0.43$, and $t = 15$.

is independent of time (i.e. the size of the structure). In this model, the length and diameter of an MU have been assumed to grow identically from the same initial value (eqn (18.1j)), and thus to a good approximation

$$\lambda_m = \delta_m(\mu_m/\beta_m); \tag{18.3b}$$

this can be verified in Table 18.2.

Also, denoting the mean age of branches of order m by age_m, we note that

$$\text{age}_m = \text{age}_1 + (m - 1)(1 + a)\,\Delta t; \tag{18.3c}$$

the mean age of branches of order m increases linearly with order number—this result could be useful physiologically, for example when considering the performance of fruit trees which have been characterized using Strahler ordering.

These structures grow exponentially, and therefore the specific growth rate (SGR) is a well-defined quantity. This can be calculated approximately by

$$\text{SGR} = \ln(\text{total number of MUs at time } t)/t. \tag{18.3d}$$

As the delay in bud outgrowth increases, the SGR decreases, as shown in Fig. 18.4.

18.2.4 Comparison with observational data Some data obtained by applying Strahler's analysis to two types of botanical tree are given in Table 18.3: the first three examples are for a fungal mycelium and the next two examples for ordinary trees—an apple tree and a birch tree. We discuss the performance of the delayed bud outgrowth model in relation to these data.

In the first two fungal mycelia two branches and three branches respectively of the highest order are present. The delayed bud outgrowth model is not able to describe such data, since it necessarily gives a single highest-order branch. The branching ratios of the second sample of *Mucor hiemalis*, the apple tree, and the

Table 18.3. Strahler analyses of some botanical tree structures

	Strahler branch order	No. of branches	Total length	Mean length	Mean diameter
Itersonia perplexans	1	59	5731	97	—
	2	17	2739	161	—
	3	6	797	133	—
	4	2	225	112	—
Mucor hiemalis	1	32	324	10.0	—
	2	8	179	22.4	—
	3	3	54.8	18.3	—
Mucor hiemalis	1	67	1034	15.4	—
	2	13	399	30.7	—
	3	4	82.9	20.7	—
	4	1	31.3	31.3	—
Apple tree	1	492	10 820	22	2.9
	2	67	11 320	169	3.8
	3	14	9900	707	9.3
	4	5	1500	299	16.0
	5	1	217	217	35.0
Birch tree	1	944	177 000	187	1.6
	2	213	53 900	253	3.3
	3	60	13 300	221	5.5
	4	14	3370	241	9.9
	5	3	1100	367	23.7
	6	1	86.0	86.0	45.0

The first three examples are of fungal mycelia; the Strahler analyses are reported by Park (1985) on the basis of drawings given by Brady (1960) (*Itersonilia perplexans*) and Hutchinson, Sharma, Clarke, and MacDonald (1980) (*Mucor hiemalis*). The collection and analysis of the apple and birch tree data are reported by Barker, Cumming, and Horsfield (1973). For the mycelia, the lengths are given in microns; for the apple and birch trees, the lengths and diameters are in millimetres.

birch tree are approximately 4.1, 4.7, and 3.9 respectively. The results of attempts to predict the observed branching structures of *Mucor hiemalis* (second sample) and the apple tree are given in Table 18.4.

Owing to limitations in array size in the computer it was not possible to simulate the whole structure of the apple tree, and two simulations are reported which can be compared with the observed structure in Table 18.3. In both cases the mean diameters fail to show the continued diameter growth that is observed. This is a consequence of the growth equation (eqn (18.1j)) applied to diameter growth which gives an asymptotic diameter value of 10; a better physiological assumption might be that the activity in the cambium gives slow but continuing

Table 18.4. Simulation of two of the botanical trees of Table 18.3 using the delayed bud outgrowth model

m	β_m	β_m/β_{m+1}	μ_m	μ_m/β_m	Age_m	λ_m	δ_m
Simulated Mucor hiemalis: $t = 15, a = 3, k_0 = 5, k_1 = 2.17$							
1	69	3.63	181	2.62	1.12	15.5	5.91
2	19	3.80	50	2.63	5.10	26.3	10.0
3	5	5.0	14	2.80	9.14	28.0	10.0
4	1	—	4	4.00	13.5	40.0	10.0
Simulated apple tree: $(a)\ t = 17, a = 4, k_0 = 1, k_1 = 0.43$							
1	60	4.00	185	3.08	1.48	9.43	3.06
2	15	3.75	45	3.00	6.44	25.56	8.52
3	4	4.00	11	2.75	11.27	26.97	9.81
4	1	—	3	3.00	16.00	29.93	9.98
Stimulated apple tree: $(b)\ t = 19, a = 5, k_0 = 1, k_1 = 0.43$							
1	55	4.58	196	3.56	1.82	12.54	3.52
2	12	4.00	43	3.58	7.91	32.72	9.13
3	3	3.00	9	3.00	13.78	29.78	9.93
4	1	—	2	2.00	18.50	19.98	9.99

The parameters of the model are $\Delta t = 1$, $t_1 = 0$, $l_0 = 1$, $d_0 = 1$; k_0 and k_1 are as shown, giving a maximum length and diameter of 10; a and t are as shown. The columns in the table are as follows: m is the Strahler order number; β_m is the number of branches of order m; the third column is the ratio of the branch numbers in successive orders, i.e. the branching ratio R_B of eqn (18.3a); μ_m is the number of MUs in the branches of order m; the ratio in the fifth column gives the number of MUs per Strahler branch; age_m denotes the mean age of the branches in that order; λ_m and δ_m are the mean length and diameter from eqns (18.21).

exponential diameter growth, which, since the mean age age_m varies linearly with order (eqn (18.3c)), will lead to results similar to those given in Table 18.3. Considering the branch lengths, there is about the same number of MUs per Strahler branch in simulation (a), and this leads to an increase in mean branch length with order which is not reflected in the observational data. In simulation (b) the highest-order branch has only two MUs, and as a consequence the predicted mean lengths are slightly closer to reality. However, it must be emphasized that some of the results of this computer model (e.g. the number of MUs in the branch of highest Strahler order) can change quite abruptly with small changes in the growth time t or other parameters (Exercise 18.1 applies the method of this section to a trifurcating system).

18.3 Apical bifurcation model of branching

A model which generates branching structures by continued bifurcation at the apex when it reaches a certain size is described in this section. The bifurcation can be asymmetric, and the asymmetry parameter λ is the critical parameter of the model, not dissimilar to the way in which the delay a in bud outgrowth was the critical parameter of the last model. This model has been described by Thornley (1977). Branching by apical bifurcation is uncommon amongst trees, but does occur, for example, in some algae, liverworts, and ferns, although we are not aware that these have been analysed by Strahler's method.

18.3.1 *The model*

Following the notation of Section 18.2.1 and Fig. 18.2, a morphological unit (MU) is assigned the number of the bifurcation which terminates (or would terminate) it. The Nth MU has daughters $2N$ and $2N + 1$. The MU with decimal number N is also represented by a set of k binary digits (cf. eqn (18.1c)):

$$N(i_1, i_2, \ldots, i_{k-1}, i_k), \tag{18.4a}$$

with k given by

$$k = 1 + \text{integer part}(\log_2 N). \tag{18.4b}$$

Growth It is assumed that both the volume and dry mass of the material increase with a specific growth rate ρ, and that each of the three linear dimensions increases at a specific rate $\rho/3$. This assumption is peripheral to the performance of the model, and it disagrees with a report by McMahon and Kronauer (1976) that the radius of an MU is proportional to its length to the power of 3/2; McMahon and Kronauer's results imply that larger branches bending under gravity bend with a smaller radius of curvature than smaller branches (Section 18.6.1, p. 528).

Branching The apical radius of a terminal MU increases due to growth, and it is assumed that when this radius reaches a critical value r the apex divides into two daughters, with basal radii R_a and R_b, and apical radii r_a and r_b (Fig. 18.5). A parameter λ is defined as the ratio of the radii of the daughter morphological units, with

$$\lambda = R_a/R_b. \tag{18.4c}$$

If $\lambda = 1$, then branching is symmetric, and as λ departs from 1 there is increasing asymmetry. It is also assumed that

$$r/R = r_a/R_a = r_b/R_b, \tag{18.4d}$$

where R is the basal radius of the MU just before bifurcation occurs and the apical radius is r.

It is also necessary to assume some relationship between the apical radius r just before bifurcation and the basal radii R_a and R_b of the daughters. Murray (1927a) observed that the weight of a part of a tree above a given cross-section is proportional to the cube of the radius of the cross-section, suggesting that (see

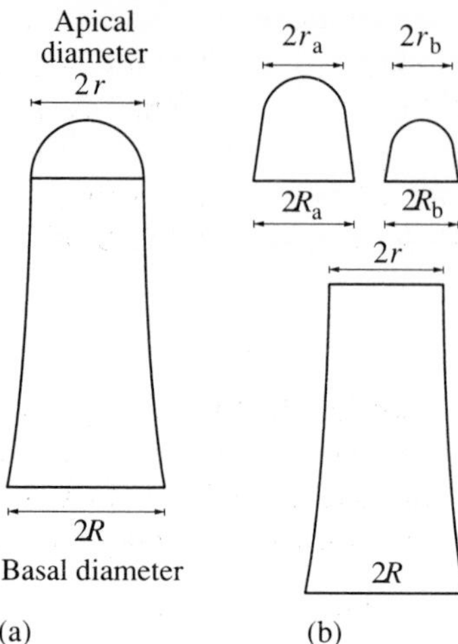

Fig. 18.5. Bifurcation at the apex: (a) before bifurcation; (b) after bifurcation. The daughter units resulting from the bifurcationare denoted by the subscripts a and b. R denotes the basal radius of a unit and r the apical radius.

Fig. 18.5(b))

$$r^3 = R_a^{\,3} + R_b^{\,3}. \tag{18.4e}$$

This equation seems reasonable in terms of conserving the material in the dividing apex, and a similar relationship is found to apply to many dividing bacteria. Combining eqns (18.4c) and (18.4e) gives

$$R_a = \frac{\lambda}{(1 + \lambda^3)^{1/3}} r \qquad R_b = \frac{1}{(1 + \lambda^3)^{1/3}} r. \tag{18.4f}$$

Times between bifurcations Let Δt_a and Δt_b be the times elapsing between successive bifurcations. If a constant specific growth rate $\rho/3$ is applicable to the apical radius,

$$r = r_a \exp\left(\frac{\rho \Delta t_a}{3}\right) \qquad r = r_b \exp\left(\frac{\rho \Delta t_b}{3}\right) \tag{18.4g}$$

Using eqns (18.4f)

$$\begin{aligned} \Delta t_a &= \frac{1}{\rho} \ln\left[\left(\frac{R_a}{r_a}\right)^3 \left(\frac{1 + \lambda^3}{\lambda^3}\right)\right] \\ \Delta t_b &= \frac{1}{\rho} \ln\left[\left(\frac{R_b}{r_b}\right)^3 (1 + \lambda^3)\right]. \end{aligned} \tag{18.4h}$$

In order to solve this equation, it is necessary to assume a value for the ratio R/r

of the basal radius to the apical radius; because of the growth assumptions made above this ratio does not change with time. It is assumed that

$$\gamma = \frac{R}{r} = \frac{R_a}{r_a} = \frac{R_b}{r_b}. \tag{18.4i}$$

A more general form of eqn (18.4e) is considered in Exercise 18.3, and a generalization of eqns (18.4h) is derived.

Generation of the tree structure Let t_1 be the time of the first bifurcation and t_N be the time of the Nth bifurcation. The Nth bifurcation is represented by a set of binary digits as in eqn (18.4a). The time Δt_a (eqn (18.4h)) is associated with the interval between the Nth and $2N$th bifurcations; Δt_b is associated with the interval between the Nth and the $(2N + 1)$th bifurcations. If Fig. 18.2 is used to represent the bifurcations of the present model, time interval Δt_a is associated with moving straight up in the diagram, and time interval Δt_b is associated with moving to the right. Thus the time when the Nth bifurcation occurs is given by

$$t_N = t_1 + \sum_{j=2}^{k} [(1 - i_j)\Delta t_a + i_j \Delta t_b], \tag{18.4j}$$

where j is a summation index and the i_j are the binary digits of (18.4a).

When bifurcation occurs, by assumption each MU has the same basal and apical radii R and r; at time t these radii are given by

$$\begin{aligned} r_N(t) &= r \exp\left[\frac{\rho(t - t_N)}{3}\right] \\ R_N(t) &= R \exp\left[\frac{\rho(t - t_N)}{3}\right]. \end{aligned} \tag{18.4k}$$

Five parameters are required in order to generate the tree structure at a given time t. These are the asymmetry parameter λ, the SGR ρ, the time t_1 of the first bifurcation, the apical radius r at bifurcation, and the basal radius R at bifurcation. λ is important in determining the type of structure; ρ, t, and t_1 determine the time scale and the number of events that occur; r and R determine the shape of the morphological units (for some purposes, it might be necessary to make additional assumptions about the relation of R and r to the length of a unit, and the nature of the radius–length profile).

Given these parameters, R_a and R_b are obtained from eqns (18.4f), and then r_a and r_b are obtained from eqns (18.4d). Next Δt_a and Δt_b are calculated from eqns (18.4h). The time t_N for the Nth bifurcation is given by eqn (18.4j). If $t_N > t$, then the Nth bifurcation has not taken place, and the $2N$th and $(2N + 1)$th morphological units do not exist. If $t_N \leqslant t$, then the $2N$th and $(2N + 1)$th units do exist, and their dimensions are given by eqns (18.4k).

Ordering of the structure by Strahler's method proceeds as outlined in Section 18.2.2.

18.3.2 *Results and discussion*

The model has only one significant parameter, λ of eqn (18.4c), which determines the symmetry of the bifurcations. Three simulations of tree growth for different values of λ are shown in Fig. 18.6. Variations in λ are clearly able to give rise to very different branching patterns.

The application of Strahler's analysis to structures generated by the current algorithm leads to well-defined values of the branching ratio R_B (the ratio of the numbers of Strahler-enumerated branches in adjacent orders; see Section 18.2.2 and eqn (18.3a)) and the diameter ratio R_D (the ratio of the mean diameters of the branches in adjacent orders). With symmetric branching $\lambda = 1$ and R_B has its minimum value of 2; as λ deviates from unity, R_B increases. Similarly, the

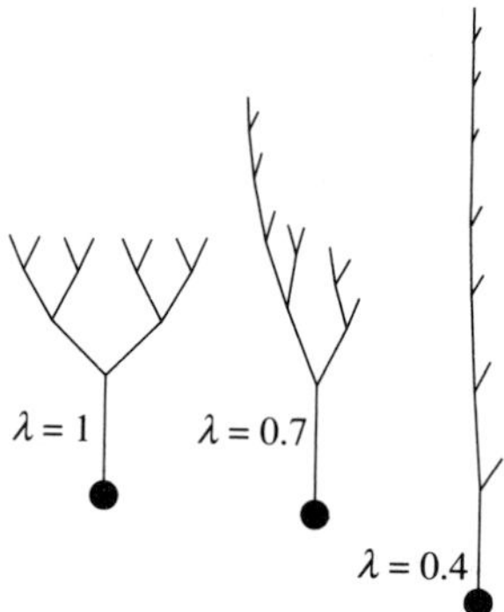

Fig. 18.6. Apical bifurcation model: simulation of tree growth for three values of λ (eqn (18.4c)). The values of the other parameters are $t = 4.4$, $t_1 = 1$, $\rho = 0.693$, $r = 0.10$, and $R = 0.11$. The angles shown above have no particular significance.

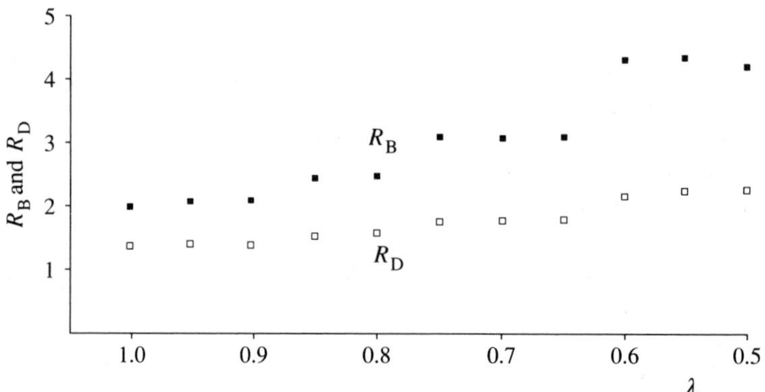

Fig. 18.7. Apical bifurcation model of branching. Dependence of the branching ratio R_B (■) and the diameter ratio R_D (□) on the branching parameter λ. Remaining model parameters are $\rho = 0.693$, $t = 9.5$, $t_1 = 1$, $r = 0.1$, and $R = 0.11$.

diameter ratio R_D takes its minimum value of 1.39 (this is $(R/r) \times 2^{1/3}$ with $R = 0.11$ and $r = 0.10$] for $\lambda = 1$, and again R_D increases as λ deviates from unity. These relationships are shown in Fig. 18.7.

As in Fig. 18.4(a), Fig. 18.7 does not shown completely smooth changes in the branching ratios; this is because of the completely deterministic nature of the two models, and can occur where the maximum Strahler order number suddenly changes by unity.

Both this and the preceding model are essentially single-parameter models, and when compared with natural structures using Strahler's ordering rules they appear capable of realistic simulation.

18.4 Optimization of different functions

Branching patterns and the resulting shape of the plant can be viewed teleonomically in many different ways. In this approach it is assumed (p. 11) that the result of evolution has been to produce, within the constraints set by the environment and within the range of possibilities allowed by physicochemical mechanism, a plant which has optimum response characteristics to perform some particular function. Since several different plant functions seem likely to contribute to the success of the essential teleonomic project, the assumption that a particular plant function is dominant is not likely to be totally satisfactory. However, this viewpoint can lead to interesting relationships between certain aspects of form and function, and may be relevant in restricted circumstances.

Some plant functions that have been considered in this way include leaf area (Honda and Fisher 1978), plant shape in relation to photosynthesis and water (Paltridge 1973), and mechanical design considerations (McMahon and Kronauer 1976). In a fine paper, Warner and Wilson (1976) have investigated the three-dimensional distribution of end-points in a branching network and its dependence on branching angle and a branch length decay parameter; this has an obvious relevance to the exploitation of a soil volume by a root branching system. Also, considering the problem of arterial branching, Zamir (1976) gives an excellent account of the application of four different optimality principles to the angles and radii of branches meeting at a node. In this section, we give an account of some of this work.

18.4.1 *Exploration of space and the distribution of end-points*

Consider a one-dimensional branching process, which begins at time $t = 0$ with a stem of unit length, so that there is a single end-point at $y = 1$ (Fig. 18.8(a)). After a single generation Δt the end-point at $y = 1$ is replaced by two end-points at a distance γ from $y = 1$, i.e.

$$y = 1 \pm \gamma \qquad n = 1 \qquad t = n\,\Delta t = \Delta t, \tag{18.5a}$$

where n is the number of generations and γ is a decay ratio. After two generations

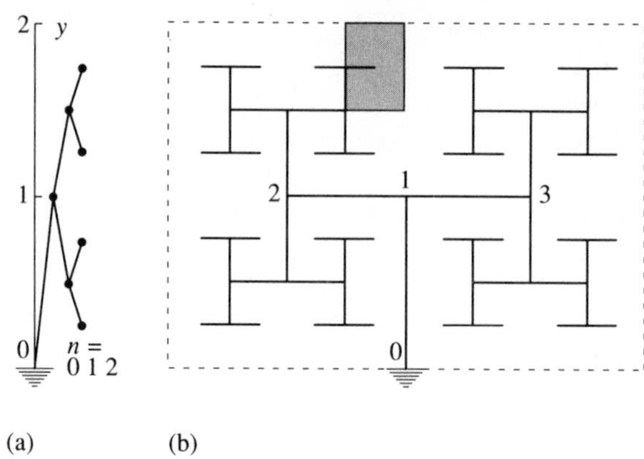

Fig. 18.8. End-point distributions in one and two dimensions. (a) A one-dimensional branching process leading to a distribution of points along the y axis (decay ratio $\gamma = 1/2$). (b) A two-dimensional branching process with symmetric branching at angle $\theta = 90°$ and decay ratio $\gamma = 1/\sqrt{2}$. These parameters give rise to uniform end-point distributions and the exploration of a finite space. The shaded area denotes an area that may be associated with each end-point, and will cover the space enclosed by the broken line.

there are four end-points:

$$y = 1 \pm \gamma \pm \gamma^2 \qquad n = 2 \qquad t = 2\Delta t. \tag{18.5b}$$

The region occupied by these four end-points is

$$1 - \gamma - \gamma^2 \quad \text{to} \quad 1 + \gamma + \gamma^2. \tag{18.5c}$$

This region is larger for increasing values of the decay ratio γ, but a uniform distribution of end-points is only obtained for $\gamma = 2^{-1}$, in which case the four points are

$$1/4 \quad 3/4 \quad 5/4 \quad 7/4 \tag{18.5d}$$

and these are equally spaced. For large n with $\gamma = 2^{-1}$ the region between 0 and 2 becomes uniformly and densely covered with end-points.

In two dimensions there is an extra degree of freedom (Fig. 18.8(b)), and so, in addition to the decay ratio γ (the ratio of successive branch lengths), there is a branching angle θ. This is the angle between 01 and 12 or 13 in Fig. 18.8(b), where $\theta = 90°$ and symmetric branching is assumed. Figure 18.8(b) shows the pattern obtained with $\gamma = 2^{-1/2} = 0.707$ and $\theta = 90°$, which gives a uniform end-point distribution within a region of space and corresponds to (18.5d) in the one-dimensional case. If the decay ratio γ is greater than $2^{-1/2}$ and $\theta = 90°$, there will be some crossing of the branch members; this would be wasteful, and space can be explored more efficiently by reducing θ below 90° so that crossing does not occur. Clearly, for values of the decay ratio γ below unity, the space filled by the end-points approaches a finite limit as the number of bifurcations becomes very large. The functional considerations are concerned with matters such as the

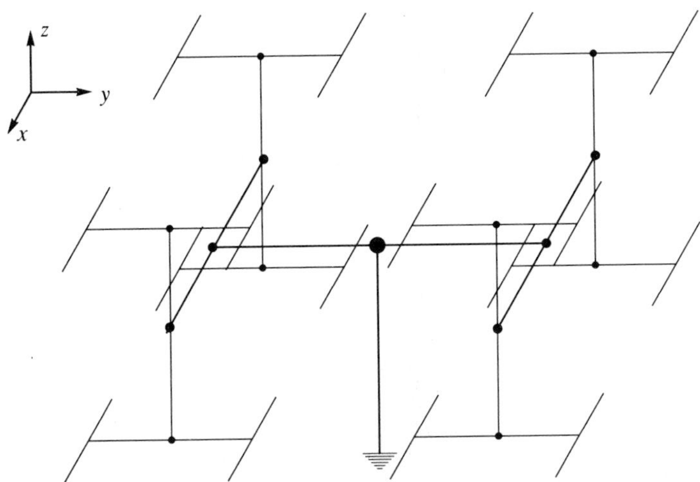

Fig. 18.9. Branching system in three dimensions. A uniform end-point distribution is achieved with a decay ratio of $2^{-1/3} = 0.794$ and a branching angle of 90°.

magnitude of the space explored, the uniformity of the end-point distribution, and the investment in the branching network.

In three dimensions there are two angular degrees of freedom. Figure 18.9 shows the pattern obtained by applying a branching angle of 90° successively in different directions with a branch decay ratio of $2^{-1/3} = 0.794$; this leads to a uniform end-point distribution. While some limited progress with these problems can be made by analytical means, the properties of these distributions are best explored by computer. Warner and Wilson (1976) have investigated the characteristics of the end-point distributions as they depend on the decay ratio γ and the branching angles.

Warner and Wilson (1976) were primarily concerned with the bronchial tree, in which the regular way in which the dimensions of the airways change has given rise to much teleonomic speculation about the advantages of the arrangement. The observed length decay ratio is very constant at about 0.84, whereas the diameter decay ratio is smaller at about 0.79 but increases in later generations. The branching angle is about 35°, with a tendency to increase in later generations. Discussion has been concerned with items such as the costs of moving gas through the network against viscous forces and the dead space associated with the network, some of which are considered below. However, these arguments are strongly dependent on assumptions about constraints and cost parameters.

18.4.2 *Optimality in a single bifurcation*

Consider the bifurcation shown in Fig. 18.10 (this section is largely based on Zamir (1976)). AJ represents the mother branch and JB and JC the daughter branches, with coordinates and angles as shown. A, J, B, and C lie in a plane: it

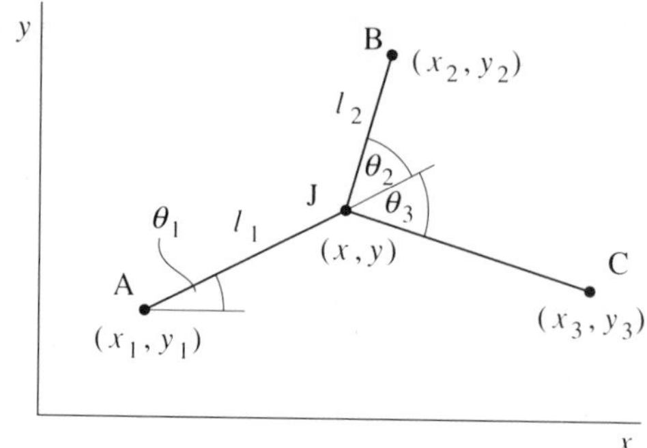

Fig. 18.10. Single bifurcation defined in the xy plane. AJ is the parent branch, JB and JC are the daughter branches, and J is the point where the bifurcation is situated. Lengths l_1, l_2, and l_3, angles θ_1, θ_2, and θ_3, and coordinates are as shown.

is assumed that in establishing communication between A and B, and between A and C, there is no advantage in placing the junction J outside the plane ABC. The lengths l_1, l_2, and l_3 of the three branches are given by

$$\begin{aligned} l_1 &= [(x - x_1)^2 + (y - y_1)^2]^{1/2} \\ l_2 &= [(x - x_2)^2 + (y - y_2)^2]^{1/2} \\ l_3 &= [(x - x_3)^2 + (y - y_3)^2]^{1/2}. \end{aligned} \tag{18.5e}$$

The angles and coordinates are related by

$$\begin{aligned} \cos(\theta_2 + \theta_1) &= \frac{x_2 - x}{[(x - x_2)^2 + (y - y_2)^2]^{1/2}} \\ \cos(\theta_3 - \theta_1) &= \frac{x_3 - x}{[(x - x_3)^2 + (y - y_3)^2]^{1/2}} \\ \cos\theta_1 &= \frac{x - x_1}{[(x - x_1)^2 + (y - y_1)^2]^{1/2}} \\ \sin\theta_1 &= \frac{y - y_1}{[(x - x_1)^2 + (y - y_1)^2]^{1/2}}. \end{aligned} \tag{18.5f}$$

General conditions for optimization of the branching geometry It is assumed that the geometry of the junction is such that some property H is minimized. H may be related to the flow of material through the junction, to the materials of which the junction is constructed, or to a combination of these things; some particular optimality principles are discussed below. It is assumed here that H can be calculated additively by means of an integral

$$H = \int h \, dl, \tag{18.5g}$$

taken along the elements of the junction, where h is the value of H per unit length. For a single element of length l, therefore,

$$H = hl. \tag{18.5h}$$

For the junction shown in Fig. 18.10

$$\begin{aligned} H &= h_1 l_1 + h_2 l_2 + h_3 l_3 \\ &= h_1[(x - x_1)^2 + (y - y_1)^2]^{1/2} \\ &\quad + h_2[(x - x_2) + (y - y_2)^2]^{1/2} \\ &\quad + h_3[(x - x_3)^2 + (y - y_3)^2]^{1/2}. \end{aligned} \tag{18.5i}$$

Next, H is minimized with respect to the position of the junction point (x, y) or with respect to the two angles θ_2 and θ_3. Differentiating eqn (18.5i) with respect to x and then y, therefore, we have

$$\begin{aligned} \frac{\partial H}{\partial x} &= \frac{h_1(x - x_1)}{l_1} + \frac{h_2(x - x_2)}{l_2} + \frac{h_3(x - x_3)}{l_3} \\ \frac{\partial H}{\partial y} &= \frac{h_1(y - y_1)}{l_1} + \frac{h_2(y - y_2)}{l_2} + \frac{h_3(y - y_3)}{l_3}. \end{aligned} \tag{18.5j}$$

Equations (18.5j) can be written

$$\begin{aligned} \frac{\partial H}{\partial x} &= h_1 \cos\theta_1 - h_2 \cos(\theta_2 + \theta_1) - h_3 \cos(\theta_3 - \theta_1) \\ \frac{\partial H}{\partial y} &= h_1 \sin\theta_1 - h_1 \sin(\theta_2 + \theta_1) + h_3 \sin(\theta_3 - \theta_1). \end{aligned} \tag{18.5k}$$

H is a minimum, maximum, or point of inflexion of zero slope if

$$\frac{\partial H}{\partial x} = 0 \quad \text{and} \quad \frac{\partial H}{\partial y} = 0. \tag{18.5l}$$

Combining eqns (18.5k) and (18.5l) to give

$$\begin{aligned} 0 &= h_1 \cos\theta_1 - h_2 \cos(\theta_2 + \theta_1) - h_3 \cos(\theta_3 - \theta_1) \\ 0 &= h_1 \sin\theta_1 - h_2 \sin(\theta_2 + \theta_1) + h_3 \sin(\theta_3 - \theta_1) \end{aligned} \tag{18.5m}$$

eliminating θ_2 or θ_3 by taking the term containing θ_2 or θ_3 across to the left-hand side and squaring and adding the two equations, and then simplifying (using the expansion for $\cos(A + B)$), we obtain

$$\begin{aligned} \cos\theta_2 &= \frac{h_1{}^2 + h_2{}^2 - h_3{}^2}{2h_1 h_2} \\ \cos\theta_3 &= \frac{h_1{}^2 + h_3{}^2 - h_2{}^2}{2h_1 h_3}. \end{aligned} \tag{18.5n}$$

By taking the first term in eqns (18.5m) across to the left-hand side, and squaring, adding, and simplifying (using the expansions of cos A cos B and sin A sin B), we obtain

$$\cos(\theta_2 + \theta_3) = \frac{h_1^{\,2} - h_2^{\,2} - h_3^{\,2}}{2h_2 h_3}. \tag{18.5o}$$

To show that the solutions given in eqns (18.5n) correspond to a minimum in H, rather than a maximum or a point of inflexion, it is strictly necessary to demonstrate that the higher derivatives satisfy

$$\frac{\partial^2 H}{\partial x^2} > 0 \qquad \frac{\partial^2 H}{\partial y^2} > 0 \qquad \left(\frac{\partial^2 H}{\partial x^2}\right)\left(\frac{\partial^2 H}{\partial y^2}\right) - \left(\frac{\partial^2 H}{\partial x \partial y}\right)^2 > 0. \tag{18.5p}$$

Zamir (1976) gives explicit expressions for these quantities, which are in fact satisfied by eqns (18.5n); this result seems intuitively acceptable from Fig. 18.10 (see Exercise 18.4).

Three optimality principles Next we examine the consequences of making three different assumptions for the property H which is to be minimized (eqns (18.5g)–(18.5i)). The three hypotheses examined are as follows.

1. The junction is in an optimum state if the total surface area S of the branches is minimum. S is given by

$$S = sl \tag{18.6a}$$

   where $s = 2\pi r$, s is the surface area per unit length corresponding to h of eqns (18.5g) and (18.5h), and r is the branch radius.
2. The junction is in an optimum state if the total volume V of the branches is minimum. V is given by

$$V = vl \tag{18.6b}$$

   where $v = \pi r^2$ is the volume per unit length.
3. The junction is in an optimum state if the power W required to drive fluid through the junction is minimum. W is given by

$$W = wl, \tag{18.6c}$$

   where $w = 8\eta q^2/\pi r^4$ is the power required per unit length assuming Poiseuille flow, η is the viscosity, and q is the volume flow rate. This result follows immediately from Poiseuille's equation for the volume flow rate q:

$$q = \frac{\Delta p}{\Delta l}\frac{\pi r^4}{8l\eta} \tag{18.6d}$$

   where Δp is the pressure difference across length element Δl and the rate of working in this element is $q\Delta p$.

Zamir (1976) considers a fourth optimality principle: a minimum in the total drag force acting on the walls of the system. However, we think that this is unlikely to be a consideration in the present context.

Applying these three principles (eqns (18.6a)–(18.6c)) to eqns (18.5n) and (18.5o) for the branching angles gives

$$S_{\min}: \quad \cos\theta_2 = \frac{r_1^2 + r_2^2 - r_3^2}{2r_1 r_2}$$

$$\cos\theta_3 = \frac{r_1^2 + r_3^2 - r_2^2}{2r_1 r_3} \tag{18.6e}$$

$$\cos(\theta_2 + \theta_3) = \frac{r_1^2 - r_2^2 - r_3^2}{2r_2 r_3}$$

$$V_{\min}: \quad \cos\theta_2 = \frac{r_1^4 + r_2^4 - r_3^4}{2r_1^2 r_2^2}$$

$$\cos\theta_3 = \frac{r_1^4 + r_3^4 - r_2^4}{2r_1^2 r_3^2} \tag{18.6f}$$

$$\cos(\theta_2 + \theta_3) = \frac{r_1^4 - r_2^4 - r_3^4}{2r_2^2 r_3^2}$$

$$W_{\min}: \quad \cos\theta_2 = \frac{q_1^4/r_1^8 + q_2^4/r_2^8 - q_3^4/r_3^8}{2(q_1^2 q_2^2/r_1^4 r_2^4)}$$

$$\cos\theta_3 = \frac{q_1^4/r_1^8 + q_3^4/r_3^8 - q_2^4/r_2^8}{2(q_1^2 q_3^2/r_1^4 r_3^4)} \tag{18.6g}$$

$$\cos(\theta_2 + \theta_3) = \frac{q_1^4/r_1^8 - q_2^4/r_2^8 - q_3^4/r_3^8}{2(q_2^2 q_3^2/r_2^4 r_3^4)}.$$

Consequences for branching Equations (18.6e)–(18.6g) for the branching angles with the three optimality conditions of eqns (18.6a)–(18.6c) are of similar structure, and thus only one of them need be considered in detail; the corresponding results for the other cases are obtained by replacing r of eqns (18.6e) by r^2 for eqns (18.6f) and by q^2/r^4 for eqns (18.6g). We therefore investigate the properties of eqns (18.6e), since this is the simplest case.

It is assumed that

$$r_1 \geqslant r_2, r_3. \tag{18.6h}$$

By substitution for r_3 in the first two of eqns (18.6e), it can be shown that

$$\text{If } r_1 = r_2 + r_3, \text{ then } \cos\theta_2 = \cos\theta_3 = 1 \quad \text{and} \quad \theta_2 = \theta_3 = 0. \tag{18.6i}$$

Thus the branching angles are zero, and B and C in Fig. 18.10 cannot be simultaneously attained.

$$\text{If } r_1 > r_2 + r_3, \text{ then } \cos\theta_2, \cos\theta_3 > 1, \tag{18.6j}$$

and there is no optimum solutions. Lastly,

$$\text{if } r_1 < r_2 + r_3, \text{ then } \cos\theta_2, \cos\theta_3 < 1, \tag{18.6k}$$

and there are solutions for θ_2 and θ_3, although for these solutions to be applicable it is required that $\theta_2 + \theta_3$ is greater than the angle $B\widehat{A}C$ in Fig. 18.10. The three possibilities given above can be verified by inserting the three radius triplets (r_1, r_2, r_3) of (10, 7, 3), (10, 7, 1), and (10, 7, 5) into eqns (18.6e).

The solutions of eqns (18.6e) are particularly simple for the special case

$$r_1{}^2 = r_2{}^2 + r_3{}^3. \tag{18.6l}$$

Equations (18.6e) become

$$\cos\theta_2 = \frac{r_2}{r_1} \qquad \cos\theta_3 = \frac{r_3}{r_1} \qquad \theta_2 + \theta_3 = 90°. \tag{18.6m}$$

Equations (18.6m) show many of the qualitative features applicable to all the optimality principles. A large branch of similar size to the parent branch continues in the same direction as its parent with little change; however, a small branch may be deflected by almost a right angle.

A symmetrical bifurcation In this simplified case the three optimality principles can be directly compared. The assumption of symmetry requires that

$$r_2 = r_3 \qquad q_2 = q_3 = q_1/2 \qquad \theta_2 = \theta_3. \tag{18.7a}$$

Equations (18.6e)–(18.6g) for the total angle between the daughter branches now become

$$S_{\min}: \qquad \cos(2\theta_2) = \frac{r_1{}^2 - 2r_2{}^2}{2r_2{}^2} \tag{18.7b}$$

$$V_{\min}: \qquad \cos(2\theta_2) = \frac{r_1{}^4 - 2r_2{}^4}{2r_2{}^4} \tag{18.7c}$$

$$W_{\min}: \qquad \cos(2\theta_2) = \frac{8r_2{}^8 - r_1{}^8}{r_1{}^8}. \tag{18.7d}$$

The total bifurcation angle now depends on the radii alone. We define the ratio ρ of daughter branch to parent branch radii by

$$\rho = \frac{r_2}{r_1}, \tag{18.7e}$$

and substituting in eqns (18.7b)–(18.7d) we obtain

$$S_{\min}: \qquad \cos(2\theta_2) = \frac{1 - 2\rho^2}{2\rho^2} \tag{18.7f}$$

$$V_{\min}: \qquad \cos(2\theta_2) = \frac{1 - 2\rho^4}{2\rho^4} \tag{18.7g}$$

$$W_{\min}: \qquad \cos(2\theta_2) = 8\rho^8 - 1. \tag{18.7h}$$

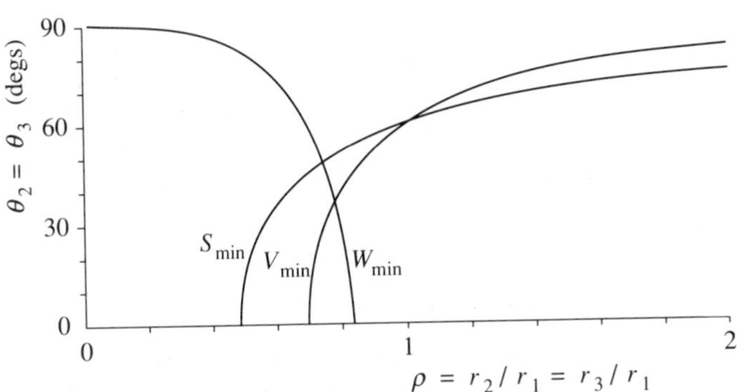

Fig. 18.11. Optimum branching angle for a symmetrical bifurcation. θ_2 and θ_3 are defined in Figure 18.10, the r_i denote the radii of the branches, and ρ is the radius ratio as given. S min, V min and W min denote the application of the three optimality principles of minimum surface, volume, and power. The curves represent eqns (18.7f)–(18.7h).

These are plotted out in Fig. 18.11. It can be seen that it is possible to satisfy two of the optimality principles simultaneously for certain values of the radius ratio ρ and the bifurcation angle θ_2, but it is not possible to satisfy more than two at the same time. The results tentatively suggest that a bifurcation angle in the range 30°–60° and a radius ratio ρ in the range 0.6–0.8 may produce a realistic compromise between conflicting requirements.

A more satisfactory approach is to minimize some weighted sum of the various physiological functions, for example by writing eqn (18.5h) as

$$H = hl = (g_A h_A + g_B h_B + g_C h_C)l \tag{18.7i}$$

where the g_x are weighting factors for the physiological functions h_x. Thus, with eqns (18.6a)–(18.6c), h is given by

$$h = g_S(2\pi r) + g_V(\pi r^2) + g_W\left(\frac{8\eta q^2}{\pi r^4}\right), \tag{18.7j}$$

using an obvious notation. This equation can then be substituted into eqns (18.5n) and (18.5o) to provide equations equivalent to eqns (18.6e) for the optimum angles.

18.5 Geometrical models of branching

The geometical models of branching are essentially empirical, and are based on a rule which relates a branch to its daughter branches in terms of relative length and angular disposition. The rule is applied repeatedly to generate a branching structure. Sometimes the branching rules are modified by a set of developmental rules in order to obtain a more realistic simulation of tree structures (Borchert and Honda 1984). Sometimes branching rules are applied to a morphological unit or shoot unit which is repeated (Bell, Roberts, and Smith 1979). Honda and colleagues have been largely responsible for developing this approach (e.g. Honda 1971, Fisher and Honda 1977, 1979, Honda, Tomlinson, and Fisher 1981, 1982), although our account of the method rests also on the analysis by Warner and Wilson (1976). We shall consider only the case of simple bifurcations, although other studies consider both whorls and bifurcations (Fisher and Honda 1977, 1979).

18.5.1 *Angular relations and transformations*

The branches are numbered as shown in Figs 18.2 and 18.3. The vector $\boldsymbol{r}_N$ represents the Nth branch in direction and magnitude, and the distal end of the Nth branch is at position $\boldsymbol{R}_N$. Since the Nth branch has daughter branches numbered $2N$ and $2N+1$,

$$\boldsymbol{R}_{2N} = \boldsymbol{R}_N + \boldsymbol{r}_{2N},$$
$$\boldsymbol{R}_{2N+1} = \boldsymbol{R}_N + \boldsymbol{r}_{2N+1}. \tag{18.8a}$$

Assuming that the basal end of the first branch $\boldsymbol{r}_1$ is placed at the origin (Fig. 18.12), we find

$$\boldsymbol{R}_1 = \boldsymbol{r}_1. \tag{18.8b}$$

The next two branches, denoted by the vectors $\boldsymbol{r}_2$ and $\boldsymbol{r}_3$, are obtained from $\boldsymbol{r}_1$ by a change in magnitude λ_A or λ_B and a rotation, denoted by a rotation matrix A' or B'. Hence

$$\boldsymbol{r}_2 = \lambda_A A'\boldsymbol{r}_1 \qquad \boldsymbol{r}_3 = \lambda_B B'\boldsymbol{r}_1. \tag{18.8c}$$

All vectors are represented with respect to a coordinate system (x, y, z) that is fixed in space, whose origin is at the beginning of the first branch, and where the first branch lies along the z axis ((Fig. 18.12(a)). Associated with each branch

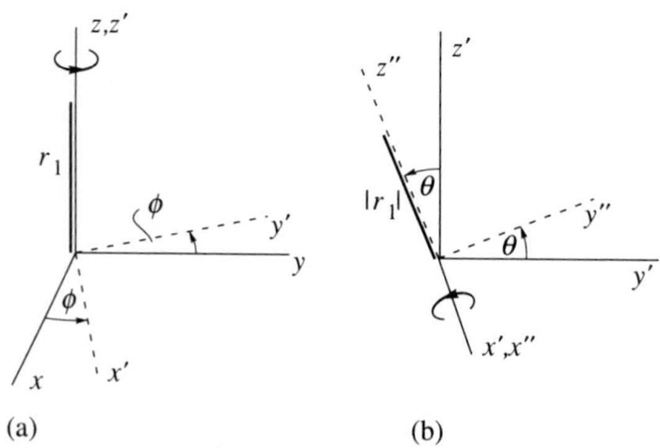

Fig. 18.12. Transformations between coordinate systems (x, y, z), (x', y', z'), and (x'', y'', z''): (a) rotation $G_z(\phi)$ through angle ϕ about Oz as given by eqn (18.8d); (b) rotation $G_x(\theta)$ through angle θ about the local x axis Ox' as given by eqn (18.8e). The thick line represents the first branch, which is considered here as being embedded in the coordinate axis system and transforming with it.

is a local set of coordinate axes, shown by x'', y'', and z'' in Fig. 18.12(b), and with z'' along the branch direction; these are related to the local axes of the parent branch (shown by x, y, and z in Fig. 18.12(a)) by two rotations, denoted by G_z and G_x, indicating rotations about the *local* z and x axes as shown.

In the $Oxyz$ system of axes, a vector $\boldsymbol{r}$ has components (a, b, c). We imagine that the vector is embedded in a system of local axes, as in rigid body. When this is rotated, the vector itself becomes a different vector $\boldsymbol{r}'$ with the same length but pointing in a different direction; however, the components (a, b, c) are unchanged relative to its local axes, which move with the vector and are labelled (x', y', z') in Fig. 18.12(a). We need to know the components of the rotated vector in the original (and invariant) system of axes $Oxyz$.

The first rotation shown in Fig. 18.12(a) is $G_x(\phi)$; this relates the coordinate axes (x', y', z') to (x, y, z). From the geometry, it can be seen that

$$(x', y', z') = \begin{pmatrix} \cos\phi & \sin\phi & 0 \\ -\sin\phi & \cos\phi & 0 \\ 0 & 0 & 1 \end{pmatrix}\begin{pmatrix} x \\ y \\ z \end{pmatrix} = G_z(\phi)\begin{pmatrix} x \\ y \\ z \end{pmatrix}. \tag{18.8d}$$

The second operation, which is rotation through angle θ about the local x axis (x') in Fig. 18.12(b)) and is denoted by $G_x(\theta)$, can be represented by the matrix in the equation

$$(x'', y'', z'') = \begin{pmatrix} 1 & 0 & 0 \\ 0 & \cos\theta & \sin\theta \\ 0 & -\sin\theta & \cos\theta \end{pmatrix}\begin{pmatrix} x' \\ y' \\ z' \end{pmatrix} = G_x(\theta)\begin{pmatrix} x' \\ y' \\ z' \end{pmatrix}. \tag{18.8e}$$

Substitution of eqn (18.8d) for (x', y', z') in eqn (18.8e) gives the total rotation, represented by the matrix G, so that

$$(x'', y'', z'') = G\begin{pmatrix} x \\ y \\ z \end{pmatrix} = G_x(\theta)G_z(\phi)\begin{pmatrix} x \\ y \\ z \end{pmatrix}. \tag{18.8f}$$

On multiplying the two matrices together, $G(\theta, \phi)$ is given by

$$G(\theta, \phi) = \begin{pmatrix} \cos\phi & \sin\phi & 0 \\ -\cos\theta\sin\phi & \cos\theta\cos\phi & \sin\theta \\ \sin\theta\sin\phi & -\sin\theta\cos\phi & \cos\theta \end{pmatrix}. \tag{18.8g}$$

Note that in the equation

$$(x'', y'', z'') = G\begin{pmatrix} x \\ y \\ z \end{pmatrix} \tag{18.8h}$$

the transformed vector in the local coordinate system is (x'', y'', z'') and this is, by definition, $(0, 0, 1)$ (in the absence of any change in length); we require the components of the vector (x, y, z) in the invariant coordinate system. Inverting eqn (18.8h) gives

$$(x, y, z) = G^{-1}\begin{pmatrix} x'' \\ y'' \\ z'' \end{pmatrix}, \tag{18.8i}$$

where G^{-1} is the inverse matrix of G. The inverse matrix can be obtained by transposing the elements of G (reflecting them in the diagonal) to give

$$G^{-1} = \begin{pmatrix} \cos\phi & -\cos\theta\sin\phi & \sin\theta\sin\phi \\ \sin\phi & \cos\theta\cos\phi & -\sin\theta\cos\phi \\ 0 & \sin\theta & \cos\theta \end{pmatrix}. \tag{18.8j}$$

Thus, taking $(x'', y'', z'') = (0, 0, 1)$ and substituting this and eqn (18.8j) into eqn (18.8i) gives the components of the branch in the invariant coordinate system as

$$x = \sin\theta\sin\phi, \qquad y = -\sin\theta\cos\phi, \qquad z = \cos\theta. \tag{18.8k}$$

We can regard the components of the branch as being given either by the third column of matrix G^{-1} or, more conveniently, by the third row of matrix G. This is more convenient for the following reason. Suppose that two transformations are applied, going from the branch to a daughter branch with G_1 and thence to a granddaughter branch with coordinate axes (x', y', z') with G_2. Thus

$$(x', y', z') = G\begin{pmatrix} x \\ y \\ z \end{pmatrix} = G_2 G_1\begin{pmatrix} x \\ y \\ z \end{pmatrix}. \tag{18.8l}$$

Inverting this gives, as in eqn (18.8i),

$$(x, y, z) = G^{-1}\begin{pmatrix} x' \\ y' \\ z' \end{pmatrix}. \tag{18.8m}$$

Since

$$G^{-1} = (G_2 G_1)^{-1} = G_1{}^{-1} G_2{}^{-1}, \tag{18.8n}$$

we can either use the inverse matrices of the G matrices and multiply by successive matrices from the right, as in

$$G_1^{-1} G_2^{-1} G_3^{-1} \ldots \tag{18.8o}$$

or we can use the G matrices as they are and multiply by successive matrices from the left, as in eqn (18.8l) and in

$$\ldots G_4 G_3 G_2 G_1. \tag{18.8p}$$

We choose the latter procedure. Note that matrices do not in general 'commute', i.e. multiplication from the left, as in $G_2 G_1$, is not usually the same as in multiplication from the right, as in $G_1 G_2$.

Consider a parent branch with index i and its two daughter branches with indices $2i$ and $2i + 1$. If it is assumed that taking $\theta = \pm\alpha$ gives rise to the two daughters, the coordinate systems are related by

$$\begin{aligned}(x_{2i}, y_{2i}, z_{2i}) &= G(\alpha, \phi)\begin{pmatrix} x_i \\ y_i \\ z_i \end{pmatrix} \\ (x_{2i+1}, y_{2i+1}, z_{2i+1}) &= G(-\alpha, \phi)\begin{pmatrix} x_i \\ y_i \\ z_i \end{pmatrix}.\end{aligned} \tag{18.8q}$$

Equation (18.8g) gives

$$\begin{aligned} x_{2i} &= x_i \cos\phi + y_i \sin\phi \\ x_{2i+1} &= x_i \cos\phi + y_i \sin\phi. \end{aligned} \tag{18.8r}$$

It can be seen that these expressions do not depend on θ and the local x axes of the two daughter branches are the same. The next ϕ rotation can be thought of as either rotating the yz plane about z (Fig. 18.12(a)) or, since this plane contains the branch and its twin, rotating the plane containing the two branches about the branch lying along the local z axis.

Examples Before discussing the general problem of generating a tree structure from the geometrical rules, we describe two elmentary examples of the application of the above considerations. For a bifurcating system in which each parent has two daughters, the matrix G of eqn (18.8g) has two possibilities A and B.

If $\phi_A = \phi_B = 0$, the branching system lies entirely within the Oyz plane: with, for instance, $\theta_A = \pi/6$ and $\theta_B = -\pi/6$, substituting these values into G (eqn (18.8g)) gives

$$A = \begin{pmatrix} 1 & 0 & 0 \\ 0 & \sqrt{3}/2 & 1/2 \\ 0 & -1/2 & \sqrt{3}/2 \end{pmatrix} \qquad B = \begin{pmatrix} 1 & 0 & 0 \\ 0 & \sqrt{3}/2 & -1/2 \\ 0 & 1/2 & \sqrt{3}/2 \end{pmatrix}. \tag{18.9a}$$

When these are applied to the vector $\boldsymbol{r}_1 = (0, 0, 1)$ as discussed in eqns (18.8i)–(18.8k), with $\lambda_A = \lambda_B = 1$ (eqn (18.8c)), $\boldsymbol{r}_2$ and $\boldsymbol{r}_3$ are given by the third rows of A and B respectively, so that the two second-generation branches are

$$\boldsymbol{r}_2 = (0, -1/2, \sqrt{3}/2) \qquad \boldsymbol{r}_3 = (0, 1/2, \sqrt{3}/2). \tag{18.9b}$$

The compound matrices relevant to $\boldsymbol{r}_4$, $\boldsymbol{r}_5$, $\boldsymbol{r}_6$, and $\boldsymbol{r}_7$ are AA, BA, AB, and BB, which are given by

$$\begin{aligned}
\boldsymbol{r}_4: \quad & AA = \begin{pmatrix} 1 & 0 & 0 \\ 0 & 1/2 & \sqrt{3}/2 \\ 0 & -\sqrt{3}/2 & 1/2 \end{pmatrix} \\
\boldsymbol{r}_5: \quad & BA = \begin{pmatrix} 1 & 0 & 0 \\ 0 & 1 & 0 \\ 0 & 0 & 1 \end{pmatrix} \\
\boldsymbol{r}_6: \quad & AB = \begin{pmatrix} 1 & 0 & 0 \\ 0 & 1 & 0 \\ 0 & 0 & 1 \end{pmatrix} \\
\boldsymbol{r}_7: \quad & BB = \begin{pmatrix} 1 & 0 & 0 \\ 0 & 1/2 & -\sqrt{3}/2 \\ 0 & \sqrt{3}/2 & 1/2 \end{pmatrix}.
\end{aligned} \tag{18.9c}$$

The bottom rows of these four matrices give the four third-generation branches with vectors

$$\begin{aligned}
\boldsymbol{r}_4 &= (0, -\sqrt{3}/2, 1/2) \\
\boldsymbol{r}_5 &= (0, 0, 1) \\
\boldsymbol{r}_6 &= (0, 0, 1) \\
\boldsymbol{r}_7 &= (0, \sqrt{3}/2, 1/2).
\end{aligned} \tag{18.9d}$$

These are the branch vectors. The end-points of the four terminal vectors are given by

$$\begin{aligned}
\boldsymbol{R}_4 &= \boldsymbol{r}_4 + \boldsymbol{r}_2 + \boldsymbol{r}_1 \\
\boldsymbol{R}_5 &= \boldsymbol{r}_5 + \boldsymbol{r}_2 + \boldsymbol{r}_1 \\
\boldsymbol{R}_6 &= \boldsymbol{r}_6 + \boldsymbol{r}_3 + \boldsymbol{r}_1 \\
\boldsymbol{R}_7 &= \boldsymbol{r}_7 + \boldsymbol{r}_3 + \boldsymbol{r}_1.
\end{aligned} \tag{18.9e}$$

With $\phi_A = \phi_B = \pi/2$, the branching occurs alternately in the Oxz plane and the Oyz plane. Again taking $\theta_A = -\theta_B = \pi/6$, eqn (18.8g) gives

$$A = \begin{pmatrix} 0 & 1 & 0 \\ -\sqrt{3}/2 & 0 & 1/2 \\ 1/2 & 0 & \sqrt{3}/2 \end{pmatrix} \qquad B = \begin{pmatrix} 0 & 1 & 0 \\ -\sqrt{3}/2 & 0 & -1/2 \\ -1/2 & 0 & \sqrt{3}/2 \end{pmatrix}. \tag{18.9f}$$

With $\boldsymbol{r}_1 = (0, 0, 1)$, the bottom rows of these matrices give the two daughter branches, again with $\lambda_A = \lambda_B = 1$, so that

$$\boldsymbol{r}_2 = (1/2, 0, \sqrt{3}/2) \qquad \boldsymbol{r}_3 = (-1/2, 0, \sqrt{3}/2). \tag{18.9g}$$

These two branches lie in the Oxz plane, and are separated in the x direction only. The matrices AA, BA, AB, and BB relevant to $\boldsymbol{r}_4$–$\boldsymbol{r}_7$ are

$$\begin{aligned}
\boldsymbol{r}_4: \quad AA &= \begin{pmatrix} -\sqrt{3}/2 & 0 & 1/2 \\ 1/4 & -\sqrt{3}/2 & \sqrt{3}/4 \\ \sqrt{3}/4 & 1/2 & 3/4 \end{pmatrix} \\
\boldsymbol{r}_5: \quad BA &= \begin{pmatrix} -\sqrt{3}/2 & 0 & 1/2 \\ -1/4 & -\sqrt{3}/2 & -\sqrt{3}/4 \\ \sqrt{3}/4 & -1/2 & 3/4 \end{pmatrix} \\
\boldsymbol{r}_6: \quad AB &= \begin{pmatrix} -\sqrt{3}/2 & 0 & -1/2 \\ -1/4 & -\sqrt{3}/2 & \sqrt{3}/4 \\ -\sqrt{3}/4 & 1/2 & 3/4 \end{pmatrix} \\
\boldsymbol{r}_7: \quad BB &= \begin{pmatrix} -\sqrt{3}/2 & 0 & -1/2 \\ 1/4 & -\sqrt{3}/2 & -\sqrt{3}/4 \\ -\sqrt{3}/4 & -1/2 & 3/4 \end{pmatrix}.
\end{aligned} \tag{18.9h}$$

Comparing eqns (18.9h) with (18.9c), in this case $BA = AB$ no longer holds. The bottom rows of these matrices give the four branch vectors, with

$$\begin{aligned}
\boldsymbol{r}_4 &= (\sqrt{3}/4, 1/2, 3/4) \\
\boldsymbol{r}_5 &= (\sqrt{3}/4, -1/2, 3/4) \\
\boldsymbol{r}_6 &= (-\sqrt{3}/4, 1/2, 3/4) \\
\boldsymbol{r}_7 &= (-\sqrt{3}/4, -1/2, 3/4).
\end{aligned} \tag{18.9i}$$

The terminal points of the four branches can be obtained from eqns (18.9e). It can be seen that $\boldsymbol{R}_4$ and $\boldsymbol{R}_5$ are separated in the y direction only, and similarly with $\boldsymbol{R}_6$ and $\boldsymbol{R}_7$.

18.5.2 *Generation of the geometry of the branching structure*

Following the terminology used in Section 18.2 and Fig. 18.2, we number branches in decimal integers from 1, using N to denote the decimal integer. The first branch of the tree is represented by the vector $\boldsymbol{r}_1 = (0, 0, 1)$ in the invariant coordinate system (x_1, y_1, z_1) to which everything is referred. Since the system branches by bifurcation, the matrix G of eqns (18.8g) and (18.8h), which relates the coordinate system of a branch to that of one of its daughter branches, has two possibilities denoted by the matrices A and B. The branch $\boldsymbol{r}_N$ has the two daughters $\boldsymbol{r}_{2N}$ and $\boldsymbol{r}_{2N+1}$. The coordinate systems of these branches are related by

$$(x_{2N}, y_{2N}, z_{2N}) = A\begin{pmatrix} x_N \\ y_N \\ z_N \end{pmatrix}$$

$$(x_{2N+1}, y_{2N+1}, z_{2N+1}) = B\begin{pmatrix} x_N \\ y_N \\ z_N \end{pmatrix}. \tag{18.10a}$$

Thus A relates the coordinate systems of $\boldsymbol{r}_1$ to $\boldsymbol{r}_2$ to $\boldsymbol{r}_4$ etc., and B relates those of $\boldsymbol{r}_1$ to $\boldsymbol{r}_3$ to $\boldsymbol{r}_7$ etc.

The decimal integer N, which labels the branch $\boldsymbol{r}_N$ and its coordinate system (x_N, y_N, z_N), can be represented by the binary integer

$$N(i_i, i_2, \ldots, i_{k-1}, i_k) \tag{18.10b}$$

where the i_j are binary digits (0 or 1) and the length of the binary number k is given by

$$k = 1 + \text{integer part}(\log_2 N). \tag{18.10c}$$

Branch N belongs to the kth generation of branches; branch 1 belongs to the first generation, branches 2 and 3 to the second generation, branches 4, 5, 6, and 7 to the third generation, etc. There are 2^{k-1} branches or potential branches in the kth generation; the total number of branches or potential branches in all generations up to the kth is $2^k - 1$.

In moving from the Nth branch to the $2N$th branch, a zero is added to the binary representation on the right and this is associated with the matrix A, as in the first of eqns (18.10a); in moving from the Nth branch to the $(2N + 1)$th branch, a 1 is added to the binary representation and this is associated with matrix B. Using G_j to denote the coordinate axis transformation associated with binary digit i_j, we obtain

$$G_j = A^{1-i_j} B^{i_j}, \tag{18.10d}$$

which gives the desired result of $G_j = A$ when $i_j = 0$ and $G_j = B$ when $i_j = 1$. We can now write

$$\boldsymbol{r}_N = (x_N, y_N, z_N) = G_k G_{k-1} \ldots G_3 G_2 \begin{pmatrix} x_1 \\ y_1 \\ z_1 \end{pmatrix} = \Gamma_N \begin{pmatrix} x_1 \\ y_1 \\ z_1 \end{pmatrix}, \tag{18.10e}$$

where the matrix Γ_N denotes the total transformation. As discussed after eqn (18.8k), when $\boldsymbol{r}_1 = (0, 0, 1)$ the components of branch r_N in the invariant (x_1, y_1, z_1)

system are (ignoring any change in length for the time being)

$$\boldsymbol{r}_N = (\Gamma_{31}, \Gamma_{32}, \Gamma_{33})_N, \tag{18.10f}$$

where the subscripts within the parentheses indicate the components of the matrix.

To take account of the change in length on moving from branch to branch, it is assumed that α is associated with A and the N to $2N$ parentage, and β is associated with B and the N to $2N + 1$ parentage. Analogously with eqn (18.10d), λ is defined by

$$\lambda_j = \alpha^{1-i_j}\beta^{i_j}, \tag{18.10g}$$

and analogously with eqn (18.10e), a scaling factor Λ_N, for the Nth branch is defined by

$$\Lambda_N = \lambda_k\lambda_{k-1}\ldots\lambda_3\lambda_2. \tag{18.10h}$$

Equation (18.10f) is then modified to

$$\boldsymbol{r}_N = \Lambda_N(\Gamma_{31}, \Gamma_{32}, \Gamma_{33})_N. \tag{18.10i}$$

Finally, the position of the end of the Nth branch, denoted by the vector $\boldsymbol{R}_N$, is given by

$$\boldsymbol{R}_N = \boldsymbol{r}_N + \boldsymbol{r}_{[N/2]} + \boldsymbol{r}_{[N/4]} + \cdots + \boldsymbol{r}_1. \tag{18.10j}$$

The square brackets indicate the integer part of the enclosed quantity. From eqn (18.10j) it can be seen that

$$\boldsymbol{R}_N = \boldsymbol{R}_{[N/2]} + \boldsymbol{r}_N; \tag{18.10k}$$

that is, branch $\boldsymbol{r}_N$ originates at $\boldsymbol{R}_{[N/2]}$ (the terminal of its parent) and terminates at $\boldsymbol{R}_N$ (Exercise 18.5).

18.5.3 *Discussion*

Equations (18.10i) and (18.10j) provide an explicit means of calculating the length, direction, and positions of the basal and distal ends of each branch. Unfortunately, further analytical progress with the problem is difficult and can only be made with highly simplifying assumptions.

Warner and Wilson (1976) considered three stochastic models for the transformation matrices of eqns (18.10a) (see Fig. 18.12). In the first model ϕ is assigned a fixed value between $-\pi/2$ and $\pi/2$, whereas θ is given either of the values $\pm\theta_0$ with a probability of 1/2. The second model makes the same assumption for θ but assumes that ϕ is a random variable uniformly distributed on $(-\pi/2, \pi/2)$. The third model fixes θ at a constant value θ_0 and distributes ϕ uniformly on $(-\pi, \pi)$. After a given number of branches, all three models lead to the same position of the centroid of the end-point distribution and to the same relative variance perpendicular and parallel to the parent stem.

If the branch-length scaling parameters α and β introduced in eqn (18.10g) are less than unity, then, even with $N \to \infty$, the spatial extent of the tree is finite. For a finite structure, let N_{max} be the highest branch number present. Not all the branches with numbers up to N_{max} may exist, depending on the growth rules that are assigned to the system. As applied in eqn (18.1h), we use $e_N = 1$ to denote the existence of the Nth branch and $e_N = 0$ to denote its non-existence.

The terminal end-points are those $\boldsymbol{R}_N$ of eqn (18.10j) where $e_N = 1$ and $e_{2N} = e_{2N+1} = 0$. The three components of $\boldsymbol{R}_N$ are denoted by

$$\boldsymbol{R}_N = (X_N, Y_N, Z_N). \tag{18.10l}$$

For different purposes we might be concerned with height in the z direction (competition), area in the xy plane (light interception), and volume (soil exploration). A simple measure of these quantities is to determine the maximum values of X_N, Y_N, and Z_N for the terminal end-points:

$$X_{max} = \max_N (X_N) \qquad Y_{max} = \max_N (Y_N) \qquad Z_{max} = \max_N (Z_N). \tag{18.10m}$$

In general these three maxima will be found at three different N values. We can then calculate approximately

$$\text{height} = Z_{max} \qquad \text{area} = \pi X_{max} Y_{max} \qquad \text{volume} = \frac{2\pi X_{max} Y_{max} Z_{max}}{3}. \tag{18.10n}$$

It is also possible to compute mean values using equations such as

$$\begin{aligned} \overline{(X^2 + Y^2)} \sum_{i=1}^{N} e_i &= \sum_{i=1}^{N} e_i (X_i^2 + Y_i^2) \\ \bar{Z} \sum_{i=1}^{N} e_i &= \sum_{i=1}^{N} e_i Z_i. \end{aligned} \tag{18.10o}$$

These may be more meaningful for particular circumstances and applications than the maximum values of (18.10m).

18.6 Some mechanical considerations

Trees and other branching structures often have some mechanical functions to fulfill. For some structures and in some situations, the performance of these tasks may be so important for the continued existence of the organism that evolution may have optimized some of the mechanical characteristics of the structure (McMahon and Kronauer 1976). Some of these aspects are considered in this section.

18.6.1 *Bending of branches*

Consider the branch shown in Fig. 18.13. This is idealized as a cylindrical rod of radius r (m) and length AB $= l$ (m), which is held rigidly at one end. Assume that

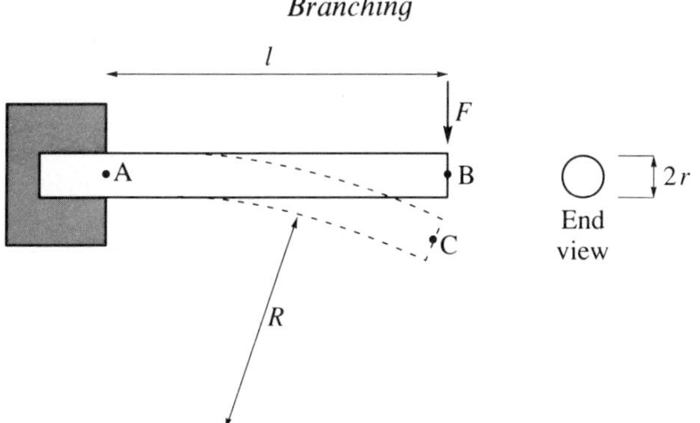

Fig. 18.13. Bending of branches. AB represents a branch of length l and circular cross-section with diameter $2r$. The branch is held rigidly at end A. A force F applied at B deflects B to C, where the branch AC assumes a constant radius of curvature R.

a force F (kg m s^{-2}) is applied to the distal end at B, to give a couple C (kg m^2 s^{-2}) where

$$C = lF. \tag{18.11a}$$

It can be shown that the resulting radius of curvature R (m) is given by (Newman and Searle 1948, p. 105)

$$CR = EAk^2 \tag{18.11b}$$

where E is the elastic modulus (kg m^{-1} s^{-2}). A is the cross-sectional area (m^2), and k is the radius of gyration (m). For a cylindrical rod of circular cross-section

$$A = \pi r^2 \qquad k^2 = r^2/2, \tag{18.11c}$$

so that

$$R = \frac{E\pi r^4}{2C}. \tag{18.11d}$$

Note that the radius of curvature R depends on the fourth power of the radius divided by the magnitude of the couple C.

Murray (1927a, p. 838) reported that if a tree is cut through at any point, the weight of the whole part distal to the cut is proportional to the cube of the radius at the cut. Thus, in making cuts just below and above a branch point, the relationship between the radii of the parent and daughter branches given in eqn (18.4e) (p. 508) holds. In a later paper Murray (1927b, p. 727) reported a 2.5 power law instead of a cubic power law. If it is assumed that branch length is proportional to cross-sectional radius, then, with weight proportional to r^3, this leads to a couple C proportional to r^4. From eqn (18.11d) this gives a constant radius of curvature R independent of cross-sectional radius r. A 2.5 power law leads instead to the radius of curvature varying with the square root of the cross-sectional radius, so that small branches bend more than larger branches.

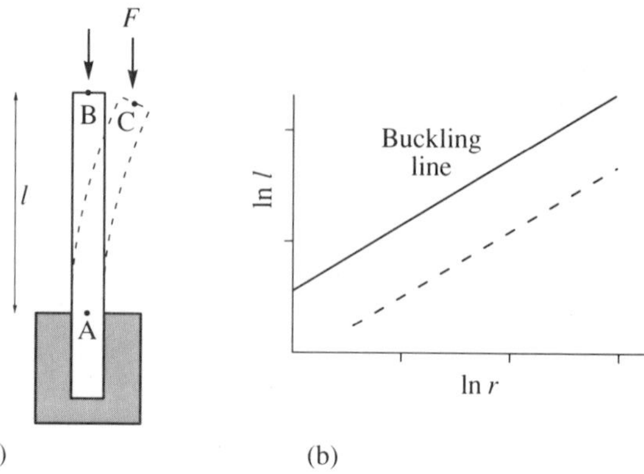

Fig. 18.14. Buckling in trees. In (a) a downward compressive force F is applied to a cylindrical rod AB of length l and radius r which is held rigidly at end A. AC indicates the buckling deformation. In (b) the solid line denotes the relationship between length l and radius r for self-buckling, as given by eqn (18.11h); actual tree data are scattered around the broken line which is on the stable side of the buckling line.

18.6.2 *Buckling of trunks or branches*

A body such as a tree trunk is usually subject to an axial compressive load due to the weight of the rest of the tree, and it may fail by buckling. In this section the criteria for the stability of a cylindrical rod under a compressive load are considered.

Figure 18.14(a) shows a rod held rigidly at one end on which an axial force F is acting downwards at the distal end. It can be shown (Temperley 1953, p. 105, writing $J = \pi r^4/2$ for a circular cross-section in his third equation) that

$$F = \frac{E\pi^3 r^4}{8l^2}. \tag{18.11e}$$

F is the force at which the configuration AB in Fig. 18.14(a) becomes unstable and deforms to AC. This expression assumes an isotropic elastic modulus which is not a good approximation for most woods.

If we consider self-buckling of AB due to its own weight, proceeding approximately we can write

$$F = \pi r^2 l \rho g, \tag{18.11f}$$

where ρ is the density of the wood and g is the acceleration due to gravity. Combining eqns (18.11e) and (18.11f) leads to the relationship

$$l^3 = \frac{cEr^2}{\rho g}, \tag{18.11g}$$

where c is a constant. Taking logarithms gives

$$3 \ln l = 2 \ln r + \text{constant}. \tag{18.11h}$$

A buckling line can be drawn as in Fig. 18.14(b). The observational data for forest trees always lie to the right of this line, and are scattered about the broken line shown in Fig. 18.14(b). McMahon and Kronauer (1976, Fig. 5) show such data, and they observe that trees generally limit their overall height to about a quarter of the critical height for self-buckling.

18.6.3 *Wind*

Wind forces often exert considerable influence on tree structure. In this section we describe how the wind force can be calculated for a wind of speed v at right angles to a cylindrical branch of diameter d and length l. First we note that the Reynolds number Re is given by

$$Re = \frac{v\rho d}{\eta}, \tag{18.11i}$$

where v (m s^{-1}) is the wind speed, ρ (kg m^{-3}) is the density of air, d (m) is the branch diameter, and η (kg m^{-1} s^{-1}) is the viscosity of air. Re is dimensionless. If it is assumed that (at 0 °C and atmospheric pressure)

$$\begin{aligned} &v = 10\ \text{m s}^{-1}\ (36\ \text{km h}^{-1}) \qquad d = 0.01\ \text{m} \\ &\eta = 1.71 \times 10^{-5}\ \text{kg m}^{-1}\ \text{s}^{-1} \qquad \rho = 1.29\ \text{kg m}^{-3}, \end{aligned} \tag{18.11j}$$

then

$$Re = 7.5 \times 10^3. \tag{18.11k}$$

The Reynolds number represents the ratio of impinging fluid momentum ρv^2 per unit area and time to the frictional viscous force $\eta v/d$ per unit area which balances the fluid momentum. It is the Reynolds number that determines the nature of the flow pattern round the body. If Re is below about 2100, the motion is streamline; above this value there are eddies and vortices, and above about 3000 there is steady turbulent flow (Fishenden and Saunders 1950, p. 78). Thus we are considering the turbulent flow region with an Re of some 7000.

It is convenient to make use of another dimensionless number, the Nusselt number Nu defined by

$$Nu = \frac{Hd}{k\theta} \tag{18.11l}$$

where H (J m^{-2} s^{-1}) is the heat flux density, k is the thermal conductivity (J m^{-1} s^{-1} K^{-1}), and θ (K) is the temperature difference between the surface and the main stream of the fluid. We can then employ an accurate empirical relationship between Re and Nu (see eqn (18.11o) below) and the simple proportionality between the heat flux density H and the momentum flux density M (kg m^{-1} s^{-1}) (Fishenden and Saunders 1950, p. 81)

$$\frac{M}{H} = \frac{v}{c\theta}, \tag{18.11m}$$

where c (J kg^{-1} K^{-1}) is the coefficient of specific heat at constant volume of air. Combining eqns (18.11l) and (18.11m) gives

$$M = \frac{vk}{dc} Nu. \tag{18.11n}$$

For gases with Re within the region 1000–100 000, the Nusselt and Reynolds numbers, Nu and Re, are related by (Fishenden and Saunders 1950, p. 130)

$$Nu = 0.24Re^{0.6}. \tag{18.11o}$$

From eqns (18.11i), (18.11n), and (18.11o) the momentum flux density M is given by

$$M = 0.24\left(\frac{k}{c}\right)\left(\frac{\rho}{\eta}\right)^{0.6} v^{1.6}d^{-0.4}. \tag{18.11p}$$

The force f (kg s^{-2}) per unit length of branch is

$$f = dM = 0.24\left(\frac{k}{c}\right)\left(\frac{\rho}{\eta}\right)^{0.6} v^{1.6}d^{0.6}. \tag{18.11q}$$

For the thermal conductivity k and specific heat c of air at 0 °C and atmospheric pressure, we take

$$k = 2.41 \times 10^{-2}\ \text{J m}^{-1}\ \text{s}^{-1}\ \text{K}^{-1} \qquad c = 718\ \text{J kg}^{-1}\ \text{K}^{-1}, \tag{18.11r}$$

and with eqns (18.11j), eqn (18.11q) becomes

$$f = 0.0068v^{1.6}d^{0.6}. \tag{18.11s}$$

This is the force per unit length. The force F (kg m s^{-2}) on a branch of length l (m) is

$$F = lf. \tag{18.11t}$$

It can be seen that doubling the diameter d and reducing the branch length l by 4 so as to retain the same branch volume, reduces the force F by multiplying it by a factor of $2^{-1.4} = 0.38$. Also, the force on a branch of length $l = 1$ m and diameter $d = 0.01$ m at a wind speed $v = 10$ m s^{-1} is

$$F = 0.017\ \text{kg m s}^{-2}. \tag{18.11u}$$

If it is assumed that wood has a density of 1000 kg m^{-3} and that the acceleration due to gravity is 10 m s^{-2}, eqn (18.11u) is comparable with a gravitational force of 0.8 kg m s^{-2}.

Exercises

18.1. Assume a trifurcating system with delays in bud outgrowth of a, b, and c with $a \leqslant b \leqslant c$. Using the method described in Section 18.2 derive the equivalent expressions for eqns (18.1a)–(18.1i).

18.2. Devise an algorithm equivalent to the Strahler ordering algorithm in Section 18.2.2 (p. 500) for ordering a structure according to Shreve's ordering rules (Fig. 18.1(b)). Assume that the branching structure is as defined in the last paragraph of Section 18.2.1 (p. 500).

18.3. In deriving eqns (18.4f)–(18.4h) it was assumed that eqn (18.4e) holds, and also that the linear specific growth rate is one-third of the mass specific growth rate. Derive a more general form of eqn (18.4h) assuming that eqn (18.4e) is replaced by $r^x = R_a{}^x + R_b{}^x$, where x is a constant (e.g. Murray (1927b) reported an x of about 2.5)) and with a specific rate of growth ρ_L of the apical radius.

18.4. Derive the second derivatives:

$$\frac{\partial^2 H}{\partial x^2} \qquad \frac{\partial^2 H}{\partial y^2} \qquad \frac{\partial^2 H}{\partial x \partial y}.$$

from eqns (18.5j). The conditions for H to be a minimum are

$$\frac{\partial^2 H}{\partial x^2} > 0 \qquad \frac{\partial^2 H}{\partial y^2} > 0 \qquad \frac{\partial^2 H}{\partial x^2}\frac{\partial^2 H}{\partial y^2} - \left(\frac{\partial^2 H}{\partial x \partial y}\right)^2 > 0.$$

Prove that these are satisfied.

18.5. For branch number $N = 19$, express N in binary form, and write out eqns (18.10e), (18.10h), and (18.10j) explicitly.

19
Phyllotaxis

19.1 Introduction

For centuries scientists have been fascinated by biological structure and pattern. Within the plant world, the most striking manifestation of such phenomena is in the area of phyllotaxis. This includes several related matters such as the positioning of leaves round a vegetative plant stem, the arrangement of florets or seeds in a flower, such as sunflower, and the patterns of the scales in a pine cone. Research into the subject can be largely categorized under four headings: observation and experiment; empirical mathematical modelling; mechanistic mathematical modelling; functional consequences at the next hierarchical level (p. 12) (i.e. what are the consequences of Fibonacci phyllotaxis for light interception, photosynthesis, and crop yield?). Our concern in this chapter is with the second and third of these: empirical modelling and mechanistic modelling.

Compared with most topics in the plant and crop sciences, phyllotaxis is often considered to be a 'difficult' subject. At least some of this difficulty arises from the many jargon words associated with the phenomenon, which are not defined with sufficient clarity. However, there is no doubt that a mathematical treatment is unavoidable for a more than superficial understanding. Our aim here is to give, as far as possible, a simple but complete account; our literature references are highly selective, and the long history of work in the area is largely ignored—these deficiencies can be partly remedied by reference to the monograph by Jean (1984), which has a large bibliography.

To help the reader find his way through the ensuing material, we begin with a summary of our view of the current status of theory and models in phyllotaxis.

19.1.1 *Summary of the status of phyllotaxis theory*

1. The theoretical problem posed by phyllotaxis has essentially been solved. We use τ to denote the golden ratio ($\tau = (1 + \sqrt{5})/2$) and $2\pi\alpha$ (rad) to denote the divergence angle (the angle between successive primordia projected onto a horizontal plane). Then an explanation can be given of why the observed α are given by $\alpha = 1/\tau^2$ (the Fibonacci angle—the commonest type of phyllotaxis) or $\alpha = (a + b\tau)/(c + d\tau)$ where a, b, c, and d are positive integers, $a < b$, $c < d$, and $ad - bc = \pm 1$ (giving the less common types of spiral phyllotaxis).
2. Two sorts of explanation are required, and these are on different levels: (i) mechanistic models which lead to values of α related to the golden ratio above are needed; (ii) it is also desirable that teleonomic considerations (such as

efficient light interception by leaves on a plant or optimal packing of seeds on a capitulum) should favour the observed values of α.

3. There are two main mechanistic approaches to phyllotaxis: packing or contact theories (Adler 1974, 1977, Ridley 1982a,b, Williams and Brittain 1984), and reaction–diffusion models (Thornley 1975a, Veen and Lindenmayer 1977). Both theories are to be regarded as candidate theories, as neither has been conclusively demonstrated to operate in nature. However, both methods are able to generate the required values of α. This satisfies 2(i).
4. Ridley (1982a) demonstrated that values of α as in 1 give the most efficient packing of primordia placed on a cyclotron spiral. Similarly, Marzec and Kappraff (1983) showed that optimum spacing of points on a circle is obtained for these values of α. These results satisfy 2(ii).
5. Marzec and Kappraff (1983) proved by analysis that, in general, reaction–diffusion models could give rise to the special angles of phyllotaxis, which had previously only been demonstrated numerically for a special case (Thornley 1975a).
6. The juxtaposition of plausible mechanistic models that can (although do not necessarily) produce the observed behaviour with the demonstrated teleonomic advantages of the angles related to the golden ratio, is able to provide a satisfactory theoretical basis for phyllotaxis through the mechanisms of evolution and natural selection.

19.1.2 Some preliminaries

On observing a sunflower head, a pineapple, or a pine cone, we can generally see conspicuous spirals winding in different directions. Usually there is one highly visible set of n_A spirals (where n_A is an integer) winding in one direction, and a second highly visible set of n_B spirals (n_B is an integer) winding in the oppositive direction (e.g. see Fig. 19.5, p. 542). These spirals are called *parastichies*, or sometimes *contact* parastichies, because the elements making up the spirals (the seeds on the sunflower head or the scales on the pine cone) are in contact with one another. The components of the integer number pair (n_A, n_B) are sometimes called the parastichy numbers.

It was noticed many centuries ago that the parastichy numbers (n_A, n_B) are generally members of the Fibonacci series. The Fibonacci series is an infinite set of integers, denoted by $f_0, f_1, f_2, \ldots, f_i, \ldots$; the first 12 numbers in the series are

i	0	1	2	3	4	5	6	7	8	9	10	11
f_i	0	1	1	2	3	5	8	13	21	34	55	89.

(19.1a)

A given term is the sum of the two preceding terms, with

$$f_i = f_{i-1} + f_{i-2} \qquad i = 2, 3, 4, \ldots. \tag{19.1b}$$

An alternative method of generating the series is given in Exercise 19.1. Note also that the ratio of any two adjacent terms approaches a constant on moving to higher terms in the series, the so-called golden mean or Fibonacci ratio τ. τ is given by (Exercises 19.2 and 19.3)

$$\tau = \lim_{i\to\infty}\left(\frac{f_i}{f_{i-1}}\right) = \frac{1+\sqrt{5}}{2} = 1.6180. \tag{19.1c}$$

It is easily shown that

$$\frac{1}{\tau} = \frac{(\sqrt{5}-1)}{2} = 0.6180 = \tau - 1, \tag{19.1d}$$

and that

$$\frac{1}{\tau^2} = \frac{3-\sqrt{5}}{2} = 0.3820 = 1 - \frac{1}{\tau}. \tag{19.1e}$$

Primordia are initiated in a temporal sequence, and the time interval elapsing between the initiation of successive primordia is called a plastochron, denoted here by t_p (day). The whole system of primordia may exist on what is approximately a flat disc, as with a sunflower head, or on some surface of revolution, such as a paraboloid, in a vegetative shoot apex. The system can be projected on to a horizontal plane, and can be analysed in terms of plane polar coordinates (r, θ) (see Fig. 19.1). Considering the primordia as points in the horizontal plane, temporally adjacent primordia subtend an angle β (rad) at the centre of the apical circle. β is called the *divergence* angle.

It has long been observed that in a developing primordial system the divergence angle usually settles down to a constant value, given frequently (but not always) by

$$\beta_0 = 2\pi/\tau^2 \tag{19.1f}$$

This angle is referred to as the Fibonacci angle, and is 137.5°. A high-order Fibonacci system is where the parastichy numbers (n_A, n_B), are high in the Fibonacci series (19.1a). Sometimes great stress is placed on the precision with which the divergence angle must be defined in nature in order to give rise to high-order Fibonacci systems; while this is the case for reproductive apices and capitula such as the sunflower head, with vegetative apices and leaf primordia the divergence angle is usually far more variable.

Much effort has been expended in considering the relationships between the parastichy numbers (n_A, n_B), the corresponding Fibonacci or Fibonacci-like series (defined by giving the first two terms), and the divergence angle β. This is the problem addressed later, but first it is necessary to consider the radial growth and outward movement of the primordia.

19.2 Radial growth

In a flower head the seeds exist on the surface of a capitulum. In the sunflower this is almost flat, giving a disc-like surface; more generally, the seeds are displayed on a surface of revolution, which may be a parabola or other conic section. In the vegetative shoot apex, the developing and newly initiated primordia are

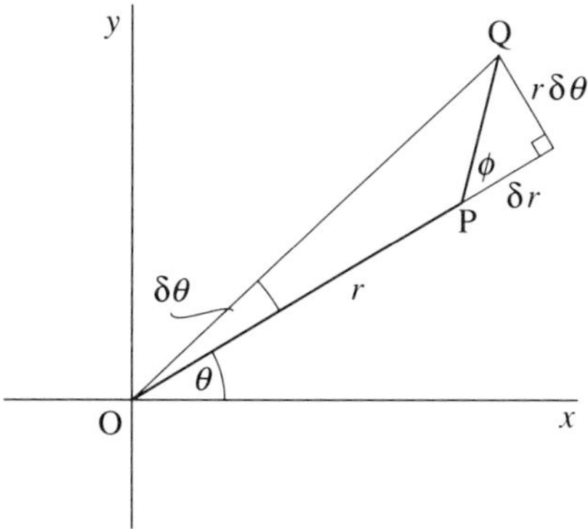

Fig. 19.1. Relationship between coordinates (x, y) and polar coordinates (r, θ). Point P is at (r, θ); PQ is an element of a curve being traced out by P; ϕ is the angle between PQ and the radius vector OP.

similarly displayed on a surface of revolution, which may approximate to a cylindical surface at one extreme. It is assumed here that all these systems can be represented on a disc, which is easily visualized and drawn; some transformation may be necessary to do this, which could distort geometrical considerations. It is convenient to represent the disc using the plane polar coordinates (r, θ); these are related to the Cartesian coordinates (x, y) by the equations (Fig. 19.1)

$$x = r\cos\theta \qquad y = r\sin\theta. \tag{19.2a}$$

Referring to Fig. 19.1, let P be a point on a curve with coordinates (r, θ) and Q be another point on the curve a short distance away from P with coordinates $(r + \delta r, \theta + \delta\theta)$, where δr and $\delta\theta$ are small increments in r and θ. From Fig. 19.1

$$\tan\phi = r\frac{\mathrm{d}\theta}{\mathrm{d}r}, \tag{19.2b}$$

where ϕ denotes the angle between PQ and the radius vector OP and $\delta\theta/\delta r$ has been replaced by $\mathrm{d}\theta/\mathrm{d}r$ for small PQ.

The successive primordia (or seeds) in the system are referred to as $P_0, P_1, P_2, \ldots$; P_0 is the youngest primordium, and $P_1, P_2, \ldots$ are successively older. The position of each primordium on the disc is defined by a point P_i, $i = 0, 1, 2, \ldots$. The points P_i lie on a spiral, which can be defined by an (r, θ) equation. Different authors have used different (r, θ) equations to define this spiral, and we now consider the three main candidates.

19.2.1 *Archimedes spiral*

This is described by

$$r = a\theta, \tag{19.3a}$$

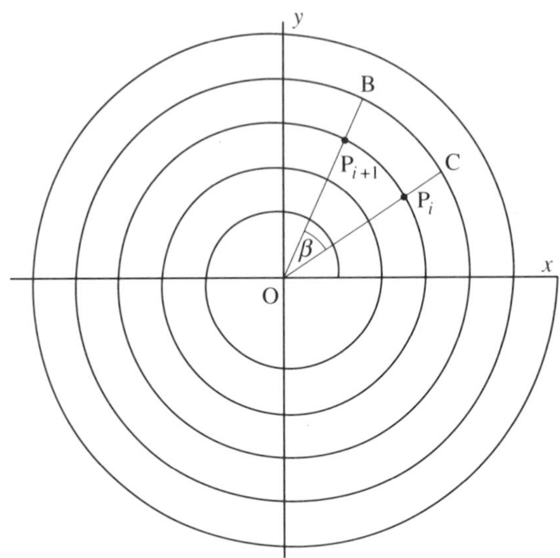

Fig. 19.2. Archimedes spiral (eqn (19.3a)).

where a is a constant. Note that $dr/d\theta = a$, and from eqn (19.2b)

$$\tan \phi = r/a. \qquad (19.3b)$$

This is drawn in Fig. 19.2, where it can be seen that the radii of successive turns of the spiral increase by the same amount, i.e. $r(\theta + 2\pi) - r(\theta) = 2\pi a = \text{constant}$.

Now assume that primordia are placed at a constant angular separation of β along the spiral. Let A_i be the area associated with primordium P_i in Fig. 19.2. Then

$$A_i \approx P_iCBP_{i+1} = (2\pi a)r_i\beta. \qquad (19.3c)$$

The youngest primordium P_0 has the coordinates (r_0, θ_0), and the ith primordium is at

$$r_i = a(\theta_0 + i\beta), \qquad \theta_i = \theta_0 + i\beta. \qquad (19.3d)$$

Substituting for r_i in (19.3c), therefore, we have

$$A_i = 2\pi a(a\theta_0 + ia\beta)\beta. \qquad (19.3e)$$

Assume that each new primordium is produced at the same radius a on the apex (without loss of generality we can take θ_0 as a constant), and that primordia are generated at a constant rate, so that

$$i = ct, \qquad (19.3f)$$

where t is the time variable and c is a constant. The area associated with a given primordium (whose integer index increases by unity each time a new primordium is formed) is, from (19.3e),

$$A_i = 2\pi a(a\theta_0 + cta\beta)\beta. \tag{19.3g}$$

This area increases linearly with time (although the shape changes), and the Archimedes spiral may thus be applicable to primordia which have a constant absolute growth rate. Most primordia grow exponentially in the early stages of growth; however, a constant growth rate might apply to some seeds or florets (e.g. cauliflower).

19.2.2 *Exponential spiral*

The exponential spiral is shown in Fig. 19.3, and is described by

$$r = a\exp(\theta\cot\phi), \tag{19.4a}$$

where a and ϕ are constants. Differentiating gives

$$\frac{\mathrm{d}r}{\mathrm{d}\theta} = a\cot\phi\exp(\theta\cot\phi). \tag{19.4b}$$

Substituting eqns (19.4a) and (19.4b) in eqn (19.2b), it follows immediately that the exponential spiral makes a constant angle ϕ with the radius vector (see Fig. 19.1).

As in the last section, primordia are placed along the spiral at a constant angular separation β, and A_i denotes the area associated with primordium P_i (Fig. 19.3). Since

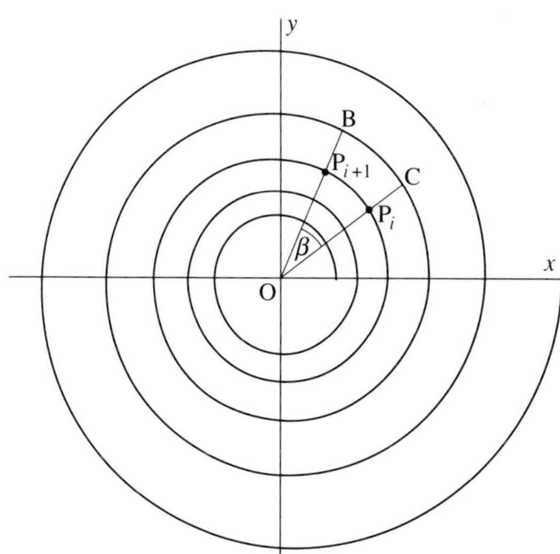

Fig. 19.3. Exponential spiral (eqn (19.4a)).

$$\mathrm{P}_i\mathrm{C} = \mathrm{OC} - \mathrm{OP}_i = a\exp(\theta_i \cot\phi)[\exp(2\pi\cot\phi) - 1], \tag{19.4c}$$

where θ_i is the angular coordinate of P_i,

$$A_i \approx \mathrm{P}_i\mathrm{CBP}_{i+1} = r_i{}^2\beta[\exp(2\pi\cot\phi) - 1]. \tag{19.4d}$$

Next substitute

$$\begin{aligned} r_i &= a\exp(\theta_i \cot\phi) \\ \theta_i &= \theta_0 + i\beta \qquad \text{with } \theta_0 = 0 \\ i &= ct, \end{aligned} \tag{19.4e}$$

where the first two equations give the coordinates of P_i and the last equation generates primordia at a constant rate. Equation (19.4d) becomes

$$A_i = \beta[\exp(2\pi\cot\phi) - 1]a^2\exp(2\cot\phi\,\beta ct). \tag{19.4f}$$

Note that in this case, in contrast with Archimedes spiral and eqn (19.3g), the area associated with each primordium increases exponentially with time. Because exponential growth is often observed in young growing tissues, the exponential spiral has frequently been applied to problems of phyllotaxis.

19.2.3 *Cyclotron spiral*

The cyclotron is a device used for accelerating charged particles, and the path described by a non-relativistic charged particle in a constant magnetic field is a spiral which is referred to as a cyclotron spiral. It is shown in Fig. 19.4 and has

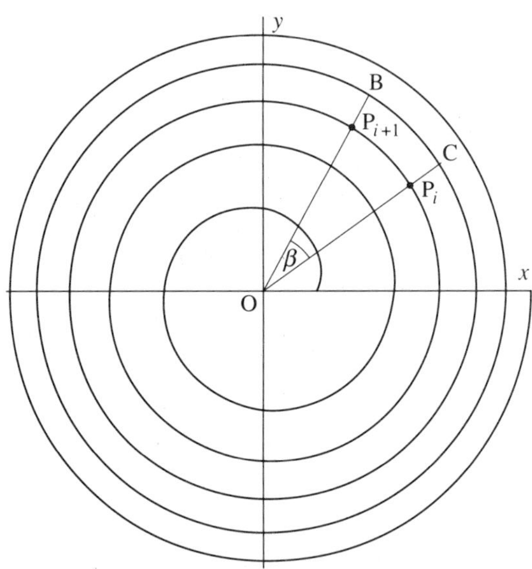

Fig. 19.4. Cyclotron spiral (eqn (19.5a)).

the equation

$$r = a\theta^{1/2} \tag{19.5a}$$

where a is a constant. Differentiating gives

$$\frac{\mathrm{d}r}{\mathrm{d}\theta} = \frac{1}{2}a\theta^{-1/2} = \frac{1}{2}\frac{a^2}{r}. \tag{19.5b}$$

From eqn (19.2b), we obtain

$$\tan\phi = \frac{2r^2}{a^2}. \tag{19.5c}$$

Now place primordia at a constant angular separation β along the spiral, and let primordium P_i have coordinates (r_i, θ_i). An area A_i equal to $P_i CBP_{i+1}$ (Fig. 19.4) can be associated with P_i. As

$$P_i P_{i+1} = a\theta_i^{1/2}\beta \qquad P_i C = a[(\theta_i + 2\pi)^{1/2} - \theta_i^{1/2}]. \tag{19.5d}$$

A_i is given by

$$A_i \approx a^2\beta\left[\theta_i\left(1 + \frac{2\pi}{\theta_i}\right)^{1/2} - \theta_i\right], \tag{19.5e}$$

and for $\theta_i \gg 2\pi$, this becomes

$$A_i = a^2\pi\beta. \tag{19.5f}$$

This element of area is independent of θ_i, and is therefore constant as i increases owing to the formation of new primordia (see eqn (19.3f) and the second of eqns (19.3d)).

The cyclotron spiral was introduced to phyllotaxis studies by Vogel (1979), who considered the sunflower head and the requirement that there should be equal areas available per seed for seeds at the periphery of the capitulum and for those in more central regions. The cyclotron spiral fulfills this requirement, whereas the Archimedes and exponential spirals do not. (Exercise 19.4).

19.3 Packing efficiency in phyllotaxis

Ridley (1982a), in an excellent paper, discusses the packing efficiency of primordia centred at points along a cyclotron spiral. While teleonomic considerations of efficiency are unlikely to lead immediately to the mechanisms that determine primordial positioning, Ridley's analysis gives valuable insights into the problem, and this section owes much to his work (Ridley 1982a,b).

In Figs 19.5, 19.6, and 19.7, 1001 primordia are drawn on spirals that are cyclotron, Archimedes, and exponential respectively. In all three cases we use an angular separation between successive primordia equal to the Fibonacci angle (eqn (19.1f)). Only for the cyclotron spiral are the primordia uniformly distributed over the disc, and it is seen that the visibility of the parastichy systems varies according to the radial growth pattern.

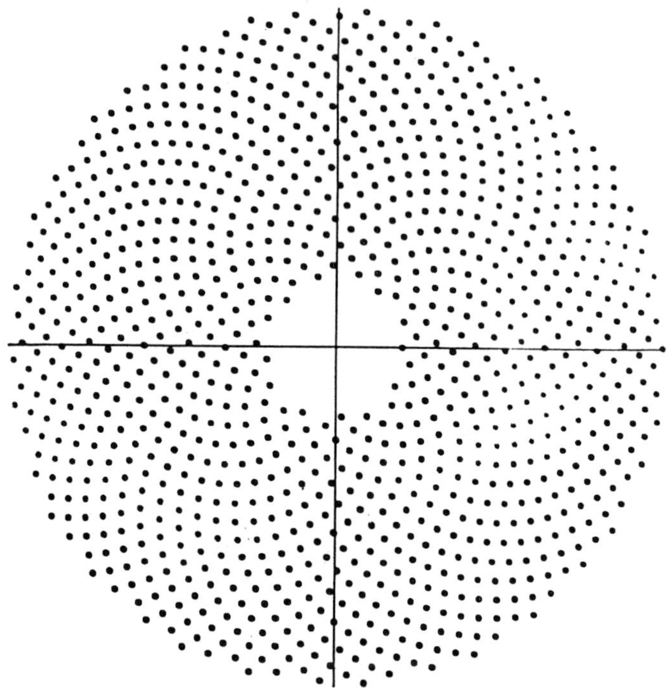

Fig. 19.5. Primordia spaced at the Fibonacci angle along a cyclotron spiral.

Considerations of packing efficiency on the Archimedes or exponential spirals would make sense if the primordia were of increasing size moving outwards (linearly or exponentially; see eqns (19.3g) and (19.4f)). Following Ridley, we consider primordia of equal area, distributed at a constant angular separation β along a cyclotron spiral. We denote the 'fundamental' Fibonacci angle by β_0 (eqn (19.1f). An infinite set of angles β_f related to the Fibonacci angle can be constructed using

$$\beta_f = 2\pi \frac{a + b\tau}{c + d\tau}, \tag{19.6a}$$

where a, b, c, and d are positive integers satisfying $0 < a < b$, $0 < c < d$, and

$$|ad - bc| = 1. \tag{19.6b}$$

For example, β_0 is a member of the β_f with

$$\beta_0 = 2\pi \frac{\tau}{1 + 2\tau}, \tag{19.6c}$$

since $2 + 1/\tau = \tau^2$. Ridley showed that the packing efficiency of primordia is optimized if the divergence angle β is equal to the Fibonacci angle β_0, or one of

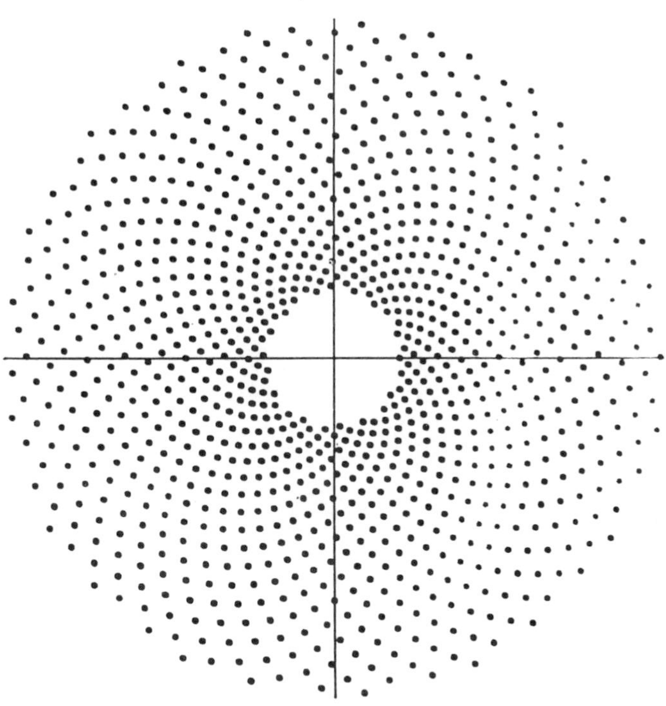

Fig. 19.6. Primordia spaced at the Fibonacci angle along an Archimedes spiral.

the related angles β_f, although the latter approach the same limiting efficiency as β_0, but more slowly.

Consider a region of the capitulum of area A containing N primordia. The average area per point a is given by

$$a = A/N. \tag{19.6d}$$

The packing efficiency η is defined by

$$\eta = \frac{\text{mean square distance between nearest neighbours}}{\text{average area per point } a} \tag{19.6e}$$

The more uniformly the points are distributed, the greater η will be. For hexagonal close packing (h.c.p), it can be shown that

$$\eta(\text{h.c.p}) = \frac{1}{\sin(\pi/3)} = \frac{2}{\sqrt{3}} = 1.155. \tag{19.6f}$$

Ridley demonstrated that, for a cyclotron spiral, the packing efficiency is maximized at the Fibonacci angle, with

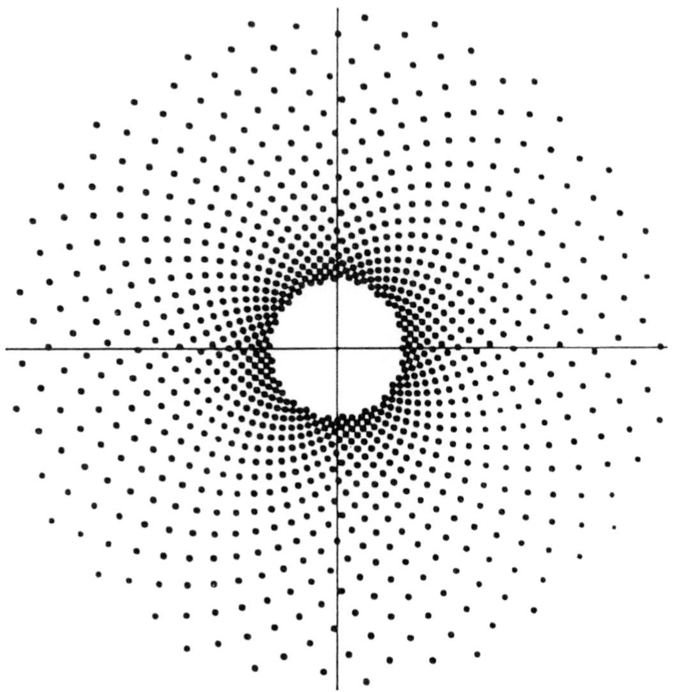

Fig. 19.7. Primordia spaced at the Fibonacci angle along an exponential spiral.

$$\eta(\beta_0) = \frac{5 - 4\cos(6\pi/\tau^2)}{\pi} = 0.8169. \tag{19.6g}$$

Spiral Fibonacci packing is not as efficient as hexagonal close packing.

The main features of Ridley's analysis are illustrated in Figs. 19.8–19.10. In Fig. 19.8, $\beta = (2\pi)34/89$, which is a close rational approximation to the ideal angle β_0. Although the more central regions are evenly covered, in the outer regions the points are grouped on 89 radii (parastichies) and in fact the points on a given radius become arbitrarily close on moving further away from the origin. Although the area per point is still the same, the packing efficiency falls to zero on moving outwards. In Fig. 19.9, an irrational value has been given to β, namely

$$\beta = 2\pi(\sqrt{2} - 1). \tag{19.6h}$$

Although the primordia accumulate along parastichy spirals, these behave differently than with the Fibonacci and Fibonacci-like irrational angles. In the inner region there are 29 anticlockwise spirals with no conspicuous clockwise spirals; further out these are replaced by 70 clockwise parastichies with no conspicuous anticlockwise spirals; eventually these are replaced by 169 (70 + 70 + 29) clockwise parastichies, and so on. In this case the points never approach each other

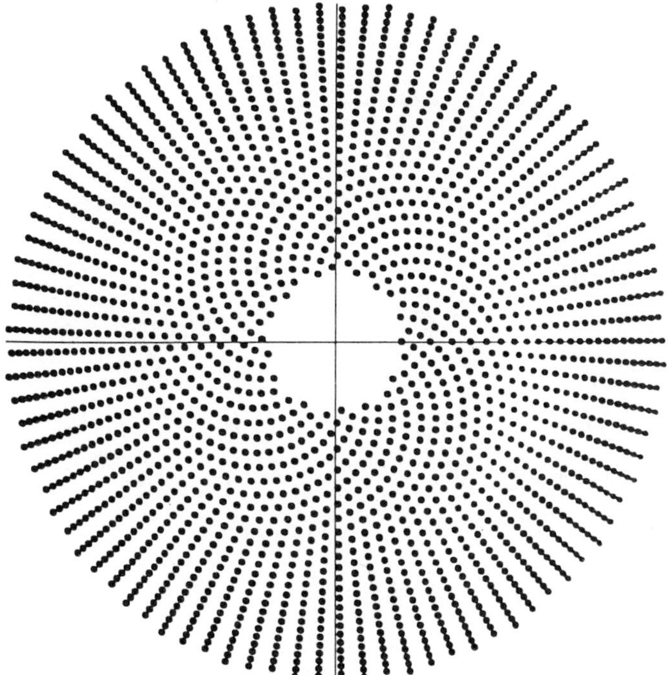

Fig. 19.8. Primordia are spaced at an angle of $2\pi(34/89)$ rad (a rational approximation to the Fibonacci angle) along a cyclotron spiral.

arbitrarily closely, as in Fig. 19.8; on following a given parastichy, it can be seen that the adjacent points first come closer together, and then move further apart. The fact that opposing parastichy systems (clockwise and anticlockwise) are not equally conspicuous suggests that the packing is less efficient, as the four nearest neighbours (two per parastichy) are not approximately equidistant. Ridley (1982a) has shown that for non-Fibonacci-like irrational numbers the packing efficiency is less than for Fibonacci angles, which give an optimum packing, as mentioned above. In Fig. 19.10, one of the Fibonacci-like irrational angles of eqn (19.6a) is taken, that is

$$\beta = 2\pi \frac{\tau}{1 + 4\tau}, \tag{19.6i}$$

which is an angle (77.96°) that is sometimes observed in primordial systems. Here, conspicuous opposed parastichy pairs are seen, and (for large arrays) the packing efficiency approaches the value given in eqn (19.6g).

Considering the regions that lie within and outside a given radius r_0 separately, it can also be shown (Ridley 1982a) that in the outer region the most efficient spiral packing is still given by $\beta = 2\pi/\tau^2$, with the efficiency η approaching

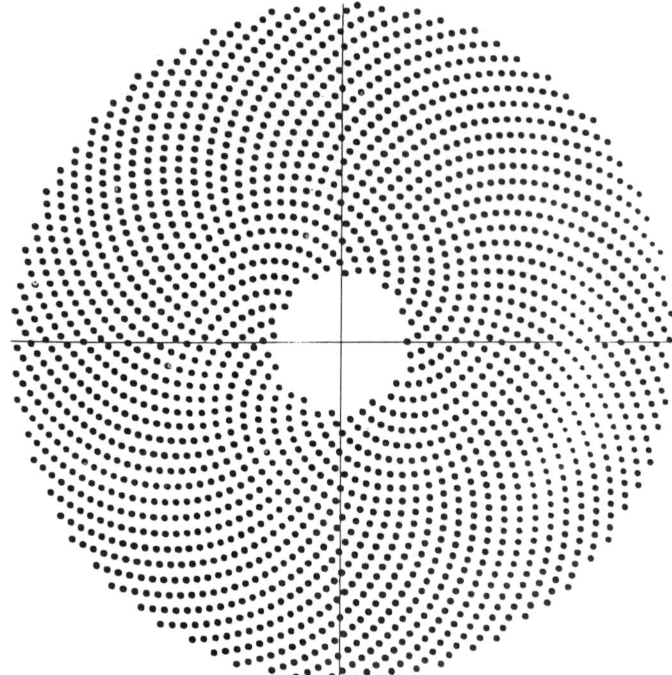

Fig. 19.9. Primordia are spaced at an irrational angle of $2\pi(\sqrt{2} - 1)$ rad along a cyclotron spiral.

$2/\sqrt{5} = 0.8944$ as $r_0 \to \infty$, which also applies to all β given by eqn (19.6a), although the latter approach this limiting value more slowly than with β_0. For other irrational angles (non-Fibonacci-like) the limiting efficiency at large r_0 is at best $1/\sqrt{2}$. Within the interior of the capitulum, with $r < r_0$ and for smaller numbers of primordia, there are some more efficient spiral arrangements than with β_0 or the β_f (this is clearly so: two primordia are best placed π rad apart; three primordia are usually best placed $2\pi/3$ rad apart).

These considerations of packing efficiency are interesting and important, but cannot be viewed as providing a complete account of phyllotaxy. It is not sufficient to demonstrate that the mathematical properties of the golden ratio τ (the most irrational of the irrational numbers, in a sense to be made plain in Section 19.5) make $2\pi/\tau^2$ an optimum angle. Bearing in mind the extent of variation present in most biological phenomena, it is almost inconceivable that genetic mechanisms alone can determine the divergence angle to the accuracy observed—for example, in most sunflower heads, the 55 and 89 parastichies are conspicuous, and this requires a relative accuracy of 1 part in 1869 (21×89) in fixing β, which must lie between 21/55 and 34/89. What is needed are mechanisms by which divergence angles are adjusted during the growth and development of primordia, so as to satisfy more nearly the requirements of efficient packing.

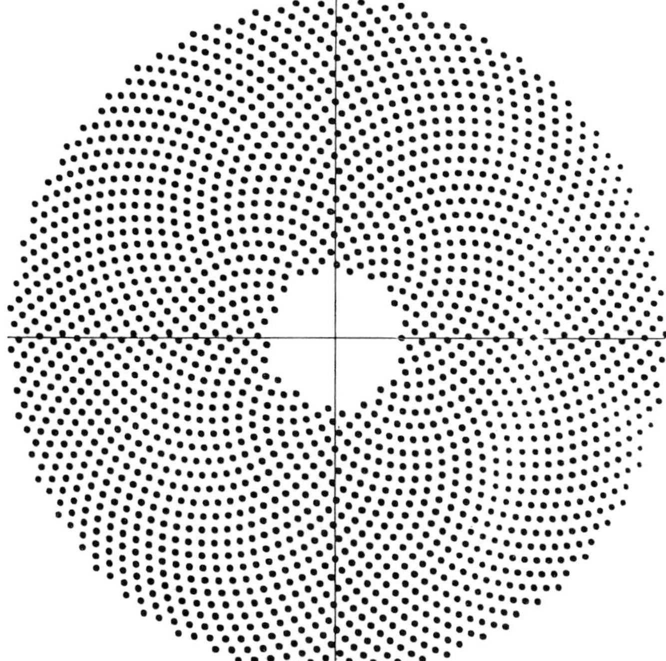

Fig. 19.10. Primordia are spaced at a Fibonacci-related angle of $2\pi\tau/(1 + 4\tau)$ rad along a cyclotron spiral.

Ridley (1982b) describes a computer model in which primordia are originally positioned at various angles, sometimes with a random component. After the genesis of each primordium, a jostling procedure is allowed to take place, in which the existing primordia can re-adjust their positions to take better advantage of the available space. The results are shown in Table 19.1, and largely speak for themselves. The conclusion is that an approximate original positioning mechanism plus subsequent adjustment, perhaps by differential growth, is sufficient to give Fibonacci phyllotaxis and other Fibonacci-like angles.

19.4 Continued fractions

This section provides some background material for subsequent sections.

The golden ratio τ can be obtained from the simple continued fraction (Exercise 19.5)

$$\tau = 1 + \cfrac{1}{1 + \cfrac{1}{1 + \cdots}}. \tag{19.7a}$$

The theory of continued fractions is highly relevant to phyllotaxis and to different

Table 19.1. Mean (standard deviation) of divergence angles (in degrees) of primordia in simulated phyllotaxis systems where readjustment through a jostling procedure can take place

Initial values	After 10 primordia	After 50 primordia	After 100 primordia
126.0 (0.0)	135.9 (3.8)	137.0 (1.4)	136.6 (6.3)
135.0 (0.0)	137.2 (3.8)	137.4 (0.9)	137.5 (0.7)
149.8 (0.0)	139.6 (7.0)	137.9 (1.6)	138.0 (3.0)
137.5 (3.6)	137.8 (4.3)	137.4 (1.0)	137.5 (0.6)
137.5 (10.8)	138.0 (5.9)	137.6 (1.3)	137.5 (0.8)
137.5 (27.0)	135.2 (13.7)	131.5 (23.7)	134.4 (24.4)
133.9 (7.2)	136.9 (6.2)	137.4 (0.9)	137.5 (0.7)
141.1 (7.2)	138.6 (6.8)	137.5 (1.2)	137.5 (0.9)
97.2 (0.0)	101.2 (4.7)	99.7 (1.1)	99.6 (0.7)
101.5 (0.0)	101.9 (4.2)	100.0 (0.6)	100.4 (4.6)
99.5 (1.8)	101.4 (4.7)	99.8 (1.3)	99.7 (0.8)
99.5 (3.6)	101.7 (4.9)	99.8 (1.5)	99.7 (1.0)
99.5 (10.8)	101.8 (5.9)	99.7 (1.5)	99.6 (0.8)

After Ridley 1982b.
$99.5 = 360\tau/(1 + 3\tau)$; see (19.6a).

theories of phyllotaxis, whether these are contact theories (Adler 1974, 1977), packing theories (Ridley 1982a, b), or reaction–diffusion morphogen theories (Marzec and Kappraff 1983). We therefore outline theorems that have a bearing on phyllotaxis; detailed derivations are given in standard texts (e.g. Niven and Zuckerman 1960).

Every positive real number α can be written uniquely in the form

$$\alpha = a_0 + \frac{1}{a_1 + 1/(a_2 + \cdots)} \tag{19.7b}$$

where $a_0, a_1, a_2, \ldots,$ are positive integers. To carry this out, write $\alpha = a_0 + d_0$, where a_0 is the integer part of α and d_0 is the decimal (non-integer) part of α; write $1/d_0 = a_1 + d_1$ where a_1 and d_1 are the integer and decimal parts of $1/d_0$; continue repeating this process (Exercise 19.6). Equation (19.7b) is abbreviated to

$$\alpha = [a_0, a_1, a_2, a_3, \ldots, a_n]. \tag{19.7c}$$

If α is rational, then the continued fraction terminates and n is finite; if α is irrational, then the continued fraction does not terminate and n is infinite. Whether n is finite or infinite, a continued fraction can be truncated at some term k to give an approximation to α. Equation (19.7c) can be written

$$\alpha = [a_0, a_1, a_2, \ldots, a_k + d_k], \tag{19.7d}$$

where d_k is the decimal remainder. α_k is known as the kth convergent to α with

$$\alpha_k = [a_0, a_1, a_2, a_3, \ldots, a_k]. \tag{19.7e}$$

From the form of (19.7b), it can be seen that the even-order convergents

$$\alpha_0 = [a_0] = a_0, \qquad \alpha_2 = [a_0, a_1, a_2] = a_0 + \cfrac{1}{a_1 + \cfrac{1}{a_2}}, \cdots, \tag{19.7f}$$

are smaller than α. The odd-order convergents

$$\begin{aligned} \alpha_1 &= [a_0, a_1] = a_0 + \frac{1}{a_1}, \\ \alpha_3 &= [a_0, a_1, a_2, a_3] = a_0 + \cfrac{1}{a_1 + \cfrac{1}{a_2 + \cfrac{1}{a_3}}}, \cdots, \end{aligned} \tag{19.7g}$$

are larger than α. Thus the even convergents α_{2i}, $i = 0, 1, 2, \ldots$, are an increasing sequence converging on α from below, and the odd convergents α_{2i+1}, $i = 0, 1, 2, \ldots$, are a decreasing sequence converging on α from above.

Every convergent α_k is a rational number (for finite k), and can be written in the form

$$\alpha_k = [a_0, a_1, a_2, \ldots, a_k] = \frac{p_k}{q_k} \tag{19.7h}$$

where p_k and q_k are integers. By direct calculation it can be shown that

$$\begin{aligned} &p_0 = a_0, \quad q_0 = 1; \quad p_1 = 1 + a_0 a_1, \quad q_1 = a_1; \\ &p_2 = a_0 + a_2(1 + a_0 a_1), \quad q_2 = 1 + a_1 a_2; \\ &p_3 = 1 + a_0 a_1 + a_3(a_0 + a_2 + a_0 a_1 a_2), \quad q_3 = a_1 + a_3(1 + a_1 a_2). \end{aligned} \tag{19.7i}$$

It can be verified that the p_k and q_k obey

$$p_k = a_k p_{k-1} + p_{k-2}, \quad q_k = a_k q_{k-1} + q_{k-2}. \tag{19.7j}$$

Additionally the p_k and q_k obey

$$p_{k-1} q_k - p_k q_{k-1} = (-1)^k \tag{19.7k}$$

$$p_{k-2} q_k - p_k q_{k-2} = (-1)^{k-1} a_k. \tag{19.7l}$$

These relations can be used and verified by working through Exercise 19.7.

Consider three successive convergents:

$$\frac{p_{k-2}}{q_{k-2}}, \quad \frac{p_{k-1}}{q_{k-1}}, \quad \frac{p_k}{q_k}. \tag{19.7m}$$

A sequence of intermediate fractions can be constructed, finishing at p_k/q_k by applying eqns (19.7j), namely

$$\frac{p_{k-2}}{q_{k-2}}, \frac{p_{k-2}+p_{k-1}}{q_{k-2}+q_{k-1}}, \frac{p_{k-2}+2p_{k-1}}{p_{k-2}+2q_{k-1}}, \ldots, \frac{p_{k-2}+a_k p_{k-1}}{q_{k-2}+a_k q_{k-1}} = \frac{p_k}{q_k}. \tag{19.7n}$$

With any two fractions a/b and c/d, it can be shown that $(a + c)/(b + d)$ lies between a/b and c/d. Therefore all these intermediate fractions lie between p_{k-2}/q_{k-2} and p_{k-1}/q_{k-1}. Note, however, that if $a_k = 1$, there are no intermediate fractions between these two convergents (Exercise 19.8).

Given a real number α and a rational fraction p/q, then p/q is said to be a best approximation to α if every other rational fraction with a denominator equal to or less than q differs from α by a greater amount. Thus, with $\alpha = 1.42$, 7/5 is a best approximation to α, since with $q < 5$, there is no p/q that lies closer to α; similarly 4/3 and 17/12 are also best approximations to α. Every best approximation to a number α is a convergent or an intermediate convergent of the continued fraction representing that number.

19.5 Spacing of points round a circle

We now give a result that is discussed extensively by Marzec and Kappraff (1983), who show that the Fibonacci angle leads to the most uniform distribution of points on a circle. Let α be any irrational number and use $\{x\}$ to denote the fractional part of x so that $\{x\}$ lies in the interval $0 \leqslant \{x\} \leqslant 1$. The n points $\{\alpha\}$, $\{2\alpha\}, \ldots, \{n\alpha\}$ lie in the interval $[0, 1]$, and divide the line segment $[0, 1]$ into $n + 1$ pieces. These $n + 1$ pieces have lengths that fall into at most three different length categories (Exercise 19.9). The next point $\{(n + 1)\alpha\}$ always falls into one of the largest segments, splitting it into two. This largest segment may be split by a 'bad break', i.e. it is split so unequally that one of the subsegments is twice or more than twice as large as the other. For all numbers α, except $1/\tau$ and $1/\tau^2$ (which are equivalent), there are $\{n\alpha\}$ which give a bad break. In fact, $\alpha = 1/\tau$ or $\alpha = 1/\tau^2$ divide each largest section according to the golden ratio τ. Considering the distribution of the n points $\{\alpha\}, \{2\alpha\}, \ldots, \{n\alpha\}$ along the line segment $[0, 1]$, it can be shown that $\alpha = 1/\tau$ or $\alpha = 1/\tau^2$ lead to the most uniform distribution of the n points.

While it is easy to understand that any rational α will not lead to a uniform distribution of points, it is less apparent that, amongst the irrational numbers, the golden ratio and related numbers are singled out and are, in a definable sense, the most irrational of the irrational numbers (cf. Figs. 19.9 and 19.10). The most uniform distribution of points arises from irrational numbers that, when expressed as continued fractions, are free from intermediate convergents, i.e. all the a_k in eqn (19.7b) are equal to unity. Whenever α has an intermediate convergent and some $a_k \geqslant 2$, a bad break occurs and the spacing of points round the circle is less than optimal. It can also be shown that irrationals of the form

$$\alpha = [a_0, a_1, a_2, \ldots, a_k, 1, 1, 1, \ldots] \tag{19.8a}$$

also lead, eventually, to a most uniform distribution of points. In fact, any α of the form

$$\alpha = \frac{a + b\tau}{c + d\tau}, \tag{19.8b}$$

where a, b, c, and d are integers satisfying

$$ad - bc = \pm 1, \qquad a < b, c < d, \tag{19.8c}$$

fulfils this criterion.

19.6 Parastichy numbers and spiral systems

In this section, which stands alone and may be omitted, we discuss some of the purely descriptive aspects of parastichy spiral systems considered by Thornley (1975b), who analysed the use of intersecting spirals to describe primordia lying along an exponential spiral. His analysis is equally applicable to Archimedes or cyclotron spirals.

Phyllotaxis systems have frequently been described in terms of a number pair (n_A, n_B), the parastichy numbers. Wherever the n_A A spirals intersect the n_B B spirals, a primordium is positioned (Fig. 19.11). It can be seen from Figs. 19.5–19.10 that there are conspicuous opposed spirals only if the divergence angle β is equal to the Fibonacci angle β_0 (eqn (19.1f)) or if it is equal to one of the related angles β_f (eqn (19.6a)). The ease with which opposed spirals can be discerned depends upon the presence of approximately equidistant nearest neighbours moving outwards in either a clockwise or an anticlockwise direction; this is equivalent to an efficient packing criterion. In Fig. 19.8 the rational angle used is only able to sustain efficient packing, and conspicuous opposed parastichy pairs, for a while. In Fig. 19.9 there is always one highly visible spiral although the opposed spiral is not so readily seen, which implies inefficient packing. However, in Fig. 19.10 conspicuous opposed spirals are easily seen at all radii, owing to the efficient packing, and the parastichy numbers (n_A, n_B) change systematically on moving outwards. The radial growth equation (Section 19.2) does of course affect the packing of the primordia; which particular primordia are in fact nearest neighbours depends upon both the divergence angle and the radial growth equation. Thus in Figs 19.5, 19.6, and 19.7 the divergence angle is the same in all cases (equal to the Fibonacci angle), but the different radial distributions affect packing, nearest neighbours, and the conspicuousness of parastichy systems.

A $(5, 8)$ parastichy system is sketched in Fig. 19.11, which is relevant to the following analysis. Let the most recently produced primordium be P_0 with coordinates $r_0 = 1$, $\theta_0 = 0$, and let the successively older primordia be situated at (r_1, θ_1), (r_2, θ_2), $\ldots$, (r_i, θ_i), $\ldots$. The angle between successive primordia is β proceeding to younger primordia; proceeding to older primordia, the angular separation is β', where

$$\beta' = 2\pi - \beta. \tag{19.9a}$$

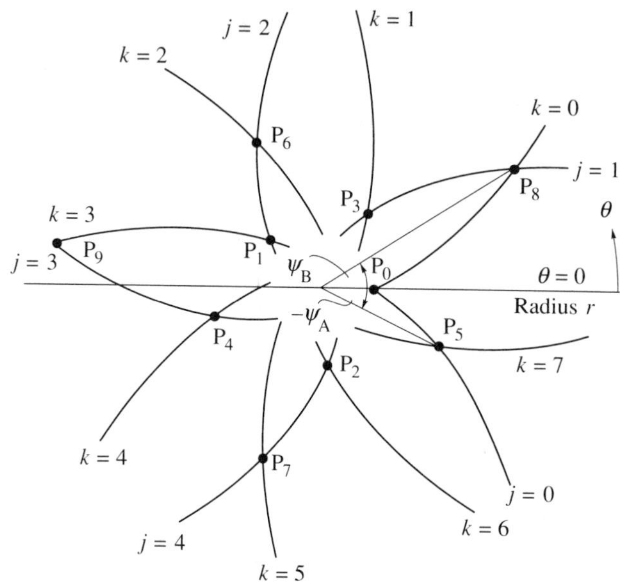

Fig. 19.11. A (5, 8) parastichy system drawn through primordia P_0, P_1, P_2, ... located on an exponential spiral with a divergence angle of 221°. An infinite set of (5, 8) spirals can be drawn through this set of primordia, and those shown here are characterized by $m_A = 2$, $m_B = 3$ (eqns (19.9i) and (19.9j)). The five A spirals are indicated by j values of 0, 1, 2, 3, 4, and the eight B spirals by k values of 0, 1, ..., 6, 7. ψ_A and ψ_B are the angles traversed along a given A spiral or B spiral in going from one primordium to the next on that spiral; these angles are shown only for the $j = 0$ A spiral and the $k = 0$ B spiral. In this case $\psi_A = -25°$ and $\psi_B = 32°$.

The angle of the ith primordium is

$$\theta_i = i\beta'. \tag{19.9b}$$

When the three radial growth equations discussed earlier (eqns (19.3a), (19.4a), and (19.5a)) are used, r_i for primordium P_i is given by

$$r_i = a\theta_i + 1 \tag{19.9c}$$

or

$$r_i = \exp(c\theta_i) \tag{19.9d}$$

or

$$r_i = b\theta_i^{1/2} + 1, \tag{19.9e}$$

where a, b, and c are parameters.

Consider the parastichy system (n_A, n_B) with $n_A < n_B$. The n_A A spirals can be obtained from each other by an angular displacement of $2\pi/n_A$. Let (r_A, θ_A) be a point on an A spiral; then for the three radial growth equations, the n_A A spirals can be written

$$r_A = a_A\left(\theta_A - j\frac{2\pi}{n_A}\right) + 1 \tag{19.9f}$$

or

$$r_A = \exp\left[c_A\left(\theta_A - j\frac{2\pi}{n_A}\right)\right] \tag{19.9g}$$

or

$$r_A = b_A\left(\theta_A - j\frac{2\pi}{n_A}\right)^{1/2} + 1 \tag{19.9h}$$

with $j = 0, 1, 2, \ldots, n_A - 1$. a_A, b_A, and c_A are parameters; the $j = 0$ spirals all pass through the point $r = 1$, $\theta = 0$, where the primordium P_0 is situated. The n_B B spirals, numbered by the integer k with $k = 0, 1, 2, \ldots, n_B - 1$, are given by equations similar to eqns (19.9f)–(19.9h) but with the A subscripts changed to B; (r_B, θ_B) is a point on a B spiral. The $k = 0$ B spiral passes through P_0 at the point $r = 1$, $\theta = 0$. The primordia $P_0, P_{n_A}, P_{2n_A}, \ldots, P_{n_B n_A}, \ldots$, lie on the A spiral with $j = 0$; the primordia $P_0, P_{n_B}, P_{2n_B}, \ldots, P_{n_A n_B}, \ldots$, lie on the B spiral with $k = 0$.

Let ψ_A be the angle travelled along the A spiral with $j = 0$ on going from P_0 to P_{n_A} (Fig. 19.11), and similarly ψ_B is the angle travelled along the B spiral with $k = 0$ from P_0 to P_{n_B}. In Fig. 19.11, the A and B spirals are shown opposed (rotating in different senses) and ψ_A is negative. The angles ψ_A and ψ_B are related to the divergence angle β', and to n_A and n_B by

$$n_A\beta' = m_A 2\pi + \psi_A \tag{19.9i}$$

and

$$n_B\beta' = m_B 2\pi + \psi_B \tag{10.9j}$$

where m_A and m_B are integers. It is assumed that primordia are present at all the points where the A spirals and the B spirals intersect. One can travel from P_0 to $P_{n_A n_B}$ along the A spiral ($j = 0$) through angle $n_B\psi_A$ or along the B spiral ($k = 0$) through angle $n_A\psi_B$. Thus

$$n_B\psi_A - n_A\psi_B = \pm 2\pi. \tag{19.9k}$$

Combining eqns (19.9i), (19.9j), and (19.9k) gives

$$m_B n_A - m_A n_B = \pm 1. \tag{19.9l}$$

Δn and Δm are now defined by

$$\Delta n = n_B - n_A \qquad \Delta m = m_B - m_A, \tag{19.9m}$$

and eqn (19.9l) becomes

$$n_A\Delta m - m_A\Delta n = \pm 1$$

which can be written as

$$m_A = \frac{n_A \Delta m \pm 1}{\Delta n} \tag{19.9n}$$

Note again that n_A and n_B are positive and ordered so that $n_A < n_B$. This excludes $n_A = n_B = 1$ which is not of interest; also n_A and n_B are relatively prime, i.e. without a factor in common—there is only one primordium for a given value of radius r (if in Fig. 19.11 $n_A = 3$ and $n_B = 9$, there would be three points of intersection and three primordia at $r = 1$).

Given (n_A, n_B) and thus Δn, different values of Δm can be tried in eqn (19.9n) (i.e. 0, 1, 2, ...,) to obtain an integer m_A. There are two number pairs (m_A, m_B) which satisfy $0 \leqslant \Delta m \leqslant \Delta n$. (m_A, m_B) are also relatively prime. From any given (m_A, m_B) an infinite series of such numbers can be generated by

$$m_A + in_A, \qquad m_B + in_B, \tag{19.9o}$$

where i is any integer; there are two such infinite series. For example, for $(n_A, n_B) = (13, 20)$ (this is an arbitrary choice; (13, 20) are not Fibonacci numbers but they are relatively prime) the two series are

$$\ldots, (-24, -37), (-11, -17), (2, 3), (15, 23), (28, 43), \ldots$$

$$\ldots, (-15, -23), (-2, -3), (11, 17), (24, 37), (37, 57), \ldots.$$

The above properties of the parastichy spiral systems only depend upon the angular relations, and not at all on the radial growth equation. If we wish to determine the parameters of the equations of the parastichy spirals (eqns (19.9f)–(19.9h)), then this can be achieved by equating the radial position of primordium P_{n_A} given by the primary spiral (eqns (19.9c)–(19.9e)) to that given by the parastichy spiral. For example, for the Archimedes spiral put $\theta_i = n_A \beta'$ in eqn (19.9c) and put $\theta_A = \psi_A$ in eqn (19.9f) with $j = 0$, and equate the results to give

$$an_A \beta' = a_A \psi_A.$$

Hence it is easily shown that

$$a_A = \frac{an_A \beta'}{\psi_A} \qquad c_A = \frac{cn_A \beta'}{\psi_A} \qquad b_A = b\left(\frac{n_A \beta'}{\psi_A}\right)^{1/2} \tag{19.9p}$$

for the three radial equations. The parameters for the B spirals are given by changing the subscripts.

Summary of procedure The number pair (n_A, n_B) can be chosen almost arbitrarily and for each choice an infinite number of spirals can be constructed, although these are of course not all conspicuous and opposed. The steps involved in doing this are outlined below.

1. Specify a unijugate system of primordia by the divergence angle β (hence β' by eqn (19.9a)), and a radial growth equation (eqns (19.9c)–(19.9e) or some other).

2. Choose two numbers (n_A, n_B) to define the spiral system; $n_A < n_B$, and (n_A, n_B) are relatively prime.
3. Generate auxiliary numbers (m_A, m_B) which satisfy eqn (19.9l); use eqns (19.9n) and (19.9o) or otherwise.
4. Use eqns (19.9i) and (19.9j) to give ψ_A and ψ_B.
5. Use eqns (19.9p) to give the parameters of the spiral equations, which can then be drawn.

19.7 Simple reaction–diffusion theory of phyllotaxis

In the preceding sections only static aspects of phyllotaxis have been considered —for instance, the packing efficiency and spacing of given arrangements, and the ways in which systems of primordia can be described. We now wish to turn to the dynamic mechanistic problem—that is, the construction of a dynamic mechanistic model which is able to generate a time sequence of events in phyllotaxis. Two approaches have been used, and although these are not entirely distinct we shall treat them separately. The packing or contact theories have been pioneered by Adler (1974, 1977), and have been pursued independently by others (Williams and Brittain, 1984). The specific application of reaction–diffusion theory and morphogen fields to phyllotaxis was initiated by Thornley (1975a), and has been considered further by, for example, Veen and Lindenmayer (1977) and Roberts (1978). While many elaborations of the reaction–diffusion approach are possible (consideration of two or three dimensions, apices of a particular shape, polar transport), the single-morphogen one-dimensional analysis of Thornley (1975a) contains the essential elements; it is relatively easily described and understood, it is able to produce the Fibonacci and related angles (eqns (19.1f) and (19.6a)), and therefore it is discussed below.

A field theory using diffusion–reaction mechanisms assumes that there is a chemical field, and that cells or groups of cells are able to respond to this field by differentiating and developing in a directed manner. Of course a packing or contact constraint could be expressed in terms of chemical interactions, although a field theory description of such constraints might not be convenient. It should also be remembered that the mechanisms that determine primordium initiation may differ from those that are important in the subsequent stages of primordial outgrowth. Turing (1952) demonstrated how a combination of diffusion and chemical reaction could give rise to steady state patterns; the morphogen field theories of phyllotaxis are of this type.

19.7.1 *The model*

Consider a flat disc of radius r. New primordia are initiated at a constant radius r from the origin, and it is assumed that only the apical tissue in this zone is competent to initiate primordia. Apical shape is not further considered. The first primordium P_0 is placed at radius r and is taken to be the origin of the angular coordinate θ. The distance x of a point Q from P_0 is measured along the circle radius r and is given by

$$x = r\theta. \tag{19.10a}$$

Also, x has a maximum value of x_m given by

$$x_m = 2\pi r \qquad 0 \leqslant x < x_m. \tag{19.10b}$$

It is assumed that each primordium acts as a point source of a morphogen M, which then diffuses round the circle and is degraded at a rate kM, where k is a constant. The differential equation governing M is (cf. (4.5g), p. 93)

$$\frac{\partial M}{\partial t} = D\frac{\partial^2 M}{\partial x^2} - kM, \tag{19.10c}$$

where t is the time variable and D is a diffusion constant. In the steady state this becomes

$$0 = \frac{d^2 M}{dx^2} - \gamma^2 M \tag{19.10d}$$

where $\gamma^2 = k/D$. This has solutions of the general form

$$M = Ae^{\gamma x} + Be^{-\gamma x}, \tag{19.10e}$$

where A and B are constants whose values are determined by the boundary conditions.

Primordium P_0 at $x = 0$ is assigned a source strength S_0. For a point source S_0 placed at $x = 0$, the boundary conditions are

$$M(x = 0) = M(x = x_m) \qquad \frac{S_0}{2} = -D\frac{dM}{dx}(x = 0+). \tag{19.10f}$$

The gradient is discontinuous at $x = 0$ (Fig. 19.12), half of S_0 diffuses in the positive x direction and half in the negative x direction, and the expression $dM/dx(x = 0+)$ indicates the gradient on the positive side of the origin where the gradient is negative.

Substituting eqn (19.10e) into eqns (19.10f) gives

$$A + B = A\exp(\gamma x_m) + B\exp(-\gamma x_m)$$

$$S_0/2 = -D\gamma(A - B).$$

Solving for A and B and substituting back into eqn (19.10e) gives

$$M_0 = \frac{S_0}{2D\gamma \sinh(\gamma x_m/2)}\cosh\left[\gamma\left(x - \frac{x_m}{2}\right)\right]. \tag{19.10g}$$

M_0 is the morphogen field arising from the source S_0, and it is plotted in Fig. 19.12 for three values of γ. The mean value $\overline{M_0}$ of M_0 is

$$\overline{M_0} = \frac{S_0}{kx_m}. \tag{19.10h}$$

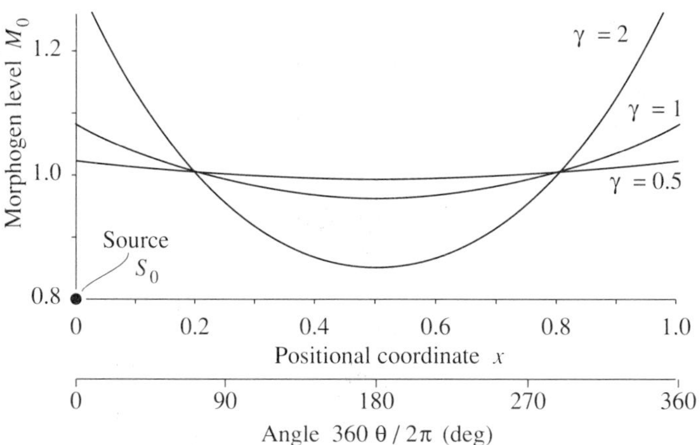

Fig. 19.12. Morphogen field from a source S_0 at the origin ($x = 0$) in the steady state. The curves represent eqn (19.10g) with $S_0 = 1$, $x_m = 1$, $k = 1$; D is given the values 4, 1, and 0.25, so that γ is 0.5, 1, or 2 (eqn (19.10d)). The mean value $\overline{M}_0$ of the field is unity in all three cases (eqn (19.10h)).

Differentiating eqn (19.10g) gives

$$\frac{dM_0}{dx} = \frac{S_0}{2D\sinh(\gamma x_m/2)}\sinh\left[\gamma\left(x - \frac{x_m}{2}\right)\right]. \tag{19.10i}$$

At $x = x_m/2$, dM_0/dx is zero and M_0 has a minimum value of $M_0(\text{min})$ opposite the position of the source S_0. Although this is an obvious consequence of restricting diffusion to the circumference of the disc of radius r, it is also true for diffusion across the disc for reasonable values of the diffusion constants.

The maximum value $M_0(\text{max})$ of M_0 occurs at the source position itself. From eqn (19.10g)

$$\begin{aligned} M_0(\text{min}) &= \frac{S_0}{2D\gamma\sinh(\gamma x_m/2)} \\ M_0(\text{max}) &= \frac{S_0\cosh(\gamma x_m/2)}{2D\gamma\sinh(\gamma x_m/2)}. \end{aligned} \tag{19.10j}$$

The effect of x_m, which is a measure of apex size, on these quantities is to be noted. With increasing x_m, both $\overline{M}_0$ and $M_0(\text{min})$ decrease and approach zero for large x_m; $M_0(\text{max})$ decreases but approaches an asymptote of $S_0/2D\gamma$. If primordium initiation takes place when the morphogen field falls to a certain value, this occurs more rapidly in an apex whose size is increasing if other factors remain the same.

It is important to realize that eqns (19.10g) and (19.10i) are linear: if M_0 is the morphogen field from S_0 and M_1 is the field from source S_1, then $M_0 + M_1$ is the field from sources $S_0 + S_1$. Each source makes its own contribution to the field independently of the other sources present.

For further use, eqns (19.10g) and (19.10i) can be generalized for a source at any position. Let M_i be the morphogen field from the ith source S_i at position x_i. s_i is defined as

$$s_i = \frac{S_i}{2D\gamma \sinh(\gamma x_{\mathrm{m}}/2)}. \tag{19.10k}$$

The morphogen field and its gradient are given by

$$M_i(x) = s_i \cosh\left[\gamma\left(x - x_i - \frac{f_i x_{\mathrm{m}}}{2}\right)\right], \tag{19.10l}$$

$$\frac{\mathrm{d}M_i}{\mathrm{d}x}(x) = \gamma s_i \sinh\left[\gamma\left(x - x_i - \frac{f_i x_{\mathrm{m}}}{2}\right)\right], \tag{19.10m}$$

where f_i is defined by

$$f_i = -1, \quad 0 \leqslant x < x_i; \quad f_i = +1, \quad x_i \leqslant x < x_{\mathrm{m}}. \tag{19.10n}$$

19.7.2 *The initiation of two, three, and four primordia*

It is assumed that primordium initiation occurs when the morphogen field M falls to a threshold value—this will occur at one of the minimum points in the field. It is assumed also that the minimum of the steady-state field gives the position of the next primordium. From Fig. 19.12, it is seen that the second primordium S_1 will arise opposite the first primordium S_0 at $x = 0$, i.e. at $x = x_{\mathrm{m}}/2$. Once initiated, the second primordium acts as a point source of morphogen of strength S_1, giving rise to a field $M_1(x)$. Therefore by eqns (19.10l) and (19.10m)

$$M(x) = s_0 \cosh\left[\gamma\left(x - \frac{x_{\mathrm{m}}}{2}\right)\right] + s_1 \cosh[\gamma(x - x_{\mathrm{m}})] \tag{19.11a}$$

and

$$\frac{\mathrm{d}M(x)}{\mathrm{d}t} = \gamma s_0 \sinh\left[\gamma\left(x - \frac{x_{\mathrm{m}}}{2}\right)\right] + \gamma s_1 \sinh[\gamma(x - x_{\mathrm{m}})] \tag{19.11b}$$

for $x_{\mathrm{m}}/2 \leqslant x < x_{\mathrm{m}}$. Equating the gradient to zero, using $\sinh(A - B) = \sinh A \cosh B - \cosh A \sinh B$, the minimum of the field is at x_2 given by

$$\tanh(\gamma x_2) = \frac{s_0 \sinh(\gamma x_{\mathrm{m}}/2) + s_1 \sinh(\gamma x_{\mathrm{m}})}{s_0 \cosh(\gamma x_{\mathrm{m}}/2) + s_1 \cosh(\gamma x_{\mathrm{m}})} \tag{19.11c}$$

is the region $x_{\mathrm{m}}/2 < x < x_{\mathrm{m}}$ (there is symmetry about $x = x_{\mathrm{m}}/2$; see Fig. 19.13).

The minimum of the morphogen field in eqn (19.11c) depends upon the relative strengths S_0 and S_1 of the two sources. A parameter λ is defined by

$$\lambda = \frac{s_0}{s_1} = \frac{S_0}{S_1}. \tag{19.11d}$$

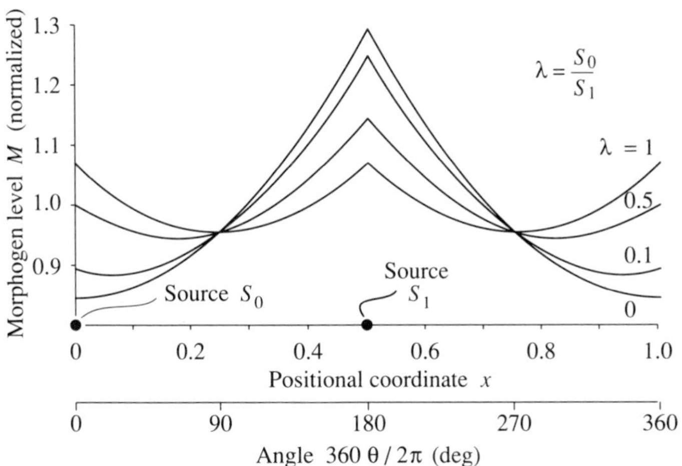

Fig. 19.13. Morphogen field from two sources, S_0 at $x = 0$ and S_1 at $x = x_m/2$, in the steady state. The total field is normalized to have a mean value of unity. The curves are calculated from eqn (19.11e) with λ values as shown.

Equations (19.11a) and (19.11c) become

$$M(x) = s_1 \left\{ \cosh[\gamma(x - x_m)] + \lambda \cosh\left[\gamma\left(x - \frac{x_m}{2}\right)\right]\right\} \tag{19.11e}$$

$$\tanh(\gamma x_2) = \frac{\sinh(\gamma x_m) + \lambda \sinh(\gamma x_m/2)}{\cosh(\gamma x_m) + \lambda \cosh(\gamma x_m/2)}. \tag{19.11f}$$

Variation of λ in eqn (19.11f) permits the competing effects of two morphogen sources of different strengths to be examined, as shown in Fig. 19.13. The limiting cases given in Fig. 19.13 are for $\lambda = 0$ and $\lambda = 1$. With $\lambda = 0$, the first primordium S_0 has ceased the production of morphogen entirely, and the field is due to the second primordium S_1 alone; such a system simply continues to produce opposite primordia. With $\lambda = 1$, S_0 and S_1 produce morphogen equally, and there are two identical minima lying midway between S_0 and S_1 at $x = x_m/4$ and $x = 3x_m/4$. The third primordium arises at one of these points. Intermediate values of λ produce minima that lie in intermediate positions.

In Table 19.2 the angle $360(x_2 - x_1)/x_m$ between primordia P_1 and P_2 is shown for a range of γ and λ, so that the relative importance of these parameters in determining x_2 through eqn (19.11f) can be seen. From eqn (19.11f), it can be shown that (for $\lambda \neq 0$)

$$\begin{aligned} \gamma \to 0 \qquad & \frac{x_2}{x_m} = \frac{1 + \lambda/2}{1 + \lambda} \\ \gamma \to \infty \qquad & \frac{x_2}{x_m} = \frac{3}{4} - \frac{1}{2\gamma x_m} \ln \lambda \end{aligned} \tag{19.11g}$$

Table 19.2. Calculated angle between the second and third primordia $P_1\hat{O}P_2$

λ	$\gamma = 0$	$\gamma = 0.1$	$\gamma = 0.2$	$\gamma = 0.5$	$\gamma = 1.0$	$\gamma = 2.0$	$\gamma = 5.0$	$\gamma = 10.0$	$\gamma = \infty$
0	180	180	180	180	180	180	180	180	
0.01	178.2	178.2	178.2	178.2	178.1	177.9	175.9	163.6	90
0.1	163.6	163.6	163.6	163.5	163.1	161.6	151.6	130.3	90
0.2	150.0	150.0	150.0	149.8	149.3	147.3	136.2	118.4	90
0.5	120.0	120.0	120.0	119.9	119.5	118.0	110.9	102.3	90
0.7	105.9	105.9	105.9	105.8	105.6	104.7	100.9	96.3	90
0.9	94.7	94.7	94.7	94.7	94.6	94.4	93.2	91.9	90
1.0	90	90	90	90	90	90	90	90	90

The competing effects of the first and second primordia are shown for a range of the parameters γ and λ. See Fig. 19.13 and eqn (19.11f).

For small λ, and writing $x_2 = x_m - \Delta x_2$, it can be proved that

$$\gamma \Delta x_2 \approx \lambda \sinh\left(\frac{\gamma x_m}{2}\right). \tag{19.11h}$$

Varying λ from zero to unity (Table 19.2) causes the minimum in the field to move smoothly from $x_2 = x_m$ (180°) to $x_2 = 3x_m/4$ (90°). Varying γ from zero to unity causes very little change in x_2, although further increases in γ cause the angle to fall rapidly towards its limiting value of 90° ($x_2 = 3x_m/4$). If γ is large (eqn (19.11g)) sources of unequal magnitude compete as through they were equal.

The combined effects of a large number of primordia in determining the position of the next primordium are considered later. The older primordia may contribute weakly to the field, and not influence the position of the next primordium significantly. Much computing time can be saved by neglecting weak primordia, and eqn (19.11h) allows us to estimate the positional error incurred by this approximation. With $\gamma = 2$, $x_m = 1$, and $\lambda = 0.0005$

$$\Delta x_2 \approx 0.6\lambda = 0.0003. \tag{19.11i}$$

This is equivalent to an angle of 0.1°. If angular errors of 0.1° are tolerated, then sources that are less than 0.05 per cent of the strongest source can be ignored for $\gamma = 2$.

The third primordium P_2 can be placed at either of the two equivalent positions shown in Fig. 19.13; we select $x_m/2 < x_2 < x_m$. From the first three primordia, the morphogen field and its gradient are (eqns (19.10l) and (19.10m))

$$M(x) = s_0 \cosh\left[\gamma\left(x - \frac{x_m}{2}\right)\right] + s_1 \cosh(\gamma x)$$
$$+ s_2 \cosh\left[\gamma\left(x - x_2 + \frac{x_m}{2}\right)\right] \tag{19.11j}$$

and

$$\frac{dM}{dx}(x) = \gamma s_0 \sinh\left[\gamma\left(x - \frac{x_m}{2}\right)\right] + \gamma s_1 \sinh(\gamma x)$$

$$+ \gamma s_2 \sinh\left[\gamma\left(x - x_2 + \frac{x_m}{2}\right)\right] \tag{19.11k}$$

for $0 < x < x_m/2$. Equating the gradient of the field to zero defines the position x_3 of the fourth primordium according to

$$\tanh(\gamma x_3) = \frac{s_0 \sinh(\gamma x_m/2) + s_1 + s_2 \sinh[\gamma(x_2 - x_m/2)]}{s_0 \cosh(\gamma x_m/2) + s_1 + s_2 \cosh[\gamma(x_2 - x_m/2)]}. \tag{17.11l}$$

As x_2 was chosen to lie in the range $x_m/2 < x_2 < x_m$, x_3 lies in the range $0 < x_3 < x_m/2$.

19.7.3 *The initiation of higher primordia*

Let there be j primordia, $P_0, P_1, \ldots, P_{j-1}$, at positions $x_0, x_1, \ldots, x_{j-1}$; let s_i be the morphogen source strength of primordium P_i. From eqns (19.10l) and (19.10m), the field and its gradient are

$$M(x) = \sum_{i=0}^{j-1} s_i \cosh\left[\gamma\left(x - x_i - \frac{f_i x_m}{2}\right)\right] \tag{19.12a}$$

and

$$\frac{dM(x)}{dx} = \sum_{i=0}^{j-1} \gamma s_i \sinh\left[\gamma\left(x - x_i - \frac{f_i x_i}{2}\right)\right]. \tag{19.12b}$$

Let $x = x'$ be a minimum point of the field M. From eqn (19.12b)

$$\tanh(\gamma x') = \frac{\sum_{i=0}^{j-1} s_i \sinh[\gamma(x_i + f_i x_m/2)]}{\sum_{i=0}^{j-1} s_i \cosh[\gamma(x_i + f_i x_m/2)]} \tag{19.12c}$$

From eqns (19.12a) and (19.12c) the morphogen concentration at $x = x'$ is

$$M(x') = \cosh(\gamma x') \sum_{i=0}^{j-1} s_i \cosh\left[\gamma\left(x_i + \frac{f_i x_m}{2}\right)\right]$$

$$- \sinh(\gamma x') \sum_{i=0}^{j-1} s_i \sinh\left[\gamma\left(x_i + \frac{f_i x_m}{2}\right)\right]. \tag{19.12d}$$

The morphogen field may have a number of minima with different x' values; the lowest minimum is assumed to be where the initiation of the $(j + 1)$th primordium P_j occurs. It can be shown that, with j primordia present, there are at most j minima, with zero or one minimum lying in each of the j intervals between the j primordia.

To evaluate eqns (19.12c) and (19.12d) the source strengths s_i must be assigned values. Equation (19.11d) is extended, with

$$s_{j-1} = 1, \quad s_{j-2} = \lambda, \ldots, s_i = \lambda^{j-1-i}, \ldots, s_0 = \lambda^{j-1}. \tag{19.12e}$$

The older is a primordium, the weaker it becomes as a source of morphogen. Note that by making each primordium source become weaker with the passage of time, a degree of time dependence is put into the system; however, we still calculate the morphogen field using the steady state assumption (eqn (19.10d)), which implies that the sources are of constant strength. This is valid if the rate at which the field equilibrates is fast compared with the rates at which the sources decline in strength. This point is further discussed later.

19.7.4 *Simulation of phyllotaxis*

Two methods were used for obtaining numerical solutions to the model. The first of these consisted of simply generating successive primordia, starting from the first primordium P_0. The divergence angle (the angle between successive primordia) generally settled down quite quickly to a constant value (Table 19.3, first four rows), although for some parameter values a stable solution was not obtained (Table 19.3, last row). However, commonsense tells us that solutions corresponding to a constant divergence angle should exist, although the definition of such solutions must become increasingly tenuous as $\lambda \to 1$, $\gamma \to 0$, or $\gamma \to \infty$. Convergence to a steady state solution might not be given by the use of a deterministic method for generating primordia sequentially for certain parameter values; also, the presence of computer rounding errors may be a contributing factor. To overcome this difficulty, a self-consistent method was employed to find solutions corresponding to a constant divergence angle. This latter method allows the evaluation of all steady state solutions, although many of these may not be accessible to the first method; presumably all such solutions might be obtained with certain probabilities if stochastic elements were introduced.

Sequential generation of primordia This method is applied essentially as outlined in eqns (19.12), and some typical solutions are shown in Table 19.3. As previously discussed, the potential field may have several minima, and the angles in Table 19.3 are obtained by placing the next primordium at the lowest of these minima. This process can be interpreted physiologically as follows. A primordium is initiated when the morphogen field falls to a certain threshold value; this will be reached first for the lowest of the several minima in the field. When a primordium has been initiated, it then acts as a source of morphogen, raising the morphogen field level and preventing the immediate initiation of further primordia. A new steady state morphogen field is set up, and finally the next primordium is initiated when the lowest of the minima in the new field falls to the critical value.

The minima in the morphogen field may not be very different, and it is possible that some irregularity in the system may cause primordial initiation to occur at a minimum which is not the lowest present. This sort of effect can result in a

Table 19.3. Sequential generation of primordia

	Angle $P_j\hat{O}P_{j+1}$ subtended by primordia P_j and P_{j+1} at the origin O (deg)											Limiting angle (deg)
λ	0–1	1–2	2–3	3–4	4–5	5–6	6–7	7–8	8–9	9–10	10–11	
0.1	180.0	151.6	159.2	157.5	157.9	157.8	157.9	157.9	157.9	157.9	157.9	157.9
0.2	180.0	136.2	152.8	148.4	149.7	149.5	149.4	149.5	149.5	149.5	149.5	149.5
0.5	180.0	110.9	156.0	141.5	149.6	133.2	154.7	130.7	140.6	140.8	138.8	139.9
0.7	180.0	100.9	165.1	139.5	160.1	118.0	168.4	117.3	138.2	141.3	134.9	138.2
0.8	180.0	96.8	170.1	138.2	166.7	108.8	174.2	244.3	222.6	202.5	210.8	unstable

The angles between successive primordia are given for the first 12 primordia. The parameter γ is 5 throughout. The values of the parameter λ are given in the far left-hand column, and the limiting angles are given in the far right-hand column; the latter can be compared with the ideal Fibonacci angle of 137.5°. With $\lambda = 0.8$, the angle showed no sign of settling down even after initiating the 80th primordium (but see Marzec and Kappraff 1983, p. 226).

different limiting angle being approached, and an example of this is given in Table 19.4. Thus it seems possible for a model, embodying the same physiological and biochemical features, to give different behaviour at the macroscopic level as a result of a change that might be barely discernible at the microscopic level. Schwabe (1971) demonstrated that it is possible to modify phyllotaxis by means of chemical treatment, and this supports the idea that a viable model of phyllotaxis should contain implicitly several possible phyllotaxis types, some of which will be more favoured than others.

A self-consistent method for steady state solutions A constant divergence angle of β is assumed, corresponding to $\Delta x = x_{j+1} - x_j$ of $x_m\beta/2\pi$. Consider a large number of primordia, say 101: $P_0, P_1, \ldots, P_{100}$. Place the last primordium at the origin, so that $x(P_{100}) = 0$, the next to last primordium at $x(P_{99}) = x_m - \Delta x$, and so on, with $x(P_i) = (\text{integer})x_m - (100 - i)\Delta x$. The integer is chosen so that $x(P_i)$ lies in the range $0 \leqslant x(P_i) < x_m$. The source strengths of the primordia are assigned values according to eqn (19.12e) with $j = 101$. Using eqns (19.12), the resulting field is calculated and the lowest minimum is obtained: this determines the position for the 102nd primordium, $x(P_{101})$. If the angle between P_{101} and P_{100} is equal to the constant divergence angle β originally assumed, then β must be one of the required solutions to the problem.

In order to keep the computing time involved within acceptable limits, primordia whose source strengths were less than 0.05 per cent of the most recent primordia were neglected, giving an error of about 0.2° for $\gamma = 5$, decreasing to about 0.1° as γ decreases (eqn (19.11i)). Steady state solutions are given in Table 19.5 for a range of values of the two parameters γ and λ.

It will be noticed that the solutions lie in a number of quite well-defined hands: 78°–85°, 97°–101°, 105°–110°, 132°–134°, 136°–143°, and 146°–180°. One of these bands covers the Fibonacci angle of 137.51° and, bearing in mind the computational approximations that have been made, one steady state solution appears to approach the Fibonacci angle as $\lambda \to 1$. An angle close to the Fibonacci angle becomes a possible solution as soon as λ exceeds about 0.5. Also, solutions close to the other Fibonacci-related angles of 99.50° and 77.96° (eqn (19.6a)) are generated at the higher values of the parameters γ and λ. The model suggests that there are zones in which the divergence angles are unlikely to lie. In Table 19.5 the steady state solutions that can also be obtained by the sequential generation method with the same parameter values are italicized, and it is apparent that most solutions are not accessible to this method. It has so far been assumed that a primordium can arise at any point, as it is solely dependent on the level of the morphogen field. Later, a spatial requirement is postulated. This extra constraint might be regarded as ruling out many of the solutions of Table 19.5 on spatial grounds.

Marzec and Kappraff (1983) have carried out a more extensive simulation of phyllotaxis using this reaction–diffusion model. In general they corroborated the above results, which give rise to the Fibonacci angle β_0 (eqn (19.1f)) and the

Table 19.4. Sequential generation of primordia

4	Angle $P_j\hat{O}P_{j+1}$ subtended by primordia P_j and P_{j+1} at the origin O (deg)											Limiting angle (deg)
	0–1	1–2	2–3	3–4	4–5	5–6	6–7	7–8	8–9	9–10	10–11	
A	180.0	104.7	163.2	140.5	159.7	120.5	165.3	149.9	147.6	150.7	148.5	148.8
B	180.0	104.7	163.2	140.5	159.7	120.5	135.6	142.9	138.9	137.9	139.6	139.2

The effect of a small perturbation of the system in producing a different limiting divergence angle is shown. In both cases $\gamma = 2$ and $\lambda = 0.7$. In A the lowest of the minima is always chosen as giving the position of the next primordium, with a limiting divergence angle of 148.8°. In B it was assumed that some irregularity caused the eighth primordium (P7) to be initiated at a minimum adjacent to the lowest minimum; the resulting divergence angle is 139.2°.

Table 19.5. Steady state solutions obtained using a self-consistent method

λ	Angle (deg) $\gamma = 0.2$			$\gamma = 0.5$			$\gamma = 1.0$			$\gamma = 2.0$			$\gamma = 5.0$		
0.1	*165.3*			*165.2*			*164.9*			*163.9*			*157.9*		
0.2	*156.0*			*155.9*			*155.7*			*154.7*			*149.4*		
0.5	142.3	*148.1*		142.2	*148.1*		142.1	*148.0*		141.7	*147.7*		*139.9*	146.7	
0.7	109.7	139.3	*148.9*	109.6	139.3	*148.9*	109.6	139.3	*148.9*	109.3	139.2	*148.8*	106.7	*138.2*	148.7
													82.4	83.9	
0.9	100.7	106.8	109.4	101.0	106.6	109.3	101.0	106.5	109.3	101.0	106.5	109.3	99.8	100.5	
	133.6	137.6	139.2	133.6	137.7	139.2	133.8	137.7	139.2	133.5	137.6	139.2	105.6	106.4	109.1
	149.1	150.8	156.2	149.1	150.8	156.2	149.1	150.8	156.2	149.1	150.8	156.2	133.5	137.6	139.1
													149.1	150.8	156.2
0.95	99.7	101.0	105.5	82.5	99.7	101.0	82.5	99.7	100.6	82.5	84.2	99.7	78.9	82.2	83.9
	106.5	109.3	111.3	105.5	106.5	106.9	105.5	106.5	106.8	100.6	105.5	106.5	85.0	97.4	99.7
	133.1	137.0	137.5	109.3	109.7	111.3	109.3	109.8	111.3	106.8	109.3	111.3	100.4	105.5	106.5
	139.2	140.3	147.9	133.1	137.0	137.5	133.0	133.5	136.8	133.1	133.5	136.9	109.2	132.3	133.1
	149.1	150.8	156.2	139.2	140.3	147.8	137.5	139.2	140.3	137.5	139.2	139.5	133.5	136.9	137.5
				149.1	150.8	156.2	147.8	149.1	150.1	140.3	147.9	149.1	139.1	139.5	147.9
							151.8	156.2	156.8	150.8	151.8	156.2	149.1	151.1	151.8
							157.8			156.8	157.8		156.2	156.8	157.8

The angles are tabulated for a range of the parameters γ and λ. Solutions obtained using the sequential generation method without perturbations are given in italics.

related angles β_f (eqn (19.6a)). They also observed other phenomena which are worth mentioning here. Stable and unstable limit cycles of divergence angles were found—a period-n limit cycle is a sequence of n intervals (angles) which repeats itself. Various lengths of the period were seen. When the parameter λ was varied, complex bifurcation–stability relationships were observed. For large λ ($0.95 < \lambda < 1$) with the generation of sequential solutions, the system could oscillate aperiodically for several hundred leaves but always settled down into a stable limit cycle before 1000 leaves had been placed. For $\lambda = 1$, the system was non-dissipative—no leaf was ever 'forgotten', and the solution never converged to a stable orbit.

19.7.5 *Time-dependent solutions*

The steady state solution in eqn (19.10l) is only valid for a primordium with a constant source strength. It is likely that the rate of production of morphogen will change with time, owing to factors such as growth, senescence, or increasing remoteness from the zone of primordial initiation. The temporal behaviour may be comparatively complex in the case of, for example, non-spiral phyllotaxis, and the production of floral primordia. Time-dependent solutions are briefly considered in this section, and it is shown that, for simple exponential time dependence of the morphogen sources, the steady state theory can still be applied but with modified values of the parameters γ and λ.

It can be verified that the equation

$$M_i(x,t) = \frac{S_i e^{-\mu t}}{2D\gamma \sinh(\gamma x_m/2)} \cosh\left[\gamma\left(x - x_i - \frac{f_i x_m}{2}\right)\right] \tag{19.13a}$$

satisfies eqn (19.10c) if

$$\gamma = \left(\frac{k-\mu}{D}\right)^{1/2} \tag{19.13b}$$

for a morphogen source of strength $S_i e^{-\mu t}$ at position $x = x_i$. For a number of sources, each of which decays at the specific rate μ_i, eqn (19.12a) becomes

$$M(x,t) = \sum_{i=0}^{j-1} s_i \exp(-\mu_i t) \cosh\left[\gamma_i\left(x - x_i - \frac{f_i x_m}{2}\right)\right] \tag{19.13c}$$

where γ_i is given by eqn (19.13b) with μ_i replacing μ. If the constants μ_i are all the same and equal to μ, comparison of eqn (19.13c) with eqns (19.12a) and (19.12e) shows that the steady state formalism is valid if γ is calculated by eqn (19.13b) instead of eqn (19.10d), and if λ is replaced by $\lambda \exp(-\mu \Delta t_p)$ where Δt_p is the time interval between the initiation of successive primordia.

19.7.6 *Physiological applications*

Apex size and spatial competence The parameter x_m, which is the circumference of the zone of primordial initiation, is a measure of apex size. As discussed after eqn (19.10j), if the size of the apex increases, the morphogen level is depressed and primordial initiation will occur earlier than it would have done in an apex of constant size, other things being equal. The model predicts that primordial initiation should take place at an increasing rate in an apex whose size is increasing.

The model, developed primarily for the case where single primordia are produced at approximately equal time intervals, does not easily describe situations where primordia are produced in bursts, as in some types of non-spiral phyllotaxis or in certain stages of flowering. To include such phenomena, an additional hypothesis, called *spatial competence*, is introduced.

Spatial competence is postulated to prevent indiscriminate production of primordia whenever the morphogen level falls below the threshold value. Three requirements must now be satisfied for primordial initiation: first, the tissue must be competent in the usual sense for primordial development; second, the morphogen level must be below the threshold value; lastly, spatial competence requires that there must be a sufficient amount of competent tissue so that self-organization into a primordium is possible. This is equivalent to saying that in its early stages a primordium has a certain minimum size, and a smaller primordium may not exist. There are sound precedents for this type of behaviour in physics with what are known as cooperative phenomena, examples of which are ferromagnetism and superconductivity. In these cases, bulk matter may exhibit unique properties which are not manifest at the atomic or molecular levels; a minimum amount of material must be present in order that the interactions, which give rise to the special bulk properties, are effective.

Non-spiral phyllotaxis Here a qualitative account is given of how the model, supplemented by the idea of spatial competence, is able to account for more complex forms of phyllotaxis. A whorled system, where whorls of six leaves are produced, is first considered. Let us take a growing apex in which there is a morphogen field due to some fairly distant primordia. Suppose that the apex is large enough to support six primordia on the same horizontal level, but that the morphogen field is above the threshold level where primordial initiation may occur. The apex continues to grow and the morphogen level falls. Eventually the threshold value is reached and a primordium is initiated. The morphogen field is still falling and allows the initiation of a further five primordia in rapid succession before the spatial competence requirement prevents further primordial initiation. Production of morphogen by the six new primordia gradually builds up and the morphogen field goes above its threshold value. At the same time the apex (more strictly the part of the apex that is competent to initiate primordia), which is probably considerably reduced in size by the production of the six primordia, continues to grow. Eventually it becomes large enough to fulfil the spatial requirements for a new primordium, but now the morphogen field prevents initiation from occurring. A comparatively long period of time elapses before the morphogen field approaches the threshold level, and in this time the apex grows to an extent where the spatial competence requirement once again allows the initiation of six primordia. The cycle of events is then gone through anew.

The leaves in a whorl generally appear to be symmetrically arranged with respect to the azimuth. However, the present model implies that the primordia, at least in their earlier stages of development, might not lie at exactly equal angles but may be disposed according to the first five angles in Table 19.3. It seems reasonable to assume that outgrowth could cause the primordia to take up their final configuration (cf. Ridley's jostling procedure, p. 547).

The reaction–diffusion model account easily for the phyllotaxes referred to as spiral, distichous, and spirodistichous; the model, extended by the spatial competence concept, is able to describe decussate, bijugate, and whorled phyllotaxes.

Bract formation and flowering When a vegetative apex becomes reproductive, there is an almost discontinuous change to a much higher rate of primordial initiation. There are a number of ways in which this could be achieved in a model, although it might be unwise to develop speculative hypotheses too far without support and guidance from experimental data. For instance, it would be possible to build switch mechanisms into a model as suggested in Chapter 6 (particularly Sections 6.4 and 6.5). Alternatively, the phyllotaxis model could be extended by interfacing it with the models of apical growth and stem extension. For example, apex size is an important parameter of the phyllotaxis model, since it affects both the morphogen level and the spatial competence requirement; quite subtle changes in apex size or growth rate could possibly cause the behaviour that characterizes reproductive growth.

Parameters γ and λ The parameter γ is the ratio of a rate of a chemical degradation to a diffusion constant (19.10d); as such it might be expected that γ should not vary widely as growth occurs, or indeed from plant to plant if the diffusible morphogen is the same. In contrast, λ (19.11d) is a conglomerate quantity that represents many aspects of the problem about which there is little or no information. λ describes the relative effectivness of temporally adjacent primordia in contributing to the morphogen field. The vertical separation of primordia through stem growth may be important; so might the radial separation, which is affected by apex growth. The morphogen source strength of a growing primordium may vary, and this could also influence λ. The diffusion constant appearing in the model refers to tangential movement; radial and vertical diffusion might be significant, and it is possible that they could be regarded as contributing to the parameter λ.

19.8 Minimization properties of morphogen fields

In Section 19.5 (p. 550) we discussed how the set of Fibonacci angles β_f of eqn (19.6a) gives rise to the most uniform spacing of points round a circle (Marzec and Kappraff 1983, pp. 215–18). Here we describe further aspects of the work of

Marzec and Kappraff (1983, pp. 218–24) who showed that certain measures of the morphogen field $M(x)$ have a minimum when sequential primordia are separated by the set of angles β_f.

Using the angular coordinate θ and an angle β between successive primordia, consider an infinite lattice of primordia at positions $\theta = 0, -\beta, -2\beta, -3\beta, \ldots, -n\beta, \ldots$. The primordium at the origin is called P_0, and it is assumed that primordium n contributes to the morphogen concentration with a weighting factor of λ^n $(0 < \lambda < 1)$. Let $M_\beta(\theta)$ be the morphogen field arising from this disposition of primordia with infinite n.

From symmetry considerations, the lattice of primordia $-1, -2, \ldots, -\infty$ produces a morphogen field of

$$\lambda M_\beta(\theta + \beta). \tag{19.14a}$$

Therefore

$$M_\beta(\theta) = \lambda M_\beta(\theta + \beta) + m(\theta), \tag{19.14b}$$

where $m(\theta)$ denotes the field from a primordium of unit strength at $\theta = 0$. $M_\beta(\theta)$ has a period of 2π, and can be written as a Fourier series:

$$M_\beta(\theta) = \sum_{q=-\infty}^{\infty} A_q \exp(iq\theta), \tag{19.14c}$$

where the A_q are Fourier coefficients, q is an integer index, and $i = (-1)^{1/2}$ (Churchill 1963). Thus

$$M_\beta(\theta + \beta) = \sum_{q=-\infty}^{\infty} A_q \exp\{iq(\theta + \beta)\}. \tag{19.14d}$$

Define the field $m(\theta)$ from a single primordium at the origin to be

$$m(\theta) = \sum_{q=-\infty}^{\infty} B_q \exp(iq\theta), \tag{19.14e}$$

where the B_q are also Fourier coefficients; $m(\theta)$ has period 2π and $m(\theta) = m(2\pi - \theta)$. Therefore substituting eqns (19.14c)–(19.4e) into eqn (19.14b) gives

$$A_q = \lambda A_q \exp(iq\beta) + B_q, \tag{19.14f}$$

whence

$$A_q = \frac{B_q}{1 - \lambda \exp(iq\beta)}. \tag{19.14g}$$

Substituting eqn (19.14g) in eqn (19.14c) gives

$$M_\beta(\theta) = \sum_{q=-\infty}^{\infty} \frac{B_q \exp(iq\theta)}{1 - \lambda \exp(iq\beta)}. \tag{19.14h}$$

Equation (19.14h) can be represented differently. The primordia, which are the sources of the morphogen field, can be treated as an infinite series of weighted delta functions ((1.5f), p. 18). Consider the source function

$$\sigma(\theta) = \sum_{j=0}^{\infty} \lambda^j \delta(\theta + j\beta). \tag{19.14i}$$

This gives a spike of unit area at $\theta = 0$ with $j = 0$, a spike of area λ at $\theta = -\beta$ with $j = 1$, a spike of area λ^2 at $\theta = -2\beta$ with $j = 2$, and so on. The part $\sigma(\theta')$ of the source function at θ' gives rise to a morphogen field of $\sigma(\theta')\, m(\theta - \theta')$; the total morphogen field arises from all part of the source functions and is

$$M_\beta(\theta) = \int_0^{2\pi} \sigma(\theta') m(\theta - \theta')\, \mathrm{d}\theta'. \tag{19.14j}$$

Returning to eqn (19.14h), the conditions on eqn (19.14e) for $m(\theta)$ require that the B_q are real and $B_q = B_{-q}$. Equation (19.14h) can therefore be developed to give

$$\begin{aligned} M_\beta(\theta) &= \frac{B_0}{1-\lambda} + \sum_{q=1}^{\infty} \left[\frac{B_q \exp(\mathrm{i}q\theta)}{1 - \lambda \exp(\mathrm{i}q\beta)} + \frac{B_q \exp(-\mathrm{i}q\theta)}{1 - \exp(-\mathrm{i}q\beta)} \right] \\ &= \frac{B_0}{1-\lambda} + 2 \sum_{q=1}^{\infty} \frac{B_q \{\cos(q\theta) - \lambda \cos[q(\theta - \beta)]\}}{1 + \lambda^2 - 2\lambda \cos(q\beta)}, \end{aligned} \tag{19.14k}$$

using

$$\begin{aligned} e^{\mathrm{i}x} &= \cos x + \mathrm{i} \sin x \\ e^{-\mathrm{i}x} &= \cos x - \mathrm{i} \sin x. \end{aligned} \tag{19.14l}$$

Consider now the morphogen concentration at distance β beyond the most recent primordium at $\theta = 0$. From eqn (19.14k) this is given by

$$M_\beta(\theta = \beta) = \frac{B_0}{1-\lambda} + \frac{1}{\lambda} \sum_{q=1}^{\infty} B_q \left[-1 + \frac{1 - \lambda^2}{1 + \lambda^2 - 2\lambda \cos(q\beta)} \right]. \tag{19.14m}$$

The particular reaction–diffusion scheme assumed will determine the coefficients $B_0, B_1, B_2, \ldots,$ in the Fourier expansion of the morphogen field from a single primordium. In general the B_q decrease with increasing q. Marzec and Kappraff (1983) made assumptions for B_q (e.g. $B_q = \exp(-q), \exp(-0.1q^2), (9 + q^2)^{-1}$) and for λ (e.g. $\lambda = 0.7$) and evaluated eqn (19.14m) numerically. They found that minima occur for values of β close to β_0 (eqn (19.1f)) and for some members of the set of the related angles β_f of eqn (19.6a). This behaviour can be seen to be a consequence of the form of the denominator in eqn (19.14m). This is a minimum (giving a maximum in the function) if $q\beta = m2\pi$, where m is an integer. However, if β is perturbed from this value of $2\pi m/q$ to

$$\beta \approx \beta_f = 2\pi \frac{m\tau + a}{q\tau + b}, \tag{19.14n}$$

with $|aq - bm| = 1$, then, as discussed in Section 19.5 (p. 550), the qth and bth terms of the sum in eqn (19.14m) represent the $(k-1)$th and kth denominators of the convergents of $\beta_f/(2\pi)$ with $a_{k+1} = 1$; these values of $q\beta$ avoid the origin, giving large denominators in eqn (19.14m) and resulting in minimum values in the sum. Which of the β_f give a minimum in $M_{\beta_f}(\theta = \beta_f)$ depends upon the values of λ and the B_q; Marzec and Kappraff (1983) report the results of several numerical experiments which demonstate that the behaviour of the simple reaction–diffusion model (p. 555) is of wider significance in that more complex reaction–diffusion models can also be expected to lead to similar results.

19.9 Cylindrical model of contact pressure in phyllotaxis

Adler (1974, 1977) has developed this approach to phyllotaxis in two substantial and important papers. Contact pressure occurs when the position of a primordium is determined by the positions of neighbouring primordia as a result of physical contact. If primordium P_n is in contact with primordia P_{n+p} and P_{n+q} then the contact parastichy numbers are (p, q). Adler showed that if (p, q) are the parastichy numbers when contact pressure first occurs, then the continuation and intensification of contact pressure can cause the parastichy numbers to change successively, as in a Fibonacci series, to $(q, p+q), (p+q, p+2q), (p+2q, 2p+3q), \ldots$. If the divergence angle initially has a value $2\pi\alpha$ and α can be expressed as the continued fraction $[a_0, a_1, a_2, \ldots, a_k + d_k]$ (eqns (19.7b)–(19.7d)), then contact pressure will ensure that α converges on the value $[a_0, a_1, a_2, \ldots, a_k + 1/\tau]$ where τ is the golden ratio (eqn (19.1c)). Thus a necessary and sufficient condition for the divergence angle to converge on the ideal angle of 137.5° is that the parastichy numbers (p, q), where contact pressure first occurs, should be consecutive members of the Fibonacci series (19.1a). An outline of this theory is given in this section. The cylindrical coordinates (r, θ, z) are employed, and primordium P_0 is at the position $r = a$, $\theta = 0$, $z = 0$.

Consider a helix wound around a cylinder of constant radius a as in Fig. 19.14. Let $i_{n\alpha}$ be the integer nearest to $n\alpha$. Consider that the primordia are projected onto the circumference of the disc at $z = 0$ and let $c_{i,j}$ be the distance round the circumference from primordium P_i to primordium P_j. Hence

$$c_{0,n} = 2\pi a|(n\alpha - i_{n\alpha})|. \tag{19.15a}$$

Let $z_{i,j}$ be the z-axis separation between primordia P_i and P_j, and suppose that successive primordia are separated by h along the z axis. Then

$$z_{0,n} = nh. \tag{19.15b}$$

Then, by Pythagoras,

$$d_{0,n}{}^2 = (nh)^2 + (2\pi a)^2(n\alpha - i_{n\alpha})^2,$$

where $d_{i,j}$ denotes the distance between primordia i and j measured on the surface of the cylinder. To simplify the equations, distances are measured in units of $2\pi a$, so that this equation becomes

$$d_{0,n}{}^2 = (nh)^2 + (n\alpha - i_{n\alpha})^2. \tag{19.15c}$$

This has a minimum value with respect to h when $h = 0$, and

$$d_{0,n}{}^2 = c_{0,n}{}^2 = (n\alpha - i_{n\alpha})^2 \tag{19.15d}$$

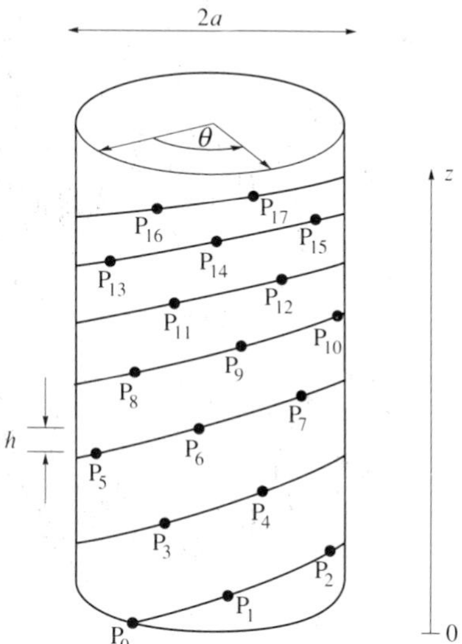

Fig. 19.14. Primordia P_0, P_1, P_2, ... are placed along a helix wrapped round a cylinder of radius a; the angular separation shown here is the Fibonacci angle 137.5°; h is the vertical distance between successive primordia; the cylinder surface is shown unwrapped.

In Fig. 19.15, ${c_{0,n}}^2$ is plotted against α over the range $0 \leqslant \alpha \leqslant 0.5$ and for $n =$ 1, 2, 3. Over this range of α

$$\begin{aligned} {c_{0,1}}^2 &= \alpha^2 \\ {c_{0,2}}^2 &= (2\alpha)^2 \quad \text{or} \quad (2\alpha - 1)^2 \\ {c_{0,3}}^2 &= (3\alpha)^2 \quad \text{or} \quad (3\alpha - 1)^2. \end{aligned} \tag{19.15e}$$

The range of α is divided into three segments: 0–0.25, 0.25–0.4, and 0.4–0.5, where P_1, P_3, or P_2 is *potentially* the closest primordium to P_0, depending upon the value of h. In fact, with just the three primordia P_1, P_2, and P_3 over the range $0 < \alpha < 0.25$, P_1 is always the closest primordium to P_0, irrespective of the value of h; however, over the rest of the range of α, the value of h determines whether P_1, P_2, or P_3 is nearest to P_0. This can be seen from Fig. 19.14, where collapsing the spiral (reducing h) can cause P_3, P_5, P_8, or P_{13} to approach P_0 closely. It is clear from eqns (19.15a) and (19.15d) or Fig. 19.15 that the higher the primordial number n, the more frequently $c_{0,n}$ returns to zero as α varies.

It Fig. 19.16 the $d_{0,n}$ for $n = 1, 2, \ldots, 9$ are drawn over a restricted range of α and with $h = 0.01$. The principal effect of varying h is to move these curves around vertically. With the help of this diagram and Fig. 19.14, the contact pressure mechanisms can now be outlined. It is assumed that primordium P_0 is the primordium that is first involved in the effects of contact pressure. Primordia are fed into the system at P_0 (Fig. 19.14). In the absence of contact pressure, α (equal to the divergence angle divided by 2π) and vertical separation h between successive primordia have values determined by other mechanisms; the primordia

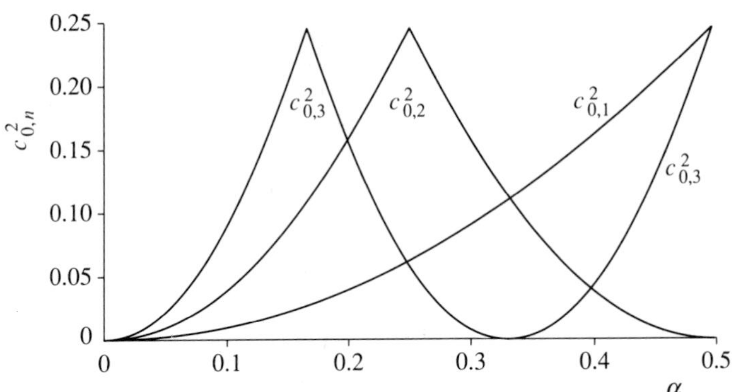

Fig. 19.15. The distance round the circumference between primordia P_1, P_2, and P_3 and the primordium P_0 at the origin is shown as a function of $\alpha(=$ divergence angle$/2\pi)$, using eqns (19.15e).

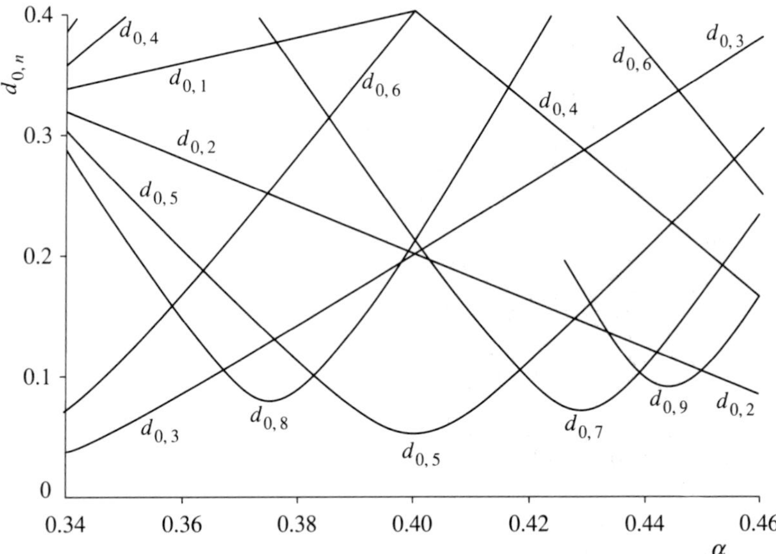

Fig. 19.16. Distance $d_{0,n}$ between primordium P_n and the primordium P_0 at the origin calculated using eqn (19.15c) with $h = 0.01$, as a function of $\alpha(=$ divergence angle$/2\pi)$.

expand at some rate. At some instant primordium P_{n_c} makes physical contact with P_0. There are two possibilities: either the higher primordia which might be closer to P_0 than P_{n_c} do not yet exist (for $d_{0,n} < d_{0,n_c}$, $n > n_c$, the P_n are not present), or h is such that these potentially closer primordia (with $c_{0,n} < c_{0,n_c}$) are in fact more distant ($d_{0,n} > d_{0,n_c}$). In Fig. 19.16 (which is drawn for $h = 0.01$) suppose that $\alpha = 0.36$ and that there are four primordia present when contact pressure occurs from $d_{0,3}$. It is assumed that α readjusts its value to allow more space; the $d_{0,3}$ curve is followed to the right until we come to another contact pressure point, in this case $d_{0,2}$, and a (2, 3) parastichy system is established. With the passage of time more primordia are added: if h remains at 0.01, α will have to readjust again from 0.4 (the intersection of $d_{0,3}$ and $d_{0,2}$) to 0.376 (the intersection of $d_{0,2}$ and $d_{0,5}$) when P_5 is present, and again at a later time P_8 causes α to move to 0.382 (the intersection of $d_{0,5}$ and $d_{0,8}$) with $d_{0,5} = d_{0,8} \approx 0.1$. If h remains at 0.01 no further adjustment is necessary since $d_{0,13} \geqslant 13 \times 0.01$ is always larger then this value. If it is assumed that h falls in the developing apex, then eventually the $d_{0,13}$ curve (not drawn in Fig. 19.16) descends and compels a shift from the (5, 8) point to the (8, 13) point.

The phyllotaxis system (p, q) is established, and whether this changes with time depends upon α, the quantitative temporal characteristics of h, and primordial expansion. Adler (1974, 1977) has established rigorously the ability of the contact pressure mechanism to furnish an explanation for Fibonacci phyllotaxy; many of the results he derives can be seen intuitively from consideration of Figs 19.14, 19.15, and 19.16.

19.10 Next-available-space model of phyllotaxis on a disc

Primordia can be treated as systems of circles arranged in two dimensions on a disc which represents the developing apex. It can be postulated that the circles are in contact with one another; each additional primordium is placed in the next available space which is of sufficient size. This geometrical approach has been pursued numerically (rather than analytically) by Williams and Brittain (1984), and their results are in general agreement with the more theoretical considerations discussed in Sections 19.3 (p. 541), 19.5 (p. 550), and 19.9 (p. 572). We give a brief account of their work, which provides many fine diagrams of different phyllotaxis systems and much historical and botanical background which has been largely omitted from this chapter.

Primordia of a constant radius r_P are initiated at a constant radius r_A from the centre of the apex which is represented by a disc. The primordium : apex ratio ϕ is defined by

$$\phi = r_P/r_A. \tag{19.16a}$$

The plastochrone ratio p is the ratio of the distances from the centre of the system to the centres of successive primordia, and

$$p = r_{n-1}/r_n \tag{19.16b}$$

where primordium n is younger than primordium $n-1$; therefore $p > 1$. The divergence angle β between successive primordia is not constant, but is determined by geometrical criteria of space. New primordia arise at approximately equal units of time, and the radius of a primordium and also the distance of its centre from the centre of the disc expand exponentially relative to this time scale; r_P and r_A remain constant.

Williams and Brittain (1984) describe static models of the ideal phyllotactic systems found in nature. More interestingly, they develop dynamic models in which primordia are added to the system sequentially using the rules outlined above.

The simple dynamic models with a constant plastochrone ratio were found only to give rise to low-order parastichy systems (1, 2) and (2, 3). To obtain transitions to higher-order systems, it was necessary to assume that the plastochrone ratio decreased with time. It will be remembered that in the contact theory model using the cylindrical representation of the last section, it is necessary to assume that the vertical separation h between primordia falls with time to obtain transitions to higher-order systems and convergence on the ideal angle. While the simulation of some natural systems, such as whorls, needed the addition of secondary constraints, the authors interpreted their results as demonstrating that simple packing considerations on a disc could give rise to Fibonacci and related systems of phyllotaxis.

Exercises

19.1. Another way of generating the Fibonacci numbers is by means of a branching or reproductive process. Suppose that we have an organism with two states: young (y) and mature (m). It takes one time unit for a young organism (y) to become mature (m), and one time unit for a mature organism (m) to produce one offspring (y); the mature organisms live for ever. Given that at time $t = 1$ there is one young organism (y), generate the total numbers of organisms at successive time intervals.

19.2. By considering three successive terms of the Fibonacci series (eqns (19.1a) and (19.1b)) derive eqn (19.1c), i.e.

$$\tau = \lim_{i \to \infty} \left(\frac{F_i}{F_{i-1}} \right) = \frac{\sqrt{5} + 1}{2}.$$

19.3. The Fibonacci ratio τ can also be considered geometrically or trigonometrically. Consider the construction within the pentagon in Fig. 19.17, and show that U divides the line QS so that QS/QU = QU/US = τ. By considering triangle QUR, show that τ = QU/UR = $2\cos(\pi/5)$.

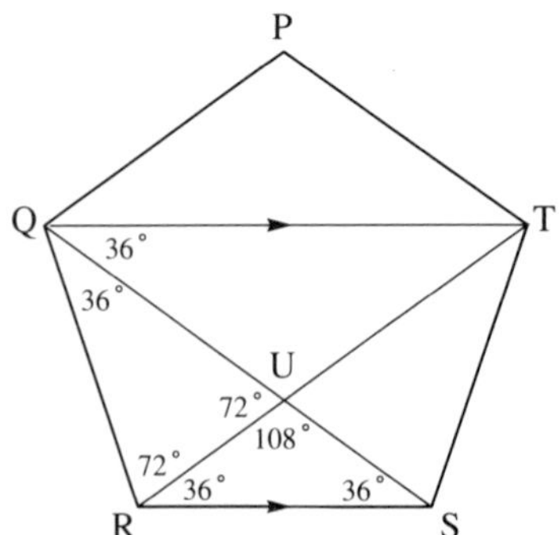

Fig. 19.17. PQRST is a regular pentagon. Various angles of $\pi/10$, $\pi/5$, and $3\pi/10$ have been marked. Note that QT $\parallel$ RS, QT = QS, QU = QR = RS, and that the triangles SRQ and SRU are isosceles and congruent, as are triangles QUR and QRT.

19.4. In plane polar coordinates (r, θ), the element of area $\mathrm{d}A$ is given by $\mathrm{d}A = \mathrm{d}r\, r\mathrm{d}\theta$ (see Fig. 19.1, p. 537). For the three spirals $r = a\theta$, $r = a\exp(\theta \cot \phi)$, and $r = a\theta^{1/2}$, eliminate r and $\mathrm{d}r$ from the expression for $\mathrm{d}A$ to show explicitly the dependence of $\mathrm{d}A$ on θ and $\mathrm{d}\theta$.

19.5. Demonstrate numerically, and also analytically, that the Fibonacci ratio τ can also be obtained from the continued fraction

$$\tau = 1 + \cfrac{1}{1 + \cfrac{1}{1 + \cfrac{1}{1 + \cdots}}}.$$

19.6. Express the number 1.42 as a continued fraction, as in eqn (19.7b) and the subsequent remarks.

19.7. Evaluate the convergents in the form $\alpha_k = p_k/q_k$ to the continued fraction found in the preceding exercise (S19.6a).

19.8. Find the intermediate convergents to the number 1.42, using the solution of the preceding exercise and (19.7n).

19.9. For $\alpha = 0.42$, corresponding to a divergence angle of 151.2°, place successive primordia at the points $0, 0.42, 0.84, \ldots, \{n\alpha\}, \ldots$, where $\{x\}$ denotes the fractional part of x, and investigate the way in which the line segment $[0, 1]$ is subdivided. Note that bad breaks and intermediate sized intervals occur, and are connected with the existence of intermediate convergents in 0.42 expressed as a continued fraction (Exercises 19.6–19.8).

19.10. Verify that the irrational number $\alpha = \sqrt{2} - 1$ can be written as the continued fraction

$$\cfrac{1}{2 + \cfrac{1}{2 + \cdots}}.$$

Find the convergents and the intermediate convergents to this number. As in Exercise 19.9, investigate the way in which the points $\{n\alpha\}$, $n = 0, 1, 2, \ldots$, subdivide the interval $[0, 1]$.

Solutions to exercises

Chapter 1

1.1. Define a second state variable μ by

$$\mu = \mu_0 e^{-Dt}. \tag{S1.1a}$$

Differentiating once and eliminating t gives

$$\frac{d\mu}{dt} = -D\mu, \tag{S1.1b}$$

with $\mu = \mu_0$ at $t = 0$. The two rate-state equations defining the system are now

$$\frac{dW}{dt} = \mu W \qquad \frac{d\mu}{dt} = -D\mu, \tag{S1.1c}$$

with $W = W_0$ and $\mu = \mu_0$ at $t = 0$. A biological interpretation of these equations is that the growth machinery, represented by the dry matter specific growth rate μ, is decaying at a specific rate D, possibly as a result of differentiation or senescence.

1.2. The integral equation can be written as

$$Z(t) = e^{-kt} \int_{-\infty}^{t} kX(t')\exp(kt')\,dt' = \exp(-kt)[u(t) - u(-\infty)], \tag{S1.2a}$$

where u is the indefinite integral of $kX(t')\exp(kt')$, so that

$$\frac{du}{dt} = kX(t)e^{kt}. \tag{S1.2b}$$

Differentiating both sides with respect to t, therefore, gives

$$\frac{dZ}{dt} = -ke^{-kt}[u(t) - u(-\infty)] + e^{-kt}\frac{du}{dt}. \tag{S1.2c}$$

The first term on the right is, from eqn (S1.2a), equal to $-kZ$; the second term, with eqn (S1.2b), is kX. Thus eqn (S1.2c) becomes

$$\frac{dZ}{dt} = k(X - Z). \tag{S1.2d}$$

Z can be described as a memory function since Z 'follows' the value of X but with a time delay of $1/k$; we could say that Z is taking a running average of X over the last $1/k$ time units.

1.3. Euler's method of eqn (1.7d) is

$$x(t + \Delta t) = x(t) + \Delta t \frac{\mathrm{d}x}{\mathrm{d}t}. \tag{S1.3a}$$

Integration of $\mathrm{d}x/\mathrm{d}t = -x$ with $x = 1$ at time $t = 0$ gives the analytic solution

$$x = \mathrm{e}^{-t}. \tag{S1.3b}$$

With $\Delta t = 0.1$, a table can be constructed, working line by line from left to right:

t	x	$\mathrm{d}x/\mathrm{d}t$	$\Delta t\, \mathrm{d}x/\mathrm{d}t$	e^{-t}
0	1.000	−1.000	−0.1000	1.000
0.1	0.900	−0.900	−0.0900	0.905
0.2	0.810	−0.810	−0.0810	0.819
0.3	0.729	−0.729	−0.0729	0.741

Comparison of the second column (from Euler's method) with the last column (from the analytic solution) shows the truncation error.

1.4. For $\mathrm{d}x/\mathrm{d}t = f(x) = -x$, define

$$f_1 = f(x) = -x \qquad x_1 = x + \Delta t \frac{\mathrm{d}x}{\mathrm{d}t}(x)$$
$$f_2 = \frac{\mathrm{d}x}{\mathrm{d}t}(x_1) \qquad g = \tfrac{1}{2}(f_1 + f_2) \qquad \Delta x = g\Delta t. \tag{S1.4a}$$

Updating is by means of

$$t \to t + \Delta t \quad \text{and} \quad x \to x + \Delta x. \tag{S1.4b}$$

As in the last example, a table is constructed, working line by line from left to right. We just give the first two rows:

t	x	f_1	x_1	f_2	g	Δx
0	1.000	−1.000	0.900	−0.900	−0.850	−0.095
0.1	0.905	−0.905	0.815	−0.815	−0.860	−0.086
0.2	0.819	...				

Comparison of the second column with the analytic solutions in the fifth column in the table in the solution to Exercise 1.3 shows that the accuracy is much improved.

1.5. From eqn (1.11q) the degrees of freedom are 84 − 4 = 80. The mean residual sum of squares (MRSS) is obtained by dividing the residual sum of squares ($R = 0.8$) by the degrees of freedom ($m - n = 84 - 4$):

$$\text{MRSS} = \frac{R}{m-n} = \frac{0.8}{80} = 0.01. \qquad \text{(S1.5a)}$$

Thus $\ln(y/Y) = (0.01)^{1/2} = \pm 0.1$, and $y/Y \approx 0.9$ or 1.1; on average there is about a 10 per cent difference between predicted and observed values.

The error variance (eqn (1.11i)) is

$$\sigma^2 = 0.4/50 = 0.008, \qquad \text{(S1.5b)}$$

and hence the variance ratio F is given by

$$F = \text{MRSS}/\sigma^2 = 0.01/0.008 = 1.25 \qquad \text{(S1.5c)}$$

with $\upsilon_1 = 80$ and $\upsilon_2 = 50$. The 10 per cent probability level is given by the 95 per cent point of the F distribution (e.g. France and Thornley 1984, p. 274), since F has been constrained to be greater than unity and lie in one half of the distribution. Interpolation in the F distribution table gives the 95 per cent point at about 1.55. Thus the model gives an acceptable fit to the observational data at the 10 per cent level.

1.6. The units of f_C are kg C (kg total dry matter)$^{-1}$. These units cannot be simplified, and it is not permissible to cancel out the kilograms for instance. The units of leaf area index are m^2 leaf (m^2 ground)$^{-1}$, and again m^2 cannot be cancelled out.

Chapter 2

2.1. (a) Proceeding as for the Michaelis–Menten equation, in steady state the concentration of $[ES_2]$ is constant, so that

$$k_{+1}[\text{E}][\text{S}]^2 = (k_{-1} + k_{+2})[\text{ES}_2]. \qquad \text{(S2.1a)}$$

The total concentration E_0 of enzyme is constant, which means that

$$E_0 = [\text{E}] + [\text{ES}_2]. \qquad \text{(S2.1b)}$$

Combining eqns (S2.1a) and (S2.1b) gives

$$[\text{ES}_2] = \frac{k_{+1}[\text{S}]^2 E_0}{k_{+1}[\text{S}]^2 + k_{-1} + k_{+2}}, \qquad \text{(S2.1c)}$$

which is equivalent to eqn (2.8b) with υ_m and K given by eqns (2.7g).

(b) Differentiating eqn (2.8b) twice with respect to $[S]$ gives

$$\frac{d^2\upsilon}{d[S]^2} = \frac{2\upsilon_m K^2(K^2 - 3[S]^2)}{(K^2 + [S]^2)^3}, \qquad \text{(S2.1d)}$$

and equating this to zero for a point of inflexion gives the required result.

2.2. (a) Proceeding as for the previous exercise,

$$k_{+1}[\mathrm{E}][\mathrm{S}]^n = (k_{-1} + k_{+2})[\mathrm{ES}_2] \tag{S2.2a}$$

and

$$E_0 = [\mathrm{E}] + [\mathrm{ES}_n]. \tag{S2.2b}$$

Combining these gives

$$[\mathrm{ES}_n] = \frac{k_{+1}[\mathrm{S}]^n E_0}{k_{+1}[\mathrm{S}]^n + k_{-1} + k_{+2}} \tag{S2.2c}$$

and the result follows.

(b) Differentiating eqn (2.8d) twice with respect to [S] gives (with $n \geqslant 2$)

$$\frac{\mathrm{d}^2 v}{\mathrm{d}[\mathrm{S}]^2} = \frac{n v_m K^2 [\mathrm{S}]^{n-2}[(n-1)K^n - (n+1)[\mathrm{S}]^n]}{(K^n + [\mathrm{S}]^n)^3}, \tag{S2.2d}$$

and hence the result.

(c) Differentiating eqn (2.8f) twice with respect to [S] gives (with $n \geqslant 2$)

$$\frac{\mathrm{d}^2 v}{\mathrm{d}[\mathrm{S}]^2} = -n v_\mathrm{m} K^2 [\mathrm{S}]^{n-2}\left[\frac{(n-1)K^n - (n+1)[\mathrm{S}]^n}{(K^n + [\mathrm{S}]^n)^3}\right], \tag{S2.2e}$$

and hence the result.

2.3. (a) Write eqn (2.9j) as

$$y = \frac{1}{2\theta}\left\{(\alpha x + y_\mathrm{m}) - (\alpha x + y_\mathrm{m})\left[1 - \frac{4\theta\alpha x y_\mathrm{m}}{(\alpha x + y_\mathrm{m})^2}\right]^{1/2}\right\}. \tag{S2.3a}$$

For small values of θ this can be written, using the binomial expansion (eqn (E2.3a)),

$$y = \frac{1}{2\theta}\left\{(\alpha x + y_\mathrm{m}) - (\alpha x + y_\mathrm{m})\left[1 - \frac{2\theta\alpha x y_\mathrm{m}}{(\alpha x + y_\mathrm{m})^2} + \cdots\right]\right\}$$

$$= \frac{\alpha x y_\mathrm{m}}{\alpha x + y_\mathrm{m}} + \text{terms of order } \theta \text{ and higher.} \tag{S2.3b}$$

Thus as $\theta \to 0$ eqn (2.9j) reduces to the rectangular hyperbola (eqn (2.9g)).

(b) When $\theta = 1$, eqn (2.9j) is

$$y = \tfrac{1}{2}\{(\alpha x + y_\mathrm{m}) - [(\alpha x - y_\mathrm{m})^2]^{1/2}\}$$

$$= \tfrac{1}{2}[(\alpha x + y_\mathrm{m}) - |\alpha x - y_\mathrm{m}|], \tag{S2.3c}$$

where $|z|$ is the modulus of z. Consideration of the separate cases where $\alpha x \leqslant y_\mathrm{m}$ and $\alpha x > y_\mathrm{m}$ leads to the required result.

(c) Differentiate eqn (2.9j) with respect to x:

$$\frac{dy}{dx} = \frac{1}{2\theta}\left\{\alpha - \frac{1}{2}[(\alpha x + y_m)^2 - 4\theta\alpha x y_m]^{-1/2}[2\alpha(\alpha x + y_m) - 4\theta\alpha y_m]\right\}. \quad \text{(S2.3d)}$$

Setting $x = 0$ gives the required result.

(d) Retaining only the terms in x, as $x \to \infty$, eqn (2.9f) becomes

$$-y\alpha x + \alpha x y_m = 0, \quad \text{(S2.3e)}$$

and the result follows.

2.4. (a) With

$$[E][A][B] = [E][B][A], \quad \text{(S2.4a)}$$

and using (2.10e) and (2.10g) on the left and (2.10f) and (2.10h) on the right gives the required result.

(b) Substitute for [E] from eqn (2.10k) in eqn (2.10e) to obtain

$$(E_0 - [EA] - [EB] - [EAB])[A] = K_A[EA]. \quad \text{(S2.4b)}$$

Dividing by [EAB] leads to

$$\left(\frac{E_0}{[EAB]} - \frac{[EA]}{[EAB]} - \frac{[EB]}{[EAB]} - 1\right)[A] = \frac{K_A[EA]}{[EAB]}. \quad \text{(S2.4c)}$$

Using eqns (2.10g) and (2.10h) this becomes

$$\left(\frac{E_0}{[EAB]} - \frac{K'_B}{[B]} - \frac{K'_A}{[A]} - 1\right) = \frac{K_A K'_B}{[A][B]}, \quad \text{(S2.4d)}$$

from which the result follows.

2.5. (a) Substitute for [E] from eqn (2.11g) in (2.11c) and (2.11d) to obtain

$$(E_0 - [ES] - [EI])[S] = K_S[ES] \quad \text{(S2.5a)}$$

and

$$(E_0 - [ES] - [EI])[I] = K_I[EI]. \quad \text{(S2.5b)}$$

Substituting for [EI] from eqn (S2.5a) in (S2.5b) leads to

$$[ES]\left(K_I + \frac{K_I K_S}{[S]} + \frac{K_S[I]}{[S]}\right) = K_I E_0, \quad \text{(S2.5c)}$$

and the result follows.

(b) The total concentration of enzyme is

$$E_0 = [E] + [ES] + [EI] + [ESI], \quad \text{(S2.5d)}$$

so that eqn (2.12e) becomes

$$(E_0 - [\mathrm{ES}] - [\mathrm{EI}] - [\mathrm{ESI}])[\mathrm{S}] = K_\mathrm{S}[\mathrm{ES}]. \tag{S2.5e}$$

This can be combined with eqns (2.12h) and (2.12i) to give

$$[\mathrm{ES}]\left(1 + \frac{K_\mathrm{S}}{[\mathrm{S}]}\right)\left(1 + \frac{[\mathrm{I}]}{K_\mathrm{I}}\right) = E_0, \tag{S2.5f}$$

which leads to the required result.

2.6. (a) The constraint is

$$\phi_0 \int_0^{\tau_\mathrm{d}} \exp(-\lambda v \tau)\,\mathrm{d}\tau = 1 \tag{S2.6a}$$

which is

$$-\frac{\phi_0}{\lambda v}[\exp(-\lambda v \tau)]_0^{\tau_\mathrm{d}} = 1 \tag{S2.6b}$$

and hence

$$\phi_0 = \frac{\lambda v}{1 - \exp(-\lambda v \tau_\mathrm{d})}. \tag{S2.6c}$$

Using eqn (2.13i) for $v\tau_\mathrm{d}$ gives the required result.

(b) $\phi(\tau_\mathrm{d})$ is, from eqn (2.14d) and using eqn (2.13i) for $v\tau_\mathrm{d}$,

$$\phi(\tau_\mathrm{d}) = \phi_0 \exp(-\lambda \ln 2) \tag{S2.6d}$$

which, substituted in eqn (2.14f), gives

$$\phi_0 = 2\theta\phi_0 \exp(-\lambda \ln 2) \tag{S2.6e}$$

and eqn (2.14g) follows.

(c) As $\phi_0 \to 0$, eqn (2.14e) for ϕ_0 becomes, using the series expansion for $\exp(-\lambda \ln 2)$,

$$\phi_0 = \frac{\lambda v}{1 - [1 - \lambda \ln 2 + (\lambda \ln 2)^2/2! + \cdots]} \tag{S2.6f}$$

and hence

$$\lim_{\lambda \to 0} \phi_0 = \frac{v}{\ln 2} = \frac{1}{\tau_\mathrm{d}} \tag{S2.6g}$$

using eqn (2.13i) for v, as required.

2.7. This is quite a difficult problem, and only an outline of the solution is presented here. First note that eqn (2.16b) can be written

$$\frac{1}{M_0}\frac{\mathrm{d}N}{\mathrm{d}t} = \xi t \exp\left[-\frac{v}{2t_\mathrm{m}}(t - t_\mathrm{m})^2\right], \tag{S2.7a}$$

where ξ is defined as

$$\xi = \frac{v}{t_m}\exp\left(\frac{1}{2}vt_m\right). \tag{S2.7b}$$

Since $N(t=0)=0$, eqn (S2.7a) integrates to

$$\frac{N}{M_0} = \xi\int_0^t t'\exp\left[-\frac{v}{2t_m}(t'-t_m)^2\right]dt', \tag{S2.7c}$$

where the prime denotes the dummy argument. Now

$$\frac{d}{dt}\left\{\exp\left[-\frac{v}{2t_m}(t-t_m)^2\right]\right\} = -\frac{v}{t_m}(t-t_m)\exp\left[-\frac{v}{2t_m}(t-t_m)^2\right], \tag{S2.7d}$$

so that eqn (S2.7c) becomes

$$\frac{N}{M_0} = \frac{\xi t_m}{v}\left\{\exp\left(-\frac{1}{2}vt_m\right) - \exp\left[-\frac{v}{2t_m}(t-t_m)^2\right]\right\}$$
$$+ \xi t_m\int_0^t \exp\left[-\frac{v}{2t_m}(t'-t_m)^2\right]dt'. \tag{S2.7e}$$

We must now evaluate the integral in this equation. Write it as

$$\int_0^t \exp\left[-\frac{v}{2t_m}(t'-t_m)^2\right]dt' = I_1 + I_2, \tag{S2.7f}$$

where

$$I_1 = \int_0^{t_m} \exp\left[-\frac{v}{2t_m}(t'-t_m)^2\right]dt' \tag{S2.7g}$$

and

$$I_2 = \int_{t_m}^{t} \exp\left[-\frac{v}{2t_m}(t'-t_m)^2\right]dt' \tag{S2.7h}$$

First consider I_1. Let

$$\left(\frac{v}{2t_m}\right)^{1/2}(t'-t_m) = -z \qquad z \geqslant 0 \tag{S2.7i}$$

so that

$$dt' = -\left(\frac{2t_m}{v}\right)^{1/2}dz, \tag{S2.7j}$$

and hence I_1 can be evaluated as

$$I_1 = \left(\frac{\pi t_m}{2v}\right)^{1/2}\text{erf}\left(\frac{v}{2t_m}\right)^{1/2}. \tag{S2.7k}$$

For I_2, set

$$\frac{v}{2t_{\rm m}}(t' - t_{\rm m})^2 = z^2 \qquad z \geqslant 0, \tag{S2.7l}$$

from which

$$z = \left(\frac{v}{2t_{\rm m}}\right)^{1/2} |t' - t_{\rm m}| \tag{S2.7m}$$

giving

$$dt' = \left(\frac{2t_{\rm m}}{v}\right)^{1/2} \operatorname{sgn}(t' - t_{\rm m})\, dz. \tag{S2.7n}$$

Thus, noting that $\operatorname{sgn}(t' - t_{\rm m})$ is constant over the range of integration and hence

$$\operatorname{sgn}(t' - t_{\rm m}) = \operatorname{sgn}(t - t_{\rm m}), \tag{S2.7o}$$

I_2 can be evaluated as

$$I_2 = \left(\frac{\pi t_{\rm m}}{2v}\right)^{1/2} \operatorname{sgn}(t - t_{\rm m}) \operatorname{erf}\left[\left(\frac{v}{2t_{\rm m}}\right)^{1/2} |t - t_{\rm m}|\right]. \tag{S2.7p}$$

Finally, substituting for I_1 and I_2 in eqn (S2.7f) and then combining this with eqns (S2.7b) and (S2.7e) leads to the required result.

Chapter 3

3.1. Equation (3.3c) can be written

$$\int_{W_0}^{W} \frac{dW'}{W_f - W'} = k \int_0^t dt', \tag{S3.1a}$$

where the prime denotes the dummy variable of integration. Integrating gives

$$[-\ln(W_f - W')]_{W_0}^{W} = kt,$$

which leads to

$$-\ln(W_f - W) + \ln(W_f - W_0) = kt. \tag{S3.1b}$$

This can be rearranged to give the required result.

To show that there is no point of inflexion, differentiate eqn (3.3c):

$$\frac{d^2W}{dt^2} = -k\frac{dW}{dt}. \tag{S3.1c}$$

Substituting from eqns (3.3c) and (3.3d) successively, this can be written

$$\frac{d^2W}{dt^2} = -k^2(W_f - W) = -k^2(W_f - W_0)e^{-kt}. \tag{S3.1d}$$

It is clear from this equation that there is no point of inflexion, since $d^2W/dt^2 \neq 0$ for finite t.

3.2. (a) Equation (3.4d) can be written as

$$\int_{W_0}^{W} \frac{dW'}{W'(1 - W'/W_f)} = \int_0^t \mu\, dt', \tag{S3.2a}$$

where the prime denotes the dummy variable of integration. Using partial fractions, this becomes

$$\mu t = \int_{W_0}^{W} \left(\frac{1}{W'} + \frac{1}{W_f - W'}\right) dW' = \left[\ln\left(\frac{W'}{W_f - W'}\right)\right]_{W_0}^{W}. \tag{S3.2b}$$

Substituting the limits of the integration, and removing the logarithm, leads to

$$\left(\frac{W}{W_f - W}\right)\left(\frac{W_f - W_0}{W_0}\right) = e^{\mu t}, \tag{S3.2c}$$

which, upon rearrangement, gives the required result.

(b) Equation (3.5d) can be written

$$\int_{W_0}^{W} \frac{dW'}{W'} = \mu_0 \int_0^t \exp(-Dt')\, dt'. \tag{S3.2d}$$

Integrating gives

$$\ln\left(\frac{W}{W_0}\right) = \frac{\mu_0}{D}(1 - e^{-Dt}), \tag{S3.2e}$$

and removing the logarithm leads to the required result.

3.3. (a) Equation (3.6a) can be written

$$\int_{W_0}^{W} \frac{dW'}{W'(1 - W'/B)} = \int_0^t \mu \exp(-Dt')\, dt', \tag{S3.3a}$$

that is

$$\int_{W_0}^{W} \left(\frac{1}{W'} + \frac{1}{B - W'}\right) dW' = \frac{\mu}{D}(1 - e^{-Dt}) \tag{S3.3b}$$

giving

$$\ln\left[\left(\frac{W}{B - W}\right)\left(\frac{B - W_0}{W_0}\right)\right] = \frac{\mu}{D}(1 - e^{-Dt}). \tag{S3.3c}$$

Removing the logarithm and rearranging leads to

$$W = \frac{W_0 B}{W_0 + (B - W_0)\exp\{-[\mu(1 - e^{-Dt})/D]\}} \tag{S3.3d}$$

as required.

In early growth when t is small, $1 - e^{-Dt} \simeq Dt$ and, since $B \geqslant W_f$ (eqn (3.6f)), $W_0/B \ll 1$ and hence it follows immediately from this equation that

$$W \approx W_0 e^{\mu t}, \tag{S3.3e}$$

thus demonstrating that μ is the specific growth rate during early growth.

(b) Differentiating eqn (3.6a) with respect to time gives

$$\frac{d^2W}{dt^2} = \mu e^{-Dt}\left[\frac{dW}{dt}\left(1 - \frac{2W}{B}\right) - DW\left(1 - \frac{W}{B}\right)\right], \tag{S3.3f}$$

which, substituting for dW/dt from eqn (3.6a), becomes

$$\frac{d^2W}{dt^2} = \mu W e^{-Dt}\left(1 - \frac{W}{B}\right)\left[\mu e^{-Dt}\left(1 - \frac{2W}{B}\right) - D\right], \tag{S3.3g}$$

so that, since $W/B < 1$, there is a point of inflexion when

$$\mu e^{-Dt}\left(1 - \frac{2W}{B}\right) - D = 0. \tag{S3.3h}$$

From eqn (S3.3c), e^{-Dt} is given by

$$e^{-Dt} = 1 - \frac{D}{\mu}\ln\left[\left(\frac{W}{B - W}\right)\left(\frac{B - W_0}{W_0}\right)\right], \tag{S3.3i}$$

so that the required equation is

$$\mu\left(1 - \frac{2W}{B}\right)\left\{1 - \frac{D}{\mu}\ln\left[\left(\frac{W}{B - W}\right)\left(\frac{B - W_0}{W_0}\right)\right]\right\} - D = 0. \tag{S3.3j}$$

This equation does not have an analytical solution and so must be solved numerically using, for example, the Newton–Raphson method ((S7.8c) *et seq.*, p. 604). However, it is readily shown that it reduces to the relevant equations for the logistic and Gompertz equations.

3.4. (a) If growth is a linear function of time, then W can be written

$$W = at + b \tag{S3.4a}$$

where a and b are constants. From (S3.4a) it follows immediately that

$$\frac{1}{W}\frac{dW}{dt} = \frac{a}{at + b}. \tag{S3.4b}$$

Using eqn (S3.4a) in (E3.4a) with (E3.4c) leads also to eqn (S3.4b).

(b) The differential equation for simple exponential growth (eqn (3.2a)) can be written

$$\frac{1}{W}\frac{dW}{dt} = \frac{d}{dt}(\ln W) = \mu. \tag{S3.4c}$$

Integrating between t_1 and t_2 gives

$$\mu = \frac{\ln W_2 - \ln W_1}{t_2 - t_1} \tag{S3.4d}$$

as required.

(c) Substituting the equation for simple exponential growth (eqn (3.2b)) in eqn (E3.4a) with eqn (E3.4c) gives

$$\frac{1}{W}\frac{\mathrm{d}W}{\mathrm{d}t} \simeq \frac{1}{\delta}\left(\frac{\mathrm{e}^{\mu\delta} - \mathrm{e}^{-\mu\delta}}{\mathrm{e}^{\mu\delta} + \mathrm{e}^{-\mu\delta}}\right) = \frac{1}{\delta}\tanh(\mu\delta), \tag{S3.4e}$$

where tanh is the hyperbolic tangent. Using the series expansion for $\tanh(\mu\delta)$, or expanding $\mathrm{e}^{\mu\delta}$ and $\mathrm{e}^{-\mu\delta}$, gives

$$\frac{1}{W}\frac{\mathrm{d}W}{\mathrm{d}t} \simeq \mu - \frac{\mu^3\delta^2}{3} \qquad \text{for small } \mu\delta. \tag{S3.4f}$$

The exact value of the specific growth rate is simply μ (eqn (3.2a)), so that the fractional error from eqn (S3.4f) is approximately $\mu^2\delta^2/3$. Taking $\mu = 0.1$, this value will be less than 0.1 provided that $\delta < 5$. Thus the linear approximation (E3.4a) is only likely to give an accurate approximation to exponentially growing plants or crops if the measurements are made quite close together.

Chapter 4

4.1. The flux is given by eqn (4.6d) multiplied by a factor of 86 400, the number of seconds in a day, giving

$$T = 86\,400AD\frac{S_1 - S_2}{x_2 - x_1} \tag{S4.1a}$$

$$= 86\,400 \times 10^{-4} \times 100 \times 5.2 \times 10^{-10}/1$$

$$= 0.449 \times 10^{-6} \text{ kg sucrose day}^{-1}. \tag{S4.1b}$$

4.2. In the steady state, the diffusion equation takes the form (cf. eqn (4.7b))

$$0 = D\frac{\mathrm{d}^2S}{\mathrm{d}x^2} - k_\mathrm{p}S, \tag{S4.2a}$$

which gives the general solution

$$S = B\mathrm{e}^{\alpha x} + C\mathrm{e}^{-\alpha x}, \tag{S4.2b}$$

where $\alpha = (k_\mathrm{p}/D)^{1/2}$, and B and C are constants. The two boundary conditions that enable B and C to be determined are first the substrate concentration at $x = 0$ so that

$$S_0 = B + C, \tag{S4.2c}$$

and second, matching the transport flux $-DA\,dS/dx$ at $x = L$ to the sink utilization flux $k_1 S$ at $x = L$, giving

$$-DA\alpha(Be^{\alpha L} - Ce^{-\alpha L}) = k_1(Be^{\alpha L} + Ce^{-\alpha L}).$$

With $\beta = DA\alpha/k_1$ (dimensionless), this becomes

$$-\beta(Be^{\alpha L} - Ce^{-\alpha L}) = Be^{\alpha L} + Ce^{-\alpha L}. \qquad \text{(S4.2d)}$$

Solving eqns (S4.2c) and (S4.2d) simultaneously, therefore, gives

$$B = \frac{S_0(1-\beta)e^{-\alpha L}}{(1-\beta)e^{-\alpha L} - (1+\beta)e^{\alpha L}} \qquad \text{(S4.2e)}$$

$$C = \frac{-S_0(1+\beta)e^{\alpha L}}{(1-\beta)e^{-\alpha L} - (1+\beta)e^{\alpha L}}. \qquad \text{(S4.2f)}$$

The flux F_S of substrate into the sink is

$$F_S = k_1 S(x = L) = k_1(Be^{\alpha L} + Ce^{-\alpha L}). \qquad \text{(S4.2g)}$$

The pathway loss F_p is given by

$$F_p = \int_0^L Ak_p S\,dx = \frac{Ak_p}{\alpha}[B(e^{\alpha L} - 1) + C(1 - e^{-\alpha L})]. \qquad \text{(S4.2h)}$$

In these equations B and C are given by eqns (S4.2e) and (S4.2f). Summing F_S and F_p gives the total flux of substrate that must be supplied at $x = 0$ to maintain S_0 at a constant value.

4.3. The equation equivalent to eqn (4.9a) is

$$T = \frac{S_1 - S_2}{r_d} = \frac{k_1 S_2}{K_1 + S_2} + \frac{k_2 S_2}{K_2 + S_2}. \qquad \text{(S4.3a)}$$

S_2 can be eliminated by writing $S_2 = S_1 - r_d T$ to give

$$T = \frac{k_1(S_1 - r_d T)}{K_1 + S_1 - r_d T} + \frac{k_2(S_1 - r_d T)}{K_2 + S_1 - r_d T}. \qquad \text{(S4.3b)}$$

Multiplying out and collecting like terms leads to

$$\begin{aligned} 0 = {} & S_1[k_1(K_2 + S_1) + k_2(K_1 + s_1)] \\ & - T\{r_d[k_1(K_2 + 2S_1) + k_2(K_1 + 2S_1)] + (K_1 + S_1)(K_2 + S_1)\} \\ & + T^2 r_d[r_d(k_1 + k_2) + K_1 + K_2 + 2S_1] - T^3 r_d^{\,2}. \end{aligned} \qquad \text{(S4.3c)}$$

Only one root of this cubic equation is biologically acceptable. The equation can be reduced to eqn (4.9c) for a single biochemical sink by putting $k_2 = 0$ and dividing by the factor $K_2 + S_1 - r_d T$.

The asymptotic value of T is obtained by taking the terms in $S_1^{\,2}$ (i.e. the highest power of S_1) to give

$$T(S_1 \to \infty) = k_1 + k_2. \tag{S4.3d}$$

Since $T = 0$ when $S_1 = 0$, the initial slope of the response is found by considering the first two terms of eqn (S4.3c) so that

$$\frac{\mathrm{d}T}{\mathrm{d}S_1}(S_1 = 0) = \frac{k_1 K_2 + k_2 K_1}{r_\mathrm{d}(k_1 K_2 + k_2 K_1) + K_1 K_2}. \tag{S4.3e}$$

4.4. Using eqn (4.10l), we obtain

$$T_\mathrm{d} = 3.1 \times 10^{-15}\ \mathrm{kg\ mol\ s^{-1}}. \tag{S4.4a}$$

Using eqn (4.10q), and noting that $S_1 \gg K$ and $S_2 \ll K$ gives

$$T_\mathrm{fd} = 5.2 \times 10^{-15}\ \mathrm{kg\ mol\ s^{-1}}. \tag{S4.4b}$$

The facilitated flux is about twice the ordinary diffusive flux. For S_2 values of K, $2K$, $4K$, and $9K$, the facilitated flux is reduced by factors of 2, 3, 5, and 10 respectively; as $S_2 \ll S_1$, even when $S_2 = 10K$, the ordinary diffusive flux is not reduced appreciably.

4.5. When substrate S and carrier C are in equilibrium,

$$k_1 C S^q = k_2 X; \tag{S4.5a}$$

eqns (4.10o) and (4.10p) are replaced by

$$X_1 = C_\mathrm{t}\frac{S_1 q}{K^q + S_1{}^q} \qquad X_2 = C_\mathrm{t}\frac{S_2{}^q}{K^q + S_2{}^q}. \tag{S4.5b}$$

The flux due to facilitated diffusion is then (cf. eqn (4.10q))

$$T_\mathrm{fd} = \frac{A}{L} D_\mathrm{X} C_\mathrm{t}\frac{K^q(S_1{}^q - S_2{}^q)}{(K^q + S_1{}^q)(K^q + S_2{}^q)}. \tag{S4.5c}$$

For the case $K \ll S_1$ and $S_2 \ll S_1$, this becomes

$$T_\mathrm{fd} = \frac{A}{L} D_\mathrm{X} C_\mathrm{t}\frac{K^q}{K^q + S_2{}^q}. \tag{S4.5d}$$

With a biochemical process is series with facilitated diffusion, in the steady state

$$\frac{A}{L} D_\mathrm{X} C_\mathrm{t}\frac{K^q}{K^q + S_2{}^q} = \frac{kS_2}{K' + S_2}. \tag{S4.5e}$$

This equation could be solved for S_2, but the essence of the solution can be seen by sketching the two sides of eqn (S4.5e) and considering the points of intersection (Fig. S4.1). It can be seen that over a wide range of activity k of the biochemical sink the change in the substrate concentration at the biochemical sink S_2 is comparatively small. A higher value of q would steepen the sigmoidal response on the left-hand side of eqn (S4.5e) (p. 56) and lead to an even more precisely

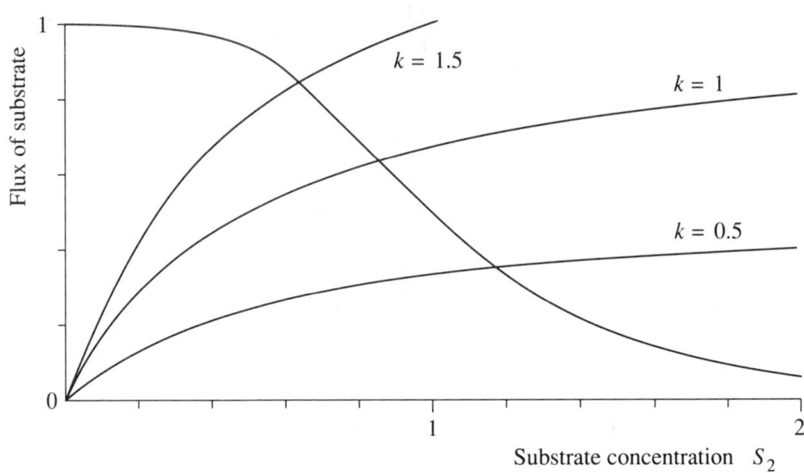

Fig. S4.1. Graphical solution to eqn (S4.5e). Each side of this equation is plotted against substrate concentration S_2 for parameter values of $AD_X C_t/L = 1$, $K = 1$, $q = 4$, $K' = 0.5$, and k values as shown. The points of intersection denote solutions to eqn (S4.5e) (Exercise 4.5).

controlled S_2 level. Such control may be biologically advantageous in situations where an enzyme system such as the nitrogenase system in nitrogen fixation may be damaged by high levels of a molecule like oxygen.

4.6. In eqns (4.11f) and (4.12l), make the substitutions $S_1 = N_{soil}$, $c_{ATP} = c_{glu}$, and $k_3 = k_{3m}/(1 + N_{plant}/J)$ to give

$$T = \frac{AC_t k_1 k_{3m} N_{soil} c_{glu}}{k_2 + k_1 N_{soil} + k_2 N_{plant}/J + k_{3m} c_{glu} + k_1 N_{soil} N_{plant}/J} \tag{S4.6a}$$

for the case of active transport with fast diffusion across the membrane, and

$$T = AC_t k_1 k_{3m} N_{soil} c_{glu} \left[k_2 + k_1 N_{soil} + \frac{k_2 N_{plant}}{J} + k_{3m}(1 + k_2 r_1) c_{glu} + \frac{k_1 N_{soil} N_{plant}}{J} + \frac{1}{2} k_1 k_{3m}(r_1 + r_2) N_{soil} c_{glu} \right]^{-1} \tag{S4.6b}$$

for the case where the diffusion resistances across the membrane are finite. Note that both these uptake functions are rectangular hyperbolic in character, with eqn (S4.6b) giving limiting behaviour for large N_{soil} and c_{glu}.

4.7. In order to apply eqn (4.14c), we need

$$S_1 = 100/342.3 = 0.292 \text{ kg mol m}^{-3}, \tag{S4.7a}$$

for $\theta = 20 + 273.15 = 293.15$ K and $R = 8314.32$ J (kg mol)$^{-1}$ K^{-1},

$$R\theta = 2.437 \times 10^6 \text{ J (kg mol)}^{-1}; \tag{S4.7b}$$

the area A for the 50 bundles is

$$A = 50 \times 0.12 \times 10^{-8} = 6 \times 10^{-8} \text{ m}^2.$$

The mass flux T (kg day^{-1}) is given by

$$T = 342.3 \times 86\,400 \frac{R\theta A^2 S_1{}^2}{8L\eta\pi}$$

$$= \frac{342.3 \times 86\,400 \times 2.437 \times 10^6 \times 36 \times 10^{-16} \times 0.292^2}{8 \times 0.1 \times 1.15 \times 0.001 \times 3.142}$$

$$= 7.65 \text{ kg day}^{-1}. \tag{S4.7c}$$

With these assumptions, it is clear that convective flow driven by osmotic pressure is able to transport very large quantities of substrate indeed.

Chapter 5

5.1. Differentiating eqn (E5.1a) gives

$$\frac{\mathrm{d}k}{\mathrm{d}T} = \frac{AE_\mathrm{a}}{RT^2}\exp\left(-\frac{E_\mathrm{a}}{RT}\right) \tag{S5.1a}$$

and differentiating again leads to

$$\frac{\mathrm{d}^2k}{\mathrm{d}T^2} = A\left[\left(\frac{E_\mathrm{a}}{RT^2}\right)^2 - \frac{2E_\mathrm{a}}{RT^3}\right]\exp\left(-\frac{E_\mathrm{a}}{RT}\right). \tag{S5.1b}$$

There is a point of inflexion at (equating (S5.1b) to zero)

$$T_\mathrm{i} = E_\mathrm{a}/2R. \tag{S5.1c}$$

To sketch eqn (E5.1a), note that at $T = 0$, $k = 0$, and $\mathrm{d}k/\mathrm{d}t = 0$; at $T = \infty$, $k = A$, and $\mathrm{d}k/\mathrm{d}T = 0$. The maximum slope is at T_i given by eqn (S5.1a) and eqn (S5.1c) (see Fig. S5.1).

Differentiating eqn (5.1b), which we write (putting $A = 1$ for simplicity)

$$k = \frac{\exp(-E_\mathrm{a}/RT)}{1 + \exp(-E_\mathrm{a}/RT)} = \frac{1}{1 + \exp(E_\mathrm{a}/RT)}, \tag{S5.1d}$$

gives

$$\frac{\mathrm{d}k}{\mathrm{d}T} = \frac{(E_\mathrm{a}/RT^2)\exp(E_\mathrm{a}/RT)}{[1 + \exp(E_\mathrm{a}/RT)]^2} \tag{S5.1e}$$

As before (eqn (S5.1a)), when $T = 0$, $k = 0$ and $\mathrm{d}k/\mathrm{d}T = 0$; at $T = \infty$, $k = \frac{1}{2}$ and $\mathrm{d}k/\mathrm{d}T = 0$. Differentiating eqn (S5.1e) again gives a transcendental equation for

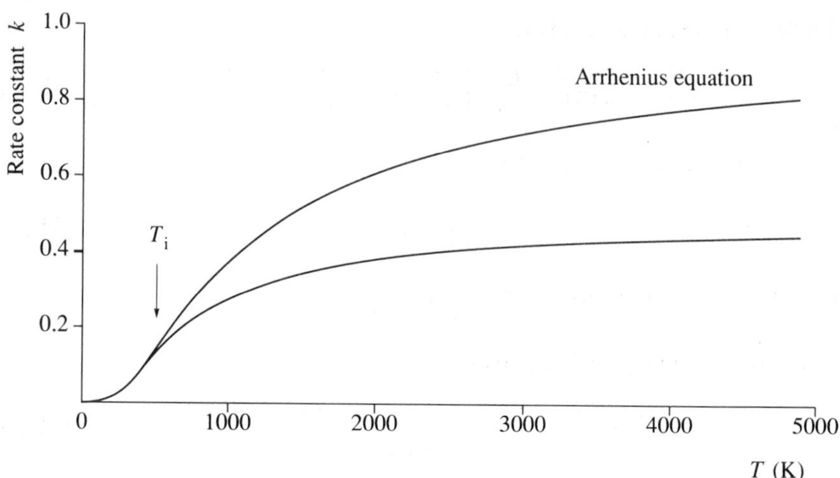

Fig. S5.1. Sketch of the Arrhenius eqn (E5.1a) with $A = 1$ and $E_a/R = 1000$ K. The arrow indicates the point of inflexion (eqn (S5.1c)) of the Arrhenius equation. The lower curve represents eqn (S5.1d), from which the Arrhenius equation is derived by approximation (p. 122).

the inflexion temperature T_i. The resulting curve is sketched in Fig. S5.1; the asymptote is now at $k = \frac{1}{2}$ rather than at $k = 1$.

5.2. Differentiating eqn (E5.2a) twice (with $B = 1$) leads first to

$$\frac{dk}{dT} = \left(\frac{1}{2T^{1/2}} + \frac{E_a}{RT^{3/2}}\right)\exp\left(-\frac{E_a}{RT}\right) \tag{S5.2a}$$

and then to

$$\frac{d^2k}{dT^2} = \exp\left(-\frac{E_a}{RT}\right)\left[\frac{E_a}{RT^2}\left(\frac{1}{2T^{1/2}} + \frac{E_a}{RT^{3/2}}\right) - \frac{1}{4T^{3/2}} - \frac{3E_a}{2RT^{5/2}}\right]$$
$$= \exp\left(-\frac{E_a}{RT}\right)\frac{E_a{}^2/R^2 - (E_a/R)T - T^2/4}{T^{7/2}}. \tag{S5.2b}$$

This has a positive root at

$$T_i = (E_a/R)2(\sqrt{2} - 1). \tag{S5.2c}$$

Although eqn (E5.2a) does not possess an asymptote, it does exhibit a point of inflexion.

Differentiating eqn (E5.2b) gives (with $C = 1$)

$$\frac{dk}{dT} = \exp\left(-\frac{E_a}{RT}\right)\left(1 + \frac{E_a}{RT}\right), \tag{S5.2d}$$

and differentiating again leads to

$$\frac{d^2k}{dT^2} = \exp\left(-\frac{E_a}{RT}\right)\left[\frac{E_a}{RT^2}\left(1 + \frac{E_a}{RT}\right) - \frac{E_a}{RT^2}\right]. \quad \text{(S5.2e)}$$

Equating eqn (S5.2e) to zero gives

$$\frac{E_a}{RT_i} = 0, \quad \text{(S5.2f)}$$

so that the point of inflexion is at $T_i = \infty$.

5.3. Differentiating eqn (E5.3a) gives

$$\frac{dk}{dx} = \frac{A[-ae^{-ax}(1 + be^{-cx}) + bce^{-cx}e^{-ax}]}{(1 + be^{-cx})^2} \quad \text{(S5.3a)}$$

Equating this to zero we obtain

$$a(1 + be^{-cx}) = bce^{-cx},$$

giving

$$e^{-cx} = \frac{a}{b(c - a)} \quad \text{(S5.3b)}$$

and thus

$$T_{max} = \frac{-c}{\ln\{a/[b(c - a)]\}}. \quad \text{(S5.3c)}$$

Substituting into eqn (E5.3a) gives the desired result.

5.4. In the steady state, the scheme of (E5.4a) gives

$$S_1(k_2 + k_3 + k_4) = Sk_1. \quad \text{(S5.4a)}$$

Therefore

$$\frac{dP}{dt} = S_1(k_3 + k_4) = \frac{Sk_1(k_3 + k_4)}{k_2 + k_3 + k_4}. \quad \text{(S5.4b)}$$

Using the Arrhenius eqn (5.1e) leads to

$$\frac{dP}{dt} = \frac{SA_1\exp(-E_1/RT)[A_3\exp(-E_3/RT) + A_4\exp(-E_4/RT)]}{A_2\exp(-E_2/RT) + A_3\exp(-E_3/RT) + A_4\exp(-E_4/RT)}. \quad \text{(S5.4c)}$$

5.5. Taking logarithms of eqn (E5.5a) gives

$$\ln D = \ln c + x\ln T. \quad \text{(S5.5a)}$$

For a reference temperature T_r, this can be written

$$\ln D = \ln c + x \ln [T_r + (T - T_r)]$$

$$= \ln c + x \ln \left[T_r \left(1 + \frac{T - T_r}{T_r} \right) \right]$$

$$= \ln c + x \ln T_r + x \ln \left(1 + \frac{T - T_r}{T_r} \right). \tag{S5.5b}$$

For $T - T_r \ll T_r$, this can be approximated by

$$\ln D = \ln c + x \ln T_r + x \frac{T - T_r}{T_r}. \tag{S5.5c}$$

Equating the coefficients of $T - T_r$ in eqns (S5.5c) and (5.1i) gives

$$Q_{10} = e^{10x/T_r}. \tag{S5.5d}$$

For CO_2 and oxygen, taking $T_r = 290$ K and putting x equal to 1.97 and 1.79 (eqns (5.5a) and (5.5b)) leads to

$$Q_{10}(CO_2) = 1.07 \qquad Q_{10}(O_2) = 1.06. \tag{S5.5e}$$

5.6. The inflexion point of the Arrhenius equation is, from eqns (S5.1c) and (E5.1a), at

$$T_i = E_a/2R \qquad k_i = Ae^{-2}. \tag{S5.6a}$$

The slope at the point (T_i, k_i) is, from eqn (S5.1a),

$$\frac{dk}{dT}(T_i, k_i) = 4\left(\frac{AR}{E_a}\right)e^{-2}. \tag{S5.6b}$$

Using the result that, in (x, y) coordinates, a line of slope m through the point (x_0, y_0) has the equation

$$y - y_0 = m(x - x_0); \tag{S5.6c}$$

then applying this to eqns (S5.6a) and (S5.6b) gives

$$k - Ae^{-2} = 4\left(\frac{AR}{E_a}\right)e^{-2}\left(T - \frac{E_a}{2R}\right),$$

which simplifies to

$$k = 4\left(\frac{AR}{E_a}\right)e^{-2}\left(T - \frac{E_a}{4R}\right). \tag{S5.6d}$$

Comparison of this equation with eqn (E5.6a) gives

$$c = 4\left(\frac{AR}{E_a}\right)e^{-2} \qquad T_0 = \frac{E_a}{4R}. \tag{S5.6e}$$

Chapter 6

6.1. Equating dx/dt to zero, $x = 0$ or

$$k_1(K_2 + x) = k_2(K_1 + x). \tag{S6.1a}$$

Therefore

$$x = \frac{k_2 K_1 - k_1 K_2}{k_1 - k_2}. \tag{S6.1b}$$

Since x is required to be positive,

$$\text{if } k_1 < k_2, \text{ then } k_1/K_1 > k_2/K_2 \tag{S6.1c}$$

or vice versa. This result is easily understood graphically: if the asymptote of one curve is greater than that of the other, the initial slope must be less in order to ensure a second intersection.

6.2. Substituting $g = \frac{1}{2}$ in eqn (6.7e) gives

$$16x^5 - 16x^4 + 8x^3 - 8x^2 + 5x - 1 = 0. \tag{S6.2a}$$

This equation factorizes to

$$(8x^3 - 12x^2 + 6x - 1)(2x^2 + x + 1) = 0,$$

which further factorizes to

$$(2x - 1)^3(2x^2 + x + 1) = 0. \tag{S6.2b}$$

This last equation has three real positive roots at

$$x = \tfrac{1}{2}. \tag{S6.2c}$$

6.3. Differentiating eqn (6.8a) twice gives

$$\begin{aligned} \frac{dU_1}{dx} &= 4bx^3 - 2ax \\ \frac{d^2U_1}{dx^2} &= 12bx^2 - 2a. \end{aligned} \tag{S6.3a}$$

Equating the first equation to zero gives the three roots

$$x = 0 \qquad x = \pm(a/2b)^{1/2}. \tag{S6.3b}$$

The second derivative is $-2a$ at $x = 0$, giving a maximum, and is $4a$ at the other two roots, giving two minima. The values of U_1 at these turning points are

$$U_1 = 0 \qquad U_1 = -\frac{a^2}{2b}\,\text{(twice)}. \tag{S6.3c}$$

6.4. Differentiating eqn (6.8b) gives

$$\frac{dU_2}{dx} = 4bx^3 - 2ax + \gamma y. \tag{S6.4a}$$

The cubic equation

$$0 = \xi^3 + p\xi + q \tag{S6.4b}$$

has only one real root if

$$-\frac{4}{27}p^3 < q^2. \tag{S6.4c}$$

Applying this result to eqn (S6.4a) equated to zero leads directly to eqn (6.8c) and also to eqn (6.8d).

Chapter 7

7.1. Equations (7.4b) are

$$\frac{dX_2}{dt} = -k_2X_2 \quad \frac{dX_3}{dt} = k_2X_2 - k_3X_3 \quad \frac{dX_4}{dt} = k_3X_3, \tag{S7.1a}$$

with

$$X_2 = 1 \qquad X_3 = X_4 = 0 \text{ at } t = 0. \tag{S7.1b}$$

Solving for X_2 gives

$$X_2 = \exp(-k_2t). \tag{S7.1c}$$

Substituting eqn (S7.1c) into the second of eqns (7.1a) gives

$$\frac{dX_3}{dt} = k_2\exp(-k_2t) - k_3X_3. \tag{S7.1d}$$

According to the standard theory of differential equations (Piaggio 1952, p. 29), the complementary function is the general solution to

$$\frac{dX_3}{dt} = -k_3X_3,$$

which is

$$X_3 = A\exp(-k_3t), \tag{S7.1e}$$

where A is a constant. If we write $D = d/dt$ and rewrite eqn (7.1d) as

$$(D + k_3)X_3 = k_2\exp(-k_2t),$$

a particular integral (any solution to this equation) is

$$X_3 = \frac{1}{k_3 + D} k_2 \exp(-k_2 t) = \frac{k_2 \exp(-k_2 t)}{k_3 - k_2}. \tag{S7.1f}$$

Adding eqns (S7.1e) and (S7.1f) gives

$$X_3 = A \exp(-k_3 t) + \frac{k_2 \exp(-k_2 t)}{k_3 - k_2}.$$

The constant A is determined by requiring $X_3 = 0$ at $t = 0$, giving

$$X_3 = \frac{k_2}{k_3 - k_2} [\exp(-k_2 t) - \exp(-k_3 t)], \tag{S7.1g}$$

which is equivalent to the second of eqns (7.4c).

7.2. Differentiating X_3 of eqns (7.4c) with $X_1(0) = 1$ and $\tau = 0$ gives

$$\frac{dX_3}{dt} = \frac{k_2}{k_3 - k_2} [k_2 \exp(-k_2 t) - k_3 \exp(-k_3 t)], \tag{S7.2a}$$

and equating this to zero at $t = t_{max}$ gives

$$k_2 \exp(-k_2 t_{max}) = k_3 \exp(-k_3 t_{max}), \tag{S7.2b}$$

whence

$$t_{max} = \frac{\ln(k_3/k_2)}{k_3 - k_2}. \tag{S7.2c}$$

Differentiating X_4 of eqns (7.4c) once gives

$$\frac{dX_4}{dt} = \frac{k_2 k_3}{k_3 - k_2} [\exp(-k_2 t) - \exp(-k_3 t)]. \tag{S7.2d}$$

This is proportional to eqn (S7.2a), so that differentiating again and equating d^2X_4/dt^2 to zero gives eqn (7.2b) with t_{max} replaced by t_i, so that

$$t_i = \frac{\ln(k_3/k_2)}{k_3 - k_2}, \tag{S7.2e}$$

which is the desired result.

7.3. Equations (7.4b) with $k_2 = k_3 = k$, $\tau = 0$, and $X_1(0) = 1$ become

$$\frac{dX_2}{dt} = -kX_2 \qquad \frac{dX_3}{dt} = kX_2 - kX_3 \qquad \frac{dX_4}{dt} = kX_3, \tag{S7.3a}$$

with

$$X_2 = 1 \qquad X_3 = X_4 = 0 \text{ at } t = 0. \tag{S7.3b}$$

Solving for X_2 gives

$$X_2 = e^{-kt}, \tag{57.3c}$$

and the equation for X_3 becomes

$$\frac{dX_3}{dt} = ke^{-kt} - kX_3. \tag{S7.3d}$$

As before, in eqn (S7.1e), the complementary function is

$$X_3 = Ae^{-kt}, \tag{S7.3e}$$

where A is a constant. Writing $D = d/dt$, eqn (S7.3d) becomes

$$(D + k)X_3 = ke^{-kt},$$

whence the particular integral is

$$X_3 = \frac{k}{D + k} e^{-kt} = ke^{-kt} \frac{1}{D} = kte^{-kt}. \tag{S7.3f}$$

Next, add eqns (S7.3e) and (S7.3f), and require $X_3 = 0$ at $t = 0$ to give $A = 0$, and thus

$$X_3 = kte^{-kt}. \tag{S7.3g}$$

Substitution into the third of eqns (S7.3a) gives

$$\frac{dX_4}{dt} = k^2te^{-kt}, \tag{S7.3h}$$

which integrates to

$$X_4 = 1 - (1 + kt)e^{-kt}. \tag{S7.3i}$$

Therefore differentiating eqn (S7.3g) gives

$$\frac{dX_3}{dt} = ke^{-kt}(1 - kt), \tag{S7.3j}$$

and thus X_3 is a maximum when

$$\tau = t_{max} = 1/k. \tag{S7.3k}$$

Differentiating eqn (S7.3i) leads to

$$\frac{dX_4}{dt} = k(1 + kt)e^{-kt} - ke^{-kt} = k^2te^{-kt}$$

and

$$\frac{d^2X_4}{dt^2} = k^2e^{-kt}(1 - kt),$$

giving a point of inflexion at

$$t = t_i = 1/k. \tag{S7.3l}$$

This completes the derivation. To use the second suggested method, let $k_3 = k$, $k_2 = k + x$, and, for small x, expand eqns (7.4c) and (7.4d) in terms of x.

7.4.

$$X_4 = k_1 k_2 k_3 \left[\frac{\exp(-k_1 t)}{(k_2 - k_1)(k_3 - k_1)(k_4 - k_1)} + \frac{\exp(-k_2 t)}{(k_3 - k_2)(k_4 - k_2)(k_1 - k_2)} + \frac{\exp(-k_3 t)}{(k_4 - k_3)(k_1 - k_3)(k_2 - k_3)} + \frac{\exp(-k_4 t)}{(k_1 - k_4)(k_2 - k_4)(k_3 - k_4)} \right]. \tag{S7.4a}$$

With $n = 1$, eqns (7.5e) and (7.5f) give

$$X_2 = k_1 \int_0^t \exp(-k_1 t)\,\mathrm{d}t = 1 - \exp(-k_1 t). \tag{S7.4b}$$

With $n = 2$, eqns (7.5e) and (7.5g) give

$$X_3 = k_1 k_2 \int_0^t \left[\frac{\exp(-k_1 t)}{k_2 - k_1} + \frac{\exp(-k_2 t)}{k_1 - k_2} \right] \mathrm{d}t = 1 - \frac{k_2 \exp(-k_1 t)}{k_2 - k_1} - \frac{k_1 \exp(-k_2 t)}{k_1 - k_2}. \tag{S7.4c}$$

With $n = 3$ eqns (7.5e) and (7.5h) give, for a terminal X_4 compartment,

$$X_4 = k_1 k_2 k_3 \int_0^t \left[\frac{\exp(-k_1 t)}{(k_2 - k_1)(k_3 - k_1)} + \frac{\exp(-k_2 t)}{(k_1 - k_2)(k_3 - k_2)} + \frac{\exp(-k_3 t)}{(k_1 - k_3)(k_2 - k_3)} \right] \mathrm{d}t$$

which gives

$$X_4 = 1 - \frac{k_2 k_3 \exp(-k_1 t)}{(k_2 - k_1)(k_3 - k_1)} - \frac{k_3 k_1 \exp(-k_2 t)}{(k_3 - k_2)(k_1 - k_2)} - \frac{k_1 k_2 \exp(-k_3 t)}{(k_1 - k_3)(k_2 - k_3)}. \tag{S7.4d}$$

We now use eqn (7.5j) with $n = 4$ to write down the equation for a terminal X_5:

$$X_5 = 1 - \frac{k_2 k_3 k_4 \exp(-k_1 t)}{(k_2 - k_1)(k_3 - k_1)(k_4 - k_1)} - \frac{k_3 k_4 k_1 \exp(-k_2 t)}{(k_3 - k_2)(k_4 - k_2)(k_1 - k_2)} - \frac{k_4 k_1 k_2 \exp(-k_3 t)}{(k_4 - k_3)(k_1 - k_3)(k_2 - k_3)} - \frac{k_1 k_2 k_3 \exp(-k_4 t)}{(k_1 - k_4)(k_2 - k_4)(k_3 - k_4)}. \tag{S7.4e}$$

7.5. From eqn (7.8a) for the nth pool of Fig. 7.10,

$$X_n = \frac{(kt)^{n-1}}{(n-1)!} e^{-kt}. \tag{S7.5a}$$

From eqn (7.8b), the differential equation for X_{n+1} is

$$\frac{dX_{n+1}}{dt} = kX_n - k_{n+1}X_{n+1},$$

which becomes

$$\frac{dX_{n+1}}{dt} + k_{n+1}X_{n+1} = k\frac{(kt)^{n-1}}{(n-1)!} e^{-kt}. \tag{S7.5b}$$

Multiply both sides by $\exp(k_{n+1}t)$, which is an integrating factor, to give

$$\frac{d}{dt}[\exp(k_{n+1}t)X_{n+1}] = k\frac{(kt)^{n-1}}{(n-1)!} \exp[(k_{n+1} - k)t],$$

and therefore

$$X_{n+1} = \exp(-k_{n+1}t) \int_0^t \frac{k(kt)^{n-1} \exp[(k_{n+1} - k)t]}{(n-1)!} dt. \tag{S7.5c}$$

Substitute for t and dt using

$$z = (k - k_{n+1})t \qquad dz = (k - k_{n+1})\,dt \tag{S7.5d}$$

to give

$$X_{n+1} = \exp(-k_{n+1}t)\left(\frac{k}{k - k_{n+1}}\right)^n \frac{1}{(n-1)!} \int_0^{(k-k_{n+1})t} z^{n-1} \exp(-z)\,dz.$$

With eqn (7.7e), this becomes

$$X_{n+1} = \exp(-k_{n+1}t)\left(\frac{k}{k - k_{n+1}}\right)^n \frac{1}{(n-1)!} \gamma[n, (k - k_{n+1})t], \tag{S7.5e}$$

which is the same as eqn (7.8d).

Equation (7.8c) is

$$\frac{dX_{n+2}}{dt} = k_{n+1}X_{n+1}. \tag{S7.5f}$$

Substituting with eqn (S7.5e), therefore, gives

$$\frac{dX_{n+2}}{dt} = c \exp(-k_{n+1}t)\gamma[n, (k - k_{n+1})t], \tag{S7.5g}$$

where

$$c = k_{n+1}\left(\frac{k}{k - k_{n+1}}\right)^n \frac{1}{(1-n)!}. \tag{S7.5h}$$

Thus

$$X_{n+2} = c\int_0^t \exp(-k_{n+1}t)\gamma[n,(k-k_{n+1})t]\,\mathrm{d}t.$$

Integrating by parts gives

$$X_{n+2} = \left\{c\frac{\exp(-k_{n+1}t)}{-k_{n+1}}\gamma[n,(k-k_{n+1})t]\right\}_0^t$$
$$+\,c\int_0^t \frac{\exp(-k_{n+1}t)}{k_{n+1}}[(k-k_{n+1})t]^n\exp[-(k-k_{n+1})t](k-k_{n+1})\,\mathrm{d}t,$$

where eqn (7.7j) has been used. With eqn (7.7h), and simplification of the integral, this becomes

$$X_{n+2} = -c\frac{\exp(-k_{n+1}t)}{k_{n+1}}\gamma[n,(k-k_{n+1})t]$$
$$+\frac{c}{k_{n+1}}\left(\frac{k-k_{n+1}}{k}\right)^n\int_0^{kt}\exp(-kt)(kt)^{n-1}\,\mathrm{d}(kt). \qquad \text{(S7.5i)}$$

The integral is $\gamma(n, kt)$ by eqn (7.7e), and simplifying we obtain

$$X_{n+2} = \frac{1}{(n-1)!}\left\{\gamma(n,kt) - \left(\frac{k}{k-k_{n+1}}\right)^n\exp(-k_{n+1}t)\gamma[n,(k-k_{n+1})t]\right\}, \qquad \text{(S7.5j)}$$

which proves eqn (7.8e).

7.6. With $k_{n+1} \to k$, applying eqn (7.7k) gives

$$\gamma[n,(k-k_{n+1})t] \approx \frac{1}{n}[(k-k_{n+1})t]^n. \qquad \text{(S7.6a)}$$

Substituting this in eqn (7.8d) or (S7.5e) gives

$$X_{n+1} = \frac{(kt)^n\mathrm{e}^{-kt}}{n!}, \qquad \text{(S7.6b)}$$

agreeing with eqn (7.6j) with $j = n + 1$.

Substituting eqn (S7.6a) into eqn (7.8e) or (S7.5j) leads to

$$X_{n+2} = \frac{1}{(n-1)!}\left[\gamma(n,kt) - \frac{(kt)^n\mathrm{e}^{-kt}}{n}\right].$$

The recurrence relationship of eqn (7.7l) then gives

$$X_{n+2} = \frac{1}{n!}\gamma(n+1,kt), \qquad \text{(S7.6c)}$$

equivalent to eqn (7.6m) with n increased by unity.

7.7. The constant term in eqn (7.9d) necessarily satisfies eqn (7.9b) and, dropping the constant factors, the problem is that of showing that I, defined by

$$I = \int_0^z \exp(-u^2)\,\mathrm{d}u, \tag{S7.7a}$$

with

$$z = \frac{x}{2(Dt)^{1/2}} \tag{S7.7b}$$

satisfies

$$\frac{\partial I}{\partial t} = D\frac{\partial^2 I}{\partial x^2}. \tag{S7.7c}$$

Let

$$I = f(z) - f(0), \tag{S7.7d}$$

where f denotes a function. Then

$$\frac{\mathrm{d}f}{\mathrm{d}z} = \exp(-z^2). \tag{S7.7e}$$

Thus

$$\frac{\partial I}{\partial t} = \frac{\mathrm{d}f}{\mathrm{d}z}\frac{\partial z}{\partial t} = -\frac{x}{4(Dt^3)^{1/2}}\exp(-z^2) \tag{S7.7f}$$

$$\frac{\partial I}{\partial x} = \frac{\mathrm{d}f}{\mathrm{d}z}\frac{\partial z}{\partial x} = \frac{1}{2(Dt)^{1/2}}\exp(-z^2) \tag{S7.7g}$$

$$\frac{\partial^2 I}{\partial x^2} = \frac{\mathrm{d}f}{\mathrm{d}z}\frac{\partial^2 z}{\partial x^2} + \frac{\mathrm{d}^2 f}{\mathrm{d}z^2}\left(\frac{\partial z}{\partial x}\right)^2$$

$$= -\frac{2z\exp(-z^2)}{4Dt} = -\frac{x}{4D(Dt^3)^{1/2}}\exp(-z^2). \tag{S7.7h}$$

Equations (S7.7f) and (S7.7h) satisfy eqn (S7.7c).

7.8. Equation (7.12o) can be written more simply as

$$\tfrac{1}{2}(b - a) = b\mathrm{e}^{-ax} - a\mathrm{e}^{-bx}, \tag{S7.8a}$$

where

$$a = k_1 + k_{1\mathrm{d}} \qquad b = k_2 + k_{2\mathrm{d}} \qquad x = t_h - \tau. \tag{S7.8b}$$

The equation whose solution is sought is

$$f(x) = 0, \tag{S7.8c}$$

where the function f is given by

$$f(x) = \tfrac{1}{2}(b - a) - b\mathrm{e}^{-ax} + a\mathrm{e}^{-bx}. \tag{S7.8d}$$

Differentiating eqn (S7.8d) gives

$$\frac{\mathrm{d}f}{\mathrm{d}x} = ab(\mathrm{e}^{ax} - \mathrm{e}^{-bx}). \tag{S7.8e}$$

The Newton–Raphson method states that, if $x = a$ is an approximation to the solution of eqn (S7.8c), then a better approximation is given by

$$x = a + h \tag{S7.8f}$$

where

$$h = -\frac{f(a)}{\mathrm{d}f/\mathrm{d}x(x = a)}. \tag{S7.8g}$$

Equation (S7.8g) can be derived from the Taylor expansion for $f(a + h)$:

$$f(a + h) \simeq f(a) + h\frac{\mathrm{d}f}{\mathrm{d}x}(a). \tag{S7.8h}$$

Equating $f(a + h)$ to zero leads directly to (S7.8g). Equation (S7.8g) is applied iteratively, replacing a by $a + h$ on each iteration, to generate a solution to eqn (S7.8c) to any desired degree of accuracy.

For the parameters given, $a = 1, b = 2$, and $x = t_h$. A first guess at the solution is $x = 1$:

$$f(1) = \tfrac{1}{2}(2 - 1) - 2\mathrm{e}^{-1} + 1\mathrm{e}^{-2} = -0.1004$$

and

$$\frac{\mathrm{d}f}{\mathrm{d}x}(x = 1) = 2(\mathrm{e}^{-1} - \mathrm{e}^{-2}) = 0.4651.$$

Applying eqn (S7.8g) gives

$$h = \frac{0.1004}{0.4651} = 0.2159.$$

A better approximation to t_h is therefore $t_h = 1.2159$. A further round of the method gives

$$f(1.2159) = -0.0050 \qquad \frac{\mathrm{d}f}{\mathrm{d}x}(1.2159) = 0.4171 \qquad h = 0.0120.$$

The next approximation is $t_h = 1.2279$, and $f(1.2279) = -0.00002$.

Chapter 8

8.1. Substituting eqn (8.4a) in eqn (8.3e) and performing the ϕ integration gives

$$\begin{aligned} I_0 &= 2\pi \int_0^{\pi/2} B_m[\mu \cos\theta + (1 - \mu)\cos^2\theta] \sin\theta \,\mathrm{d}\theta \\ &= -2\pi \int_0^{\pi/2} B_m[\mu \cos\theta + (1 - \mu)\cos^2\theta]\, \mathrm{d}(\cos\theta). \end{aligned} \tag{S8.1a}$$

This can be integrated directly to give

$$I_0 = 2\pi\left[\mu\frac{\cos^2\theta}{2} + (1-\mu)\frac{\cos^3\theta}{3}\right]_0^{\pi/2}, \tag{S8.1b}$$

and substituting the limits leads to the required result.

8.2. With ζ given by eqn (E8.2a), eqn (8.5e) becomes

$$\mathrm{d}I = -(1-m)I\left[(\zeta_m - \zeta_0)\frac{l}{l+K} + \zeta_0\right]\mathrm{d}l. \tag{S8.2a}$$

Rearranging gives

$$\mathrm{d}I = -(1-m)I\left[\zeta_m - (\zeta_m - \zeta_0)\frac{K}{l+K}\right]\mathrm{d}l, \tag{S8.2b}$$

which can be integrated from $l = 0$ to give

$$\ln\left(\frac{I}{I_0}\right) = -(1-m)\left[\zeta_m l - (\zeta_m - \zeta_0)K\ln\left(\frac{l+K}{K}\right)\right], \tag{S8.2c}$$

so that I is

$$I = I_0\exp\left\{-(1-m)\left[\zeta_m l - (\zeta_m - \zeta_0)K\ln\left(\frac{l+K}{K}\right)\right]\right\}. \tag{S8.2d}$$

As $K \to 0$,

$$K\ln\left(\frac{l+K}{K}\right) \to 0 \tag{S8.2e}$$

and hence

$$I \to I_0\exp[-(1-m)\zeta_m l]. \tag{S8.2f}$$

Now, as $K \to 0$, $\zeta \to \zeta_m$, and eqn (S8.2f) is consistent with eqn (8.5g) with ζ replaced by ζ_m.

8.3. Using eqn (8.6i) we obtain

$$\int_0^L mI_l\,\mathrm{d}l = \frac{m}{1-m}I_0(1 - e^{-kL}). \tag{S8.3a}$$

As $L \to \infty$ this becomes

$$\frac{m}{1-m}I_0, \tag{S8.3b}$$

which, taking the series expansion of the factor $m/(1-m)$, can be written

$$I_0(m + m^2 + m^3 + \cdots). \tag{S8.3c}$$

The interpretation of this equation is that all the irradiance is intercepted once, twice, three times, and so on.

8.4. To derive eqn (8.11g), we must evaluate the integral

$$\int_{-\pi/2}^{\pi/2} [\mu + (1-\mu)\cos\theta]\cos^2\theta\,d\theta \tag{S8.4a}$$

in eqn (8.11f). This, in turn, requires evaluating the integrals of $\cos^2\theta$ and $\cos^3\theta$. For the first of these, use the identity

$$\cos^2\theta = \frac{\cos(2\theta)+1}{2}, \tag{S8.4b}$$

from which

$$\int_{-\pi/2}^{\pi/2}\cos^2\theta\,d\theta = \left\{\frac{1}{2}\left[\frac{\sin(2\theta)}{2}+\theta\right]\right\}_{-\pi/2}^{\pi/2} = \frac{\pi}{2}. \tag{S8.4c}$$

Now consider $\int_{-\pi/2}^{\pi/2}\cos^3\theta\,d\theta$, which can be written

$$\begin{aligned}\int_{-\pi/2}^{\pi/2}\cos^3\theta &= \int_{-\pi/2}^{\pi/2}(1-\sin^2\theta)\cos\theta\,d\theta\\ &= \int_{-\pi/2}^{\pi/2}(1-\sin^2\theta)\,d(\sin\theta)\\ &= \left(\sin\theta - \frac{\sin^3\theta}{3}\right)_{-\pi/2}^{\pi/2} = \frac{4}{3}.\end{aligned} \tag{S8.4d}$$

Using these results leads to eqn (8.11g) for T_b.

Chapter 9

9.1. Using Fig. 9.1 and eqns (9.3a)–(9.3l) gives

$$\frac{d[CO_2]_i}{dt} = \frac{[CO_2]_a - [CO_2]_i}{r_{dc}} - k_{CO_2}[RuBP][CO_2]_i + \tfrac{1}{2}k_{glycol}[\text{P-glycolate}][O_2]_i[ATP][NADPH] \tag{S9.1a}$$

$$\begin{aligned}\frac{d[O_2]_i}{dt} = \frac{([O_2]_a - [O_2]_i)}{r_{do}} &- k_{O_2}[RuBP][O_2]_i\\ &- \tfrac{1}{2}k_{glycol}[\text{P-glycolate}][O_2]_i[ATP][NADPH]\\ &+ \tfrac{1}{2}k_{NADPH}([NP]_o - [NADPH])I_1;\end{aligned} \tag{S9.1b}$$

$$\begin{aligned}\frac{d[ATP]}{dt} = k_{ATP}([AP]_o - [ATP])I_1 &- k_{glycol}[\text{P-glycolate}][O_2]_i[ATP][NADPH]\\ &- k_{glycerate}[\text{P-glycerate}][ATP][NADPH]\\ &- k_{\text{triose-P,RuBP}}[\text{triose-P}][ATP]\end{aligned} \tag{S9.1c}$$

$$\frac{d[\mathrm{NADPH}]}{dt} = k_{\mathrm{NADP}}([\mathrm{NP}]_o - [\mathrm{NADPH}])\mathrm{I}_l$$
$$- \tfrac{1}{2}k_{\mathrm{glycol}}[\text{P-glycolate}][\mathrm{O}_2]_i[\mathrm{ATP}][\mathrm{NADPH}]$$
$$- k_{\mathrm{glycerate}}[\text{P-glycerate}][\mathrm{ATP}][\mathrm{NADPH}]. \qquad \text{(S9.1d)}$$

With these four equations supplementing the five equations (9.3h)–(9.3l), the nine-state variable problem is now fully defined, and steady state or transient solutions can be obtained in the usual way.

9.2. First eliminate P from eqns (9.7i) and (9.8a):

$$P_n + R_d = \frac{\alpha I_l C_i/r_x}{\alpha I_l + C_i/r_x} \qquad \text{(S9.2a)}$$

Equation (9.8b) can be written

$$C_i = C_a - r_d P_n, \qquad \text{(S9.2b)}$$

which when substituted into (S9.2a) leads to eqn (9.9a).

Differentiating eqn (9.9a) with respect to I_l gives

$$2r_d P_n \frac{\partial P_n}{\partial I_l} - \frac{\partial P_n}{\partial I_l}[\alpha I_l(r_x + r_d) + C_a - R_d r_d]$$
$$- P_n\alpha(r_x + r_d) + \alpha(C_a - R_d r_x) = 0. \qquad \text{(S9.2c)}$$

As $P_n(I_l = 0) = -R_d$ (eqn (9.9c)),

$$(-2r_d R_d - C_a + R_d r_d)\frac{\partial P_n}{\partial I_l}(I_l = 0) + \alpha[R_d(r_x + r_d) + C_a - R_d r_x] = 0, \qquad \text{(S9.2d)}$$

from which the first of eqns (9.9d) follows immediately.

The second of eqns (9.9d) is readily obtained by letting $I_l \to \infty$ and $P_n = 0$ respectively in eqn (9.9a). Similarly, eqn (9.9e) is derived by considering $C_a \to \infty$ in eqn (9.9a). Differentiating eqn (9.9a) with respect to C_a gives

$$2r_d P_n \frac{\partial P_n}{\partial C_a} - \frac{\partial P_n}{\partial C_a}[\alpha I_l(r_x + r_d) + C_a - R_d r_d] - P_n + \alpha I_l - R_d = 0 \qquad \text{(S9.2e)}$$

and, with $C_a = 0$, eqn (9.9g) follows.

Equation (9.9f) is obtained by setting $C_a = 0$ in eqn (9.9a) and solving the quadratic. This is straightforward , but care is needed with the algebra. The binomial expansion is

$$(1 + \xi)^{1/2} = 1 + \frac{1}{2}\xi + \frac{(1/2)(-1/2)}{(1)(2)}\xi^2 + \frac{(1/2)(-1/2)(-3/2)}{(1)(2)(3)}\xi^3 + \cdots$$

provided that $|\xi| < 1$. With $0 < |\xi| \ll 1$, this approximates to

$$(1 + \xi)^{1/2} \approx 1 + \tfrac{1}{2}\xi \qquad \text{(S9.2f)}$$

(a) This is first applied to eqn (9.9f) with $r_x/r_d \to 0$ and $\alpha I_l \gg R_d$. $P_n(C_a = 0)$ becomes

$$P_n(C_a = 0) = \frac{1}{2}(\alpha I_l - R_d)\left\{1 - \left[1 + \frac{2R_d \alpha I_l r_x/r_d}{(\alpha I_l - R_d)^2}\right]\right\},$$

and hence

$$P_n(C_a = 0) \approx -\frac{R_d r_x}{r_d} \to 0. \tag{S9.2g}$$

Substituting in eqn (9.9g) gives

$$\frac{\partial P_n}{\partial C_a}(C_a = 0) \approx \frac{1}{r_d}\left(1 - \frac{R_d}{\alpha I_l}\frac{r_x}{r_d}\right) \to \frac{1}{r_d}. \tag{S9.2h}$$

The only source of CO_2 within the leaf is through the respiration term R_d, and this may be lost to the atmosphere or utilized through refixation within the leaf. $r_x/r_d \to 0$ implies a greater propensity for refixation than for diffusion to the atmosphere. Thus, in this limit, since light is non-limiting ($\alpha I_l \gg R_d$), virtually all the respired CO_2 is refixed and $P_n(C_a = 0)$ approaches zero. With CO_2 applied at low levels, eqn (S9.2h) simply means that net photosynthesis is limited by the resistance to diffusion of this CO_2 to the site of photosynthesis.

(b) We again apply the binomial expansion (eqn (S9.2f)) with $r_x/r_d \gg 1$ and $\alpha I_l \gg R_d$, and eqn (9.9f) reduces to

$$P_n(C_a = 0) \approx -R_d. \tag{S9.2i}$$

Equation (9.9g) then becomes

$$\frac{\partial P_n}{\partial C_a}(C_a = 0) \approx \frac{1}{r_x}. \tag{S9.2j}$$

In this case the tendency is for CO_2 to diffuse readily between the atmosphere and the sites of photosynthesis rather than be fixed by photosynthesis. Thus virtually all the respired CO_2 is lost to the atmosphere. Furthermore, at low levels of atmospheric CO_2, eqn (S9.2j) implies that P_n is limited by the resistance to CO_2 fixation at the sites of photosynthesis.

9.3. Combining eqns (9.8a) and (9.9a), we can readily obtain eqn (9.10a). The initial slope and asymptote of $P(I_l)$ can be obtained similarly to those for $P_n(I_l)$ in Exercise 9.2, or directly from the results in Solution 9.2, noting that

$$\frac{\partial P}{\partial P_l}(I_l = 0) = \frac{\partial P_n}{\partial I_l}(I_l = 0) \tag{S9.3a}$$

and

$$\lim_{I_l \to \infty} P(I_l) = R_d + \lim_{I_l \to \infty} P_n(I_l). \tag{S9.3b}$$

Equation (9.10i) is derived directly from the quadratic equation (9.10f). Consider the limit $\theta \to 0$ in eqn (9.10i). Rewriting (9.10i) as

$$P = \frac{1}{2\theta}\left\{\alpha P_1 + P_m - (\alpha I_1 + P_m)\left[1 - \frac{4\theta\alpha I_1 P_m}{(\alpha I_1 + P_m)^2}\right]^{1/2}\right\} \tag{S9.3c}$$

and applying the binomial eqn (S9.2f) as $\theta \to 0$, this becomes

$$P \to \frac{1}{2\theta}(\alpha I_1 + P_m)\left\{1 - \left[1 - \frac{2\theta\alpha I_1 P_m}{(\alpha I_1 + P_m)^2}\right]\right\} \tag{S9.3d}$$

and eqn (9.10g) follows immediately.

When $\theta = 1$, eqn (9.10f) becomes

$$P = \tfrac{1}{2}\{\alpha I_1 + P_m - [(\alpha I_1 - P_m)^2]\}^{1/2},$$

and as, by definition, the square root term is positive, this can be written

$$P = \tfrac{1}{2}(\alpha I_1 + P_m - |\alpha I_1 - P_m|), \tag{S9.3e}$$

which leads directly to eqn (9.10h).

9.4. Differentiating eqn (9.12n) with respect to I_1 gives

$$\begin{aligned} 0 = {} & 2P_n\frac{\partial P_n}{\partial I_1}(r_p r_{dc} - r_x r_{do}) - P_n\alpha(r_x r_p + r_p r_{dc} + r_x r_{do}) \\ & - \frac{\partial P_n}{\partial I_1}[r_p C_a + r_x O_a + \alpha I_1(r_x r_p + r_p r_{dc} + r_x r_{do}) - R_d(r_p r_{dc} - r_x r_{do})] \\ & + \alpha[(r_p C_a - r_x O_a) - R_d r_x r_p]. \end{aligned} \tag{S9.4a}$$

Note that $I_1 = 0$ and $P_n = -R_d$ is a solution of eqn (9.12n). Next, solve (S9.4a) for $\partial P_n/\partial I_1$, and put $I_1 = 0$ and $P_n = -R_d$ to give

$$\frac{\partial P_n}{\partial I_1}(I_1 = 0) = \alpha\left[\frac{r_p C_a - r_x O_a + R_d(r_p r_{dc} + r_x r_{do})}{r_p C_a + r_x O_a + R_d(r_p r_{dc} - r_x r_{do})}\right]. \tag{S9.4b}$$

9.5. Eliminating P from eqns (9.16m) and (9.16o) gives

$$P_n + R_d = \frac{\alpha I_1(C_{bs}/r_x - O_{bs}/r_p)}{\alpha I_1 + C_{bs}/r_x + O_{bs}/r_p + C_m/r_4}. \tag{S9.5a}$$

From (9.16p) and (9.16q)

$$C_{bs} = C_a - P_n r_{dcbs} + \frac{(C_a - C_m) r_{dcbs}}{r_{dcm}} \tag{S9.5b}$$

$$O_{bs} = O_a + P_n r_{do}. \tag{S9.5c}$$

Substituting (S9.5b) and (S9.5c) in (S9.5a) gives

$$(P_\text{n} + R_\text{d})\left[\alpha I_\text{l} + \frac{C_\text{a}(1 + (r_\text{dcbs}/r_\text{dcm}))}{r_\text{x}} + C_\text{m}\left(\frac{1}{r_4} - \frac{r_\text{dcbs}}{r_\text{dcm}r_\text{x}}\right)\right.$$
$$\left. + \frac{O_\text{a}}{r_\text{p}} - P_\text{n}\left(\frac{r_\text{dcbs}}{r_\text{x}} - \frac{r_\text{do}}{r_\text{p}}\right)\right]$$
$$= \alpha I_\text{l}\left[\frac{C_\text{a}(1 + (r_\text{dcbs}/r_\text{dcm})}{r_\text{x}} - \frac{C_\text{m}r_\text{dcbs}}{r_\text{dcm}r_\text{x}} - \frac{O_\text{a}}{r_p}\right.$$
$$\left. - P_\text{n}\left(\frac{r_\text{dcbs}}{r_\text{x}} + \frac{r_\text{do}}{r_\text{p}}\right)\right]. \qquad \text{(S9.5d)}$$

Equation (9.16r) is

$$\frac{C_\text{a} - C_\text{m}}{r_\text{dcm}} = \frac{\alpha I_\text{l} C_\text{m}/r_4}{\alpha I_\text{l} + C_\text{bs}/r_\text{x} + O_\text{bs}/r_\text{p} + C_\text{m}/r_4}, \qquad \text{(S9.5e)}$$

and substituting for C_bs and O_bs from (S9.5b) and (S9.5c) gives

$$\frac{\alpha I_\text{l} C_\text{m} r_\text{dcm}}{r_4} = (C_\text{a} - C_\text{m})\left[\alpha I_\text{l} + \frac{C_\text{a}(1 + (r_\text{dcbs}/r_\text{dcm}))}{r_\text{x}} + C_\text{m}\left(\frac{1}{r_4} - \frac{r_\text{dcbs}}{r_\text{dcm}r_\text{x}}\right)\right.$$
$$\left. + \frac{O_\text{a}}{r_\text{p}} - P_\text{n}\left(\frac{r_\text{dcbs}}{r_\text{x}} - \frac{r_\text{do}}{r_\text{p}}\right)\right]. \qquad \text{(S9.5f)}$$

Next, solve (S9.5d) for C_m:

$$C_\text{m}\left[(P_\text{n} + R_\text{d})\left(\frac{1}{r_4} - \frac{r_\text{dcbs}}{r_\text{dcm}r_\text{x}}\right) + \frac{\alpha I_\text{l} r_\text{dcbs}}{r_\text{dcm}r_\text{x}}\right]$$
$$= \alpha I_\text{l}\left[\frac{C_\text{a}(1 + (r_\text{dcbs}/r_\text{dcm}))}{r_\text{x}} - \frac{O_\text{a}}{r_\text{p}} - P_\text{n}\left(\frac{r_\text{dcbs}}{r_\text{x}} + \frac{r_\text{do}}{r_\text{p}}\right)\right]$$
$$- (P_\text{n} + R_\text{d})\left[\alpha I_\text{l} + \frac{C_\text{a}(1 + (r_\text{dcbs}/r_\text{dcm}))}{r_\text{x}} + \frac{O_\text{a}}{r_\text{p}} - P_\text{n}\left(\frac{r_\text{dcbs}}{r_\text{x}} - \frac{r_\text{do}}{r_\text{p}}\right)\right]. \qquad \text{(S9.5g)}$$

Note that $I_\text{l} = 0$ and $P_\text{n} = -R_\text{d}$ satisfies this equation for all values of C_m. Note also that with $r_4 \to \infty$ and $r_\text{dcm} \to \infty$, the term on the left of the equation in C_m becomes zero, and the equation becomes identical with (9.12n) for C_3 photosynthesis and photorespiration.

Now rewrite (S9.5f) as

$$0 = C_\text{m}^2\left(\frac{1}{r_4} - \frac{r_\text{dcbs}}{r_\text{dcm}r_\text{x}}\right) + C_\text{m}\left[\alpha I_\text{l}\left(1 + \frac{r_\text{dcm}}{r_4}\right) + \frac{O_\text{a}}{r_\text{p}}\right.$$
$$\left. - P_\text{n}\left(\frac{r_\text{dcbs}}{r_\text{x}} - \frac{r_\text{do}}{r_\text{p}}\right) + C_\text{a}\left(\frac{1}{r_\text{x}} - \frac{1}{r_4} + \frac{2r_\text{dcbs}}{r_\text{dcm}r_\text{x}}\right)\right]$$
$$- C_\text{a}\left[\alpha I_\text{l} + \frac{C_\text{a}(1 + (r_\text{dcbs}/r_\text{dcm}))}{r_\text{x}} + \frac{O_\text{a}}{r_\text{p}} - P_\text{n}\left(\frac{r_\text{dcbs}}{r_\text{x}} - \frac{r_\text{do}}{r_\text{p}}\right)\right]. \qquad \text{(S9.5h)}$$

To simplify notation, (S9.5g) is written as

$$C_{\mathrm{m}} = \frac{aP_{\mathrm{n}}^{2} - bP_{\mathrm{n}} + c}{P_{\mathrm{n}} + d}, \tag{S9.5i}$$

where

$$\begin{aligned}
a\left(\frac{1}{r_4} - \frac{r_{\mathrm{dcbs}}}{r_{\mathrm{dcm}} r_{\mathrm{x}}}\right) &= \frac{r_{\mathrm{dcbs}}}{r_{\mathrm{x}}} - \frac{r_{\mathrm{do}}}{r_{\mathrm{p}}} \\
b\left(\frac{1}{r_4} - \frac{r_{\mathrm{dcbs}}}{r_{\mathrm{dcm}} r_{\mathrm{x}}}\right) &= \alpha I_1\left(\frac{r_{\mathrm{dcbs}}}{r_{\mathrm{x}}} + \frac{r_{\mathrm{do}}}{r_{\mathrm{p}}} + 1\right) + \frac{C_{\mathrm{a}}(1 + (r_{\mathrm{dcbs}}/r_{\mathrm{dcm}}))}{r_{\mathrm{x}}} + \frac{O_{\mathrm{a}}}{r_{\mathrm{p}}} \\
&\quad - R_{\mathrm{d}}\left(\frac{r_{\mathrm{dcbs}}}{r_{\mathrm{x}}} - \frac{r_{\mathrm{do}}}{r_{\mathrm{p}}}\right) \\
c\left(\frac{1}{r_4} - \frac{r_{\mathrm{dcbs}}}{r_{\mathrm{dcm}} r_{\mathrm{x}}}\right) &= \alpha I_1\left[\frac{C_{\mathrm{a}}(1 + (r_{\mathrm{dcbs}}/r_{\mathrm{dcm}}))}{r_{\mathrm{x}}} - \frac{O_{\mathrm{a}}}{r_{\mathrm{p}}} - R_{\mathrm{d}}\right] \\
&\quad - R_{\mathrm{d}}\left[\frac{C_{\mathrm{a}}(1 + (r_{\mathrm{dcbs}}/r_{\mathrm{dcm}}))}{r_{\mathrm{x}}} + \frac{O_{\mathrm{a}}}{r_{\mathrm{p}}}\right] \\
d\left(\frac{1}{r_4} - \frac{r_{\mathrm{dcbs}}}{r_{\mathrm{dcm}} r_{\mathrm{x}}}\right) &= \alpha I_1 \frac{r_{\mathrm{dcbs}}}{r_{\mathrm{dcm}} r_{\mathrm{x}}} + R_{\mathrm{d}}\left(\frac{1}{r_4} - \frac{r_{\mathrm{dcbs}}}{r_{\mathrm{dcm}} r_{\mathrm{x}}}\right).
\end{aligned} \tag{S9.5j}$$

Similarly simplifying (S9.5h), we write

$$0 = C_{\mathrm{m}}^{2} + C_{\mathrm{m}}(e - fP_{\mathrm{n}}) - g + hP_{\mathrm{n}}, \tag{S9.5k}$$

where

$$\begin{aligned}
e\left(\frac{1}{r_4} - \frac{r_{\mathrm{dcbs}}}{r_{\mathrm{dcm}} r_{\mathrm{x}}}\right) &= \alpha I_1\left(1 + \frac{r_{\mathrm{dcm}}}{r_4}\right) + \frac{O_{\mathrm{a}}}{r_{\mathrm{p}}} + C_{\mathrm{a}}\left(\frac{1}{r_{\mathrm{x}}} - \frac{1}{r_4} + \frac{2r_{\mathrm{dcbs}}}{r_{\mathrm{dcm}} r_{\mathrm{x}}}\right) \\
f\left(\frac{1}{r_4} - \frac{r_{\mathrm{dcbs}}}{r_{\mathrm{dcm}} r_{\mathrm{x}}}\right) &= \frac{r_{\mathrm{dcbs}}}{r_{\mathrm{x}}} - \frac{r_{\mathrm{do}}}{r_{\mathrm{p}}} \\
g\left(\frac{1}{r_4} - \frac{r_{\mathrm{dcbs}}}{r_{\mathrm{dcm}} r_{\mathrm{x}}}\right) &= C_{\mathrm{a}}\left[\alpha I_1 + \frac{C_{\mathrm{a}}(1 + (r_{\mathrm{dcbs}}/r_{\mathrm{dcm}}))}{r_{\mathrm{x}}} + \frac{O_{\mathrm{a}}}{r_{\mathrm{p}}}\right] \\
h\left(\frac{1}{r_4} - \frac{r_{\mathrm{dcbs}}}{r_{\mathrm{dcm}} r_{\mathrm{x}}}\right) &= C_{\mathrm{a}}\left(\frac{r_{\mathrm{dcbs}}}{r_{\mathrm{x}}} - \frac{r_{\mathrm{do}}}{r_{\mathrm{p}}}\right).
\end{aligned} \tag{S9.5l}$$

These coefficients are not all independent; in particular $a = f$, and we replace f by a in (S9.5k). Now, eliminate C_{m} between eqns (S9.5i) and (S9.5k):

$$\begin{aligned}
0 &= (aP_{\mathrm{n}}^{2} - bP_{\mathrm{n}} + c)^{2} + (e - aP_{\mathrm{n}})(P_{\mathrm{n}} + d)(aP_{\mathrm{n}}^{2} - bP_{\mathrm{n}} + c) \\
&\quad + (P_{\mathrm{n}} + d)^{2}(hP_{\mathrm{n}} - g).
\end{aligned} \tag{S9.5m}$$

This becomes, successively,

$$0 = (aP_n^2 - bP_n + c)^2 + [-aP_n^2 + (e - ad)P_n + de](aP_n^2 - bP_n + c)$$
$$+ (hP_n - g)(P_n^2 + 2dP_n + d^2)$$
$$0 = P_n^4(a^2 - a^2) + P_n^3[-2ab + ab + a(e - ad) + h]$$
$$+ P_n^2[b^2 + 2ac - ac - b(e - ad) + ade + 2hd - g]$$
$$+ P_n[-2bc + c(e - ad) - bde + hd^2 - 2dg]$$
$$+ c^2 + cde - gd^2$$
$$0 = P_n^3(-ab + ae - da^2 + h) + P_n^2(b^2 + ac - be + abd + ade + 2hd - g)$$
$$+ P_n(-2bc + ce - acd - bde + hd^2 - 2dg) + c^2 + cde - gd^2. \quad \text{(S9.5n)}$$

Solutions to this cubic can be written down directly, but this can be quite complicated, and it is easier here to generate the solutions by means of the Newton–Raphson method (p. 605). Let

$$y = u_3 P_n^3 + u_2 P_n^2 + u_1 P_n + u_0 \quad \text{(S9.5o)}$$

where

$$\begin{aligned} u_3 &= -ab + ae - da^2 + h \\ u_2 &= b^2 + ac - be + abd + ade + 2hd - g \\ u_1 &= -2bc + ce - acd - bde + hd^2 - 2dg \\ u_0 &= c^2 + cde - gd^2. \end{aligned} \quad \text{(S9.5p)}$$

Differentiating (S9.5o)

$$y'(P_n) = \frac{dy}{dP_n} = 3u_3 P_n^2 + 2u_2 P_n + u_1.$$

If $P_n = v$ is a first guess at the solution to $y = 0$, then a better approximation is

$$P_n = v + w, \quad \text{(S9.5q)}$$

where $w = -y(v)/y'(v)$. This is repeated until the desired accuracy is obtained. A first guess that converges on the physiologically acceptable solution is $P_n = 0$.

An expression for $P_n(I_1 \to \infty)$ can be derived. First pick out the terms in (S9.5n) containing the highest power of I_1, namely I_1^3, giving

$$P_n(-bde) + cde - gd^2 = 0.$$

Therefore

$$P_n = \frac{ce - gd}{be}.$$

Substituting the coefficients, carrying only the terms in αI_1, and cancelling out $(\alpha I_1)^2$ gives

$$\lim_{I_l \to \infty} P_n = \frac{[C_a(1 + r_{dcbs}/r_{dcm})/r_x - O_a/r_p - R_d](1 + r_{dcm}/r_4) - C_a(r_{dcbs}/r_{dcm}r_x)}{(r_{dcbs}/r_x + r_{do}/r_p + 1)(1 + r_{dcm}/r_4)}. \tag{S9.5r}$$

In the limit $r_4 \to \infty$ (no C_4 photosynthesis), (S9.5r) reduces to eqn (9.13b) for C_3 photosynthesis and photorespiration.

Finding the initial slope of the light response curve $dP_n/dI_l(I_l = 0)$ is algebraically tedious (although straightforward), and here we merely indicate the method. First differentiate (S9.5o) with $y = 0$:

$$0 = u_3' P_n^{\,3} + u_2' P_n^{\,2} + u_1' P_n + u_0' + 3u_3 P_n P_n' + 2u_2 P_n P_n' + u_1 P_n',$$

where the prime indicates differentiation with respect to I_l. Therefore

$$P_n' = -\frac{u_3' P_n^{\,3} + u_2' P_n^{\,2} + u_1' P_n + u_0'}{3u_3 P_n + 2u_2 P_n + u_1}. \tag{S9.5s}$$

Equation (S9.5s) is evaluated at $I_l = 0$ with $P_n = -R_d$.

The procedure is now to use (S9.5j) and (S9.5l) to substitute for the coefficients a–h in (S9.5p) to obtain expressions for u_i, $i = 0, 1, 2, 3$. Simplify these and then substitute $I_l = 0$ in these expressions; differentiate with respect to I_l to give u_i' and also put $I_l = 0$. These are substituted in (S9.5s) to give the initial slope $dP_n/dI_l(I_l = 0)$.

Chapter 10

10.1. (a) Taking the Taylor series expansion (eqn (1.7e), p. 20) for $F(x)$, eqn (10.5d), with respect to θ, it can be shown after some involved but straightforward analysis that, as $\theta \to 0$,

$$F(x) \to P_m\{1 + 2\ln[2(\alpha x + P_m)]\}\theta \tag{S10.1a}$$

and (10.5e) follows immediately.

(b) First differentiate eqn (10.5c) with respect to I_0 to obtain

$$\frac{dP}{dI_0}(I_0 = 0) = \frac{1}{2\theta(1 - m)}[1 - e^{-kL}]\frac{dF}{dx}(x = 0). \tag{S10.1b}$$

To evaluate this again involves expanding $F(x)$, but now with respect to x rather than θ. This can either be done directly using the Taylor series, as in part (a), or by showing that, as $x \to 0$,

$$[(\alpha x)^2 + 2P_m(1 - 2\theta)\alpha x + P_m^{\,2}]^{1/2} \to P_m + (1 - 2\theta)\alpha x, \tag{S10.1c}$$

from which

$$F(x) \to -P_m + 2\theta\alpha x - P_m(1 - 2\theta)\ln[2(1 - \theta)(P_m + \alpha x)] + P_m \ln\{2[P_m + (1 - 2\theta)\alpha x]\}. \tag{S10.1d}$$

Differentiating and taking the limit $x \to 0$ gives

$$\frac{dF}{dx}(x = 0) = 2\theta\alpha, \tag{S10.1e}$$

and substituting in (S10.1b) gives the required result. This result is to be expected since at low light the leaf photosynthesis approximates to αI_l and the canopy photosynthesis becomes the product of α and the total irradiance intercepted, as given by eqn (8.7a).

For the asymptote, note that as $x \to \infty$

$$F(x) \to P_m(1 - 2\theta) + P_m \ln(1 - \theta) + 2\theta P_m \ln 2\alpha x, \tag{S10.1f}$$

and the result follows by substituting in eqn (10.5c) with (10.1b) and taking the limit $I_0 \to \infty$. The asymptote is physiologically realistic in that all the leaves in the canopy are photosynthesizing at their maximum rate.

10.2. The curves are shown in Fig. S10.1. The obvious feature is that as L increases P increases. Also of interest is the fact that at low leaf area index (LAI) and high irradiance, increasing k causes P to increase, whereas at high LAI this is reversed. This latter observation may be quite surprising, but is a result of most of the irradiance being intercepted by leaves which are already at saturating irradiance. By reducing k, this allows leaves lower in the canopy to intercept more irradiance and thus increase the canopy photosynthetic rate.

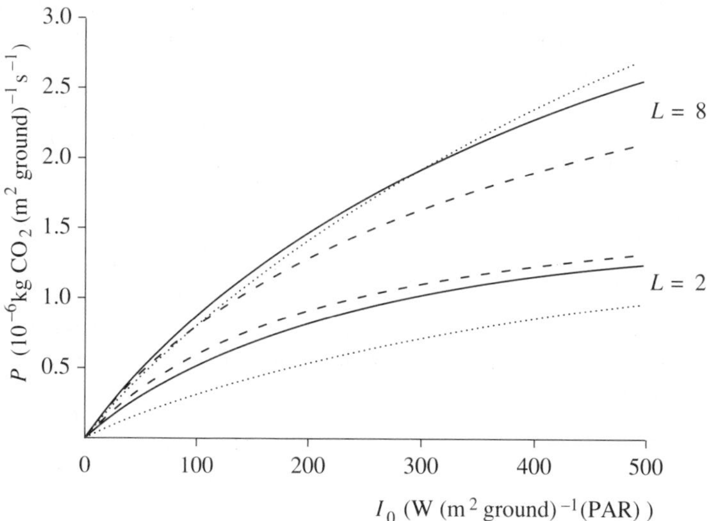

Fig. S10.1. Instantaneous canopy gross photosynthetic rate as required in Exercise 10.2: $\cdots$, $k = 0.2$; ——, $k = 0.5$; – – –, $k = 0.8$.

10.3. (a) Differentiating eqn (10.9g) with respect to I_l gives, after some algebra

$$\frac{\partial f}{\partial I_l} = -\frac{2P_m{}^2\alpha^2 I_l(1-\theta)}{k}[(\alpha I_l + P_m)^2 - 4\theta\alpha I_l P_m]^{-3/2}, \tag{S10.3a}$$

so that, with $0 \leqslant \theta < 1$, this implies that

$$\frac{\partial f}{\partial I_l} < 1 \quad \text{for all values of } I_l \tag{S10.3b}$$

and the result follows.

(b) Equation (10.9g) for $f(I_l, P_m)$ can be written as

$$f(I_l, P_m) = \frac{P_m}{2\theta k}\left(1 - \frac{2\theta\alpha I_l}{\alpha I_l + P_m}\right)\left[1 - \frac{2\theta\alpha I_l P_m}{(\alpha I_l + P_m)^2}\right]^{-1/2}, \tag{S10.3c}$$

and taking the leading terms in the expansion with respect to θ gives

$$f(I_l, P_m) = \frac{P_m}{2\theta k}\left[1 - 2\theta\left(\frac{\alpha I_l}{\alpha I_l + P_m}\right)^2\right] \tag{S10.3d}$$

and the result follows.

10.4. These calculations simply require care with the integrations and the algebra. The following identities may be helpful:

$$\cos(2x) = 1 - 2\sin^2 x$$

$$\sin x \sin y = \tfrac{1}{2}[\cos(x-y) - \cos(x+y)].$$

10.5. In Chapter 8 we saw that kL represents the projection of L on to the horizontal plane. Note that it incorporates the transmission of the leaves and so kL is the equivalent leaf area index of horizontal non-transmitting leaves that absorbs the same irradiance as L. By interpreting k in this way, the following conclusions can be drawn:

k_e can be interpreted as the mean extinction coefficient of the canopy (note that this is not the same as the mean of the canopy extinction coefficients);

L_i^* is that leaf area index of a canopy with extinction coefficient k_i that would have the same projection onto the horizontal plane as does the mixed canopy;

$L_i k_i/(L_1 k_1 + L_2 k_2)$ is the ratio of the projected leaf area index of the ith component to that of the total canopy.

Chapter 11

Equation (11.3o) can be written

$$R = \frac{1 - Y_g}{Y_g}\frac{dw}{dt} + mw. \tag{S11.1a}$$

Substituting for R in eqn (E11.1a) and simplifying, therefore, gives

$$\frac{dw}{dt} = Y_g(P - mw). \tag{S11.1b}$$

Next, substituting for photosynthetic rate P with eqn (E11.1c) gives

$$\frac{1}{w}\frac{dw}{dt} = Y_g(\alpha I \lambda f_{leaf} - m). \tag{S11.1c}$$

Numerically

$$\frac{1}{w}\frac{dw}{dt} = 0.7(0.3 \times 10^{-8} \times 2 \times 10^6 \times 50 \times 0.5 - 0.01)$$

$$= 0.7(0.15 - 0.01) = 0.10 \text{ day}^{-1}. \tag{S11.1d}$$

Note the reasonable size of this very simplistic estimate of crop specific growth rate, and that maintenance has contributed a 7 per cent reduction to specific growth rate.

11.2. Equation (11.3r) is

$$r = \left(\frac{1 - Y_g}{Y_g}\right)\frac{1}{w}\frac{dw}{dt} + m. \tag{S11.2a}$$

Substituting eqn (E11.2a) for dw/dt gives

$$r = \mu_0\left(\frac{1 - Y_g}{Y_g}\right)\left(1 - \frac{w}{w_f}\right) + m. \tag{S11.2b}$$

This is a straight line of negative slope passing through the points ($w = 0$, $r = \mu_0(1 - Y_g)/Y_g + m$) and ($w = w_f, r = m$).

From eqn (3.5l) the differential form of the Gompertz growth equation can be written as

$$\frac{1}{w}\frac{dw}{dt} = D(\ln w_f - \ln w), \tag{S11.2c}$$

where D (day^{-1}) is a parameter and w_f is as above. With eqn (S11.2a), therefore, we obtain

$$r = D\left(\frac{1 - Y_g}{Y_g}\right)(\ln w_f - \ln w) + m. \tag{S11.2d}$$

Note that any time-course data for dry mass and specific respiration rate, i.e. $w:t$ and $r:t$ relationships, can be assessed for compatibility with the substrate balance analysis by means of eqn (S11.2a).

11.3. Working directly from Fig. E11.1, we obtain

$$\frac{dw_{st}}{dt} = Y_{st}P - k_{st}w_{st} \qquad \text{(S11.3a)}$$

$$\frac{dw_{sol}}{dt} = (1 - Y_{st})P + k_{st}w_{st} - k_g w_{sol} + k_d w_d \qquad \text{(S11.3b)}$$

$$\frac{dw_d}{dt} = Y_d Y_g k_g w_{sol} - k_d w_d \qquad \text{(S11.3c)}$$

$$\frac{dw_n}{dt} = (1 - Y_d) Y_g k_g w_{sol}. \qquad \text{(S11.3d)}$$

Adding these gives (cf. eqn (11.5e))

$$\frac{dw}{dt} = P - k_g(1 - Y_g)w_{sol}, \qquad \text{(S11.3e)}$$

so that eqn (11.5f) becomes

$$R = k_g(1 - Y_g)w_{sol}. \qquad \text{(S11.3f)}$$

Equation (11.5g) takes the form (use eqns (E11.3a) and (E11.3b))

$$\frac{dw}{dt} - \frac{dw_S}{dt} = \frac{dw_d}{dt} + \frac{dw_n}{dt} = Y_g k_g w_{sol} - k_d w_d. \qquad \text{(S11.3g)}$$

Substitution of $k_g w_{sol}$ in eqn (S11.3f) gives eqns (11.5h)–(11.5k) in unaltered form.

11.4. With $P = kw_d$, and for balanced exponential growth with all the state variables $w_i \approx \exp(\lambda t)$, eqns (S11.3a)–(S11.3d) become

$$\lambda w_{st}(0) = Y_{st}kw_d(0) - k_{st}w_{st}(0) \qquad \text{(S11.4a)}$$

$$\lambda w_{sol}(0) = (1 - Y_{st})kw_d(0) + k_{st}w_{st}(0) + k_d w_d(0) - k_g w_{sol}(0) \qquad \text{(S11.4b)}$$

$$\lambda w_d(0) = Y_d Y_g k_g w_{sol}(0) - k_d w_d(0) \qquad \text{(S11.4c)}$$

$$\lambda w_n(0) = (1 - Y_d) Y_g k_g w_{sol}(0). \qquad \text{(S11.4d)}$$

$w_i(0)$ denotes the value of w_i at time $t = 0$ (i = st, sol, d, n). From eqns (S11.4a)–(S11.4c), $w_{sol}(0)$, $w_{st}(0)$, and $w_d(0)$ can be eliminated. For example, derive an expression for $w_{sol}(0)$ from eqn (S11.4b) and substitute into eqn (S11.4c). The resulting equation is (omitting the (0)s for convenience)

$$(\lambda + k_g)\lambda w_d + (1 + k_g)k_d w_d = Y_d Y_g k_g[k(1 - Y_{st}) + k_d]w_d + Y_d Y_g k_g k_{st} w_{st}. \qquad \text{(S11.4e)}$$

Next substitute for w_{st} from eqn (S11.4a) and simplify to give

$$\lambda^3 + \lambda^2(k_{st} + k_g + k_d) + \lambda\{k_{st}(k_g + k_d) + k_d k_g - Y_d Y_g k_g[k(1 - Y_{st}) + k_d]\} + k_{st}k_g k_d - Y_d Y_g k_g k_{st}(k + k_d) = 0. \qquad \text{(S11.4f)}$$

If (see Fig. E11.1) k_{st} becomes very large, the starch pool becomes negligibly small and the scheme of Fig. E11.1 reduces to that of Fig. 11.2. In eqn (S11.4f), we collect the terms that have the factor k_{st}, since these dominate the equation, and drop the k_{st} factor to give

$$\lambda_2 + \lambda(k_g + k_d) + k_g k_d - Y_d Y_g k_g(k + k_d) = 0. \qquad \text{(S11.4g)}$$

This is identical to eqn (11.7e).

11.5. For balanced exponential growth the specific growth rate is given by the larger (positive) of the two roots of eqns (11.7f) and (11.7g). We denote this by Λ, where

$$\Lambda = \tfrac{1}{2}\{-(k_g + k_d) + [(k_g - k_d)^2 + 4Y_g Y_d k_g(k + k_d)]^{1/2}\}. \qquad \text{(S11.5a)}$$

From eqn (11.5c) (or from eqns (11.7b) and (11.7c)

$$(\Lambda + k_d)w_d = Y_d Y_g k_g w_s, \qquad \text{(S11.5b)}$$

giving the ratio of storage to degradable structure in balanced exponential growth. Without loss in generality, we can take $t = 0$ as the time when the plants are placed in the dark, and assume that at this time

$$w_s(0) + w_d(0) = 1, \qquad \text{(S11.5c)}$$

so that

$$w_d(0) = \frac{Y_d Y_g k_g}{\Lambda + k_d + Y_d Y_g k_g}$$
$$w_s(0) = \frac{\Lambda + k_d}{\Lambda + k_d + Y_d Y_g k_g} \qquad \text{(S11.5d)}$$

In the dark, eqns (11.7f) and (11.7g) are solved for the two roots λ_1 and λ_2, which are both negative. Using eqns (11.7j) and (11.7k), we can now obtain A_1 and A_2. Substitution into eqn (11.7h) gives w_s, and the respiration rate R in the dark, given by

$$R = k_g(1 - Y_g)w_s, \qquad \text{(S11.5e)}$$

follows immediately with

$$R = k_g(1 - Y_g)[A_1 \exp(\lambda_1 t) + A_2 \exp(\lambda_2 t)]. \qquad \text{(S11.5f)}$$

The two terms in this equation will dominate the solution at different times. Taking logarithms, and assuming that the first term and then the second term are dominant, we obtain

$$\ln R = \ln[k_g A_1(1 - Y_g)] + \lambda_1 t$$
$$\ln R = \ln[k_g A_2(1 - Y_g)] + \lambda_2 t. \qquad \text{(S11.5g)}$$

These two straight lines with negative slopes give the limiting dark respiration : time relations. The intercepts depend on the values of A_1 and A_2, which themselves depend on the conditions before the plants are placed in the dark through eqns (11.7j), (11.7k), (S11.5d), and (S11.5a).

Chapter 12

12.1. For the first case (12.1j)

$$(1 - 1/18)\text{ glucose} \rightarrow 12\text{NADH}. \qquad \text{(S12.1a)}$$

For the second case

$$\text{glucose} \rightarrow 12\text{NADH}. \qquad \text{(S12.1b)}$$

For the pentose phosphate pathway (12.1l)

$$(1 + 1/36)\text{ glucose} \rightarrow 12\text{NADPH}. \qquad \text{(S12.1c)}$$

For our estimates of glucose requirements we take (S12.1b), as outlined in (12.1c). However, comparison of (S12.1b) with (S12.1a) and (S12.1c) gives an idea of the magnitude of possible errors.

12.2. The stoichometric equation is

$$(5 + 5/36)\text{ glucose} \rightarrow 6\text{ ribose}. \qquad \text{(S12.2a)}$$

This leads to a glucose requirement of

$$1.028\text{ kg glucose} \rightarrow 1\text{ kg ribose}. \qquad \text{(S12.2b)}$$

The biochemical efficiency is given by

$$\text{efficiency} = \frac{6 \times 2339}{(5 + 5/36) \times 2803} = 0.974. \qquad \text{(S12.2c)}$$

This process is marginally more efficient than (12.3i)–(12.3k) for ribose synthesis.

12.3. Per galacturonate residue, (12.6b) is

$$\text{glucose} + 2\text{ATP} \rightarrow (\text{galacturonate}) + 2\text{NADH}. \qquad \text{(S12.3a)}$$

Equating NADPH, NADH and $FADH_2$, and neglecting water, (12.2m) becomes

$$0.5\text{ glucose} + \text{tetrahydrofolate} + 0.5O_2$$
$$\rightarrow \text{methyl tetrahydrofolate} + 2CO_2 + 2\text{NADH}. \qquad \text{(S12.3b)}$$

Additionally, we need

$$\text{homocysteine} + \text{methyl tetrahydrofolate}$$
$$\rightarrow \text{methionine} + \text{tetrahydrofolate} \qquad \text{(S12.3c)}$$

and

(galacturonate) + methionine

$$\rightarrow \text{(methyl galacturonate)} + \text{homocysteine.} \quad \text{(S12.3d)}$$

Summing (S12.3a)–(S12.3d) gives

$$1.5\ \text{glucose} + 0.5O_2 + 2ATP \rightarrow \text{(methyl galacturonate)} + 2CO_2 + 4NADH. \quad \text{(S12.3e)}$$

With (12.1c), the stoichiometric equation is

$$(1.5 + 2/36 - 4/12)\ \text{glucose} \rightarrow \text{(methyl galacturonate).} \quad \text{(S12.3f)}$$

Assuming a relative molecular mass (Table 12.1) of $(176 - 1 + 15) = 190$ leads to

$$1.158\ \text{kg glucose} \rightarrow 1\ \text{kg fully methylated galacturonan.} \quad \text{(S12.3g)}$$

12.4. Equation (12.7e) for malate synthesis is

$$0.5\ \text{glucose} + CO_2 + NADPH \rightarrow \text{malate} + ATP + NADH. \quad \text{(S12.4a)}$$

Assuming that NADH + ATP is equivalent to NADPH, then the stoichiometric equations and biochemical efficiency (cf. eqns (12.7f)–(12.7h)) are

$$0.5\ \text{glucose} \rightarrow \text{malate} \quad \text{(S12.4b)}$$

giving a glucose requirement of

$$0.672\ \text{kg glucose} \rightarrow 1\ \text{kg malate} \quad \text{(S12.4c)}$$

and a biochemical efficiency of

$$\text{efficiency} = \frac{1328}{0.5 \times 2803} = 0.948. \quad \text{(S12.4d)}$$

12.5. We work through the answers for glycerol tristearate and trilinolenate. For glycerol tristearate, with $n_c = 18$ and $n_{db} = 0$, (12.13k) becomes

$$14\ \text{glucose} + 4ATP + 48NADPH \rightarrow \text{glycerol tristearate} + 27CO_2 + 24H_2O + 53NADH; \quad \text{(S12.5a)}$$

for glycerol trilinolenate with $n_c = 18$ and $n_{db} = 3$, (12.13k) becomes

$$14\ \text{glucose} + 3O_2 + 4ATP + 57NADPH \rightarrow \text{glycerol trilinolenate} + 27CO_2 + 42H_2O + 53NADH. \quad \text{(S12.5b)}$$

Note again that if we wish to check the balance of these equations for C, H and O atoms, we must remember that NADH and NADPH correspond to transfers of two H atoms.

The stoichiometric relations are (using (12.1c))

$$(14 + 4/36 - 5/12)\text{ glucose} \rightarrow \text{glycerol tristearate} \quad \text{(S12.5c)}$$

$$(14 + 4/36 + 4/12)\text{ glucose} \rightarrow \text{glycerol trilinolenate.} \quad \text{(S12.5d)}$$

Thus

$$2.770\text{ kg glucose} \rightarrow 1\text{ kg glycerol tristearate} \quad \text{(S12.5e)}$$

$$2.982\text{ kg glucose} \rightarrow 1\text{ kg glycerol trilinolenate.} \quad \text{(S12.5f)}$$

12.6. The combustion reaction occurs according to

$$NH_3 + \tfrac{3}{4}O_2 \rightarrow \tfrac{1}{2}N_2 + \tfrac{3}{2}H_2O. \quad \text{(S12.6a)}$$

The heat of combustion of NH_3 is given by

$$\begin{aligned} &-(3/2) \times \text{heat of formation of } H_2O \\ &+ 1 \times \text{heat of formation of } NH_3 \\ &= (3/2) \times 285.9 - 46.2 = 383\text{ MJ (kg mol)}^{-1}. \end{aligned} \quad \text{(S12.6b)}$$

One way of understanding this is to relate both sides of (S12.6a) back to the elements:

$$\tfrac{1}{2}N_2 + \tfrac{3}{2}H_2 + \tfrac{1}{2}O_2 \quad \text{(S12.6c)}$$

with a zero energy level, and then to consider the energy changes occurring when the compounds on the left and right of (S12.6a) are formed.

12.7. Combining (E12.7a) and (12.15h) leads to

$$\begin{aligned} &2.5\text{ glucose} + O_2 + 5HNO_3 + 11ATP + 12NADH \\ &\quad \rightarrow AMP + 5CO_2 + 3NADPH + 2FADH_2 + 11ADP + 10P_i. \end{aligned} \quad \text{(S12.7a)}$$

The glucose requirement is now (cf. (12.15i) and (12.15j))

$$(2.5 + 11/36 + 7/12)\text{ glucose} \rightarrow AMP \quad \text{(S12.7b)}$$

$$1.758\text{ kg glucose} \rightarrow 1\text{ kg AMP.} \quad \text{(S12.7c)}$$

This result checks against (12.14m). Other results are listed in Table 12.5.

12.8. Combining (E12.8a) and (E12.8b) gives

$$\begin{aligned} &0.5\text{ glucose} + NH_3 + H_2SO_4 + 4NADPH + 3ATP \\ &\quad \rightarrow \text{cysteine} + NADH. \end{aligned} \quad \text{(S12.8a)}$$

With (12.1c), this gives

$$(0.5 + 3/36 + 3/12) \text{ glucose} \rightarrow \text{cysteine}, \quad \text{(S12.8b)}$$

and therefore

$$1.240 \text{ kg glucose} \rightarrow 1 \text{ kg cysteine}. \quad \text{(S12.8c)}$$

With nitrate replacing ammonia as the nitrogen source (12.14a):

$$(0.5 + 3/36 + 7/12) \text{ glucose} \rightarrow \text{cysteine} \quad \text{(S12.8d)}$$

$$1.736 \text{ kg glucose} \rightarrow 1 \text{ kg cysteine}. \quad \text{(S12.8e)}$$

For the cysteine residue in a polypeptide chain, there is an extra cost of four ATP per molecule (12.20a), so that (S12.8b)–(S12.8e) become (using relative molecular mass of 121 − 18 = 103)

$$(0.5 + 7/36 + 3/12) \text{ glucose} \rightarrow \text{cysteine residue} \quad \text{(S12.8f)}$$

$$1.650 \text{ kg glucose} \rightarrow 1 \text{ kg cysteine residue} \quad \text{(S12.8g)}$$

and for the nitrate source

$$(0.5 + 7/36 + 7/12) \text{ glucose} \rightarrow \text{cysteine residue} \quad \text{(S12.8h)}$$

$$2.33 \text{ kg glucose} \rightarrow 1 \text{ kg cysteine residue}. \quad \text{(S12.8i)}$$

Note the large difference in glucose requirement for the cysteine residue formed from NH_3 and H_2S (0.922) and from nitrate and sulphate (2.233).

12.9. Working from the first AMP row of Table 12.5, using (12.1d) and (12.1e), we obtain for synthesis from NH_3:

$$(5 + 11/6 - 13/2)CO_2 \text{ results from 1 AMP formed}, \quad \text{(S12.9a)}$$

giving a CO_2 production coefficient of

$$44 \times (5 + 11/6 - 13/2)/347 = 0.042 \text{ kg } CO_2 \text{ (kg AMP)}^{-1}. \quad \text{(S12.9b)}$$

For synthesis from nitrate, (S12.9b) is replaced by

$$44 \times (5 + 11/6 + 7/2)/347 = 1.310 \text{ kg } CO_2 \text{ (kg AMP)}^{-1} \quad \text{(S12.9c)}$$

The oxygen requirement factor is

$$\frac{32 \times (1 + 11/6)}{347} = 0.261 \text{ kg } O_2 \text{ (kg AMP)}^{-1}. \quad \text{(S12.9d)}$$

The oxygen requirement factor is not changed by the change in nitrogen source.

For RNA with equal molar proportions of the four bases, we work from the row labelled 'sum' in Table 12.5. Then (S12.9b)–(S12.9d) become, first from ammonia,

$$\frac{44 \times (10 + 44/6 - 33/2)}{1357 - 4 \times 18} = 0.029 \text{ kg } CO_2 \text{ (kg RNA)}^{-1} \quad \text{(S12.9e)}$$

and second from nitrate,

$$\frac{44 \times (10 + 44/6 + 27/2)}{1357 - 4 \times 18} = 1.056 \text{ kg } CO_2 \text{ (kg RNA)}^{-1} \quad \text{(S12.9f)}$$

$$\frac{32 \times (2 + 44/6)}{1357 - 4 \times 18} = 0.232 \text{ kg } O_2 \text{ (kg RNA)}^{-1} \quad \text{(S12.9g)}$$

12.10. Working from the second row of Table 12.6, first the CO_2 production for arginine formation from ammonia is

$$\frac{44 \times (-3 + 6/6 + 5/2)}{174} = 0.126 \text{ kg } CO_2 \text{ (kg arginine)}^{-1} \quad \text{(S12.10a)}$$

and from nitrate it is

$$\frac{44 \times (-3 + 6/6 + 21/2)}{174} = 2.149 \text{ kg } CO_2 \text{ (kg arginine)}^{-1}; \quad \text{(S12.10b)}$$

the oxygen requirement is

$$\frac{32 \times (6/6)}{174} = 0.184 \text{ kg } O_2 \text{ (kg arginine)}^{-1} \quad \text{(S12.10c)}$$

for both ammonia and nitrate as the nitrogen source.

For the arginine residue in the polypeptide chain, the ATP requirement is increased by 4 and the relative molecular mass becomes 174 − 18 = 156. Thus (E12.10a)–(E12.10c) are replaced by, from ammonia,

$$\frac{44 \times (-3 + 10/6 + 5/2)}{156} = 0.329 \text{ kg } CO_2 \text{ (kg arginine residue)}^{-1} \quad \text{(S12.10d)}$$

and from nitrate,

$$\frac{44 \times (-3 + 10/6 + 21/2)}{156} = 2.585 \text{ kg } CO_2 \text{ (kg arginine residue)}^{-1}; \quad \text{(S12.10e)}$$

the oxygen requirement is

$$\frac{32 \times (10/6)}{156} = 0.342 \text{ kg } O_2 \text{ (kg arginine residue)}^{-1}. \quad \text{(S12.10f)}$$

12.11. The answer is constructed in Table S12.1.

Table S12.1. Efficiency of formation of grass at different stages of maturity

Stage of maturity	Category	Fraction	Glucose requirement	CO_2 production	O_2 requirement
Young vegetative	Protein†	0.33	0.532 (0.774)	0.126 (0.481)	0.083
	Lipid (glycerol trioleate)	0.10	0.284	0.133	0.013
	Sugars (sucrose)	0.10	0.110	0.006	0.005
	Minerals (11.21d)–(11.21i)	0.12	0.015	0.022	0.016
	Hemicellulose (xylan)	0.14	0.170	0.016	0.011
	Cellulose	0.18	0.211	0.016	0.012
	Lignin (polyconiferyl alcohol)	0.03	0.054	0.013	0.006
Total		1.00	1.376 (1.618)	0.332 (0.687)	0.146
Mature flowering	Protein†	0.07	0.113 (0.164)	0.027 (0.102)	0.018
	Lipid (glycerol trioleate)	0.03	0.085	0.040	0.004
	Sugars (sucrose)	0.25	0.274	0.016	0.012
	Minerals (11.21d)–(11.21i)	0.05	0.006	0.009	0.007
	Hemicellulose (xylan)	0.23	0.279	0.026	0.019
	Cellulose	0.30	0.352	0.032	0.020
	Lignin (polyconiferyl alcohol)	0.07	0.127	0.031	0.014
Total		1.00	1.236 (1.287)	0.181 (0.437)	0.094

The fractions are expressed on a mass basis. The glucose, CO_2, and oxygen requirements and production are all in kg (kg total dry matter)$^{-1}$; these are obtained from Tables 12.2, 12.4, and 12.6.

† Equal molar amounts of 20 amino acids (not cystine) (see Table 12.6); from NH_3 and H_2S; values from nitrate are in parentheses.

12.12. Equating hydrogen and oxygen atoms in (E12.12b) gives

$$\text{H:} \quad 2c + n + 2s + r = h + 2w \tag{S12.12a}$$

and

$$\text{O:} \quad 2c + 3n + 4s = x + w. \tag{S12.12b}$$

Eliminating w gives

$$r(HNO_3, H_2SO_4) = h - 2x + 5n + 6s. \qquad (S12.12c)$$

Replacing the n HNO_3 term in (E12.12b) by $(n/2)N_2$, and then equating hydrogen and oxygen atoms gives

$$H: \quad 2 + 2s + r = h + 2w \qquad (S12.12d)$$

and

$$O: \quad 1 + 4s = x + w. \qquad (S12.12e)$$

Elimination of w leads to

$$r(N_2, H_2SO_4) = h - 2x + 6s. \qquad (S12.12f)$$

12.12. Equating hydrogen and oxygen atoms in (E12.13b) gives

$$H: \quad 2c + 3n + 2s + 3p + r = h + 2w \qquad (S12.13a)$$

$$O: \quad c + 4p = x + w. \qquad (S12.13b)$$

Eliminating w leads to

$$r = h - 2x - 3n - 2s + 5p. \qquad (S12.13c)$$

12.14. For AMP, from Table 12.1,

$$c_{\text{stoich}} = 10. \qquad (S12.14a)$$

Using eqn (E12.14a)

$$r_{\text{stoich}} = 14 - 14 + 25 + 5 = 30. \qquad (S12.14b)$$

Substituting in eqn (12.25a) gives

$$R_{\text{glc, stoich}} = 10/6 + 30/24 = 2.917. \qquad (S12.14c)$$

From Table 12.5,

$$c_{\text{pathway}} = 6 \times 2.5 = 15 \qquad (S12.14d)$$

$$r_{\text{pathway}} = 2 \times (12 - 3 - 2) = 14 \qquad (S12.14e)$$

$$a_{\text{pathway}} = 11. \qquad (S12.14f)$$

Therefore with eqn (12.25b)

$$R_{\text{glc, pathway}} = 3.389. \qquad (S12.14g)$$

The percentage difference between the stoichiometric and pathway glucose requirements is −13.9 per cent, compared with an error of −29.7 per cent when ammonia is the nitrogen substrate (Table 12.9).

Protein with equal molar amino acids has the stoichiometric formula

$$C_{10}H_{157}O_{29}N_{29}S_2. \qquad (S12.14h)$$

Thus

$$c_{\text{stoich}} = 107 \quad \text{(S12.14i)}$$

$$r_{\text{stoich}} = 157 - 2 \times 29 + 5 \times 29 - 2 \times 2 = 240. \quad \text{(S12.14j)}$$

Therefore

$$R_{\text{glc, stoich}} = 107/6 + 240/24 = 27.833. \quad \text{(S12.14k)}$$

From Table 12.6,

$$c_{\text{pathway}} = 6 \times 19 = 114 \quad \text{(S12.14l)}$$

$$r_{\text{pathway}} = 2 \times (97 + 13 - 2) = 216 \quad \text{(S12.14m)}$$

$$a_{\text{pathway}} = 107. \quad \text{(S12.14n)}$$

Therefore, with eqn (12.25b),

$$R_{\text{glc, pathway}} = 30.972. \quad \text{(S12.14o)}$$

The percentage difference between the stoichiometric and the pathway estimates is -10.1 per cent, compared with -14.6 per cent in Table 12.9 (bottom row).

Chapter 13

13.1. If P_n is the rate of instantaneous single-leaf net photosynthesis (kg CO_2 m^{-2} (leaf) s^{-1}) then, averaged over the day,

$$P_n = \frac{\sigma_c}{20}\frac{44}{12}\frac{1}{h}, \quad \text{(S13.1a)}$$

where h is the day length (s day^{-1}) and the factor 44/12 converts carbon to CO_2. If a 16 day is assumed, $h = 57\,600$ s day^{-1}, and so

$$P_n = 0.48 \times 10^{-6} \text{ kg } CO_2 \text{ m}^{-2} \text{ (leaf) s}^{-1}. \quad \text{(S13.1b)}$$

This is a typical rate for single-leaf net photosynthesis.

13.2. Differentiate eqns (13.5g)–(13.5i) with respect to f_{sh}:

$$\frac{d\mu}{df_{sh}} = \mu\left(\frac{1}{C}\frac{dC}{df_{sh}} + \frac{1}{N}\frac{dN}{df_{sh}}\right) \quad \text{(S13.2a)}$$

$$\frac{d\mu}{df_{sh}}C + \mu\frac{dC}{df_{sh}} = \sigma_C - f_C\frac{d\mu}{df_{sh}} \quad \text{(S13.2b)}$$

$$\frac{d\mu}{df_{sh}}N + \mu\frac{dN}{df_{sh}} = -\sigma_N - f_N\frac{d\mu}{df_{sh}}. \quad \text{(S13.2c)}$$

Substituting dN/df_{sh} and dC/df_{sh} from eqns (S13.2b) and (S13.2c) in (S13.2a) yields

$$\left(\frac{f_C}{C} + \frac{f_N}{N} + 3\right)\frac{d\mu}{df_{sh}} = \frac{\sigma_C}{C} - \frac{\sigma_N}{N}. \tag{S13.2d}$$

Thus for $d\mu/df_{sh} = 0$

$$\frac{\sigma_C}{C} = \frac{\sigma_N}{N}; \tag{S13.2e}$$

this is a necessary and sufficient condition for $d\mu/df_{sh} = 0$. To show that this maximizes μ, differentiate eqn (S13.2d):

$$\left(\frac{f_C}{C} + \frac{f_N}{N} + 3\right)\frac{d^2\mu}{df_{sh}{}^2} + \frac{d\mu}{df_{sh}}\frac{d}{df_{sh}}\left(\frac{f_C}{C} + \frac{f_N}{N} + 3\right) = -\frac{\sigma_C}{C^2}\frac{dC}{df_{sh}} + \frac{\sigma_N}{N^2}\frac{dN}{df_{sh}}. \tag{S13.2f}$$

Now, when $d\mu/df_{sh} = 0$ it follows from eqns (S13.2b) and (S13.2c) that

$$\frac{dC}{df_{sh}} = \frac{\sigma_C}{\mu} \qquad \frac{dN}{df_{sh}} = -\frac{\sigma_N}{\mu}, \tag{S13.2g}$$

so that eqn (S13.2f) becomes

$$\left(\frac{f_C}{C} + \frac{f_N}{N} + 3\right)\frac{d^2\mu}{df_{sh}{}^2} = -\frac{1}{\mu}\left[\left(\frac{\sigma_C}{C}\right)^2 + \left(\frac{\sigma_N}{N}\right)^2\right] < 0. \tag{S13.2h}$$

This proves the desired result.

Note that, since

$$\left(\frac{\sigma_C}{C}\right)^2 + \left(\frac{\sigma_N}{N}\right)^2 = \left(\frac{\sigma_C}{C} - \frac{\sigma_N}{N}\right)^2 + \frac{4\sigma_C\sigma_N}{CN}, \tag{S13.2i}$$

using eqns (S13.2d) and (13.5g), eqn (S13.2h) can alternatively be written

$$\left(\frac{f_C}{C} + \frac{f_N}{N} + 3\right)\frac{d^2\mu}{df_{sh}{}^2} = -\frac{2k\sigma_C\sigma_N}{\mu^2} < 0. \tag{S13.2j}$$

13.3. With eqn (E13.3a), μ is given by (in balanced exponential growth) (cf. eqn (13.5g))

$$\mu = \frac{k_{CN}}{2(1 + C/K_C + N/K_N + CN/K_{CN})}, \tag{S13.3a}$$

which can be differentiated to give, after some algebra,

$$\frac{d\mu}{df_{sh}} = \frac{\mu}{1 + C/K_C + N/K_N + CN/K_{CN}}\left[\frac{1}{C}\left(1 + \frac{N}{K_N}\right)\frac{dC}{df_{sh}} + \frac{1}{N}\left(1 + \frac{C}{K_C}\right)\frac{dN}{df_{sh}}\right]. \tag{S13.3a}$$

Now, using eqns (S13.2b) and (S13.2c), from Solution 13.2, for dC/df_{sh} and dN/df_{sh} this becomes

$$\frac{d\mu}{df_{sh}}\left[1+\frac{C}{K_C}+\frac{N}{K_N}+\frac{CN}{K_{CN}}+\frac{1}{C}(f_C+C)\left(1+\frac{N}{K_N}\right)+\frac{1}{N}(f_N+N)(1+\frac{C}{K_C}\right)\right]$$
$$=\frac{\sigma_C}{C}\left(1+\frac{N}{K_N}\right)-\frac{\sigma_N}{N}\left(1+\frac{C}{K_C}\right). \qquad \text{(S13.3c)}$$

Thus a necessary and sufficient condition for $d\mu/df_{sh} = 0$ is

$$\frac{\sigma_C}{C(1+C/K_C)}=\frac{\sigma_N}{N(1+N/K_N)}, \qquad \text{(S13.3d)}$$

as required. Substituting this into eqn (13.51), we obtain the partitioning function

$$P=\frac{f_r}{f_{sh}}\frac{N(1+N/K_N)/(N+f_N)}{C(1+C/K_C)/(C+f_C)}, \qquad \text{(S13.3e)}$$

with the constraint $P = 1$ in balanced exponential growth, as required.

13.4. In balanced exponential growth

$$\mu = k_{sh}CN\lambda_{sh} \qquad \text{(S13.4a)}$$

$$\mu = k_rCN\lambda_r, \qquad \text{(S13.4b)}$$

and, with $\lambda_{sh} + \lambda_r = 1$, $k_{sh}\lambda_{sh} = k_r\lambda_r$ is satisfied by

$$\lambda_{sh}=\frac{k_r}{k_{sh}+k_r} \qquad \lambda_r=\frac{k_r}{k_{sh}+k_r}. \qquad \text{(S13.4c)}$$

Thus, in general, by defining

$$\lambda_{sh}=\frac{k_rP}{k_{sh}+k_rP} \qquad \lambda_r=\frac{k_{sh}}{k_{sh}+k_rP} \qquad \text{(S13.4d)}$$

these equations are consistent with the particular case $k_{sh} = k_r = k$ and eqn (13.4d); in balanced exponential growth $P = 1$.

13.5. Differentiating eqn (13.16c) with respect to f_{sh} yields

$$\mu\frac{dE}{df_{sh}}+\frac{d\mu}{df_{sh}}E=-\sigma_E-f_E\frac{d\mu}{df_{sh}}. \qquad \text{(S13.5a)}$$

Now, again for balanced exponential growth, eqn (13.15a) or (13.15b) gives

$$\mu = \tfrac{1}{2}kCNE, \qquad \text{(S13.5b)}$$

and differentiating this with respect to f_{sh} gives

$$\frac{d\mu}{df_{sh}}=\mu\left(\frac{1}{C}\frac{dC}{df_{sh}}+\frac{1}{N}\frac{dN}{df_{sh}}+\frac{1}{E}\frac{dE}{df_{sh}}\right). \qquad \text{(S13.5c)}$$

Substituting for $\mathrm{d}C/\mathrm{d}f_{\mathrm{sh}}$ and $\mathrm{d}N/\mathrm{d}f_{\mathrm{sh}}$ from Solution 13.2 and eqns (S13.2b) and (S13.2c), and for $\mathrm{d}E/\mathrm{d}f_{\mathrm{sh}}$ from eqn (S13.5a) gives

$$\left(\frac{f_{\mathrm{C}}}{C}+\frac{f_{\mathrm{N}}}{N}+\frac{f_{\mathrm{E}}}{E}+4\right)\frac{\mathrm{d}\mu}{\mathrm{d}f_{\mathrm{sh}}}=\frac{\sigma_{\mathrm{C}}}{C}-\frac{\sigma_{\mathrm{N}}}{N}-\frac{\sigma_{\mathrm{E}}}{E}, \tag{S13.5d}$$

so that for μ to be maximized in balanced exponential growth

$$\frac{\sigma_{\mathrm{C}}}{C}=\frac{\sigma_{\mathrm{N}}}{N}+\frac{\sigma_{\mathrm{E}}}{E}. \tag{S13.5e}$$

To derive the partitioning function, eqns (13.5h), (13.5i), and (13.16c) can be combined to give

$$\sigma_{\mathrm{N}}f_{\mathrm{r}}(C+f_{\mathrm{C}})=\sigma_{\mathrm{C}}f_{\mathrm{sh}}(N+f_{\mathrm{r}}) \tag{S13.5f}$$

$$\sigma_{\mathrm{E}}f_{\mathrm{r}}(C+f_{\mathrm{C}})=\sigma_{\mathrm{C}}f_{\mathrm{sh}}(E+f_{\mathrm{E}}) \tag{S13.5g}$$

$$\sigma_{\mathrm{E}}(N+f_{\mathrm{N}})=\sigma_{\mathrm{N}}(E+f_{\mathrm{E}}). \tag{S13.5h}$$

Combining these with eqn (S13.5e) gives

$$\frac{C+f_{\mathrm{C}}}{Cf_{\mathrm{sh}}}=\frac{N+f_{\mathrm{N}}}{Nf_{\mathrm{r}}}+\frac{E+f_{\mathrm{E}}}{Ef_{\mathrm{r}}}, \tag{S13.5i}$$

so that the required partitioning function is (cf. eqn (13.5n))

$$P=\frac{f_{\mathrm{r}}}{f_{\mathrm{sh}}}\frac{[(N+f_{\mathrm{N}})/N+(E+f_{\mathrm{E}})/E)]^{-1}}{C/(C+f_{\mathrm{C}})}, \tag{S13.5j}$$

with $P=1$ in balanced exponential growth.

13.6. First we rewrite eqns (13.12a)–(13.12f) directly in terms of the state variables, by substituting from eqns (13.11c), (13.11d), (13.11h), and (13.11i) to give

$$\frac{\mathrm{d}W_{\mathrm{sh}}}{\mathrm{d}t}=k\frac{W_{\mathrm{C,sh}}W_{\mathrm{N,sh}}}{W_{\mathrm{sh}}} \qquad \frac{\mathrm{d}W_{\mathrm{r}}}{\mathrm{d}t}=k\frac{W_{\mathrm{C,r}}W_{\mathrm{N,r}}}{W_{\mathrm{r}}} \tag{S13.6a}$$

$$\begin{aligned}
\frac{\mathrm{d}W_{\mathrm{C,sh}}}{\mathrm{d}t}&=\sigma_{\mathrm{C}}W_{\mathrm{sh}}-\frac{W_{\mathrm{sh}}+W_{\mathrm{r}}}{r_{\mathrm{C}}}\left(\frac{W_{\mathrm{C,sh}}}{W_{\mathrm{sh}}}-\frac{W_{\mathrm{C,r}}}{W_{\mathrm{r}}}\right)-f_{\mathrm{C}}k\frac{W_{\mathrm{C,sh}}W_{\mathrm{N,sh}}}{W_{\mathrm{sh}}}\\
\frac{\mathrm{d}W_{\mathrm{N,sh}}}{\mathrm{d}t}&=\frac{W_{\mathrm{sh}}+W_{\mathrm{r}}}{r_{\mathrm{N}}}\left(\frac{W_{\mathrm{N,r}}}{W_{\mathrm{r}}}-\frac{W_{\mathrm{N,sh}}}{W_{\mathrm{sh}}}\right)-f_{\mathrm{N}}k\frac{W_{\mathrm{C,sh}}W_{\mathrm{N,sh}}}{W_{\mathrm{sh}}}\\
\frac{\mathrm{d}W_{\mathrm{C,r}}}{\mathrm{d}t}&=\frac{W_{\mathrm{sh}}+W_{\mathrm{r}}}{r_{\mathrm{C}}}\left(\frac{W_{\mathrm{C,sh}}}{W_{\mathrm{sh}}}-\frac{W_{\mathrm{C,r}}}{W_{\mathrm{r}}}\right)-f_{\mathrm{C}}k\frac{W_{\mathrm{C,r}}W_{\mathrm{N,r}}}{W_{\mathrm{r}}}\\
\frac{\mathrm{d}W_{\mathrm{N,r}}}{\mathrm{d}t}&=\sigma_{\mathrm{N}}W_{\mathrm{r}}-\frac{W_{\mathrm{sh}}+W_{\mathrm{r}}}{r_{\mathrm{N}}}\left(\frac{W_{\mathrm{N,r}}}{W_{\mathrm{r}}}-\frac{W_{\mathrm{N,sh}}}{W_{\mathrm{sh}}}\right)-f_{\mathrm{N}}k\frac{W_{\mathrm{C,r}}W_{\mathrm{N,r}}}{W_{\mathrm{r}}}.
\end{aligned} \tag{S13.6b}$$

Substitution of eqns (E13.6a)–(E13.6c) into the above equations gives

$$\frac{dw_{sh}}{d\tau} = \frac{w_{C,sh}w_{N,sh}}{w_{sh}} \qquad \frac{dw_r}{d\tau} = \frac{w_{C,r}w_{N,r}}{w_r} \tag{S13.6c}$$

$$\frac{dw_{C,sh}}{d\tau} = \alpha_C w_{sh} - \sigma_C(w_{sh} + w_r)\left(\frac{w_{C,sh}}{w_{sh}} - \frac{w_{C,r}}{w_r}\right) - \frac{w_{C,sh}w_{N,sh}}{w_{sh}}$$

$$\frac{dw_{N,sh}}{d\tau} = \gamma_N(w_{sh} + w_r)\left(\frac{w_{N,r}}{w_r} - \frac{w_{N,sh}}{w_{sh}}\right) - \frac{w_{C,sh}w_{N,sh}}{w_{sh}}$$

$$\frac{dw_{N,r}}{d\tau} = \gamma_C(w_{sh} + w_r)\left(\frac{w_{C,sh}}{w_{sh}} - \frac{w_{C,r}}{w_r}\right) - \frac{w_{C,r}w_{N,r}}{w_r} \tag{S13.6d}$$

$$\frac{dw_{N,r}}{d\tau} = \alpha_N w_r - \gamma_N(w_{sh} + w_r)\left(\frac{w_{N,r}}{w_r} - \frac{w_{N,sh}}{w_{sh}}\right) - \frac{w_{C,r}w_{N,r}}{w_r}.$$

The parameter space of the four parameters α_C, α_N, γ_C, and γ_N can be investigated much more easily than with the model in its unreduced form.

Chapter 14

14.1. 1 kg water m^{-2} occupies a volume above 1 m^2 of ground given by $1 \times 1 \times h$ m^3 where h (m) is the height of water. Thus, denoting the density of water by ρ_W kg water m^{-3}

$$1 \text{ kg water m}^{-2} = \rho_W h \text{ kg water m}^{-2}, \tag{S14.1a}$$

and hence

$$\frac{1 \text{ kg water m}^{-2}}{\rho_W} = h \text{ m water.} \tag{S14.1b}$$

Therefore, with $\rho_W = 1000$ kg water m^{-3}, it follows that

$$1 \text{ kg water m}^{-2}\text{ s}^{-1} \equiv 1 \text{ mm water s}^{-1} \tag{S14.1c}$$

$$1 \text{ kg water m}^{-2}\text{ s}^{-1} \equiv 3600 \text{ mm water hour}^{-1} \tag{S14.1d}$$

$$1 \text{ kg water m}^{-2}\text{ s}^{-1} \equiv 86\,400 \text{ mm water day}^{-1}. \tag{S14.1e}$$

With $\rho_W = 998.2$ kg water m^{-3}, eqn (S14.1c) becomes

$$1 \text{ kg water m}^{-2}\text{ s}^{-1} \equiv 1.0018 \text{ mm water s}^{-1}, \tag{S14.1f}$$

so that the error in using (S14.1c) is

$$\left(\frac{1.0018 - 1}{1}\right) \times 100 = 0.18 \text{ per cent.} \tag{S14.1g}$$

Thus the conversion factors given by eqns (S14.1c)–(S14.1e) are unlikely to give any appreciable error over practical temperature ranges.

14.2. (This analysis follows Campbell, 1977, p. 28.) Differentiating eqn (E14.2d) gives

$$\mathrm{d}V = -\frac{nRT}{p^2}\mathrm{d}p, \tag{S14.2a}$$

so that in eqn (E14.2b)

$$\mathrm{d}U = \frac{nRT}{p}\mathrm{d}p. \tag{S14.2b}$$

Thus, the change in energy from the reference state p' is

$$U = \int_{p'}^{p} \frac{nRT}{p}\mathrm{d}p = nRT\ln\left(\frac{p}{p'}\right). \tag{S14.2c}$$

Now, when this is applied to water vapour, the mass of water is simply $\rho_v V$, so that

$$\psi = \frac{U}{\rho_v V} \tag{S14.2d}$$

which, with eqns (E14.2c) and (S14.2c), yields

$$\psi = \frac{RT}{M}\ln\left(\frac{p_v}{p_v'}\right), \tag{S14.2e}$$

and hence

$$h_r = \frac{p_v}{p_v'} = \exp\left(\frac{\psi M}{RT}\right) \tag{S14.2f}$$

14.3. (a) The density of dry air is simply

$$\rho = \rho_{N_2} + \rho_{O_2} + \rho_{Ar} + \rho_{CO_2} \tag{S14.3a}$$

which is, using eqns (E14.3c) and (E14.3d),

$$\rho = (M_{N_2}f_{N_2} + M_{O_2}f_{O_2} + M_{Ar}f_{Ar} + M_{CO_2}f_{CO_2}) \times 0.0446158 \tag{S14.3b}$$

where f with the appropriate subscript denotes fractional content. Using Table E14.1, this gives

$$\rho = 1.292 \text{ kg air m}^{-3}. \tag{S14.3c}$$

Applying eqn (E14.3c) to air, combined with eqn (S14.3b), gives

$$M_a = M_{N_2}f_{N_2} + M_{O_2}f_{O_2} + M_{Ar}f_{Ar} + M_{CO_2}f_{CO_2} = 28.95 \tag{S14.3d}$$

and, since $M_v = 18.02(= 2.02 + 16)$, it follows that

$$\varepsilon = \frac{M_v}{M_a} = 0.622. \tag{S14.3e}$$

(b) The partial pressures of water vapour and air are

$$p_v = \frac{\rho_v}{M_v} RT \tag{S14.3f}$$

and

$$p_a = \frac{\rho_a}{M_a} RT, \tag{S14.3g}$$

with the pressure of the mixture being

$$P = p_a + p_v. \tag{S14.3h}$$

Eliminating the factor RT from eqns (S14.3f) and (S14.3g) gives

$$p_a = \frac{\rho_a}{\rho_v} \varepsilon p_v, \tag{S14.3i}$$

and substituting for p_a from eqn (S14.3h) yields

$$\rho_v = \rho_a \varepsilon \frac{p_v}{P - p_v}. \tag{S14.3j}$$

Assuming that ρ_a can be approximated by the density of dry air ρ gives the required result.

(c) Equation (S14.3j), with ρ_a replaced by ρ, can be rewritten

$$p_v = \frac{\rho_v P}{\rho\varepsilon + \rho_v}, \tag{S14.3k}$$

so that

$$\frac{p_v}{p'_v} = \frac{\rho_v}{\rho'_v}\left(\frac{\rho\varepsilon + \rho'_v}{\rho\varepsilon + \rho_v}\right). \tag{S14.3l}$$

This can be rearranged to give

$$\frac{p_v}{p'_v} = \frac{\rho_v}{\rho'_v}\left(1 - \frac{\Delta\rho_v}{\rho\varepsilon + \rho'_v}\right)^{-1} \tag{S14.3m}$$

which, since

$$\frac{\Delta\rho_v}{\rho\varepsilon + \rho'_v} \ll 1, \tag{S14.3n}$$

approximates to

$$\frac{p_v}{p'_v} = \frac{\rho_v}{\rho'_v}\left(1 + \frac{\Delta\rho_v}{\rho\varepsilon + \rho'_v}\right) \tag{S14.3o}$$

by taking the first term in the binomial expansion. (For a brief discussion of the binomial theorem, see Exercise 2.3, p. 72) Thus the percentage error in using eqn (E14.3g) is

$$\frac{\Delta\rho_v}{\rho\varepsilon + \rho'_v} \times 100 \text{ per cent.} \tag{S14.3p}$$

If the air is saturated, then $\Delta\rho_{va} = 0$ and there is no error. The error will be maximum when the vapour content of the air is zero. Using the values of ρ and ε from eqns (S14.3c) and (S14.3e), and taking ρ'_v from Table 14.3, the maximum errors at (10, 20, 30) °C are (1.2, 2.1, 3.6) per cent. Since, in practice, $\rho_v \neq 0$, the errors will generally be less than this, and it is apparent that the approximation is unlikely to introduce an appreciable error.

14.4. Differentiating eqn (14.4k) with respect to g_a gives

$$\frac{\partial E}{\partial g_a} = \frac{\gamma[g_c\lambda\Delta\rho_{va}(s+\gamma) - s\phi_N]}{\lambda g_c[s+\gamma(1+g_a/g_c)]^2}, \tag{S14.4a}$$

which is zero when

$$g_c = \frac{s\phi_N}{\lambda\Delta\rho_{va}(s+\gamma)} \tag{S14.4b}$$

so that with this value E is independent of variation in g_a. Substituting in eqn (14.4k) yields

$$E = \frac{s\phi_N}{\lambda(s+\gamma)} \tag{S14.4c}$$

which is independent of g_a.

Chapter 15

15.1. Equation (E15.1b) can easily be derived by considering the mechanical stability of the liquid column of volume $\pi r^2 h$, supported by the surface tension force acting upwards round its circumference against the downward gravitational force, so that

$$\pi r^2 \rho g h = 2\pi r T, \tag{S15.1a}$$

leading to (E15.1b).

The gravitational potential per unit mass of the liquid just inside the meniscus is hg (J kg^{-1}), from which it can be deduced that the water potential inside a concave meniscus (relative to a free surface at the same height) is $-hg$, giving

$$\psi = -\frac{2T}{\rho r}. \tag{S15.1b}$$

15.2. Combining eqns (E15.2a) and (E15.2b) gives

$$\theta_s = n\pi \int_{z_s}^{\infty} r_0{}^2 \left(\frac{z}{z_0}\right)^{-2q} dz = \frac{n\pi r_0{}^2 z_0 (z_s/z_0)^{-2q+1}}{2q-1}. \tag{S15.2a}$$

Using the inverse of (E15.2a), i.e.

$$\frac{z_s}{z_0} = \left(\frac{r_s}{r_0}\right)^{-1/q}, \tag{S15.2b}$$

we can rewrite (S15.2a) as

$$\theta_s = \frac{n\pi z_0 r_0{}^2}{2q-1}\left(\frac{r_s}{r_0}\right)^{(2q-1)/q}. \tag{S15.2c}$$

This is the desired expression for θ_s versus r_s, which is used in the text in (15.3i).

15.3. Note that from (15.16f) and (15.16h)

$$\frac{V - V_0}{V_0} = \frac{cQ_{pl}}{W_G} - 1. \tag{S15.3a}$$

Then (15.19d) becomes

$$\varepsilon = \varepsilon_0 + \varepsilon_1\left(\frac{cQ_{pl}}{W_G} - 1\right). \tag{S15.3b}$$

Combining (15.16g), (S15.3b), (15.16i), and (15.15b) gives

$$\psi_{pl} = \left[\varepsilon_0 + \varepsilon_1\left(\frac{cQ_{pl}}{W_G} - 1\right)\right]\left(\frac{cQ_{pl}}{W_G} - 1\right) - \frac{RTf_0 W_S}{\mu Q_{pl}}. \tag{S15.3c}$$

This cubic equation replaces the quadratic (15.17a), and the analysis then proceeds as outlined after (15.17a).

Chapter 16

16.1. Differentiating (16.1a), (16.1b), and (16.1c) gives

$$\frac{dY}{d\rho} = a - 2b\rho \qquad a - \frac{b}{2\rho^{1/2}} \qquad a(1 - b\rho)e^{-b\rho} \tag{S16.1a}$$

respectively. The initial slopes $dY/d\rho\,(\rho = 0)$ of the responses are a, ∞, and a respectively. Equating (S16.1a) to zero gives the maximum yield at planting densities of

$$\rho_{max} = \begin{cases} a/2b \\ b^2/4a^2 \\ 1/b. \end{cases} \tag{S16.1b}$$

Substituting these into (16.1a), (16.1b), and (16.1c) gives the maximum yields Y_{max}.

16.2. Differentiating (16.2b) and simplifying gives

$$\frac{dY}{d\rho} = \frac{a(1 - c\rho^2)}{1 + b\rho + c\rho^2)}. \tag{S16.2a}$$

Equating this to zero gives

$$\rho_{max} = 1/c^{1/2}. \tag{S16.2b}$$

16.3. Differentiating (16.3h) gives

$$\begin{aligned}\frac{dY}{d\rho} &= w_m\left[\frac{1 + g_1\rho + g_2\rho^{3/2} + g_3\rho^{5/2} - \rho(g_1 + \frac{3}{2}g_2\rho^{1/2} + \frac{5}{2}g_3\rho^{3/2})}{(1 + g_1\rho + g_2\rho^{3/2} + g_3\rho^{5/2})^2}\right]\\ &= w_m\left[\frac{1 - (g_2/2)\rho^{3/2} - (3g_3/2)\rho^{5/2}}{(1 + g_1\rho + g_2\rho^{3/2} + g_3\rho^{5/2})^2}\right].\end{aligned} \tag{S16.3a}$$

On equating this to zero, ρ_{max} is given by the solution of

$$2 = g_2\rho^{3/2} + 3g_3\rho^{5/2},$$

which can be written as

$$4 = g_2{}^2\rho^3 + 6g_2g_3\rho^4 + 9g_3{}^2\rho^5. \tag{S16.3b}$$

16.4. Equating (16.4a) to zero with $w = w_f$ gives ($w_f \neq 0$)

$$k/m = 1 + gw_f{}^n,$$

whence

$$w_f = \left(\frac{k/m - 1}{g}\right)^{1/n}. \tag{S16.4a}$$

This equation is only meaningful for $k > m$. Differentiating (16.4a) gives

$$\frac{d^2w}{dt^2} = \frac{k(1 + gw^n) - kw(ngw^{n-1})}{(1 + gw^n)^2} - m. \tag{S16.4b}$$

Equating this to zero and rearranging gives

$$mg^2w^{2n} + [2mg + kg(n - 1)]w^n + m - k = 0. \tag{S16.4c}$$

Solving this quadratic in w^n gives $w = w^*$ at the inflexion point.

16.5. Taking Y/N as $N \to 0$ gives an initial slope of b_N. For the asymptote, as $N \to \infty$,

$$Y(N \to \infty) = \frac{Y_{max}b_Pb_KPK}{b_Pb_KPK + Y_{max}(b_KK + b_PP)}. \tag{S16.5}$$

Chapter 17

17.1. In the steady state, (E17.1a) becomes

$$0 = Dw_r\frac{d^2w_r}{dz^2} - m_rw_r. \tag{S17.1a}$$

Cancelling w_r ($w_r \neq 0$), and with $\lambda = m_r/D$,

$$0 = \frac{d^2 w_r}{dz^2} - \lambda. \tag{S17.1b}$$

At $z = 0$, $w_r = w_0$ and we define $w_0' = dw_r/dz(z = 0)$; then (S17.1b) can be written

$$\int_{w_0'}^{dw_r/dz} d\left(\frac{dw_r}{dz}\right) = \int_0^z \lambda \, dz. \tag{S17.1c}$$

Integrating gives

$$\frac{dw_r}{dz} = w_0' + \lambda z. \tag{S17.1d}$$

As, by assumption, at $z = z_0$, $dw_r/dz = 0$, we have

$$w_0' = -\lambda z_0 \qquad \frac{dw_r}{dz} = -\lambda(z_0 - z). \tag{S17.1e}$$

Note that w_0' and dw_r/dz are negative quantities. A further integral,

$$\int_{w_0}^{w_r} dw_r = \int_0^z -\lambda(z_0 - z)\, dz,$$

leads to

$$w_r = w_0 - \frac{\lambda(2z_0 z - z^2)}{2}. \tag{S17.1f}$$

At $z = z_0$, $w_r = 0$, and hence

$$w_0 = \frac{\lambda z_0}{2}. \tag{S17.1g}$$

Using (S17.1g), (S17.1f) can be written

$$w_r = \frac{\lambda(z_0 - z)^2}{2}. \tag{S17.1h}$$

The total root dry mass W_r, is obtained by integrating (S17.1h) from $z = 0$ to $z = z_0$, to give

$$W_r = \frac{\lambda z_0^{\,3}}{6}. \tag{S17.1i}$$

Thus, for a plant with a constant shoot : root ratio, a doubling in plant size increases the rooting depth by only $2^{3/2}$.

17.2. First, we notice that the first term in square brackets is just the carbon substrate concentration gradient near the soil surface. This is an intensive quantity, which does not change (necessarily) as plant size changes. The general problem is one of considering how the rate of some transport process P, which occurs between two organs w_1 and w_2 (say) of dry matter, varies with the size of the organs. That is, how does P depend on w_1 and w_2? Assume that we can write

$$P = f_i g_e, \tag{S17.2a}$$

where f_i denotes a function of the intensive variables of the two organs and g_e denotes a function of the extensive variables. In our example, f_i corresponds to the first term in square brackets in (E17.2a), and g_e corresponds to the second term in square brackets. In (E17.2a), we have taken

$$g_e = \frac{w_1 w_2}{w_1 + w_2}. \tag{S17.2b}$$

(S17.2b) scales such that if w_1 and w_2 are both increased by a factor s, then g_e increases by the same factor s. The smaller of w_1 and w_2 dominates the behaviour of g_e. If, in (E17.2a), we allow $\Delta z \to 0$, the equation becomes

$$F_C(1) = \zeta_C \left[\frac{C_{sh} - C_r(1)}{\Delta z} \right] w_{Gr}(1)\Delta z. \tag{S17.2c}$$

As $\Delta z \to 0$, the term in square brackets becomes constant, so that the substrate carbon flux density $F_C(1)$ approaches zero. This does not seem satisfactory. Note that, for large Δz, (E17.2a) becomes

$$F_C(1) = \zeta_C \left[\frac{C_{sh} - C_r(1)}{\Delta z} \right] W_{Gsh}. \tag{S17.2d}$$

This is well behaved, and this probably accounts for the good results obtained with this approach. The difficulty noted in (S17.2c) does not arise in (17.11a), and this illustrates the problems that can arise from interfacing a single compartment shoot with a distributed root.

An alternative expression to (S17.2b) is

$$g_e = (w_1 w_2)^{1/2}. \tag{S17.2e}$$

This gives the two organs equal importance regardless of their actual size; that is, a 10 per cent change in w_1 changes g_e by 5 per cent regardless of the magnitude of w_1.

Chapter 18

18.1. The number n of generations at time t is given by the value of n that satisfies

$$t_1 + (n + 1)a > t > t_1 + na. \tag{S18.1a}$$

There are potentially

$$\begin{aligned} &3^n \text{ terminal MUs} \\ &M_{max} = 3^{n+1} - 1 \text{ total MUs (terminal + non-terminal).} \end{aligned} \tag{S18.1b}$$

The morphological units (MUs) are numbered with a decimal integer N, starting with 1, and which can be represented by k digits $(0, 1, 2)$ counting to base 3 (trinary):

$$N(i_1, i_2, \ldots, i_{k-1}, i_k), \tag{S18.1c}$$

where

$$j = \text{integer part}(\log_3 N) \qquad k = j + 1. \tag{S18.1d}$$

The time delays a, b and c in outgrowth are associated respectively with

$$N \to 3N \qquad N \to 3N+1 \qquad N \to 3N+2. \tag{S18.1e}$$

With a trifurcating system, the counting of the MUs is not quite as convenient as with a bifurcating system. Applying eqn (S18.1e) gives successive generations of MUs with decimal integer numbers of 1; 3, 4, 5; 9, 10, ..., 16, 17; 27, 28, ..., 52, 53; .... Numbering the branches in this way gives gaps where some numbers are missing; this is because in the trinary representation we only count numbers beginning with 1; thus, the trinary representation of the above numbers is (1); (1, 0), (1, 1), (1, 2); (1, 0, 0), (1, 0, 1), ..., (1, 2, 1), (1, 2, 2); (1, 0, 0, 0), (1, 0, 0, 1), ..., (1, 2, 2, 1), (1, 2, 2, 2); .... All the trinary numbers beginning with 2 (e.g. 6, 7, 8; 18, 19, ..., 25, 26; 54, 55, ..., 79, 80; ....) are excluded. This requirement can be expressed as follows: MU N is allowed if

$$\text{integer part}(N/3^j) = 1, \tag{S18.1f}$$

where j is given by eqn (S18.1d).

The time of formation t_N of the Nth MU is

$$t_N = t_1 + \sum_{j'=2}^{k} (a\delta_{0i_{j'}} + b\delta_{1i_{j'}} + c\delta_{2i_{j'}}) \tag{S18.1g}$$

where $\delta_{pi_{j'}}$ is zero unless $p = i_{j'}$, in which case it is unity. The age τ_N of the Nth MU is then calculated from

$$\tau_N = t - t_N. \tag{S18.1h}$$

18.2. Following the procedure of Section 18.2.2, we define the following quantities:

S_N, Shreve order of MU N (S18.2a)

β_m, number of Shreve-ordered branches (not MUs) of Shreve order m; (S18.2b)

μ_m, number of MUs of Shreve order m; (S18.2c)

L_m, total length of all the branches (or MUs) of Shreve order m; (18.2d)

D_m, sum of the diameters of all the MUs of Shreve order m. (S18.2e)

The β_m, μ_m, L_m, and D_m are initialized to zero for a range of m ($m = 1, 2, 3, \ldots$) sufficient to accommodate the structure to be described.

The algorithm to sort and count the MUs according to the Shreve-ordering rules is as follows.

N is successively assigned integer values form $N_{\max}$ down to unity; i.e.

$$N := N_{\max}, N_{\max} - 1, \ldots, 2, 1. \tag{S18.2f}$$

Here and elsewhere the symbol := denotes 'is assigned the value of'. If MU N does not exist, then N is simply decremented so that

$$\text{if } e_N = 0, \text{ then decrement } N. \tag{S18.2g}$$

If an MU is terminal (without daughters) it necessarily has a Shreve order of unity. Thus

$$\text{if } 2N > N_{\max} \quad \text{or} \quad e_{2N} = e_{2N+1} = 0, \text{ then}$$
$$S_N := 1, \quad \beta_1 := \beta_1 + 1, \quad \mu_1 := \mu_1 + 1,$$
$$L_1 := L_1 + l_N, \quad D_1 := D_1 + d_N. \tag{S18.2h}$$

If the MU is not terminal, but has a dormant bud, then its Shreve order is the same as that of its daughter (2N), and it belongs to the same Shreve-ordered branch. Note that, in the case of Shreve ordering as contrasted with Strahler ordering, only the terminal branches have the possibility of having more than one MU. Hence

$$\text{if } e_{2N} = 1 \quad \text{and} \quad e_{2N+1} = 0, \text{ then}$$
$$S_N := S_{2N} \quad (\beta_{sN} \text{ is not incremented}),$$
$$\mu_{SN} := \mu_{sN} + 1, \quad L_{sN} := L_{sN} + l_N,$$
$$D_{sN} := D_N + d_N. \tag{S18.2i}$$

The two-letter subscript sN denotes the value of s_N.

If the bud on the MU has grown out to give a lateral shoot, then MU N has two daughter MUs, numbered $2N$ and $2N + 1$. In this case, it does not matter whether these have equal or unequal Shreve order numbers.

$$\text{if } e_{2N} = e_{2N+1} = 1, \text{ then}$$
$$S_N := S_{2N} + S_{2N+1}, \quad \beta_{sN} := \beta_{2N} + 1,$$
$$\mu_{sN} := \mu_{sN} + 1, \quad L_{sN} := L_{sN} + l_N,$$
$$D_{sN} := D_{sN} + d_N. \tag{S18.2j}$$

Note that with the conventions shown in Fig. 18.2, $S_{2N} \geqslant S_{2N+1}$ always.

Finally, the mean length λ_m and mean diameter δ_m of the branches of Shreve order m are given by

$$\lambda_m = L_m/\beta_m \qquad \delta_m = D_m/\mu_m. \tag{S18.2k}$$

18.3. Equation (18.4c) and (18.4d) are

$$\lambda = R_a/R_b, \tag{S18.3a}$$

$$r/R = r_a/R_a = r_b/R_b. \tag{S18.3b}$$

Equation (18.4e) is replaced by

$$r^x = R_a{}^x + R_b{}^x. \tag{S18.3c}$$

Combining eqns (S18.3a), (S18.3b), and (S18.3c) gives

$$R_a = \frac{\lambda}{(1 + \lambda^x)^{1/x}} r \qquad R_b = \frac{1}{(1 + \lambda^x)^{1/x}} r. \tag{S18.3d}$$

If we denote the time intervals between successive bifurcations by Δt_a and Δt_b, then, since this is the time required for the apical radius (r_a or r_b) just after a bifurcation to grow to the critical value of r, eqns (18.4g) can be replaced by

$$r = r_a \exp(\rho_L \Delta t_a)$$
$$r = r_b \exp(\rho_L \Delta t_a) \qquad \text{(S18.3e)}$$

Then combining eqns (S18.3d) and (S18.3e) gives

$$\Delta t_a = \frac{1}{\rho_L} \ln\left[\frac{R_a}{r_a}\frac{(1+\lambda^x)^{1/x}}{\lambda}\right]$$
$$\Delta t_b = \frac{1}{\rho_L} \ln\left[\frac{R_b}{r_b}\frac{(1+\lambda^x)^{1/x}}{\lambda}\right]. \qquad \text{(S18.3f)}$$

These equations replace eqns (18.4h).

18.4. Differentiating the first of eqns (18.5j) with respect to x gives

$$\frac{\partial^2 H}{\partial x^2} = \frac{h_1}{l_1} + \frac{h_2}{l_2} + \frac{h_3}{l_3} - h_1(x-x_1)\frac{x-x_1}{l_1{}^3}$$
$$- h_2(x-x_2)\frac{x-x_2}{l_2{}^3} - h_3(x-x_3)\frac{x-x_3}{l_3{}^3}$$
$$= \frac{h_1}{l_1}\left[1 - \frac{(x-x_1)^2}{l_1{}^2}\right] + \frac{h_2}{l_2}\left[1 - \frac{(x-x_2)^2}{l_2{}^2}\right] + \frac{h_3}{l_3}\left[1 - \frac{(x-x_3)^2}{l_3{}^2}\right]$$

which, with eqns (18.5f), becomes

$$\frac{\partial^2 H}{\partial x^2} = \frac{h_1}{l_1}(1-\cos^2\theta_1) + \frac{h_2}{l_2}\left[1-\cos^2(\theta_2+\theta_1)\right] + \frac{h_3}{l_3}\left[1-\cos^2(\theta_3-\theta_1)\right]$$
$$= \frac{h_1}{l_1}\sin^2\theta_1 + \frac{h_2}{l_2}\sin^2(\theta_2+\theta_1) + \frac{h_2}{l_3}\sin^2(\theta_3-\theta_1). \qquad \text{(S18.4a)}$$

In an analogous manner it can be shown that

$$\frac{\partial^2 H}{\partial y^2} = \frac{h_1}{l_2}\cos^2\theta_1 + \frac{h_2}{l_2}\cos^2(\theta_2+\theta_1) + \frac{h_3}{l_3}\cos^2(\theta_3-\theta_1). \qquad \text{(S18.4b)}$$

The mixed second derivative is given by differentiating the first of eqns (18.5j) with respect to y, or the second of these equations with respect to x:

$$\frac{\partial^2 H}{\partial x \partial y} = -\frac{h_1}{l_1}\frac{(x-x_1)(y-y_1)}{l_1{}^2} - \frac{h_2}{l_2}\frac{(x-x_2)(y-y_2)}{l_2{}^2} - \frac{h_3}{l_3}\frac{(x-x_3)(y-y_3)}{l_3{}^2}$$
$$= -\frac{h_1}{l_1}\cos\theta_1\sin\theta_1 - \frac{h_2}{l_2}\cos(\theta_2+\theta_1)\sin(\theta_2+\theta_1)$$
$$-\frac{h_3}{l_3}\cos(\theta_3-\theta_1)\sin(\theta_3-\theta_1). \qquad \text{(S18.4c)}$$

Cf. Zamir (1976, eqns (17)).

The first two inequalities are clearly satisfied for positive h. The last inequality is also satisfied: the terms in $(h_1/l_1)^2$, $(h_2/l_2)^2$, and $(h_3/l_3)^2$ cancel identically. Considering the coefficient of just one of the cross-terms, that in $h_1 h_2/l_1 l_2$, this is

$$\cos^2\theta_1 \sin^2(\theta_2+\theta_1) + \sin^2\theta_1 \cos^2(\theta_2+\theta_1) - 2\cos\theta_1 \sin\theta_1 \cos(\theta_2+\theta_1)\sin(\theta_2+\theta_1).$$

This is a perfect square and is therefore positive.

18.5. Since $19 = 1(16) + 0(8) + 0(4) + 1(2) + 1(1)$, the binary representation of N is

$$19(1 \quad 0 \quad 0 \quad 1 \quad 1). \tag{S18.5a}$$

Using eqn (18.10d) with $j = 2, 3, 4, 5$ gives

$$G_2 = A \qquad G_3 = A \qquad G_4 = B \qquad G_5 = B. \tag{S18.5b}$$

Substituting into eqn (18.10e) gives

$$\Gamma_{19} = BBAA. \tag{S18.5c}$$

Similarly from eqns (18.10g) and (18.10h) gives

$$\Lambda_{19} = \alpha^2\beta^2. \tag{S18.5d}$$

Finally, applying eqn (18.10j) to find the position of the tip of the 19th branch, we obtain

$$R_{19} = r_{19} + r_9 + r_4 + r_2 + r_1. \tag{S18.5e}$$

Chapter 19

19.1. Applying the rules gives

t	Organisms	Total
1	y	1
2	m	1
3	m y	2
4	m y m	3
5	m y m m y	5
6	m y m m y m y m	8
7	m y m m y m y m m y m m y	13
.	.	.

19.2. Calculation of the ratios of adjacent terms of (19.1a), such as 21/13, 34/21, and 55/34, indicates that this ratio is approaching a limit. Let this limiting value be τ, so that the relative values of five high-order Fibonacci terms are

$$\cdots, \frac{1}{\tau^2}, \frac{1}{\tau}, 1, \tau, \tau^2, \cdots. \tag{S19.2a}$$

Since any term is the sum of the two preceding terms,

$$\tau^2 = \tau + 1. \tag{S19.2b}$$

Thus

$$\tau = \frac{\sqrt{5}+1}{2} = 1.6180 \quad \text{or} \quad \tau = \frac{-\sqrt{5}+1}{2} = -0.6180. \tag{S19.2c}$$

19.3. In Fig. 19.17, denote the ratio QU/US by τ. Now

$$\frac{\text{QU}}{\text{US}} = \frac{\text{QT}}{\text{RS}} \qquad \frac{\text{QT}}{\text{RS}} = \frac{\text{QS}}{\text{RS}}. \tag{S19.3a}$$

As RS = QR = QU,

$$\frac{\text{QS}}{\text{RS}} = \frac{\text{QS}}{\text{QU}}. \tag{S19.3b}$$

We have shown that

$$\tau = \frac{\text{QU}}{\text{US}} = \frac{\text{QS}}{\text{QU}}. \tag{S19.3c}$$

As

$$\text{QS} = \text{QU} + \text{US}, \tag{S19.3d}$$

$$\tau = 1 + 1/\tau, \tag{S19.3e}$$

which is the same as eqn (S19.2b) defining the Fibonacci ratio.

We can also write

$$\tau = \frac{\text{QU}}{\text{US}} = \frac{\text{QU}}{\text{UR}} = \frac{\sin(\pi/5)}{\sin(\pi/10)} = \frac{2\sin(\pi/10)\cos(\pi/10)}{\sin(\pi/10)} = 2\cos\left(\frac{\pi}{10}\right). \tag{S19.3f}$$

19.4. Differentiating gives

$$\frac{\mathrm{d}r}{\mathrm{d}\theta} = 1 \qquad a\cot\phi\exp(\theta\cot\phi) \qquad \tfrac{1}{2}a\theta^{-1/2}. \tag{S19.4a}$$

Substituting in the expression for dA gives

$$\mathrm{d}A = a^2\theta(\mathrm{d}\theta)^2 \qquad a^2\cot\phi\exp(2\theta\cot\phi)(\mathrm{d}\theta)^2 \qquad \tfrac{1}{2}a^2(\mathrm{d}\theta)^2. \tag{S19.4b}$$

Note that for Archimedes spiral dA increases linearly with θ, for the exponential spiral dA increases exponentially with θ, and for the cyclotron spiral dA is independent of θ.

19.5. In the continued fraction, delete successive + signs and everything following, to give the fractions

$$\frac{1}{1}, \frac{1}{2}, \frac{2}{3}, \frac{3}{5}, \frac{5}{8}, \ldots . \tag{S19.5a}$$

These are the ratios of successive terms in (19.1a), which converge to $\tau - 1$. This is connected with the fact that in any Fibonacci-like series

$$\ldots, i, j, i + j, \ldots, \tag{S19.5b}$$

where i and j are integers, the ratios of successive terms, i.e. i/j and $j/(i + j)$, are related by

$$\frac{1}{1 + (i/j)} = \frac{i}{i + j}. \tag{S19.5c}$$

Analytically, in the continued fraction, the first denominator is equal to τ so that we can write

$$\tau = 1 + (1/\tau) \tag{S19.5d}$$

which is identical with eqn (S19.2b).

19.6. Repeatedly take the integer part, and the reciprocal of the remainder, to give

$$1.42 = [1, 2, 2, 1, 1, 1, 2]. \tag{S19.6a}$$

19.7. By applying eqns (19.7i) and (19.7j) to (S19.6a), a table can be constructed:

i	0	1	2	3	4	5	6
a_i	1	2	2	1	1	1	2
p_i	1	3	7	10	17	27	71
q_i	1	2	5	7	12	19	50
α_i		1.500		1.429		1.421	
	1.000		1.400		1.417		1.420

Note that the p_i/q_i are always relatively prime (without a common factor); the way in which the even convergents α_0, α_2, α_4, and α_6, and the odd convergents α_1, α_3, and α_5 converge to $\alpha = 1.420$.

19.8. Using (19.7n), an intermediate convergent (int. conv.) can be added at $i = 2$, $a_i = 2$, and at $i = 6$, $a_i = 2$, to give an expanded set of convergents:

p_i	1	3	4	7	10	17	27	44	71
q_i	1	2	*3*	5	7	12	19	*31*	50
		1.500			1.429		1.421		
	1.000			1.400		1.417			1.420
Int. conv.			1.333					1.419	

The q_i of the intermediate convergents are in italics.

19.9. A table can be constructed as in Table S19.1. After 49 primordia have been placed on [0, 1] the pattern repeats itself. For each intermediate convergent (see Solution 19.8) there is a bad break and an intermediate interval is created. The smallest intervals are always created by the denominator q of a convergent so that the number of points $n = q$.

Table S19.1. Subdivision of the interval [0, 1] by sequential placing of primordia at a rational angle

Number of points	Position of last point	Lengths and numbers of intervals			Comments
		Smallest	Intermediate	Largest	
1	0.42	0.42 (1)	0.58 (1)	—	
2	0.84	0.16 (1)	0.42 (2)	—	Bad break
3	0.26	0.16 (2)	0.26 (1)	0.42 (1)	Intermediate interval created
4	0.68	0.16 (3)	0.26 (2)	—	
5	0.10	0.10 (1)	0.16 (4)	0.26 (1)	
6	0.52	0.10 (2)	0.16 (5)	—	
7	0.94	0.06 (1)	0.10 (3)	0.16 (4)	
8	0.36	0.06 (2)	0.10 (4)	0.16 (3)	
9	0.78	0.06 (3)	0.10 (5)	0.16 (2)	
10	0.20	0.06 (4)	0.10 (6)	0.16 (1)	
11	0.64	0.06 (5)	0.10 (7)	—	
12	0.06	0.04 (1)	0.06 (6)	0.10 (6)	
⋮					
18	0.56	0.04 (7)	0.06 (12)	—	
19	0.98	0.02 (1)	0.04 (8)	0.06 (11)	Bad break
⋮					
30	0.60	0.02 (12)	0.04 (19)	—	
31	0.02	0.02 (14)	0.04 (18)	—	Intermediate interval created
49	0.58	0.02 (50)	—	—	

19.10. Write

$$x = 1 + \cfrac{1}{2 + \cfrac{1}{2 + \cdots}} \tag{S19.10a}$$

This can be rewritten as

$$x = 1 + \frac{1}{x + 1} \tag{S19.10b}$$

which has roots

$$x = \pm\sqrt{2}. \tag{S19.10c}$$

$\alpha = \sqrt{2} - 1$ can therefore be represented by the continued fraction $[0, 2, 2, 2, \ldots,]$.

The convergents of $\sqrt{2} - 1$ are found to be

$$\frac{1}{2}, \frac{2}{5}, \frac{5}{12}, \frac{12}{29}, \frac{29}{70}, \ldots. \qquad \text{(S19.10d)}$$

After the first two convergents, there is one intermediate convergent between every two convergents. The intermediate convergents can be found from (19.7n) to give

$$\frac{3}{7}, \frac{7}{17}, \frac{17}{41}, \frac{41}{99}, \ldots. \qquad \text{(S19.10e)}$$

The reader should evaluate all these convergents in decimal form to reveal their convergence properties. It is easily verified that there is no intermediate convergent between 1/2 and 2/5, as it is not possible to construct a fraction $i/3$ or $i/4$ (i is integer) lying between 1/2 and 2/5.

Similarly to Table S19.1, which dealt with a rational number, Table S19.2 can be constructed to reveal a series of bad breaks.

Table S19.2. Subdivision of the interval [0, 1] by sequential placing of primordia at an irrational angle with intermediate convergents

Number of points	Position of last point	Lengths and numbers of intervals			Comments
		Smallest	Intermediate	Largest	
1	0.4142	0.41 (1)	0.59 (1)	—	
2	0.8284	0.17 (1)	0.41 (2)		Bad break
3	0.2426	0.17 (2)	0.24 (1)	0.41 (1)	Intermediate interval created
4	0.6569	0.17 (3)	0.24 (2)	—	
5	0.0711	0.07 (1)	0.17 (4)	0.24 (1)	Bad break
6	0.4853	0.07 (2)	0.17 (5)	—	
7	0.8995	0.07 (3)	0.10 (1)	0.17 (4)	Intermediate interval created

Bibliography

Chapter 1. Dynamic modelling

Further reading and references

Mitchell and Gauthier Associates (1987). *Advanced Continuous Simulation Language (ACSL)* Mitchell and Gauthier Associates, Concord, Ma 01742.

Cvitanovic, P. (1984). *Universality in chaos.* Hilger, Bristol.

De Wit, C. T. (1970). Dynamic concepts in biology. In *Prediction and measurement of photosynthetic producivity* (ed. I. Setlik), pp. 17–23. Pudoc, Wageningen.

France, J. and Thornley, J. H. M. (1984). *Mathematical models in agriculture.* Butterworths, London.

Gear, C. W. (1971). *Numerical initial value problems in ordinary differential equations.* Prentice-Hall, Englewood Cliffs, NJ.

Lock, S. (1986). *A difficult balance: editorial peer review in medicine.* ISI Press, Philadelphia, PA.

Monod, J. (1972). *Chance and necessity.* Collins, London.

Popper, K. R. (1958). *The logic of scientific discovery.* Hutchinson, London.

Popper, K. R. (1982). *The open universe: an argument for indeterminism.* Hutchinson, London.

Speckhart, F. H. and Green, W. L. (1976). *A Guide to using CSMP–the continuous system modelling program.* Prentice-Hall, Englewood Cliffs, NJ.

Thornley, J. H. M. (1976). *Mathematical models in plant physiology.* Academic Press, London.

Thornley, J. H. M. (1980). Research strategy in the plant sciences. *Plant, Cell and Environment* **3**, 233–236.

Thornley, J. H. M. and Doyle, C. J. (1984). The management of publicly-funded agricultural research and development in the UK: questions of autonomy and accountability. *Agricultural Systems* **15**, 195–208.

Thornley, J. H. M., Hurd, R. G., and Pooley, A. (1981). A model of growth of the fifth leaf of tomato. *Annals of Botany* **48**, 327–340.

Chapter 2. Some subjects of general importance

Dixon, M. and Webb, E. C. (1979). *Enzymes* (3rd edn, assisted by C. J. R. Thorne and K. F. Tipton). Longmans, London.

Powell, E. O. (1956). Growth rate and generation time of bacteria, with special reference to continuous culture. *Journal of General Microbiology* **15**, 492–511.

Royal Society (1975). *Quantities, Units and symbols.* Symbols Committee, Royal Society of London.

Thornley, J. H. M. (1981). Organogenesis. In *Mathematics and plant physiology* (eds D. A. Rose and D. A. Charles-Edwards), pp. 49–65. Academic Press, London.

Chapter 3. Plant growth functions

Brouwer, R. (1962). Nutritive influences on the distribution of dry matter in the plant. *Netherlands Journal of Agricultural Science*, **10**, 399–408.

Chanter, D. O. (1976). Mathematical models in mushroom research and production. D.Phil. Thesis, University of Sussex, UK.

France, J. and Thornley, J. H. M. (1984). *Mathematical models in agriculture*. Butterworths, London.

Hunt, R. (1982). *Plant growth curves*. Arnold, London.

Chapter 4. Transport processes

Aikman, D. P. and Anderson, W. P. (1971). A quantitative investigation of a peristaltic model for phloem translocation. *Annals of Botany* **35**, 761–772.

Canny, M. J. (1973). *Phloem translocation*. Cambridge University Press, Cambridge.

Crank, J. (1975). *Mathematics of diffusion* (2nd edn). Oxford University Press, Oxford.

Curran, P. F. and Schultz, S. G. (1968). Transport across membranes: general principles. In *Handbook of Physiology*, Section 6: *Alimentary Canal*, Volume III, *Intestinal Absorption* (Section Editor, C. F. Code; Executive Editor, W. Heidel), pp. 1217–1243. American Physiological Society, Washington, DC.

Ho, L. C. and Thornley, J. H. M. (1978). Energy requirements for assimilate translocation from mature tomato leaves. *Annals of Botany* **42**, 481–483.

Katchalsky, A. and Curran, P. F. (1965). *Nonequilibrium thermodynamics in biophysics*. Harvard University Press, Cambridge, Ma.

Lang, A. (1978). A model of mass flow in the phloem. *Australian Journal of Plant Physiology* **5**, 535–546.

Mason, T. G. and Maskell, E. J. (1928). Studies on the transport of carbohydrates in the cotton plant. II. The factors determining the rate and direction of movement of sugars. *Annals of Botany* **42**, 571–636.

Moorby, J. (1981). *Transport systems in plants*. Longmans, London.

Munch, E. (1930). *Die Stoffbewegungen in der Pflanze*. Fischer, Jena.

Murray, J. D. (1977). *Lectures on nonlinear-differential-equation models in biology*. Clarendon Press, Oxford.

Nubar, J. (1971). Blood flow, slip, and viscometry. *Biophysical Journal* **11**, 252–264.

Peel, A. J. (1974). *Transport of nutrients in plants*. Butterworths, London.

Pippard, A. B. (1985). *Response and stability*. Cambridge University Press, Cambridge.

Richmond, P. and Wardlaw, I. F. (1976). On the translocation of sugar: van der Waals' forces and surface flow. *Australian Journal of Plant Physiology* **3**, 545–549.

Ross, S. M. and Tyree, M. T. (1979). Mason and Maskell's diffusion analogue reconciled with a translocation theory. *Annals of Botany* **44**, 637–640.

Spanner, D. C. (1958). The translocation of sugar in sieve tubes. *Journal of Experimental Botany* **9**, 332–342.

Spanner, D. C. (1970). The electro-osmotic theory of phloem transport in the light of recent measurements on *Heracleum* phloem. *Journal of Experimental Botany* **21**, 325–334.

Stein, W. D. (1986). *Transport and diffusion across cell membranes*. Academic Press, Orlando, FL.

Sutcliffe, J. F. and Collins, O. D. G. (1975). A mechanism of phloem transport based on interfacial flow controlled by solute potential gradients. *Annals of Botany* **39**, 627–629.

Thornley, J. H. M. (1976). *Mathematical models in plant physiology*. Academic Press, London.

Thornley, J. H. M. (1977). Root : shoot interactions. *Symposia of the Society for Experimental Biology* **31**, 367–389.

Thornley, J. H. M. (1987). The magnitude of the temperature-driven contribution to within-plant transport. *Plant, Cell and Environment* **10**, 699–700.

Walker, A. J. and Thornley, J. H. M. (1977). The tomato fruit: import, growth, respiration and carbon metabolism at different fruit sizes and temperatures. *Annals of Botany* **41**, 977–985.

Chapter 5. Temperature effects on plant and crop processes

Barrow, G. M. (1961). *Physical chemistry*. McGraw-Hill, New York.

Dixon, M. and Webb, E. C. (1964). *Enzymes* (2nd ed). Longmans, London.

Ehrlinger, J. and Björkman, O. (1977). Quantum yields for CO_2 uptake in C_3 and C_4 plants. *Plant Physiology* **59**, 86–90.

Eyring, H. (1935). The activated complex in chemical reactions. *Journal of Chem. Physics* **3**, 107–115.

Geiger, D. R. and Sovonick, S. A. (1970). Temporary inhibition of translocation velocity and mass transfer rates by petiole cooling. *Plant Physiology* **46**, 847–849.

Gordon, A. J. and Flood, A. E. (1979). Effect of 2,4-dichlorophenoxyacetic acid on invertases in chicory root. *Phytochemistry* **18**, 405–408.

Hearon, J. Z. (1952). Rate behaviour of metabolic systems. *Physiological Review* **32**, 499–523.

Helms, K. and Wardlaw, I. F. (1977). Effect of temperature on symptoms of tobacco mosaic virus and movement of photosynthate in Nicotiniana glutinosa. *Phytopathology* **67**, 344–350.

Ingraham, J. L. (1958). *Journal of Bacteriology* **76**, 75–80.

Johnson, I. R. and Thornley, J. H. M. (1984). A model of instantaneous and daily canopy photosynthesis. *Journal of Theoretical Biology* **107**, 531–545.

Johnson, I. R. and Thornley, J. H. M. (1985). Temperature dependence of plant and crop processes. *Annals of Botany* **55**, 1–24.

Jost, W. (1960). *Diffusion in solids, liquids, gases*. Academic Press, New York.

Lang, A. (1974). The effect of petiolar temperature upon the translocation rate of ^{137}Cs in the phloem of *Nymphoides peltata*. *Journal of Experimental Botany* **25**, 71–80.

Lyons, J. M. (1973). Chilling injury in plants. *Annual Review of Plant Physiology* **24**, 445–466.

Lyons, J. M. and Asmundson, C. M. (1965). Solidification of unsaturated saturated fatty acid mixtures and its relationship to chilling sensitivity in plants. *Journal of the American Oil Chemical Society* **42**, 1056–1058.

Robertson, G. W. (1968). A biometeorological time scale for a cereal crop involving day and night temperatures and photoperiod. *International Journal of Biometeorology* **12**, 191–223.

Sharpe, P. J. H. and Demichele, D. W. (1977). Reaction kinetics of poikilotherm development. *Journal of Theoretical Biology* **64**, 649–670.

Simon, E. W., Minchin, A., McMenamin, M. M. and Smith, J. M. (1976). The low temperature limit for seed germination. *New Phytologist* **77**, 301–311.

Thornley, J. H. M. (1977). Growth, maintenance and respiration: a re-interpretation. *Annals of Botany* **41**, 1191–1203.

Thornley, J. H. M., Gifford, R. M. and Bremner, P. M. (1981). The wheat spikelet—growth response to light and temperature—experiment and hupothesis. *Annals of Botany* **47**, 713–725.

Watson, B. T. (1975). The influence of low temperature on the rate of translocation in the phloem of *Salix viminalis* L. *Annals of Botany* **39**, 889–900.

Wolfe, J. (1978). Chilling injury in plants—the role of membrane lipid fluidity. *Plant, Cell and Environment* **1**, 241–247.

Chapter 6. Biological switches

Bellairs, R., Goodwin, B. and Mackley, M. R. (1977). In support of catastrophe theory. *Nature* (*London*) **270**, 381–382.

Charles-Edwards, D. A., Cockshull, K. E., Horridge, J. S., and Thornley, J. H. M. (1979). A model of flowering in chrysanthemum. *Annals of Botany* **44**, 557–566.

Horridge, J. S. and Cockshull, K. E. (1979). Chrysanthemum shoot apex in relation to inflorescence initiation and development. *Annals of Botany* **44**, 547–556.

Johnson, I. R. and Parsons, A. J. (1985). A theoretical analysis of grass growth under grazing. *Journal of Theoretical Biology* **112**, 345–367.

Lyndon, R. F. (1977). Interacting processes in vegetative development and in the transition to flowering at the shoot apex. *Symposia of the Society for Experimental Biology* **31**, 221–250.

Miller, M. B. and Lyndon, R. F. (1976). Rates of growth and cell division in the shoot apex of *Silene* during the transition to flowering. *Journal of Experimental Botany* **27**, 1142–1153.

Poston, T. and Stewart, I. (1978). *Catastrophe theory and its applications*. Pitman, London.

Thornley, J. H. M. (1972). A model of a biochemical switch and its application to flower initiation. *Annals of Botany* **36**, 861–871.

Thornley, J. H. M. (1975). Phyllotaxis. I. A mechanistic model. *Annals of Botany* **39**, 491–507.

Thornley, J. H. M. (1976). *Mathematical models in plant physiology*. Academic Press, London.

Thornley, J. H. M. (1981). Organogenesis. In *Mathematics and plant physiology* (eds D. A. Rose and D. A. Charles-Edwards), pp. 49–65. Academic Press, London.

Thornley, J. H. M. (1987). Modelling flower initiation. In *Manipulation of Flowering* (ed. J. G. Atherton), pp. 67–79. Butterworths, London.

Thornley, J. H. M. and Cockshull, K. E. (1980). A catastrophe model for the switch from vegetative to reproductive growth in the shoot apex. *Annals of Botany* **46**, 331–341.

Wilson, A. G. and Kirkby, M. J. (1980). *Mathematics for geographers and planners* (2nd edn). Oxford University Press, Oxford.

Zahler, R. S. and Sussmann, H. J. (1977). Claims and accomplishments of applied catastrophe theory. *Nature* (*London*) **269**, 759–763.

Chapter 7. Development

Angus, J. F., Cunningham, R. B., Moncur, M. W., and Mackenzie, D. H. (1981a). Phasic development in field crops. I. Thermal response in the seedling phase. *Field Crops Research* **3**, 365–378.

Angus, J. F., Mackenzie, D. H., Morton, R., and Schafer, C. A. (1981b). Phasic development in field crops. II. Thermal and photoperiodic responses of spring wheat. *Field Crops Research* **4**, 269–283.

Crank, J. (1956). *The mathematics of diffusion*. Oxford University Press, London.

France, J., Thornley, J. H. M., Dhanoa, M. S., and Siddons, R. C. (1985). On the mathematic of digesta flow kinetics. *Journal of Theoretical Biology* **113**, 743–758.

Garcia-Huidobro, J., Monteith, J. L., and Squire, G. R. (1982). Time, temperature and germination of pearl millet (*Pennisetum typhoides* S. & H.). I. Constant temperature. *Journal of Experimental Botany* **33**, 288–296.

Godfrey, K. (1983). *Compartmental models and their application*. Academic Press, London.

Goloff, A. A. and Bazzaz, F. A. (1975). A germination model for natural seed populations. *Journal of Theoretical Biology* **52**, 259–283.

Hageseth, G. T. (1978). Kinetic and thermodynamic parameters that describe isothermal seed germination. *Journal of* Experimental Botany **29**, 281–293.

Hageseth, G. T. and Joyner, R. D. (1975). Kinetics and thermodynamics of isothermal seed germination. *Journal of Theoretical Biology* **53**, 51–65.

Hsu, F. H., Nelson, C. J. and Chow, W. S. (1984). A mathematical model to utilize the logistic function in germination and seedling growth. *Journal of Experimental Botany* **35**, 1629–1640.

Labouriau, L. G. and Osborn, J. H. (1984). Temperature dependence of the germination of tomato seeds. *Journal of Thermal Biology* **9**, 285–294.

Nuttonson, M. Y. (1955). *Wheat–climate relationship and the use of phenology in ascertaining the thermal and photo-thermal requirements for wheat.* American Institute of Crop Ecology, Washington, DC.

Piaggio, H. T. H. (1952). *An elementary treatise on differential equations and their applications.* Bell, London.

Robertson, G. W. (1968). A biometeorological time scale for a cereal crop involving day and night temperatures and photoperiod. *International Journal of Biometeorology* **12**, 191–223.

Thornley, J. H. M. (1977). Germination of seeds and spores. *Annals of Botany* **41**, 1363–1365.

Thornley, J. H. M. (1986). A germination model: responses to time and temperature. *Journal of Theoretical Biology* **123**, 481–492.

Washitani, I. and Takenaka, A. (1984). Mathematical description of the seed germination dependency on time and temperature. *Plant, Cell and Environment* **7**, 359–362.

Chapter 8. Light relations in canopies

Bell, C. J. and Rose, D. A. (1981). Light measurement and the terminology of flow. *Plant, Cell and Environment* **4**, 89–96.

Brown, R. H. and Blaser, R. E. (1968). Leaf area index in pasture growth. *Herbage Abstracts* **38**, 1–9.

Charles-Edwards, D. A. and Thornley, J. H. M. (1973). Light interception by an isolated plant: a simple model. *Annals of Botany* **37**, 919–928.

Charles-Edwards, D. A. and Thorpe, M. R. (1976). Interception of diffuse and direct-beam radiation by a hedgerow apple orchard. *Annals of Botany* **40**, 603–613.

Davidson, I. A., Robson, M. J., and Denis, W. D. (1982). The effect of nitrogenous fertilizer on the composition, canopy structure and growth of a mixed grass-clover sward. *Grass and Forage Science* **37**, 178–179.

Jackson, J. E. and Palmer, J. W. (1979). A simple model of light transmission and interception by discontinuous canopies. *Annals of Botany* **44**, 381–383.

Johnson, I. R., Parsons, A. J., and Ludlow, M. M. (1989). Modelling photosynthesis in monocultures and mixtures. *Australian Journal of Plant Physiology* (In press).

Ludlow, M. M. (1983). External factors influencing photosynthesis and respiration. In *The growth and functioning of leaves* (eds J. E. Dale and F. L. Milthorpe), pp. 347–380. Cambridge University Press, Cambridge.

Monsi, M. and Saeki, T. (1953). Uber den Lichtfaktor in den Pflanzengesellschaften und seine Bedeutung für die Stoffproduktion. *Japanese Journal of Botany* **14**, 22–52.

Palmer, J. W. (1977). Diurnal light interception and a computer model of light interception by hedgerow apple orchards. *Journal of Applied Ecology* **14**, 601–614.

Robson, M. J. and Sheehy, J. E. (1981). Leaf area and light interception. In *Sward measurement handbook* (eds J. Hodgson, R. D. Baker, A. Davies, A. S. Laidlaw, and J. D. Leaver), pp. 115–139. British Grassland Society, Hurley.

Ross, P. J., Henzell, E. F., and Ross, D. R. (1972). Effects of nitrogen and light in grass–legume pastures—a systems analysis approach. *Journal of Applied Ecology* **9**, 535–556.

Sheehy, J. E. and Peacock, J. M. (1975). Canopy photosynthesis and crop growth rate of eight temperate forage grasses. *Journal of Experimental Botany* **26**, 679–691.

Simons, S (1970) *Vector analysis for mathematicians, scientists and engineers.* Pergamon Press, Oxford.

Stern, W. R. and Donald, C. M (1962). Light relationships in grass–clover swards. *Australian Journal of Agricultural Research* **13**, 599–614.

Whitfield, D. M. (1980). Interaction of single tobacco plants with direct beam light. *Australian Journal of Plant Physiology* **7**, 435–447.

Whitfield, D. M. and Connor, D. J. (1980). Architecture of individual plants in a field-grown tobacco crop. *Australian Journal of Plant Physiology* **7**, 415–433.

Wilson, G. L. and Ludlow, M. M. (1983). The distribution of leaf photosynthetic activity in a mixed grass-legume pasture canopy. *Photosynthesis Research* **4**, 137–144.

Chapter 9. Leaf photosynthesis

Edwards, G. and Walker, D. (1983). *C_3, C_4: mechanisms and cellular and environmental regulation of photosynthesis.* Blackwell Scientific Publications, Oxford.

Hahn, B. D. (1987). A mathematical model of photosynthesis and photorespiration. *Annals of Botany* **60**, 157–169.

Johnson, I. R. and Thornley, J. H. M. (1985). Dynamic model of the response of a vegetative grass crop to light, temperature and nitrogen. *Plant, Cell and Environment* **8**, 485–499,

Jones, H. G. (1983). *Plants and microclimate.* Cambridge University Press, Cambridge.

Ku, S.-B. and Edwards, G. E. (1978). Oxygen inhibition of photosynthesis. III. Temperature dependence of quantum yield and its relation to O_2/CO_2 solubility ratio. *Planta* **140**, 1–6.

Marshall, B. and Biscoe, P. V. (1980). A model for C_3 leaves describing the dependence of net photosynthesis on irradiance. I. Derivation. *Journal of Experimental Botany* **31**, 29–39.

Thornley, J. H. M. (1974). Light fluctuations and photosynthesis. *Annals of Botany* **38**, 363–373.

Thornley, J. H. M. (1976). *Mathematical models in plant physiology.* Academic Press, London.

Woledge, J. and Dennis, W. D. (1982). The effect of temperature on photosynthesis of ryegrass and white clover leaves. *Annals of Botany* **50**, 15–35.

Chapter 10. Canopy photosynthesis

Acock, B., Charles-Edwards, D. A., Fitter, D. J., Hand, D. W., Ludwig, L. J., Warren-Wilson, J., and Withers, A. C. (1978). The contribution of leaves from different levels within a tomato crop to canopy net photosynthesis: an experimental examination of two canopy models. *Journal of Experimental Botany* **29**, 815–827.

Charles-Edwards, D. A. and Acock, B. (1977). Growth response of a *Chrysanthemum* crop to the environment. II. A mathematical analysis relating photosynthesis and growth. *Annals of Botany* **41**, 49–58.

Gradshteyn, I. S. and Ryzhik, J. W. (1980). Tables of integrals, series and products. Academic Press, New York.

Johnson, I. R. and Thornley, J. H. M. (1984). A model of instantaneous and daily canopy photosynthesis. *Journal of Theoretical Biology* **107**, 531–545.

Johnson, I. R., Parsons, A. J., and Ludlow, M. M. (1989). Modelling photosynthesis in monocultures and mixtures. *Australian Journal of Plant Physiology* (In press).

Ludlow, M. M. and Charles-Edwards, D. A. (1980). Analysis of the regrowth of a tropical grass/legume sward subjected to different frequencies and intensities of defoliation. *Australian Journal of Agricultural Research* **31**, 673–692.

Ludlow, M. M. and Wilson, G. L. (1971). Photosynthesis of tropical pasture plants. II. Temperature and illuminance history. *Australian Journal of Biological Science* **24**, 1065–1075.

Monteith, J. L. (1965). Light distribution and photosynthesis in field crops. *Annals of Botany* **29**, 17–37.

Monteith, J. L. (1981). Does light limit crop production. In *Physiological processes limiting plant productivity* (ed. C. B. Johnson), pp. 23–28. Butterworth, London.

Prioul, J.-L., Brangeon, J., and Reyss, A. (1980a). Interaction between external and internal conditions in the development of photosynthetic features in a grass leaf. I. Regional responses along a leaf during and after low-light or high-light acclimation. *Plant Physiology* **66**, 762–769.

Prioul, J.-L., Brangeon, J., and Reyss, A. (1980b). Interaction between external and internal conditions in the development of photosynthetic features in a grass leaf. II. Reversibility of light-induced responses as a function of development stages. *Plant Physiology* **66**, 770–774.

Thornley, J. H. M. (1976). Mathematical models in plant physiology. Academic Press, New York.

Woledge, J. (1971). The effect of light intensity during growth on the subsequent rate of photosynthesis. *Annals of Botany* **35**, 311–322.

Woledge, J. and Dennis, W. D. (1982). The effect of temperature on photosynthesis of ryegrass and white clover leaves. *Annals of Botany* **50**, 25–35.

Chapter 11. Whole-plant respiration and growth energetics

Amthor, J. S. (1984). The role of maintenance in plant growth. *Plant, Cell and Environment* **7**, 561–569.

Barnes, A. and Hole, C. C. (1978). A theoretical basis of growth and maintenance respiration. *Annals of Botany* **42**, 1217–1221.

Breeze, V. and Elston, J. (1983). Examination of a model and data describing the effect of temperature on the respiration rate of crop plants. *Annals of Botany* **51**, 611–616.

Hansen, G. K. and Jensen, C. R. (1977). Growth and maintenance respiration in whole plants, tops and roots of *Lolium multiflorum. Physiologia Plantarum* **39**, 155–164.

Johnson, I. R. (1983). Nitrate uptake and respiration in roots and shoots. *Physiologia Plantarum* **58**, 145–147.

Johnson, I. R. (1987). Models of respiration. In Plant growth modeling and resource management, Vol. I (eds K. Wisiol and J. D. Hesketh), pp. 89–108. CRC Press, Boca Raton, FL.

Loehle, C. (1982). Growth and maintenance respiration: a reconciliation of Thornley's model and the traditional view. *Annals of Botany* **51**, 741–77.

McCree, K. J. (1970). An equation for the respiration of white clover plants grown under controlled conditions. In *Prediction and Measurement of Photosynthetic Productivity* (ed. I. Setlik), pp. 221–229. Pudoc, Wageningen.

McCree, K. J. (1974). Equations for the rate of dark respiration of white clover and grain sorghum. *Crop Science* **14**, 509–514.

McCree, K. J. (1982). Maintenance requirements of white clover at high and low growth rates. *Crop Science* **22**, 345–351.

Pirt, S. J. (1965). The maintenance energy of bacteria in growing cultures. *Proceedings of the Royal Society of London, Series B* **163**, 224–231.

Ryle, G. J. A., Cobby, J., and Powell, C. E. (1976). Synthetic and maintenance respiratory losses of $^{14}CO_2$ in uniculm barley and maize. *Annals of Botany* **40**, 571–586.

Szaniawski, R. K. and Kielkiewicz, M. (1982). Maintenance and growth respiration in shoots and roots of sunflower plants grown at different root temperatures. *Physiologia Plantarum* **54**, 500–504.

Thornley, J. H. M. (1970). Respiration, growth and maintenance in plants. *Nature (London)* **227**, 304–305.

Thornley, J. H. M. (1971). Energy, respiration and growth in plants. *Annals of Botany* **35**, 721–728.

Thornley, J. H. M. (1976). *Mathematical models in plant physiology*. Academic Press, London.

Thornley, J. H. M. (1977). Growth, maintenance and respiration: a re-interpretation. *Annals of Botany* **41**, 1191–1203.

Thornley, J. H. M. (1982). Interpretation of respiration coefficients. *Annals of Botany* **49**, 257–259.

Thornley, J. H. M. and Hesketh, J. D. (1972). Growth and respiration in cotton bolls. *Journal of Applied Ecology* **9**, 315–317.

Treharne, K. J. and Nelson, C. J. (1975). Effect of growth temperature on photosynthetic and photorespiratory activity in tall fescue. In *Environmental and biological control of photosynthesis* (ed. R. Marcelle), pp. 61–69. Junk, The Hague.

Veen, B. W. (1981). Relation between root respiration and root activity. *Plant and Soil* **63**, 73–76.

Woledge, J. and Dennis, W. D. (1982). The effect of temperature on photosynthesis of ryegrass and white clover leaves. *Annals of Botany* **50**, 25–35.

Chapter 12. Biochemical and chemical approaches to plant growth

Amthor, J. S. (1984). The role of maintenance respiration in plant growth. *Plant, Cell and Environment* **7**, 561–569.

Bauer, W. D., Talmadge, K. W., Keegstra, K., and Albersheim, P. (1973). The structure of plant cell walls. II. The hemicellulose of the walls of suspension-cultured sycamore cells. *Plant Physiology* **51**, 174–187.

Beevers, H. (1970). Respiration in plants and its regulation. In *Prediction and measurement of photosynthetic productivity* (ed. I. Setlik), pp. 209–214. Pudoc, Wageningen.

Blaxter, K. L. (1962). *The energy metabolism of ruminants*. Hutchison, London.

Dagley, S. and Nicholson, D. E. (1970). *An introduction to metabolic pathways*. Blackwell Scientific Publications, Oxford.

Gill, M., Thornley, J. H. M., Black, J. L., Oldham, J. D., and Beever, D. E. (1984). Simulation of the metabolism of absorbed energy-yielding nutrients in young sheep. *British Journal of Nutrition* **52**, 621–649.

Goodwin, T. W. and Mercer, E. I. (1972) *Introduction to plant biochemistry*. Pergamon, Oxford.

Harwood, J. L. (1975). Fatty acid biosynthesis. In *Recent advances in the chemistry and biochemistry of plant lipids* (eds T. Galliard and E. I. Mercer), pp. 44–93. Academic Press, London.

Hue, L. (1982). Futile cycles and regulation of metabolism. In *Metabolic compartmentation* (ed. H. Sies), pp. 71–97. Academic Press, New York.

Lainson, R. A. and Thornley, J. H. M. (1982). A model for leaf expansion in cucumber. *Annals of Botany* **50**, 407–425.

Lehninger, A. L. (1975). *Biochemistry*. Worth, New York.

Long, C. (ed.) (1968). *Biochemist's handbook*. Spon, London.

Luckner, M. (1984). *Secondary metabolism in microorganisms, plants, and animals*. Springer-Verlag, Berlin.

McDermitt, D. K. and Loomis, R. S. (1981a). Elemental composition of biomass and its relation to energy content, growth efficiency, and growth yield. *Annals of Botany* **48**, 275–290.

McDermitt, D. K. and Loomis, R. S. (1981b). A new approach to the analysis of reductive and dissipative costs in nitrogen assimilation. In *Genetic engineering of symbiotic nitrogen fixation and conservation of fixed nitrogen* (eds J. M. Lyons, R. C. Valentine, D. A. Phillips, D. W. Rains, and R. C. Huffaker), pp. 639–650. Plenum, New York.

Mahler, H. R. and Cordes, E. H. (1966). *Biological chemistry*. Harper & Row, New York.

Michal, G. (1982). *Biochemical pathways* (wallchart and booklet). Boehringer Mannheim GmbH Biochemica, FRG.

Miller, E. C. (1938). *Plant physiology*. McGraw-Hill, New York.

Newsholme, E. A. and Start, C. (1973). *Regulation in metabolism*. Wiley, Chichester.

Osbourn, D. F. (1980). The feeding value of grass. In *Grass: its production and utilization* (ed. W. Holmes), pp. 70–124. Blackwell, Oxford, for British Grassland Society.

Penning de Vries, F. W. T. (1975a). Use of assimilates in higher plants. In *Photosynthesis and productivity in different environments* (ed. J. P. Cooper), pp. 459–480. Cambridge University Press, Cambridge.

Penning de Vries, F. W. T. (1975b). The cost of maintenance processes in plant cells. *Annals of Botany* **39**, 77–92.

Penning de Vries, F. W. T., Brunsting, A. H. M., and van Laar, H. H. (1974). Products, requirements and efficiency of biosynthesis: a quantitative approach. *Journal of Theoretical Biology* **45**, 339–377.

Preiss, J. and Kosuge, T. (1976). Regulation of enzyme activity in metabolic pathways. In *Plant biochemistry* (3rd edn) (eds J. Bonner and J. E. Varner), pp. 277–336. Academic Press, New York.

Rabkin, M. and Blum, J. J. (1985). Quantitative analysis of intermediary metabolism in hepatocytes incubated in the presence and absence of glucagon with a substrate mixture containing glucose, ribose, fructose, alanine and acetate. *Biochemical Journal* **225**, 761–786.

Schulz, A. R. (1978). Simulation of energy metabolism in the simple-stomached animal. *British Journal of Nutrition* **39**, 235–254.

Simpson, F. B. (1987). Hydrogen reactions of nitrogenase. *Physiological Plantarum* **69**, 187–190.

Stein, R. B. and Blum, J. J. (1978). On the analysis of futile cycles in metabolism. *Journal of Theoretical Biology* **72**, 487–522.

Stumpf, P. K. (1977). Lipid biosynthesis in developing seeds. In *Lipids and lipid polymers in higher plants* (ed. M. Tevini and H. K. Lichtenthaler), pp. 75–84. Springer, Berlin.

Sutcliffe, J. F.(1962). *Mineral salts absorption in plants*. Pergamon, Oxford.

Weast, R. C. (ed.) (1980). *CRC Handbook of Chemistry and Physics* (61st edn). CRC Press, Boca Raton, Fl.

Williams, K., Percival, F., Merino, J., and Mooney, H. A. (1987). Estimate of tissue construction cost from heat of combustion and organic nitrogen content. *Plant, Cell and Environment* **10**, 725–734.

Wilson, L. G. and Reuveny, Z. (1976). Sulfate reduction. In *Plant biochemistry* (3rd edn) (eds J. Bonner and J. E. Varner), pp. 599–632. Academic Press, New York.

Chapter 13. Partitioning during vegetative growth

Bhat, K. K. S., Nye, P. H. and Brereton, A. J. (1979). The possibility of predicting solute uptake and plant growth response from independently measured soil and plant characteristics. IV. The growth and uptake of rape in solutions of constant nitrate concentration. *Plant and Soil* **53**, 137–167.

Brouwer, R. (1962). Distribution of dry matter in the plant. *Netherlands Journal of Agricultural Science* **10**, 361–376.

Clement, C. R., Hopper, M. J., Jones, L. H. P., and Leafe, E. L. (1978). Uptake of nitrate by *Lolium perenne* from flowing nutrient solution. II. Effect of light, defoliation, and relationship to CO_2 flux. *Journal of Experimental Botany* **29**, 1173–1183.

Davidson, R. L. (1969). Effect of root/leaf temperature differentials on root/shoot ratios in some pasture grasses and clover. *Annals of Botany* **33**, 561–569.

France, J. and Thornley, J. H. M. (1984). *Mathematical models in agriculture*. Butterworths, London.

Johnson, I. R. (1985). A model of the partitioning of growth between the shoots and roots of vegetative plants. *Annals of Botany* **55**, 421–431.

Johnson, I. R. and Thornley, J. H. M. (1987). A model of shoot : root partitioning with optimal growth. *Annals of Botany* **60**, 133–142.

Mäkelä, A. A. and Sievänen, R. P. (1987). Comparison of two shoot–root partitioning models with respect to substrate utilization and functional balance. *Annals of Botany*, **59**, 129–140.

Reynolds, J. F. and Thornley, J. H. M. (1982). A shoot : root partitioning model. *Annals of Botany* **49**, 585–597.

Robson, M. J. (1973). The growth and development of simulated swards of perennial ryegrass. I. Leaf growth and dry weight changes as related to ceiling yield of a seedling sward. *Annals of Botany* **37**, 487–500.

Thornley, J. H. M. (1972a). A model to describe the partitioning of photosynthate during vegetative plant growth. *Annals of Botany* **36**, 419–430.

Thornley, J. H. M. (1972b). A balanced quantitative model for root : shoot ratios in vegetative plants. *Annals of Botany* **36**, 431–441.

Thornley, J. H. M. (1976). *Mathematical models in plant physiology*. Academic Press, London.

Thornley, J. H. M. (1977). Root : shoot interactions. *Symposium of the Society for Experimental Biology* **31**, 367–389.

White, H. L. (1937). The interaction of factors in the growth of *Lemma* XII. The interaction of nitrogen and light intensity in relation to root length. *Annals of Botany* **1**, 649–654.

White, R. E. (1973). Studies on mineral ion absorption by plants. II. The interaction between metabolic activity and the rate of phosphorus uptake. *Plant and Soil* **38**, 509–523.

Chapter 14. Transpiration by a crop canopy

Campbell, G. S. (1977). *An introduction to environmental biophysics*. Springer-Verlag, New York.

Jones, H. G. (1983). *Plants and microclimate*. Cambridge University Press, Cambridge.

Monteith, J. L. (1965). Evaporation and environment. *Symposium of the Society for Experimental Biology* **19**, 205–234.

Monteith, J. L. (1973). *Principles of environmental physics*. Edward Arnold, London.

Penman, H. L. (1948). Natural evaporation from open water, bare soil and grass. *Proceedings of the Royal Society of London, Series A* **193**, 120–145.

Thom, A. S. (1975). Momentum, mass and heat exchange. In *Vegetation and the atmosphere*, Vol. 1, *Principles* (ed. J. L. Monteith), pp. 57–109. Academic Press, London.

Chapter 15. Crop water relations

Barlow, E. W. R. (1986). Water relations of expanding leaves. *Australian Journal of Plant Physiology* **13**, 45–58.

Boyer, J. S., Armond, P. A., and Sharp, R. E. (1987). Light stress and leaf water relations. In *Photoinhibition* (eds D. J. Kyle, C. B. Osmond, and C. J. Arntzen), pp. 111–122. Elsevier, New York.

Campbell, G. S. (1974). A simple method for determining unsaturated conductivity from moisture retention data. *Soil Science* **117**, 311–315.

Campbell, G. S. (1985). *Soil physics with basic. Transport models for soil–plant systems.* Elsevier, Amsterdam.

Childs, E. C. (1969). *An introduction to the physical basis of soil water phenomena.* Wiley, London.

Dainty, J. (1972). Plant cell-water relations: the elasticity of the cell wall. *Proceeding of the Royal Society of Edinburgh, Series A* **70**, 89–93.

Dainty, J. (1976). Water relations of plant cells. In *Transport in plant II*, Part A, *Cells* (eds U. Luttge and M. G. Pitman), pp. 12–35. Springer-Verlag, Berlin.

Gregson, K., Hector, D. J. and McGowan, M. (1987). A one-parameter model for the soil water characteristic. *Journal of Soil Science* **38**, 483–486.

Hsiao, T. C. and Bradford, K. J. (1983). Physiological consequences of cellular water deficits. In *Limitations to efficient water use in crop production* (eds H. M. Taylor, W. R. Jordan, and T. M. Sinclair), pp. 227–265. American Society of Agronomy, Crop Science Society of America, Soil Science Society of America, Madison, WI.

Johnson, I. R. and Thornley, J. H. M. (1985). Dynamic model of the response of a vegetative grass crop to light, temperature and nitrogen. *Plant, Cell and Environment* **8**, 484–499.

Lainson, R. A. and Thornley, J. H. M. (1982). A model for leaf expansion in cucumber. *Annals of Botany* **50**, 407–425.

Ludlow, M. M. (1987). Defining root water status in the most meaningful way to relate to physiological processes. In *Measurement of soil and plant water status*, Vol. 2 (eds R. J. Hanks and R. W. Brown). pp. 47–53. Utah Agricultural Experiment Station, Logan, UT.

Sinell, H. J. (1980). Interacting factors affecting mixed populations. In *Microbiology of foods*, Vol. I (eds J. H Silliker, R. P. Elliott, A. C. Baird-Parker, F. L. Bryan, J. H. B. Christian, D. S. Clark, J. C. Olson, and T. A. Roberts), pp. 215–231, Academic Press, New York.

Taylor, H. M. and Klepper, B. (1978). The role of rooting characteristics in the supply of water to plants. *Advances in Agronomy* **30**, 99–128.

Taylor, H. M., Jordan, W. R., and Sinclair, T. M. (eds) (1983). *Limitations to efficient water use in crop production.* American Society of Agronomy, Crop Science Society of America, Soil Science Society of America, Madison, WI.

Tyree, M. T. and Hammett, H. T. (1972). The measurement of the turgor pressure and the water relations of plants by the pressure-bomb technique. *Journal of Experimental Botany* **23**, 267–282.

Zur, B. and Jones, J. W. (1981). A model for the water relations, photosynthesis, and expansive growth of crops. *Water Resources Research* **17**, 311–320.

Chapter 16. Crop responses

France, J. and Thornley, J. H. M. (1984). *Mathematical models in agriculture.* Butterworths, London.

Harper, J. L. (1977). *Population biology of plants.* Academic Press, London.

Johnson, I. R. and Thornley, J. H. M. (1985). Dynamic model of the response of a vegetative grass crop to light, temperature and nitrogen. *Plant, Cell and Environment* **8**, 485–499.

Pickard, W. F. (1983). Three interpretations of the self-thinning rule. *Annals of Botany* **51**, 749–757.

Thornley, J. H. M. (1983). Crop yield and planting density. *Annals of Botany* **52**, 257–259.

Yoda, K., Kira, T., Ogawa, H., and Hozumi, K. (1963). Self thinning in overcrowded pure stands under cultivated and natural conditions. *Journal of Biology, Osaka City University* **14**, 107–129.

Chapter 17. Root growth

Brouwer, R. (1962). Distribution of dry matter in the plant. *Netherlands Journal of Agricultural Science* **10**, 361–376.

Brugge, R. (1985). A mechanistic model of grass root growth and development dependent upon photosynthesis and nitrogen uptake. *Journal of Theoretical Biology* **116**, 443–467.

Brugge, R and Thornley, J. H. M. (1985). A growth model of root mass and vertical distribution, dependent on carbon substrate from photosynthesis and with non-limiting soil conditions. *Annals of Botany* **55**, 563–577.

Cooper, A. J. and Thornley, J. H. M. (1976). Response of dry matter partitioning, growth, and carbon and nitrogen levels in the tomato plant to changes in root temperature: experiment and theory. *Annals of Botany* **40**, 1139–1152.

Dunphy, E. J. and Hanway, J. J. (1976). Water-soluble carbohydrate accumulation in soybean plants. *Agronomy Journal* **68**, 697–700.

Garwood, E. A. (1967). Seasonal variation in appearance and growth of grass roots. *Journal of the British Grassland Society* **22**, 121–130.

Gerwitz, A. and Page, E. R. (1974). An empirical mathematical model to describe plant root systems. *Journal of Applied Ecology* **11**, 773–781.

Hackett, C. and Rose, D. A. (1972a). A model of the extension and branching of a seminal root of barley and its use in studying relations between root dimensions. I. The model. *Australian Journal of Biological Sciences* **25**, 669–679.

Hackett, C. and Rose, D. A. (1972b). A model of the extension and branching of a seminal root of barley and its use in studying relations between root dimensions. II. Results and inferences from manipulations of the model. *Australian Journal of Biological Sciences* **25**, 681–690.

Johnson, I. R., Ameziane, T. E., and Thornley, J. H. M. (1983). A model of grass growth. *Annals of Botany* **51**, 599–609.

Johnson, I. R. and Thornley, J. H. M. (1983). Vegetative crop growth model incorporating leaf area expansion and senescence, and applied to grass. *Plant, Cell and Environment* **6**, 721–729.

Lungley, D. R. (1973). The growth of root systems—a numerical computer simulation model. *Plant and Soil* **38**, 145–159.

Silk, M. W. K. and Erickson, R. O. (1979). The kinematics of plant growth. *Journal of Theoretical Biology* **76**, 481–501.

Silk, W. K., Walker, R. C., and Labavitch, J. (1984). Uronide deposition rates in the primary root of *Zea mays*. *Plant Physiology* **74**, 721–726.

Troughton, A. (1978). The influence of reproductive development upon the root system of perennial ryegrass and some effects on herbage production. *Journal of Agricultural Science, Cambridge* **91**, 427–431.

Waite, R. and Boyd, J. (1953). The water-soluble carbohydrates of grasses. I. Changes occurring during the normal life cycle. *Journal of the Science of Food and Agriculture* **4**, 197–204.

Chapter 18. Branching

Barker, S. B., Cumming, G. and Horsfield, K. (1973). Quantitative morphometry of the branching structure of trees. *Journal of Theoretical Biology* **40**, 33–43.

Bell, A. D., Roberts, D., and Smith, A. (1979). Branching patterns: simulation of plant architecture. *Journal of Theoretical Biology* **81**, 351–375.

Borchert, R. and Honda, H. (1984). Control of development in the bifurcating branch system of *Tabebuia rosae*: a computer simulation. *Botanical Gazette* **145**, 184–195.

Brady, B. L. (1960). Occurrence of *Itersonilia* and *Tilletiopsis* on lesions caused by *Entyloma*. *Transactions of the British Mycological* **43**, 31–50.

Fishenden, M. and Saunders, O. A. (1950). *An introduction to heat transfer*. Oxford University Press, Oxford.

Fisher, J. B. and Honda, H. (1977). Computer simulation of branching pattern and geometry in *Terminalia* (*combretaceae*), a tropical tree. *Botanical Gazette* **138**, 377–384.

Fisher, J. B. and Honda, H. (1979). Branch geometry and effective leaf area: a study of *Terminalia*-branching pattern. I. Theoretical trees. *American Journal of Botany* **66**, 633–644.

Honda, H. (1971). Description of the form of trees by the parameters of the tree-like body: effects of the branching angle and the branch length on the shape of the tree-like body. *Journal of Theoretical Biology* **31**, 331–338.

Honda, H. and Fisher, J. B. (1978). Tree branch angle: maximizing effective leaf area. *Science* **199**, 888–890.

Honda, H., Tomlinson, P. B., and Fisher, J. B. (1981). Computer simulation of branch interaction and regulation by unequal flow rates in botanical trees. *American Journal of Botany* **68**, 569–585.

Honda, H., Tomlinson, P. B., and Fisher, J. B. (1982). Two geometrical models of branching of botanical trees. *Annals of Botany* **49**, 1–11.

Horton, R. E. (1945). Erosional development of streams and their drainage basins: hydrophysical approach to quantitive morphology. *Bulletin of the Geological Society of America* **56**, 275–370.

Hutchinson, S. A., Sharma, P., Clarke, K. R., and Macdonald, I. (1980). Control of hyphal orientation in colonies of *Mucor hiemalis*. *Transactions of the British Mycological Society* **75**, 171–191.

Macdonald, N. (1983). *Trees and networks in biological models*. Wiley, Chichester.

McMahon, T. A. and Kronauer, R. E. (1976). Tree structures: deducting the principle of mechanical design. *Journal of Theoretical Biology* **59**, 443–466.

Murray, C. D. (1927a). The physiological principle of minimum work applied to the angle of branching of arteries. *Journal of General Physiology* **9**, 835–841.

Murray, C. D. (1927b). A relationship between circumference and weight in trees and its bearing on branching angles. *Journal of General Physiology* **10**, 725–729.

Newman, F. H. and Searle, V. H. L. (1948). *The general properties of matter* (4th edn): Edward Arnold, London.

Paltridge, G. W. (1973). On the shape of trees. *Journal of Theoretical Biology* **38**, 111–137.

Park, D. (1985). Does Horton's law of branch length apply to open branching systems? *Journal of Theoretical Biology* **112**, 299–313.

Rose, D. A. (1983). The description of the growth of root systems. *Plant and Soil* **75**, 405–415.

Shreve, R. L. (1966). Statistical law of stream numbers. *Journal of Geology* **74**, 17–37.

Strahler, A. N. (1952). Hypsometric (area-altitude) analysis of erosional topography. *Bulletin of the Geological Society of America* **63**, 1117–1142.

Temperley, H. N. V. (1953). *Properties of matter.* University Tutorial Press, London.

Thornley, J. H. M. (1977). A model of apical bifurcation applicable to trees and other organisms. *Journal of Theoretical Biology* **64**, 165–176.

Warner, W. H. and Wilson, T. A. (1976). Distribution of end-points of a branching network with decaying branch length. *Bulletin of Mathematical Biology* **38**, 219–237.

Zamir, M. (1976). Optimality principles in arterial branching. *Journal of Theoretical Biology* **62**, 227–251.

Chapter 19. Phyllotaxis

Adler, I. (1974). A model of contact pressure in phyllotaxis. *Journal of Theoretical Biology* **45**, 1–79.

Adler, I. (1977). The consequences of contact pressure in phyllotaxis. *Journal of Theoretical Biology* **65**, 29–77.

Churchill, R. V. (1963). *Fourier series and boundary value problems.* McGraw-Hill, New York.

Jean, R. V. (1984). *Mathematical approach to pattern and form in plant growth.* Wiley–Interscience, New York.

Marzec, C. and Kappraff, J. (1983). Properties of maximal spacing on a circle related to phyllotaxis and to the golden mean. *Journal of Theoretical Biology* **103**, 201–226.

Niven, I. and Zuckerman, H. S. (1960). *An introduction to the theory of numbers* (2nd edn). Wiley, New York.

Ridley, J. N. (1982a). Packing efficiency in sunflower heads. *Mathematical Biosciences* **58**, 129–139.

Ridley, J. N. (1982b). Computer simulation of contact pressure in capitula. *Journal of Theoretical Biology* **95**, 1–11.

Roberts, D. W. (1978). The origin of Fibonacci phyllotaxis—an analysis of Adler's contact pressure model and Mitchison's expanding apex model. *Journal of Theoretical Biology* **74**, 217–233.

Schwabe, W. W. (1971). Chemical modifications of phyllotaxis and its implications. *Symposia of the Society of Experimental Biology* **25**, 301–322.

Thornley, J. H. M. (1975a). Phyllotaxis. I. A mechanistic model. *Annals of Botany* **39**, 491–507.

Thornley, J. H. M. (1975b). Phyllotaxis. II. A description in terms of intersecting logarithmic spirals. *Annals of Botany* **39**, 509–524.

Turing, A. M. (1952). The chemical basis of morphogenesis. *Philosophical Transactions of the Royal Society of London* **237**, 37–72.

Veen, A. H. and Lindenmayer, A. (1977). Diffusion mechanism for phyllotaxis. *Plant Physiology* **60**, 127–139.

Vogel, H. (1979). A better way to construct the sunflower head. *Mathematical Biosciences* **44**, 179–189.

Williams, R. F. and Brittain, E. G. (1984). A geometrical model of phyllotaxis. *Australian Journal of Botany* **32**, 43–72.

Index